GENERAL CHEMISTRY

GENERAL CHEMISTRY

LINUS PAULING

Research Professor
Linus Pauling Institute
of Science and Medicine

DOVER PUBLICATIONS, INC., New York

Published in Canada by General Publishing Company, Ltd., 30 Lesmill Road, Don Mills, Toronto, Ontario.
Published in the United Kingdom by Constable and Company, Ltd.

This Dover edition, first published in 1988, is an unabridged, slightly altered and corrected republication of the work published by W. H. Freeman and Company, San Francisco in 1970. The endpaper tables have been moved to the front section of the book and the color illustrations have been moved to the inside covers.

Manufactured in the United States of America
Dover Publications, Inc., 31 East 2nd Street, Mineola, N.Y. 11501

Library of Congress Cataloging-in-Publication Data

Pauling, Linus, 1901–
General chemistry / Linus Pauling.
p.   cm.
Reprint. Originally published: 3rd ed. San Francisco : W.H. Freeman, 1970.
Bibliography: p.
Includes index.
ISBN 0-486-65622-5 (pbk.)
1. Chemistry. I. Title.
QD33.P34   1988
540—dc19   87-32165
                                             CIP

Preface

In the first edition of this book, published 22 years ago, I attempted to simplify the teaching of general chemistry by the correlation of the facts of descriptive chemistry, the observed properties of substances, to as great an extent as possible with theoretical principles, especially the theory of atomic and molecular structure. This correlation with theory was extended in the second edition, and is extended still further in the present edition.

The theories of greatest value in modern chemistry are the theories of atomic and molecular structure, quantum mechanics, statistical mechanics, and thermodynamics. In this book I have tried to present in a sound and logical way the development of these theories, in their relation to chemistry. The principles of quantum mechanics are discussed on the basis of the de Broglie wavelength of the electron. The quantized energy levels of a particle in a box are derived by means of a simple assumption about the relation of the de Broglie waves to the walls of the box. No attempt is made to solve the Schrödinger wave equation for other systems, but the wave functions of hydrogen-like electrons are presented and discussed in some detail, and the quantum states for other systems are also discussed.

I have found that an understanding of statistical mechanics (especially in its quantum-mechanical form) is more easily obtained by the beginning student than an understanding of chemical thermodynamics. It is for this reason that I have introduced statistical mechanics before thermodynamics, and have based the discussion of thermodynamics on it.

A simple derivation of the Boltzmann distribution law is given in Chapter 9. The principles of chemical thermodynamics are then developed in a reasonably straightforward way from this law in Chapters 10 and 11. It has been customary for many years to include some parts of chemical thermodynamics, especially in relation to chemical equilibrium, in general chemistry, without relating the equations to the basic principles of thermodynamics and statistical mechanics. I think that it will be helpful to the student to have available in his textbook a discussion of the basic theory, even though it may not be possible for him to master the material in the time during which the course is presented.

The amount of descriptive chemistry has been decreased somewhat in this edition, and the presentation of this subject, especially in relation to the nonmetals, has been revised in such a way as to permit greater correlation with the electronic structure of atoms, especially electronegativity. It is not possible to add new material to the one-year course of general chemistry without omitting some of the old material. It is important for the student to know some of the facts of descriptive chemistry, and I have attempted to make a reasonable decision about the extent to which the elimination of descriptive chemistry should be carried out.

There is, of course, somewhat more material in this book than can be mastered by even a very good student in one year. I have assumed that some of the chapters and sections will be omitted in formal study. The inclusion in the book of this extra material makes it available for the clarification of questions that might arise and for additional reading by the interested student.

In this book extensive use is made of the International System of units. Many students have already become acquainted with this system (or the closely related MKS system) through their study of physics. The acceptance of the IS system by most countries in the world and the inherent advantages of the system (the elimination of arbitrary conversion factors) seem to me to justify the transition to the IS units at the present time. The principal change is the use of the joule in place of the calorie as the unit of energy.

This edition, like the preceding editions, is designed especially for use by first-year college university students who plan to major in chemistry or in closely related fields, and by other well-prepared students with a special interest in chemistry. I have assumed that the students have some background of knowledge of physics and mathematics.

I am glad to acknowledge the general assistance of Gustav Albrecht, Barclay Kamb, Peter Pauling, Arthur B. Robinson, and Fred Wall, and also thank John Ricci for preparation of the stereoscopic drawings.

LINUS PAULING

9 December 1969

Contents

15 Oxidation-Reduction Reactions. Electrolysis 512

16 The Rate of Chemical Reactions 551

17 The Nature of Metals and Alloys 577

GENERAL
CHEMISTRY

The Periodic System of the Elements

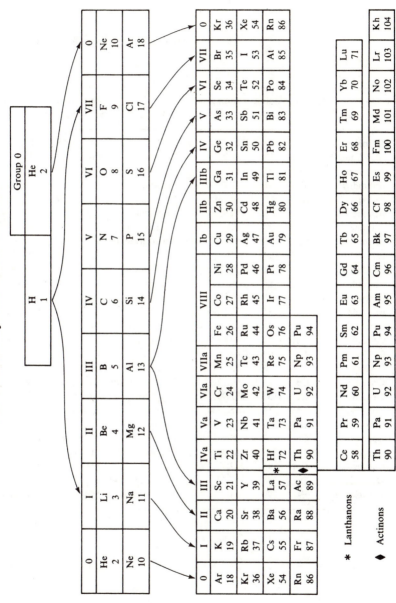

International Atomic Weights

Element	Symbol	Atomic Number	Atomic Weight	Element	Symbol	Atomic Number	Atomic Weight
Actinium	Ac	89	[227]*	Erbium	Er	68	167.26
Aluminum	Al	13	26.9815	Europium	Eu	63	151.96
Americium	Am	95	[243]	Fermium	Fm	100	[253]
Antimony	Sb	51	121.75	Fluorine	F	9	18.9984
Argon	Ar	18	39.948	Francium	Fr	87	[223]
Arsenic	As	33	74.9216	Gadolinium	Gd	64	157.25
Astatine	At	85	[210]	Gallium	Ga	31	69.72
Barium	Ba	56	137.34	Germanium	Ge	32	72.59
Berkelium	Bk	97	[247]	Gold	Au	79	196.967
Beryllium	Be	4	9.0122	Hafnium	Hf	72	178.49
Bismuth	Bi	83	208.980	Helium	He	2	4.0026
Boron	B	5	10.811 †	Holmium	Ho	67	164.930
Bromine	Br	35	79.909 ‡	Hydrogen	H	1	1.00797 †
Cadmium	Cd	48	112.40	Indium	In	49	114.82
Calcium	Ca	20	40.08	Iodine	I	53	126.9044
Californium	Cf	98	[249]	Iridium	Ir	77	192.2
Carbon	C	6	12.01115 †	Iron	Fe	26	55.847 ‡
Cerium	Ce	58	140.12	Khurchatovium	Kh	104	[260]
Cesium	Cs	55	132.905	Krypton	Kr	36	83.80
Chlorine	Cl	17	35.453 ‡	Lanthanum	La	57	138.91
Chromium	Cr	24	51.996 ‡	Lawrencium	Lr	103	[257]
Cobalt	Co	27	58.9332	Lead	Pb	82	207.19
Copper	Cu	29	63.54	Lithium	Li	3	6.939
Curium	Cm	96	[247]	Lutetium	Lu	71	174.97
Dysprosium	Dy	66	162.50	Magnesium	Mg	12	24.312
Einsteinium	Es	99	[254]	Manganese	Mn	25	54.9380

Element	Symbol	Atomic Number	Atomic Weight	Element	Symbol	Atomic Number	Atomic Weight
Mendelevium	Md	101	[256]	Ruthenium	Ru	44	101.07
Mercury	Hg	80	200.59	Samarium	Sm	62	150.35
Molybdenum	Mo	42	95.94	Scandium	Sc	21	44.956
Neodymium	Nd	60	144.24	Selenium	Se	34	78.96
Neon	Ne	10	20.183	Silicon	Si	14	28.086†
Neptunium	Np	93	[237]	Silver	Ag	47	107.870‡
Nickel	Ni	28	58.71	Sodium	Na	11	22.9898
Niobium	Nb	41	92.906	Strontium	Sr	38	87.62
Nitrogen	N	7	14.0067	Sulfur	S	16	32.064†
Nobelium	No	102	[256]	Tantalum	Ta	73	180.948
Osmium	Os	76	190.2	Technetium	Tc	43	[97]
Oxygen	O	8	15.9994†	Tellurium	Te	52	127.60
Palladium	Pd	46	106.4	Terbium	Tb	65	158.924
Phosphorus	P	15	30.9738	Thallium	Tl	81	204.37
Platinum	Pt	78	195.09	Thorium	Th	90	232.038
Plutonium	Pu	94	[242]	Thulium	Tm	69	168.934
Polonium	Po	84	[210]	Tin	Sn	50	118.69
Potassium	K	19	39.102	Titanium	Ti	22	47.90
Praseodymium	Pr	59	140.907	Tungsten	W	74	183.85
Promethium	Pm	61	[147]	Uranium	U	92	238.03
Protactinium	Pa	91	[231]	Vanadium	V	23	50.942
Radium	Ra	88	[226]	Xenon	Xe	54	131.30
Radon	Rn	86	[222]	Ytterbium	Yb	70	173.04
Rhenium	Re	75	186.2	Yttrium	Y	39	88.905
Rhodium	Rh	45	102.905	Zinc	Zn	30	65.37
Rubidium	Rb	37	85.47	Zirconium	Zr	40	91.22

*A value given in brackets is the mass number of the most stable known isotope.

†The atomic weight varies because of natural variations in the isotopic composition of the element. The observed ranges are boron, ±0.003; carbon, ±0.00005; hydrogen, ±0.00001; oxygen, ±0.0001; silicon, ±0.001; sulfur, ±0.003.

‡The atomic weight is believed to have an experimental uncertainty of the following magnitude: bromine, ±0.002; chlorine, ±0.001; chromium, ±0.001; iron, ±0.003; silver, ±0.003. For other elements the last digit given is believed to be reliable to ±0.05.

Energy Level Diagram of Electron Shells and Subshells of the Elements

Shell

Ra
88

Q   7s ⇅   6d ◹◹◹◹◹

Cm Bk Cf Es Fm Md No
96 97 98 99 100 101 102

5f ⇅⇅⇅⇅⇅⇅⇅

Fr
87

Po At Rn
84 85 86

6p ⇅⇅⇅⇅

Ac Th Pa U Np Pu Am
89 90 91 92 93 94 95

Ba
56

6s ⇅

Tl Pb Bi
81 82 83

Os Ir Pt Au Hg
76 77 78 79 80

5d ⇅⇅⇅⇅⇅

Gd Tb Dy Ho Er Tm Yb
64 65 66 67 68 69 70

4f ⇅⇅⇅⇅⇅⇅⇅

Cs
55

Te I Xe
52 53 54

5p ⇅⇅⇅⇅

Lu Hf Ta W Re
71 72 73 74 75

La Ce Pr Nd Pm Sm Eu
57 58 59 60 61 62 63

Sr
38

5s ⇅

In Sn Sb
49 50 51

Ru Rh Pd Ag Cd
44 45 46 47 48

4d ⇅⇅⇅⇅⇅

Rb
37

Se Br Kr
34 35 36

4p ⇅⇅⇅⇅

Y Zr Nb Mo Tc
39 40 41 42 43

Ca
20

4s ⇅

Ga Ge As
31 32 33

Fe Co Ni Cu Zn
26 27 28 29 30

3d ⇅⇅⇅⇅⇅

K
19

S Cl Ar
16 17 18

3p ⇅⇅⇅⇅

Sc Ti V Cr Mn
21 22 23 24 25

Mg
12

Al Si P
13 14 15

3s ⇅

M

Na
11

O F Ne
8 9 10

2p ⇅⇅⇅⇅

Be
4

B C N
5 6 7

2s ⇅

L

Li
3

He
2

K   1s ⇅   Most stable orbital

H
1

Energy

↑ = Electron with positive
orientation of spin

↓ = Electron with negative
orientation of spin

N

O

P

Q

1

The Nature and Properties of Matter

1-1. Matter and Chemistry

The universe is composed of matter and radiant energy. Matter (from the Latin *materia*, meaning wood or other material) may be defined as any kind of mass-energy (see Section 1-2) that moves with velocities less than the velocity of light, and radiant energy as any kind of mass-energy that moves with the velocity of light.

The different kinds of matter are called *substances*. Chemistry is the science of substances—their structure, their properties, and the reactions that change them into other substances.

This definition of chemistry is both too narrow and too broad. It is too narrow because the chemist in his study of substances must also study radiant energy, in its interaction with substances. He may be interested in the color of substances, which is produced by the absorption of light. Or he may be interested in the atomic structure of substances, as determined by the diffraction of x-rays (Section 3-7 and Appendix IV) or by the absorption or emission of radiowaves by the substances.

On the other hand, the definition is too broad, in that almost all of science could be included within it. The astrophysicist is interested in the substances that are present in stars and other celestial bodies, or that are distributed, in very low concentration, through interstellar space. The nuclear physicist studies the substances that constitute the nuclei of atoms. The biologist is interested in the substances that are present in living organisms. The geologist is interested in the substances, called minerals, that make up the earth. It is hard to draw a line between chemistry and other sciences.

1-2. Mass and Energy

Matter has mass, and any portion of matter on the earth is attracted toward the center of the earth by the force of gravity; this attraction is called the weight of the portion of matter. For many years scientists thought that matter and radiant energy could be distinguished through the possession of mass by matter and the lack of possession of mass by energy. Then, early in the present century (1905), it was pointed out by Albert Einstein (1879–1955) that energy also has mass, and that light is accordingly attracted by matter through gravitation. This was verified by astronomers, who found that a ray of light traveling from a distant star to the earth and passing close by the sun is bent toward the sun by its gravitational attraction. The observation of this phenomenon was made during a solar eclipse, when the image of the star could be seen close to the sun.

The amount of mass associated with a definite amount of energy is given by an important equation, the *Einstein equation*, which is an essential part of the theory of relativity:

$$E = mc^2 \qquad\qquad (1\text{-}1)$$

In this equation E is the amount of energy (J), m is the mass (kg), and c is the velocity of light (m s^{-1}).* The velocity of light, c, is one of the fundamental constants of nature;† its value is 2.9979×10^8 meters per second.

Until the present century it was also thought that matter could not be created or destroyed, but could only be converted from one form into another. In recent years it has, however, been found possible to convert matter into radiant energy, and to convert radiant energy into matter. The mass m of the matter obtained by the conversion of an amount E of radiant energy or convertible into this amount of radiant energy is given by the Einstein equation. Experimental verification of the Einstein equation has been obtained by the study of processes involving nuclei of atoms. The nature of these processes will be described in later chapters in this book.

---

*See Section 1–3 for a discussion of the units.

†The symbol c represents the velocity of light in empty space.

Until early in the present century scientists made use of a law of conservation of matter and a law of conservation of energy. These two conservation laws were then combined into a single one, the *law of conservation of mass*, in which the mass to be conserved includes both the mass of the matter in the system and the mass of the radiant energy in the system.

1-3. The International System of Units

The metric system of units of length, mass, force, and other physical quantities was developed during the French Revolution. Because of their greater convenience and simplicity, metric units have replaced native units (such as the foot and the pound) in scientific work everywhere and have been formally accepted for practical use in many countries (all except the United States, Canada, and some African countries). An extended and improved form of the metric system, called the International System (IS, or sometimes SI, for *Systéme International*), was formally adopted by the General Conference of Weights and Measures in 1960.

The symbols of the basic IS units and of the prefixes for fractions and multiples and those for some derived IS units are given in Appendix I. If you have made use of the MKS system (meter-kilogram-second system) in your study of physics the IS system will be familiar to you, for the most part, but if you have made use of the cgs system (centimeter-gram-second system) you will have to learn some new units.

The IS unit of mass, the *kilogram*, is defined as the mass of a standard object made of a platinum-iridium alloy and kept in Paris. One pound is equal approximately to 453.59 g, and hence 1 kg is equal approximately to 2.205 lb. (Note that it has become customary for the abbreviation of units in the metric system to be written without periods.) There is at the present time a flaw in the International System, in that the name for the unit of mass involves a prefix, kilo. This flaw will remain until agreement about a new name and symbol has been reached. In the meantime we must remember that 1 milligram (symbol 1 mg, not 1 μkg) is one millionth of the unit of mass, not one thousandth, as indicated by the prefix milli.

The IS unit of length, the *meter* (m), is equal to about 39.37 inches (1 inch equals exactly 2.54 cm). The meter was formerly defined as the distance between two engraved lines on a standard platinum-iridium bar kept in Paris by the International Bureau of Weights and Measures; in 1960 it was redefined, by international agreement, as 1,650,763.73 wavelengths of the orange-red spectral line* of krypton 86.

---

*This line corresponds to the transition of the atom from the state designated $5d_5$ to the state designated $2p_{10}$, each of which has $J = 1$. In this transition an electron drops from a $6d$ orbital to a $5p$ orbital (see Section 5-3).

The IS unit of time is the *second* (s). It is defined as the interval occupied by 9,192,631,770 cycles of the microwave line* of cesium 133 with wavelength about 3.26 cm. The second was formerly defined as 1/86400th of the mean solar day.

The IS unit of volume is the *cubic meter*, m^3. In chemistry a unit that is much used† is the liter, symbol l, which is 1×10^{-3} m^3. The milliliter, 1×10^{-3} l, is equal to the cubic centimeter: 1 ml = 1 cm^3.

The IS unit of force is the *newton* (N), which is defined as the force needed to accelerate a mass of 1 kg by 1 m s^{-2}. The newton is 10^5 dyne (the dyne, the unit of force in the cgs system, is the force that accelerates 1 g by 1 cm s^{-2}). The IS unit of energy, the *joule*‡ (J), is the work done by 1 newton in the distance 1 meter: 1 J = 1 N m = 10^7 erg = 10^7 dyne cm.

In chemistry the *calorie* has been extensively used as the unit of energy. The thermochemical calorie, defined as 4.184 J (Appendix I), is approximately the amount of energy needed to raise the temperature of 1 g of water by 1°C. The large calorie (kcal or Cal) is 10^3 cal. In this book we shall use the joule in most of the tables and discussions. Since most thermochemical reference books use the calorie or kilocalorie, you will find it worth while to remember the conversion factor:

$$1 \text{ cal} = 4.184 \text{ J}$$
$$1 \text{ kcal} = 1 \text{ Cal} = 4.184 \text{ kJ}$$

Example 1-1. Niagara Falls (Horseshoe) is 160 feet high. How much warmer is the water at the bottom than at the top, as the result of the conversion of potential energy into thermal energy? The standard acceleration of gravity is 9.80665 m s^{-2}.

Solution. The gravitational force on a mass of 1 kg at the earth's surface is 9.80665 N. The change is potential energy of 1 kg over a vertical distance h (in meters) is $9.80665 \times h$ J. In this problem h has the value $0.3048 \times 160 = 48.77$ m (conversion factor from Appendix I); hence the change in potential energy produces $9.80665 \times 48.77 = 478$ J of thermal energy. The energy required to raise the temperature of 1 kg of water by 1°C is given above as 1 kcal = 4.184 kJ = 4184 J. Hence the increase in temperature of the water is 478/4184 = 0.114°C.

---

*The normal state of cesium involves one unpaired electron, in a 6*s* orbital, symbol $^2S_{1/2}$. The nucleus cesium 133 has a spin, the spin quantum number having the value $I = 7/2$. The nuclear spin angular momentum and the electron spin angular momentum combine to a resultant angular momentum corresponding either to the value $F = 4$ or to the value $F = 3$ of the total angular momentum quantum number. The 3.26-cm line involves the transition between these two levels (see Section 26-7).

†The liter was formerly defined in such a way that there was a small difference between the milliliter and the cubic centimeter (1 old milliliter = 1.000027 cm^3).

‡Usually pronounced to rhyme with howl.

Example 1-2.   When 2 kg of uranium 235 undergoes nuclear fission (as in the detonation of the Hiroshima atomic bomb on 6 August 1945), 1.646×10^{14} J of radiant energy and thermal energy is liberated. What is the mass of the material products of the reaction?

Solution.   We can calculate the mass of the liberated energy by the use of the Einstein equation (1-1). Rewriting this equation by dividing each side by c^2 and introducing the values of E and c, we obtain

$$m = \frac{E}{c^2} = \frac{1.646 \times 10^{14} \text{ J}}{(2.998 \times 10^8)^2 \text{ m}^2 \text{ s}^{-2}} = 0.183 \times 10^{-2} \text{ kg}$$

Thus, the material mass of 2 kg has decreased by 0.00183 kg (that is, by 0.0915%), leaving material products of the reaction with mass 1.99817 kg.

The Einstein relation between mass and energy has been verified by the direct measurement of the mass of the products and of the energy emitted in nuclear reactions.

Example 1-3.   It is found by experiment that when 1 kg of glyceryl trinitrate (nitroglycerine) is exploded, the amount 8.0×10^6 J of energy is liberated. What is the mass of the products of the explosion?

Solution.   This example is to be solved in exactly the same way as the preceding one. The mass of the radiant energy that is produced by the explosion is obtained by dividing the energy, E, by the square of the velocity of light:

$$m = \frac{E}{c^2} = \frac{8.0 \times 10^6 \text{ J}}{(2.998 \times 10^8)^2 \text{ m}^2 \text{ s}^{-2}} = 0.89 \times 10^{-10} \text{ kg}$$

Thus we calculate that the mass of the products of the explosion is 0.999999999911 kg.

We see that the mass of the products of this chemical reaction differs very slightly from the mass of the reactant—so slightly that it is impossible to detect the change in a direct way. The change, less than one part in ten billion (1 in 10^{10}), is so small that for practical purposes we may say that there is conservation of mass in ordinary chemical reactions.

1-4.   Temperature

If two objects are placed in contact with one another, thermal energy may flow from one object to the other one. *Temperature* is the quality that determines the direction in which thermal energy flows—it flows from the object at higher temperature to the object at lower temperature.

Temperatures are ordinarily measured by means of a thermometer, such as the ordinary mercury thermometer, consisting of a quantity of mercury in a glass tube. The temperature scale used by scientists is the *centigrade* or *Celsius scale*; it was introduced by Anders Celsius (1701–1744), a Swedish professor of astronomy, in 1742. On this scale the temperature of freezing water saturated with air is 0°C and the temperature of boiling water is 100°C at 1 atm pressure.

On the *Fahrenheit scale*, used in everyday life in English-speaking countries, the freezing point of water is 32°F and the boiling point of water is 212°F. On this scale the freezing point and the boiling point differ by 180°, rather than the 100° of the centigrade scale.*

To convert temperatures from one scale to another, you need only remember that the Fahrenheit degree is $\frac{100}{180}$ or $\frac{5}{9}$ of the centigrade degree, and that 0°C is the same temperature as 32°F.

The Kelvin Temperature Scale

About 200 years ago scientists noticed that a sample of gas that is cooled decreases in volume in a regular way, and they saw that if the volume were to continue to decrease in the same way it would become zero at about −273°C. The concept was developed that this temperature, −273°C (more accurately, −273.15°C), is the minimum temperature, the *absolute zero*. A new temperature scale was then devised by Lord Kelvin, a great British physicist (1824–1907). The *Kelvin scale*† is defined in such a way as to permit the laws of thermodynamics to be expressed in simple form (see Chapter 10).

The IS temperature scale is the Kelvin scale with a new definition of the degree. The absolute zero is taken to be 0°K and the triple point of water is taken to be 273.16°K. (The triple point of water, the temperature at which pure liquid water, ice, and water vapor are in equilibrium, is discussed in Section 11-9.) With this definition of the degree, the boiling

---

*The Fahrenheit scale was devised by Gabriel Daniel Fahrenheit (1686–1736), a natural philosopher who was born in Danzig and settled in Holland. He invented the mercury thermometer in 1714; before then alcohol had been used as the liquid in thermometers. As the zero point on his scale he took the lowest temperature he could obtain, produced by mixing equal quantities of snow and ammonium chloride. His choice of 212° for the boiling point of water was made in order that the temperature of his body should be 100°F. The normal temperature of the human body is 98.6°F; perhaps Fahrenheit had a slight fever while he was calibrating his thermometer.

†Another absolute scale, the *Rankine scale*, is sometimes used in engineering work in the English-speaking countries. It uses the Fahrenheit degree, and has 0°R at the absolute zero.

point of water at one atmosphere pressure is 373.15°K and the freezing point of water saturated with air at one atmosphere pressure* is 273.15°K. Hence the IS Kelvin temperature is 273.15°K greater than the centigrade temperature.

1-5.   Kinds of Matter

We shall first distinguish between objects and kinds of matter. An object, such as a human being, a table, a brass doorknob, may be made of one kind of matter or of several kinds of matter. The chemist is primarily interested not in the objects themselves, but in the kinds of matter of which they are composed. He is interested in the alloy brass, whether it is in a doorknob or in some other object; and his interest may be primarily in those properties of the material that are independent of the nature of the objects containing it.

Materials

The word *material* is used in referring to any kind of matter, whether homogeneous or heterogeneous.

A *heterogeneous* material is a material that consists of parts with different properties. A *homogeneous* material has the same properties throughout.

Wood, with soft and hard rings alternating, is obviously a hetero-geneous material, as is also granite, in which grains of three different substances (the minerals quartz, mica, and feldspar) can be seen.

A *mineral* is any chemical element, compound, or other homogeneous material (such as a liquid solution or a crystalline solution) occurring naturally as a product of inorganic processes. Most minerals are solids. Water and mercury are examples of liquid minerals, and air and helium (from rocks or helium wells) are examples of gaseous minerals. Amalgam (mercury containing dissolved silver and gold) is an example of a solution occurring as a mineral. Rocks are simple minerals (limestone consists of the mineral calcite, which is calcium carbonate) or mixtures of minerals (granite is such a mixture).

---

*The definition of 0°C (and 32°F) is that it is the freezing temperature of water saturated with air at one atmosphere pressure. This temperature is 0.010°K below the triple-point temperature. Increase of pressure from the triple-point value 0.00603 atm to 1 atm decreases the freezing point by 0.0075°K and the presence of dissolved air (nitrogen and oxygen) decreases it further by 0.0024°K.

Substances

A *substance* is usually defined by chemists as a homogeneous species of matter with reasonably definite chemical composition.

By this definition, pure salt, pure sugar, pure iron, pure copper, pure sulfur, pure water, pure oxygen, and pure hydrogen are representative substances. On the other hand, a solution of sugar in water is not a substance; it is, to be sure, homogeneous, but it does not satisfy the second part of the above definition, inasmuch as its composition is not definite but is widely variable, being determined by the amount of sugar that happens to have been dissolved in a given amount of water. Similarly, the gold of a gold ring or watchcase is not a pure substance, even though it is apparently homogeneous. It is an alloy of gold with other metals, and it usually consists of a crystalline solution of copper in gold. The word *alloy* is used to refer to a metallic material containing two or more elements: the intermetallic compounds are substances, but most alloys are crystalline solutions or mixtures.

Sometimes (as in the first section of this chapter) the word "substance" is used in a broader sense, essentially as equivalent to material. For the sake of clarity, the chemist's more restricted meaning may be indicated by the phrase "pure substance."

Our definition is not precise, in that it says that a substance has reasonably definite chemical composition. Most materials that the chemist classifies as substances (pure substances) have definite chemical composition; for example, pure salt consists of the two elements sodium and chlorine in exactly the ratio of one atom of sodium to one atom of chlorine. Others, however, show a small range of variation of chemical composition; an example is the iron sulfide that is made by heating iron and sulfur together. This substance has a range in composition of a few percent.

Kinds of Definition

Definitions may be either precise or imprecise. The mathematician may define the words that he uses precisely; in his further discussion he then adheres rigorously to the defined meaning of each word. We have given some precise definitions above. One of them is the definition of the kilogram as the mass of a standard object, the prototype kilogram, that is kept in Paris. Similarly, the gram is defined rigorously and precisely as 1/1000 the mass of the kilogram.

On the other hand, the words that are used in describing nature, which is itself complex, may not be capable of precise definition. In giving a definition for such a word the effort is made to describe the accepted usage.

Mixtures and Solutions

A specimen of granite, in which grains of three different species of matter can be seen, is obviously a *mixture*. An emulsion of oil in water (a suspension of droplets of oil in water) is also a mixture. The heterogeneity of a piece of granite is obvious to the eye. The heterogeneity of an emulsion containing large drops of oil suspended in water is also obvious; the emulsion is clearly seen to be a mixture. But as the oil droplets in the emulsion are made smaller and smaller, it may become impossible to observe the heterogeneity of the material, and uncertainty may arise as to whether the material should be called a mixture or a solution.

An ordinary *solution* is homogeneous; it is not usually classified as a substance, however, because its composition is variable. A solution of liquids, such as alcohol and water, or of gases, such as oxygen and nitrogen (the principal constituents of air), may also be called a mixture. The word "mixture" may thus be used to refer to a homogeneous material that is not a pure substance or to a heterogeneous aggregate of two or more substances.

A homogeneous crystalline material is not necessarily a pure substance. Thus natural crystals of sulfur are sometimes deep yellow or brown in color, instead of light yellow. The dark crystals contain some selenium, distributed at random throughout the crystals in place of some of the sulfur, the crystals being homogeneous, and with faces as well formed as those of pure sulfur. These crystals are a *crystalline solution* (or *solid solution*). The gold-copper alloy used in jewelry is another example of a crystalline solution. It is a homogeneous material, but its composition is variable.

Phases

A material system (that is, a limited part of the universe) may be described in terms of the *phases* constituting it. A phase is a homogeneous part of a system, separated from other parts by physical boundaries. For example, if a flask is partially full of water in which ice is floating, the system comprising the contents of the flask consists of three phases: ice (a solid phase); water (a liquid phase); and air (a gaseous phase). A piece of malleable cast iron can be seen with a microscope to be a mixture of small grains of iron and particles of graphite (a form of carbon); it hence consists of two phases, iron and graphite (Figure 1-1).

A phase in a system comprises all of the parts that have the same properties and composition. Thus, if there were several pieces of ice in the system discussed above, they would constitute not several phases, but only one phase, the ice phase.

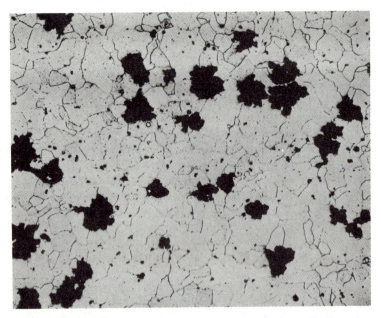

FIGURE 1-1
A photomicrograph (linear magnification 100 times—that is, magnified by a linear
factor of 100) of a polished and etched surface of a specimen of malleable
cast iron, showing small grains of iron and roughly spherical particles of graphite
(carbon). The grains of iron look somewhat different from one another
because of different illumination. (From Malleable Founders' Society.)

Constituents and Components

Chemists use the words *constituent* and *component* in special ways.

The constituents of a system are the various phases that constitute the
system.

A set of components of a system is a set of substances (the minimum
number of substances) from which the phases (constituents) of the system
could be made.

The constituents of the system discussed above are the three phases air,
water, and ice. The components of the system may be taken to be either
air and water, or air and ice, because the water phase and the ice phase
can both be made from a single substance, water (or ice).* In this case the

---

*In the above discussion air has been described as a component of the system. In dis-
cussing changes in state of the system in which the air behaves in the same way that
nitrogen would behave, this would not lead to any difficulty, but in a rigorous treatment
air might have to be described as involving several components (nitrogen, oxygen, argon,
and others).

number of components is less than the number of phases. It may be greater; for example, a system consisting just of a solution of sugar in water is constituted of one phase, the solution, but it has two components, sugar and water.

1-6.   The Physical Properties of Substances

Properties of substances are their characteristic qualities.

Sodium chloride (common salt) may be selected as an example of a substance. We have all seen this substance in what appear to be different forms—table salt, in fine grains; salt in the form of crystals 2 or 3 mm in diameter, for use in regenerating water-softening minerals; and natural crystals of rock salt 5 cm or more across. Despite their obvious difference all of these samples of salt have the same fundamental properties. In each case the crystals, small or large, are bounded by square or rectangular faces, of different sizes, but with each face always at right angles to each adjacent face. The possession of different properties in different directions —in particular the *formation of faces, edges, and vertices*—is character-istic of crystals. The *cleavage* of the different crystals of salt is the same: when crushed, the crystals always break (cleave) along planes parallel to the original faces, producing smaller crystals similar to the larger ones. The different samples have the same salty *taste*. Their *solubility* is the same: at room temperature (18°C) 35.86 g of salt can be dissolved in 100 g of water. The *density* (ratio of mass to volume) of the salt is the same, 2.163 g cm^{-3}.

Properties of this sort, which are not affected appreciably by the size of the sample or its state of subdivision, are called the *intrinsic properties**
of the substance represented by the samples.

There are other properties besides density and solubility that can be measured precisely and expressed in numbers. Another such property is the *melting point*, the temperature at which a crystalline substance melts to form a liquid. The *electric conductivity* and the *thermal conductivity* are similar properties. On the other hand, there are also interesting physical properties of a substance that are not so simple in nature. One such property is the *malleability* of a substance—the ease with which the sub-stance can be hammered out into thin sheets. A related property is the *ductility*—the ease with which the substance can be drawn into a wire. *Hardness* is a similar property: we say that one substance is less hard than a second substance when it is scratched by the second substance, but this test provides only qualitative information about the hardness. A discussion of hardness is presented in Section 7-1. The *color* of a substance is an im-

---

*Also called *specific properties*.

portant physical property. It is interesting to note that the apparent color of a substance depends upon its state of subdivision: the color becomes lighter as large particles are ground up into smaller ones, because the distance through which the light penetrates before it is reflected back from the interfaces (surfaces) becomes less as the particles become smaller.

It is customary to say that under the same external conditions all specimens of a particular substance have the same specific physical properties (density, hardness, color, melting point, crystalline form, and others). Sometimes, however, the word "substance" is used in referring to a material without regard to its state of aggregation; for example, ice, liquid water, and water vapor may be referred to as the same substance. Moreover, a specimen containing crystals of rock salt and crystals of table salt may be called a mixture, even though the specimen may consist entirely of the one chemical substance sodium chloride. This lack of definiteness in usage seems to cause no confusion in practice.

The concept "substance" is, of course, an idealization; all actual substances are more or less impure. It is a useful concept, however, because we have learned through experiment that the properties of various specimens of impure substances with the same major component and different impurities are usually nearly the same if the impurities are present in only small amounts. These properties are accepted as the properties of the ideal substance.

1-7.   The Chemical Properties of Substances

The *chemical properties* of a substance are those properties that relate to its participation in chemical reactions. *Chemical reactions* are the processes that convert substances into other substances.

Thus sodium chloride has the property of changing into a soft metal (sodium) and a greenish-yellow gas (chlorine) when it is decomposed by electrolysis. It also has the property, when it is dissolved in water, of producing a white precipitate when a solution of silver nitrate is added to it; and it has many other chemical properties. Iron has the property of combining readily with the oxygen in moist air, to form iron rust; whereas an alloy of iron with chromium and nickel (stainless steel) is found to resist this process of rusting. It is evident from this example that the chemical properties of materials are important in engineering.

Most substances have the power to enter into many chemical reactions. The study of these reactions constitutes a large part of the study of chemistry.

The properties of *taste* and *odor* are closely correlated with the chemical nature of substances, and are to be considered as chemical properties; the senses of smell and taste possessed by animals are the *chemical senses*.

There is still complete lack of knowledge as to the way in which the molecules of tasty and odorous substances interact with the nerve endings in the mouth and nose to produce the sensations of taste and odor; this problem, like the problem of the molecular basis of the action of drugs, is one that awaits solution by the younger generation of chemists.

1-8. The Scientific Method

An important reason for studying science is to learn the scientific method of attack on a problem. The method may be valuable not only in the field of science but also in other fields—of business, of law, of government, of sociology, of international relations.

It is not possible to present a complete account of the scientific method in a few paragraphs. At this point there is given a partial account, which is amplified at the beginning of the following chapter, and in later chapters. Here I may say that one part of the scientific method consists in the application of the principles of rigorous argument that are developed in mathematics and in logic, the deduction of sound conclusions from a set of accepted postulates. In a branch of mathematics the basic postulates are accepted as axioms, and the entire subject is then derived from these postulates. In science, and in other fields of human activity, the basic postulates (principles, laws) are not known, but must be discovered. The process of discovering these laws is called *induction*.

The first step in applying the scientific method consists in being curious about the world, about the facts that have been found by observation and experiment. In our science these are the facts of descriptive chemistry. The next step is the classification and correlation of many facts by one statement. Such a general statement, which includes within itself a number of facts, is called a *law*—sometimes a *law of nature*. Such an effort to classify facts often suggests new experiments, which uncover new facts.

For example, when it was discovered, early in the nineteenth century, that water could be decomposed into hydrogen and oxygen by electrolysis (the action of an electric current), quantitative measurements of the amounts of hydrogen and oxygen were made. It was found in one experiment that 9 g of water on electrolysis produced 1 g of hydrogen and 8 g of oxygen. This fact, for a particular specimen of water, was then amplified by additional facts, all showing that the same amount of hydrogen, 1 g, and the same amount of oxygen, 8 g, were obtained by the electrolysis of 9 g of water from other sources. After many more experiments had been made, all giving the same result, the facts were summarized in a law: all samples of water give on electrolysis the same relative amounts of hydrogen and oxygen. When similar results were obtained for some other substances, this law was generalized into the law of constant composition (or law of definite proportions): in every pure sample of a given compound, the elements are present in the same proportion by weight.

It must be pointed out that the process of induction is never completely reliable. If one hundred analyses of water are made (by weighing the amounts of hydrogen and oxygen obtained by electrolysis of samples of water obtained from different sources), and the same proportion by weight of hydrogen to oxygen is found—to within the limits of accuracy of the experiments—it would seem to be reasonably well justified to state that all samples of water have the same ratio of hydrogen to oxygen by weight. However, if a single reliable analysis were then to be made that gave a different ratio, the law would have to be modified. It might turn out that the law is valid if the weighings of the gases are made with an accuracy of 0.1%, but not if the weighings are made with greater accuracy. This has, in fact, been found to be the case for water. In 1929 William F. Giauque (born 1895) discovered that there are three different kinds of oxygen atoms, with different masses (these atoms are called isotopes; see Chapter 4), and shortly thereafter Harold C. Urey (born 1893) discovered that there are two kinds of hydrogen atoms, with different masses. Water consisting of molecules made with these different kinds of hydrogen atoms and oxygen atoms contains hydrogen and oxygen in different ratios by weight, and it has been found that the composition by weight of pure water from different natural sources is, in fact, slightly different. It has accordingly become necessary to revise the law of constant composition in such a way as to take into account the existence of these isotopic forms of atoms. The way in which this is done is described in Chapter 4.

One important way in which progress has been made in science is through a process of *successive approximations*. Some measurements are made with a certain precision, such as the measurements of the compositions of substances with an accuracy of 1%, and a rough law is formulated that encompasses all of these measurements. It may then happen that, when more precise measurements are made, deviations from the first law are found to exist. A second, more refined but more complicated law may then be formulated to include these deviations. This procedure may have been carried out several times in the formulation of a law of nature in its now accepted state.

It is wise to remember that a law obtained by the process of induction may at any time be found to have limited validity. Conclusions that are reached from such a law by the process of deduction should be recognized as having a probability of being correct that is determined by the probability that the original law is correct.

The application of the scientific method does not consist solely of the routine use of logical rules and procedures. Often a generalization that encompasses many facts has escaped notice until a scientist with unusual insight has discovered it. Intuition and imagination play an important part in the scientific method.

As more and more people gain a sound understanding of the nature of the scientific method and learn to apply it in the solution of the problems of everyday life, we may hope for an improvement in the social, political, and international affairs of the world. Technical progress represents one way in which the world can be improved through science. Another way is through the social progress that results from application of the scientific method—through the development of "moral science." I believe that the study of science, the learning of the scientific method by all people, will ultimately help the people of the world in the solution of our great social and political problems.

Exercises

1-1. What is the difference between matter and radiant energy?

1-2. What is the Einstein relation between mass and energy? Indicate the IS units of the terms in this relation.

1-3. Approximately how much energy, in IS units, is needed to raise 1 liter (1 kg) of liquid water from 273.15°K to 373.15°K? (See the discussion of the calorie, Section 1-3.)

1-4. Verify the following method of converting temperature on the Celsius scale to temperature on the Fahrenheit scale (or the reverse): add 40, multiply by $\frac{9}{5}$ (or by $\frac{5}{9}$), subtract 40.

1-5. Mercury freezes at $-40°C$. What is its freezing point on the Fahrenheit scale?

1-6. For each of the following systems (*i*) state how many phases are present in the system; (*ii*) state for each phase whether it is a pure substance or a mixture; (*iii*) give the constituents of the system; (*iv*) give a set of components for the system:
   (*a*) A flask containing a saturated aqueous solution of salt and several crystals of salt.
   (*b*) An evacuated, sealed quartz tube of 100-ml volume containing 10 g of pure zinc heated until about one half the zinc is melted.
   (*c*) As (*b*), but containing 10 g of a copper-gold alloy instead of 10 g of zinc.

1-7. What is meant by "intrinsic property of a substance"? Are odor, shape, density, color, weight, taste, luster, area, magnetic susceptibility, and heat capacity intrinsic properties? Which of these are properties that can be quantitatively measured?

2

The Atomic and Molecular
Structure of Matter

The properties of any kind of matter are most easily and clearly learned and understood when they are correlated with its structure, in terms of the molecules, atoms, and still smaller particles that compose it. This subject, the atomic structure of matter, will be taken up in this chapter.

2-1. Hypotheses, Theories, and Laws

When it is first found that an idea explains or correlates a number of facts, the idea is called a *hypothesis*. A hypothesis may be subjected to further tests and to experimental checking of deductions that may be made from it. If it continues to agree with the results of experiment the hypothesis is called a *theory* or *law*.

A theory, such as the atomic theory, usually involves some idea about the nature of some part of the universe, whereas a law may represent a summarizing statement about observed experimental facts. For example, there is a law of the constancy of the angles between the faces of crystals. This law states that whenever the angles between corresponding faces of various crystals of a pure substance are measured they are found to have the same value, whether the crystal is a small one or a large one. It does not explain this fact. An explanation of the fact is given by the atomic theory of crystals, the theory that in crystals the atoms are arranged in a regular order (as described later in this chapter.)

Chemists and other scientists use the word theory in two somewhat different senses. The first meaning of the word is that described above— namely, a hypothesis that has been verified. The second use of the word theory is to represent a systematic body of knowledge, compounded of facts, laws, theories in the limited sense described above, deductive arguments, and so on. Thus by the atomic theory we mean not only the idea that substances are composed of atoms, but also all the facts about substances that can be explained and interpreted in terms of atoms and the arguments that have been developed to explain the properties of substances in terms of their atomic structure.

2-2. The Atomic Theory

In 1805 the English chemist and physicist John Dalton (1766–1844) stated the hypothesis that all substances consist of small particles of matter, of several different kinds, corresponding to the different elements. He called these particles atoms, from the Greek word *atomos*, meaning indivisible. This hypothesis gave a simple explanation or picture of previously observed but unsatisfactorily explained relations among the weights of substances taking part in chemical reactions with one another. As it was verified by further work in chemistry and physics, Dalton's atomic hypothesis became the atomic theory.

The rapid progress of our science during the current century is well illustrated by the increase in our knowledge about atoms. In a popular textbook of chemistry written in the early years of the twentieth century atoms were defined as the "imaginary units of which bodies are aggregates." The article on "Atom" in the 11th edition of the *Encyclopaedia Britannica*, published in 1910, ends with the words "The atomic theory has been of priceless value to chemists, but it has more than once happened in the history of science that a hypothesis, after having been useful in the discovery and the coordination of knowledge, has been abandoned and replaced by one more in harmony with later discoveries. Some distinguished chemists have thought that this fate may be awaiting the atomic theory. . . . But modern discoveries in radioactivity are in favor of the existence of the atom, although they lead to the belief that the atom is not so eternal and unchangeable a thing as Dalton and his predecessors had imagined." Now, only half a century later, we have precise knowledge of the structure and properties of atoms and molecules. Atoms and molecules can no longer be considered "imaginary."

Dalton's Arguments in Support of the Atomic Theory

The Greek philosopher Democritus (about 460–370 B.C.), who had adopted some of his ideas from earlier philosophers, stated that the universe is composed of void (vacuum) and atoms. The atoms were considered to be everlasting and indivisible—absolutely small, so small that their size could not be diminished. He considered the atoms of different substances, such as water and iron, to be fundamentally the same, but to differ in some superficial way; atoms of water, being smooth and round, could roll over one another, whereas atoms of iron, being rough and jagged, would cling together to form a solid body.

The atomic theory of Democritus was pure speculation, and was much too general to be useful. Dalton's atomic theory, however, was a hypothesis that explained many facts in a simple and reasonable way.

In 1785 the French chemist Antoine Laurent Lavoisier (1743–1794) showed clearly that there is no measurable change in mass during a chemical reaction: the mass of the products is equal to the mass of the reacting substances.

In 1799 another general law, the *law of constant proportions*, was enunciated by the French chemist Joseph Louis Proust (1754–1826). The law of constant proportions states that different samples of a substance contain its elementary constituents (elements) in the same proportions. For example, it was found by analysis that the two elements hydrogen and oxygen are present in any sample of water in the proportion by weight 1:8.

Dalton stated the hypothesis that elements consist of atoms, all of the atoms of one element being identical, and that compounds result from the combination of atoms of two or more elements, each in definite number. In this way he could give a simple explanation of the law of conservation of mass and of the law of constant proportions.

A *molecule* is a group of atoms bonded to one another. If a molecule of water is formed by the combination of two atoms of hydrogen with one atom of oxygen, the mass of the molecule should be the sum of the masses of two atoms of hydrogen and an atom of oxygen, in accordance with the law of conservation of mass. The definite composition of a compound is then explained by the definite ratio of atoms of different elements in the molecules of the compound.

Dalton also formulated another law, the *law of simple multiple proportions*.* This law states that when two elements combine to form more than one compound, the weights of one element that combine with the same weight of the other are in the ratios of small integers.

It is found by experiment that, whereas one oxide of carbon consists of

---

*The discovery of the law of simple multiple proportions was the first great success of Dalton's atomic theory. This law was not induced from experimental results, but was derived from the theory, and then tested by experiments.

carbon and oxygen in the weight ratio 3:4, another consists of carbon and oxygen in the ratio 3:8. The weights of oxygen combined with the same weight of carbon, 3 g, in the two substances are 4 g and 8 g; that is, they are in the ratio of the small integers 1 and 2. This ratio can be explained by assuming that twice as many atoms of oxygen combine with the same number of atoms of carbon in the second substance as in the first.

Dalton had no way of determining the correct formulas of compounds, and he arbitrarily chose formulas to be as simple as possible; for example, he assumed that the molecule of water consisted of one atom of hydrogen and one atom of oxygen, whereas in fact it consists of two atoms of hydrogen and one of oxygen.

2-3. Modern Methods of Studying Atoms and Molecules

During the second half of the nineteenth century chemists began to discuss the properties of substances in terms of assumed structures of the molecules. Precise information about the atomic structure of molecules and crystals of many substances was finally obtained during the recent period, beginning in 1912. Physicists have developed many powerful methods of investigating the structure of matter. One of these methods is the interpretation of the spectra of substances (see Figure 21-1). A flame containing water vapor, for example, emits light that is characteristic of the water molecule. The spectrum of water vapor may be observed when this light passes through a spectroscope. Measurements of the lines in the water spectrum have been made and interpreted, and it has been found that the two hydrogen atoms in the molecule are about 97 pm (that is, 97×10^{-12} m) from the oxygen atom. Moreover, it has been shown that the two hydrogen atoms are not on opposite sides of the oxygen atom, but that the molecule is bent, the angle formed by the three atoms being 104.5°. The distances between atoms and the angles formed by the atoms in many simple molecules have been determined by spectroscopic methods.

Also, the structures of many substances have been determined by the methods of diffraction of electrons and diffraction of x-rays. In the following pages we shall describe many atomic structures that have been determined by these methods. The x-ray diffraction method of determining the structure of crystals is discussed in Appendix IV.

It is stated above that in the water molecule the two hydrogen atoms are 97 pm from the oxygen atom. This usage is in accordance with the International System. In the past, however, it has been customary to describe interatomic distances with use of the Ångström* as the unit of

---

*The Ångström was named in honor of a Swedish physicist, Anders Jonas Ångström (1814–1874), who in 1868 had published his measured values, to six significant figures, of the wavelengths of 1000 lines in the solar spectrum.

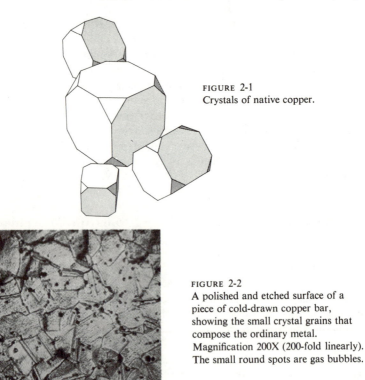

FIGURE 2-1
Crystals of native copper.

FIGURE 2-2
A polished and etched surface of a
piece of cold-drawn copper bar,
showing the small crystal grains that
compose the ordinary metal.
Magnification 200X (200-fold linearly).
The small round spots are gas bubbles.

length; 1 Å $= 10^{-8}$ cm $= 10^{-10}$ m $= 100$ pm. Because of the convenience
of the Ångström in the discussion of interatomic distances and because of
its use in reference books, we shall use it in this book. The hydrogen-
oxygen distance in the water molecule is 0.97 Å, and most bond lengths in
molecules and crystals lie between 1 Å and 4 Å.

2-4.   The Arrangement of Atoms in a Crystal

Most solid substances are crystalline in nature. Sometimes the particles of
a sample of solid substance are themselves single crystals, such as the cubic
crystals of sodium chloride in table salt. Sometimes these single crystals
are very large; occasionally crystals of minerals several meters in diameter
are found in nature.

In our discussion we shall use *copper* as an example. Crystals of copper
as large as a centimeter on edge, as shown in Figure 2-1, are found in

deposits of copper ore. An ordinary piece of the metal copper does not consist of a single crystal of copper, but of an aggregate of crystals. The crystal grains of a specimen of a metal can be made clearly visible by polishing the surface of the metal, and then etching the metal slightly with an acid. Often the grains are small, and can be seen only with the aid of a microscope (Figure 2-2), but sometimes they are large, and can be easily seen with the naked eye, as in some brass doorknobs.

It has been found by experiment (Section 2-5) that *every crystal consists of atoms arranged in a three-dimensional pattern that repeats itself regularly.* In a crystal of copper all of the atoms are alike, and they are arranged in the way shown in Figures 2-3 and 2-4. This is a way in which spheres of uniform size may be packed together to occupy the smallest volume. This structure, called the *cubic closest-packed structure*, was assigned to the copper crystal by W. L. Bragg in 1913.

It is the regularity of arrangement of the atoms in a crystal that gives to the crystal its characteristic properties, in particular the property of growing in the form of polyhedra. (A polyhedron is a solid figure bounded by plane faces.) The faces of crystals are defined by surface layers of atoms, as shown in Figures 2-3 and 2-4. These faces lie at angles to one another that have definite characteristic values, the same for all specimens of the same substance. The principal surface layers for copper correspond to the six faces of a cube (cubic faces or cube faces); these faces are always orthogonal to (at right angles with) one another. The eight smaller surface layers, obtained by cutting off the corners of the cube, are called "octahedral faces." Native copper, found in deposits of copper ore, often is in the form of crystals with cubic and octahedral faces (Figure 2-1).

Atoms are not hard spheres, but are soft, so that by increased force they may be pushed more closely together (be compressed). This compression occurs, for example, when a copper crystal becomes somewhat smaller in volume under increased pressure. The sizes that are assigned to atoms correspond to the distances between the center of one atom and the center of a neighboring atom of the same kind in a crystal under ordinary circumstances. The distance from a copper atom to each of its twelve nearest neighbors in a copper crystal at room temperature and atmospheric pressure is 2.55 Å; this is called the diameter of the copper atom in metallic copper. The radius of the copper atom is half this value.

2-5.   The Description of a Crystal Structure

Chemists often make use of the observed shapes of crystals to help in the identification of substances. The description of the shapes of crystals is the subject of the science of *crystallography*. The method of studying the structure of crystals by the diffraction of x-rays, which was discovered by

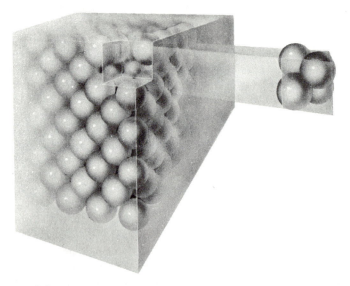

FIGURE 2-3
The arrangement of atoms in a crystal of copper. The small cube, containing four copper atoms, is the unit of structure; by repeating it the entire crystal is obtained.

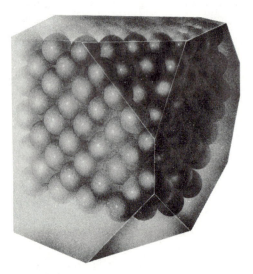

FIGURE 2-4
Another atomic view of a copper crystal, showing small octahedral faces and large cube faces.

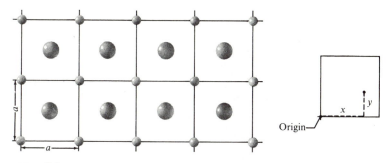

FIGURE 2-5
Arrangement of atoms in a plane. The unit of structure is a square. Small atoms have the coordinates 0, 0 and large atoms the coordinates $\frac{1}{2}$, $\frac{1}{2}$.

the German physicist Max von Laue (1879–1960) in 1912 and developed by the British physicists W. H. Bragg (1862–1942) and W. L. Bragg (born 1890), has become especially valuable in recent decades. Much of the information about molecular structure that is given in this book has been obtained by the x-ray diffraction technique.

The basis of the description of the structure of a crystal is the unit of structure, often called the *unit cell*. For cubic crystals the unit cell is a small cube, which, when repeated parallel to itself in such a way as to fill space, reproduces the entire crystal.

The way in which this is done can be seen from a two-dimensional example. In Figure 2-5 there is shown a portion of a structure based on a square lattice. The unit of structure of this square lattice is a square; when this square is repeated parallel to itself in such a way as to fill the plane, we obtain a sort of two-dimensional crystal. In this case there are present a lattice of atoms of one sort, represented by small spheres at the intersections of the lattice lines, and a lattice of atoms of another sort, represented by larger spheres at the centers of the unit squares. It is customary to describe the structure by the use of coordinates x and y, giving the position of the atoms relative to an origin at the corner of the unit cell, with x and y taken as fractions of the edges of the unit of structure, as indicated in the figure. The atom represented by the small sphere has the coordinates $x = 0$, $y = 0$, and the atom at the center of the square has the coordinates $x = \frac{1}{2}$, $y = \frac{1}{2}$.

Similarly, the unit cell of a cubic crystal is a cube which when reproduced in parallel orientation would fill space to produce a cubic lattice, as shown in Figure 2-6. The unit cell of a given cubic crystal could be described by giving the value of the edge of the unit, a, and the values of the coordinates x, y, and z for each atom, as fractions of the edge of the unit. Thus, for metallic copper, a cubic structure, the unit of structure is

FIGURE 2-6
The simple cubic arrangement of atoms. The unit of structure is a cube, with one atom per unit, its coordinates being 0, 0, 0.

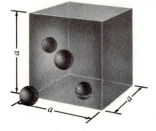

FIGURE 2-7
The cubic unit of structure for the face-centered cubic arrangement, corresponding to cubic closest packing of spheres. There are four atoms in the unit, with coordinates $0, 0, 0$; $0, \frac{1}{2}, \frac{1}{2}$; $\frac{1}{2}, 0, \frac{1}{2}$; $\frac{1}{2}, \frac{1}{2}, 0$.

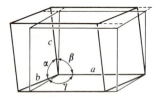

FIGURE 2-8
The parallelepiped representing a general unit of structure. It is determined by the lengths of its three edges, and by the three angles between the edges. A rectangular parallelepiped with its *ab* face in the same plane is also shown, light lines.

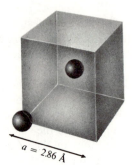

FIGURE 2-9
The unit of structure corresponding to the cubic body-centered arrangement. There are two atoms in the unit, with coordinates $0, 0, 0$ and $\frac{1}{2}, \frac{1}{2}, \frac{1}{2}$.

$a = 2.86 \text{ Å}$

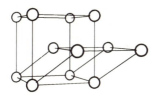 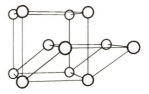

FIGURE 2-10

Stereoscopic view of the body-centered cubic structure, found for iron and some other metals. [The stereo drawings in this book can be viewed by looking at the right drawing with the right eye and the left drawing with the left eye from a distance of a few inches. It may help to hold a stiff piece of paper upright between the drawings. A moment or two may be required before the viewer learns to integrate the two images.]

a cube with edge $a = \sqrt{2} \times 2.55$ Å and with four atoms per unit, with coordinates $x = 0$, $y = 0$, $z = 0$; $x = 0$, $y = \frac{1}{2}$, $z = \frac{1}{2}$; $x = \frac{1}{2}$, $y = 0$, $z = \frac{1}{2}$; and $x = \frac{1}{2}$, $y = \frac{1}{2}$, $z = 0$, as shown in Figure 2-7. Often these coordinates are written without giving the symbols x, y, and z; it is then said that there are four copper atoms in the unit, at 0, 0, 0; 0, $\frac{1}{2}$, $\frac{1}{2}$; $\frac{1}{2}$, 0, $\frac{1}{2}$; $\frac{1}{2}$, $\frac{1}{2}$, 0.

In the unit cube shown in Figure 2-7 the coordinates place an atom at only one of the eight corners. Of course, when this unit cube is surrounded by other unit cubes, atoms are placed at the seven other corners, these atoms being formally associated with the adjacent unit cubes.

The unit of structure of the most general sort of crystal, a triclinic crystal (see Appendix III and Figure III-4), is a general parallelepiped, as shown in Figure 2-8. It can be described by giving the values a, b, and c, the lengths of the three edges, and the values of α, β, and γ, the angles between pairs of edges.

Example 2-1. The metal iron is cubic, with $a = 2.86$ Å, and with two iron atoms in the unit cube, at 0, 0, 0 and $\frac{1}{2}$, $\frac{1}{2}$, $\frac{1}{2}$. How many nearest neighbors does each iron atom have, and how far away are they?

Solution. We draw a cubic unit of structure, with edge 2.86 Å, as shown in Figure 2-9, and we indicate in it the positions 0, 0, 0 and $\frac{1}{2}$, $\frac{1}{2}$, $\frac{1}{2}$. When cubes of this sort are reproduced parallel to one another, we see that we obtain the structure shown in Figure 2-10; this is called the *body-centered arrangement*. It is seen that the atom at $\frac{1}{2}$, $\frac{1}{2}$, $\frac{1}{2}$ is surrounded by eight atoms, the atom at 0, 0, 0 and seven similar atoms. Also, the atom at 0, 0, 0 is surrounded by eight atoms. In each case the surrounding atoms are at the corners of a cube. This situation is described by saying that each atom in the body-centered arrangement has *ligancy* 8 (or *coordination number* 8).

To calculate the interatomic distance, we note that the square of this distance is equal to $(a/2)^2 + (a/2)^2 + (a/2)^2$, and hence the distance itself is equal to $\sqrt{3}a/2$. Thus the distance between each iron atom and its neighbors is found to be 1.732×2.86 Å$/2 = 2.48$ Å. The metallic radius of iron is hence 1.24 Å.

Example 2-2. The English mathematician and astronomer Thomas Harriot (1560–1621), who was tutor to Sir Walter Raleigh and who traveled to Virginia* in 1585, was interested in the atomic theory of substances. He believed that the hypothesis that substances consist of atoms was plausible, and capable of explaining some of the properties of matter. His writings contain the following propositions:

" 9. The more solid bodies have Atoms touching on all Sydes.
"10. Homogeneall bodies consist of Atoms of like figure, and quantitie.
"11. The waight may increase by interposition of lesse Atoms in the vacuities betwine the greater.
"12. By the differences of regular touches (in bodies more solid), we find that the lightest are such, where euery Atom is touched with six others about it, and greatest (if not intermingled) where twelve others do touch euery Atom."

Assuming that the atoms can be represented as hard spheres in contact with one another, what difference in density would there be between the two structures described in the above proposition 12?

Solution. The structure in which every atom is in contact with six others about it that Harriot had in mind is probably the simple cubic arrangement, shown in Figure 2-11. In this arrangement of atoms the unit of structure is a cube that contains one atom, which can be assigned the coordinates 0, 0, 0. Each atom is then in contact with six other atoms, which are at the distance d from it. The volume of the unit cube is accordingly d^3. If the mass of the atom is M, the density for this arangement is Md^{-3}.

The denser structure referred to by Harriot, where twelve atoms are in contact with each atom, is the cubic closest-packed arrangement described in the preceding section. (Harriot had apparently discovered that there is no way of packing equal hard spheres in space that gives a greater density than is given by this arrangement.) The cubic unit of structure for this arrangement contains four atoms. Its edge, a, is equal to $2^{1/2}d$, and its volume to $2^{3/2}d^3$. The mass contained in the unit cube is $4M$, and the density is accordingly $4M/2^{3/2}d^3$, or $2^{1/2}M/d^3$. We have thus found that the dense structure described by Harriot has density $2^{1/2} = 1.414$ times that of the less dense structure; it is accordingly 41.4% denser than the less dense structure.

---

*He took the potato and tobacco back to Europe, and was perhaps the first smoker known to have died of lung cancer. (R. Taton (ed.), *The Beginnings of Modern Science*.)

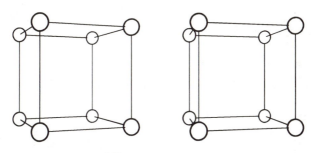

FIGURE 2-11
Simple cubic packing of spheres (stereo).

2-6.   Crystal Symmetry; the Crystal Systems

The principal classification of crystals is on the basis of their symmetry. An object has symmetry if some operation can be carried out on it such that, when the operation is completed, the object appears unchanged. For example, a three-bladed propeller can be rotated about its axis through 120° ($\frac{1}{3}$ of a revolution), and it is then indistinguishable from its original condition, provided that the three blades are exactly identical with one another. Similarly, it can be rotated through 240° ($\frac{2}{3}$ of a revolution), again becoming indistinguishable from its original condition. These operations of rotation by $\frac{1}{3}$ of a revolution and rotation by $\frac{2}{3}$ of a revolution, together with the original operation involving no change, constitute the symmetry of a threefold axis of symmetry. Other examples of symmetry are given in Appendix III.

Only a few symmetry elements are shown by crystals. They can be combined in 32 different ways, called *point groups*, corresponding to the 32 crystal classes. These crystal classes are divided into six crystal systems, as described in Appendix III.

An infinite crystal also has translations and other operations involving translations as identity operations. There are 230 possible combinations of these identity operations, which are called *space groups*. A discussion of the 230 space groups is given in Appendix III.

2-7.   The Molecular Structure of Matter

Molecular Crystals

The only building stone of a copper crystal is the copper atom; in this crystal there do not exist any discrete groups of atoms smaller than the crystal specimen itself. On the other hand, crystals of many other substances contain discrete groups of atoms, which are called molecules. An example of a molecular crystal is shown in the upper left part of Figure

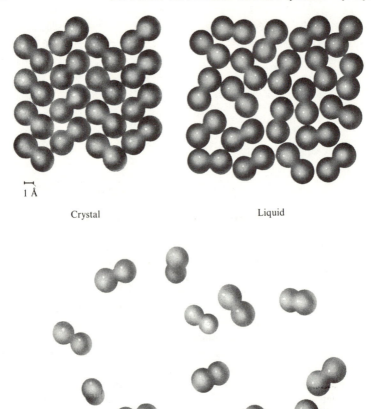

Crystal   Liquid

1 Å

Gas

FIGURE 2-12
Crystalline, liquid, and gaseous iodine, showing diatomic molecules I_2.

2-12, which is a drawing representing the structure of a crystal of the blackish-gray solid substance iodine, as determined by the x-ray diffraction method. The iodine atoms are grouped together in pairs, to form diatomic molecules (molecules containing two atoms each).

The distance between the two atoms in the same molecule of a molecular crystal is smaller than the distances between atoms in different molecules. In the iodine crystal the two iodine atoms in each molecule are only 2.68 Å apart, whereas the small distances between iodine atoms in different molecules range from 3.56 to 4.40 Å.

Moreover, the forces acting between atoms within a molecule are much stronger than those acting between molecules. As a result, it is much harder to cause the molecule to change its shape than to change the positions of the molecules relative to one another. For example, when a crystal of iodine is put under pressure it decreases in size: the molecules can be pushed together until the intermolecular distances have decreased by several percent; but the molecules themselves retain their original size, with no comparable change in interatomic distance within the molecule. Also, when a crystal of iodine at low temperature is heated it expands, so that each of the molecules occupies a larger space in the crystal; but the iodine–iodine distance within the molecule stays very close to the normal 2.68 Å.

The molecules of different chemical substances contain varying numbers of atoms, bonded together. An example of a molecule more complicated than iodine is shown in Figure 2-13; this molecule, of the substance cyanuric triazide, contains three carbon atoms and twelve nitrogen atoms. Within the molecule the small interatomic distances have values between 1.11 Å and 1.38 Å. The smallest interatomic distances between molecules are 3.12 Å and 3.16 Å. It is found that in most molecular crystals the intermolecular distances between atoms are about 1.60 Å greater than the intramolecular distances (bond lengths) for the same kinds of atoms.

Crystals Containing Giant Molecules

Immense numbers of molecules of different kinds are found in the complicated structures of plants and animals. Many of these molecules are very large, containing tens of thousands of atoms. Scientists are beginning to gather detailed information about the structure of these giant organic molecules.

The *viruses* are aggregates of giant molecules with very interesting properties. Some diseases, such as measles, smallpox, infantile paralysis (poliomyelitis), and the common cold, are caused by viruses. Virus particles have the power of self-duplication—that is, the power of causing other particles identical with themselves to be formed when they are in the right environment. A disease such as measles results from the formation of a great many measles-virus particles in a human body that has been infected by a few of these particles.

Another property that virus particles have, in common with ordinary small molecules, is the ability to form crystals. The particles of a virus are all essentially alike in size and shape and can order themselves in the regular arrangement that constitutes a crystal.

In recent years it has become possible to photograph virus particles. They are too small to be seen with a microscope using ordinary visible light, which cannot permit objects much smaller in diameter than the

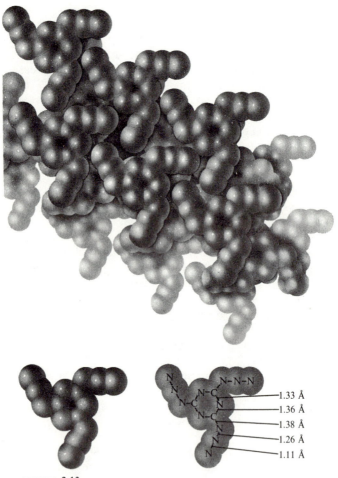

FIGURE 2-13

Above, portion of crystal of cyanuric triazide, C_3N_{12}, showing how the molecules pack together. Below, separate molecules.

wavelength of light, about 5000 Å, to be seen. However, electron microscopes, which use beams of electrons in place of beams of light, permit objects as small as 5 Å in diameter to be seen.

A photograph made with the electron microscope, reproduced here as Figure 2-14, shows the virus that causes a disease in tomato plants. Each virus particle is about 200 Å in diameter. It is made of about 450,000 atoms. In the photograph the individual particles can be clearly seen, and the regular way in which they arrange themselves in the crystal is evident.

FIGURE 2-14
Electron micrograph of crystals of necrosis virus protein, showing particles in ordered arrangement. Linear magnification 65,000. (From R. W. G. Wyckoff.)

The Evaporation of a Molecular Crystal

At a low temperature the molecules in a crystal of iodine lie rather quietly in their places in the crystal. As the temperature increases, the molecules become more and more agitated; each one bounds back and forth more and more vigorously in the little space left for it by its neighbors, and each one strikes its neighbors more and more strongly as it rebounds from them. This increase in molecular motion with increase in temperature causes the crystal to expand somewhat, giving each molecule a larger space in which to carry out its thermal oscillation.

A molecule on the surface of the crystal is held to the crystal by the forces of attraction that its neighboring molecules exert on it. Attractive forces of this kind, which are operative between all molecules when they are close together, are called *van der Waals attractive forces*, this name being used because it was the Dutch physicist J. D. van der Waals (1837–1923) who first gave a thorough discussion of intermolecular forces in relation to the nature of gases and liquids.

These attractive forces are quite weak. Hence occasionally a molecule will become so agitated as to break loose from its neighbors, and to fly off into the surrounding space. If the crystal is in a vessel, there will soon be present in the space within the vessel, through this process of evaporation, a large number of these free molecules, each moving in a straight-line path, and occasionally colliding with another molecule or with the walls of the vessel to change the direction of its motion. These free molecules constitute iodine vapor, or iodine gas (Figure 2-12). The gas molecules are very much like the molecules in the crystal, their interatomic distance being practically the same; it is the distances between molecules that are much larger in a gas than in a crystal.

Iodine vapor is violet in color and has a characteristic odor. The odor of tincture of iodine (a solution of iodine in alcohol, used as an antiseptic) is a combination of the odor of iodine vapor and the odor of ethyl alcohol.

It may seem surprising that molecules on the surface of a crystal should evaporate directly into a gas, but in fact the process of slow evaporation of a crystalline substance is not uncommon. Solid pieces of camphor or of naphthalene (as used in moth balls, for example) left out in the air slowly decrease in size because of the evaporation of molecules from the surface of the solid. In the same manner, snow may disappear from the ground without melting if the ice crystals evaporate at a temperature below that of their melting point. Evaporation is accelerated if a wind is blowing, to take the water vapor away from the immediate neighborhood of the snow crystals and to prevent the vapor from condensing again on the crystals.

The Nature of a Gas

The characteristic feature of a gas is that its molecules are not held together, but are moving about freely in a volume rather large compared with the volume of the molecules themselves. The van der Waals attractive forces between the molecules still operate whenever two molecules come close together, but usually these forces are negligibly small because the molecules are far apart. Because of this freedom of molecular motion, a specimen of gas does not have either definite shape or definite size. A gas shapes itself to its container. A quantitative study of the properties of gases will be taken up later (Chapter 9).

Gases at ordinary pressure are very dilute: the molecules themselves constitute only about one one-thousandth of the total volume of the gas, the rest being empty space. Thus 1 g of solid iodine has a volume of about 0.2 cm³ (its density is 4.93 g cm^{-3}), whereas 1 g of iodine gas at 1 atm pressure and at the temperature 184°C (its boiling point) has a volume of 148 cm³, over 700 times greater. The volume of all the molecules in a gas at ordinary pressure is accordingly small compared with the volume of the gas itself. On the other hand, the diameter of a gas molecule is not extremely small compared with the distance between molecules; in a gas at room temperature and 1 atm pressure the average distance from a molecule to its nearest neighbors is about ten times its molecular diameter, as indicated in Figure 2-12.

The Vapor Pressure of a Crystal

A crystal of iodine in an evacuated vessel will gradually change into iodine gas by the evaporation of·molecules from its surface. Occasionally one of these free gas molecules will again strike the surface of the crystal, and it may stick to the surface, held by the van der Waals attraction of the other crystal molecules. This is called *condensation* of the gas molecules.

The rate at which molecules evaporate from a crystal surface is proportional to the area of the surface, but is essentially independent of the pressure of the surrounding gas, whereas the rate at which gas molecules strike the crystal surface is proportional to the area of the surface and also is proportional to the concentration of the molecules in the gas phase (the number of gas molecules in a unit of volume).

If some iodine crystals are put into a flask, which is then stoppered and allowed to stand at room temperature, it will soon be evident that a quantity of iodine has evaporated, because the gas phase in the flask will become violet in color. After a while the intensity of coloration of the gas phase will no longer increase, but will remain constant. This steady state is reached when the concentration of gas molecules becomes so great that the rate at which gas molecules strike the crystal surface and stay there is just equal to the rate at which molecules leave the crystal surface. The gas pressure at this steady state is called the *vapor pressure* of the crystal.

A steady state of this sort is an example of *equilibrium.* The study of physical equilibrium, such as this equilibrium between the crystal and its vapor, and of chemical equilibrium, representing a steady state among various substances that are reacting chemically with one another, constitutes an important part of general chemistry. It must be recognized that equilibrium is not a situation in which nothing is happening, but rather a situation in which opposing reactions are taking place at the same rate, so as to result in no over-all change.

The vapor pressure of every crystal increases as temperature increases. The vapor pressure of iodine has the value 0.26×10^{-3} atm at 20°C and 0.118 atm at 114°C, the melting point of a crystal. The crystals of iodine that are heated to a temperature only a little below the melting point evaporate rapidly, and the vapor may condense into crystals in a cooler part of the vessel. The complete process of evaporation of a crystal and recondensation of a gas directly as crystals, without apparently passing through the liquid state, is called *sublimation*. Sublimation is often a valuable method of purifying a substance.

Units of Pressure

In the preceding paragraph the vapor pressure of iodine is given in atmospheres. *One atmosphere* (standard atmospheric pressure) is equal to 101.325 kN m^{-2}. It is not an approved IS unit of pressure, but it is especially important in chemistry because many properties of substances have been measured and recorded at 1 atm pressure.

Another unit of pressure that has been much used in the past is the *torr*, which is the height in millimeters of a column of mercury that balances the pressure. The symbol mm Hg is sometimes used for torr. The name of the unit is taken from the name of the inventor of the mercury barometer, the Italian physicist Evangelista Torricelli (1608–1647). One atmosphere is equal to 760 torr.

The Nature of a Liquid

When iodine crystals are heated to 114°C they melt, forming liquid iodine. The temperature at which the crystals and the liquid are in equilibrium—that is, at which there is no tendency for the crystals to melt or for the liquid to freeze—is called the *melting point* of the crystals, and the *freezing point* of the liquid. This temperature is 114°C for iodine.

Liquid iodine differs from solid iodine (crystals) mainly in its fluidity. Like the solid, and unlike the gas, it has a definite volume, 1 g occupying about 0.2 cm^3, but it does not have a definite shape; instead, it fits itself to the shape of the bottom part of its container.

From the molecular viewpoint the process of melting can be described in the following way. As a crystal is heated, its molecules are increasingly agitated, and move about more and more vigorously; but at lower temperatures this thermal agitation does not carry any one molecule any significant distance away from the position fixed for it by the arrangement of its neighbors in the crystal. At the melting point the agitation finally becomes so great as to cause the molecules to slip by one another and to change somewhat their location relative to one another. They continue to stay close together, but do not continue to retain a regular fixed arrangement; instead, the grouping of molecules around a given molecule changes continually, sometimes being much like the close packing of the crystal, in which each iodine molecule has twelve near neighbors, and sometimes considerably different, the molecule having only ten or nine or eight near neighbors, as shown in Figure 2-12. Thus a liquid, like a crystal, is a *condensed phase*, as contrasted with a gas, the molecules being piled rather closely together; but whereas a crystal is characterized by regularity of atomic or molecular arrangement, a liquid is characterized by randomness of structure. The randomness of structure usually causes the density of a liquid to be somewhat less than that of the corresponding crystal.

The Vapor Pressure and Boiling Point of a Liquid

A liquid, like a crystal, is, at any temperature, in equilibrium with its own vapor when the vapor molecules are present in a certain concentration. The pressure corresponding to this concentration of gas molecules is called the *vapor pressure of the liquid* at the given temperature.

The vapor pressure of every liquid increases with increasing temperature. The temperature at which it reaches a standard value (usually 1 atm) is called the *boiling point* (standard boiling point) of the liquid. At this temperature it is possible for bubbles of the vapor to appear in the liquid and to escape to the surface.

The vapor pressure of liquid iodine at its freezing point, 114°C, is 0.118 atm. This is exactly the same as the vapor pressure of iodine crystals at this temperature, as described in the section before the preceding one. That is, iodine gas at a pressure of 0.118 atm is in equilibrium with liquid iodine at 114°C, the freezing point of the liquid, and this gas is also in equilibrium with iodine crystals at this temperature, their melting point. The crystals and the liquid are in equilibrium at the freezing point (melting point), and they then have exactly the same vapor pressure. If the two phases had different vapor pressures, the phase with the larger pressure would continue to evaporate, and the vapor would continue to condense as the other phase, until the first phase had disappeared.

The vapor pressure of liquid iodine reaches 1 atm at 184°C, which is the boiling point of iodine.

Other substances undergo similar changes in phase when they are heated. When copper melts, at 1083°C, it forms liquid copper, in which the arrangement of the copper atoms shows the same sort of randomness as that of the molecules in liquid iodine. Under 1 atm pressure copper boils at 2310°C to form copper gas; the gas molecules are single copper atoms. The vapor pressure of a substance can be measured by various techniques, such as that illustrated in Figure 2-15. In this experiment the barometric pressure is measured by measuring the height of a column of mercury in an evacuated tube. If the barometric pressure happened to be exactly normal, the height of the column would be 760.0 mm. The column of mercury is supported by the pressure of the air on the exposed surface of the mercury. Now if a substance, such as a drop of water, were to be introduced into the space in the tube above the column of mercury, by slipping it underneath the open bottom end of the tube and releasing it to permit it to rise

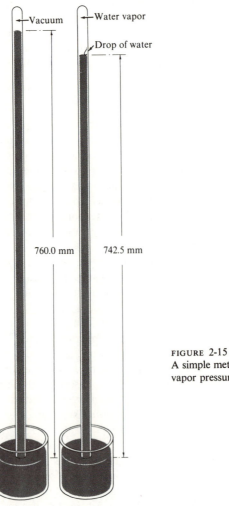

Vacuum

Water vapor

Drop of water

760.0 mm

742.5 mm

FIGURE 2-15
A simple method of measuring the
vapor pressure of a liquid.

to the top of the column, molecules of this substance would evaporate into
the space above the column of mercury, until equilibrium between the
vapor and the condensed phase was reached. The top of the column of
mercury would then be subjected to the pressure of this vapor, equal to the
vapor pressure of the substance, and the column of mercury would decrease in height. The results of measurements of this sort on iodine
crystals and liquid iodine are shown in Figure 2-16.

The vapor pressure of liquids and crystals is discussed in greater detail
in Sections 11-2 and 11-3.

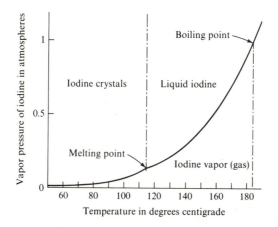

FIGURE 2-16

A graph showing the vapor-pressure curve of iodine
crystal and the vapor-pressure curve of liquid iodine.
The melting point of the crystal is the temperature
at which the crystal and the liquid have the same
vapor pressure, and the boiling point of the liquid
(at 1 atm pressure) is the temperature at which the
vapor pressure of the liquid equals 1 atm.

Exercises

2-1. What are the differences between a hypothesis, a theory, a law, and a fact?
Classify the following statements as hypotheses, theories, laws, or facts:
(a) The interior of the moon consists of granite and similar silicate rocks.
(b) Hydrogen, nitrogen, oxygen, and neon are all gases under ordinary
conditions.
(c) The force f acting on a body with mass m causes it to be accelerated
by the amount $f m^{-1}$.
(d) The properties of gases can be explained by considering the motion
of the molecules composing them.
(e) All crystals contain atoms or molecules arranged in a regular way.

2-2. Discuss some of the evidence for the atomic nature of matter.

2-3. The metal indium forms tetragonal crystals. The unit of structure is a rectangular parallelepiped, with edges a = 3.24 Å, b = 3.24 Å, and c = 4.94 Å. There are two atoms in the unit of structure, with coordinates 0, 0, 0 and $\frac{1}{2}, \frac{1}{2}, \frac{1}{2}$.

    (*a*) Calculate distances from each atom to its twelve nearest neighbors; note that four are at one distance, and eight at another distance.

    (*b*) Show that if the axial ratio c/a were equal to 1.414, the twelve distances would be equal.

    (*c*) What is the relation between the unit of structure with this axial ratio (1.414) and the cubic unit of structure described in the text for cubic closest packing?

2-4. Diamond has a cubic unit of structure, with a = 3.56 Å. There are eight atoms in the unit, with coordinates 0, 0, 0; 0, $\frac{1}{2}, \frac{1}{2}$; $\frac{1}{2}$, 0, $\frac{1}{2}$; $\frac{1}{2}, \frac{1}{2}$, 0; $\frac{1}{4}, \frac{1}{4},$ $\frac{1}{4}$; $\frac{1}{4}, \frac{3}{4}, \frac{3}{4}$; $\frac{3}{4}, \frac{1}{4}, \frac{3}{4}$; $\frac{3}{4}, \frac{3}{4}, \frac{1}{4}$. How many nearest neighbors does each atom have? What is the distance to each of these neighbors? (Answer: 4; 1.54 Å.)

2-5. The sodium chloride crystal has a cubic unit of structure, with a = 5.628 Å. There are four sodium atoms (sodium ions) in the unit of structure, with coordinates 0, 0, 0; 0, $\frac{1}{2}, \frac{1}{2}$; $\frac{1}{2}$, 0, $\frac{1}{2}$; $\frac{1}{2}, \frac{1}{2}$, 0; and also four chlorine atoms (chloride ions), with coordinates $\frac{1}{2}, \frac{1}{2}, \frac{1}{2}$; $\frac{1}{2}$, 0, 0; 0, $\frac{1}{2}$, 0; 0, 0, $\frac{1}{2}$. Make a drawing showing the cubic unit of structure and the positions of the atoms. How many nearest neighbors does each atom have? What is the inter-atomic distance for these neighbors? What polyhedron is formed by them (at the corners)? This arrangement of atoms, called the sodium chloride arrangement, is a common one for salts.

2-6. The cesium chloride crystal is cubic with a = 4.11 Å, Cs at 0 0 0 and Cl at $\frac{1}{2}$ $\frac{1}{2}$ $\frac{1}{2}$. How many nearest neighbors does each atom have? At what distance are they? What polyhedron do they define?

2-7. The vapor pressure of solid carbon dioxide at its melting point, $-56.5°C$, is 5 atm. How do you explain the fact that solid carbon dioxide when used for packing ice cream does not melt to form liquid carbon dioxide? If you wanted to make some liquid carbon dioxide, what would you have to do?

2-8. Define vapor pressure of a crystal, and also vapor pressure of a liquid. Can you think of an argument showing that these two vapor pressures of a substance must be equal at the melting point?

2-9. What is the effect of increase in pressure on the boiling point of a liquid? Estimate the boiling point of liquid iodine at a pressure of 0.5 atm (see Figure 2-16).

2-10. Carbon dioxide (dry ice) consists of CO_2 molecules. These molecules are linear, with the carbon atom in the center. Make three drawings, representing your concepts of carbon dioxide gas, carbon dioxide liquid, and carbon dioxide crystal, and describe these structures.

3

The Electron
the Nuclei of Atoms
and the Photon

During the period of fourteen years beginning in 1897 it was discovered that atoms are composed of smaller particles, and that these smaller particles carry electric charges. Photons, the units of radiant energy, were discovered at the same time. The discovery of the components of atoms and the investigation of the structure of atoms constitute one of the most interesting stories in the history of science. Moreover, knowledge about the electronic structure of atoms has permitted the facts of chemistry to be systematized in a striking way, making the subject easier to understand and to remember: it has been discovered that the bonds that hold atoms together in molecules consist of pairs of electrons held jointly by two atoms. The student of chemistry can be helped greatly in mastering his subject by first obtaining a good understanding of the structure of the atom.

3-1. The Nature of Electricity

The ancient Greeks knew that when a piece of amber is rubbed with wool or fur it achieves the power of attracting light objects, such as feathers or bits of straw. The phenomenon was studied by William Gilbert (1540-1603), Queen Elizabeth I's physician, who invented the adjective *electric* to describe the force of attraction, after the Greek word *elektron*, meaning amber. Gilbert and many other scientists, including Benjamin Franklin (1706–1790), investigated electric phenomena, and during the nineteenth century many discoveries were made about the nature of electricity and of magnetism (which is closely related to electricity).

It was found that if a rod of sealing wax, which behaves in the same way as amber, is rubbed with a woolen cloth, and a rod of glass is rubbed with a silken cloth, an electric spark will pass between the sealing-wax rod and the glass rod when they are brought near one another. Moreover, it was found that a force of attraction operates between them. If the sealing-wax rod that has been electrically charged by rubbing with the woolen cloth is suspended from a thread and the charged glass rod is brought near one end of it, this end will turn toward the glass rod. An electrified sealing-wax rod is repelled, however, by a similar sealing-wax rod, and also an electrified glass rod is repelled by a similar glass rod. The ideas were developed that there are two kinds of electricity, and that opposite kinds of electricity attract one another, whereas similar kinds repel one another. Franklin assumed that when a glass rod is rubbed with a silken cloth something is transferred from the cloth to the glass rod, and he described the glass rod as positively charged, meaning that it had an excess of the electric fluid that had been transferred to it by the rubbing, with the cloth having a deficiency, and hence being negatively charged. He pointed out that he did not really know whether the electric fluid had been transferred from the silken cloth to the glass rod or from the glass rod to the silken cloth, and that accordingly the decision to describe electricity on the charged glass rod as positive (involving an excess of electric fluid) was an

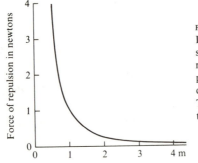

FIGURE 3-1
Diagram illustrating the inverse-square law of electrostatic repulsion: the force is inversely proportional to the square of the distance between the charges. The numbers correspond to two 1-stoney charges.

arbitrary one. We now know, in fact, that when the glass rod is rubbed with a silken cloth negatively charged particles, the electrons, are transferred from the glass rod to the silken cloth, and that Franklin thus made the wrong decision.

Careful experiments carried out by Joseph Priestley (1733–1804), Henry Cavendish (1731–1810), and the French physicist Charles Augustin Coulomb (1736–1806) led to the discovery that the force of attraction between two opposite charges of electricity is inversely proportional to the square of the distance between them, and directly proportional to the magnitudes of the charges—that is, to the amount of positive electricity, $+e_1$, and the amount of negative electricity, $-e_2$. This relation can be expressed by the following equation:

$$\text{force of attraction} = \frac{k \times e_1 \times e_2}{r^2} \qquad (3\text{-}1)$$

In this equation r is the distance between the two charges, considered to be located at points, and k is a constant that has the value 1 for a suitably chosen unit of electric charge, described in the following paragraphs.

Units of Electric Charge

The International System unit of electric charge is the *coulomb* (C). The coulomb is one ampere-second, and the ampere is defined as that current in each of two infinitely long parallel wires 1 m apart that causes an electromagnetic force of 2×10^{-7} N per meter of its length to act on each wire. (This unit, which is the practical unit of electric current, is retained in the International System, even though it is less rational than a unit ten times as large.)

A unit of electric charge that is especially useful in discussing atomic and molecular structure is the electrostatic unit, defined in such a way that the electrostatic force between two such units one meter apart is 1 newton. We shall use the symbol S for this unit, and the name "stoney."* One stoney is $\sqrt{10} \times 10^3 c^{-1}$ C $= 1.054822 \times 10^{-5}$ coulomb. (Here c is the velocity of light.)

Force and Potential Energy

The force of repulsion between two positive charges, each of one stoney, as a function of the distance apart of the charges is represented in Figure 3-1. With use of Coulomb's law for the force as a function of distance, as represented in this figure, we can calculate the amount of work that must

---

*The name "stoney" for this unit is suggested in honor of G. Johnstone Stoney (Section 3-2). The corresponding unit in the c.g.s. system is the statcoulomb. The electrostatic force between two 1-statcoulomb point charges 1 cm apart is 1 dyne. One coulomb is equal to $c/10$ statcoulombs.

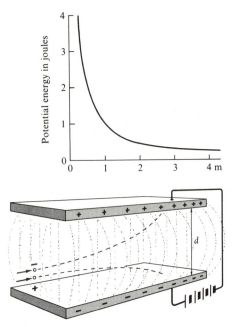

FIGURE 3-2
Diagram showing the dependence of the potential energy of two electric charges (each 1 S) on the distance between them; the potential energy is inversely proportional to the first power of the distance.

FIGURE 3-3
The motion of an electrically charged particle in the uniform electric field between charged plates.

be done to bring the two unit charges from an infinite distance apart to the distance r apart. The relation between work and force is

$$\text{work} = \text{force} \times \text{distance} \tag{3-2}$$

It is found that the work that must be done to bring two unit charges (each 1 S) from infinity to the distance r is $1/r$ J. This work is stored in the system of the two charges the distance r apart; by making use of the force of repulsion between the charges, they would do the amount of work $1/r$ J as they separate from one another. The stored-up capacity for doing work is called the *potential energy* of the system. It is represented in Figure 3-2.

We see that the potential energy of two unit charges 1 m apart is 1 J, and potential energy of the unit charges $\frac{1}{2}$ m apart is 2 J. Also, it is noteworthy that the slope of the curve of potential energy against distance between the charges, as given in Figure 3-2, is equal to the force of repulsion between the charges;

$$\text{force} = -\frac{d\,(\text{potential energy})}{dr}.$$

The general expression for the mutual potential energy of two electric charges q_1 and q_2 is

$$\text{potential energy} = \frac{q_1 q_2}{r} \text{ J} \tag{3-3}$$

where q_1, q_2 are in stoneys and r is in meters. The potential energy is positive if the charges q_1 and q_2 have the same sign, and negative if they have opposite signs. If q_1 and q_2 were given in coulombs, we would have to write

$$\text{potential energy} = 8.9876 \times 10^9 \frac{q_1 q_2}{r} \text{ J} \qquad (3\text{-}4)$$

We can see the advantage of using the charge unit S rather than C in calculation of electrostatic interactions in the relative simplicity of Equation 3-3.

The Interaction of an Electric Charge with Electric and Magnetic Fields

An electric charge is said to be surrounded by an *electric field*, which exercises a force on other electric charges in its neighborhood. (Sometimes the electric field is called the *electrostatic field*.) The strength of an electric field is measured by determining the force that operates on a unit electric charge; the strength of the field in electrostatic units is equal to the force in newtons that operates on a charge of 1 stoney. The field due to a charge q at a distance r from the center of the charge is equal to q/r^2, and the field is directed away from the charge (for a positive charge).

The field can be conveniently indicated by a method invented by Michael Faraday (1791–1867), who made many discoveries in electricity and magnetism as well as in chemistry. Faraday assumed that *lines of force* emanate from every charged body. The direction of the lines of force at any point coincides with the direction of the electric field at that point, and the number of lines of force per square centimeter of a cross section perpendicular to this direction is proportional to the strength of the field.

An important situation is that shown in Figure 3-3, in which two large parallel plates of metal are held a small, constant distance from one another. One of these parallel plates is charged positively and the other is charged negatively. Except near the edges, the lines of force are straight lines between the plates, and with uniform density. Accordingly, in the region between the plates the electric field is constant.

The amount of work required to move a unit positive charge from the negative plate to the positive plate, through the distance d, is equal to the product of the field strength and the distance d, in meters. This quantity is called the *potential difference* of the upper plate and the lower plate.

The International System unit of electric potential difference is the volt (V). Its value is such that the energy involved in moving a charge of one coulomb through a potential difference of one volt is one joule: $1 \text{ C V} = 1 \text{ J}$.

The force exerted on a charged particle by a constant electric field has the same effect as the force exerted on a mass by the gravitational field of the earth, which is fairly constant in the neighborhood of the surface of the earth. A positively charged particle projected into the region between

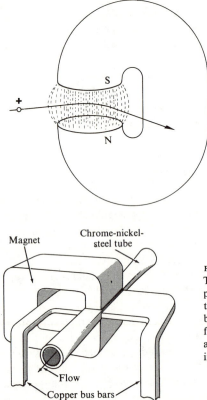

FIGURE 3-4
The path of a moving electric
charge in a magnetic field.

FIGURE 3-5
The construction of an electromagnetic
pump. When electric current flows
through the liquid alloy in the region
between the poles of the magnet, a
force operates on the moving charges,
and hence on the alloy, in the direction
indicated.

the plates, as indicated in Figure 3-3, would fall to the bottom plate along
the parabolic course represented by the dashed line, in the same way that
a projectile fired horizontally would fall toward the surface of the earth.

No force is exerted on a stationary electric charge by a constant *magnetic field*. The direction of the force exerted on an electric charge that is
traversing a magnetic field is at right angles to the direction of motion of
the charged particle and also at right angles to the direction of the magnetic
field. This is illustrated in Figure 3-4. The poles of the magnet are marked
N (north-seeking pole) and *S* (south-seeking pole); the lines of force are
indicated as going from the north-seeking pole to the south-seeking pole.
A positively charged particle is shown as moving through the field from
the left to the right. The nature of electricity and magnetism is such that a
force operates that is proportional to the strength of the magnetic field, the
quantity of electric charge on the particle, and the speed of the particle;
the direction of this force is at right angles to the plane formed by the

direction of motion of the moving particle and the direction of the lines of force of the magnetic field, its sense being out of the plane of the paper. This causes the moving positively charged particle to be deflected to the front, as indicated in the drawing.

The Explanation of Electricity and Magnetism

You may well ask how scientists explain the fact that an electron is repelled by another electron, or that in general two electrically charged bodies repel or attract one another in accordance with Coulomb's law; or how scientists explain the still more extraordinary and unexpected fact that an electric charge that is moving through a magnetic field is pushed to one side by interaction of its charge with the field. The answer to these questions is that there is no explanation. These properties of electric charges and of electric and magnetic fields are a part of the world in which we live.

It has been found that observed electric and magnetic phenomena are compatible with a general theory that is expressed in a set of equations, called Maxwell's equations, formulated in 1873 by the British physicist James Clerk Maxwell (1831–1879). These equations, like Newton's laws of motion, summarize a tremendous number of phenomena in a beautiful and satisfying way, but they lie outside the scope of this book.

The Electromagnetic Pump

An interesting device that makes use of the force that acts on a moving electric charge in a magnetic field is the electromagnetic pump, which has been devised to pump the liquid metal (such as sodium-potassium alloy) that is used as a heat transfer agent to carry the heat from a nuclear reactor to an external boiler. The liquid metal is in a pipe (made of a nickel-chromium alloy steel) between the poles of a large permanent magnet. An electric current of about 20,000 amperes, at 1 volt potential difference, passes through the liquid metal, as indicated in Figure 3-5. The sideways force that is exerted on the electrons moving through the magnetic field causes the metal to flow in the indicated direction, at right angles to the direction of the electric current and the direction of the magnetic field.

3-2. The Discovery of the Electron

It was discovered in the early part of the nineteenth century that compounds can be decomposed by an electric current, and the quantity of electricity needed to liberate a certain amount of an element from one of its compounds was measured. It was found that, for example, 96,490 C of

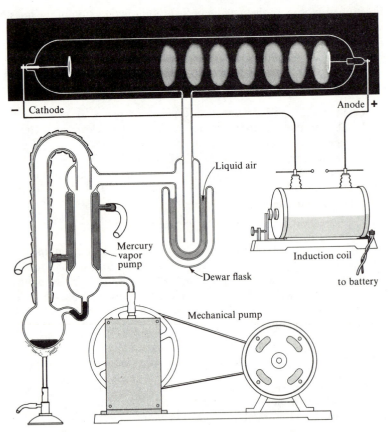

FIGURE 3-6
Apparatus used to observe the discharge of electricity in a gas at low pressure.
The dark space around the cathode is called the Crookes dark space; at still lower
pressures the Crookes dark space fills the whole tube.

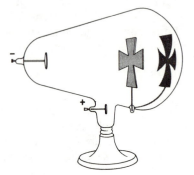

FIGURE 3-7
Experiment showing that cathode rays,
starting from the cathode at the left,
move through the Crookes tube in
straight lines.

electricity is required to liberate 1 g of hydrogen from water (we shall discuss these matters in detail in Chapter 15). After thinking about these phenomena, Dr. G. Johnstone Stoney (1826–1911), an English scientist, stated as early as 1874 that they indicate that electricity exists in discrete units, and that the units are associated with atoms, which also carry an equal charge of opposite sign. In 1891 he emphasized this point and suggested the name *electron* for the postulated unit of electricity.

At that time experiments were being carried on by physicists on the conduction of electricity through gases; these experiments after some years (in 1897) led Joseph John Thomson (1856–1940), then director of the Cavendish Laboratory at Cambridge University, to the firm conclusion that the electron exists and to the determination of some of its properties.

If a glass tube about 50 cm long is fitted with electrodes and a potential difference of about 10,000 volts is applied between the two electrodes, as indicated in Figure 3-6, no electric discharge occurs until part of the air or other gas has been pumped out of the tube. Discharge begins to occur when the pressure has become about 0.01 atm. The nature of the discharge, which is associated with the emission of light by the gas in the tube, changes as the pressure becomes lower. When the pressure is somewhat less than 0.001 atm a dark space appears in the neighborhood of the cathode, and striations are observed in the rest of the tube. At still lower pressures the dark space increases in size, until at about 10^{-5} atm it fills the whole tube. At this pressure no light is given out by the residual gas within the tube, but the glass of the tube itself glows (fluoresces) with a faint greenish light.

It was discovered that the greenish light is due to bombardment of the glass by rays liberated at the cathode (called *cathode rays*), and traveling in straight lines. This was revealed by the experiment illustrated in Figure 3-7; it was observed that an object placed within the discharge tube casts a shadow on the glass, which fluoresces except in the region of this shadow.

It was shown by the French physicist Jean Perrin (1870-1942) in 1895 that these cathode rays consist of negatively charged particles. His experiment is illustrated in Figure 3-8. He introduced in the tube a shield with a slit, such as to form a beam of cathode rays. He also placed a fluorescent screen in the tube, so that the path of the beam could be followed by the trace of the fluorescence. When a magnet was placed near the tube, in such a way that the lines of force of the magnetic field were perpendicular to the direction of motion of the cathode rays, the beam was observed to be deflected in the direction corresponding to the presence of a negative charge on the particles.

The amount of deflection could not be used to determine the ratio of charge to mass of the particles, however, because of lack of knowledge of

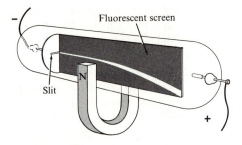

FIGURE 3-8
Experiment showing that the cathode rays
have a negative charge.

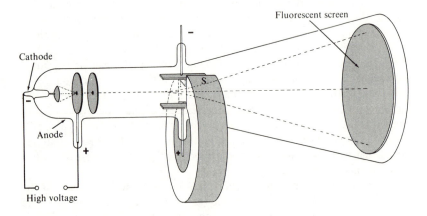

FIGURE 3-9
The apparatus used by J. J. Thomson to determine the ratio of electric charge to
mass of the cathode rays, through the simultaneous deflection of the rays by an
electric field and a magnetic field.

the speed with which they were moving. J. J. Thomson then carried out an
experiment that permitted him to determine the speed. This was done
with the use of the apparatus shown in Figure 3-9, in which the cathode
rays could be affected simultaneously by a magnetic field and an electric
field. For a fixed value of the magnetic field strength, the strength of the
electric field could be varied until the beam struck the center of the
fluorescent screen, the same position at which it would strike the screen
in the absence of both the magnetic field and the electric field. Under these
conditions the force due to the magnetic field is just balanced by the force
due to the electrostatic field. The force due to the magnetic field is, how-

ever, equal in magnitude to *Hev*, in which *H* is the strength of the magnetic field, *e* the electric charge, and *v* the velocity. The force due to the electric field is *Ee*, in which *E* is the field strength. When these are equal, we have

$$Hev = Ee$$

The electric charge of the particles, *e*, cancels out, and *v*, the velocity of the particles, is then given by the equation

$$v = \frac{E}{H}$$

By inserting the values of *E* and *H* Thomson was able to calculate the velocity *v*. The values that he obtained were found to be dependent only upon the voltage at which the Crookes tube was operated; they were of the order of 6×10^7 m s^{-1}—that is, about $\frac{1}{5}$ the velocity of light.

Knowing the velocity of the cathode rays, he could then determine the ratio of charge to mass of the particles by measuring the deflection of the beam produced by either the electrostatic field alone or the magnetic field alone. From these experiments he obtained the value $e/m = 1 \times 10^8$ coulombs per gram. To within his experimental error (involving a possible factor of about 2) the same value was obtained whether the tube had been filled originally with air, hydrogen, carbon dioxide, or methyl iodide; also, the same value was obtained whether the cathodes were made of platinum, aluminum, copper, iron, lead, silver, tin, or zinc.

The value of e/m obtained from this experiment convinced Thomson that the particles constitute a form of matter different from the ordinary forms of matter. His argument involved comparison with the value of e/m for hydrogen or other substances as obtained by electrolysis. It was mentioned above that 1 g of hydrogen is produced in the electrolysis of water by 96,490 coulombs; hence, if Stoney's suggestion that one elementary charge is to be associated with each atom of hydrogen is correct, the ratio e/M for hydrogen is 96,490 coulombs per gram. If the same electric charge is involved in both cases, then the mass of the particles present in the cathode rays must be about 1000 times smaller than the mass of a hydrogen atom. Later experiments showed that in his original work in 1897 Thomson had obtained a value too small by a quotient of nearly two, and accordingly that the mass of the cathode-ray particle (the electron) is approximately 1/2000 of the mass of the hydrogen atom (actually 1/1837).

Although other investigators had carried out important experiments on cathode rays, the quantitative experiments by Thomson provided the first convincing evidence that these rays consist of particles (electrons) much lighter than atoms, and Thomson is hence given the credit for discovering the electron.

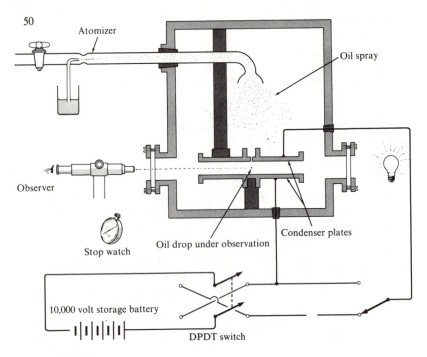

FIGURE 3-10
A diagram of the apparatus used by R. A. Millikan in determining the charge of the electron by the oil-drop method.

Measurement of the Charge of the Electron

After the discovery of the electron by J. J. Thomson, many investigators worked on the problem of determining either e or m separately. The American physicist R. A. Millikan (1868–1953), who began his experiments in 1906, was the most successful; by means of his oil-drop experiment he determined the value of e in 1909 to within 1%.

The apparatus that he used is illustrated in Figure 3-10. Small drops of oil are formed by a sprayer, and some of them attach themselves to ions that have been produced in the air by irradiation with a beam of x-rays. The experimenter watches one of the small oil drops through a microscope and measures the rate at which it falls in the earth's gravitational field. Because of the large frictional force of the air, the drops reach a terminal velocity at which the frictional force is just equal to the gravitational force. This terminal velocity depends on the size and mass of the drop, and the mass can be calculated if the density of the oil and the viscosity of air are known. When the electric field is turned on, by charging the plates above and below the region where the oil drops are suspended, some of the drops (which carry no electric charge) continue to fall as before; others, carrying electric charges, change their speed and may rise, being pulled up by the attraction of the electric charge by the oppositely charged upper plate in the apparatus. The rate of rise of a drop that has been watched falling is then observed. From this rate, and its rate of fall when the electric field was turned off, the magnitude of the electric charge on the drop can be calculated. Millikan was able to measure the rates of rise and fall of a single drop many times. In various experiments with different drops, he obtained values such as the following:

$$e = 1.6 \times 10^{-19} \text{ C}$$
$$e = 3.2 \times 10^{-19} \text{ C} = 2 \times 1.6 \times 10^{-19} \text{ C}$$
$$e = 8.0 \times 10^{-19} \text{ C} = 5 \times 1.6 \times 10^{-19} \text{ C}$$

All of these values have a common factor, 1.6×10^{-19} C. Millikan accordingly concluded that this is the smallest electric charge that can occur under these conditions. The average value of his measurements was 1.591×10^{-19} C. This was accepted as the charge of the electron. It was found later (1935) to be nearly 1% low, mainly because of an error in the value used for the viscosity of air.

Since Millikan carried out his work, a number of other methods have been developed for determining the charge of the electron, and its value is now known to an accuracy of about 0.001% (1 in 10^5).

It is convenient to have available the value of the electronic charge in two different units, and to use two symbols for the electronic charge. The value in coulombs, $e = 0.160206 \times 10^{-18}$ C, is useful in that eV is the energy in joules. The value in stoneys, $\epsilon = 15.1880 \times 10^{-15}$ S, is useful in that ϵ^2/r is the electrostatic energy in joules.*

The mass of the electron is 0.91083×10^{-30} kg. It is about 1/1837 the mass of the hydrogen atom.

Example 3-1.   If, in one·experiment, J. J. Thomson accelerated the electrons in his apparatus by applying a potential difference of 10,000 V between the cathode and the anode, and if an electron that escaped from the cathode was accelerated by falling through the entire field between cathode and anode, with what velocity would it be moving?

Solution.   The strength of the electric field between the cathode and the anode is $10{,}000/d$ V m^{-1}, where d is the distance between the cathode and the anode in meters. The force acting on the charge e is $10{,}000\ e/d$, and the kinetic energy imparted to the electron when this force operates over the distance d is $d \times 10{,}000\ e/d = 10{,}000\ e$ J. (Note that in general,

---

*We have followed the convention of writing e as a positive quantity. The charge of the electron is $-e$. Also, we usually shall express values of constants with a factor $10^{\pm n}$, with n a multiple of three, in accordance with the convention of the International System.

whether the field is uniform or nonuniform, the energy of moving a charge e through the potential difference V is eV.) From the value of e we see that the electron has been given the kinetic energy $10,000 \times 0.1602 \times 10^{-18}$ J $= 0.1602 \times 10^{-14}$ J. We place this quantity equal to $\frac{1}{2}mv^2$, with $m = 0.91083 \times 10^{-30}$ kg, and solve for v, obtaining the value 0.593×10^8 m sec^{-1}. Thus we see that by falling through a potential difference of 10,000 V the electron has been accelerated to a velocity about one-fifth the velocity of light.

J. J. Thomson was surprised that all of the particles in the cathode-ray beam were moving with practically the same velocity. The reason for this is that all of the electrons had been liberated at the cathode and had fallen through the same potential difference. If there had been a large amount of gas in the tube many of the electrons might have collided with gas molecules and imparted some of their energy to them, in which case they would not have had the same energy.

The Relativistic Correction

There is a small error in the foregoing calculation, resulting from use of Newton's laws of motion rather than the relativistic laws of motion. The special theory of relativity states that a particle with rest-mass m_0 moving with velocity v relative to the observer has momentum $m_0v/\sqrt{1 - \beta^2}$, rather than mv, relative to the observer. Here $\beta = v/c$ is the ratio of the velocity of the particle to the velocity of light. The kinetic energy of the particle is given by the equation

$$\text{kinetic energy} = m_0 c^2 \left(\frac{1}{\sqrt{1 - \beta^2}} - 1 \right) \tag{3-5}$$

The velocity of a 10,000-V electron is calculated by this equation to be 0.585×10^8 m s^{-1}, 1.4% less than given by the nonrelativistic calculation. The effective inertial mass of a 10,000-V electron,

$$m = \frac{m_0}{\sqrt{1 - \beta^2}} \tag{3-6}$$

is 1.96% greater than m_0. One of the early experimental tests of the theory of relativity was the measurement of the effective mass of fast electrons under the influence of electric and magnetic fields.

The Flow of Electricity in a Metal

A direct current of electricity passing along a copper wire is a *flow of electrons* along the wire. In a metal or similar conductor of electricity there are electrons that have considerable freedom of motion and that move along between the atoms of the metal when an electric potential difference is applied.

The flow of electricity along a wire may be compared with the flow of water in a pipe. *Quantity* of water is measured in liters; quantity of electricity is usually measured either in coulombs or in stoneys. *Rate of flow*, or *current*, of water, the quantity passing a given point of the pipe in a unit of time, is measured in liters per second; current of electricity is measured in *amperes* (coulombs per second). The rate of flow of water in a pipe depends on the *pressure difference* at the ends of the pipe, with atmospheres or kilograms per square meter as units. The current of electricity in the wire depends on the *electric pressure difference* or *potential difference* or *voltage drop* between its ends, which is usually measured in volts.

An electric generator is essentially an electron pump, which pumps electrons under pressure out of one wire and into another. A generator of direct current pumps electrons continually in the same direction, and one of alternating current reverses its pumping direction regularly, thus building up electron pressure first in one direction and then in the other. A 60-cycle generator reverses its pumping direction 120 times per second.

3-3. The Discovery of X-rays and Radioactivity

The period of sixteen years beginning in 1895 was a period of great discovery. X-rays were discovered in 1895; radioactivity was discovered in 1896, and the new radioactive elements polonium and radium were isolated in the same year; the electron was discovered in 1897; the quantum theory was discovered in 1900; the photon was discovered in 1905; and atomic nuclei were discovered in 1911.

Wilhelm Konrad Röntgen (1845–1923), professor of physics in the University of Würzburg, reported in 1895 that he had discovered a new kind of rays, which he called x-rays. He began his paper with a sentence that read essentially as follows: "If the discharge of a large induction coil is allowed to pass through a Crookes tube or similar apparatus, and the tube is covered with a fairly closely fitting mantle of thin, black cardboard, it is seen that light is emitted by a fluorescent screen in the neighborhood." He showed that the x-rays that he had discovered could penetrate matter that is impervious to ordinary light, and could produce fluorescence in various substances, such as glass and calcite. He found that a photographic plate is blackened by the radiation, that the rays are not deflected by a magnet, and that they appear to come from the place in the vacuum tube that is struck by the cathode rays. Within a few weeks after the announcement of this great discovery x-rays were being used by physicians for the investigation of patients.

For nearly twenty years there was doubt as to the nature of x-rays; some scientists thought that their properties could be accounted for best on a corpuscular basis, whereas others thought x-rays to be similar to

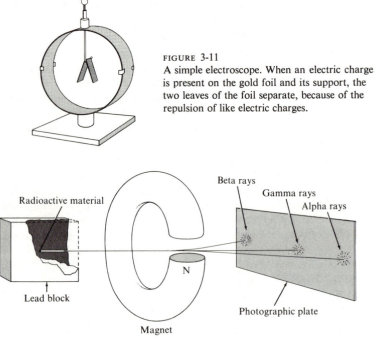

FIGURE 3-11
A simple electroscope. When an electric charge is present on the gold foil and its support, the two leaves of the foil separate, because of the repulsion of like electric charges.

FIGURE 3-12
The deflection of alpha rays and beta rays by a magnetic field.

ordinary light, but with very short wavelengths. The second alternative was shown to be correct when, in 1912, the diffraction of x-rays by crystals was observed (Section 3-7).

Soon after the discovery of x-rays the French mathematician Henri Poincaré suggested at a meeting of the French Academy of Sciences that the x-radiation might be connected with the fluorescence shown by the glass of the Crookes tube at the place where the x-rays are emitted. This suggestion stimulated the French physicist Henri Becquerel (1852–1908) to investigate some fluorescent minerals. He was professor of physics at the Museum of Natural History in Paris, as successor to his father and grandfather. His father had collected many fluorescent minerals, which were available in the museum. Becquerel selected a uranium salt, exposed it to sunlight until it showed strong fluorescence, and then placed it against a photographic plate wrapped in black paper. He found upon developing the plate that it had been blackened, which seemed to confirm Poincaré's idea. Becquerel then found, however, that the salts of uranium would blacken a photographic plate wrapped in paper even though they had not been exposed to sunlight to make them fluorescent, and he showed that the effect could be observed with any compound of uranium. He also found that the radiation produced by the uranium compounds could, like x-rays, discharge an electroscope, by ionizing the air and making it conductive.

Marie Sklodowska Curie (1867–1934) then began a systematic investigation of "Becquerel radiation," using the electroscope technique (Figure 3-11), to see if substances other than uranium exhibited similar properties; this work was the subject of her doctoral dissertation. She found that natural pitchblende, an ore of uranium, is several times more active than purified uranium oxide, and, with her husband, Professor Pierre Curie (1859–1906), she began to separate pitchblende into fractions and to determine their activity. She isolated a bismuth sulfide fraction that was 400 times more active than uranium. Since pure bismuth sulfide is not radioactive, she assumed that a new, strongly radioactive element, similar in chemical properties to bismuth, was present as a contaminant. This element, which she named polonium, was the first element discovered through its properties of radioactivity. In the same year, 1896, the Curies isolated an active barium chloride fraction containing another new element, which they named radium.

Becquerel also continued to study the properties of his new radiation, and he was able to make use of the strongly radiating preparations produced by the Curies. In 1899 Becquerel showed that the radiation from radium could be deflected by a magnet. Also in 1899, a young physicist from New Zealand, Ernest Rutherford (1871–1937), working in the Cavendish Laboratory in Cambridge under J. J. Thomson, reported that the radiation from uranium is of at least two distinct types, which he called α (alpha) radiation and β (beta) radiation. A French investigator, Paul Villard (1860–1934), soon reported that a third kind of radiation, γ (gamma) radiation, is also emitted.

Alpha, Beta, and Gamma Rays

The experiments showing the presence of three kinds of rays emitted by natural radioactive materials are illustrated by Figure 3-12. The rays, collimated by passing along a narrow hole in a lead block, traverse a strong magnetic field. They are differently affected, showing them to have different electric charges. Alpha rays are positively charged; further studies, made by Rutherford, showed them to be the positive parts of helium atoms, moving at high speed. Beta rays are electrons, also moving at high speeds. Gamma rays are similar to visible light, but with very short wavelengths—they are identical with x-rays produced in an x-ray tube operated at very high voltage.

Rutherford identified the positively charged alpha particles with helium atoms by an experiment in which he allowed the alpha particles to be shot

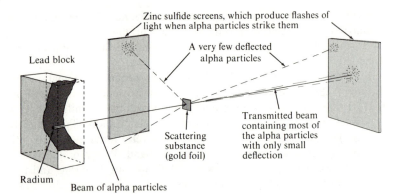

FIGURE 3-13

A diagram representing the Rutherford-Geiger-Marsden experiment, which showed that atoms contain very small, heavy atomic nuclei.

through a thin metal foil into a chamber, and later was able to show the presence of helium in the chamber. He could, moreover, show that the amount of helium in the chamber equaled the number of alpha particles penetrating the foil.

The nature of the nuclear processes that occur in radioactive elements will be discussed in Chapter 26.

3-4.   The Nuclei of Atoms

In 1906 Ernest Rutherford, then professor of physics in McGill University (after 1907 in the University of Manchester, England), carried out some experiments on the scattering of alpha rays by a thin sheet of mica. Similar experiments with gold foil were then carried out by H. Geiger and E. Marsden, in Rutherford's Manchester laboratory, and the results were published in 1909. Two years later, in 1911, Rutherford published his interpretation of the experimental results as showing that most of the mass of an atom is concentrated in a single particle that is very small compared with the atom itself.

The experimental method is indicated by the drawing in Figure 3-13. A piece of radium emits alpha particles in all directions. A set of slits is constructed to define a beam of the alpha particles. This beam of alpha particles then passes through the metal foil (labeled "scattering substance" in the figure), and the direction in which the alpha particles continue their motion is observed. The direction of motion of the alpha particles can be detected by use of a screen coated with zinc sulfide. When an alpha particle strikes this screen a flash of light is sent out.

If the atoms bombarded with alpha particles were solid throughout their volume, nearly every alpha particle should bounce off an atom and change direction slightly. It was observed that most of the particles passed through a foil with only very small deflection, about 1°, with a distribution in angle corresponding to the idea that each particle had suffered a large number of encounters, each of which produced a small deflection in a random direction. However, a very small number, one in 100,000 for a foil 0.5 μm thick, showed a great deflection, often more than 90°, which could not be accounted for in this way. When foil twice as thick was taken, it was found that about twice as many alpha particles showed deflection through large angles, with most of them still passing nearly straight through.

It is clear that these experimental results can be understood if the assumption is made that most of the mass of the atom is concentrated into a very small particle, which Rutherford called the *atomic nucleus*. If the alpha particle were also very small, then the chance of collision of these two small particles as the alpha particle passed through the atom would be small. Rutherford concluded that the heavy nucleus has a cross-sectional area less than 10^{-8} as great as the cross-sectional area of the atoms, and hence that the ratio of the maximum diameter of the nucleus to the diameter of the atom is the square root of this factor—$(10^{-8})^{1/2} = 10^{-4}$. Since the diameter of the gold atom is about 3×10^{-10} m, the diameter of the nucleus is indicated to be less than 3×10^{-14} m.

The picture of the atom that has been developed from this experiment and similar experiments is indeed an extraordinary one. If we could magnify a piece of gold leaf by the linear factor 10^9, we would see it as an immense pile of atoms about 30 cm in diameter, each atom thus being somewhat larger than a basketball. Nearly the entire mass of each atom would, however, be concentrated in a single particle, the nucleus, less than 0.003 cm in diameter, like an extremely small grain of sand. This nucleus would be surrounded by electrons, moving very rapidly about. Rutherford's experiment would consist in shooting through a pile of these basketball atoms a stream of minute grains of sand, each of which would continue in a straight line unless it happened to collide with one of the minute grains of sand representing the nuclei of the atoms. It is obvious that the chance of such a collision would be very small. (Note that in the Rutherford experiment the alpha particles are not deflected by the electrons because they are very much heavier than the electrons.)

The proton, which is the nucleus of the hydrogen atom, has the same electric charge as the electron, but with opposite sign (positive instead of negative). The nuclei of other atoms have positive charges that are multiples of this fundamental charge. The structure of these atomic nuclei will be discussed in the following chapter.

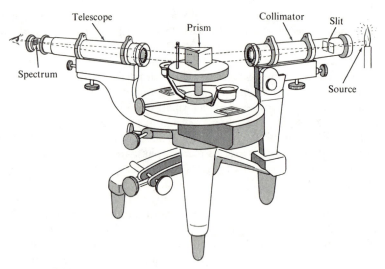

FIGURE 3-14
A simple spectroscope. The light from the source is refracted into a spectrum by use of a glass prism; it could instead be diffracted into a spectrum by use of a ruled grating, in place of the prism.

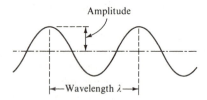

FIGURE 3-15
Diagram representing wave motion.

3-5. The Birth of the Quantum Theory

The quantum theory is a child of the twentieth century: it was announced, in its primitive form, in 1900, by its discoverer, Max Planck (1858–1947), a professor in the University of Berlin. He was led to the theory by the consideration of the nature of the radiation given out by a hot solid body.

Waves and Their Interference

During the second half of the 17th century, Isaac Newton, with the use of a glass prism (Figure 3-14), resolved a fine beam of sunlight into its component colors: violet, indigo, blue, green, yellow, orange, red. He found that with a second prism he could recombine the whole spectrum into a beam of white light, but that if he selected one of the colors no further treatment could change it. He also studied the colors of soap bubbles and of a slightly convex lens in contact with a plane glass surface (Newton's rings). Newton recognized that these colors (interference colors) could be explained by a wave theory of light, but he felt that the observed rectilinear propagation of light was most simply explained by the assumption that light consists of particles (corpuscles). He tried, but without success, to explain interference phenomena by attributing suitable properties to the corpuscles. Other investigators, especially Christian Huygens (1629–1695), Augustin Jean Fresnel (1788–1827), and Thomas Young (1773–1829), developed a thoroughly sound foundation for the wave nature of light. James Clerk Maxwell in 1873 deduced from his electromagnetic field equations that electromagnetic waves with the properties of light would be produced by moving an electric charge back and forth in an oscillatory manner, and confirmatory experiments were carried out in 1888 by the German physicist Heinrich Hertz (1857–1894). The electromagnetic waves are described as involving an oscillating electric field and an oscillating magnetic field.

The nature of the wave motion is represented by the sine curve shown in Figure 3-15. This curve might represent, for example, the instantaneous contour of waves on the surface of the ocean. The distance between one crest and an adjacent crest is called the *wavelength*, usually represented by the symbol λ (Greek letter lambda). The height of the crest, which is also the depth of the trough, with reference to the average level is called the *amplitude* of the wave. If the waves are moving with the velocity c m s^{-1}, the frequency of the waves, represented by the symbol ν (Greek letter nu), is equal to c/λ; that is, it is the number of waves that pass by a fixed point in unit time (1 s). The dimensions of the wavelength are those of length. The dimensions of frequency, number of waves per second, are [time^{-1}]. We see that the product of wavelength and frequency has the dimensions [length][time^{-1}]— that is, the dimensions of velocity. The equation connecting wavelength λ, frequency ν, and velocity c is

$$\lambda\nu = c \qquad (3\text{-}7)$$

In the case of a light wave the sine curve shown in Figure 3-15 is considered to represent the magnitude of the electric field in space. The electric field of a light wave is perpendicular to the direction of motion of the beam of light.

The phenomenon of interference of waves is used to determine the wavelength of light waves. This phenomenon can be illustrated by Figures 3-16 and 3-17. In 3-16 there is shown a set of water waves approaching a

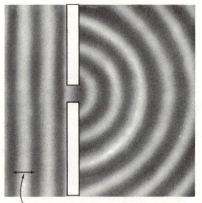

Wavelength λ

FIGURE 3-16
Diagram representing waves on
the surface of water, from the
left, striking a pier; waves
propagated through the opening
in the pier then spread out in
circles.

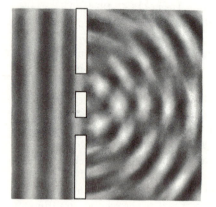

FIGURE 3-17
The interference and reinforcement of two sets of circular
waves, from two openings.

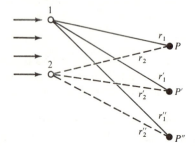

FIGURE 3-18
Diagram illustrating the conditions for interference and
reinforcement of circular waves
from two points, 1 and 2. The
incident waves are moving
toward the right, as shown by
the arrows.

jetty in which there is a small opening. The waves that strike the jetty dissipate their energy among the rocks of the jetty, but the part of the waves that strikes the opening causes a disturbance on the other side of the jetty. This disturbance is in the form of a set of circular waves, that spread out from the opening of the jetty. The wavelength of these circular waves is the same as the wavelength of the incident water waves. When light or x-rays strike atoms, part of the energy of the incident light is scattered by the atoms. Each atom scatters a set of circular waves. If two atoms that are excited by the same incident waves scatter light, as illustrated in Figure 3-17, there are certain directions in which the circular waves (spherical waves in the case of atoms in three-dimensional space) from the two scattering centers reinforce one another, producing waves with twice the amplitude of either set, and other directions in which the trough of one set of waves coincides with the crest of the other set, and interference occurs. The directions of reinforcement and interference for two sets of circular waves are shown in the figure.

It is easy to calculate the angles at which reinforcement and interference would occur, in terms of the distance between the two scattering centers and the wavelength of the waves. The way in which the calculation is made is shown in Figure 3-18. Here r_1 is the distance from the first scattering center, and r_2 the distance from the second scattering center. At all points in the median plane these two distances are equal. Accordingly, a crest of a wave of the first set will reach a distant point P at the same time as a crest of a wave of the second set, and there will be reinforcement at this point. The point P'' lies at such distances r_1'' and r_2'' that $r_1'' - r_2''$ is just equal to one wavelength of the waves. Accordingly, the crest of a wave from the first scattering center will reach P'' at the same time as the crest of the preceding wave from the second scattering center, and again there will be reinforcement. At the intermediate point P', however, the difference $r_1' - r_2'$ is just one-half of a wavelength. The crest of a wave from one scattering center will coincide with the trough of a wave from another scattering center, and there will be interference.

A diffraction grating spectroscope can be made by replacing the glass prism shown in Figure 3-14 by thin film of plastic with a set of parallel raised lines on it (a replica grating, formed by casting the film on a metal plate in which grooves have been cut with a diamond point). If the plane of the replica is placed perpendicular to the incident beam of light, the angles of reinforcement φ for light with wavelength λ are found by the method of the preceding paragraph to be given by the equation

$$n\lambda = d \sin \varphi \qquad (3\text{-}8)$$

In this equation d is the distance between rulings, φ the angle between the incident beam and the diffracted beam, and n is an integer (the order of

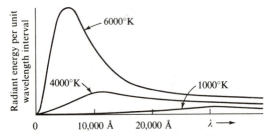

FIGURE 3-19
Curves showing the distribution of energy as a function
of wavelength in the light in equilibrium with a hot body,
at three different temperatures. Through the analysis
of experimental curves of this sort Max Planck was led
to the discovery of the quantum theory, in 1900.

diffraction, which is the number of wavelengths difference in path length
for adjacent rulings the distance *d* apart). Wavelengths found in this way
range from about 400 nm for violet to 700 nm for red light (see Figure 21-1).

The Emission of Light

When gases are heated or are excited by the passage of an electric spark,
the atoms and molecules in the gases emit light of definite wavelengths.
The light that is emitted by an atom or molecule under these conditions is
said to constitute its *emission spectrum*. The emission spectra of the alkali
metals, mercury, and neon are shown in Figure 21-1. The emission
spectra of elements, especially of the metals, can be used for identifying
them, and *spectroscopic chemical analysis* is an important technique of
analytical chemistry. When a solid body is heated it emits light with a
distribution of intensity with wavelength characteristic of the chemical
composition of the body. During the second half of the 19th century it was
found, however, that the light that emerges through a small hole from the
hollow center of a hot body does not show characteristic emission lines,
but has a smooth distribution of intensity with wavelength, characteristic
of the temperature but independent of the nature of the hot body. This
distribution is indicated, for three temperatures, in Figure 3-19. It is seen
that at low temperatures, below 4000°K, most of the energy is in the
infrared region and only a small amount is in the visible region, between
400 nm and 800 nm. At 6000°K the wavelength with the maximum
amount of energy is about 500 nm, and a large fraction of the emitted
energy is in the visible region. This is the temperature of the surface of
the sun.

The Discovery of Planck's Constant

During the years before 1900 the theoretical physicists who were interested in the problem of the emission of light by hot bodies found themselves unable to account for the curves shown in Figure 3-19 on the basis of the emission and absorption of light by vibrating molecules in the hot body, making use of the kinetic theory of molecular motion. Max Planck then discovered that a satisfactory theory could be formulated if the assumption were made that the hot body cannot emit or absorb light of a given wavelength in an arbitrarily small amount, but must emit or absorb a certain quantum of energy of light of that wavelength. Although Planck's theory did not require that the light itself be considered as consisting of bundles of energy—*light quanta* or *photons*—it was soon pointed out by Einstein (in 1905) that other evidence supports this concept.

The amount of light energy of wavelength λ absorbed or emitted by a solid body in a single act was found by Planck to be proportional to the frequency ν (equal to c/λ):

$$E = h\nu \qquad (3\text{-}9)$$

In this equation E is the amount of energy of light with frequency ν emitted or absorbed in a single act, and h is the constant of proportionality. This constant h is a very important constant; it is one of the fundamental constants of nature, and the basis of the whole quantum theory. It is called *Planck's constant*. Its value is

$$h = 0.66252 \times 10^{-33} \text{ J s}$$

(The units of h, J s, have the dimensions of energy times time, as is required by Equation 3-9.)

We see that light of short wavelength consists of large bundles of energy and light of long wavelength of small bundles of energy. Some of the experiments in which these bundles of energy express their magnitudes will be discussed in the following section.

3-6.   The Photoelectric Effect and the Photon

In 1887 Heinrich Hertz, who discovered radiowaves, observed that a spark passes between two metal electrodes at a lower voltage when ultraviolet light is shining on the electrodes than when they are not illuminated. It was then discovered by J. J. Thomson in 1898 that negative electric charges are emitted by a metal surface on which ultraviolet light impinges. A simple experiment to show this effect is represented in Figure 3-20. An electroscope is negatively charged, and ultraviolet light is allowed to fall on the zinc plate in contact with it. The leaves of the electroscope fall,

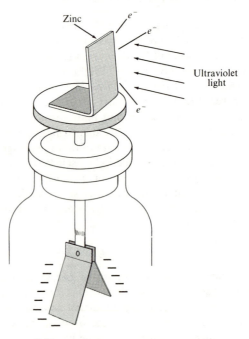

FIGURE 3-20
A simple experiment showing the photoelectric
effect. A negative charge is emitted by a zinc plate
upon which ultraviolet light impinges

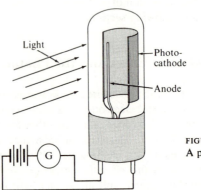

FIGURE 3-21
A photoelectric cell.

showing that the negative electric charge is being removed under the action of the ultraviolet light. If the electroscope has a large positive charge the leaves do not fall, showing that positive charges are not emitted under similar conditions. An uncharged electroscope becomes charged when the metal plate is illuminated with ultraviolet light, and the charge that remains on the leaves of the electroscope is a positive charge, showing that negative charges have left the metal.

J. J. Thomson was able to show that the negative electric charge that leaves the zinc plate under the influence of ultraviolet light consists of electrons. The emission of electrons by action of ultraviolet light or x-rays is called the *photoelectric effect*. The electrons that are given off by the metal plate are called *photoelectrons*; they are not different in character from other electrons.

A great deal was learned by the study of the photoelectric effect. It was soon found that visible light falling on a zinc plate does not cause the emission of photoelectrons, but ultraviolet light with a wavelength shorter than about 350 nm does cause their emission. The maximum wavelength that is effective is called the *photoelectric threshold*.

Substances differ in their photoelectric thresholds: the alkali metals are especially good photoelectric emitters, and their thresholds lie in the visible region; that for sodium is about 650 nm, so that visible light is effective with this metal except at the red end of the spectrum.

It was discovered that the photoelectrons are emitted with an amount of kinetic energy that depends upon the wavelength of the light. An apparatus somewhat like the photoelectric cell shown in Figure 3-21 can be used for this purpose. In this apparatus the photoelectrons that are emitted when the metal is illuminated are collected by a collecting electrode, and the number of them that strike the electrode can be found by measuring the current that flows along the wire to the electrode. A potential difference can be applied between the electrode and the emitting metal. If the collecting electrode is given a slight negative potential, which requires work to be done on the electrons to transfer them from the emitting metal to the collecting electrode, the flow of photoelectrons to the collecting electrode is stopped if the incident light has a wavelength close to the threshold, but it continues if the incident light has a wavelength much shorter than the threshold wavelength. By increasing the negative charge on the collecting electrode the potential difference can be made great enough to stop the flow of photoelectrons to the electrode.

These observations were explained by Einstein in 1905, by means of his theory of the photoelectric effect. He assumed that the light that impinges on the metal plate consists of *light quanta*, or *photons*, with energy $h\nu$, and that when the light is absorbed by the metal all of the energy of one photon is converted into energy of a photoelectron. However, the electron must use a certain amount of energy to escape from the metal. This may

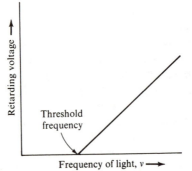

FIGURE 3-22

Curve representing the result of experiments on the retarding voltage necessary to prevent the flow of photoelectrons to the anode, as a function of the frequency of light producing the photoelectrons. The threshold frequency is the frequency of light such that one quantum has just enough energy to remove an electron from the metal; a quantum of light with larger frequency is able to remove the electron from the metal and to give it some kinetic energy.

be represented by the symbol E_i (the energy of ionizing the metal). The remaining energy is kinetic energy of the photoelectron. The *Einstein photoelectric equation* is

$$h\nu = E_i + \tfrac{1}{2}mv^2 \tag{3-10}$$

This famous equation states that the energy of the light quantum, $h\nu$, is equal to the energy required to remove the electron from the metal, E_i, plus the kinetic energy imparted to the electron, $\tfrac{1}{2}mv^2$. The success of this equation in explaining the observations of the photoelectric effect was largely responsible for the acceptance of the idea of light quanta.

It is difficult to measure the velocity of the electrons directly. Instead, the energy quantity $\tfrac{1}{2}mv^2$ is measured by measuring the potential difference, V, which is necessary to keep the photoelectrons from striking the collecting electrode; the product of the potential difference V and the charge of the electron, e, is the amount of work done against the electrostatic field, and when V has just the value required to prevent the electrons from reaching the collecting plate the following relation is satisfied:

$$eV = \tfrac{1}{2}mv^2$$

Introducing this in the preceding equation, we obtain

$$eV = h\nu - E_i$$

or

$$V = \frac{h\nu}{e} - \frac{E_i}{e} \tag{3-11}$$

In Figure 3-22 this equation is plotted. The equation expresses a linear relation between the retarding voltage and the frequency of the light. The experimental observations lie on a line of just this sort. The intercept on the frequency axis, ν_0, corresponds to the photoelectric threshold for the metal. The slope of the curve is seen from Equation 3-11 to be equal to h/e—that is, to the ratio of Planck's constant, h, to the charge of the electron, e. R. A. Millikan in 1912 carried out careful measurements of the retarding voltage, in order to verify the Einstein equation. He used an apparatus similar to that shown in Figure 3-21, and his measurements led to a value for Planck's constant; this value for many years was the most accurate known.

The Photoelectric Cell

The photoelectric cell is used in talking motion pictures, television, automatic door-openers, and many other practical applications. The cell may be made by depositing a thin layer of an alkali metal on the inner surface of a small vacuum tube, as shown in Figure 3-21. The collecting electrode is positively charged, so that the photoelectrons are attracted to it. Illumination of the metal surface by any radiation with wavelength shorter than the photoelectric threshold causes the emission of photoelectrons, and a consequent flow of electric current through the circuit. The current may be registered on an ammeter. It is found that the magnitude of the current is proportional to the intensity of the light.

Example 3-2.   How much energy is there in one quantum of light with wavelength 650 nm?

Solution.   The amount of energy in a light quantum is $h\nu$, where h is Planck's constant and ν is the frequency of the light. The frequency of light of wavelength λ is c/λ; hence

$$\nu = \frac{3 \times 10^8}{650 \times 10^{-9}} = 4.62 \times 10^{14} \text{ Hz (Hertz, cycles per sec)}$$

Thus we obtain

$$\text{energy of photon} = h\nu = 0.66252 \times 10^{-33} \times 4.62 \times 10^{14}$$
$$= 3.06 \times 10^{-19} \text{ J}$$

The Electron-Volt, a Unit of Energy.

Physicists have found it convenient to use the *electron-volt* (eV) as a unit of energy. It is the energy acquired by an electron accelerated by a potential difference of 1V. Its value is

$$1 \text{ eV} = 0.160206 \times 10^{-18} \text{ J} \tag{3-12}$$

The MeV (mega-electron-volt = 10^6 eV) and GeV (giga-electron-volt = 10^9 eV) are used in nuclear physics and elementary-particle physics.

Example 3-3.   What retarding voltage would be required to stop the flow of photoelectrons produced by light of wavelength 650 nm from a sodium metal surface?

Solution.   The photoelectric threshold of sodium metal is 650 nm. Accordingly, the photoelectrons that are produced have no kinetic energy: the amount of energy in the light quantum is just enough to remove the electron from the metal. Hence an extremely small retarding potential would stop the flow of photoelectrons under these conditions.

Example 3-4.   What retarding potential would be required to stop the flow of photoelectrons in a photoelectric cell with sodium metal illuminated with light of wavelength 325 nm?

Solution.   If, using the method of the solution of Example 3-2, we calculate the energy of a light quantum with wavelength 325 nm, we obtain the value 6.12×10^{-19} J. This result can be obtained, in fact, with little calculation by noting that this wavelength is just half that of the photoelectric threshold, 650 nm; hence the frequency ν is twice as great, and the energy $h\nu$ is also twice as great (Example 3-2).

Of this total amount of energy in the light quantum, the amount 3.06×10^{-19} J is used to remove the electron from the metal. The remaining amount, 3.06×10^{-19} J, is kinetic energy of the photoelectron. The retarding potential that would slow the electron down to zero speed is such that its product with the charge of the electron is equal to this amount of energy:

$$eV = 3.06 \times 10^{-19} \text{ J}$$
$$V = \frac{3.06 \times 10^{-19}}{0.1602 \times 10^{-18}} = 1.91 \text{ V}$$

Hence the retarding potential necessary to prevent the flow of photoelectrons in the sodium photoelectric cell illuminated with wavelength 325 nm is 1.91 volts.

The Production of X-rays

In an x-ray tube electrons from a hot filament are accelerated by a voltage V and then brought to rest by striking a solid target. Much or all of the kinetic energy of such an electron is converted into a photon. This phenomenon is called the inverse photoelectric effect. If all the energy eV of an electron is converted into a photon, the frequency of the photon (the x-ray) is given by the photoelectric equation $eV = h\nu$ (the ionization energy of the metal, E_i, can be neglected in this case, because it is a small energy quantity in comparison with the other). If the electron is not slowed down completely the frequency of the x-ray quantum that is emitted will be somewhat smaller than the limiting value.

Example 3-5. An x-ray tube is operated at 50,000 volts. What is the short-wavelength limit of the x-rays that are produced?

Solution. The energy of an electron that strikes the anode in the x-ray tube is eV. The value of eV is accordingly 8.01×10^{-15} J (Equation 3-12). This is equal to $h\nu$; hence for ν we have

$$\nu = \frac{8.01 \times 10^{-15}}{0.6625 \times 10^{-33}} = 1.209 \times 10^{19} \text{ Hz}$$

The wavelength λ is obtained by dividing the velocity of light by this quantity:

$$\lambda = \frac{c}{\nu} = \frac{3 \times 10^8}{1.209 \times 10^{19}} = 2.48 \times 10^{-11} \text{ m} = 0.248 \text{ Å}$$

Hence the short-wavelength limit of an x-ray tube operated at 50,000 volts is calculated to be 0.248 Å.

It is interesting to note that the preceding calculation can be simplified by combining all the steps into a single equation:

$$\text{wavelength (in Å)} = \frac{12,398}{\text{accelerating potential (in volts)}} \quad (3\text{-}13)$$

This equation states that a photon in the near infrared, with wavelength 12,398 Å, has the same energy as an electron that has been accelerated by a potential difference of 1 volt.

Example 3-6. A beam of light with wavelength 6500 Å and carrying the energy 0.01 W = 0.01 J per second falls on a photoelectric cell, and is completely used in the production of photoelectrons. (This is about the energy of the light from the sun and sky on a bright day that strikes an area of 1 cm².) What is the magnitude of the photoelectric current that then flows in the circuit of which the photoelectric cell is a part?

Solution. The energy of one quantum of light with wavelength 6500 Å is 3.06×10^{-19} J. Hence there are $0.01/3.06 \times 10^{-19} = 0.327 \times 10^{17}$ photons in the amount of light carrying 0.01 J of radiant energy, and this number of photons impinges on the metal of the photoelectric cell every second. The same number of photoelectrons would be produced. Multiplying by the charge of the electron, 0.1602×10^{-18} C, we obtain 5.24×10^{-3} C as the number of coulombs transferred per second. One ampere is a flow of electricity at the rate of 1 coulomb per second; hence the current that is produced by the beam of light is 5.24×10^{-3} A—that is, 5.24 mA.

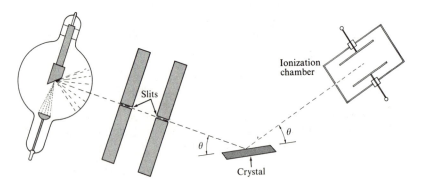

FIGURE 3-23
The Bragg ionization-chamber technique of investigating the diffraction of x-rays by crystals.

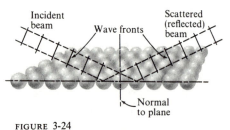

FIGURE 3-24
Diagram showing the equality of path lengths when the conditions for specular reflection from a layer are satisfied.

3-7. The Diffraction of X-rays by Crystals

During the decade after the discovery of x-rays an attempt was made to obtain a diffraction pattern by passing x-rays through a very narrow slit. The experiments showed that if these rays were similar to ordinary light their wavelength must be of the order of magnitude of 1 Å; that is, about 1/5000 of the wavelength of visible light. Then the German physicist Max von Laue (1879–1960) had the idea that crystals, in which atoms are arranged in a regular lattice with interatomic distances around 300 pm, might serve to produce diffraction effects with x-rays. The experiment was immediately carried out by two experimental physicists, W. Friedrich and P. Knipping, with use of a crystal of copper sulfate pentahydrate. A narrow beam of x-rays from an x-ray tube was passed through the crystal, and photographic plates were placed around the crystal. It was found that on the photographic plate behind the crystal there was a blackened spot representing the position where the direct beam of x-rays had struck the plate, and also several other spots, showing the preferential scattering of the beam of x-rays in certain directions, corresponding to diffraction maxima. This experiment showed at once that x-rays are similar to light, in having a wave nature, and that the wavelength of the x-rays produced by the x-ray tube that was used was of the order of 1 Å.

W. Lawrence Bragg (born 1890), who was a student at Cambridge University, then developed the theory of x-ray diffraction (the Bragg equation, described below) and used it in November, 1912, to determine the structure of sphalerite, the cubic form of zinc sulfide, by analyzing the x-ray diffraction photographs of sphalerite that had been published by Laue, Friedrich, and Knipping. His father, William H. Bragg (1862–1942), then devised the x-ray spectrometer (Figure 3-23), and within a year W. L. and W. H. Bragg had determined the exact atomic arrangements for many crystals and also the wavelengths of the characteristic x-ray lines emitted by several elements serving as targets in the x-ray tubes. In the Bragg technique (Figure 3-23) a beam impinges on the face of a crystal, such as the cleavage face of a salt crystal. An instrument for detecting x-rays (in the case of the original experiments by the Braggs an *ionization chamber*, but in modern work a *Geiger counter* or *scintillation counter** may be used) was then placed as shown in the figure. The simple theory of diffraction of x-rays by crystals developed by Lawrence Bragg is illustrated in Figures 3-24 and 3-25. He pointed out that if the beam of rays incident on a plane of atoms and the scattered beam are in the same vertical plane and at the same angle with the plane, as shown in Figure 3-24, the conditions for reinforcement are satisfied. This sort of scattering is called *specular reflection*—it is similar to reflection from a mirror. He then formulated the conditions for reinforcement of the beam specularly reflected from one plane of atoms and the beam specularly reflected from another plane of atoms separated from it by the *interplanar distance d*. This situation is illustrated in Figure 3-25. We see that the difference in path is equal to $2d \sin \theta$, in which θ is the *Bragg angle* (the angle between the

---

*The ionization chamber, Geiger counter, and proportional counter are detectors for energetic charged particles or photons. Gas molecules in these detectors are ionized (see Section 5-2) by the absorption of the particle or photon. In the ionization chamber this ionization discharges a charged electroscope. In a Geiger or proportional counter the original gaseous ionization is amplified by subsequent ionization of more gas molecules for which the energy comes from a large potential difference between the central wire of the counter and the case. In a scintillation counter the energetic particle or photon is converted to a flash of light by a *scintillator*. The light is converted to electrons at the photocathode of a photomultiplier and the electrons are amplified by a high potential difference across a series of plates.

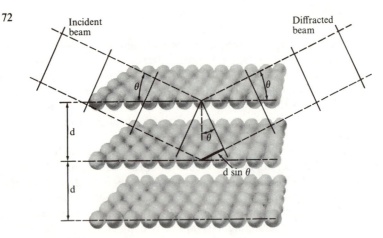

Diagram illustrating the derivation of the Bragg equation for the diffraction of x-rays by crystals.

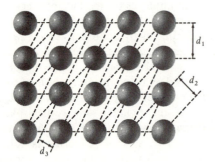

FIGURE 3-26
Spacings between different rows of atoms in a two-dimensional crystal.

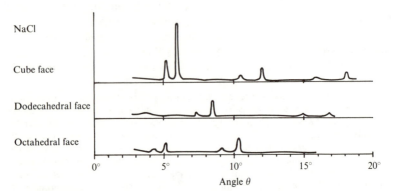

NaCl

Cube face

Dodecahedral face

Octahedral face

Angle θ

FIGURE 3-27
Experimental data obtained by the Braggs for the diffraction of x-rays by the sodium chloride crystal.

incident beam and the plane of atoms). In order to have reinforcement, this difference in path $2d \sin \theta$ must be equal to the wavelength λ or an integral multiple of this wavelength—that is, to $n\lambda$, in which n is an integer. We thus obtain the *Bragg equation* for the diffraction of x-rays:

$$n\lambda = 2d \sin \theta \qquad (3\text{-}14)$$

The way in which the structure of crystals was then determined is illustrated in Figure 3-26. Here we show a simple cubic arrangement of atoms, as seen along one of the cube faces. It is evident that there are layers of atoms, shown by their traces in the plane of the paper, with spacings $d_1, d_2, d_3, \ldots$, which are in the ratios $1:2^{-1/2}:5^{-1/2}, \ldots$. Since the relative values of the spacings $d_1, d_2, d_3, \ldots$ could be determined without knowledge of the wavelengths of x-rays, but simply as inversely proportional to their values of $\sin \theta$, the nature of the atomic arrangement could be discovered by the Bragg experiments.

A reproduction of some of the first experimental measurements made by the Braggs is shown as Figure 3-27. It is seen that there occurs a pattern of reflections that is repeated at values of $\sin \theta$ representing the values 1, 2, and 3 for the integer n, which is called the *order of the reflection*. The pattern that is repeated consists of a weak scattered beam, at the smaller angle, and a stronger scattered beam, at the larger angle. This shows that there were present in the beam of x-rays produced by the x-ray tube a shorter wavelength and a longer wavelength, with greater intensity of the x-rays of longer wavelength than of shorter wavelength.

Within a few months the Braggs had succeeded in determining the wavelengths of x-rays to an accuracy of about 1%, and also in determining the structures of about twenty different crystals. We have already discussed the results of their discovery in the preceding chapter, in which the structures of the copper crystal and of the iodine crystal were described.

A further discussion of x-ray diffraction and crystal structure is given in Appendix IV.

Example 3-7. The Braggs calculated from the density of salt and Millikan's value of the charge of the electron (see Exercise 4-18) that the spacing d_1 for the cube face of the sodium chloride crystal has the value 2.81 Å. Using the experimental data given in Figure 3-27, find the wavelengths of the two radiations produced by the x-ray tube that they used.

Solution. Let us first calculate the wavelength of the x-rays with shorter wavelength. This is called the $K\beta$ line. The Bragg angle θ for this line in the first order is seen from the figure to be $5°18'$. The value of $\sin \theta$ is accordingly 0.0924. The Bragg equation for $n = 1$ is

$$\lambda = 2d \sin \theta$$

We insert the value 2.81 Å for d and 0.0924 for sin θ and obtain

$$\lambda = 2 \times 2.81 \times 0.0924 = 0.519 \text{ Å}$$

This is the value of the wavelength for the $K\beta$ line. In the same way we calculate for the other line, using the observed value of θ of 6°0', the value

$$\lambda = 0.587 \text{ Å}$$

This is the wavelength of the $K\alpha$ line. These two wavelengths correspond to the metal palladium, which was the target (the anode) of the x-ray tube used by the Braggs in this experiment. The two lines are part of the *characteristic x-radiation* of the element palladium.

3-8.   Electron Wave Character and Electron Spin

Until 1924 the observed properties of the electron were considered to justify describing it as a small electrically charged particle. In that year the wave character of the electron was discovered by the French physicist Louis de Broglie (born 1892). While making a theoretical study of the quantum theory, to serve as his thesis for the doctor's degree from the University of Paris, he found that a striking analogy between the properties of electrons and the properties of photons could be recognized if a moving electron were to be assigned a wavelength. This wavelength is now called the *de Broglie wavelength* of the electron.

The equation for the wavelength of the electron is

$$\lambda = \frac{h}{mv} \tag{3-15}$$

In this equation λ is the wavelength of the electron, h is Planck's constant, m is the mass of the electron, and v is the velocity of the electron. The product mv is the translational momentum of the electron.* It is seen that according to this equation a stationary electron has infinite wavelength, and the wavelength decreases with increase in the velocity of the electron.

Example 3-8. What is the wavelength of an electron with 13.6 eV of kinetic energy?

Solution. The energy of a 13.6-eV electron is

$$E = 13.6 \times 0.1602 \times 10^{-18} = 2.18 \times 10^{-18} \text{ J}$$

This is equal to $\frac{1}{2}mv^2$, the kinetic energy of an electron moving with velocity v; accordingly,

$$mv^2 = 4.36 \times 10^{-18} \text{ J}$$

---

*In Newtonian mechanics the momentum of translational motion of a body with mass m moving with velocity v is defined as mv.

Multiplying both sides of this equation by the mass of the electron, $m = 0.9108 \times 10^{-30}$ kg, we obtain

$$m^2v^2 = 4.36 \times 10^{-18} \times 0.9108 \times 10^{-30} = 3.97 \times 10^{-48} \text{ kg}^2 \text{ m}^2 \text{ s}^{-2}$$

By taking the square root of each side of this equation we obtain

$$mv = 1.99 \times 10^{-24} \text{ kg m s}^{-1}$$

By use of the de Broglie equation we can now obtain the value for the wavelength:

$$\lambda = \frac{h}{mv} = \frac{0.6625 \times 10^{-33} \text{ kg m}^2 \text{ s}^{-1}}{1.99 \times 10^{-24} \text{ kg m s}^{-1}} = 0.332 \times 10^{-9} \text{ m}$$

Accordingly, we have found that the de Broglie wavelength of an electron that has been accelerated by a potential difference of 13.6 V is 3.32 Å.

We can now calculate easily the wavelength of an electron with 100 times as much kinetic energy—that is, an electron that has been accelerated by the potential difference of 1360 V. Since the energy is proportional to the square of the velocity, such an electron has a velocity ten times that of a 13.6-eV electron, and, according to the de Broglie equation, its wavelength is $\frac{1}{10}$ as great. Thus the wavelength of a 1360-eV electron is 0.332 Å.

The Analogy Between the Photon and the Electron

One part of the argument carried out by de Broglie in his discovery of the wavelength of the electron can be easily presented. The energy of a photon with frequency ν is $h\nu$. The mass of the photon is related to the energy by the Einstein equation

$$mc^2 = h\nu$$

where m represents the mass of the photon. Dividing each side of this equation by c, we obtain for the momentum mc of the photon the value

$$mc = \frac{h\nu}{c}$$

In this equation ν/c can be replaced by $1/\lambda$, giving

$$mc = \frac{h}{\lambda}$$

or

$$\lambda = \frac{h}{mc}$$

De Broglie pointed out that the same equation might be applied to an electron, by using m for the mass of the electron instead of the mass of the photon, and replacing c, the velocity of the photon, by v, the velocity of the electron. In this way the de Broglie equation is obtained.

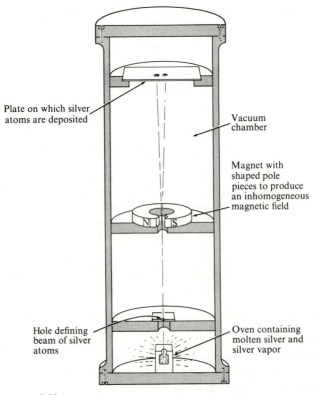

Plate on which silver
atoms are deposited

Vacuum
chamber

Magnet with
shaped pole
pieces to produce
an inhomogeneous
magnetic field

N S

Hole defining
beam of silver
atoms

Oven containing
molten silver and
silver vapor

FIGURE 3-28
Diagram illustrating the Stern-Gerlach experiment, showing quantized
orientation of the magnetic moment of an atom in a magnetic field.

The Direct Experimental Verification
of the Wavelength of the Electron

The wave character of moving electrons was established beyond question by the work of the American physicist C. J. Davisson (1881–1958) and the English physicist G. P. Thomson (born 1892). These investigators found that electrons that are scattered by crystals produce a diffraction pattern similar to that produced by x-rays scattered by crystals, and, moreover, that the diffraction pattern, interpreted by the Bragg law, corresponds to the wavelength given by the de Broglie equation.

The penetrating power of electrons through matter is far less than that of x-rays with the same wavelength. It is accordingly necessary to reflect the beam of electrons from the surface of a crystal (as was done by Davisson and his collaborators, using a single crystal of nickel), or to shoot a stream of high-speed electrons through a very thin crystal or layer of crystalline powder (as was done by Thomson).

The structure of crystals can be investigated by the electron-diffraction method as well as by the x-ray-diffraction method. The electron-diffraction method has been especially useful in studying the structure of very thin films on the surface of a crystal. For example, it has been shown that when argon is absorbed on a clean face of a nickel crystal the argon atoms occupy only one-quarter of the positions formed by triangles of nickel atoms (in the octahedral face of the cubic closest-packed crystal; Figure 2-4). The structure of very thin films of metal oxide that are formed on the surface of metals, and that protect them against further corrosion, has been studied by this method.

The electron-diffraction method is also very useful for determining the structure of gas molecules. The way in which the diffraction pattern is formed is illustrated by Figure 3-17, which corresponds to the scattering of waves by a diatomic molecule. The molecules in a gas have different orientations, and the diffraction pattern is accordingly somewhat blurred. It consists of a series of rings. Knowledge of the wavelength of the electrons and measurement of the diameters of these rings permit calculation of the interatomic distances in the molecules. Since the discovery of the electron-diffraction method the structures of several hundred molecules have been determined in this way.

The Spin of the Electron

It was discovered in 1925 by two Dutch physicists, George E. Uhlenbeck (born 1900) and Samuel A. Goudsmit (born 1902), that the electron has properties corresponding to its having a spin; it can be described as rotating about an axis in a way that can be compared with the rotation of the earth about an axis through its north pole and south pole. The amount of the spin (angular momentum) is the same for all electrons, but the orientation of the axis can change. With respect to a specified direction, such as the direction of the earth's magnetic field, a free electron can orient itself in either one of only two ways: either it lines up parallel to the field, or antiparallel (with the opposite orientation).

The experimental observations that led to the discovery of the spin of the electron were mainly those of the fine structure of spectral lines, discussed briefly in a later chapter (Chapter 5). One experiment of significance was the Stern-Gerlach experiment, suggested by the German physicist Otto Stern (1888–1969) in 1921 and carried out by him and W. Gerlach the same year. The experiment is illustrated in Figure 3-28. Silver is vaporized from the vessel at the bottom into a high vacuum. A narrow beam of

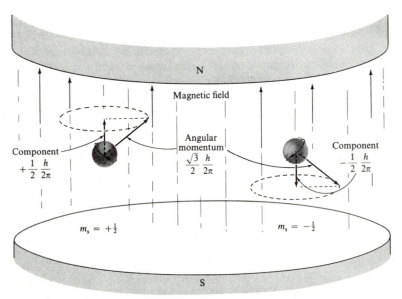

FIGURE 3-29
Diagram illustrating spatial quantization of the spin of an electron. The spin quantum number is $s = \frac{1}{2}$, leading to angular momentum $\frac{\sqrt{3}}{2} \frac{h}{2\pi}$. The magnetic quantum number m_s can be either $+\frac{1}{2}$ or $-\frac{1}{2}$, with values $+\frac{1}{2} \frac{h}{2\pi}$ and $-\frac{1}{2} \frac{h}{2\pi}$, respectively, for the component of the angular momentum in the direction of the magnetic field.

atoms, segregated by a slit, is passed through a highly inhomogeneous magnetic field produced by the shaped pole pieces of a magnet. The deflected beams impinge on a plate, and their traces are later made evident by a process of chemical development. It was found that the beam of silver atoms was split into two beams. This observation was explained by the assumptions that the silver atom has a magnetic moment (later found to be just that associated with the spin of one electron), that the moment can be oriented in either one of two ways relative to the lines of force of a magnetic field, and that the inhomogeneity (the gradient) of the field produces a force accelerating the atoms to the right or the left.

Quantized Angular Momentum

The angular momentum of a rigid body rotating with angular velocity ω (radians per second) about an axis through its center of mass is $I\omega$, where I is the moment of inertia (equal to $\Sigma_i m_i \rho_i$, with m_i the mass of the *i*th particle in the body and ρ_i its distance from the axis of rotation). Angular momentum is conventially represented by a vector along the axis of rotation. The dimensions of angular momentum are seen to be mass $\times$ (length)^2 $\times$ time^{-1}. These are also the dimensions of Planck's constant, h, which suggested that the angular momenta of atoms and molecules might be quantized in units h or $h/2\pi$. Unsuccessful efforts in 1912 by the British mathematician J. W. Nicholson and the Danish chemist Niels Bjerrum were followed in 1913 by the successful attack on the hydrogen atom by Niels Bohr, as described in Chapter 5.

Angular Momentum Quantum Numbers

In general it is found that the angular momentum of a particle or system of particles has a value (or values) given by the following equation:

$$\text{angular momentum} = \sqrt{J(J+1)}\,\frac{h}{2\pi}$$

This equation may also be written as

$$\text{angular momentum} = \sqrt{J(J+1)}\,\hbar \tag{3-16}$$

with $\hbar = h/2\pi$. (The quantity $\hbar$, read "h bar," has the value 0.105443×10^{-33} J s.) The quantum number J, called the angular momentum quantum number, has for some systems half-integral values ($\frac{1}{2}, \frac{3}{2}, \frac{5}{2}, \ldots$), and for other systems integral values ($0, 1, 2, \ldots$).

The component of angular momentum in the direction of a magnetic field has the values $M_J \hbar$, with $M_J = -J, -J+1, \ldots, +J$. Thus the values of M_J are half-integral if J is half-integral, and are integral if J is integral. There are $2J+1$ values of M_J for a given J; that is, there are $2J+1$ ways of spatial quantization (orientiation in space) for a system with angular momentum described by the quantum number J.

The angular momentum quantum number for the spin of the electron is given the symbol s. Its value is $\frac{1}{2}$. The magnitude of the spin angular momentum of the electron is accordingly $\sqrt{3}\hbar/2 = 91.32 \times 10^{-30}$ J s.

The values of m_s, the magnetic quantum number for the spin of an electron, are $+\frac{1}{2}$ and $-\frac{1}{2}$. The component of angular momentum in the field direction accordingly has one or the other of the two values $+\hbar/2$ and $-\hbar/2$, as shown in Figure 3-29.

Magnetic Moment

In 1820 the Danish physicist Hans Christian Oersted (1777–1851) discovered that a wire carrying an electric current exerts a force on a magnet. The current may be said to produce a magnetic field. If the current flows in a ring, the magnetic field is similar to that of a small magnet—that is, of a magnetic dipole. The value of the dipole moment for a current i (in amperes) is $i \times$ area $\times 4\pi \times 10^{-7}$ weber meter (here area is the area of the ring around which the current is flowing and $4\pi \times 10^{-7}$ weber $A^{-1} m^{-1}$ is the permeability of empty space). If we consider the current i to result from the flow in a circle of particles with mass m and charge e, we calculate that the angular momentum of the current has the value $i \times$ area $\times (2m/e)$, and that the magnetic moment is related to the angular momentum in the following way:

$$\text{magnetic moment} = \text{angular momentum} \times \frac{e}{2m} \times 4\pi \times 10^{-7} \text{ Wb m} \qquad (3\text{-}17)$$

A quantized unit component of angular momentum, with magnitude $\hbar = h/2\pi$, that results from the motion of electrons accordingly has associated with it the magnetic moment μ_B, given by the equation

$$\mu_B = \hbar \times \frac{e}{2m} \times 4\pi \times 10^{-7} = 11.653 \times 10^{-30} \text{ Wb m} \qquad (3\text{-}18)$$

The quantity μ_B is called the Bohr magneton.

The spin angular momentum of the electron is $\sqrt{3}\hbar/2$. The spin magnetic moment is not $\sqrt{3}/2$ Bohr magnetons, as given by Equation 3-18, but is $\sqrt{3}\,\mu_B$. This fact is described by saying that there is a factor 2 to be introduced (called the g-factor for spin). The g-factor 2 for electron spin is a requirement of the theory of relativity.

In the Stern-Gerlach experiment with silver atoms the observed paths of the atoms correspond to a component $\pm 1\mu_B$ of magnetic moment in the field direction.

Stern-Gerlach experiments with hydrogen molecules and experiments of other kinds have shown that the proton has a spin, with spin quantum number $s = \frac{1}{2}$. The associated magnetic moment is not $\sqrt{3}/2$ nuclear Bohr magnetons (the nuclear Bohr magneton, 6.347×10^{-33} weber meter, is given by Equation 3-18 with use of the mass of the proton), nor is it twice this great (g-factor 2); it is in fact observed to be $2.79275\sqrt{3}$ nuclear magnetons, corresponding to the g-factor 5.5855. This surprising value constitutes some of the evidence that the proton is not a simple particle, but has a complex structure.

3-9.   What Is Light? What Is an Electron?

During recent years many people have asked the following questions: Does light *really* consist of waves, or of particles? Is the electron *really* a particle, or is it a wave?

These questions cannot be answered by one of the two stated alternatives. Light is the name that we have given to a part of nature. The name refers to all of the properties that light has, to all of the phenomena that are observed in a system containing light. Some of the properties of light resemble those of waves, and can be described in terms of a wavelength. Other properties of light resemble those of particles, and can be described in terms of a light quantum, having a certain amount of energy, $h\nu$, and a certain mass, $h\nu/c^2$. A beam of light is neither a sequence of waves nor a stream of particles; it is both.

In the same way, an electron is neither a particle nor a wave, in the ordinary sense. In many ways the behavior of electrons is similar to that expected of small spinning particles, with mass m, electric charge $-e$, and certain values of angular momentum and magnetic moment. But electrons differ from ordinary particles in that they also behave as though they had wave character, with wavelength given by the de Broglie equation. The electron, like the photon, has to be described as having the character both of a particle and of a wave.

After the first period of adjustment to these new ideas about the nature of light and of electrons, scientists became accustomed to them, and found that they could usually predict when, in a certain experiment, the behavior of a beam of light would be determined mainly by its wavelength, and when it would be determined by the energy and mass of the photon; that is, they would know when it was convenient to consider light as consisting of waves, and when to consider it as consisting of particles, the photons. Similarly, they learned when to consider an electron as a particle, and when as a wave. In some experiments the wave character and the particle character both contribute significantly, and it is then necessary to carry out a careful theoretical treatment, using the equations of quantum mechanics, in order to predict how the light or the electron will behave.

You might ask two other questions: Do electrons exist? What do they look like?

The answer to the first question is that electrons do exist: "electron" is the name that scientists have used in discussing certain phenomena, such as the beam in the electric-discharge tube studies by J. J. Thomson, the carrier of the unit electric charge on the oil drops in Millikan's apparatus, the part that is added to the neutral fluorine atom to convert it into a fluoride ion. As to the second question—what does the electron look like?—we may say that some information has been obtained by studying the scattering of very-high-velocity electrons by protons and other atomic nuclei. These experiments have given much information about the size and structure of the nuclei (see Chapter 26), and have also shown that the electron behaves as a point particle, with no structure extending over a diameter as great as 0.1 fm (0.1×10^{-15} m).

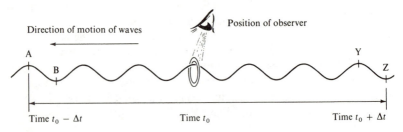

Direction of motion of waves

Position of observer

A

B

Y

Z

Time $t_0 - \Delta t$ Time t_0 Time $t_0 + \Delta t$

FIGURE 3-30
A diagram illustrating the uncertainty in determining the frequency by counting the number of waves passing a point during a period of time.

3-10.   The Uncertainty Principle

The *uncertainty principle*, an important relation that is a consequence of quantum mechanics, was discovered by the German physicist Werner Heisenberg (born 1901) in 1927. Heisenberg showed that in consequence of the wave-particle duality of matter it is impossible to carry out simultaneously a precise determination of the position of a particle and of its velocity. He also showed that it is impossible to determine exactly the energy of a system at an instant of time.

The uncertainty principle in quantum mechanics is closely related to an uncertainty equation between frequency and time for any sort of waves. Let us consider, for example, a train of ocean waves passing a buoy that is anchored at a fixed point. An observer on the buoy could measure the frequency (number of waves passing the buoy per unit time) at time t_0 by counting the number of crests and troughs passing the buoy between the times $t_0 - \Delta t$ and $t_0 + \Delta t$, dividing by 2 to obtain the number of waves in time $2\Delta t$, and by $2\Delta t$ to obtain the frequency (ν), which is defined as the number of waves in unit time:

$$\nu = \frac{\text{number of crests plus number of troughs}}{2 \times 2\Delta t}$$

This measurement is an average for a period of time $2\Delta t$ in the neighborhood of t_0; we may describe it as being for the time t_0 with uncertainty Δt. There is also an uncertainty in the frequency. The crest A (Figure 3-30) might or might not be counted, and similarly the trough Z. Hence there is an uncertainty of about 2 in the number of crests plus the number of troughs, and hence of about $1/2\Delta t$ in the frequency:

$$\Delta \nu = \frac{2}{2 \times 2\Delta t} = \frac{1}{2\Delta t}$$

We may rewrite this equation as $\Delta\nu \times \Delta t = \frac{1}{2}$. A more detailed discussion based on an error-function definition of $\Delta\nu$ and Δt leads to the *uncertainty equation for frequency and time* in its customary form:

$$\Delta\nu \times \Delta t = \frac{1}{2\pi} \tag{3-19}$$

By use of quantum theory this equation can be at once converted into the uncertainty equation for energy and time for photons. The energy of a photon with frequency ν is $h\nu$. The uncertainty in frequency $\Delta\nu$ when multiplied by h is the uncertainty in energy ΔE:

$$\Delta E = h\Delta\nu$$

By substituting this relation in Equation 3-19 we obtain the energy-time uncertainty equation:

$$\Delta E \times \Delta t = \frac{h}{2\pi} \tag{3-20}$$

It has been found by analysis of many experiments on the basis of quantum theory that this relation holds for any system. Only by making the measurement of the energy of any system over a long period of time can the error in the measured energy of the system be made small.

Example 3-9. The yellow D lines, with wavelengths 5890 and 5896 Å, are emitted by sodium gas when it is excited by an electric discharge. The rate at which the D lines are emitted has been determined by measuring the intensity of the lines as absorption lines in sodium gas. It is such as to correspond to a mean life of the excited states of 1.6×10^{-8} s. (Most excited states of atoms have about this mean life.) Does this value of the mean life lead to a broadening of the spectral lines?

Solution. We use the uncertainty principle here on the argument that the train of waves corresponding to the photon emitted by an atom is emitted during a period of time approximately equal to the mean life of the excited state of the atom. The frequency corresponding to wavelength $\lambda = 0.589 \times 10^{-6}$ m is $\nu = c/\lambda = 3 \times 10^8/0.589 \times 10^{-6} = 5.09 \times 10^{14}$ s^{-1}. The uncertainty in frequency corresponding to the uncertainty in time $\Delta t = 1.6 \times 10^{-8}$ s is given by Equation 3-19 as $\Delta\nu = 1/(2\pi\Delta t) = 1/(2\pi \times 1.6 \times 10^{-8}) = 1.00 \times 10^7$ s^{-1}. Hence the relative uncertainty in frequency is $\Delta\nu/\nu = 1.00 \times 10^7/5.09 \times 10^{14} = 1.96 \times 10^{-8}$. This is also the value of $\Delta\lambda/\lambda$, the relative uncertainty in wavelength. Hence we obtain $\Delta\lambda = 1.96 \times 10^{-8} \times 0.589 \times 10^{-6} = 1.15 \times 10^{-14}$ m $= 0.0001$ Å.

The uncertainty in frequency of these lines thus gives a line width that is only about 1/50,000,000 of the wavelength. This line broadening is usually masked by other effects. However, some excited states of atoms, with energy greater than the ionization energy, have mean life only 10^{-12} s, because of very rapid decomposition into an electron and a positive ion, and the spectral lines with one of these states as upper state are about 1 Å broad.

The uncertainty principle between position and momentum of a particle states that the product of the uncertainty in one of the coordinates x, y, z describing the position, such as Δx, and the uncertainty in the corresponding component of the momentum, $\Delta(mv_x)$, is equal to or greater than $h/2\pi$:

$$\Delta x \times \Delta(mv_x) \gtrapprox \frac{h}{2\pi} \tag{3-21}$$

Exercises

3-1. An ordinary electric light bulb is operated under conditions such that one ampere of current (1 C s^{-1}) is passing through the filament. How many electrons pass through the filament each second? (Remember that the charge of the electron is -0.160×10^{-18} C.)

3-2. According to the law of gravitation, the gravitational force of attraction between two particles with masses m_1 and m_2 a distance r apart is Gm_1m_2/r^2, where G, the constant of gravitation, has been found by experiment to have the value 0.6673×10^{-10} N m^2 kg^{-2}; that is, the force of attraction between two particles each with mass 1 kg and the distance 1 m apart is 0.6673×10^{-10} N.

    (a) Calculate the force of electrostatic attraction between an electron and a proton 10 Å apart.

    (b) Calculate the force of gravitational attraction between an electron and a proton 10 Å apart. What is the relationship of the electrostatic attraction to the gravitational attraction at this distance?

    (c) What is the dependence on distance of the ratio of electrostatic attraction and gravitational attraction of an electron and a proton? (Answer: 2.3068×10^{-10} N; 1.0165×10^{-49} N; 2.269×10^{39}.)

3-3. Calculate the velocity with which the electrons would move in the apparatus used by J. J. Thomson, operated at an accelerating voltage of 6000 V. Assume that each electron has kinetic energy equal to eV, where e is the charge of the electron in coulombs and V is the accelerating potential in volts.

3-4. Describe Millikan's oil-drop experiment. Why is a knowledge of the value of the viscosity of air needed in order for the value of the charge of the electron to be calculated from the measurements?

3-5. The principal α particles from radium have an energy of 4.79 MeV (4.79 million electron volts). With what velocity are they moving? What is the ratio of their velocity to the velocity of light? The mass of the α particle is 6.66×10^{-27} kg. (Answer: Velocity 1.5×10^7 m s^{-1}.) (Note that if the velocity of a particle is less than 10% of the velocity of light, the expression

1/2 mv^2 for its kinetic energy can be used with error less than 1%. For larger values of the velocity the theory of relativity must be used to obtain the correct answers.)

3-6. Calculate the energy, in J, of α particles of radium, which have energy 4.79 MeV. The nucleus of an atom of gold has the electric charge 79ϵ, in which ϵ is the magnitude of the charge of the electron (15.188 $\times$ 10^{-15} S). The α particle has the charge 2ϵ. At what distance (for a head-on collision) is the mutual potential energy of the α particle and the nucleus of an atom of gold (equal, in J, to the product of the two charges, in S, divided by the distance between them, in meters) equal to the original kinetic energy of the α particle? This radius may be taken as indicating how closely the α particle must approach the atom of gold in order to experience a large deflection. (Answer: 4.75 $\times$ 10^{-4} Å.)

3-7. Calculate the fraction of α particles that would be predicted to undergo very large deflection, as a result of approaching to within the distance 4.75 $\times$ 10^{-4} Å from a gold nucleus (as calculated in the preceding problem), when the α particles pass through a single layer of gold atoms. Assume the gold atoms in the layer to be in the triangular close-packed arrangement, with distance 2.88 Å between centers of adjacent atoms. (Answer: 9.87 $\times$ 10^{-8}.)

3-8. A beam of light impinges normally (perpendicularly) upon a ruled grating, with 5000 rulings per cm. Calculate the angles between which the visible spectrum appears in the first order. Make the calculations for the wavelengths 4000 Å (violet) and 7500 Å (red).

3-9. Explain, in terms of reinforcement of scattered waves, why there is reinforcement of the x-rays scattered by all of the atoms in a plane of atoms, if the incident beam and the diffracted beam lie in the same vertical plane (perpendicular to the plane of the atoms), and the two beams make the same angle with the plane of the atoms.

3-10. Assuming specular reflection from a plane of atoms (see preceding exercise), derive the Bragg equation for the diffraction of x-rays from a sequence of planes of atoms the distance d apart.

3-11. The distance between layers of atoms parallel to the cube face of a crystal of sodium chloride is 2.81 Å. Using the Bragg equation, calculate the first three angles of reflection of copper $K\alpha$ radiation (that is, of the $K\alpha$ x-ray line from an x-ray tube with a copper target). The wavelength of the $K\alpha$ line of copper is 1.54 Å.

3-12. It was shown by Selig Hecht and collaborators that a flash of light, wavelength 5500 Å, with energy 20 $\times$ 10^{-18} J passing through the pupil of the eye is detected by a young observer. About 10% of the light passing through the pupil of the eye reaches and is absorbed by the retinal receptors. From this information calculate the number of photons needed for threshold visual perception (the number of retinal receptors excited).

3-13. The low-temperature form of the element polonium crystallizes in a simple-cubic form, with one atom in the unit cell. The first four lines of the x-ray powder diagram occur at Bragg angles corresponding to the interplanar distances 3.346, 2.366, 1.932, and 1.673 Å. To what sets of indices $h\,k\,l$ do they correspond? What is the value of a, the length of edge of the unit cell? How many nearest neighbors does each polonium atom have? At what distance? (See Appendix IV.)

3-14. Potassium metal crystallizes with a cubic body-centered arrangement: 2K in the unit cell, at 0 0 0 and $\frac{1}{2}\frac{1}{2}\frac{1}{2}$. Make a drawing to show that there are layers with the same number of atoms parallel to a cube face and the distance $a/2$ apart, to explain why the reflection 100, with spacing $d = a$, does not occur in the x-ray diffraction pattern. (For any structure based on a body-centered lattice only reflections $h\,k\,l$ with $h + k + l$ even are observed.) The first two observed reflections for potassium, $h\,k\,l = 110$ and 200, have $d = 3.70$ Å and 2.62 Å, respectively, at 78°K. What is the value of a? How many nearest neighbors and next-nearest neighbors does each atom have? At what distances?

4

Elements and Compounds
Atomic and Molecular Masses

One of the most important parts of chemical theory is the division of substances into the two classes *elementary substances* and *compounds*. This division was achieved in 1787 by the French chemist Antoine Laurent Lavoisier (1743–1794), on the basis of the quantitative studies that he had made during the preceding fifteen years of the masses of the substances (reactants and products) involved in chemical reactions. Lavoisier defined a compound as a substance that can be decomposed into two or more other substances, and an elementary substance (or element) as a substance that can not be decomposed. In his *Traité Elémentaire de Chimie* [Elementary Treatise on Chemistry], published in 1789, he listed 33 elements, including 10 that had not yet been isolated as elementary substances, but were known as oxides, the compound nature of which was correctly surmised by Lavoisier. Since the discovery of the electron and the atomic nucleus the definitions of elementary substances and compounds have been revised in the ways presented in the following paragraphs.

4-1.   The Chemical Elements

A kind of matter consisting of atoms that all have nuclei with the same electric charge is called an *element*.

For example, all of the atoms that contain nuclei with the charge $+e$, each nucleus having one electron attached to it to neutralize its charge, comprise the element hydrogen, and all of the atoms that contain nuclei with the charge $+92e$ comprise the element uranium.

An *elementary substance* is a substance that is composed of atoms of one element only. An elementary substance is commonly called an element.

A *compound* is a substance that is composed of atoms of two or more different elements. These atoms of two or more different elements must be present in a definite numerical ratio, since compounds are defined as having a definite composition.

Atomic Number

The electric charge of the nucleus of an atom, in units equal to the charge of the proton, is called the atomic number of the atom. It is usually given the symbol Z, the electric charge of a nucleus with atomic number Z being Z times e, with the charge of the proton equal to e, and the charge of the electron equal to $-e$. Thus the simplest atom, that of hydrogen, has atomic number 1; it consists of a nucleus with electric charge e, and an electron with electric charge $-e$.

The Assignment of Atomic Numbers to the Elements

Soon after the discovery of the electron as a constituent of matter it was recognized that elements might be assigned atomic numbers, representing the number of electrons in an atom of each element, but the way of doing this correctly was not known until 1913. In that year H. G. J. Moseley (1887–1915), a young English physicist working in the University of Manchester, found that the atomic number of any element could be determined by the study of the x-rays emitted by an x-ray tube containing the element. By a few months of experimental work he was able to assign their correct atomic numbers to many elements. The apparatus shown in Figure 3-23 resembles that used by Moseley. The x-ray tube is drawn at the left side of this figure. Electrons that come from the cup near the bottom of the tube (as drawn) are speeded up by the electric potential (several thousand volts) applied to the two ends of the tube, and strike the target, which is near the center of the tube. The x-rays are emitted by the atoms of the target when they are struck by the fast-moving electrons.

It was found that the x-rays produced by an x-ray tube contain lines of

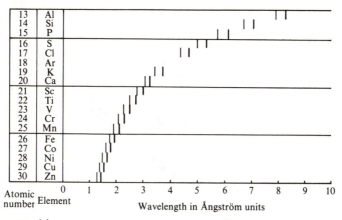

FIGURE 4-1
Diagram showing regular change of wavelength of x-ray emission lines for a
series of elements.

definite wavelengths, characteristic of the material in the target of the x-ray
tube. Moseley measured the wavelengths produced by a number of different elements, and found that they change in a regular way. The wavelengths of the two principal x-ray lines of the elements from aluminum to
zinc (omitting the gas argon) are shown in Figure 4-1.

The regularity in the wavelengths can be shown more strikingly by
plotting the square root of the reciprocals of the wavelengths of the two
x-ray lines for the various elements arranged in the proper sequence,
which is the sequence of the atomic numbers of the elements. In a graph of
this sort, called a Moseley diagram, the points for a given x-ray line lie on
a straight line. The Moseley diagram for the elements from aluminum to
zinc is shown in Figure 4-2. It was easy for Moseley to assign the correct
atomic numbers to the elements with use of a diagram of this sort (see the
discussion of the Bohr theory in Chapter 5).

Isotopes

In a partially evacuated tube through which an electric discharge is
passing (Figure 3-6) the collision of high-velocity electrons with atoms or
molecules may knock one or more electrons from an atom or molecule,
leaving it with a positive electric charge. It is called a *cation*. (Note that a
cation is attracted toward the cathode, the negatively charged terminal;
an *anion*, which is a negatively charged atom or molecule, is attracted
toward the anode, the positive terminal. These names were introduced by
Michael Faraday in 1834, in his discussion of the conduction of electricity

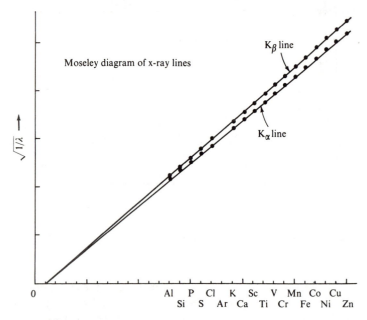

Moseley diagram of x-ray lines

K_β line

K_α line

$\sqrt{1/\lambda}$ →

0

Al	P	Cl	K	Sc	V	Mn	Co	Cu
Si	S	Ar	Ca	Ti	Cr	Fe	Ni	Zn

FIGURE 4-2

A graph of the reciprocal of the square root of the wavelengths of x-ray lines, for the K_α line and the K_β line, of elements, plotted against the order of the elements in the periodic table. This graph, called the Moseley diagram, was used by Moseley in determining the atomic numbers of the elements.

by aqueous solutions of salts.) The cations, attracted toward a perforated cathode, pass through the perforation and emerge on the other side as a beam of positively charged particles. In 1912 J. J. Thomson, using an apparatus with electric and magnetic fields (somewhat similar to that shown in Figure 3-9), found the cations of neon ($Z = 10$) to be of two kinds, one with mass about 20 times the proton mass and the other with mass about 22 times the proton mass. They are called isotopes (from the Greek *isos*, the same, and *topos*, place—in the table of elements), and are said to differ in *mass number*, which is given the symbol A. The element neon as found in nature (the atmosphere) contains 89.97% of the isotope with $A = 20$, 9.73% with $A = 22$, and 0.30% of a third isotope, with $A = 21$, not noticed by Thomson.

All known elements have two or more isotopes. In some cases (such as aluminum) only one isotope occurs naturally, the others being unstable. The maximum number of stable isotopes of any element is 10, possessed by tin.

The chemical properties of all the isotopes of an element are essentially the same. These properties are determined in the main by the atomic number of the nucleus, and not by its mass.

The word *nucleide* (sometimes written *nuclide*) is used for the kind of matter involving nuclei with given values of Z and A. Nucleides of the same element are isotopes.

The Names and Symbols of the Elements

The names of the elements are given in Table 4-1. The chemical symbols of the elements, used as abbreviations for their names, are also given in the table. These symbols are usually the initial letters of the names, plus another letter when necessary. In some cases the initial letters of Latin names are used: Na for sodium (natrium), K for potassium (kalium), Fe for iron (ferrum), Cu for copper (cuprum), Ag for silver (argentum), Au for gold (aurum), Hg for mercury (hydrargyrum), Sn for tin (stannum), Sb for antimony (stibium), and Pb for lead (plumbum). The system of chemical symbols was proposed by the Swedish chemist Jöns Jakob Berzelius (1779–1848) in 1811.

The elements are also shown in a special arrangement, the *periodic table*, at the front of the book and in Table 5-3.

A symbol is used to represent an atom of an element, as well as the element itself. The symbol I represents the element iodine, and also may be used to mean the elementary substance. However I_2 is the customary formula for the elementary substance, because it is known that elementary iodine consists of molecules containing two atoms in the solid and liquid states as well as in the gaseous state (except at very high temperature). In formulas showing composition or molecular structure the numerical subscript of the symbol of an element gives the number of atoms of the element in the molecule.

4-2. The Neutron. The Structure of Nuclei

In 1921 the American chemist W. D. Harkins described nuclei as being built of protons and neutrons; he used the word "neutron" for a hypothetical particle with mass equal to that of the proton and with no electric charge. Ernest Rutherford made a similar suggestion during the same year. In 1932 the discovery of the neutron was reported by the English physicist James Chadwick (born 1891). It had been observed in 1930 by two German investigators, W. Bothe and H. Becker, that a very penetrating radiation is produced when beryllium metal is bombarded with alpha particles from radium. Bothe and Becker considered the radiation to consist of γ-rays. Frédéric Joliot (1900–1958) and his wife Irene Joliot-

TABLE 4-1
International Atomic Weights

Element	Symbol	Atomic Number	Atomic Weight	Element	Symbol	Atomic Number	Atomic Weight
Actinium	Ac	89	[227]*	Erbium	Er	68	167.26
Aluminum	Al	13	26.9815	Europium	Eu	63	151.96
Americium	Am	95	[243]	Fermium	Fm	100	[253]
Antimony	Sb	51	121.75	Fluorine	F	9	18.9984
Argon	Ar	18	39.948	Francium	Fr	87	[223]
Arsenic	As	33	74.9216	Gadolinium	Gd	64	157.25
Astatine	At	85	[210]	Gallium	Ga	31	69.72
Barium	Ba	56	137.34	Germanium	Ge	32	72.59
Berkelium	Bk	97	[247]	Gold	Au	79	196.967
Beryllium	Be	4	9.0122	Hafnium	Hf	72	178.49
Bismuth	Bi	83	208.980	Helium	He	2	4.0026
Boron	B	5	10.811 †	Holmium	Ho	67	164.930
Bromine	Br	35	79.909 ‡	Hydrogen	H	1	1.00797 †
Cadmium	Cd	48	112.40	Indium	In	49	114.82
Calcium	Ca	20	40.08	Iodine	I	53	126.9044
Californium	Cf	98	[249]	Iridium	Ir	77	192.2
Carbon	C	6	12.01115 †	Iron	Fe	26	55.847 ‡
Cerium	Ce	58	140.12	Khurchatovium	Kh	104	[260]
Cesium	Cs	55	132.905	Krypton	Kr	36	83.80
Chlorine	Cl	17	35.453 ‡	Lanthanum	La	57	138.91
Chromium	Cr	24	51.996 ‡	Lawrencium	Lr	103	[257]
Cobalt	Co	27	58.9332	Lead	Pb	82	207.19
Copper	Cu	29	63.54	Lithium	Li	3	6.939
Curium	Cm	96	[247]	Lutetium	Lu	71	174.97
Dysprosium	Dy	66	162.50	Magnesium	Mg	12	24.312
Einsteinium	Es	99	[254]	Manganese	Mn	25	54.9380

Curie (1897–1956) then discovered that this radiation from beryllium, when passed through a block of paraffin or other substance containing hydrogen, produces a large number of protons. Because of the difficulty of understanding how protons could be produced by γ-rays, Chadwick carried out a series of experiments that led to the discovery that the rays from beryllium are in fact composed of particles with no electric charge and with mass approximately equal to that of the proton. Because they have no electric charge, neutrons interact with other forms of matter very weakly, except at very small distances, less than about 5 fm (5×10^{-15} m).

The neutron has mass 1.67470×10^{-27} kg, which is 0.14% greater than that of the proton. Like the proton, it has a spin, with the spin quantum

Element	Symbol	Atomic Number	Atomic Weight	Element	Symbol	Atomic Number	Atomic Weight
Mendelevium	Md	101	[256]	Ruthenium	Ru	44	101.07
Mercury	Hg	80	200.59	Samarium	Sm	62	150.35
Molybdenum	Mo	42	95.94	Scandium	Sc	21	44.956
Neodymium	Nd	60	144.24	Selenium	Se	34	78.96
Neon	Ne	10	20.183	Silicon	Si	14	28.086 †
Neptunium	Np	93	[237]	Silver	Ag	47	107.870 ‡
Nickel	Ni	28	58.71	Sodium	Na	11	22.9898
Niobium	Nb	41	92.906	Strontium	Sr	38	87.62
Nitrogen	N	7	14.0067	Sulfur	S	16	32.064 †
Nobelium	No	102	[256]	Tantalum	Ta	73	180.948
Osmium	Os	76	190.2	Technetium	Tc	43	[97]
Oxygen	O	8	15.9994 †	Tellurium	Te	52	127.60
Palladium	Pd	46	106.4	Terbium	Tb	65	158.924
Phosphorus	P	15	30.9738	Thallium	Tl	81	204.37
Platinum	Pt	78	195.09	Thorium	Th	90	232.038
Plutonium	Pu	94	[242]	Thulium	Tm	69	168.934
Polonium	Po	84	[210]	Tin	Sn	50	118.69
Potassium	K	19	39.102	Titanium	Ti	22	47.90
Praseodymium	Pr	59	140.907	Tungsten	W	74	183.85
Promethium	Pm	61	[147]	Uranium	U	92	238.03
Protactinium	Pa	91	[231]	Vanadium	V	23	50.942
Radium	Ra	88	[226]	Xenon	Xe	54	131.30
Radon	Rn	86	[222]	Ytterbium	Yb	70	173.04
Rhenium	Re	75	186.2	Yttrium	Y	39	88.905
Rhodium	Rh	45	102.905	Zinc	Zn	30	65.37
Rubidium	Rb	37	85.47	Zirconium	Zr	40	91.22

*A value given in brackets is the mass number of the most stable known isotope.

†The atomic weight varies because of natural variations in the isotopic composition of the element. The observed ranges are boron, ± 0.003; carbon, ± 0.00005; hydrogen, ± 0.00001; oxygen, ± 0.0001; silicon, ± 0.001; sulfur, ± 0.003.

‡The atomic weight is believed to have an experimental uncertainty of the following magnitude: bromine, ± 0.002; chlorine, ± 0.001; chromium, ± 0.001; iron, ± 0.003; silver, ± 0.003. For other elements the last digit given is believed to be reliable to ± 5.

number $s = \frac{1}{2}$. Although it has no charge, it has a magnetic moment, $\mu = -3.3137$ nuclear magnetons (the negative sign means that the magnetic moment corresponds to the rotation of negative charge).

Although the detailed structures of nuclei are not known, physicists seem to be agreed in accepting the idea that nuclei can all be described as being built up of protons and neutrons.

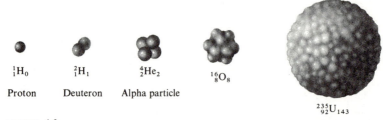

$_1^1H_0$ $_1^2H_1$ $_2^4He_2$ $_8^{16}O_8$

Proton   Deuteron   Alpha particle

$_{92}^{235}U_{143}$

FIGURE 4-3

Hypothetical structures of some atomic nuclei. We do not yet know just how these
nuclei are constructed out of elementary particles, but it is known that nuclei are
about 10^{-15} m in diameter, and are, accordingly, very small even compared with atoms.

Let us first discuss, as an example, the *deuteron*. This is the nucleus of
the deuterium atom. The deuteron has the same electric charge as the
proton, but has about twice the mass of the proton. It is thought that the
deuteron is made of one proton and one neutron, as indicated in Figure 4-3.

The nucleus of the helium atom, the alpha particle, has electric charge
twice as great as that of the proton, and mass about four times as great as
that of the proton. It is thought that the alpha particle is composed of two
protons and two neutrons.

In Figure 4-3 there is also shown a drawing representing the nucleus of
an oxygen atom, composed of eight protons and eight neutrons. The mass
of this oxygen nucleus is about 16 mass units.

There is also shown in the figure a hypothetical drawing of the nucleus
of a uranium atom. This nucleus is composed of 92 protons and 143
neutrons. The electric charge of this nucleus is 92 times that of the proton;
it would be neutralized by the negative charges of 92 electrons. The mass
of this nucleus is about 235 times the mass of the proton.

The number of neutrons in a nucleus is represented by the letter N. The
mass number A is the number of neutrons, N, plus the number of protons,
Z. As shown in Figure 4-3, the symbol for a nuclide is $_Z^A X_N$. (Formerly
the symbol $_Z X_N^A$ was used.) This symbol is, of course, redundant, in that
$A = Z + N$, and the letter X (U for uranium, for example) carries the
same information as Z (92 for U). $^A X$ would be concise; but it is often
convenient to the reader for the complete symbols to be given.

The complete symbol for the neutron is $_0^1 n_1$, and that for the proton $_1^1 p_0$.

Nuclides with the same value of the neutron number N are called
isotones. This word is not derived from a Greek root, but was formulated
by analogy with the word isotopes, through replacing the letter p (protons)
by n (neutrons). Nuclides with the same value of A are called *isobars*.

4-3.   Chemical Reactions

The formula of a compound should give as much as possible of the information known about its composition and structure. The formula of benzene is written C_6H_6, not CH; it is known that the benzene molecule contains six carbon atoms and six hydrogen atoms. The formula of crystalline copper sulfate pentahydrate is written $CuSO_4(H_2O)_5$ or $CuSO_4 \cdot 5H_2O$, to show that it contains the sulfate group SO_4 and that five molecules of water are easily removed; the name also reflects these facts.

The equations for chemical reactions can be correctly written if the nature of the products is known. For example, in the reaction of a rocket propellant made of carbon and potassium perchlorate, $KClO_4$, the products may be potassium chloride, KCl, and either carbon monoxide or carbon dioxide, or a mixture of the two. It would probably be wise to write two equations, corresponding to two reactions:

$$KClO_4 + 4C \longrightarrow KCl + 4CO$$

$$KClO_4 + 2C \longrightarrow KCl + 2CO_2$$

For each of these equations the same number of atoms of each element is shown on the right side as on the left side: the equations are balanced. Writing a balanced equation for a reaction is often the first step in solving a chemical problem.

4-4.   Nuclidic Masses and Atomic Weights

The accepted unit of mass for nuclides is the *dalton*, symbol d. The dalton is defined as exactly $\frac{1}{12}$ of the mass of the neutral atom $^{12}_6C_6$. The nuclidic mass of ^{12}C is exactly 12.00000 d. The dalton is approximately 1.66033×10^{-27} kg.

In a quantitative discussion of an ordinary chemical reaction the relative amounts of various substances that react or are formed can be calculated from the numbers of atoms of different elements and the masses of the atoms. If an element is present as a mixture of isotopes, it is the average atomic mass that is needed. It is this average atomic mass for an element that is called the *chemical atomic weight*. The unit for atomic weight is the dalton.

The name atomic mass might be adopted in place of atomic weight. There might, however, be some advantage to retaining the old name atomic weight in referring to the natural mixtures of isotopes that are involved in most chemical reactions, and it seems likely that this usage will continue for some time.

The History of the Atomic Weight Scale

John Dalton chose the value 1 for hydrogen as the base of his scale of atomic weights. The Swedish chemist J. J. Berzelius used 100 for oxygen, and the Belgian chemist J. S. Stas (1813–1891), who carried out many quantitative analyses of compounds, proposed 16 for oxygen (the natural mixture of isotopes), and this base was used for many years. For several decades nuclidic masses were expressed on a scale (called the physical scale) based on $\frac{1}{16}$th the mass of the neutral atom $^{16}_{8}O_8$; the chemical atomic-weight unit was then 1.000272 times the physical atomic-mass unit. This period of confusion was brought to an end in 1961 by the acceptance of $\frac{1}{12}$th the mass of $^{12}_{6}C_6$ as the unit for both atomic weights and nuclidic masses.

4-5.   Avogadro's Number. The Mole

Avogadro's number (N) is defined as the number of carbon-12 atoms in exactly 12 g of carbon 12. Its value is approximately 0.60229×10^{24}. It was named for the Italian physicist Amedeo Avogadro, whose work is discussed in Chapter 9.

A *mole* of a substance is defined as Avogadro's number of molecules of the substance. Thus a mole of water, H_2O, is the quantity of water containing $N\ H_2O$ molecules. The molecular weight of water (the sum of the atomic weights of the atoms in the molecule: $2 \times 1.00797 + 1 \times 15.9994$, from Table 4-1) is 18.0153. From the definitions of atomic weights and the mole we see that a mole of water is 18.0153 g.

A mole of iodine atoms is 126.9044 g of iodine, and a mole of iodine molecules (I_2) is 253.8088 g of iodine. Sometimes, to prevent confusion, a mole of atoms of an element is called a *gram-atom*. The weight of a mole of a compound for which a formula is written is sometimes taken to be the gram-formula-weight (gfw) (the number of grams equal to the sum of the atomic weights corresponding to the formula), whether or not the formula is a correct molecular formula for the substance. Thus the molar weight of liquid acetic acid, for which the formula $CH_3COOH(l)$ is written, is taken as 60.05, even though it is likely that the liquid contains some *dimers* (double molecules, $(CH_3COOH)_2$), as does the vapor.

Often, as shown above, the state of aggregation of a substance is represented by appended letters: Cu(c) refers to crystalline copper (c standing for crystal; sometimes, for solid, s is used), Cu(l) to liquid copper, and Cu(g) to gaseous copper. A substance in solution is sometimes represented by its formula followed by the name of the solvent in parentheses; (aq) is used for aqueous solutions.

4-6.   Examples of Weight-relation Calculations

The way to work a weight-relation problem is to think about the problem, in terms of atoms and molecules, and then to decide how to carry out the calculations. You should not memorize any rule about these problems—such rules are apt to confuse you and to cause you to make mistakes.

The way to work these problems is best indicated by some examples.

In general, chemical problems may be solved by using a slide rule for the numerical work. This gives about three reliable figures in the answer, which is often all that is justified by the accuracy of the data. Sometimes the data are more reliable, and logarithms or long-hand calculations might be used to obtain the answer with the accuracy required. Unless the problem requires unusual accuracy, you may round values of atomic weight off to the first decimal point.

Example 4-1.   What is the percentage of lead in galena, PbS? Calculate to 0.1%.

Solution.   The formula weight of PbS is obtained by adding the atomic weights of lead and sulfur, which we obtain from the table inside the front cover:

$$\text{Weight of one lead atom (1 Pb)} = 207.2 \text{ d}$$
$$\text{Weight of one sulfur atom (1 S)} = 32.1 \text{ d}$$
$$\text{Weight of 1 PbS} = 239.3 \text{ d}$$

Hence 239.3 d of PbS contains 207.2 d of lead. We see that 100.0 g of PbS would contain

$$\frac{207.2 \text{ d of lead}}{239.3 \text{ d of galena}} \times 100.0 \text{ g of galena} = 86.6 \text{ g of lead}$$

Hence the percentage of lead in galena is 86.6%.

Example 4-2.   A propellant for rockets can be made by mixing powdered potassium perchlorate, $KClO_4$, and powdered carbon (carbon black), C, with a little adhesive to bind the powdered materials together. What weight of carbon should be mixed with 1000 g of potassium perchlorate in order that the products of the reaction be KCl and CO_2?

Solution.   Taking the equation for the reaction as

$$KClO_4 + 2C \longrightarrow KCl + 2CO_2$$

we first calculate the formula weight of $KClO_4$:

$$\text{Weight of 1K} = 39.1$$
$$\text{Weight of 1Cl} = 35.5$$
$$\text{Weight of 4O} = 4 \times 16.0 = 64.0$$
$$\text{Weight of } KClO_4 = 138.6$$

The atomic weight of carbon is 12.0; the weight 2C is 24.0. Hence the weight of carbon required is 24.0/138.6 times the weight of potassium perchlorate:

$$\frac{24.0 \ (C)}{138.6 \ (KClO_4)} \times 1000 \ g \ (KClO_4) = 173 \ g \ (C)$$

Hence about 173 g of carbon* is required for 1000 g of potassium perchlorate.

Example 4-3.   In a determination of the atomic weight of iron, 7.59712 g of carefully purified ferric oxide, Fe_2O_3, was reduced by heating in a stream of hydrogen, and found to yield 5.31364 g of metallic iron. Given the atomic weight of oxygen, to what value of the atomic weight of iron does this result lead?

Solution.   The difference between the weight of ferric oxide and the weight of iron, 2.28348 g, is the weight of the oxygen contained in the sample of ferric oxide. From the formula we see that two thirds of this quantity is the weight of the number of oxygen atoms equal to the number of iron atoms in the sample, 5.31364 g. The atomic weight of oxygen is 15.9994 (Table 4-1). Hence the atomic weight of iron is

$$\frac{5.31364}{2/3 \times 2.28348} \times 15.9994 = 55.8457$$

This value for the atomic weight of iron, 55.8457, is from an investigation made by G. P. Baxter and C. R. Hoover fifty years ago. It agrees closely with the value in Table 4-1, 55.847.

Example 4-4.   An oxide of arsenic contains 65.2% arsenic. What is its simplest formula?

Solution.   In 100 g of this oxide of arsenic there are contained, according to the reported analysis, 65.2 g of arsenic and 34.8 g of oxygen. If we divide 65.2 g by the gram-atomic weight of arsenic, 74.9 g, we obtain 0.870 as the number of gram-atoms of arsenic. Similarly, by dividing 34.8 g by the gram-atomic weight of oxygen, 16 g, we obtain 2.17 as the number of gram-atoms of oxygen in 100 g of this oxide of arsenic. Hence the numbers of atoms of arsenic and oxygen in the compound are in the ratio 0.870 to 2.17. If we set this ratio, 0.870/2.17, on the slide rule, we see that it is very close to 2/5, being 2/4.99. Hence the simplest formula is As_2O_5.

We say that this is the simplest formula in order not to rule out the possibility that the substance contains more complex molecules, such as As_4O_{10}, in which case it would be proper to indicate in the formula the larger numbers of atoms per molecule.

---

*Note the convention in chemistry that "173 g of carbon" is singular in number; it means a quantity of carbon weighing 173 g, rather than 173 separate grams of carbon.

Example 4-5.   A substance is found by qualitative tests to consist of only carbon and hydrogen (it is a hydrocarbon). A quantitative analysis is made of the substance by putting a weighed amount, 0.2822 g, in a tube that can be strongly heated from outside, and then burning it in a stream of dry air. The air containing the products of combustion is passed first through a weighed tube containing calcium chloride, which absorbs the water vapor, and then through another weighed tube containing a mixture of sodium hydroxide and calcium oxide, which absorbs the carbon dioxide. When the first tube is weighed, after the combustion is completed, it is found to have increased in weight by 0.1598 g, this being accordingly the weight of water produced by the combustion of the sample. The second tube is found to have increased in weight by 0.9768 g. What is the simplest formula of the substance?

Solution.   It is convenient to solve this problem in steps. Let us first find out how many moles of water were produced. The number of moles of water produced is found by dividing 0.1598 g by 18.02 g, the weight of a mole of water; it is 0.00887. Each mole of water vapor contains two gram-atoms of hydrogen; hence the number of gram-atoms of hydrogen in the original sample is twice this number, or 0.01774.

Similarly, the number of moles of carbon dioxide in the products of combustion is obtained by dividing the weight of carbon dioxide, 0.9768 g, by the molar weight of the substance, 44.01 g. It is 0.02219, which is also the number of gram-atoms of carbon in the sample of substance, because each molecule of carbon dioxide contains one atom of carbon.

The original substance accordingly contained carbon atoms and hydrogen atoms in the ratio 0.02219 to 0.01774. This ratio is found on calculation to be 1.251, which is equal to $\frac{5}{4}$, to within the accuracy of the analysis. Accordingly the simplest formula for the substance is C_5H_4.

If the analyst had smelled the substance, and noticed an odor resembling moth balls, he would have identified the substance as naphthalene, $C_{10}H_8$.

4-7.   Determination of Atomic Weights by the Chemical Method

It is hard to overestimate the importance of the table of atomic weights. Almost every activity of a chemist involves the use of atomic weights in some way. For nearly two centuries successive generations of chemists have carried out experiments in the effort to provide more and more accurate values of the atomic weights, in order that chemical calculations could be carried out with greater accuracy.

Until recently almost all atomic weight determinations were made by the chemical method. This method consists in determining the amount of the element that will combine with one gram-atom of oxygen or of another element with known atomic weight. One example has been given above (Example 4-3); the following is another one, which was important in the development of the theory of radioactivity.

> **Example 4-6.** A 1.0000-g sample of lead sulfide, PbS, prepared from an ore of uranium (Katangan curite) was found to give 0.8654 g of metallic lead on reduction. Another 1.0000-g sample of lead sulfide, prepared from an ore of thorium (Norwegian thorite) was found to give 0.8664 g of metallic lead on reduction. Assuming the atomic weight of sulfur to be 32.064, calculate the two values of the atomic weight of lead.

> **Solution.**   For the curite lead sulfide we see that the Pb/S ratio of atomic weights is 0.8654/0.1346, and that the atomic weight of lead is accordingly 0.8654/0.1346 × 32.064 = 206.15. For thorite lead sulfide a similar calculation gives 0.8664/0.1336 × 32.064 = 207.94.

Results of this sort were obtained in 1914 by the American chemist Theodore William Richards (1868–1928). They confirmed the deductions about the existence of isotopes that had been made in 1913 by the English chemist Frederick Soddy (1877–1956) from the study of radioactive transformations of uranium and thorium.

4-8. Determination of Atomic Weights by Use of the Mass Spectrograph

The apparatus used by J. J. Thomson in discovering two isotopes of neon, described in Section 4-1, was a simple form of mass spectrograph. Modern mass spectrographs have been used in attacking many physical and chemical problems, including that of determining nuclidic masses and the abundance ratios of isotopes.

The principle of the modern mass spectrograph can be illustrated by the simple apparatus shown in Figure 4-4.

At the left is a chamber in which positive ions are formed by an electric discharge, and then accelerated toward the right by an electric potential. The ions passing through the first pinhole have different velocities; in the second part of the apparatus a beam of ions with approximately a certain velocity is selected, and allowed to pass through the second pinhole, the ions with other velocities being stopped. (We shall not attempt to describe the construction of the velocity selector.) The ions passing through the second pinhole then move on between two metal plates, one of which has a positive electric charge and the other a negative charge. The ions accordingly undergo an acceleration toward the negative plate, and are

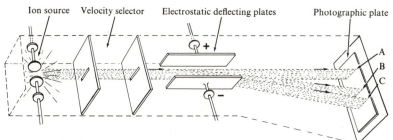

Details of vacuum chamber and velocity selector not shown

FIGURE 4-4
Diagram of a simple mass spectrograph.

deflected from the straight path A that they would pursue if the plates were not charged.

The force acting on an ion between these plates is proportional to $+ze$, its electric charge (z being the number of missing electrons), and its inertia is proportional to its mass M. The amount of deflection is hence determined by ze/M, the ratio of the charge of the ion to its mass.

Of two ions with the same charge, the lighter one will be deflected in this apparatus by the greater amount. The beam C might accordingly represent the ion C^+, with charge $+e$ and mass 12 atomic weight units (the atomic weight of carbon), and the beam B the heavier ion O^+, with the same charge but with mass 16.

Of two ions with the same mass, the one with the greater charge will be deflected by the greater amount. Beams B and C might represent O^{++} and O^{+++}, respectively.

By measuring the deflection of the beams, relative values of ze/M for different ions can be determined. Since e is constant, relative values of ze/M for different ions are also inverse relative values of M/z: therefore this method permits the direct experimental determination of the relative masses of atoms, and hence of their atomic weights.

The value of the integer z—the degree of ionization of the ions—can usually be fixed from knowledge of the substances present in the discharge tube; thus neon gives ions with $M/z = 20$ and 22 ($z = 1$), 10 and 11 ($z = 2$), and so on.

Instead of the mass spectrograph described above, others of different design, using both an electric field and a magnetic field, are usually used. These instruments are designed so that they focus the beam of ions with a given value of M/z into a sharp line on a photographic plate or other detector. An instrument of this sort, using both an electric field, with

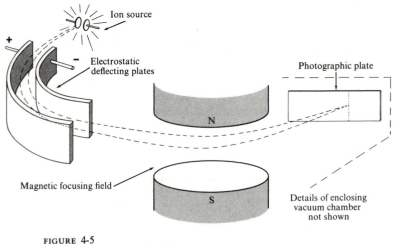

FIGURE 4-5
A focusing mass spectrograph, using both electrostatic and magnetic
deflection of the beam of ions.

curved plates, and a magnetic field, is sketched in Figure 4-5. Focusing
mass spectrographs can also be made with the direction of magnetic deflection opposite to that of electric deflection.

In the time-of-flight mass spectrograph successive pulses of ions are
accelerated through a long tube (2 m long), and the number of ions reaching the detector is displayed on a screen as a function of time after the
start of the pulse. The ions with the largest ratio of charge to mass have the
smallest time of flight.

Modern types of mass spectrographs have an accuracy of about one
part in 200,000 and a resolving power of 20,000 (that is, they are able to
separate ion beams with values of M/z differing by only one part in 20,000)
The great accuracy of modern mass spectrographs makes the mass-
spectrographic method of determining atomic weights more useful and
important at present than the chemical method.

Mass-spectrographic comparisons with ^{12}C or ^{16}O are made in the
following way. An ion source that produces ions both of carbon (or
oxygen) and of the element to be investigated is used; the lines of carbon
or oxygen and of the element in such states of ionization that their ze/M
values are nearly the same are then obtained; thus for ^{32}S, ^{33}S, and ^{34}S the
lines for the doubly ionized atoms would lie near the line for singly
ionized oxygen. Accurate relative measurements of these lines can then
be made.

Examples of the Determination of Atomic Weights with the Mass Spectograph

For a simple element, with only one isotope, the value of the atomic mass of that isotope is the atomic weight of the element. Thus for gold, which consists entirely of one stable isotope, ^{197}Au, the mass-spectrographic mass is reported to be 196.967. This value has been accepted by the International Committee on Atomic Weights.

For elements with two or more stable isotopes the atomic weight can be calculated from the masses of the isotopes and values of their relative amounts, as shown in the following example.

> **Example 4-7.**   Silver has two stable isotopes. Their masses as determined by use of the mass spectrograph were found to be 106.902 and 108.900, with relative amounts (numbers of nuclei) 51.35% and 48.65%, respectively. What is the atomic weight of silver, from this investigation?
>
> **Solution.**   The average atomic weight is $0.5135 \times 106.902 + 0.4865 \times 108.900$. We see that this can be rewritten as $106.902 + 0.4865(108.900 - 106.902) = 106.902 + 0.4865 \times 1.998 = 106.902 + 0.972 = 107.874$. The value of the atomic weight of silver is accordingly calculated to be 107.874.

4-9.   Determination of Nuclidic Masses by Nuclear Reactions

A tremendous amount of information about nuclidic masses has been obtained by studying the energy released in nuclear reactions. A discussion of nuclear reactions is given in Chapter 26. The following example illustrates the general method.

> **Example 4-8.**   The fraction 0.0118% of the element potassium in its natural occurrence is the radioactive isotope $^{40}_{19}$K$_{21}$. It undergoes beta decay:
>
> $$^{40}_{19}\text{K}_{21} \longrightarrow \, ^{40}_{20}\text{Ca}^{+}_{20} + e^{-}$$
>
> The kinetic energy of the emitted electron (the beta ray) has been measured by use of a magnetic beta-ray spectrometer (measurement of the curvature of the path of the beta ray in a magnetic field) and found to be 1.32 MeV. The mass of ^{40}Ca, which is stable and constitutes 97% of natural calcium, is 39.96259 d, as determined with the mass spectrograph. What is the mass of ^{40}K?

Solution. The energy of the reaction $Ca^+ + e^- \longrightarrow Ca$ (6 eV) is negligible. Hence the energy given out in the reaction $^{40}K \longrightarrow {}^{40}Ca$ is 1.32 MeV, which, with $e = 0.1602 \times 10^{-18}$ C, equals 0.211×10^{-12} J. We use the equation $E = mc^2$ to convert this energy to mass. Division by $c^2 = (2.9979 \times 10^8)^2$ gives 2.35×10^{-30} kg as the amount of mass changed to kinetic energy. The dalton is 1.660×10^{-27} kg. Hence the decrease in mass from ^{40}K to ^{40}Ca is $2.35 \times 10^{-30}/1.660 \times 10^{-27} = 0.00142$ d, and the mass of ^{40}K is $39.96259 + 0.00142 = 39.96401$ d.

4-10. The Discovery of the Correct Atomic Weights. Isomorphism

In the early years of the atomic theory there was no secure knowledge of the true relative weights of different elements. Dalton assigned atomic weights in such a way as to lead to simple formulas for compounds. Many chemists continued to write HO for water until 1858. In that year a principle discovered much earlier (1811) by Avogadro was applied so effectively by Stanislao Cannizzaro (1826–1910) as to convince most chemists that the atomic weights obtained by its use could be accepted as correct. This principle will be discussed in a later chapter.

Some other methods for assigning correct atomic weights were also developed during the first half of the nineteenth century. One of them (the rule of Dulong and Petit) will be discussed in Section 10-3.

In 1819 the German chemist Eilhard Mitscherlich (1794–1863) discovered the phenomenon of *isomorphism*, the existence of different crystalline substances with essentially the same crystal form, and suggested his rule of isomorphism, which states that isomorphous crystals have similar chemical formulas.

As an example of isomorphism we may consider the minerals rhodochrosite, $MnCO_3$, and calcite, $CaCO_3$. Crystals of these two substances resemble one another closely, as shown in Figure 4-6. They both belong to the hexagonal crystal system (Appendix III), and they both have excellent rhombohedral cleavage. The larger of the two angles of a rhombic face of the cleavage rhombohedron has the value $102°50'$ for rhodochrosite and $101°55'$ for calcite. These facts justified the description of the two crystals as isomorphous more than a century ago. When x-ray diffraction was discovered it was possible to verify that the crystals have the same structure.

An illustration of the use of the rule of isomorphism is given by the work of the English chemist Henry E. Roscoe in determining the correct atomic weight of vanadium. Berzelius had attributed the atomic weight 68.5 to vanadium in 1831. In 1867 Roscoe noticed that the corresponding

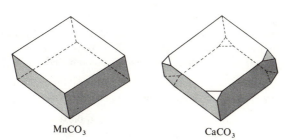

MnCO₃   CaCO₃

FIGURE 4-6
Isomorphous crystals of rhodochrosite and calcite
(hexagonal system).

formula for the hexagonal mineral vanadinite was not analogous to the
formulas of other hexagonal minerals apparently isomorphous with it:

		Axial ratio
Apatite	$Ca_5(PO_4)_3F$	$c/a = 1.363$
Hydroxyapatite	$Ca_5(PO_4)_3OH$	1.355
Pyromorphite	$Pb_5(PO_4)_3Cl$	1.362
Mimetite	$Pb_5(AsO_4)_3Cl$	1.377
Vanadinite	$Pb_5(VO_3)_3Cl$ (wrong)	1.404

The formula for vanadinite analogous to the other formulas is
$Pb_5(VO_4)_3Cl$. On reinvestigating the compounds of vanadium Roscoe
found that this latter formula is indeed the correct one, and that Berzelius
had accepted VO, vanadium monoxide, as the elementary substance. The
atomic weight of vanadium now accepted is 50.942.

There are occasional exceptions to the rule of isomorphism. An example
is provided by hydroxyapatite, which in the form of its crystals is classed
as isomorphous with apatite, but which has an extra atom in its formula.
The explanation of this deviation from the rule is that the hydrogen atom
is smaller than other atoms, and in the hydroxyapatite crystal the hydrogen
atoms occupy interstices among the larger atoms that are vacant in the
apatite crystal.

Exercises

4-1. What is an element? Is it possible to prove rigorously by chemical methods that a substance is an element? Is it possible to prove rigorously by chemical methods that a substance is a compound?

4-2. Describe two chemical experiments that would prove that water is not an element. Can you devise a chemical proof that oxygen is an element?

4-3. Gasoline burns to form water and carbon dioxide. Does this prove rigorously that gasoline is not an element?

4-4. Use the Moseley diagram (Figure 4-2) to make a prediction of the approximate value of the wavelength of the $K\alpha$ x-ray emission line of argon. Can you surmise why it was not measured by Moseley?

4-5. Define isotope, isotone, isobar, nuclide, mass number, N, A, Z, the dalton.

4-6. What are the atomic number and approximate atomic weight of the element each of whose nuclei contains 81 protons and 122 neutrons? Give the complete symbol for this nuclide, including chemical symbol, atomic number, mass number, and neutron number.

4-7. How many protons and how many neutrons are there in the nucleus of the isotope of cobalt with mass number 60? Of the isotope of nickel with mass number 60? Of the isotope of plutonium with mass number 238?

4-8. An atom of ^{90}Sr emits a beta ray. What are the atomic number and mass number of the resulting nucleus? What element is it? This nucleus also emits a beta ray. What nucleus does it produce?

4-9. Argon, potassium, and calcium all have nuclides with mass number 40. How many protons and how many neutrons constitute each of the three nuclei?

4-10. What are the relative advantages of basing the chemists' atomic weight scale on H = 1.000, ^{16}O = 16.00000, or ^{12}C = 12.00000?

4-11. What would be the implications of defining Avogadro's number as 1.00000×10^{24}? What units would have to be changed?

4-12. By counting the flashes of light produced by alpha particles when they strike a screen coated with zinc sulfide, Sir William Ramsay and Professor Frederick Soddy found that 1 g of radium gives off 13.8×10^{10} alpha particles (nuclei of helium atoms) per second. They also measured the amount of helium gas produced in this way, finding 0.158 cm³ (at 0°C and 1 atm) per year per gram of radium. At this temperature and pressure 1 l of helium weighs 0.179 g. Avogadro's number of helium atoms weighs 4.003 g (the atomic weight of helium is 4.003). From these data calculate an approximate value of Avogadro's number.

4-13. The molecule of the anesthetic agent nitrous oxide, N_2O, contains two atoms of nitrogen and one of oxygen. Using Avogadro's number and the atomic weights of nitrogen and oxygen, calculate the weight in kilograms of one atom of oxygen and that of two atoms of nitrogen. Also calculate the weight of the nitrous oxide molecule. What is the percentage composition by weight of nitrous oxide in terms of nitrogen and oxygen?

4-14. Balance the following equations of chemical reactions. The molecular formulas are correct.

$$Fe_2O_3 + C \longrightarrow Fe + CO_2$$
$$Ag + S_8 \longrightarrow Ag_2S$$
$$C_{12}H_{22}O_{11} + O_2 \longrightarrow CO_2 + H_2O$$
$$H_3PO_4 + NaOH \longrightarrow Na_3PO_4 + H_2O$$
$$Li + H_2O \longrightarrow LiOH + H_2$$
$$HCl + Ba(OH)_2 \longrightarrow BaCl_2 + H_2O$$
$$CuO + NH_3 \longrightarrow N_2 + Cu + H_2O$$
$$N_2 + H_2 \longrightarrow NH_3$$
$$H_3BO_3 \longrightarrow B_2O_3 + H_2O$$
$$Fe + F_2 \longrightarrow FeF_3$$

4-15. How much coal (assumed to be pure carbon) is needed to reduce one ton of Fe_2O_3 to iron? How much iron is produced?

4-16. Exactly what is meant by the statement that the atomic weight of samarium is 150.35?

4-17. Thallium occurs in nature as ^{203}Tl and ^{205}Tl. From the atomic weight of natural thallium, 204.39, calculate the nucleide composition of thallium. The atomic mass of ^{203}Tl is 202.97 and that of ^{205}Tl is 204.97.

4-18. The density of NaCl(c) is 2.165 g cm^{-3}. Calculate the molar volume, and, with use of Avogadro's number, the volume of the unit cube, containing four sodium atoms and four chlorine atoms, and the value of a, the edge of the unit cube. (This is the method used by the Braggs in their determination of the wavelengths of x-rays, Example 3-7.)

5

Atomic Structure and the Periodic Table of the Elements

It was discovered about one hundred years ago that many physical and chemical properties of the elements vary in a roughly periodic way with increasing atomic weight. The study of chemistry is greatly aided by continued reference to the *periodic table* (Table 5-3), an arrangement of the elements determined by their atomic numbers, in which this periodicity is made evident. Moreover, the detailed electronic structures of atoms have been determined during the last fifty years, and it has been found possible to correlate these electronic structures with the properties of the elements in a reasonably satisfactory way. These matters are discussed in the following sections.

5-1. The Bohr Theory of the Hydrogen Atom

Most of our knowledge of the electronic structure of atoms has been obtained by the study of the light given out by atoms when they are excited by high temperature or by an electric arc or spark. The light that

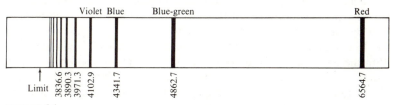

FIGURE 5-1
The Balmer series of spectral lines of atomic hydrogen. The line at the right, with the longest wavelength, is H_α. It corresponds to the transition from the state with $n = 3$ to the state with $n = 2$.

is emitted by atoms of a given substance can be refracted or diffracted into a distinctive pattern of lines of certain frequencies; such a distinctive pattern of lines is described as the *line spectrum* of the atom.

The careful study of line spectra began about 1880. Early investigators made some progress in the interpretation of spectra, in recognizing regularities in the frequencies of the lines: the frequencies of the spectral lines of the hydrogen atom, for example, show an especially simple relationship with one another, which will be discussed below. The regularity is evident in the reproduction of a part of the spectrum of hydrogen in Figure 5-1. It was not until 1913, however, that the interpretation of the spectrum of hydrogen in terms of the electronic structure of the hydrogen atom was achieved. In that year the Danish physicist Niels Bohr (1885–1962) successfully applied the quantum theory to this problem, and laid the basis for the extraordinary advance in our understanding of the nature of matter that has been made since then.

The hydrogen atom consists of an electron and a proton. The interaction of their electric charges, $-\epsilon$ and $+\epsilon$, respectively, corresponds to inverse-square attraction, in the same way that the gravitational interaction of the earth and the sun corresponds to inverse-square attraction. If Newton's laws of motion were applicable to the hydrogen atom we should accordingly expect that the electron, which is light compared with the nucleus, would revolve about the nucleus in an elliptical orbit, in the same way that the earth revolves about the sun. The simplest orbit for the electron about the nucleus would be a circle, and Newton's laws of motion would permit the circle to be of any size, as determined by the energy of the system.

After the discovery of the electron and the proton this model was considered by physicists interested in atomic structure, and it became evident that the older theories of the motion of particles (Newton's laws of motion) and of electricity and magnetism could not apply to the atom. If the electron were revolving around the nucleus it should, according to electromagnetic theory, produce light, with the frequency of the light equal to the frequency of revolution of the electron in the atom. This emission of light by the moving electron is similar to the emission of radiowaves by the electrons that move back and forth in the antennae of a radio station.* With the continued emission of energy by the atom in the form of light, however, the electron would move in a circle approaching more and more closely to the nucleus, and the frequency of its motion about the nucleus would become greater and greater. Accordingly, the older (classical) theories of motion and of electromagnetism would require that hydrogen atoms produce a spectrum of light of all wavelengths (a *continuous spectrum*). This is contrary to observation: the spectrum of hydrogen, produced in a discharge tube containing hydrogen atoms (formed by dissociation of hydrogen molecules), consists of discrete lines, as shown in Figure 5-1. Moreover, it is known that the volume occupied by a hydrogen atom, in a solid or liquid substance, corresponds to a diameter of about 1 Å, whereas the older theory of the hydrogen atom contained no mechanism for preventing the electron from approaching more and more closely to the nucleus, and the atom from becoming far smaller than 1 Å in diameter.

A hint as to the way to solve this difficulty had been given to Bohr by Planck's quantum theory of emission of light by a hot body, and by Einstein's theory of the photoelectric effect and the light quantum. Both Planck and Einstein had assumed that light of frequency v is not emitted or absorbed by matter in arbitrarily small amounts, but only in quanta of energy hv. If a hydrogen atom in which the electron is revolving about the nucleus in a large circular orbit emits a quantum of energy hv, the electron must then be in a much different (smaller) circular orbit, corresponding to an energy value of the atom hv less than its initial energy. Bohr accordingly assumed that *the hydrogen atom can exist only in certain discrete states*, which are called the *stationary states* of the atom. He assumed that one of these states, the *ground state* or *normal state*, represents the minimum energy possible for the atom; it is accordingly the most stable state of the atom. The other states, with an excess of energy relative to the ground state, are called the *excited states* of the atom.

Bohr further assumed, in agreement with the earlier work of Planck

---

*According to classical electromagnetic theory, an electric charge ϵ carrying out harmonic oscillation with frequency v and amplitude x_0 along a line (its coordinate x being given by the equation $x = x_0 \cos 2\pi v t$) loses energy by emitting light at the rate

$$-\frac{dE}{dt} = \frac{16\pi^4 v^4 \epsilon^2 x_0^2}{3c^3}$$

An electron in a circular orbit in the xy plane is equivalent to a linear oscillator along the x-axis and one along the y-axis, and hence should radiate at twice this rate.

and Einstein, that when an atom changes from a state with energy E'' to a state with energy E' the difference in energy $E'' - E'$ is equal to the energy of the light quantum that is emitted. This equation,

$$h\nu = E'' - E' \tag{5-1}$$

is called the *Bohr frequency rule*; it tells what the frequency of the light is that is emitted when an atom changes from an excited state with energy E'' to a lower state with energy E'.

The same equation also applies to the absorption of light by atoms. The frequency of the light absorbed in the transition from a lower state to an upper state is equal to the difference in energy of the upper state and the lower state, divided by Planck's constant. The equation also applies to the emission and absorption of light by molecules and more complex systems.

Example 5-1.   It is found that a tube containing hydrogen atoms in their normal state does not absorb any light in the visible region, but only in the far ultraviolet. The absorption line of longest wavelength has $\lambda = 1216$ Å. What is the energy of the excited state of the hydrogen atom that is produced from the normal state by the absorption of a quantum of this light?

Solution.   The frequency of the absorbed light is $\nu = c/\lambda = (2.998 \times 10^8$ m s$^{-1})/(1.216 \times 10^{-7}$ m$) = 2.466 \times 10^{15}$ s^{-1}. The energy of a light quantum is $h\nu$. This is just the energy of the excited state relative to the normal state of the hydrogen atom. Accordingly, the answer to our problem is

energy of excited state relative to normal state
$$= h\nu = 0.6625 \times 10^{-33} \text{ J s} \times 2.466 \times 10^{15} \text{ s}^{-1}$$
$$= 1.634 \times 10^{-18} \text{ J}$$

This can be converted into electron volts in the usual way (Equation 3-12); the answer is 10.20 eV. This result could be obtained simply by applying Equation 3-13: 12,398/1216 Å $= 10.20$ eV.

Bohr also discovered a method of calculating the energy of the stationary states of the hydrogen atom, with use of Planck's constant. He found that the correct values of the energies of the stationary states were obtained if he assumed that the orbits of the electrons are circular, and that the angular momentum of the electron has for the normal state the value $\hbar$, for the first excited state the value $2\hbar$, for the next excited state the value $3\hbar$, and so on (see the discussion of angular momentum in Section 3-8). Note that here it is convenient to use the angular-momentum quantum $\hbar$, rather than Planck's constant h.

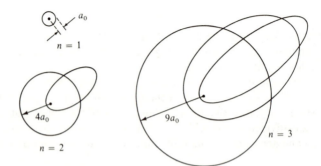

FIGURE 5-2

Bohr-Sommerfeld orbits for an electron in the hydrogen atom.
These circular and elliptical orbits were involved in the
Bohr-Sommerfeld theory. They do not provide a correct description
of the motion of the electron in the hydrogen atom. According
to the theory of quantum mechanics, which seems to be essentially
correct, the electron moves about the nucleus in the hydrogen
atom in roughly the way described by Bohr, but the motion
in the normal state ($n = 1$) is not in a circle, but is radial
(in and out, toward the nucleus and away from the nucleus).

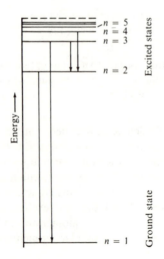

FIGURE 5-3
Energy-level diagram for the
hydrogen atom.

In general, the angular momentum of the electron in the circular orbit about the nucleus (the Bohr orbit) was represented by Bohr as having the value

$$\text{angular momentum} = n\hbar \qquad \text{with } n = 1, 2, 3, \cdots \qquad (5\text{-}2)$$

The number n introduced in this way in the Bohr theory is called the *principal quantum number* of the Bohr orbit.

It is shown below that the radius of the Bohr orbit is equal to $n^2 a_0$, in which

$$a_0 = \hbar^2/m\epsilon^2 = 0.530 \text{ Å} \qquad (5\text{-}3)$$

In this equation m is the mass of the electron and ϵ is the electronic charge. Thus the radius of the Bohr orbit for the normal state of the hydrogen atom is 0.530 Å, that for the first excited state is four times as great, that for the next excited state nine times as great, and so on, as illustrated in Figure 5-2.

For an atom with atomic number Z (nuclear charge $Z\epsilon$) the radius of the circular Bohr orbit is

$$r(Z, n) = \frac{n^2 a_0}{Z} \qquad (5\text{-}4)$$

For the same value of the quantum number the size of the electron orbit is inversely proportional to the atomic number.

The energy of the electron in the nth stationary state is shown below to be given by the equation

$$E_n = -\frac{m\epsilon^4 Z^2}{2\hbar^2 n^2} = -13.60 \text{ eV} \times \frac{Z^2}{n^2} \qquad (5\text{-}5)$$

This equation, with $Z = 1$, gives the energy levels for the hydrogen atom shown in Figure 5-3, and, with the Bohr frequency rule $E_i - E_j = h\nu_{ij}$, accounts for the spectral lines of hydrogen.

The lines of the Balmer series, shown in Figure 5-1, correspond to transitions from $n = 3, 4, 5, \ldots$ to $n = 2$. The series involving as its lower state the state with $n = 3$, and that involving the state with $n = 4$, were also known at the time that Bohr carried out his work. The series with the ground state, $n = 1$, as the lower state for the transition had not, however, been recognized. Bohr predicted that this series would exist, and he calculated the wavelengths of the lines ($\lambda = 1,216$ Å, and so on). Experimental physicists immediately began to search the far ultraviolet region for these lines, a search which involved difficult experimental technique. In 1915 Professor Theodore Lyman of Harvard University discovered the lines, which are called the Lyman series.

Let us calculate the properties of the hydrogen atom in the way first carried out by Bohr. In Figure 5-4, representing the motion of the electron about the nucleus in the circular orbit, we see that the velocity of the

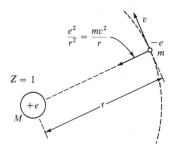

FIGURE 5-4
Diagram illustrating the calculation of acceleration to balance the centripetal force of electrostatic attraction, in the Bohr theory of the hydrogen atom.

electron, which at any instant is in a direction tangent to the circular orbit, changes as the electron proceeds around the circle. This requires that the electron be accelerated toward the nucleus. The amount of the acceleration is calculated from the geometry of the figure to be v^2/r, and hence the force required to produce it is mv^2/r. This force is the coulomb force of attraction, $Z\epsilon^2/r^2$, for a nucleus with atomic number Z. We thus have the equation

$$\frac{mv^2}{r} = \frac{Z\epsilon^2}{r^2} \qquad (5\text{-}6)$$

or, multiplying by r,

$$mv^2 = \frac{Z\epsilon^2}{r} \qquad (5\text{-}7)$$

Note that this equation gives a relation between the kinetic energy, which is $\frac{1}{2}mv^2$, and the potential energy, which is $-Z\epsilon^2/r$; namely, the kinetic energy is equal to $\frac{1}{2}$ the potential energy, with the sign changed.

The angular momentum of the electron in its circular orbit is the linear momentum times the radius; that is, it is equal to mvr. Bohr made the assumption that the angular momentum must be equal to $nh/2\pi$; that is, to $n\hbar$, with n an integer:

$$mvr = n\hbar \qquad (5\text{-}8)$$

Multiplying Equation 5-7 by mr^2, we obtain

$$m^2v^2r^2 = Z^2\epsilon mr$$

The left side of this equation is just the square of the quantity on the left side of Equation 5-8; accordingly, it is equal to the square of the right side of this equation, which is $n^2\hbar^2$. Thus we obtain the equation

$$n^2\hbar^2 = Z\epsilon^2mr$$

The equation for the radius of the orbit is given by solving this for r:

$$r = \frac{n^2\hbar^2}{Zm\epsilon^2} = \frac{n^2}{Z} \times 0.530 \text{ Å} \qquad (5\text{-}9)$$

With use of Equation 5-8, we can solve for the velocity, obtaining

$$v = \frac{Z\epsilon^2}{n\hbar} = \frac{Z}{n} \times 2.188 \times 10^6 \text{ m s}^{-1} \tag{5-10}$$

The total energy, which is the sum of the kinetic energy and the potential energy, is seen to be given by the expression quoted above (Equation 5-5). We may note that the total energy is just equal to the kinetic energy, $\frac{1}{2}mv^2$, with changed sign, and also equal to half the potential energy.*

We see that the energy of the hydrogen atom in its normal state is $-m\epsilon^4/2\hbar^2$, which equals 13.60 eV. This is the amount of energy required to separate a normal hydrogen atom into an electron and a proton. It is called the *ionization energy* of the normal hydrogen atom.

Elliptical Orbits

In 1915 the German physicist Arnold Sommerfeld extended Bohr's treatment to include certain elliptical orbits. In his treatment he introduced three quantum numbers to describe the orbit of the electron. The total quantum number n, giving the energy of the atom (Equation 5-4), also determined the semimajor axis of the ellipse as $n^2 a_0$. The angular-momentum quantum number k, equal to or smaller than n, determined the semiminor axis as nka_0. The third quantum number, m, described the component of angular momentum along the direction of an applied magnetic field (see Section 3-8). Some of the Sommerfeld elliptical orbits are shown in Figure 5-2.

The Bohr-Sommerfeld description of electrons in atoms has now been superseded by the wave-mechanical description, which, however, retains some of the features of this earlier model.

5-2. Excitation and Ionization Energies

Interesting verification of Bohr's idea about stationary states of atoms and molecules was provided by some electron-impact experiments carried out by James Franck and Gustav Hertz during the years 1914 to 1920. They were able to show that when a fast-moving electron collides with an atom or molecule it bounces off with only small loss of kinetic energy, unless it

---

*In the above treatment the electron has been discussed as moving about a stationary nucleus. In fact, the two particles should be described as carrying on similar motions on opposite sides of their center of mass. Correction for the motion of the nucleus can be made by replacing the mass of the electron by the quantity $(1/m + 1/M_p)^{-1}$, which is called the *reduced mass*. This correction changes the energy levels for hydrogen by about 0.05%, and those for He$^+$ by $\frac{1}{4}$ as much. This explanation of the displacement of certain lines of He$^+$ from those of H was one of the early successes of the Bohr theory.

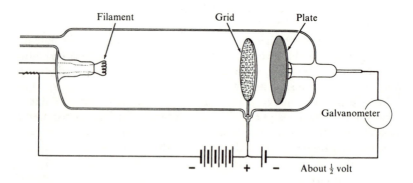

FIGURE 5-5
Apparatus for electron-impact experiments of the sort carried out by Franck and Hertz.

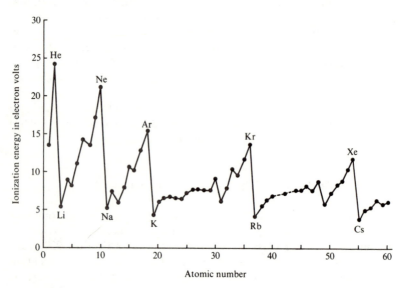

FIGURE 5-6
The ionization energy, in electron-volts, of the first electron of atoms from hydrogen, atomic number 1, to neodymium, atomic number 60. Symbols of the elements with very high and very low ionization energy are shown in the figure.

has a high enough speed to be able to raise the atom or molecule from its normal electronic state to an excited electronic state, or even to ionize the atom or molecule by knocking an electron out of it.

The apparatus that they used is indicated diagrammatically in Figure 5-5. Electrons are boiled out of the hot filament, and are accelerated toward the grid by the accelerating potential difference V_1. Many of these electrons pass through the perforations in the grid, and strike the collecting plate, which is held at a negative voltage relative to the grid. The electrons are able to move against the electrostatic field between the grid and the collecting plate because of the kinetic energy that they have gained while being accelerated from the filament to the grid. Even if there are some atoms or gas molecules in the space between the filament and the grid, the electrons may bounce off from them without much loss in energy.

If, however, the accelerating voltage V_1 is great enough that the kinetic energy picked up by the electron exceeds the excitation energy of the atom or molecule with which the electron may collide, the electron on impact with the atom or molecule may raise it from the normal state to the first excited state. The colliding electron will retain only the kinetic energy equal to its original kinetic energy minus the excitation energy of the atom or molecule with which it has collided. It may then not have enough residual kinetic energy to travel to the plate, against the opposing field, and the current registered by the galvanometer to the plate may show a decrease as the accelerating voltage is increased.

With hydrogen atoms in the tube, for example, no change in the plate current would be registered on the galvanometer until the accelerating voltage reaches 10.2 V. At this accelerating voltage the electrons obtain from the field between the filament and the grid just enough energy to raise a normal hydrogen atom to the first excited state—that is, to change the quantum number from $n = 1$ to $n = 2$. There then occurs a decrease in the plate current. The voltage 10.2 V is called a *critical voltage* or *critical potential* for atomic hydrogen. Other critical potentials occur, corresponding to the other excited states, and a large one occurs at 13.60 V. This critical voltage, 13.60 V, corresponds to the energy, 13.60 eV, required to remove an electron completely from the hydrogen atom; that is, it corresponds to the energy required to convert a normal hydrogen atom into a proton plus an electron, far removed from it. The voltage 13.60 V is called the *ionization potential* of the hydrogen atom.

Values of the first and second ionization energies for many atoms are given in Table 5-1, and the values are shown as a function of Z in Figure 5-6. Most of these values have been obtained by analysis of the spectra of the elements. Values of the third and higher ionization energies are also known for many elements; for aluminum, for example, the 13 successive ionization energies are 5.984, 18.823, 28.44, 119.96, 153.77, 190.92, 241.38, 284.53, 330.1, 398.5, 441.9, 2085, and 2298 eV.

TABLE 5-1
First and Second Ionization Energies of the Elements (*in eV*)

Z		I_1	I_2	Z		I_1	I_2
1	H	13.60		24	Cr	6.76	16.49
2	He	24.58	54.40	25	Mn	7.43	15.64
3	Li	5.390	75.62	26	Fe	7.87	16.18
4	Be	9.32	18.21	27	Co	7.86	17.05
5	B	8.30	25.15	28	Ni	7.63	18.15
6	C	11.26	24.38	29	Cu	7.72	20.29
7	N	14.53	29.59	30	Zn	9.39	17.96
8	O	13.61	35.11	31	Ga	6.00	20.51
9	F	17.42	34.98	32	Ge	7.88	15.93
10	Ne	21.56	41.07	33	As	9.81	18.63
11	Na	5.138	47.29	34	Se	9.75	21.5
12	Mg	7.64	15.03	35	Br	11.84	21.6
13	Al	5.98	18.82	36	Kr	14.00	24.56
14	Si	8.15	16.34	37	Rb	4.176	27.5
15	P	10.48	19.72	38	Sr	5.69	11.03
16	S	10.36	23.4	39	Y	6.38	12.23
17	Cl	13.01	23.80	40	Zr	6.84	13.13
18	Ar	15.76	27.62	41	Nb	6.88	14.32
19	K	4.339	31.81	42	Mo	7.10	16.15
20	Ca	6.11	11.87	43	Tc	7.28	15.26
21	Sc	6.54	12.80	44	Ru	7.36	16.76
22	Ti	6.82	13.57	45	Rh	7.46	18.07
23	V	6.74	14.65	46	Pd	8.33	19.42

A simple interpretation of these values is based on the idea that each electron is shielded somewhat from the nucleus by the other electrons in the atom. Let us represent the effective nuclear charge of an atom (for a particular electron) by $(Z - S)\epsilon$, in which S is the *shielding constant* (also called the *screening constant*). The ionization energy for an electron with total quantum number N is then given (from Equation 5-5) by the equation

$$I = \frac{(Z - S)^2}{n^2} \times 13.60 \text{ eV} \qquad (5\text{-}11)$$

Example 5-2.   What is the shielding constant of one of the two electrons in the helium atom for the other? Assume that both electrons have $n = 1$.

Solution.   The value of I_2, 54.40 eV, is just $2^2 \times 13.60$ eV, as given by Equation 5-11, with $n = 1$, $Z = 2$ (for helium), and $S = 0$, as expected. We equate I_1, 24.58 eV (Table 5-1), to $(2 - S)^2 \times 13.60$ eV, and obtain $S = 0.66$.

Z		I_1	I_2	Z		I_1	I_2
47	Ag	7.57	21.48	70	Yb	6.2	
48	Cd	8.99	16.90	71	Lu	5.0	
49	In	5.79	18.86	72	Hf	5.5	14.9
50	Sn	7.34	14.63	73	Ta	7.88	16.2
51	Sb	8.64	16.5	74	W	7.98	17.7
52	Te	9.01	18.6	75	Re	7.87	16.6
53	I	10.45	19.09	76	Os	8.7	17
54	Xe	12.13	21.2	77	Ir	9	
55	Cs	3.893	25.1	78	Pt	9.0	18.56
56	Ba	5.21	10.00	79	Au	9.22	20.5
57	La	5.61	11.43	80	Hg	10.43	18.75
58	Ce	6.6	14.8	81	Tl	6.11	20.42
59	Pr	5.8		82	Pb	7.42	15.03
60	Nd	6.3		83	Bi	8	16.68
61	Pm			84	Po	8.43	
62	Sm	6	11.4	85	At		
63	Eu	5.66	11.4	86	Rn	10.75	
64	Gd	6.2		87	Fr		
65	Tb	6.7		88	Ra	5.28	10.14
66	Dy	6.8		89	Ac	6.9	12.1
67	Ho			90	Th		
68	Er			91	Pa		
69	Tm			92	U	5	

Example 5-3. Assuming Li^+ and Al^{11+} to contain two electrons with $n = 1$, evaluate the shielding constant from the observed values $I_2 = 75.62$ eV for Li (Table 5-1) and $I_{12} = 2085$ eV for Al (given above).

Solution. Use of Equation 5-11 as in Example 5-2 leads to $S = 0.64$ for Li^+ and $S = 0.62$ for Al^{11+}. We conclude that the screening effect of one electron with $n = 1$ for another with $n = 1$ is nearly independent of Z.

Example 5-4. We assume from the trend of values of I_1 and I_2 in Table 5-1 that the lithium atom contains two inner electrons with $n = 1$ and one outer electron with $n = 2$. If the inner electrons shielded the outer electron completely from the nucleus the value of I_1 would be $(3 - 2)^2/2^2 \times 13.60 = \frac{1}{4} \times 13.60 = 3.40$ eV. If they had no shielding effect, the value of I_2 would be $3^2/2^2 \times 13.60 = 30.60$ eV. The observed value (Table 5-1) is 5.390 eV. How effective are the two inner electrons in their shielding action for the outer electron?

Solution. We equate 5.390 eV to $(3 - S)^2/4 \times 13.60$ eV and obtain $S = 1.74$. Hence each inner electron is 87% effective in shielding the outer electron from the nucleus (as compared with 64% effective (Example 5-3) is shielding the other inner electron).

Example 5-5. Estimate the relative sizes of the neutral lithium atom and the lithium ion, Li^+.

Solution. As an approximation known to us, we use Equation 5-9, with Z replaced by $Z - S$:

$$r = \frac{n^2}{Z - S} \times 0.530 \text{ Å}$$

The values $n = 2$ and $S = 1.74$ (Example 5-4) lead to $r = 1.68$ Å for the Bohr radius of the outer electron in Li, and the values $n = 1$ and $S = 0.64$ (Example 5-3) lead to $r = 0.22$ Å for the Bohr radius of Li^+. We conclude that Li is nearly 8 times as large as Li^+.

5-3. The Wave-mechanical Description of Atoms

By 1923, it was recognized that Bohr's formulation of the theory of the electronic structure of atoms needed to be improved and extended. The theory gave correct values for the energy of the hydrogen atom (also for those of the one-electron ions He^+, Li^{++}, and so on), but it did not provide a satisfactory way of calculating values of the probability of transition from one quantum state to another; that is, it did not explain satisfactorily the observed distribution of intensities of the lines of the hydrogen spectrum. Moreover, the Bohr theory did not give correct values for the energy levels of the helium atom or the hydrogen molecule-ion, H_2^+, or of any other atom with more than one electron or any molecule. The observed rotational spectra (band spectra) of diatomic molecules were seen to show that the rotational energy levels are not proportional to J^2 ($J = 0, 1, 2, \ldots$), as required by Bohr's postulate about the quantization of angular momentum, but instead are proportional to $J(J + 1)$; and many other observed properties of substances were seen to require some changes to be made in the old quantum theory. The search for a better theory was soon successful, in the discovery of the theory of quantum mechanics (also called wave mechanics, as described below).

During the two-year period 1924 to 1926 the Bohr description of electron orbits in atoms was replaced by the greatly improved description of wave mechanics, which is still in use and seems to be completely satisfactory.

The discovery by de Broglie in 1924 that an electron moving with the velocity v has a wavelength $\lambda = h/mv$ was mentioned in Section 3-8. De Broglie pointed out that the wavelength of the electron as given by his equation has just the value to give reinforcement of electron waves in the circular Bohr orbits. An example is the circular Bohr orbit with total quantum number n equal to 5. The length of the orbit, 2π times the radius, is just equal to five times the de Broglie wavelength for an electron moving with the velocity given by Bohr's theory for the electron in this orbit. Thus the electron waves could be considered to reinforce one another as the electron moves about the nucleus in this orbit, whereas in slightly smaller or slightly larger orbits the waves would interfere.

The calculation verifying this statement has been made in Example 3-8. The kinetic energy of an electron in the first Bohr orbit, the normal state of the hydrogen atom, is 13.60 eV. In the solution of this example we found the wavelength of the electron to be 3.33 Å. The radius of the first Bohr orbit is 0.530 Å. When this is multiplied by 2π, the value 3.33 Å is obtained. Hence in the first Bohr orbit there is, according to de Broglie's calculation, just one wavelength in the circumference of the orbit. According to the Bohr theory the velocity of the electron in the nth Bohr orbit is just $1/n$ times the velocity in the first Bohr orbit, and the wavelength is accordingly $n \times 3.33$ Å. The circumference of the Bohr orbit is, however, proportional to n^2 (Equation 5-4), being equal to $n^2 \times 3.33$ Å. This calculation hence shows, as de Broglie discovered, that there are n electron wavelengths in the circumference of the nth Bohr orbit.

This discovery was interesting in suggesting that the wave character of the electron was involved in determining the stationary states of the hydrogen atom, but there is no close connection between this relation of the de Broglie wavelength to the circumference of the Bohr orbits and the description of the hydrogen atom given by quantum mechanics (wave mechanics).

The theory of quantum mechanics was developed in 1925 by the German physicist Werner Heisenberg (born 1901). An equivalent theory, called wave mechanics, was independently developed early in 1926 by the Austrian physicist Erwin Schrödinger (1887–1961). Important contributions to the theory were also made by the English physicist Paul Adrien Maurice Dirac (born 1902).

The theory seems to be in complete agreement with the experimental information about the structure of atoms and molecules, but some additions probably need to be made before the theory is applicable to the discussion of the structure of nuclei. Quantum mechanics cannot be discussed in a brief elementary way, and in this book we shall have to be content with the description of some of its results, especially with respect to the electronic structure of atoms and molecules.

Quantum mechanics does not describe the motion of electrons in the atom so precisely as was done by Bohr. The properties of the atom that can be measured are, however, correctly given by the quantum-mechanical equations. These properties include, for example, the average and most probable distances of the electron from the nucleus, in a particular quantum state, and also the average speed with which the electron moves. It is found that the most probable distance of the electron from the nucleus is the same as was calculated by Bohr, and the average speed (the root-mean-square speed) is also the same. The angular momentum, however, is different, and in particular the electron in the hydrogen atom in its normal state is not moving around the nucleus in an orbit with angular momentum $\hbar$, but is moving in toward and out from the nucleus, in an orbit with zero angular momentum.

The electrons moving about a nucleus are described in quantum mechanics by certain mathematical functions, called *wave functions*. The wave function for one electron is called an *orbital wave function*, and the electron is said to occupy an *orbital* (rather than an orbit). The use of a different name indicates that the motion of the electron according to quantum mechanics is somewhat different from the motion in a Bohr orbit.

Orbital Quantum Numbers

An orbital is described by three quantum numbers, n, l, and m_l, defined as follows.

The *total quantum number*, n, is essentially equivalent to Bohr's quantum number n, as discussed above. It largely determines the energy of the electron occupying the orbital and also the size of the orbital.

The *orbital angular-momentum quantum number* l determines the value of the orbital angular momentum of the electron occupying the orbital. The value of the orbital angular momentum is $\{l(l+1)\}^{1/2}\hbar$; compare this with the spin angular momentum of the electron, $\{s(s+1)\}^{1/2}\hbar$, $s = \frac{1}{2}$ (Sec. 3-8). The values of l are $0, 1, 2, \ldots, n-1$; thus for $n = 1$ the value of l is 0; for $n = 2$ l has values 0 and 1, and so on.

The *orbital magnetic quantum number* m_l determines the value of the component of angular momentum in a specified direction in space (such as the direction of a magnetic field): this component has the value $m_l\hbar$; compare with m_s (Section 3-8). The values of m_l are $-l, -l+1, \ldots, 0, 1, \ldots, +l$. There are accordingly $2l+1$ values of m_l for a given value of l, corresponding to $2l+1$ different orientations of the orbital in space.

The common way of labeling the orbitals defined by the values $0, 1, 2, \ldots$ of the orbital angular-momentum quantum number is by use of the letters

$s, p, d, f, g, h, \ldots$, respectively. The letter s is used to represent an orbital with $l = 0$, p an orbital with $l = 1$, d an orbital with $l = 2$, and so on.* The value of principal quantum number n is written before the letter: thus $3d$ means an orbital with $n = 3$ and $l = 2$. The same nomenclature is applied to electrons: a $3d$ electron is an electron occupying the $3d$ orbital of an atom.

For a hydrogen-like atom (one electron) the energy is given by the Bohr equation, 5-5. The average value of r, the distance of the electron from the nucleus, is

$$r_{average} = 0.530 \text{ Å} \times \frac{n^2}{Z} \left\{ 1 + \frac{1}{2} \left(1 - \frac{l(l+1)}{n^2} \right) \right\} \qquad (5\text{-}12)$$

Expressions for the simplest hydrogen-like functions in polar coordinates (r, θ, φ) are given in Appendix V.

The Normal State of the Hydrogen Atom

The wave function for the normal state of the hydrogen atom, with $n = 1$, $l = 0$, $m_l = 0$, is†

$$\psi_{100} = \frac{1}{\sqrt{\pi}} \left(\frac{Z}{a_0} \right)^{3/2} \exp\left(-Zr/a_0 \right)$$

Its square, $\dfrac{Z^3}{\pi a_0^3} \exp\left(-2Zr/a_0 \right)$, is the probability that the electron is in unit volume element at distance r from the nucleus, and $4\pi r^2 \psi_{100}^2 \, dr$ is the probability that it be between the distances r and $r + dr$ from the nucleus. It is seen from Figure 5-7 that this last function has its maximum value at $r = a_0$. The most probable distance of the electron from the nucleus is thus just the Bohr radius a_0; the electron is, however, not restricted to this one distance. The speed of the electron also is not constant, but can be represented by a distribution function, such that the root-mean-square speed has just the Bohr value v_0. We can accordingly describe the normal hydrogen atom by saying that the electron moves in and out about the nucleus, remaining usually within a distance of about 0.5 Å, with a speed that is variable but is of the order of magnitude of v_0. Over a period of time long enough to permit many cycles of motion of the elec-

---

*This use of the letters s, p, d, f has a curious origin; they are the initial letters of the adjectives *sharp, principal, diffuse,* and *fundamental,* which happened to be used by spectroscopists in the period around 1890 to describe different observed spectral series of the alkali metals. These letters accordingly are not abbreviations of words that describe the orbitals in a meaningful way. A mnemonic for the letters is "So poorly did foolish Gelehrte have it."

†Note that exp x means e^x.

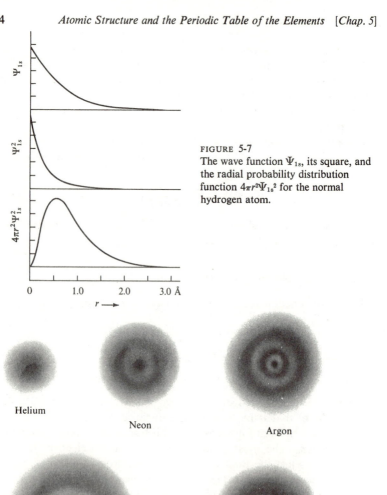

FIGURE 5-7
The wave function Ψ_{1s}, its square, and the radial probability distribution function $4\pi r^2\Psi_{1s}^2$ for the normal hydrogen atom.

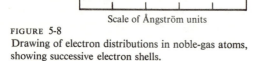

FIGURE 5-8
Drawing of electron distributions in noble-gas atoms, showing successive electron shells.

tron, the atom can be described as consisting of the nucleus surrounded by a spherically symmetrical ball of negative electricity (the electron blurred by a time-exposure of its rapid motion), as indicated by the drawing for helium in Figure 5-8 (with some increase in size).

The normal hydrogen atom, with $l = 0$, has no orbital angular momentum; the electron is not to be thought of as going around the nucleus, but rather as going in and out, in varying directions, so as to make the electron distribution spherically symmetrical. The average value of the distance of the electron from the nucleus is $\frac{3}{2} \times 0.530 \text{ Å} = 0.795 \text{ Å}$, 50% greater than the radius of the circular Bohr orbit.

There is only one orbital for $n = 1$. There are other possible orbitals for the hydrogen atom, corresponding to $n = 2$, $n = 3$, and so on. A hydrogen atom in which the electron occupies one of these other orbitals is unstable; it is said to be in an excited state. A large amount of energy is required to change the hydrogen atom from its normal state to the first excited state ($n = 2$)—three-quarters as much as to remove the electron completely. The diameter of the atom for this excited state is four times as great as for the normal state. The orbitals with $n = 2, 3, 4, \ldots$ are occupied by electrons in heavier atoms.

The Pauli Exclusion Principle

A principle of extreme importance to spectroscopy as well as to other phases of physics and chemistry is the *exclusion principle* discovered by the Austrian physicist W. Pauli (1900–1958).

Let us consider an atom in an external magnetic field so strong that the electrons orient themselves independently with respect to the field. The state of each electron is then given by fixing the values of a set of quantum numbers: for each electron we may give the values of the total quantum number n of the orbital, the orbital angular-momentum quantum number l, the orbital magnetic quantum number m_l (stating the component of orbital angular momentum in the field direction), the spin quantum number s (which has the value $\frac{1}{2}$ for every electron), and the spin magnetic quantum number m_s (which can be equal to either $+\frac{1}{2}$, corresponding to the spin oriented roughly in the direction of the field, or $-\frac{1}{2}$, corresponding to the spin oriented roughly in the opposite direction). The Pauli exclusion principle can be expressed in the following way: *there cannot exist an atom in such a quantum state that two electrons within it have the same set of quantum numbers.*

This is equivalent to saying that *two electrons can occupy the same orbital only if their spins are opposed*—that is, oriented in opposite directions.

The Electronic Structure of the Noble Gases

The distributions of electrons in atoms of the noble gases have been determined by physicists by experimental and theoretical methods that are too complex to be discussed here. The results obtained are shown in Figure 5-8. It is seen that for the atoms neon, argon, krypton, and xenon the electrons are arranged about the atomic nuclei in two or more concentric shells.

The helium atom contains two electrons, each of which carries out motion about the helium nucleus similar to that of the one electron in the hydrogen atom. These two electrons occupy the same orbital, the $1s$ orbital, and in accordance with the Pauli exclusion principle their spins are opposed.

The symbol $1s^2$ is used to express the electron configuration of the normal helium atom. The superscript 2 means that two electrons occupy the $1s$ orbital. The $1s$ orbital has $n = 1$ and $l = 0$ (also $m_l = 0$).

Two electrons with opposed spins occupying the same orbital are called an *electron pair*. The electrons can be described as forming a ball of negative electricity near the nucleus. The diameter of the helium atom is only about half that of the hydrogen atom, because of the doubled value of the nuclear charge.

These two electrons are said to constitute a completed helium shell (also called a completed K shell).

All of the atoms heavier than hydrogen have a completed helium shell, consisting of two $1s$ electrons ($1s^2$) close to the nucleus. The diameter of the helium shell is inversely proportional to the atomic number; for radon ($Z = 86$) it is only about 0.02 Å.

The neon atom has two completed shells. First, it has a helium shell of two electrons, with diameter about 0.2 Å, as shown in Figure 5-8, and around this shell an outer shell of eight electrons, called the neon shell or L shell. The diameter of this outer shell is about 2 Å.

These two shells, reduced in size, appear in argon, krypton, and xenon (Figure 5-8), together with additional shells. The nature of these shells is discussed in the following paragraphs.

Shells and Subshells of Electrons

Around 1920, while they were developing the theory of atomic spectra (line spectra and x-ray spectra of the elements), physicists discovered that the successive shells, after the helium shell, contain orbitals of more than one kind.

The K shell consists of only one orbital, the $1s$ orbital, described in the preceding section. The L shell consists of two subshells and four orbitals.

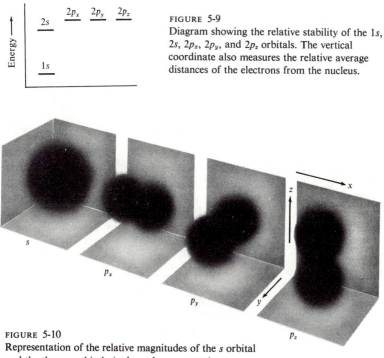

FIGURE 5-9
Diagram showing the relative stability of the $1s$, $2s$, $2p_x$, $2p_y$, and $2p_z$ orbitals. The vertical coordinate also measures the relative average distances of the electrons from the nucleus.

FIGURE 5-10
Representation of the relative magnitudes of the s orbital and the three p orbitals in dependence on angle.

The $2s$ subshell consists of only one orbital, the $2s$ orbital, which has $n = 2$, $l = 0$, and $m_l = 0$. The $2p$ subshell consists of three $2p$ orbitals, each with $n = 2$ and $l = 1$; the three correspond to the three values -1, 0, and $+1$ for m_l. An electron in a $2s$ orbital is somewhat more stable and somewhat closer to the nucleus than an electron in one of the $2p$ orbitals, as is indicated in the energy diagram, Figure 5-9. The three $2p$ orbitals have the same energy.

The $2s$ orbital, like the $1s$ orbital, corresponds to an electron distribution that is spherically symmetrical. The electron distribution for a $2p$ orbital is not spherically symmetrical, but is concentrated about an axis, as shown in Figure 5-10. The characteristic axes of the three $2p$ orbitals in an atom are orthogonal to one another, and they can be taken as the x-axis, the y-axis, and the z-axis, respectively, as indicated in Figure 5-10. The three $2p$ orbitals can be given the symbols $2p_x$, $2p_y$, and $2p_z$ (Appendix V).

In accordance with the Pauli exclusion principle, each of these orbitals can be occupied by two electrons, which must have their spins opposed.

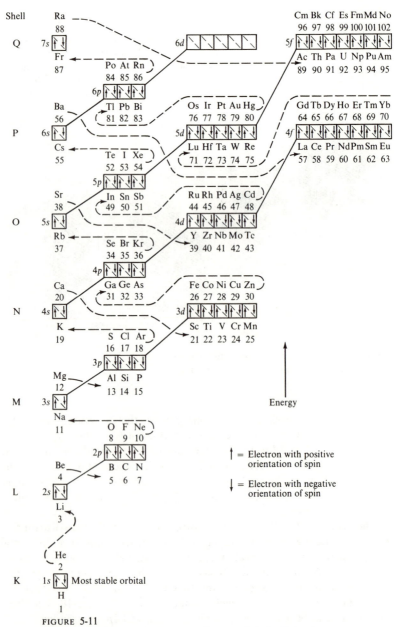

FIGURE 5-11
Energy level diagram of electron shells and subshells of the elements.

Hence the completed 2s subshell contains two electrons (one electron pair) and the complete 2p subshell contains six electrons (three electron pairs, one for each of the three 2p orbitals). The completed L shell accordingly contains eight electrons (four electron pairs).

The symbol for the completed 2s subshell is $2s^2$ and that for the completed 2p subshell is $2p_x^2 2p_y^2 2p_z^2$, which is usually simplified to $2p^6$. The symbol for the completed L shell is $2s^2 2p_x^2 2p_y^2 2p_z^2$, usually written as $2s^2 2p^6$.

The M shell, with total quantum number $n = 3$, consists of three subshells and nine orbitals. In addition to the 3s subshell (one orbital) and the 3p subshell (three orbitals), it contains a 3d subshell, with five 3d orbitals.

The next shell, the N shell, contains these subshells plus an additional one, the 4f subshell, which has seven orbitals.

The letters K, L, M, N, O, P that are used for electron shells correspond to the successive values of the principal quantum number n: K to $n = 1$, L to $n = 2$, M to $n = 3$, and so on. It is as a result of historical accident that this sequence of letters begins with K rather than A.*

The successive electron shells (K, L, and so on) described above are those used by physicists. Chemists have found it convenient to use a different classification, corresponding to a different way of grouping the subshells together. In Figure 5-11 the approximate sequence of energy values for all orbitals that are occupied by electrons in atoms in their normal states are shown. It is seen that the energy values for the physicists' shells overlap. For example, the energy of a 3d electron (in the M shell) is about the same as that of a 4p electron (in the N shell). Because of their interest in sets of electrons with the same energy, chemists have found it convenient to assign the five 3d orbitals to 4s and 4p, rather than to 3s and 3p, and to make similar assignments of 4d, 5d, 6d, 4f, and 5f, as indicated in Figure 5-11.

The successive shells are named for the noble gases in which they first are completed. The number of electrons in each shell is equal to the number of elements in the period of the periodic table that is completed by the

---

*In the period 1905 to 1910 the English physicist Charles Glover Barkla (1877–1944) measured the power of x-rays of penetrating sheets of copper and other substances and discovered that elements emit characteristic x-rays of two kinds, differing in their penetrating power. After having used the letters B and A, he decided in 1911 to assign K to the more penetrating and L to the less penetrating radiation, in order to leave other letters available for still more penetrating and less penetrating kinds of characteristic radiation, which he thought would probably be discovered. He soon discovered M and N radiation (less penetrating), but characteristic x-radiation more penetrating than the K lines is not emitted by atoms. Barkla may have taken the letters K and L from his name.

TABLE 5-2
Shells and Subshells of Electrons

Chemist's Name for Shell	Symbol for Electron Configuration of Completed Shell, Showing Subshells
Helium shell	$1s^2$
Neon shell	$2s^2 2p^6$
Argon shell	$3s^2 3p^6$
Krypton shell	$3d^{10} 4s^2 4p^6$
Xenon shell	$4d^{10} 5s^2 5p^6$
Radon shell	$4f^{14} 5d^{10} 6s^2 6p^6$
Eka-radon shell	$5f^{14} 6d^{10} 7s^2 7p^6$

corresponding noble gas: 2 for the helium shell, 8 each for the neon shell and the argon shell, 18 each for the krypton shell and the xenon shell, and 32 each for the radon shell and the eka-radon shell. The symbols for the completed shells are given in Table 5-2.

Argononic Structures

The elements helium, neon, argon, krypton, xenon, and radon are usually called the noble gases or the inert gases, and their electronic structures are called noble-gas or inert-gas structures. To simplify the nomenclature we shall call these elements the *argonons*, and their structures the *argononic structures*. The word argonon is selected to indicate low chemical reactivity (from greek *a*, privative, and *ergon*, work), with the ending *-on* to distinguish it from the element argon.

Electron-spin Multiplets

In 1925 the American astronomer Henry Norris Russell and the American physicist F. A. Saunders made an important discovery about the electronic structure of atoms. While they were trying to find the principles determining the wavelengths of the spectral lines emitted by atoms that have been excited by an electric discharge or in some other way, they discovered that the spins of the electrons in an atom may combine to form a resultant spin, which is designated by the resultant electron-spin quantum number S. Similarly, the orbital angular momenta of the several electrons can combine to form a resultant, designated by the orbital quantum number L. These two angular momentum vectors then combine to form a resultant total angular momentum vector, designated by the quantum number J. This sort of interaction of the electrons is called *Russell-Saunders coupling*.

For example, the normal boron atom, with electron configuration $1s^2 2s^2 2p$, contains two pairs of electrons, and the two electrons of each pair, occupying the same orbit, have their spins opposed (Pauli exclusion principle). Hence the total resultant spin is just that of the fifth electron, and therefore the value of S is that of one electron: $S = \frac{1}{2}$. Also, the $1s$ and $2s$ electrons have zero orbital angular momentum, and the $2p$ electron has one unit of orbital angular momentum, corresponding to the quantum number $l = 1$; hence the value of L is the same: $L = 1$.

The states of an atom with Russell-Saunders coupling are represented by *Russell-Saunders symbols*. The symbol for the normal state of the boron atom is $^2P_{1/2}$.

In this symbol the capital letter gives the value of L. The letters S, P, D, F, G, . . . correspond to $L = 0, 1, 2, 3, \ldots$, just as for one electron the letters s, p, d, f, . . . correspond to the values $0, 1, 2, 3, \ldots$ for the orbital angular momentum quantum number l.

The superscript on the left of the letter is a quantity called the multiplicity of the state. Its value is $2S + 1$, where S is the resultant electron-spin quantum number. The multiplicity represents the number of ways in which the resultant electron spin can orient itself relative to a magnetic field or to the orbital angular momentum vector. For $S = \frac{1}{2}$ there are two orientations, corresponding to the component $+\frac{1}{2}$ or $-\frac{1}{2}$ relative to L. For $L = 1$, as in the normal state of the boron atom, these two orientations of the electron spin lead to two values of J: $1 + \frac{1}{2} = \frac{3}{2}$ and $1 - \frac{1}{2} = \frac{1}{2}$. These values of J are given as a subscript in the symbol; the two states are $^2P_{3/2}$ and $^2P_{1/2}$. These two states are nearly equal in energy, with the difference (the fine-structure separation) increasing with increase in Z (0.002 eV for B, 0.014 eV for Al, 0.102 eV for Ga, 0.274 eV for In, all of which have $^2P_{1/2}$ as the normal state and $^2P_{3/2}$ as the first excited state). The two states are described as forming the two components of a *doublet*.

Similarly, the normal state of the carbon atom has the Russell-Saunders symbol $^3P_0$, corresponding to $S = 1$, $L = 1$, $J = 0$. It is a component of a triplet, the other two components being $^3P_1$ and $^3P_2$. The two p electrons of the carbon configuration $1s^2 2s^2 2p^2$ can combine their spins to the resultant $S = 0$ (which gives singlet states) or the resultant $S = 1$ (which gives triplet states, corresponding to the components $+1$, 0, and -1). Similarly, the two orbital angular momenta with value 1 (for the two p electrons) can combine to $L = 0$, 1, or 2, giving S, P, and D states. The Pauli exclusion principle operates in such a way (too complicated for discussion in an elementary course) that the actual states are ^1S, ^3P, and ^1D, and of these the triplet state, ^3P, is the most stable (Figure 5-12).

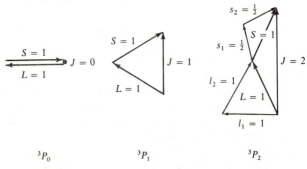

$$^3P_0 \qquad\qquad ^3P_1 \qquad\qquad ^3P_2$$

FIGURE 5-12
The three ways of combining the spin angular momentum for
$S = 1$ and the orbital angular momentum for $L = 1$ to give
the total angular momentum, $J = 0$, 1, and 2. $^3P_0$ is the normal
state for carbon, and $^3P_2$ for oxygen.

The rules (called Hund's rules) about stability of the states based upon
a given electron configuration (such as ^1S, ^3P, and ^1D for the configuration of two equivalent p electrons) are the following:

1. The states with the largest value of S (the largest multiplicity) are
the most stable. Thus for $(np)^2$ the triplet $^3P_{0,1,2}$ lies below the singlet
states $^1S_0$ and $^1D_2$.

2. For a given value of S, the state with the largest value of L is the
most stable. Thus for $(np)^2$ the state $^1D_2$ lies below $^1S_0$.

3. For a multiplet in the first half of a subshell the state with smallest J
is the most stable, and in the second half that with largest J is the most
stable. For example, the filled $2p$ subshell contains six electrons. Carbon,
with electron configuration $1s^2 2s^2 2p^2$, lies in the first half, and has $^3P_0$ as its
normal state, whereas oxygen, with electron configuration $1s^2 2s^2 2p^4$, lies
in the second half (it may be said to have two holes in a completed $2p^6$
subshell), and has $^3P_2$ as its normal state.

The importance of the first rule is evident from the drop in binding
energy (I_1) from $Z = 7$ (N) to $Z = 8$ (O), as shown in Figure 5-6. The big
drop in I from He to Li is the result of the increase in n from 1 to 2. The
smaller drop from Be to B corresponds to the change from the penetrating
orbital $2s$ to the less penetrating orbital $2p$. There are three $2p$ orbitals,
and the Pauli exclusion principle permits the three $2p$ electrons of N to
have their spins parallel, giving $S = \frac{3}{2}$ (a quartet state, $^4S_{3/2}$), in which
there are three stabilizing interactions of parallel electron spins. For
oxygen, however, the fourth $2p$ electron must have its spin opposed to the
spins of the other three, which leads to decreased stability.

5-4. The Periodic Table of the Elements

One of the most valuable parts of chemical theory is the *periodic law*. In its modern form this law states simply that *the properties of the chemical elements are not arbitrary, but depend upon the electronic structure of the atom and vary with the atomic number in a systematic way.* The important point is that this dependence involves a crude periodicity that shows itself in the periodic recurrence of characteristic properties.

For example, the elements with atomic numbers 2, 10, 18, 36, 54, and 86 are all chemically inert gases. Similarly, the elements with atomic numbers one greater—namely, 3, 11, 19, 37, 55, and 87—are all light metals that are very reactive chemically. These six metals—lithium (3), sodium (11), potassium (19), rubidium (37), cesium (55), and francium (87)—all react with chlorine to form colorless salts that crystallize in cubes and show a cubic cleavage. The chemical formulas of these salts are similar: $LiCl$, $NaCl$, KCl, $RbCl$, $CsCl$, and $FrCl$. The composition and properties of other compounds of these six metals are correspondingly similar, and different from those of other elements.

The periodicity of properties is strikingly shown by the values of the first ionization energy of atoms (Figure 5-6). The values of I_1 increase rather steadily with increase in Z until a noble gas is reached, and then drop for the next element to about one quarter the value for the noble gas. The periodicity of another property, the density of the elements in the solid state, is shown in Figure 5-13. The periodicity of properties of the elements with increasing atomic number may be effectively shown by arranging the elements in a table, called the *periodic table* or *periodic system* of the elements. Many different forms of the periodic system have been proposed and used. We shall base the discussion of the elements and their properties in this book on the simple system shown as Table 5-3.

The horizontal rows of the periodic table are called *periods*: they consist of a very short period (containing hydrogen and helium, atomic numbers 1 and 2), two short periods of 8 elements each, two long periods of 18 elements each, a very long period of 32 elements, and an incomplete period.

The vertical columns of the periodic table, with connections between the short and long periods as shown, are the *groups* of chemical elements. Elements in the same group may be called *congeners*; these elements have closely related physical and chemical properties.

The groups I, II, and III are considered to include the elements in corresponding places at the left side of all the periods, and V, VI, VII the elements at the right side. The central elements of the long periods, called the *transition elements*, have properties differing from those of the elements

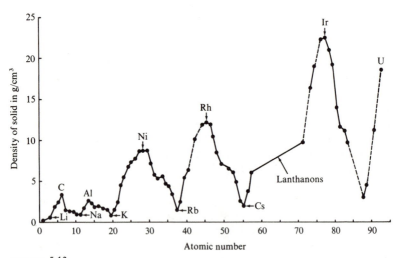

FIGURE 5-13
The density of the elements in the solid state, in g cm^{-3}. The symbols of the elements at high and low points of the jagged curve are shown.

of the short periods; these elements are discussed separately, as groups IVa, Va, VIa, VIIa, VIII (which, for historical reasons, includes three elements in each long period), Ib, IIb, IIIb, and IVb.

The very long period is compressed into the table by removing fourteen elements, the rare-earth metals, from $Z = 58$ to $Z = 71$, and representing them separately below.

The elements on the left side and in the center of the periodic table are metals. These elementary substances have the characteristic properties called metallic properties—high electric and thermal conductivity, metallic luster, the capability to be hammered into sheets (malleability) and to be drawn into wire (ductility). The elements on the right side of the periodic table are nonmetals, the elementary substances not having metallic properties.

The metallic properties are most pronounced for elements in the lower left corner of the periodic table, and the nonmetallic properties are most pronounced for elements in the upper right corner. The transition from metals to nonmetals is marked by the elements with intermediate properties, which occupy a diagonal region extending from a point near the upper center to the lower right corner. These elements, which are called metalloids, include boron, silicon, germanium, arsenic, antimony, tellurium, and polonium.

The discovery of the periodic table is discussed in the last section of this chapter.

TABLE 5-3
The Periodic System of the Elements

Group 0	
H 1	
	He 2

0	I	II	III	IV	V	VI	VII	0
								He 2
	Li 3	Be 4	B 5	C 6	N 7	O 8	F 9	Ne 10
	Na 11	Mg 12	Al 13	Si 14	P 15	S 16	Cl 17	Ar 18

0	I	II	III	IVa	Va	VIa	VIIa	VIII			Ib	IIb	IIIb	IV	V	VI	VII	0
Ar 18	K 19	Ca 20	Sc 21	Ti 22	V 23	Cr 24	Mn 25	Fe 26	Co 27	Ni 28	Cu 29	Zn 30	Ga 31	Ge 32	As 33	Se 34	Br 35	Kr 36
Kr 36	Rb 37	Sr 38	Y 39	Zr 40	Nb 41	Mo 42	Tc 43	Ru 44	Rh 45	Pd 46	Ag 47	Cd 48	In 49	Sn 50	Sb 51	Te 52	I 53	Xe 54
Xe 54	Cs 55	Ba 56	La 57 *	Hf 72	Ta 73	W 74	Re 75	Os 76	Ir 77	Pt 78	Au 79	Hg 80	Tl 81	Pb 82	Bi 83	Po 84	At 85	Rn 86
Rn 86	Fr 87	Ra 88	Ac 89 ◆	Th 90	Pa 91	U 92	Np 93	Pu 94										Kh 104

* **Lanthanons**

Ce 58	Pr 59	Nd 60	Pm 61	Sm 62	Eu 63	Gd 64	Tb 65	Dy 66	Ho 67	Er 68	Tm 69	Yb 70	Lu 71

◆ **Actinons**

Th 90	Pa 91	U 92	Np 93	Pu 94	Am 95	Cm 96	Bk 97	Cf 98	Es 99	Fm 100	Md 101	No 102	Lr 103

Electron Shells and the Periods of the Periodic Table

The successive electron shells listed in Table 5-2 involve the following numbers of electrons: 2, 8, 8, 18, 18, 32, 32. These numbers are equal to the numbers of elements in the successive periods of the periodic system (the last period incomplete).

We see that each of the two short periods, with eight elements, corresponds to the filling of two subshells, an *s* subshell (one orbital) and a *p* subshell (three orbitals).

The next two periods, the long periods, correspond to the filling not only of the next sets of these orbitals (giving $4s^24p^6$ and $5s^25p^6$), but also of sets of *d* orbitals ($3d^{10}$ and $4d^{10}$). It is the introduction of the ten *d* electrons that increases these periods to eighteen elements each.

The very long period, which ends in radon, involves not only $5d^{10}$ and $6s^26p^6$, but also $4f^{14}$.

The Octet

In every one of the noble gases except helium the outermost electrons, with the maximum value of the principal quantum number, are a set of eight (four pairs) with the symbol ns^2np^6. This set of eight electrons is called an *octet*.

It is found that many of the properties of elements close to the noble gases in the periodic table can be discussed in a simple and satisfactory way in terms of the octet and the four corresponding orbitals ns, np_x, np_y, and np_z. (For other elements, discussed for the most part in the final chapters of this book, the *d* orbitals must also be taken into consideration.)

The Electronic Structure of
the Elements of the First Short Period

Each of the elements from lithium to fluorine has an inner shell, $1s^2$. Lithium has in addition a $2s$ electron. Accordingly its electron configuration is $1s^22s$.

The electronic structure of an atom may be represented by an *electron-dot symbol*, in which the electrons of the outer shell (or outer octet) are represented by dots and the nucleus and inner electrons by the chemical symbol of the element. The electron-dot symbol of lithium, Li·, shows only the outermost electron, which is called the valence electron.

The next element, beryllium, has two valence electrons, both of which occupy the $2s$ orbital if the atom is in its normal state. The normal beryllium atom has the electron-dot symbol Be: and the electron configuration $1s^22s^2$. The two dots together represent two electrons with opposed spins in one orbital.

TABLE 5-4
Electron Configurations, Li to Ne

Atom	Number of Electrons per Orbital*				Electron Configuration*
	$2s$	$2p_x$	$2p_y$	$2p_z$	
Li	1				$2s$
Be	2				$2s^2$
B	2	1			$2s^2 2p$
C	2	1	1		$2s^2 2p^2$
N	2	1	1	1	$2s^2 2p^3$
O	2	2	1	1	$2s^2 2p^4$
F	2	2	2	1	$2s^2 2p^5$
Ne	2	2	2	2	$2s^2 2p^6$

*The inner electrons are not shown; for all of these atoms they are a $1s^2$ pair.

The electron-dot symbols for the normal states of the eight elements of the first short period are

$$\text{Li· Be: :B· :}\overset{\textstyle\cdot}{\text{C}}\text{· :}\overset{\textstyle\cdot}{\text{N}}\text{· :}\overset{\textstyle\cdot\cdot}{\underset{\cdot\cdot}{\text{O}}}\text{· :}\overset{\cdot\cdot}{\underset{\cdot\cdot}{\text{F}}}\text{· :}\overset{\cdot\cdot}{\underset{\cdot\cdot}{\text{Ne}}}\text{:}$$

Note that two or three $2p$ electrons occupy different orbitals, and remain unpaired. The normal state of carbon corresponds to the configuration $1s^2 2s^2 2p_x 2p_y$, and not to $1s^2 2s^2 2p_x^2$.

The electron configurations for the normal states of the elements lithium to neon are given in Table 5-4.

The electron configurations for the congeners of these elements are the same except for the increased value of the quantum number n. For example, sulfur, the congener of oxygen in the next period, has the configuration $3s^2 3p^4$ for its valence electrons.

The use of electronic structures in correlating the properties of substances will be illustrated in the following chapters.

An Energy-level Diagram

A diagram representing the distribution of all electrons in all atoms is given in Figure 5-11.

Each orbital is represented by a square. The most stable orbital (its electrons being held most tightly by the nucleus) is the $1s$ orbital, at the bottom of the diagram. Energy is required to lift an electron from a stable orbital to a less stable one, above it in the diagram.

The electrons are shown being introduced in sequence; the first and second in the $1s$ orbital, the next two in the $2s$ orbital, the next six in the $2p$ orbitals, and so on. The sequence is indicated by arrows. The symbol and atomic number of each element are shown adjacent to the outermost electron (least tightly held electron) in the neutral atom.

The electron configuration is indicated by the sequence along this path, up to the symbol of the element. Thus the electron configuration of nitrogen is $1s^2 2s^2 2p^3$ and that of scandium is $1s^2 2s^2 2p^6 3s^2 3p^6 4s^2 3d$.

For the heavier atoms two or more electron configurations may have nearly the same energy, and there is some arbitrariness in the diagram shown in Figure 5-11. The configuration shown for each element is either that of the most stable state of the free atom (in a gas) or of a state close to the most stable state. The electron configurations and the Russell-Saunders symbols for the normal states of atoms of the elements from 1 to 92, as determined from the atomic spectra, are given in Table 5-5. There are only occasional deviations from Figure 5-11. The first one is at chromium, $Z = 24$, for which the normal state has the configuration $3d^5 4s$, rather than $3d^4 4s^2$. The latter configuration is represented by an excited state that lies only 0.96 eV above the normal state.

5-5.   Electron Energy as the Basis of the Periodic Table

The equation for an electron in the hydrogen atom or a hydrogen-like ion (He^+, Li^{++}, . . .), Equation 5-5, shows the energy to be determined by n alone. For each value of n there are $2n^2$ sets of values of l, m_l, and m_s; hence successive hydrogen-like energy levels n could be occupied by $2n^2$ electrons (2 in the K shell, 8 in the L shell, 18 in the M shell, and so on). The periodic table would be simpler if the noble gases occurred on completion of these shells, at $Z = 2, 10, 28, 60, 110, ...$ Instead, they occur at $Z = 2, 10, 18, 36, 54,$ and 86, and the successive periods contain 2, 8, 8, 18, 18, and 32 elements. This important feature of the periodic table is now thoroughly explained. It is the result of the dependence of the orbital energy of an electron (in all atoms except hydrogen) on the quantum number l as well as on n.

Orbitals Penetrating Inner Shells

The dependence of the energy of an electron on the quantum number l as well as on the principal quantum number, n, is represented in the observed energy level diagram, Figure 5-11, where $2s$ ($l = 0$) is shown below $2p$ ($l = 1$), $3s$ below $3p$, which itself is below $3d$, and so on. It is

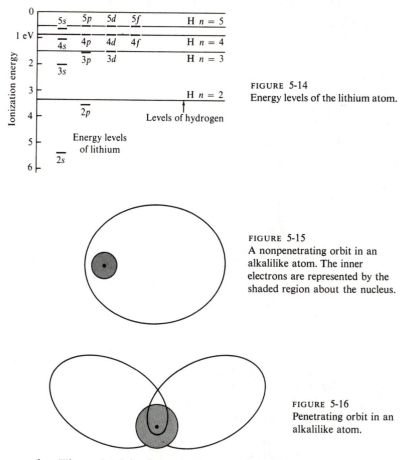

FIGURE 5-14
Energy levels of the lithium atom.

FIGURE 5-15
A nonpenetrating orbit in an alkalilike atom. The inner electrons are represented by the shaded region about the nucleus.

FIGURE 5-16
Penetrating orbit in an alkalilike atom.

seen also (Figure 5-14) in the excited states of the lithium atom,* as well as of all other atoms except hydrogen. The explanation of this behavior was suggested by Schrödinger in 1921, before the development of quantum mechanics. This explanation is indicated by the drawings in Figures 5-15 and 5-16. Schrödinger suggested that the inner electron shell of lithium might be replaced by an equivalent charge of electricity distributed uniformly over the surface of a sphere of suitable radius, which for lithium would be about 0.33 Å (Example 5-5, with the factor $\frac{3}{2}$ of Equation 5-12). The valence electron, outside this shell, would be moving in an electric field of the nucleus, with charge $+3\epsilon$, and of the two K electrons, with

---

*This energy-level diagram for Li has been obtained by combining the photon energy values corresponding to the observed lines in the lithium spectrum.

TABLE 5-5
Electron Configuration of Atoms in Their Normal States

		He	Neon		Argon		Krypton			Xenon			Radon				Eka-radon				Symbol	
		1s	2s	2p	3s	3p	3d	4s	4p	4d	5s	5p	4f	5d	6s	6p	5f	6d	7s	7p		
H	1	1																			$^2S_{1/2}$	
He	2	2																				$^1S_0$
Li	3	2	1																		$^2S_{1/2}$	
Be	4	2	2																			$^1S_0$
B	5	2	2	1																	$^2P_{1/2}$	
C	6	2	2	2																		$^3P_0$
N	7	2	2	3																	$^4S_{3/2}$	
O	8	2	2	4																		$^3P_2$
F	9	2	2	5																	$^2P_{3/2}$	
Ne	10	2	2	6																		$^1S_0$
Na	11				1																$^2S_{1/2}$	
Mg	12				2																	$^1S_0$
Al	13		10		2	1															$^2P_{1/2}$	
Si	14		Neon		2	2																$^3P_0$
P	15		core		2	3															$^4P_{3/2}$	
S	16				2	4																$^3P_2$
Cl	17				2	5															$^2P_{3/2}$	
Ar	18	2	2	6	2	6																$^1S_0$
K	19							1													$^2S_{1/2}$	
Ca	20							2														$^1S_0$
Sc	21						1	2													$^2D_{3/2}$	
Ti	22						2	2														$^3F_2$
V	23						3	2													$^4F_{3/2}$	
Cr	24						5	1														$^7S_3$
Mn	25						5	2													$^6S_{5/2}$	
Fe	26						6	2														$^5D_4$
Co	27		18				7	2													$^4F_{9/2}$	
Ni	28		Argon core				8	2														$^3F_4$
Cu	29						10	1													$^2S_{1/2}$	
Zn	30						10	2														$^1S_0$
Ga	31						10	2	1												$^2P_{1/2}$	
Ge	32						10	2	2													$^3P_0$
As	33						10	2	3												$^4S_{3/2}$	
Se	34						10	2	4													$^3P_2$
Br	35						10	2	5												$^2P_{3/2}$	
Kr	36	2	2	6	2	6	10	2	6													$^1S_0$
Rb	37										1										$^2S_{1/2}$	
Sr	38										2											$^1S_0$
Y	39									1	2										$^2D_{3/2}$	
Zr	40									2	2											$^3F_2$
Nb	41									4	1										$^6D_{1/2}$	
Mo	42									5	1											$^7S_3$
Tc	43									5	2										$^6S_{5/2}$	
Ru	44									7	1											$^5F_5$
Rh	45									8	1										$^4F_{9/2}$	
Pd	46		36							10												$^1S_0$
Ag	47		Krypton core							10	1										$^2S_{1/2}$	
Cd	48									10	2											$^1S_0$
In	49									10	2	1									$^2P_{1/2}$	
Sn	50									10	2	2										$^3P_0$
Sb	51									10	2	3									$^4S_{3/2}$	
Te	52									10	2	4										$^3P_2$
I	53									10	2	5									$^2P_{3/2}$	

		He Neon			Argon		Krypton			Xenon			Radon				Eka-radon				Symbol
		1s	2s	2p	3s	3p	3d	4s	4p	4d	5s	5p	4f	5d	6s	6p	5f	6d	7s	7p	
Xe	54	2	2	6	2	6	10	2	6	10	2	6									$^1S_0$
Cs	55														1						$^2S_{1/2}$
Ba	56														2						$^1S_0$
La	57													1	2						$^2D_{3/2}$
Ce	58												1	1	2						$^3H_4$
Pr	59												2	1	2						$^4K_{11/2}$
Nd	60												3	1	2						$^5L_6$
Pm	61												4	1	2						$^6L_{9/2}$
Sm	62												5	1	2						$^7K_4$
Eu	63												6	1	2						$^8H_{3/2}$
Gd	64												7	1	2						$^9D_2$
Tb	65							54					8	1	2						$^8H_{17/2}$
Dy	66						Xenon	core					9	1	2						$^7K_{10}$
Ho	67												10	1	2						$^6K_{19/2}$
Er	68												11	1	2						$^5L_{10}$
Tm	69												12	1	2						$^4K_{17/2}$
Yb	70												13	1	2						$^3H_6$
Lu	71												14	1	2						$^2D_{3/2}$
Hf	72												14	2	2						$^3F_2$
Ta	73												14	3	2						$^4F_{3/2}$
W	74												14	4	2						$^5D_0$
Re	75												14	5	2						$^6S_{5/2}$
Os	76												14	6	2						$^5D_4$
Ir	77												14	7	2						$^4F_{9/2}$
Pt	78												14	9	1						$^3D_3$
Au	79												14	10	1						$^2S_{1/2}$
Hg	80												14	10	2						$^1S_0$
Tl	81												14	10	2	1					$^2P_{1/2}$
Pb	82												14	10	2	2					$^3P_0$
Bi	83												14	10	2	3					$^4S_{3/2}$
Po	84												14	10	2	4					$^3P_2$
At	85												14	10	2	5					$^2P_{3/2}$
Rn	86	2	2	6	2	6	10	2	6	10	2	6	14	10	2	6					$^1S_0$
Fr	87																		1		$^2S_{1/2}$
Ra	88																		2		$^1S_0$
Ac	89							86										1	2		$^2D_{3/2}$
Th	90						Radon	core										2	2		$^3F_2$
Pa	91																	3	2		$^4F_{3/2}$
U	92																	4	2		$^5D_0$
Eka-Rn	118	2	2	6	2	6	10	2	6	10	2	6	14	10	2	6	14	10	2	6	$^1S_0$

charge -2ϵ (that is, in a field of a charge $+\epsilon$, the same as the charge of the proton). So long as the electron stayed outside the K shell it could be expected to behave in a way corresponding to a hydrogen-like electron. An orbit of this sort, shown in Figure 5-15, is called a *nonpenetrating orbit*. Reference to Figure 5-14 indicates that an f electron or a d electron in an excited lithium atom would be essentially nonpenetrating, but that surely an s electron, in an orbit that extends to the nucleus, would penetrate the K shell, and probably a p electron would also penetrate the K shell to some extent. An electron in a *penetrating orbit* (Figure 5-16) would move into the field of attraction of the nucleus with charge $+3\epsilon$ only partially shielded by the K electrons and would accordingly be stabilized by a large amount.

Example 5-6.   The spectral line corresponding to the transition $2p \longrightarrow 2s$ of Li has observed wavelength 6710 Å. What is the shielding constant of the two K electrons for the $2p$ electron?

Solution.   By the usual method (Equation 3-13) we find that a photon with $\lambda = 6710$ Å has energy 1.848 eV. The $2p$ excited state of Li accordingly lies 1.848 eV above the $2s$ (normal) state, and has ionization energy 1.848 eV less. The binding energy I for the normal state is 5.390 eV, and for the $2p$ state is $5.390 - 1.848 = 3.542$ eV. From Equation 5-11 we then obtain the result $S = 1.98$, as compared with $S = 1.74$ for $2s$. We may say that the $2p$ electron in the excited lithium atom penetrates the K shell to the extent of only 1%, whereas the $2s$ electron in the normal lithium atom penetrates it to the extent of 13%.

The First Long Period

Argon, $Z = 18$, has the electron configuration $1s^2 2s^2 2p^6 3s^2 3p^6$. The next element, potassium, adds a $4s$ electron, rather than a $3d$ electron, because the $4s$ orbital is a penetrating orbital (the electron has binding energy $I = 4.34$ eV; Table 5-1), whereas the $3d$ orbital is a nonpenetrating orbital. The $3d$ excited state of the potassium atom is found from the spectrum of the element to be 2.67 eV above the normal state, and hence to involve binding energy of the $3d$ electron, $4.34 - 2.67 = 1.67$ eV, only a little greater than that for a completely nonpenetrating orbit ($Z - S = 1$), which is $13.60/3^2 = 1.51$ eV.

It is reasonable that $3d$ should be a nonpenetrating orbital in potassium. The closest distance of approach to the nucleus for an electron in a Bohr-Sommerfeld ellipse with total quantum number n and angular momentum $\{l(l+1)\}^{1/2}\hbar$ is

$$\frac{n^2 a_0}{(Z-S)}\left\{1 + \frac{1}{2}\left(1 - \frac{l(l+1)}{n^2}\right)\right\}$$

This equation gives 2.02 Å for a 3d electron with $Z - S = 1$, whereas the core (the ion K^+) has radius about 1.4 Å.* The relation between wave-mechanics and the Bohr-Sommerfeld treatment of the electron structure of atoms is close enough for this calculation to be accepted as explaining the instability of the 3d state for potassium.

For the following elements—Ca, Sc, Ti, V, . . . —the effective nuclear charge increases rapidly, and 3d becomes about equal in energy to 4s at vanadium, $Z = 23$. The first long period (18 elements) accordingly occurs from potassium to krypton, and involves adding the subshells 3d^{10}, 4s^2, and 4p^2 to the argon structure.

Similarly, a 4d electron begins to penetrate the core at niobium, $Z = 41$, as shown by the normal-state configuration in Table 5-5, leading to the second long period. The very long period, from Cs (55) to Rn (86), results from the initiation of penetration of the core by both 5d and 4f at about $Z = 58$.

5-6. The History of the Periodic Table

The differentiation of chemical substances into two groups, elements and compounds, was achieved at the end of the eighteenth century. A long time was required for the recognition of the fact that the elements can be classified in the way now described by the periodic law. The first step was taken in 1817, when the German chemist J. W. Döbereiner (1780–1849) showed that the combining weight of strontium lies midway between the combining weights of the two related elements calcium and barium. Some years later he recognized the existence of other "triads" of similar elements (chlorine, bromine, and iodine; lithium, sodium, and potassium).

Other chemists then showed that the elements could be classified into groups consisting of more than three similar elements. Fluorine was added to the triad chlorine, bromine, and iodine, and magnesium to the triad calcium, strontium, barium. Oxygen, sulfur, selenium, and tellurium had been classed as one family, and nitrogen, phosphorus, arsenic, antimony, and bismuth as another family of elements by 1854.

In 1862 the French chemist A. E. B. de Chancourtois arranged the elements in the order of atomic weights on a helical curve in space, with corresponding points on the successive turns of the helix differing by 16 in atomic weight. He noticed that elements with similar properties appeared near corresponding points, and suggested that "the properties of elements are the properties of numbers." The English chemist J. A. R. Newlands in 1863 proposed a system of classification of the elements in order of atomic

---

*The value of I, leads, by the method of Example 5-5, to $r = 1.44$ Å. The crystal radius of K^+ (Chapter 6) is 1.33 Å.

weights, in which the elements were divided into seven groups of seven elements each. He termed his relation the *law of octaves,* by analogy with the seven intervals of the musical scale. His proposal was ridiculed, however, and he did not develop it further.

The final and most important step in the development of the periodic table was taken in 1869, when the Russian chemist Dmitri I. Mendelyeev (1834–1907) made a thorough study of the relation between the atomic weights of the elements and their physical and chemical properties, with especial attention to valence (Chapter 6). Mendelyeev proposed a periodic table containing seventeen columns, resembling in a general way the periodic Table 5-3 with the noble gases missing. In 1871 Mendelyeev revised this table, and placed a number of elements in different positions, corresponding to revised values of their atomic weights. At the same time, 1871, both he and the German chemist Lothar Meyer (1830–1895), who was working independently, proposed a table with eight columns, obtained by splitting each of the long periods into a period of seven elements, an eighth group containing the three central elements (such as Fe, Co, Ni), and a second period of seven elements. The first and second periods of seven were later distinguished by use of the letters a and b attached to the group symbols, which were the Roman numerals. This nomenclature of the periods (Ia, IIa, IIIa, IVa, Va, VIa, VIIa, VIII, Ib, IIb, IIIb, IVb, Vb, VIb, VIIb) appears, slightly revised, in the present periodic table even when it is written in the extended form.

The "zero" group was added to the periodic table after the discovery of the argonons helium, neon, argon, krypton, and xenon by Lord Rayleigh and Sir William Ramsay in 1894 and the following years. After the discovery of the electron and the development of the theory of the nuclear atom, it was suggested in 1911 by the Dutch physicist A. van den Broek that the nuclear charge of an element, which we now call its atomic number, might be equal to the ordinal number of the element in the periodic system.

The periodic law was accepted immediately after its proposal by Mendelyeev because of his success in making predictions with its use which were afterward verified by experiment. In 1871 Mendelyeev found that by changing seventeen elements from the positions indicated by the atomic weights which had been accepted for them into new positions, their properties could be better correlated with the properties of the other elements. He pointed out that this change indicated the existence of small errors in the previously accepted atomic weights of several of the elements, and large errors for several others, to the compounds of which incorrect formulas had been assigned. Further experimental work verified that Mendelyeev's revisions were correct.

Most of the elements occur in the periodic table in the order of increasing atomic weight. There still remain, however, four pairs of elements in the inverted order of atomic weight; argon and potassium (the atomic numbers of argon and potassium are 18 and 19, respectively, whereas their atomic weights are 39.948 and 39.102), cobalt and nickel, tellurium and iodine, and protactinium and thorium. The nature of the isotopes of these elements is such that the atomic weight of the naturally occurring mixture of isotopes is greater for the element of lower atomic number in each of these pairs than for the element of higher atomic number; thus argon consists almost entirely (99.6%) of the isotope with mass number 40 (18 protons, 22 neutrons), whereas potassium consists largely (93.4%) of the isotope with mass number 39 (19 protons, 20 neutrons). This inversion of the order in the periodic system, as indicated by the chemical properties of the elements, from that of atomic weight caused much concern before the atomic numbers of the elements were discovered, but has now been recognized as having little significance.

A very striking application of the periodic law was made by Mendelyeev. He was able to predict the existence of six elements which had not yet been discovered, corresponding to vacant places in his table. He named these elements eka-boron, eka-aluminum, eka-silicon, eka-manganese, dvi-manganese, and eka-tellurium (Sanskrit: *eka*, first; *dvi*, second). Three of these elements were soon discovered (they were named scandium, gallium, and germanium by their discoverers), and it was found that their properties and the properties of their compounds are very close to those predicted by Mendelyeev for eka-boron, eka-aluminum, and eka-silicon, respectively. Since then the elements technetium, rhenium, and polonium have been discovered or made artificially, and have been found to have properties similar to those predicted for eka-manganese, dvi-manganese, and eka-tellurium.

After helium and argon had been discovered, the existence of neon, krypton, xenon, and radon was clearly indicated by the periodic law, and the search for these elements in air led to the discovery of the first three of them; radon was then discovered during the investigation of the properties of radium and other radioactive substances. While studying the relation between atomic structure and the periodic law Niels Bohr pointed out that element 72 would be expected to be similar in its properties to zirconium. G. von Hevesy and D. Coster were led by this observation to examine ores of zirconium and to discover the missing element, which they named hafnium.

Exercises

5-1. By means of the Bohr frequency rule and the Bohr energy equation, show that the second line (wavelength 4862.7 Å) in the Balmer series of the atomic hydrogen spectrum (Figure 5-1) corresponds to an electron quantum change $n = 4$ to $n = 2$.

5-2. What is the meaning of the Russell-Saunders symbol $^2S_{1/2}$ for the ground state of the hydrogen atom (Table 5-5)?

5-3. The first ionization energy of helium is 24.58 eV, and the second is 54.40 eV. In each case a $1s$ electron is removed. Can you explain why the second $1s$ electron is held so much more tightly than the first?

5-4. What are the electron configurations of Li^+ and Be^+? Why does the second ionization energy have a much larger value for lithium than for beryllium?

5-5. What are the values of the quantum numbers for the initial states and the final state of the spectrum lines in the Balmer spectrum of hydrogen? Draw an energy-level diagram for the hydrogen atom, and indicate the transitions corresponding to the emission of the Balmer lines.

5-6. What is the dependence of the size of the orbits, in the Bohr theory, on the principal quantum number, and on the atomic number of the atom (the charge of the nucleus)?

5-7. Describe the electron-impact experiments carried out by Franck and Hertz in the period 1914 to 1920. What properties of atoms or molecules are measured by this experiment?

5-8. The ionization energies of hydrogen, helium, and lithium are 13.60 eV, 24.58 eV, and 5.39 eV, respectively. Discuss the relation of these values to the chemical properties of the three elements.

5-9. Describe the electronic structures of the normal hydrogen atom, helium atom, and lithium atom in terms of Bohr orbits, the spin of the electron, and the Pauli exclusion principle.

5-10. Make a drawing of each of the argonons, He, Ne, Ar, Kr, Xe, and Rn, showing by dots the proper numbers of electrons in the successive electron shells.

5-11. Satisfy yourself that you understand the way in which the periodic system is built up, by using the table of electron shells for the argonons to deduce the electronic structures of N (atomic number 7), Al (13), K (19), Ni (28), Cu (29), Ba (56), Bi (83), Ra (88). Show these electronic structures in a table like that given in the text for the noble gases.

5-12. The synthetic elements neptunium (93), plutonium (94), americium (95), and curium (96) are all reported to form tripositive ions, as does actinium (89). In what shells and subshells might the extra electrons that Np^{+++}, Pu^{+++}, Am^{+++}, and Cm^{+++} have in excess over those in Ac^{+++} be contained?

5-13. (*a*) What is the electron configuration of fluorine? (Refer to Figure 5-4, and show all nine electrons. Remember that there are three $2p$ orbitals in the subshell.)

(*b*) How many electron pairs are there in the atom? Which orbitals do they occupy?

(*c*) How many unpaired electrons are there? Which orbital does it occupy?

5-14. In an electric arc between carbon electrodes some of the carbon atoms are raised to an excited state to which the spectroscopists have assigned the electron configuration $1s^2 2s 2p_x 2p_y 2p_z$ (also written $2s2p^3$). What electron-dot symbol would you write for this state of the carbon atom?

5-15. (*a*) What are the electron configurations of beryllium and boron, as shown in Figure 5-11?

(*b*) What electron-dot symbols do they correspond to?

(*c*) What are the customary chemical electron-dot symbols for these atoms? (They show a larger number of unpaired electrons.)

5-16. Write the electron configuration for the element with $Z = 103$, showing all 103 electrons. To what orbital do you assign the last electron? Why?

5-17. What are the most important metallic properties? In what part of the periodic table are the elements with metallic properties? Classify the following elements as metals, metalloids, or nonmetals; potassium, arsenic, aluminum, xenon, bromine, silicon, phosphorus.

5-18. In the article on "Chemistry" in the Ninth Edition of the Encyclopaedia Britannica (published in 1878) the author (H. A. Armstrong) says that Mendelyeev had recently proposed that uranium be assigned the atomic weight 240 in place of the old value 120 that had been assigned to it by Berzelius, but that he himself preferred 180. Mendelyeev was right. The correct formula of pitchblende, an important ore of uranium, is U_3O_8. What formula was written for pitchblende by (*a*) Berzelius, (*b*) Armstrong?

6

The Chemical Bond

For over a century chemists have systematized the formulas of compounds by assigning certain combining powers, *valences*, to the elements. The valence of an element was formerly described as the number of valence bonds formed by an atom of the element with other atoms.

The effort to obtain a clear understanding of the nature of valence and of chemical combination in general has led in recent years to the dissociation of the concept of valence into several new concepts—especially *covalence* and *ionic valence*. We shall discuss these concepts and the related question of the nature of the chemical bond in the present chapter.

6-1. The Nature of Covalence

The atoms of most molecules are held tightly together by a very important sort of bond, the *shared-electron-pair bond* or *covalent bond*. This bond is so important, so nearly universally present in substances that Professor Gilbert Newton Lewis of the University of California (1875–1946), who discovered its electronic structure, called it *the* chemical bond.

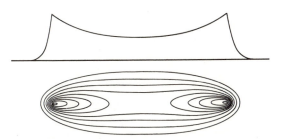

FIGURE 6-1
The electron distribution function for the hydrogen
molecule-ion. The upper curve shows the value of the
function along the line through the two nuclei, and the
lower figure shows contour lines, increasing from 0.1
for the outermost to 1 at the nuclei.

It is the covalent bond that is represented by a dash in the valence-bond

formulas, such as Br—Br and Cl—C—Cl, that have been written by

$$\text{Cl}$$
$$|$$
$$\text{Cl—C—Cl}$$
$$|$$
$$\text{Cl}$$

chemists for over a hundred years.

Modern chemistry has been greatly simplified through the development
of the theory of the covalent bond. It is now easier to understand and to
remember chemical facts, by connecting them with our knowledge of the
nature of the chemical bond and the electronic structure of molecules,
than was possible sixty years ago. It is accordingly wise for the student of
chemistry to study this chapter carefully, and to get a clear picture of the
chemical bond.

The Hydrogen Molecule-ion

The simplest molecule is the hydrogen molecule-ion, H_2^+, which consists of two protons and one electron. Its properties have been determined
from the study of the hydrogen band spectrum (molecular spectrum). The
two protons have average separation 1.06 Å, and the bond energy (the
energy required to split H_2^+ into H and H^+) is 255 kJ mole^{-1}.

The wave-mechanical treatment of H_2^+ gives values of the bond length
and the bond energy in complete agreement with experiment. The calculated electron distribution is shown in Figure 6-1. The electron is concentrated in the region between the two nuclei. Its electrostatic attraction for
the nuclei balances the mutual repulsion of the nuclei, and is the source of
the bond energy. The bond is called a *one-electron bond*.

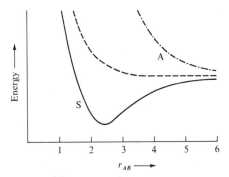

FIGURE 6-2

Curves showing the energy of interaction of a
hydrogen atom and a proton. The lower
curve corresponds to the formation of a
hydrogen molecule-ion in its stable normal
state. The scale for the internuclear distance
r_{AB} is based on the unit $a_0 = 0.530$ Å.

A simple and illuminating wave-mechanical discussion of H_2^+ can be
carried out with use of the $1s$ wave function for the hydrogen atom in its
normal state. Let us call the two protons H_a^+ and H_b^+. If the proton H_a^+
has the electron attached to it, to form a normal hydrogen atom $H_a\cdot$, we
use the symbol $1s(a)$ for the wave function. It is found, when the energy of
the system is calculated for this wave function, corresponding to a proton
H_b^+ approaching a hydrogen atom $H_a\cdot$, that no bond is formed: the
energy curve, shown by the dashed line in Figure 6-2, corresponds to
repulsion. The same result is, of course, obtained for the wave function
$1s(b)$, corresponding to approach of the proton H_a^+ to the hydrogen
atom $H_b\cdot$.

When the wave-mechanical calculation of the energy is made for the
hybrid structure

$$(H_a\cdot \quad H_b^+) + (H_a^+ \quad \cdot H_b)$$

with wave function $1s(a) + 1s(b)$, the curve labeled S (for symmetric) in
Figure 6-2 is obtained. This curve agrees with the observed properties of
the hydrogen molecule-ion.

It is customary to say that the electron *resonates* between the two nuclei,
and that the wave function is the symmetric hybrid of $1s(a)$ and $1s(b)$.
The bond energy may be called the *resonance energy* of the electron between positions about the two protons.

Another wave function can also be formed from $1s(a)$ and $1s(b)$; it is the
antisymmetric function $1s(a) - 1s(b)$. (It is called antisymmetric because
it changes sign if the nuclei a and b are interchanged.) This function gives

zero electron density midway between the nuclei. It leads to strong repulsion, as shown by curve A in Figure 6-2, and corresponds to an anti-
bonded (unstable) state of H_2^+.

Both the symmetric and the antisymmetric wave functions described
above are called *molecular orbitals*.

Bonds Involving 1s Orbitals

Let us consider the $1s$ orbitals of two nuclei, a and b. The exclusion
principle permits $1s(a)$ to be occupied by two electrons (with opposed
spins) and $1s(b)$ to be occupied by two electrons.

We consider the following cases:

			Binding electrons
I.	H_2^+	$a + b$	1
II.	H_2	$(a + b)^2$	$1 + 1$
III.	He_2^+	$(a + b)^2(a - b)$	$1 + 1 - 1$
IV.	$(He_2) = He + He$	$(a + b)^2(a - b)^2$	$1 + 1 - 1 - 1$

Because of the equivalence of the two nuclei, we use the symmetric and
antisymmetric molecular orbitals (a means $1s(a)$, b means $1s(b)$). The
resonance energy is bonding (stabilizing) for an electron in $a + b$, and
antibonding (destabilizing) for an electron in $a - b$. The exclusion principle permits only two electrons to occupy $a + b$. (They must have opposed
spins.) We conclude, as indicated above, that there are two bonding
electrons in H_2, a resultant of one in He_2^+, and a resultant of zero in
$He + He$.

This conclusion agrees with experiment. The bond energy for H_2,
429 kJ mole^{-1}, is nearly twice that for H_2^+, that for He_2^+, 243 kJ mole^{-1}, is
about the same as that of H_2^+, and two helium atoms have only very weak
attraction for one another.

The Hydrogen Molecule

Another description of the hydrogen molecule is that it involves reso-
nance between the two structures $H_a\uparrow$ $\downarrow H_b$ and $H_a\downarrow$ $\uparrow H_b$; that is, the
electron with positive spin and the electron with negative spin change
places. This wave function leads to somewhat better calculated values of
the bond energy and bond length than the molecular-orbital wave function. Still better agreement is obtained by use of an intermediate wave
function, involving a small contribution of the ionic structures $H_a^-\uparrow\downarrow$
H_b^+ and H_a^+ $\uparrow\downarrow H_b^-$, in addition to the two structures given above.

In the hydrogen molecule the two nuclei are rather firmly held at a
distance of about 0.74 Å apart; they oscillate relative to each other with
an amplitude of a few hundredths of an Ångstrom at room temperature,

and with a somewhat larger amplitude at higher temperatures. The two electrons move very rapidly about in the region of the two nuclei, their time-average distribution being indicated by the shading in Figure 6-3. It can be seen that the motion of the two electrons is largely concentrated into the small region between the two nuclei. *The two electrons held jointly by the two nuclei constitute the chemical bond between the two hydrogen atoms.*

The hydrogen atom can achieve the helium structure $1s^2$ by adding an electron, as in the salt lithium hydride, Li^+H^- (which is discussed in Section 6-8). But it can also achieve the helium structure by sharing an electron pair with another atom, as in H_2, with the shared pair being counted first for one atom and then for the other.

The Covalent Bond in Other Molecules

The covalent bond in other molecules is closely similar to that in the hydrogen molecule. For each covalent bond a pair of electrons is needed; also, two orbitals are needed, one of each atom.

The covalent bond consists of a pair of electrons shared between two atoms, and occupying two stable orbitals, one of each atom.

For example, reference to the energy-level diagram (Figure 5-11) shows that the carbon atom has four stable orbitals in its L shell, and four electrons that may be used in bond formation. Hence it may combine with four hydrogen atoms, each of which has one stable orbital (the $1s$ orbital) and one electron, forming four covalent bonds:

$$
\begin{array}{ccc}
\text{H} & & \text{H} \\
\ddot{\text{H}}\!:\!\ddot{\text{C}}\!:\!\text{H} & \text{equivalent to} & \text{H}\!-\!\text{C}\!-\!\text{H} \\
\text{H} & & \text{H}
\end{array}
$$

In this molecule each atom has achieved an argononic (noble gas) structure; the shared electron pairs are to be counted for each of the atoms sharing them. The carbon atom, with four shared pairs in the L shell and one unshared pair in the K shell, has achieved the neon structure, and each hydrogen atom has achieved the helium structure.

6-2. The Structure of Covalent Compounds

The electronic structure of molecules of covalent compounds involving the principal groups of the periodic table can usually be written by counting the number of valence electrons in the molecule and then

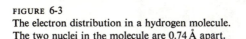

FIGURE 6-3
The electron distribution in a hydrogen molecule.
The two nuclei in the molecule are 0.74 Å apart.

distributing the valence electrons as unshared electron pairs and shared electron pairs in such a way that each atom achieves an argononic structure.

For many molecules the covalence of each atom is equal to the number of unpaired electrons in its outer shell, and is thus simply related to the position of the element in the periodic table. For other molecules and ions the covalence of the atoms is less simply related to the periodic table.

It is often necessary to have some experimental information about the way in which the atoms are bonded together. Thus there are two compounds with the composition C_2H_6O: ethyl alcohol and dimethyl ether. The chemical properties of these two substances show that one of them, ethyl alcohol, contains one hydrogen atom attached to an oxygen atom, whereas dimethyl ether does not contain such a hydroxyl group. The structures of these two molecules are shown in Figure 6-4.

Compounds of Hydrogen with Nonmetals

Let us consider first the structure expected for a compound between hydrogen and fluorine, the lightest element of the seventh group. Hydrogen has a single orbital and one electron. Accordingly, it could achieve the helium configuration by forming a single covalent bond with another element. Fluorine has seven electrons in its outer shell, the L shell. These seven electrons occupy the four orbitals of the L shell. They accordingly constitute three electron pairs in three of the orbitals and a single electron in the fourth orbital. Hence fluorine also can achieve a noble-gas con-

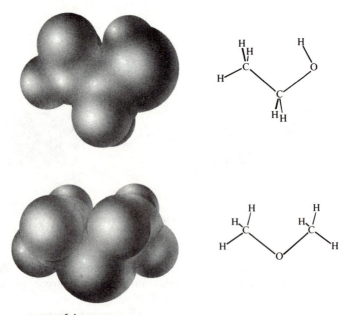

The structures of the isomeric molecules ethyl alcohol, C_2H_5OH, and dimethyl ether, $(CH_3)_2O$.

figuration by forming a single covalent bond with use of its odd electron. We are thus led to the following structure for the hydrogen fluoride molecule:

$$H:\ddot{\underset{..}{F}}:$$

In this molecule, the hydrogen fluoride molecule, there is a single covalent bond that holds the hydrogen atom and the fluorine atom firmly together.

It is often convenient to represent this electronic structure by using a dash as a symbol for the covalent bond instead of the dots representing the shared electron pair. Sometimes, especially when the electronic structure of the molecule is under discussion, the unshared pairs in the outer shell of each atom are represented, but often they are omitted:

$$H—\ddot{\underset{..}{F}}: \quad \text{or} \quad H—F$$

The other halogens form similar compounds:

$$H—\ddot{\underset{..}{Cl}}: \qquad\qquad H—\ddot{\underset{..}{Br}}: \qquad\qquad H—\ddot{\underset{..}{I}}:$$

Hydrogen chloride   Hydrogen bromide   Hydrogen iodide

Elements of the sixth group (oxygen, sulfur, selenium, tellurium) can achieve the noble-gas structure by forming two covalent bonds. Oxygen has six electrons in its outer shell. These can be distributed among the four orbitals by putting two unshared pairs in two of the orbitals and an odd electron in each of the other two orbitals. These two odd electrons can be used in forming covalent bonds with two hydrogen atoms, to give a water molecule, with the following electronic structure:

$$\begin{array}{cc} \text{H} & \text{H} \\[4pt] :\overset{..}{\underset{..}{\text{O}}}:\text{H} \quad \text{or} \quad & :\overset{..}{\underset{..}{\text{O}}}-\text{H} \end{array}$$

If a proton is removed, a hydroxide ion, OH^-, is formed:

$$[:\overset{..}{\underset{..}{\text{O}}}-\text{H}]^-$$

If a proton is added to a water molecule (attaching itself to one of the unshared electron pairs), a hydronium ion, OH_3^+, is formed:

$$\left[\begin{array}{c} \text{H} \\ | \\ :\text{O}-\text{H} \\ | \\ \text{H} \end{array} \right]^+$$

All three of the hydrogen atoms in the hydronium ion are held to the oxygen atom by the same kind of bond, a covalent bond.

In hydrogen peroxide, H_2O_2, each oxygen atom achieves the neon configuration by forming one covalent bond with the other oxygen atom and one covalent bond with a hydrogen atom:

Hydrogen sulfide, hydrogen selenide, and hydrogen telluride have the same electronic structure as water:

$$\begin{array}{ccc} \text{H} & \text{H} & \text{H} \\ | & | & | \\ :\overset{}{\underset{..}{\text{S}}}-\text{H} & :\overset{}{\underset{..}{\text{Se}}}-\text{H} & :\overset{}{\underset{..}{\text{Te}}}-\text{H} \end{array}$$

Nitrogen and the other fifth-group elements, with five outer electrons, can achieve the noble-gas configuration by forming three covalent bonds. The structures of ammonia, phosphine, arsine, and stibine are the following:

$$\begin{array}{cccc} \text{H} & \text{H} & \text{H} & \text{H} \\ | & | & | & | \\ :\text{N}-\text{H} & :\text{P}-\text{H} & :\text{As}-\text{H} & :\text{Sb}-\text{H} \\ | & | & | & | \\ \text{H} & \text{H} & \text{H} & \text{H} \end{array}$$

The ammonia molecule can attach a proton to itself, to form an ammonium ion, NH_4^+, in which all four hydrogen atoms are held to the nitrogen atom by covalent bonds:

In the ammonium ion all four of the *L* orbitals are used in forming covalent bonds. The formation of the ammonium ion from ammonia is similar to the formation of the hydronium ion from water.

The Electronic Structure of Some Other Compounds

Electronic structures of other molecules containing covalent bonds may readily be written, by keeping in mind the importance of completing the octets of atoms of nonmetallic elements. The structures of some compounds of nonmetallic elements with one another are shown below:

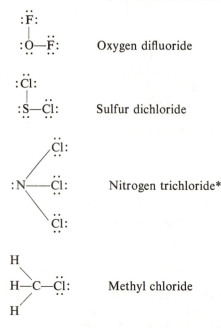

Oxygen difluoride

Sulfur dichloride

Nitrogen trichloride*

Methyl chloride

---

*Note that sometimes an effort is made in drawing the structure of a molecule to indicate the spatial configuration; the structure shown here for nitrogen trichloride is supposed to indicate that the molecule is pyramidal, with the chlorine atoms approximately at three corners of a tetrahedron about the nitrogen atom. The spatial configuration of molecules is discussed in the following section.

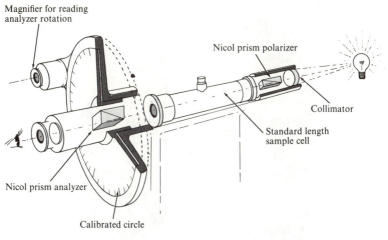

Magnifier for reading
analyzer rotation

Nicol prism polarizer

Collimator

Standard length
sample cell

Nicol prism analyzer

Calibrated circle

FIGURE 6-5
The polarimeter, an instrument used to determine the rotation of the plane of
polarization of a beam of plane-polarized light by an optically active substance.

6-3. The Direction of Valence Bonds in Space

In 1874 it was discovered that the four bonds formed by a carbon atom are
directed in space toward the four corners of a tetrahedron. This discovery
was made through the effort to explain the observed effects of some substances on polarized light, as described in the following paragraphs.

Optical Activity

When a beam of ordinary light is passed through a crystal of calcite, it
is split into two beams. Each of these beams is a beam of plane-polarized
light; the vibrating electric field of the light lies in one plane for one of the
beams and in the plane at right angles to it for the other beam.

A prism made of two pieces of calcite cut in a certain way and cemented
together has the property of permitting only one beam to pass through;
the other is reflected to the side and absorbed in the darkened side of the
prism. Such a prism (a Nicol prism) can be used to form a beam of plane-
polarized light, and also to determine the orientation of the plane of polarization. In the instrument called the polarimeter, shown in Figure 6-5, the
first prism defines the beam. If there is nothing between the two prisms to
rotate the plane of polarization, the beam will pass through the second
prism if it has the same orientation as the first, but will be absorbed if it is
oriented at right angles to the first.

Left-hand quartz   Right-hand quartz

FIGURE 6-6
Right-handed and left-handed quartz crystals.

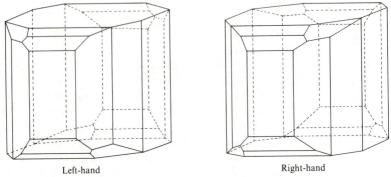

Left-hand   Right-hand

Sodium ammonium tartrate

FIGURE 6-7
Right-handed and left-handed crystals of sodium ammonium tartrate.

In 1811 the French physicist Dominique François Jean Arago (1786-1853) discovered that a quartz crystal has the power of rotating the plane of polarization of the beam of polarized light passing through it. Some quartz crystals were found to rotate the plane of polarization to the right (so that the second prism has to be turned clockwise, viewed in the direction opposite to the path of the light, in order that the beam pass through it), and others to the left. These are called dextrorotatory crystals and levorotatory crystals, respectively.

The two kinds of quartz crystals also differ in their face development, as shown in Figure 6-6. They are mirror images of one another: one can be described as a right-handed crystal and one as a left-handed crystal.

Right-handed and Left-handed Molecules

The French physicist Jean Baptiste Biot (1774–1862) then found that some liquids are optically active (that is, have the power of rotating the plane of polarization). For example, turpentine was found to be levorotatory, and an aqueous solution of sucrose (cane sugar, $C_{12}H_{22}O_{11}$) was found to be dextrorotatory. The substances that were found to be optically active in solution were all organic compounds, produced by plants or animals.

A puzzling observation was then made. It was found that two kinds of tartaric acid were deposited from wine lees. These two kinds of tartaric acid are closely similar in their properties, but they show the astounding difference that one is dextrorotatory, and the other is completely without rotatory power. How could there be two molecules with the same composition but with such greatly different power of interacting with polarized light?

The answer to this puzzle was found in 1844 by the great French chemist Louis Pasteur (1822–1895). He added sodium hydroxide and ammonium hydroxide to the solution of the optically inactive tartaric acid and allowed the solution to evaporate, so that crystals of sodium ammonium tartrate, $NaNH_4C_4H_4O_6$, were formed. On examining the crystals he first noticed that they appeared to be identical with the crystals similarly made from the optically active tartaric acid. Then, as he continued to scrutinize them carefully, he suddenly recognized that only half of them were truly identical; the others were their mirror images (Figure 6-7). He separated the two kinds of crystals by hand and dissolved them in water. One of the solutions was dextrorotatory and the other levorotatory, with the same rotatory power except for sign.

It was accordingly evident that the atoms in the tartaric acid molecule arrange themselves in a structure that does not have a plane of symmetry (mirror plane) or a center of symmetry (inversion center); hence there is a

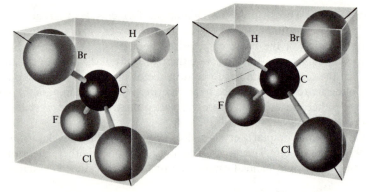

FIGURE 6-8
Right-handed and left-handed molecules of fluorochlorobromomethane.

right-handed arrangement of atoms and there also is a left-handed arrangement, the mirror image of the first.

In 1844 the possibility of discovering these arrangements (the three-dimensional structures of the molecules) was so small that neither Pasteur nor any other chemist attacked the problem. But within fifteen years the correct atomic weights were accepted and the correct formulas were assigned, the concept of the chemical bond was developed, and the quadrivalence of carbon was established. The term "chemical structure" was used for the first time in 1861 by the Russian chemist Alexander M. Butlerov (1828–1886), who wrote that it is essential to find the way in which each atom is linked to other atoms in the molecules of substances.

A few more years passed. Many students of chemistry learned about the quadrivalence of carbon, about structural formulas, and about right-handed and left-handed molecules. Then two of them, the young Dutch chemist Jacobus Hendricus van't Hoff (1852–1911) and the young French chemist Jules Achille le Bel (1847–1930), saw in 1874 that *no* structure in which the atoms lie in a single plane can lead to optical activity. A planar molecule is its own mirror image, because the plane is itself a plane of symmetry for the molecule. For example, the substance fluorochlorobromomethane, CHFClBr, can be resolved into a dextrorotatory variety and a levorotatory variety. Hence it cannot be correctly represented by the planar formula

$$
\begin{array}{c}
\text{F} \\
| \\
\text{Cl—C—H} \\
| \\
\text{Br}
\end{array}
$$

The four bonds formed by the carbon atom in this molecule cannot lie in one plane, but instead must extend toward the corners of a tetrahedron. There must be two kinds of fluorochlorobromomethane molecules, identical except for handedness, each the exact mirror image of the other (Figure 6-8).

This was the birth of the tetrahedral carbon atom and of stereochemistry (the chemistry of three-dimensional space, or structural chemistry). It led to the rapid development of chemical structure theory and of chemistry as a whole.

A pair of right-handed and left-handed molecules are called an *enantiomeric pair,* and the two substances they compose are called *enantiomers* (from the Greek *enantios*, opposite, and *meros*, part). The symbols D and L are used as prefixes to distinguish the two substances that constitute an enantiomeric pair. A pair of crystals of enantiomers is called an enantiomorphic pair.

A discussion of right-handed and left-handed molecules in living organisms is given in Chapter 24.

6-4.   Tetrahedral Bond Orbitals

In a molecule such as methane, CH_4, or carbon tetrachloride, CCl_4, in which the four bonds are equivalent, the bond angles H—C—H or Cl—C—Cl have the value 109°28′. In an asymmetric molecule such as CHFClBr the angles differ somewhat from this value, but only by a few degrees. It has been found by experiment (x-ray diffraction, electron diffraction, microwave spectroscopy) that these angles usually lie between 106° and 113°, with the average value for the six bond angles close to 109°28′.

Each of the four bonds formed by the carbon atom involves one of the four orbitals of the L shell. These orbitals are given in Chapter 5 as the $2s$ orbital and the three $2p$ orbitals. We might hence ask whether or not the bonds to the four hydrogen atoms are all alike. Would not the $2s$ electron form a bond of one kind, and the three $2p$ electrons form bonds of a different kind?

Chemists have made many experiments to answer this question, and have concluded that the four bonds of the carbon atom are alike. A theory of the tetrahedral carbon atom was developed in 1931. According to this theory, the *theory of hybrid bond orbitals*, the $2s$ orbital and the three $2p$ orbitals of the carbon atom are hybridized (combined) to form four *tetrahedral bond orbitals*. They are exactly equivalent to one another, and are directed toward the corners of a regular tetrahedron, as shown in Figure 6-9. Moreover, the nature of s and p orbitals and their hybrids is

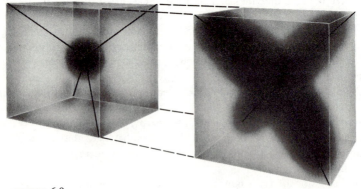

FIGURE 6-9
Diagram illustrating (left) the $1s$ orbital in the K shell of the carbon atom, and (right) the four tetrahedral orbitals of the L shell.

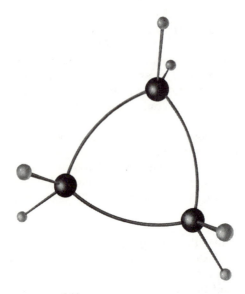

FIGURE 6-10
The molecule of cyclopropane, C_3H_6, showing the bent carbon-carbon bonds.

such that of all possible hybrid orbitals of s and p the tetrahedral orbitals are the best suited for forming strong bonds. Accordingly the tetrahedral arrangement of the bonds is the stable one (see Appendix VII).

Some molecules are known in which the bond angles are required by the molecular structure to differ greatly from the tetrahedral value. These molecules can be described as containing bent bonds, and as being strained. The substance cyclopropane, C_3H_6, is an example. The cyclopropane molecule contains a ring of three carbon atoms, as shown in Figure 6-10. Each bond is bent through nearly 50°. The molecules are less stable by about 100 kJ mole^{-1} (33 kJ mole^{-1} per bond) than corresponding unstrained molecules, such as those of cyclohexane, C_6H_{12}.

The Carbon-Carbon Double Bond

Sometimes two valence bonds of an atom are used in the formation of a double bond with another atom. There is a double bond between two carbon atoms in the molecules of ethylene, C_2H_4:

Ethylene

Such a double bond between two atoms may be represented by two tetrahedra sharing two corners—that is, sharing an edge, as shown in Figure 6-11. The amount of bending of the two bonds constituting the double bond is indicated in Figure 6-12.

It is interesting to note that the four other bonds that the two carbon atoms in ethylene can form lie in the same plane, at right angles to the plane containing the two bent bonds.

The Carbon-Carbon Triple Bond

In acetylene, C_2H_2, there is a triple bond between the two carbon atoms:

$$H\!-\!C\!\equiv\!C\!-\!H$$
Acetylene

The triple bonds between two atoms may be represented by two tetrahedra sharing a face (Figure 6-11 and 6-12). This causes the acetylene molecule to be linear.

Bond Lengths

Spectroscopic studies have led to the determination of the carbon-carbon bond length (distance between the nuclei of the two carbon atoms) in ethane, ethylene, and acetylene. The values are 1.54 Å for the single

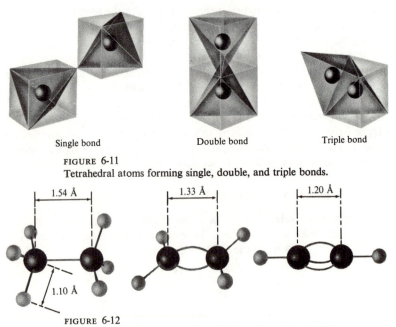

Single bond   Double bond   Triple bond

FIGURE 6-11
Tetrahedral atoms forming single, double, and triple bonds.

FIGURE 6-12
Valence-bond models of ethylene, C_2H_4, and acetylene, C_2H_2.

bond in ethane (and also in other molecules containing the C—C bond), 1.54 Å for the double bond, and 1.20 Å for the triple bond.

It is interesting that these values for C=C and C≡C are within 0.02 Å of the values that correspond to bent bonds with the normal single-bond length 1.54 Å and at tetrahedral angles, as indicated in Figure 6-12. This agreement supports the bent-bond description of the double bond and the triple bond.

6-5. Bond Orbitals with Large *p* Character

In a molecule such as ammonia, NH_3, with structural formula

the bond orbitals of the nitrogen atom are not tetrahedral orbitals, but instead have mainly the character of the three $2p$ orbitals. Quantum-mechanical calculations and nuclear magnetic resonance experiments (which measure the interaction energy of the nuclear spin magnetic moment with the valence electrons) agree in allocating the unshared electron pair to a hybrid orbital that is largely $2s$ in character (about 79%). The three bond orbitals have about 93% $2p$ character and 7% $2s$ character.

It was pointed out in Chapter 5 that a $2s$ electron is more stable than a $2p$ electron.* The nitrogen atom, $:\overset{\cdot}{\underset{\cdot}{N}}\cdot$, is about 1000 kJ mole⁻¹ more

stable if the pair of electrons is in the $2s$ orbital ($2s^2 2p_x 2p_y 2p_z$) than if it is in one of the $2p$ orbitals ($2s 2p_x^2 2p_y 2p_z$). Hence it tends to retain the $2s$ pair in forming compounds, and to use the $2p_x$, $2p_y$, and $2p_z$ orbitals for the bonding electrons.

The three $2p$ orbitals are represented in Figure 5-10. The $2p_x$ orbital extends in two opposite directions along the x-axis, and can be used in forming a bond in either direction. The $2p_y$ orbital can be used in forming a bond along the y-axis, and the $2p_z$ orbital in forming a bond along the z-axis. Hence *the bonds formed by p orbitals are approximately at 90° to one another.* With some s character added to the bond orbitals the bond angles increase, reaching 109°28' for tetrahedral orbitals, which have 25% s character.

The experimental values of bond angles for atoms with unshared electron pairs usually lie between 90° and 109°. For example, the spectroscopically determined value for NH_3 is 107°, for H_2O 104.5°, for PH_3 93°, for H_2S 92°, and for H_2Se 91°.

6-6. Molecules and Crystals of the Nonmetallic Elements

The Halogen Molecules

A halogen atom such as fluorine can achieve the noble-gas structure by forming a single covalent bond with another halogen atom:

$$:\overset{..}{\underset{..}{F}}-\overset{..}{\underset{..}{F}}: \quad :\overset{..}{\underset{..}{Cl}}-\overset{..}{\underset{..}{Cl}}: \quad :\overset{..}{\underset{..}{Br}}-\overset{..}{\underset{..}{Br}}: \quad :\overset{..}{\underset{..}{I}}-\overset{..}{\underset{..}{I}}:$$

This single covalent bond holds the atoms together into diatomic molecules, which are present in the elementary halogens in all states of aggregation—crystal, liquid, and gas.

---

*It is seen from Figure 5-14 that the $2p$ level of lithium lies about 1.85 eV above the $2s$ level. The corresponding values of the p-s separation for the other alkali metals are 2.10 eV for Na, 1.62 eV for K, 1.58 eV for Rb, and 1.44 eV for Cs. The values are about twice as great for the group II elements, three times as great for group III, and so on. We shall usually use energy values in kJ mole⁻¹ (1 eV = 96.49 kJ mole⁻¹), and as an easily remembered approximation we shall accept 200 z kJ mole⁻¹ as the p-s separation, with z equal to the group number in the periodic table. Also, the d-p separation is approximately equal to the p-s separation.

The Elements of the Sixth Group

An atom of a sixth-group element, such as sulfur, lacks two electrons of having a completed octet. It can complete its octet by forming single covalent bonds with two other atoms. These bonds may hold the molecule together either into a ring, such as an S_8 ring, or into a very long chain, with the two end atoms having an abnormal structure:

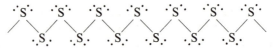

The elementary substance *sulfur* occurs in both these forms (see Section 7-1). Ordinary sulfur (orthorhombic sulfur) consists of molecules made of eight atoms. The molecule S_8 has the configuration shown in Figure 6-13. It is a staggered octagonal ring, with S—S—S bond angle 102°.

Ordinary *oxygen* consists of diatomic molecules with an unusual electronic structure. We might expect these molecules O_2 to contain a double bond:

$$:\overset{..}{O}::\overset{..}{O}: \quad \text{or} \quad :\overset{..}{O}=\overset{..}{O}:$$

Instead, only one shared pair is formed, leaving two unshared electrons:

$$:\overset{..}{\underset{.}{O}}-\overset{.}{\underset{.}{O}}:$$

These two unshared electrons are responsible for the paramagnetism of oxygen. It has been found by study of the oxygen spectrum that the force of attraction between the oxygen atoms is much greater than that expected for a single covalent bond. This shows that the unpaired electrons are really involved in the formation of bonds of a special sort. The oxygen molecule may be said to contain a single covalent bond plus two three-electron bonds, and its structure may be written as

$$:\overset{...}{\underset{...}{O-O}}:$$

Ozone, the triatomic form of oxygen, has the electronic structure

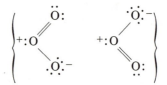

Here one of the end atoms of the molecule resembles a fluorine atom in that it completes its octet by sharing only one electron pair. It may be considered to be the negative ion, $:\overset{..}{O}\cdot^-$, which forms one covalent bond.

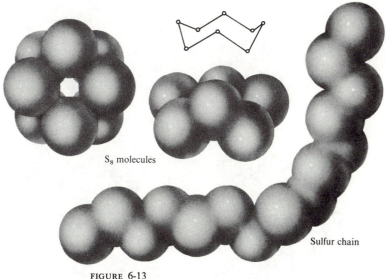

S_8 molecules

Sulfur chain

FIGURE 6-13
The S_8 ring, and a long chain of sulfur atoms.

The central oxygen atom of the ozone molecule resembles a nitrogen atom (see the following section), and may be considered to be the positive ion :$\overset{\cdot}{\underset{\cdot}{O}}\cdot^+$, which forms three covalent bonds (one double bond and one single bond). The angle between the double bond and the single bond is 116.8°, somewhat smaller than the value for tetrahedral orbitals, 125.3° (Figure 6-11).

Two structures for ozone are shown above, in braces. This indicates that the two end oxygen atoms are not different, but are equivalent. The molecule has a structure represented by the superposition of the two structures shown; that is, each bond is a hybrid of a single covalent bond and a double covalent bond (see Section 6-7, on resonance).

Nitrogen and Its Congeners

The nitrogen atom, lacking three electrons of a completed octet, may achieve the octet by forming three covalent bonds. It does this in elementary nitrogen by forming a triple bond in the molecule N_2. Three electron pairs are shared by the two nitrogen atoms:

$$:N:::N: \quad \text{or} \quad   :N\equiv N:$$

This bond is extremely strong, and the N_2 molecule is very stable.

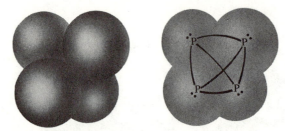

FIGURE 6-14
The P_4 molecule.

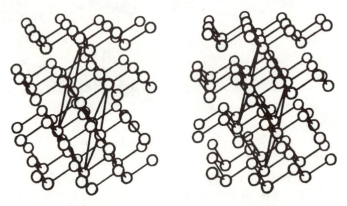

FIGURE 6-15
A stereo view of a portion of an arsenic crystal. Each atom is
attached by single bonds to three other atoms in a puckered layer.

Phosphorus gas at very high temperatures consists of similar P_2 molecules, $:P\equiv P:$. At lower temperatures the molecule P_4 is stable. This
molecule has the tetrahedral structure shown in Figure 6-14. The four
phosphorus atoms are arranged at the corners of a regular tetrahedron.
Each phosphorus atom forms covalent bonds with the three other phosphorus atoms. This P_4 molecule exists in phosphorus vapor, in solutions of
phosphorus in carbon disulfide and other nonpolar solvents, and in solid
white phosphorus. In other forms of elementary phosphorus (red phosphorus, black phosphorus) the atoms are bonded together into larger
aggregates.

Arsenic and *antimony* also form tetrahedral molecules, As_4 and Sb_4, in
the vapor phase. At higher temperatures these molecules dissociate into
diatomic molecules, As_2 and Sb_2. Crystals of these elementary substances
and of bismuth, however, contain high polymers—layers of atoms in
which each atom is bonded to three neighbors by single covalent bonds, as
shown in Figure 6-15.

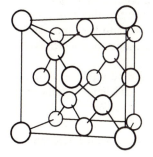
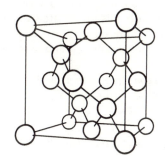

FIGURE 6-16
The structure of diamond (stereo).

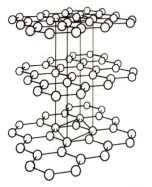
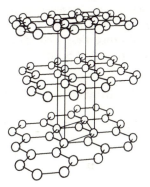

FIGURE 6-17
The structure of graphite (stereo).

Carbon and Its Congeners

Carbon, with four electrons missing from a completed octet, can form four covalent bonds. In *diamond* each atom is bonded strongly to four neighboring atoms which are held about it at the corners of a regular tetrahedron (Figure 6-16). These covalent bonds bind all of the atoms in the diamond crystal together into a single giant molecule, and since the C—C bonds are very strong the crystal is very hard. This structure helps to explain why diamond is one of the hardest substances known.

Graphite consists of layers of atoms, with the structure shown in Figure 6-17. Each atom has three near neighbors. It is bonded to two of them by

single covalent bonds, and to the third by a double bond. This completes
the octet for each atom. The double bonds are not fixed in position, but
move around so as to give each bond some double-bond character. The
covalent bonds tie the atoms very tightly together into layers; the layers,
however, are piled rather loosely on one another, and can be separated
easily, which causes graphite to be a soft substance, which is even used as a
lubricant.

Silicon, *germanium*, and gray *tin* also crystallize with the diamond
structure. Ordinary tin (white tin) and lead have metallic structures (see
Chapter 20).

The Relative Stability of Single Bonds and Multiple Bonds

From the foregoing discussion of the structures of stable molecules and
crystals of the nonmetallic elements we draw the conclusion that single
bonds are more stable than multiple bonds for all of these elements except
the first-row *p*-bonding elements, which are nitrogen and oxygen (also
fluorine, which, however, is restricted by its normal covalence 1). This
generalization is useful in explaining many observed differences in structure and properties of compounds of the first-row elements and those of
the corresponding compounds of their heavier congeners.

6-7.   Resonance

In the foregoing section it was mentioned that ozone has the structure

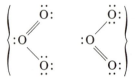

The reason for this statement is that it is known from experiment that the
two oxygen–oxygen bonds in ozone are not different, but are equivalent,
each having the length 1.278 Å. Equivalence of the bonds can be explained by the assumption of a *hybrid structure*. Each of the bonds in
ozone is a hybrid between a single bond and a double bond, and its
properties are intermediate between those of these two kinds of bonds.

The double bond may be said to *resonate* between the two positions in
ozone. The *resonance of molecules between two or more electronic structures* is an important concept. Often it is found difficult to assign to a
molecule a single electronic structure of the valence-bond type that represents its properties satisfactorily. Often, also, two or more electronic

structures seem to be about equally good. In these cases it is customary to say that the actual molecule resonates among the reasonable structures, and to represent the molecule by writing the various resonating structures together in brackets. These various structures do not correspond to different kinds of molecules; there is only one kind of molecule present, with an electronic structure that can be described as a hybrid structure of two or more valence bond structures.

The following resonating structures represent important molecules:

$$\left\{ \;:C\!-\!\ddot{O}:\qquad\qquad :C\!=\!\ddot{O}:\qquad\qquad :C\!\equiv\!O: \;\right\}\quad\begin{array}{l}\text{Carbon}\\\text{monoxide}\end{array}$$

$$\left\{ \;:\ddot{O}\!=\!C\!=\!\ddot{O}:\qquad :\ddot{O}\!-\!C\!\equiv\!O:\qquad :O\!\equiv\!C\!-\!\ddot{O}: \;\right\}\quad\begin{array}{l}\text{Carbon dioxide}\\\text{(linear molecule)}\end{array}$$

$$\left\{ \;:\ddot{S}\!=\!C\!=\!\ddot{S}:\qquad :\ddot{S}\!-\!C\!\equiv\!S:\qquad :S\!\equiv\!C\!-\!\ddot{S}: \;\right\}\quad\begin{array}{l}\text{Carbon disulfide}\\\text{(linear molecule)}\end{array}$$

$$\left\{ \;:\ddot{N}\!=\!N\!=\!\ddot{O}:\qquad :N\!\equiv\!N\!-\!\ddot{O}: \;\right\}\quad\begin{array}{l}\text{Nitrous oxide}\\\text{(linear molecule)}\end{array}$$

There is experimental evidence showing that these molecules have the resonating structures indicated above. Perhaps the simplest evidence is that given by the distances between the atoms. It has been observed that in general the distance between two atoms connected by a double bond is approximately 0.21 Å less than the distance between the same two atoms connected by a single bond, and that the distance for a triple bond is approximately 0.13 Å less than that for a double bond. For example, the single-bond distance between two carbon atoms (as in diamond or ethane, $H_3C\!-\!CH_3$) is 1.54 Å; the double-bond distance is 1.33 Å, and the triple-bond distance is 1.20 Å. The distance between a carbon atom and an oxygen atom connected by a double bond as found in compounds such as formaldehyde,

is 1.22 Å. In carbon dioxide, however, for which the structure $O\!=\!C\!=\!O$ was accepted for many years, the distance between the carbon atom and an oxygen atom has been found to be 1.16 Å. The shortening of 0.06 Å is due to the triple-bond character introduced by the two structures $O\!\equiv\!C\!-\!O$ and $O\!-\!C\!\equiv\!O$ (the shortening effect of the triple bond on the interatomic distance is greater than the lengthening effect of the single bond).

6-8. Ionic Valence

The British scientist Henry Cavendish reported that the electric conductivity of water is greatly increased by dissolving salt in it. In 1884 the young Swedish scientist Svante Arrhenius (1859–1927) published his doctor's dissertation, which included measurements of the electric conductivity of salt solutions and his ideas as to their interpretation. These ideas were rather vague, but he later made them more precise and then published a detailed paper on ionic dissociation in 1887. Arrhenius assumed that in a solution of sodium chloride in water there are present sodium ions, Na^+, and chloride ions, Cl^-. When electrodes are put into such a solution the sodium ions are attracted toward the cathode and move in that direction, and the chloride ions are attracted toward the anode and move in the direction of the anode. The motion of these ions through the solution, in opposite directions, provides the mechanism of conduction of the current of electricity by the solution.

The presence of hydrated ions such as $Na^+(aq)$, $Mg^{++}(aq)$, $Al^{+++}(aq)$, $S^{--}(aq)$, and $Cl^-(aq)$, as well as complex ions such as $SO_4^{--}(aq)$, in aqueous solutions has been verified by the study of the properties of the solutions. Many of these ions have an electric charge such as to give the atom the electron number of the nearest argonon. The number of electrons removed from or added to the atom is called its ionic valence: $+1$ for Na^+, for example, and -1 for Cl^-.

The alkali metals (in group I of the periodic table) are unipositive because their atoms contain one more electron than an argonon atom, and this electron is easily removed, to produce the corresponding cation, Li^+, Na^+, K^+, Rb^+, and Cs^+. The ease with which the outermost electron can be removed from an atom of an alkali metal is shown by the values of the first ionization energy given in Tables 5-1 (in eV) and 6-1 (in kJ mole^{-1}) and in Figure 5-13.

The halogens (in group VII of the periodic table) are uninegative because each of their atoms contains one less electron than an argonon atom, and readily gains an electron, producing the corresponding anion, F^-, Cl^-, Br^-, and I^-. The energy that is liberated when an extra electron is attached to an atom to form an anion is called the *electron affinity* of the atom. Values of electron affinities of the halogens, given in Table 6-1, are larger than those of other atoms.*

The atoms of group II of the periodic table, by losing two electrons, can also produce ions with argononic structures: these ions are Be^{++}, Mg^{++},

---

*It is surprising that the electron affinity of fluorine is less than that of chlorine. The first electron affinity of oxygen is 140 kJ mole^{-1}, and that of OH is 175 kJ mole^{-1}.

TABLE 6-1
Ionization Energy and Electron Affinity of Univalent Elements

Element	First Ionization Energy	Element	Electron Affinity
H	1312 kJ mole^{-1}	H	71 kJ mole^{-1}
Li	520	F	333
Na	496	Cl	350
K	419	Br	330
Rb	403	I	300
Cs	376		

Ca^{++}, Sr^{++}, and Ba^{++}. The alkaline-earth elements are hence bipositive in valence. The elements of group III are tripositive, those of group VI are binegative, and so on.

The formulas of binary salts of these elements can thus be written from knowledge of the positions of the elements in the periodic table:

$$Na^+F^- \qquad Na^+Br^- \qquad K^+I^- \qquad Ca^{++}(F^-)_2 \qquad Ba^{++}(Cl^-)_2$$

$$Al^{+++}(Cl^-)_3 \qquad (Na^+)_2O^{--} \qquad Ca^{++}O^{--} \qquad (Al^{+++})_2(O^{--})_3$$

Ionic compounds are formed between the strong metals in groups I and II and the strong nonmetals in the upper right corner of the periodic table. In addition, ionic compounds are formed containing the cations of the strong metals and the anions of acids, especially of the oxygen acids.

It will be pointed out later in this chapter that the description of compounds as aggregates of ions is an approximation. The electronic structure of molecules and crystals usually described as ionic involves only a partial transfer of electrons from the metal atoms to the nonmetal atoms. Nevertheless, the discussion of ionic valence in relation to the noble-gas electron configurations, as given above, is an important and useful part of chemical theory.

In 1913 the structure of NaCl(c) was determined by x-ray diffraction, and it was found that there are no discrete Na—Cl molecules in the crystal (Figure 6-18). Instead, each sodium atom is equidistant from six neighboring chlorine atoms, and each chlorine atom is similarly surrounded by six sodium atoms. It was at once recognized that the crystal can be described as an aggregate of sodium cations and chloride anions, and that each ion is bonded to each of its six neighbors by an electrostatic or ionic bond with bond-number (or strength) $\frac{1}{6}$. The alkali hydrides (LiH to CsH) and most of the alkali halides crystallize with the NaCl structure.

FIGURE 6-18
The sodium chloride structure. The cubic unit cell contains
4 Na at 000, $0\frac{1}{2}\frac{1}{2}$, $\frac{1}{2}0\frac{1}{2}$, and $\frac{1}{2}\frac{1}{2}0$, and 4 Cl at $\frac{1}{2}\frac{1}{2}\frac{1}{2}$, $\frac{1}{2}00$,
$0\frac{1}{2}0$, and $00\frac{1}{2}$. The structure is based on a face-centered cubic
lattice. This figure is from an early paper by William Barlow.

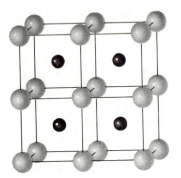

FIGURE 6-19
The cesium chloride structure. The
cubic unit cell contains 1 Cs at 000 and
1 Cl at $\frac{1}{2}\frac{1}{2}\frac{1}{2}$. The structure is based
on a simple cubic lattice.

TABLE 6-2
Crystal Radii of Some Ions

Ion	Radius	Ion	Radius	Ion	Radius	Ion	Radius	Ion	Radius	Ion	Radius
				Li^+	0.60 Å	Be^{++}	0.31 Å	B^{3+}	0.20 Å	C^{4+}	0.15 Å
O^{--}	1.40 Å	F^-	1.36 Å	Na^+	.95	Mg^{++}	.65	Al^{3+}	.50	Si^{4+}	.41
S^{--}	1.84	Cl^-	1.81	K^+	1.33	Ca^{++}	.99	Sc^{3+}	.81	Ge^{4+}	.53
Se^{--}	1.98	Br^-	1.95	Rb^+	1.48	Sr^{++}	1.13	Y^{3+}	.93	Sn^{4+}	.71
Te^{--}	2.21	I^-	2.16	Cs^+	1.69	Ba^{++}	1.35	La^{3+}	1.15	Pb^{4+}	.84

CsCl, CsBr, and CsI crystallize with a different structure, shown in Figure 6-19. In the NaCl structure each ion is said to have ligancy (coordination number) 6, and in the CsCl structure to have ligancy 8.

Ionic Radii

The electron distributions in alkali ions and halide ions are shown in Figure 6-20. It is seen that these ions are closely similar to the corresponding argonons, which are shown, on a somewhat larger scale, in Figure 5-8. With increase in nuclear charge from $+9e$ for fluoride ion to $+11e$ for sodium ion the electron shells are drawn closer to the nucleus, so that the sodium ion is about 30% smaller than the fluoride ion. The neon atom is intermediate in size.

Atoms and ions do not have a sharply defined outer surface. Instead, the electron distribution function usually reaches a maximum for the outer shell and then decreases asymptotically toward zero with increasing distance from the nucleus. It is possible to define a set of crystal radii for ions such that the radii of two ions with similar electronic structures are proportional to the relative extensions in space of the electron distribution functions for the two ions, and that the sum of two radii is equal to the contact distance of the two ions in the crystal. Figure 6-21 shows the relative sizes of various ions with noble-gas structures, chosen in this way. Some values of ionic radii are given in Table 6-2.

These radii give the observed cation-anion distance in crystals in which cation and anion have the structure of the same argonon, such as Na^+F^- (both ions with the neon structure) and K^+Cl^- (both with the argon structure). The observed distances, Na^+—$F^- = 2.31$ Å and K^+—$Cl^- = 3.14$ Å, are equal to the sums of the corresponding radii. In other crystals, in which the anions are almost in contact, the observed distance is larger than the radius sum.

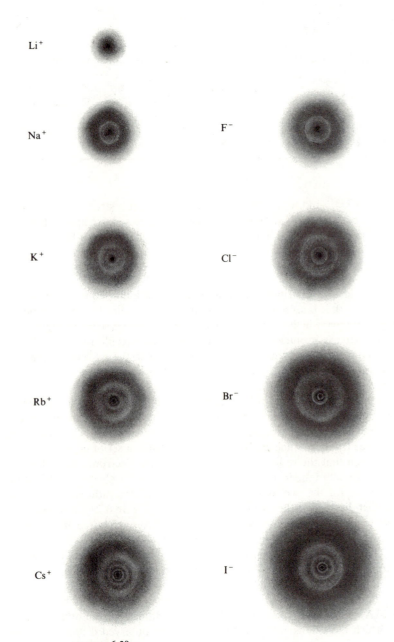

FIGURE 6-20
The electron distribution in alkali ions and halogenide ions.

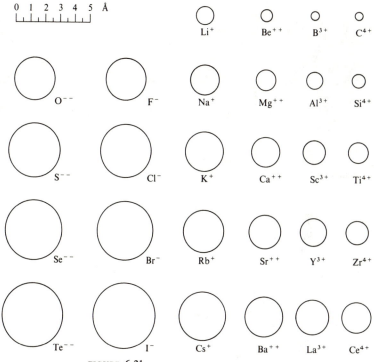

0   1   2   3   4   5   Å

Li$^+$ Be^{++} B^{3+} C^{4+}

O^{--} F$^-$ Na$^+$ Mg^{++} Al^{3+} Si^{4+}

S^{--} Cl$^-$ K$^+$ Ca^{++} Sc^{3+} Ti^{4+}

Se^{--} Br$^-$ Rb$^+$ Sr^{++} Y^{3+} Zr^{4+}

Te^{--} I$^-$ Cs$^+$ Ba^{++} La^{3+} Ce^{4+}

FIGURE 6-21
A drawing representing the ionic radii of ions.

An extreme case is that of lithium iodide. In this crystal, as shown in Figure 6-22, the iodide ions are in contact with one another, and the lithium ions are not in contact with the surrounding iodide ions. In consequence of the contact of the large iodide ions with one another the Li$^+$—I$^-$ distance in the crystal, 3.02 Å, is nearly 10% greater than the radius sum, 2.76 Å. In consequence the substance is softer than the other alkali halides, and also has lower melting point, boiling point, heat of fusion, and heat of vaporization. Figure 6-23 shows the effect on the melting point of the large ratio of radius of anion to radius of cation. The crystals containing the smallest cation, with each anion, would be expected to have the highest melting point, because of the stronger attraction. But the salts of lithium show large deviation from this expectation. Sodium iodide also has a lower melting point than expected. In this crystal both the anion-anion contact and the cation-anion contact are operating, and in consequence the crystal lattice is expanded somewhat (Figure 6-22).

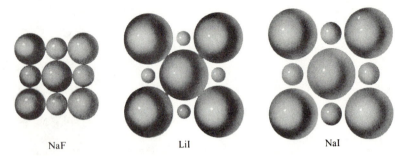

NaF   LiI   NaI

FIGURE 6-22
The arrangement of ions in three crystals. In lithium iodide the iodide ions are in contact with one another.

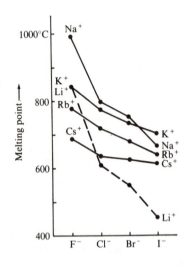

FIGURE 6-23
The observed melting points of the alkali halides, showing the effect of anion contact in the lithium salts and sodium iodide.

TABLE 6-3
Cation-anion Distances in Alkali Halide Gas Molecules

	Li^+	Na^+	K^+	Rb^+	Cs^+
Cl^-	2.03 Å	2.36 Å	2.67 Å	2.79 Å	2.91 Å
Br^-	2.17	2.50	2.82	2.95	3.07
I^-	2.39	2.71	3.05	3.18	3.32

These radius-ratio effects do not occur in the gas molecules, such as LiI. For all the alkali halide gas molecules for which the cation-anion distance has been determined (by microwave spectroscopy), as given in Table 6-3, the distance is equal to about 86% of the sum of the crystal ionic radii.

The decrease in cation-anion distance from an alkali halide crystal to the gas molecule can be explained as corresponding to an increase in the strength of the ionic bond. In the gas molecule, such as Na^+Cl^-, we may say that there is a single bond between the unipositive cation and the uninegative anion. In the crystal we may say that each cation divides its power of attraction among the six anions that surround it, and that in consequence each cation-anion contact represents one-sixth of a bond, a $\frac{1}{6}$th bond. This weaker bond is longer (by 0.4 to 0.6 Å) than the full bond in the gas molecule.

6-9. The Partial Ionic Character of Covalent Bonds

Often a decision must be made as to whether a molecule is to be considered as containing an ionic bond or a covalent bond. There is no question about a salt of a strong metal and a strong nonmetal; an ionic structure is to be written for it. Thus for lithium chloride we write

$$Li^+Cl^- \qquad \text{or} \qquad Li^+:\overset{\cdot\cdot}{\underset{\cdot\cdot}{Cl}}:^-$$

Similarly, there is no doubt about nitrogen trichloride, NCl_3, an oily molecular substance composed of two nonmetals. Its molecules have the covalent structure

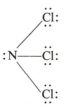

Between LiCl and NCl_3 there are the three compounds $BeCl_2$, BCl_3, and CCl_4. Where does the change from an ionic structure to a covalent structure occur?

The answer to this question is provided by the theory of resonance. *The transition from an ionic bond to a normal covalent bond in a series of compounds such as those mentioned in the preceding sentence does not occur sharply, but gradually.*

The Electric Dipole Moment of Molecules and the Partial Ionic Character of Bonds

About fifty years ago it was noticed that some liquids have a small dielectric constant (1.5 to 2.5) that is nearly independent of the temperature, and that others have a larger value that decreases rapidly with increase in the temperature. The idea was developed that the liquids of the first kind, called *nonpolar liquids*, consist of molecules with no electric dipole moment, and that the liquids of the second kind, called *polar liquids*, consist of molecules that have a dipole moment.

A molecule has an electric dipole moment if its center of positive charge does not coincide with its center of negative charge. The magnitude of the dipole moment for two charges $+q$ and $-q$ the distance d apart is qd. For example, a proton, $q = \epsilon$, and an electron, $q = -\epsilon$ ($\epsilon = 15.188 \times 10^{-15}$ S), one Ångstrom (10^{-10} m) apart, have dipole moment 1.5188×10^{-24} S m. For our purpose it is convenient to use the ϵÅ (epsilon-Ångstrom) as the unit of dipole moment.*

The water molecule is a polar molecule (Figure 6-24; note that it is customary to represent dipole moment by a vector drawn from the positive toward the negative charge).

In an electric field, as between the electrostatically charged plates of a capacitor, water molecules tend to orient themselves, pointing their positive ends toward the negative plate and their negative ends toward the positive plate (Figure 6-25). This partially neutralizes the applied field, an effect described by saying that the medium (water) has a *dielectric constant* greater than unity (about 80 for liquid water at 20°C).

The voltage required to put a given amount of electric charge on the plates of a capacitor is inversely proportional to the dielectric constant of the medium surrounding the capacitor plates. In this way the dielectric constant of the substance can be determined, and from its value, by use of a theory developed by P. Debye in 1914, the dipole moment of the molecules can be evaluated (Appendix XIII). Values of dipole moments can also be determined with great accuracy by microwave spectroscopy and molecular beam resonance methods.

The value found for the dipole moment of the water molecule in water vapor is 0.387 ϵÅ. The O—H bond length is 0.97 Å and the H—O—H angle is 104.5°, as determined from the spectrum of water vapor; hence the point midway between the two protons is 0.59 Å from the oxygen nucleus. If the bonds were completely ionic, the value of the dipole moment would be $2\epsilon \times 0.59$ Å $= 1.18$ ϵÅ. The observed value, 0.387 ϵÅ,

---

*It has been customary to report values of the dipole moment in debyes (symbol D): 1 D is 1 statcoulomb cm. Interconversion of ϵÅ and D is made with the relation 1 ϵÅ = 4.803 D.

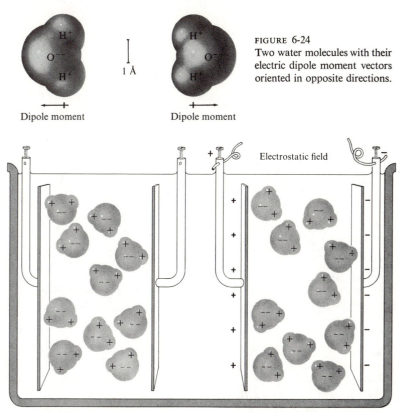

FIGURE 6-24
Two water molecules with their electric dipole moment vectors oriented in opposite directions.

Dipole moment

Dipole moment

Electrostatic field

Unoriented water molecules

Partially oriented water molecules

FIGURE 6-25
Orientation of polar molecules in an electrostatic field, producing the effect of a high dielectric constant.

suggests that each O—H bond has $0.387/1.18 \times 100 = 33\%$ ionic character (and 67% covalent character).

An extreme case is the lithium fluoride gas molecule, which has dipole moment 1.39 $_\epsilon$Å, equal to 92% of the value for Li^+ and F^- at the internuclear distance, 1.52 Å. This molecule is accordingly well represented by the symbol Li^+F^-, and the bond may be described as an ionic bond with only a small amount of covalent character (about 8%).

On the other hand, the observed dipole moment for hydrogen iodide, 0.080 $_\epsilon$Å, with the internuclear distance 1.62 Å, corresponds to only 5% of ionic character. The HI molecule is well represented by the normal covalent structure H : Ï :.

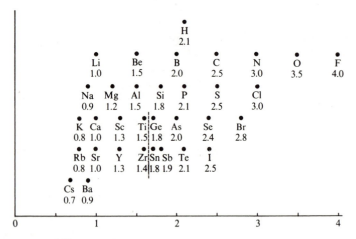

FIGURE 6-26
The electronegativity scale. The dashed line indicates approximate values
for the transition metals.

TABLE 6-4
Values of the Electronegativity of Elements

						H										
						2.1										
Li	Be	B											C	N	O	F
1.0	1.5	2.0											2.5	3.0	3.5	4.0
Na	Mg	Al											Si	P	S	Cl
0.9	1.2	1.5											1.8	2.1	2.5	3.0
K	Ca	Sc	Ti	V	Cr	Mn	Fe	Co	Ni	Cu	Zn	Ga	Ge	As	Se	Br
0.8	1.0	1.3	1.5	1.6	1.6	1.5	1.8	1.9	1.9	1.9	1.6	1.6	1.8	2.0	2.4	2.8
Rb	Sr	Y	Zr	Nb	Mo	Tc	Ru	Rh	Pd	Ag	Cd	In	Sn	Sb	Te	I
0.8	1.0	1.2	1.4	1.6	1.8	1.9	2.2	2.2	2.2	1.9	1.7	1.7	1.8	1.9	2.1	2.5
Cs	Ba	La-Lu	Hf	Ta	W	Re	Os	Ir	Pt	Au	Hg	Tl	Pb	Bi	Po	At
0.7	0.9	1.0–1.2	1.3	1.5	1.7	1.9	2.2	2.2	2.2	2.4	1.9	1.8	1.9	1.9	2.0	2.2
Fr	Ra	Ac	Th	Pa	U	Np-No										
0.7	0.9	1.1	1.3	1.4	1.4	1.4-1.3										

Hydrogen fluoride has dipole moment 0.413 ϵÅ and internuclear distance 0.92 Å. These correspond to a 45% contribution of the ionic structure H^+F^- and 55% contribution of the covalent structure $H\!:\!\ddot{F}\!:$. The H—F bond may thus be described as about midway between the ionic extreme and the covalent extreme.

The amounts of ionic character and covalent character of a bond between two atoms A and B can be estimated with use of quantity called the electronegativity. This quantity will be discussed in the following section.

6-10.   The Electronegativity Scale of the Elements

It has been found possible to assign to the elements numbers representing their power of attraction for the electrons in a covalent bond, by means of which the amount of partial ionic character of the bond may be estimated. This power of attraction for the electrons in a covalent bond is called the *electronegativity* of the element. In Figure 6-26 the elements other than the transition elements and the rare-earth metals are shown on an *electronegativity scale*. The electronegativity values are also given in Table 6-4. The symbol x is used for electronegativity.

The scale extends from cesium, with electronegativity 0.7, to fluorine, with electronegativity 4.0. Fluorine is by far the most electronegative element, with oxygen in second place, and nitrogen and chlorine in third place. Hydrogen and the metalloids are in the center of the scale, with electronegativity values close to 2. Most of the metals have values about 1.7 or less.

The electronegativity scale as drawn in Figure 6-26 is seen to be similar in a general way to the periodic table, but deformed by pushing the top to the right and the bottom to the left. In describing the periodic table we have said that the strongest metals are in the lower left corner and the strongest nonmetals in the upper right corner of the table; because of this deformation, the electronegativity scale shows the metallic or nonmetallic character of an element simply as a function of the value of the horizontal coordinate, the electronegativity.

A rough relation between the electronegativity difference $x_A - x_B$ (or $x_B - x_A$) and the amount of partial ionic character of the bond between atoms A and B can be found by plotting the partial ionic character as given by the observed values of the electric dipole moment and the internuclear distance for diatomic molecules against the electronegativity difference. Such a diagram is shown in Figure 6-27. Experimental points are shown for iodine bromide, iodine chloride, the hydrogen halides, and the lithium halides (gas molecules). The smooth curve that is drawn is

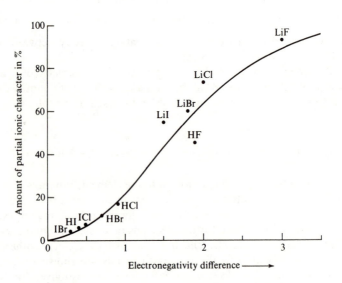

FIGURE 6-27
A diagram showing the relation between the ionic character of a
bond and the electronegativity difference of the two atoms that are
bonded together.

TABLE 6-5
Relation between Electronegativity Difference and Amount of Partial Ionic Character of Bonds

$x_A - x_B$	Amount of Ionic Character	$x_A - x_B$	Amount of Ionic Character
0.2	1%	1.8	55%
.4	4	2.0	63
.6	9	2.2	70
.8	15	2.4	76
1.0	22	2.6	82
1.2	30	2.8	86
1.4	39	3.0	89
1.6	47	3.2	92

seen to approximate the experimental points to within about $\pm 2\%$ for $x_A - x_B$ less than 1 and about $\pm 10\%$ for larger values. Numerical values corresponding to the curve are given in Table 6-5.

The dipole moment is a vector, and can be represented by an arrow. The convention has been adopted of drawing the arrow from the positive charge to the negative charge. Sometimes a special arrow, $\longmapsto$, is used.

The dipole moment of a polyatomic molecule is found to be roughly equal to the vector sum of the moments calculated from the partial ionic character of the bonds in the molecule. When the experimental value shows large disagreement with the value calculated in this way, it is likely that the electronic structure of the molecule is somewhat different from that used in making the calculation.

The farther away two elements are from one another in the electronegativity scale (horizontally in Figure 6-26), the greater is the amount of ionic character of a bond between them. When the separation on the scale is 1.7 the bond has about 50% ionic character. If the separation is greater than this, it would seem appropriate to write an ionic structure for the substance, and if it is less, to write a covalent structure. It is not necessary, however, to adhere rigorously to this rule.

Example 6-1. Describe the electronic structure of the NaCl gas molecule (the boiling point of NaCl at 1 atm is 1430°C). What is the predicted value of its electric dipole moment?

Solution. The electronegativity difference for chlorine ($x = 3.0$) and sodium ($x = 0.9$) is 2.1. Hence the molecule is largely ionic, Na^+Cl^-, with only about 33% covalent character (67% ionic character, Table 6-5). The bond length is given in Table 6-3 as 2.36 Å. Hence the dipole moment for 100% ionic character would be 2.36 ϵÅ. For 67% partial ionic character the predicted value of the dipole moment is $0.67 \times 2.36 = 1.58$ ϵÅ. (The observed value is somewhat larger (1.73 ϵÅ), corresponding to 73% of partial ionic character for the bond.)

Example 6-2. What is the electric dipole moment of acetylene, C_2H_2?

Solution. The acetylene molecule, H—C≡C—H, is linear (Section 6-4). Hence the two H—C dipole moments oppose one another,

$$\overset{\longmapsto \quad \longleftarrow}{\text{H—C≡C—H}}$$

and have the resultant zero. Thus the answer is 0.

6-11. Heats of Formation and
Relative Electronegativity of Atoms

Some chemical reactions take place with the evolution of heat, and some with the absorption of heat. The reactions that take place with the evolution of heat are called *exothermic reactions*, and those that take place with the absorption of heat are called *endothermic reactions*. Of course, any reaction that is exothermic when it takes place in one direction is endothermic when it takes place in the reverse direction.

Heat is defined as the energy that is transferred from one region to another by the thermal processes of conduction and radiation across the surface separating the regions. The heat of a chemical reaction is the quantity of heat that is evolved (that is, transferred to the environment) when the reaction takes place at constant temperature and constant pressure, under such conditions that the only work done is the pressure-volume work $P \times \Delta V$ (ΔV = volume of the products minus the volume of the reactants).

Enthalpy

It has been found by experiment that it is possible to assign to every substance at standard conditions a numerical value of a certain quantity, represented by the symbol H, such that the heat absorbed during a chemical reaction carried out at constant pressure can be found by subtracting the sum of the values of H for the reactants from the sum for the products of the reaction. The names used for the quantity H are *heat content* and *enthalpy* (from Gk. *enthalpein*, to warm). These names are equivalent to one another. The enthalpy is defined by the equation

$$H = E + PV$$

In this equation E is the internal energy of the substance.

The symbol ΔH is used to express the change in enthalpy (or heat content) of a system accompanying a change in state, such as a chemical reaction. Thus a positive value of ΔH means that heat is absorbed from the surroundings during the reaction. The symbol ΔH_{298} is used for the change in enthalpy accompanying a change (reaction) at 298.15°K (25°C) and 1 atm pressure. For example, we may write the equations for the combustion of carbon as follows:

$$C(graphite) + O_2(g) \longrightarrow CO_2(g) \qquad \Delta H_{298} = -393.5 \text{ kJ mole}^{-1}$$
$$C(graphite) + \tfrac{1}{2}O_2(g) \longrightarrow CO(g) \qquad \Delta H_{298} = -110.5 \text{ kJ mole}^{-1}$$

(The temperature 25°C has been accepted by chemists as the customary one for determination of thermochemical quantities.)

In the past it was customary to refer to the heat evolved in the reaction of the combination of elements to form a compound as the *heat of formation, Q_f*. A negative value of Q_f means that the reaction of formation of the compound from the elements is endothermic. The heat of formation is equal to $-\Delta H$ for the reaction of formation of the compound: thus $Q_f = 393.5$ kJ mole^{-1} for $CO_2(g)$ and 110.5 kJ mole^{-1} for $CO(g)$.

Values of ΔH for two reactions can be combined to give the value for a third. By combining the reactions above we obtain the enthalpy of combustion of carbon monoxide to carbon dioxide:

$$CO(g) + \tfrac{1}{2}O_2(g) \longrightarrow CO_2(g) \qquad \Delta H = -283.0 \text{ kJ mole}^{-1}$$

The value of $-\Delta H$ for a reaction of a substance with oxygen is called the *heat of combustion.*

It is often convenient to write the value of the heat of formation or combustion ($-\Delta H$ for any reaction) as though it were one of the products of the reaction; for example,

$$C(\text{graphite}) + O_2(g) \longrightarrow CO_2(g) + 393.5 \text{ kJ mole}^{-1}$$
$$CO(g) + \tfrac{1}{2}O_2(g) \longrightarrow CO_2(g) + 283.0 \text{ kJ mole}^{-1}$$

Enthalpies of reaction can be determined by use of instruments such as the bomb calorimeter shown in Figure 6-28. A weighed sample of combustible material is placed in the bomb, and oxygen is introduced under pressure. The temperature of the surrounding water is recorded, and the sample is ignited by passing an electric current through a wire embedded in it. The heat liberated by the reaction causes the entire system inside the insulating material to increase in temperature. After enough time has elapsed to permit the temperature of this material to become uniform, the temperature is again recorded. From the rise in temperature and the total water equivalent of the calorimeter (that is, the weight of water that would require the same amount of heat to cause the temperature to rise one degree as is required to cause a rise in temperature of one degree of the total material of the calorimeter inside the insulation), the amount of heat liberated in the reaction can be calculated. Correction must, of course, be made for the amount of heat introduced by the electric current that produced the ignition and for the fact that the reaction has not taken place at constant pressure.

In the standard reference book *Selected Values of Chemical Thermodynamic Properties*, Circular of the U.S. Bureau of Standards No. 500 (1952), the experimental values of the enthalpies of formation of about 5000 compounds from the elements in their standard states are given. These values may be combined to obtain the enthalpy change for any reaction involving these compounds (and elementary substances) as reactants and products.

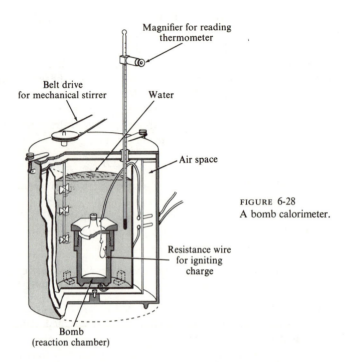

FIGURE 6-28
A bomb calorimeter.

We can, moreover, develop a general understanding of these values and related chemical properties of substances in relation to the electronegativity and other properties of atoms, as discussed in the following paragraphs and in later chapters.

Atoms of hydrogen and iodine, although quite different, are approximately equal in electronegativity. In the molecule H—I ⦂ the two atoms exert about the same attraction on the shared electron pair that constitutes the covalent bond between them. This bond is accordingly much like the covalent bonds in the elementary molecules H—H and ⦂I—I⦂. It is hence not surprising that the energy of the H—I bond is very nearly the average of the energies of the H—H bond and the I—I bond. The heat of formation of HI from the gas molecules H_2 and I_2 is only 6.3 kJ mole^{-1}:

$$\tfrac{1}{2}H_2(g) + \tfrac{1}{2}I_2(g) \longrightarrow HI(g) + 6.3 \text{ kJ mole}^{-1}$$

The other hydrogen halide molecules have larger amounts of electronegativity difference (0.7 for HBr, 0.9 for HCl, 1.9 for HF) and of partial ionic character (12%, 17%, and 45%, respectively), and their heats of formation also increase greatly in the same order:

$$\tfrac{1}{2}H_2(g) + \tfrac{1}{2}Br_2(g) \longrightarrow HBr(g) + 51 \text{ kJ mole}^{-1}$$
$$\tfrac{1}{2}H_2(g) + \tfrac{1}{2}Cl_2(g) \longrightarrow HCl(g) + 92 \text{ kJ mole}^{-1}$$
$$\tfrac{1}{2}H_2(g) + \tfrac{1}{2}F_2(g) \longrightarrow HF(g) + 269 \text{ kJ mole}^{-1}$$

This means that the bonds in these hydrogen halide molecules are stronger than the average of the bonds in the molecules of the elementary substances, and that the increased strength (bond energy) is determined by the electronegativity difference of the two atoms. *The greater the separation of two elements on the electronegativity scale, the greater is the strength of the bond between them.* The extra stability is the resonance energy between the normal covalent structure and the ionic structure.

Bond Energy

The heat of formation of a compound from the elements composing it is a measure of the difference in energy of the bonds in the compound and the bonds in the elements. Values of the bond energy for simple molecules, such as H_2, F_2, Cl_2, Br_2, and I_2, have been determined by spectroscopic methods. These bond energies are equal to the heats of formation $(-\Delta H°)$ of the diatomic molecules from atoms:

$$2H(g) \longrightarrow H_2(g) + 436 \text{ kJ mole}^{-1}$$
$$2F(g) \longrightarrow F_2(g) + 153 \text{ kJ mole}^{-1}$$
$$2Cl(g) \longrightarrow Cl_2(g) + 243 \text{ kJ mole}^{-1}$$
$$2Br(g) \longrightarrow Br_2(g) + 193 \text{ kJ mole}^{-1}$$
$$2I(g) \longrightarrow I_2(g) + 151 \text{ kJ mole}^{-1}$$

The energy required to break an H—H bond and an F—F bond is $436 + 153 = 589$ kJ mole^{-1}. The bond energy of HF is

$$H(g) + F(g) \longrightarrow HF(g) + 563 \text{ kJ mole}^{-1}$$

or

$$2H(g) + 2F(g) \longrightarrow 2HF(g) + 1126 \text{ kJ mole}^{-1}$$

We see that in forming two H—F bonds, 1126 kJ mole^{-1} is released. Of this only 589 kJ mole^{-1} is needed to dissociate H_2 and F_2 into 2H and 2F. Hence the two H—F bonds are $1126 - 589 = 537$ kJ mole^{-1} stronger than H—H and F—F. One H—F bond is 269 kJ mole^{-1} stronger than the average of H—H and F—F. This quantity is just the heat of formation of HF from $\tfrac{1}{2}H_2$ and $\tfrac{1}{2}F_2$.

A set of bond-energy values is given in Appendix VIII, with a discussion of their use.

The quantitative relation between bond energy and electronegativity difference may be expressed by an equation. For a single covalent bond between two atoms A and B the extra energy due to the partial ionic character is approximately $100(x_A - x_B)^2$ kJ mole^{-1}; that is, it is proportional to the square of the difference in electronegativity of the two atoms, and the proportionality constant has the value 100 kJ mole^{-1}. For example, chlorine and fluorine have electronegativity values differing by 1; hence the heat of formation of ClF (containing one Cl—F bond) is predicted to be 100 kJ mole^{-1}. The observed heat of formation of ClF is 107 kJ mole^{-1}. The agreement between the predicted and observed heat of formation is only approximate. For electronegativity differences of 1 or less there is usually agreement to within 5 or 10 kJ mole^{-1}, but for larger values of the difference the calculated value may show a larger error. Improved agreement for large values of Δx is obtained by including a fourth-power term, as is done in Equation 6-1 (see Exercise 6-22).

Heats of formation calculated in this way would refer to elements in states in which the atoms formed single bonds, as they do in the molecules P_4 and S_8. Nitrogen (N_2) and oxygen (O_2) contain multiple bonds, and the nitrogen and oxygen molecules are more stable (by 470 kJ mole^{-1} and 212 kJ mole^{-1}, respectively) than they would be if the molecules contained single bonds (as in P_4 and S_8). Hence we must correct for this extra stability by using the equation

$$
\begin{aligned}
Q_f &= \text{heat of formation (in kJ mole}^{-1}) \\
&= 100\Sigma(x_A - x_B)^2 - 6.5\Sigma(x_A - x_B)^4 - 235n_N - 106n_O \qquad (6\text{-}1)
\end{aligned}
$$

Here the summation indicated by Σ is to be taken over all the bonds represented by the formula of the compound. The symbol n_N means the number of nitrogen atoms in the formula, and n_O the number of oxygen atoms.

As an example, we may consider the substance nitrogen trichloride,

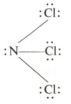

Nitrogen and chlorine have the same electronegativity; hence the first term contributes nothing. There is one nitrogen atom in the molecule. Hence $Q_f = -235$ kJ mole^{-1}. The minus sign shows that the substance is unstable, and that heat is liberated when it decomposes. Nitrogen trichloride is in fact an oil that explodes easily, with great violence:

$$\text{NCl}_3(g) \longrightarrow \text{N}_2(g) + \tfrac{3}{2}\text{Cl}_2(g) + 235 \text{ kJ mole}^{-1}$$

The instability of nitrogen trichloride is due entirely to the great stability of the triple bond in the nitrogen molecule.

Equation 6-1 may be used to calculate rough values of the heat of formation of any compound in which the atoms are held together by single bonds. Its use may be illustrated by some other examples.

Example 6-3.   Which of the two following substances is predicted to form from the elements by a strongly exothermic reaction: PI_3, PF_3?

Solution.   Phosphorus and iodine differ by only 0.4 in electronegativity; hence the formation of PI_3 is predicted to be only slightly exothermic. The extra P—I bond energy due to partial ionic character, $100(x_A - x_B)^2$ kJ mole^{-1}, is $100 \times 0.4^2 = 16$ kJ mole^{-1} per P—I bond; hence we predict $P + \frac{3}{2}I_2 \longrightarrow PI_3 + 48$ kJ mole^{-1}.

For formation of PF_3 a strongly exothermic reaction is expected, because of the large electronegativity difference, 1.9, which leads to $3 \times (100 \times 1.9^2 - 6.5 \times 1.9^4) = 829$ kJ mole^{-1}: $P(c) + \frac{3}{2}F_2(g) \longrightarrow PF_3(g) + 829$ kJ mole^{-1}.

Example 6-4.   Which compounds of nitrogen involving single bonds are predicted to be stable relative to the elements?

Solution.   In general it is assumed that a compound that is formed from the elements by an exothermic reaction is stable relative to the elements (although the heat of formation is not the only factor determining stability; see Chapter 11). The formation of a compound of nitrogen with an element with the same electronegativity (3.0) is endothermic by 235 kJ mole^{-1} per N (Equation 6-1), because of the stability of the N≡N triple bond in N_2. Therefore, to be formed by an exothermic reaction the compound must be with an element differing in electronegativity by an amount such that the extra stability of the bonds due to partial ionic character overcomes this instability. There are three N—X bonds formed per nitrogen atom. Hence we write*

$$3 \times 100 \times (x_X - x_N)^2 = 235$$

$$(x_X - x_N)^2 = \frac{235}{(3 \times 100)} = 0.78$$

$$x_X - x_N = \pm 0.78^{1/2} = \pm 0.88$$

Therefore compounds of nitrogen (involving single bonds) with elements with $x \leq 2.1$ or with $x \geq 3.9$ (fluorine) are predicted to be stable. NF_3 and NH_3 are stable; NCl_3, NBr_3, and NI_3 are unstable.

Example 6-5.   A mixture of aluminum powder and iron(III) oxide, Fe_2O_3, can be ignited; the reaction

$$2Al(c) + Fe_2O_3(c) \longrightarrow 2Fe(c) + Al_2O_3(c)$$

then takes place, producing much heat (the product is molten iron). Would you predict that magnesium could be similarly made from MgO by use of aluminum?

---

*The term $-6.5(x_A - x_B)^4$ has been left out. If it were included the answer would be ± 0.90.

Solution.   We may consider the two reactions

$$4Al(c) + 3O_2(g) \longrightarrow 2Al_2O_3(c) + Q_1$$

and

$$6Mg(c) + 3O_2(g) \longrightarrow 6MgO(c) + Q_2$$

Here Q_1 is the heat evolved in the first reaction and Q_2 is the heat evolved in the second reaction. In each reaction, as written, twelve metal-oxygen single bonds are formed. The electronegativity of Al is 1.5 and that of Mg is 1.2. Hence the electronegativity difference of Al and O, 2.0, is less than that of Mg and O, 2.3, and accordingly Q_1 is less than Q_2. By subtracting the second equation from the first we obtain the result

$$4Al(c) + 6MgO(c) \longrightarrow 6Mg(c) + 2Al_2O_3(c) + Q_1 - Q_2$$

Since Q_1 is less than Q_2, this reaction is endothermic, and probably would not take place*. Accordingly magnesium probably could not be made by igniting a mixture of aluminum and magnesium oxide.

6-12.   The Electroneutrality Principle

A useful principle in writing electronic structures for substances is the *electroneutrality principle*. This principle is that *stable molecules and crystals have electronic structures such that the electric charge of each atom is close to zero.* "Close to zero" means between -1 and $+1$.

That this principle is a reasonable one may be seen by the consideration of values of the ionization energy and electron affinity of atoms. The electron affinity of atoms of nonmetals is about 350 kJ mole^{-1} for the first electron added, to convert the atom $:\overset{\cdot\cdot}{\underset{\cdot\cdot}{F}}\cdot$ into the anion $:\overset{\cdot\cdot}{\underset{\cdot\cdot}{F}}:^-$, or the atom $:\overset{\cdot}{\underset{\cdot\cdot}{O}}\cdot$ into the anion $:\overset{\cdot\cdot}{\underset{\cdot\cdot}{O}}\cdot^-$ (Section 6-8). But there is in general no significant affinity for a second electron to convert $:\overset{\cdot\cdot}{\underset{\cdot\cdot}{O}}\cdot^-$ into $:\overset{\cdot\cdot}{\underset{\cdot\cdot}{O}}:^{--}$, even when it would complete an octet. The repulsion of the two negative charges decreases the attraction for the second electron to nearly zero. Similarly, the values of the first ionization energy of metal atoms lie between about 400 and 800 kJ mole^{-1}, but the second ionization energy is 1500 kJ mole^{-1} or more. It is accordingly unlikely that an atom in a stable molecule would have either a double negative charge or a double positive charge.

The use of the electroneutrality principle in assigning electronic structures to molecules and crystals is discussed in the following examples and in the following section and later chapters.

---

*A conclusion of this sort on the basis of the standard enthalpy of reaction is in general unjustified. It is pointed out in Section 11-10, however, that for reactions involving only crystalline substances a chemical equilibrium is largely determined by the enthalpy change.

Example 6-6.   Should the hydrogen cyanide molecule be assigned the structure HCN or the structure HNC?

Solution.   The electronic structure H—C≡N: makes the atoms nearly neutral. The partial ionic character of the bonds (4% for H—C and 7% for each C—N) leads to the charge +0.04 on H, +0.17 on C, and −0.21 on N. These charges are small, and are compatible with the electroneutrality principle. For HNC the structure H—N≡C: assigns four valence electrons to N and five to C, and hence corresponds to N^+ and C^-. The partial ionic character of the bonds then leads to the charges +0.04 on H, +0.75 on N, and −0.79 on C. These charges on N and C are much larger than for the structure H—C≡N:, and correspond to instability. Hence HCN is the preferable structure.

Example 6-7.   Methyl cyanide and methyl isocyanide have the same composition. Their heats of formation are −88 and −150 kJ mole^{-1}, respectively. Which of the two is H_3C—C≡N: ?

Solution.   Methyl cyanide is more stable than methyl isocyanide by 62 kJ mole^{-1}. The two structures H_3C—C≡N: and H_3C—N≡C: contain the same number of bonds, but the first places smaller electric charges on the atoms than the second, and is accordingly the more stable of the two. Hence methyl cyanide is to be assigned the structure H_3C—C≡N:, and methyl isocyanide the structure H_3C—N≡C:.

Example 6-8.   What is the electronic structure of carbon monoxide? Its observed electric dipole moment is very small, only 0.023 $_\epsilon$Å.

Solution.   The only electronic structure that completes the octet about both C and O is :C≡O:, which corresponds to C^- and O^+, if the shared electron pairs are divided equally between the two atoms. The electronegativity difference 1.0 corresponds to 22% partial ionic character for each bond, and hence to the charges −0.34 for C and +0.34 for O.

Another possible electronic structure, :C=Ö:, gives oxygen its normal covalence but does not complete the octet of carbon. The partial ionic character of the bonds leads to the charges +0.44 for C and −0.44 for O. We conclude that the two structures contribute about equally to a hybrid structure, which has very small resultant electric charge on each atom, in agreement with the electroneutrality principle and the observed very small dipole moment. Hence we describe the molecule as the resonance hybrid $\left\{ :C≡O:, \ :C=\ddot{O}: \right\}$.

Example 6-9.   What is the electronic structure of the anesthetic gas nitrous oxide, N_2O? Its electric dipole moment is 0.035 $_\epsilon$Å.

Solution. A ring structure is not likely, because of the strain of the bent bonds. The linear structure $^{--}:\overset{..}{\underset{..}{N}}-\overset{+}{N}\equiv\overset{+}{\underset{..}{O}}:$ completes the octet for each

atom, but we reject it because of the double negative charge on the end nitrogen atom. The two other structures that complete the octet for each atom are $:N\equiv\overset{+}{N}-\overset{..}{\underset{..}{O}}:^{-}$ and $^{-}:\overset{..}{N}=\overset{+}{N}=\overset{..}{\underset{..}{O}}:$, each of which has formal charges on two atoms as shown. These two structures look equally good, and we conclude that the molecule can best be described as the resonance hybrid with the two structures contributing about equally. Each structure alone would give the molecule a large dipole moment, but since these two dipole moments point in opposite directions they cancel one another in the hybrid, in agreement with the small observed dipole moment.

Transargononic Structures

Sometimes an atom forms so many covalent bonds as to surround itself with more than four electron pairs; it assumes a transargononic structure. An example is phosphorus pentachloride, PCl_5; in the molecule of this substance the phosphorus atom is surrounded by five chlorine atoms, with each of which it forms a covalent bond (with some ionic character):

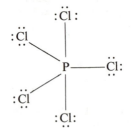

The phosphorus atom in this compound seems to be using five of the nine orbitals of the *M* shell, rather than only the four most stable orbitals, which are occupied by electrons in the argon configuration. It seems likely that of the nine or more orbitals in the *M* shell, the *N* shell, and the *O* shell four are especially stable but one or more others may occasionally be utilized.

The difference in electronegativity of chlorine and phosphorus is 0.9, which corresponds to 18% of partial ionic character. Accordingly, an alternative description of the PCl_5 molecule is that the phosphorus atom forms four covalent bonds, using only the four orbitals of the outer shell, and one ionic bond to Cl^-, and that the four covalent bonds resonate among the five positions, so that each chlorine atom is held by a bond with 80% covalent and 20% ionic character.

The oxygen acids, such as H_2SO_4, may be assigned similar transargononic structures:

The stability of transargononic structures will be discussed in Sections 7-7 and 8-1.

6-13. The Sizes of Atoms and Molecules. Covalent Radii and van der Waals Radii

Interatomic distances (bond lengths) in molecules and crystals can be determined by the methods of spectroscopy (including microwave spectroscopy), x-ray diffraction, electron diffraction, neutron diffraction, and nuclear magnetic resonance. The description of these methods is beyond the scope of this book. During the past forty years the bond lengths have been determined for many hundreds of substances, and their values have been found to be useful in the discussion of the electronic structures of molecules and crystals.

It has been found that usually the bond length for the single bond A—B is, to within about 0.03 Å, equal to the average of the bond lengths A—A and B—B. For example, the average of C—C (1.54 Å, Section 6-4) and Cl—Cl (1.98 Å) is $\frac{1}{2} \times (1.54 + 1.98) = 1.76$ Å. The value of C—Cl found by the investigation of CCl_4 by the electron diffraction method is 1.76 Å; hence the *single-bond covalent radii* 0.77 Å for C and 0.99 Å for Cl can be added together in three ways to give the three observed bond lengths C—C, Cl—Cl, and C—Cl.

Values of the single-bond covalent radii of nonmetallic elements are given in Table 6-6. The value for hydrogen is 0.30 Å, for all bonds other than H—H (the H—H bond length, 0.74 Å, corresponds to a larger radius for hydrogen than the value used for other bonds).

TABLE 6-6
Single-bond Covalent Radii

C	0.77 Å	N	0.70 Å	O	0.66 Å	F	0.64 Å
Si	1.17	P	1.10	S	1.04	Cl	0.99
Ge	1.22	As	1.21	Se	1.17	Br	1.14
Sn	1.40	Sb	1.41	Te	1.37	I	1.33

It was mentioned in Section 6-4 that the C=C and C≡C bond lengths are 0.21 Å and 0.34 Å, respectively, less than the C—C bond length. Approximately the same shortening is found for other double and triple bonds. For example, for C—N the bond length 1.47 Å is given by the sum of the radii for carbon and nitrogen (Table 6-6), and the value 1.47 − 0.34 = 1.13 Å would be expected for C≡N. The value observed in H—C≡N is 1.15 Å, in reasonably good agreement with the result of the calculation.

Bond lengths for hybrid structures have intermediate values.

> **Example 6-10.**   The bond lengths 1.13 Å for nitrogen-nitrogen and 1.19 Å for nitrogen-oxygen are observed in nitrous oxide, N_2O. What do these values indicate about the structure of the molecule?
>
> **Solution.**   Expected bond lengths (Table 6-6), with use of −0.21 Å for a double bond and −0.34 Å for a triple bond, are 1.19 Å for N=N, 1.06 Å for N≡N, 1.36 Å for N—O, and 1.15 Å for N=O. We see that the observed values indicate that the nitrogen-nitrogen bond is intermediate between a double bond and a triple bond and that the nitrogen-oxygen bond is intermediate between a single bond and a double bond. This comparison accordingly supports the conclusion reached in Example 6-9, that the structure is a resonance hybrid of :N≡N̄—Ö:⁻ and ⁻:N̈=N̄=Ö: .

Van der Waals Radii

The Dutch physicist J. D. van der Waals found that in order to explain some of the properties of gases it was necessary to assume that molecules have a well-defined size, so that two molecules begin to undergo strong repulsion when, as they approach, they reach a certain distance from one another. For example, the deviations of the argonons from ideal behavior and other properties such as viscosity lead to the assignment of effective radii between 1 Å and 2 Å to their molecules. These radii are called the *van der Waals radii* of the atoms.

It has been found that the effective sizes of molecules packed together in liquids and crystals can be described by assigning similar van der Waals radii to each atom in the molecule. Values of these radii are given in Table 6-7.

The values are seen to be about 0.8 Å larger than the corresponding single-bond covalent radii. This difference is illustrated in Figure 6-29, which represents two chlorine molecules in van der Waals contact (packed together in a crystal or colliding in the liquid or gas). Each chlorine atom is surrounded by four outer electron pairs. One pair is shared with the other chlorine atom in the same Cl_2 molecule. The point midway between

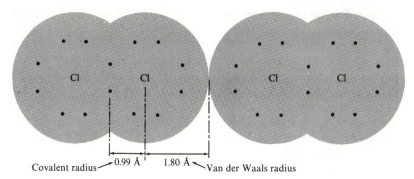

Covalent radius ←0.99 Å | 1.80 Å ↖Van der Waals radius

FIGURE 6-29
Two chlorine molecules in van der Waals contact, illustrating the difference between van der Waals radius and covalent radius.

TABLE 6-7
Van der Waals Radii of Atoms

H	1.1 Å	N	1.5 Å	O	1.40 Å	F	1.35 Å
		P	1.9	S	1.85	Cl	1.80
		As	2.0	Se	2.00	Br	1.95
		Sb	2.2	Te	2.20	I	2.15
		Half-thickness of aromatic molecule, such as benzene or naphthalene					1.70

the two nuclei, 0.99 Å from each nucleus, represents the average position of the shared pair. The three unshared pairs about each nucleus are also about the same distance (equal to the covalent radius) from the nucleus. When two nonbonded chlorine atoms are in contact there are two unshared pairs in the region between the nuclei; the van der Waals radius defines the region that includes the major part of the electron distribution function for the unshared pairs.

In making drawings of atoms or molecules, the van der Waals radii may be used in indicating the volume within which the electrons are largely contained. For ions, the ionic radii (crystal radii) discussed in Section 6-8 may be used. The van der Waals radius of an atom and the ionic radius of its negative ion are essentially the same. For example, the van der Waals radius of chlorine is 1.80 Å, and the ionic radius of the chloride ion is 1.81 Å.

The covalent radii have a different meaning and a different use. The sum of the single-bond covalent radii for two atoms is equal to the distance between the atoms when they are connected by a single covalent bond. The single-bond covalent radius of an atom may be considered to be the distance from the nucleus to the average position of the shared electron pair, whereas the van der Waals radius extends to the outer part of the region occupied by the electrons of the atom, as indicated in Figure 6-29. The effective radius of an atom in a direction that makes only a small angle with the direction of a covalent bond formed by the atom is smaller than the van der Waals radius in directions away from the bond. For example, in the carbon tetrachloride molecule the chlorine atoms are only 2.9 Å apart, and yet the properties of the substance indicate that there is no great strain, even though this distance is much less than the van der Waals diameter 3.6 Å.

6-14. Oxidation Numbers of Atoms

The nomenclature of inorganic chemistry is based upon the assignment of numbers (positive or negative) to the atoms of the elements. These numbers, called *oxidation numbers*, are defined in the following way.

The oxidation number of an atom is a number that represents the electric charge that the atom would have if the electrons in a compound were assigned to the atoms in a conventional way.

The assignment of electrons is somewhat arbitrary, but the conventional procedure, described below, is useful because it permits a simple statement to be made about the valences of the elements in a compound without considering its electronic structure in detail and because it can be made the basis of a simple method of balancing equations for oxidation-reduction reactions.

An oxidation number may be assigned to each atom in a substance by the application of simple rules. These rules, though simple, are not completely unambiguous. Although their application is usually a straightforward procedure, it sometimes requires considerable chemical insight and knowledge of molecular structure. The rules are given in the following statements.

1. The oxidation number of a monatomic ion in an ionic substance is equal to its electric charge.

2. The oxidation number of an atom in an elementary substance is zero.

3. In a covalent compound of known structure, the oxidation number of each atom is the charge remaining on the atom when each shared electron pair is assigned completely to the more electronegative of the two atoms sharing it. An electron pair shared by two atoms of the same element is usually split between them.

4. The oxidation number of an element in a compound of uncertain structure may be calculated from a reasonable assignment of oxidation numbers to the other elements in the compound.

The application of the first three rules is illustrated by the following examples; the number by the symbol of each atom is the oxidation number of that atom:

$Na^{+1}Cl^{-1}$ $Mg^{+2}(Cl^{-1})_2$ $(B^{+3})_2(O^{-2})_3$

H_2^0 O_2^0 C^0 (diamond or
                                                               graphite)

H^{+1} (hydrogen $(O^{-2}H^{+1})^-$ (hydroxide
      cation)   ion)

$N^{-3}(H^{+1})_3$ $Cl^{+1}F^{-1}$ $C^{+4}(O^{-2})_2$

$C^{+2}O^{-2}$ $C^{-4}(H^{+1})_4$ $K^{+1}Mn^{+7}(O^{-2})_4$

Fluorine, the most electronegative element, has the oxidation number -1 in all of its compounds with other elements.

Oxygen is second only to fluorine in electronegativity, and in its compounds it usually has oxidation number -2; examples are $Ca^{+2}O^{-2}$, $(Fe^{+3})_2(O^{-2})_3$, $C^{+4}(O^{-2})_2$. Oxygen fluoride, OF_2, is an exception; in this compound, in which oxygen is combined with the only element that is more electronegative than it is, oxygen has the oxidation number $+2$. Oxygen has oxidation number -1 in hydrogen peroxide, H_2O_2, and other peroxides.

Hydrogen when bonded to a nonmetal has oxidation number $+1$, as in $(H^{+1})_2O^{-2}$, $(H^{+1})_2S^{-2}$, $N^{-3}(H^{+1})_3$, $(P^{-2})_2(H^{+1})_4$. In compounds with metals, such as $Li^{+1}H^{-1}$ and $Ca^{+2}(H^{-1})_2$, its oxidation number is -1, corresponding to the electronic structure $H:^{-1}$ for a negative hydrogen ion with completed K shell (helium structure).

Oxidation Number and Chemical Nomenclature

The principal classification of the compounds of an element is made on the basis of its oxidation state. In our discussions of the compounds formed by the various elements or groups of elements in the following chapters of this book we begin by a statement of the oxidation states represented by the compounds. The compounds are grouped together in classes, representing those with the principal element in the same oxidation state. For example, in the discussion of the compounds of iron they are divided into two classes, representing the compounds of iron in oxidation state $+2$ and those in oxidation state $+3$, respectively.

The nomenclature of the compounds of the metals is also based upon their oxidation states. At the present time there are two principal nomenclatures in use. We may illustrate the two systems of nomenclature by taking the compounds $FeCl_2$ and $FeCl_3$ as examples. In the older system a compound of a metal in the lower of two important oxidation states is named by use of the name of the metal (usually the Latin name) with the suffix *ous*. Thus the salts of iron in oxidation state $+2$ are ferrous salts; $FeCl_2$ is called ferrous chloride. The compounds of a metal in the higher oxidation state are named with use of the suffix *ic*. The salts of iron in oxidation state $+3$ are called ferric salts; $FeCl_3$ is ferric chloride.

Note that the suffixes *ous* and *ic* do not tell what the oxidation states are. For copper compounds, such as CuCl and $CuCl_2$, the compounds in which copper has oxidation number $+1$ are called cuprous compounds, and those in which it has oxidation number $+2$ are called cupric compounds.

A new system of nomenclature for inorganic compounds was drawn up by a committee of the International Union of Chemistry in 1940. According to this system the value of the oxidation number of a metal is represented by a Roman numeral given in parentheses following the name (usually the English name rather than the Latin name) of the metal. Thus $FeCl_2$ is given the name iron(II) chloride, and $FeCl_3$ is given the name iron(III) chloride. These names are read simply by stating the numeral after the name of the metal: thus iron(II) chloride is read as "iron two chloride."

It may be noted that it is not necessary to give the oxidation number of a metal in naming a compound if the metal forms only one principal series of compounds. The compound $BaCl_2$ may be called barium chloride rather than barium(II) chloride, because barium forms no stable compounds other than those in which it has oxidation number $+2$. Also, if one oxidation state is represented by many compounds, and another by only a few, the oxidation state does not need to be indicated for the compounds of the important series. Thus the compounds of copper with oxidation number $+2$ are far more important than those of copper with the oxidation number $+1$, and for this reason $CuCl_2$ may be called simply copper chloride, whereas CuCl would be called copper(I) chloride.

We shall in general make use of the new system of nomenclature in the following chapters of our book, except that, for convenience, we shall use the old nomenclature for the following common metals:

> Iron: $+2$, ferrous; $+3$, ferric
>
> Copper: $+1$, cuprous; $+2$, cupric (or copper)
>
> Mercury: $+1$, mercurous; $+2$, mercuric
>
> Tin: $+2$, stannous; $+4$, stannic

Compounds of metalloids and nonmetals are usually given names in which the numbers of atoms of different kinds are indicated by prefixes. The compounds PCl_3 and PCl_5, for example, are called phosphorus trichloride and phosphorus pentachloride, respectively. We shall use the name dinitrogen trioxide for N_2O_3, and similar names for N_2O_4, N_2O_5, and other such compounds, although the prefix di is often omitted in general usage.

Exercises

6-1. Describe the wave function for the electron in the hydrogen molecule-ion, H_2^+, in terms of the $1s$ orbital wave functions about each of the two nuclei. Why is the wave function called symmetric?

6-2. The observed bond energies of H_2^+, H_2, He_2^+, and He_2 are 255, 429, 243, and 0 kJ mole^{-1}, respectively. Discuss these values in relation to the occupancy of bonding and antibonding molecular orbitals.

6-3. Calculate from information given in Exercise 6-2 and Table 5-1 the ionization energy of the hydrogen molecule-ion (to give $e^- + p + p$).

6-4. Evaluate the ionization energy of the H_2 molecule (to give $e^- + H_2^+$).

6-5. The first four excited states of the hydrogen molecule are thought to involve one electron in a molecular orbital based on $2s$ or $2p$ of the hydrogen atom. These four states lie 11.37, 11.87, 11.89, and 12.40 eV above the normal state, averaging 11.88 eV. The $2s$ and $2p$ levels of the hydrogen atom are mentioned in Chapter 5 as lying 10.20 eV above the $1s$ normal state. Can you suggest a simple explanation of the difference between the values for the molecule and the atom?

6-6. The observed internuclear distances for eleven of the thirteen excited states of H_2 for which accurate values are known are nearly the same, 1.05 ± 0.02 Å. Can you suggest a structural explanation of this fact?

6-7. Write electronic structures for the following polyatomic ions, indicating all electrons in the outer shell; assume that the various atoms of the ion are held together by covalent bonds:

O_2^{--} (peroxide ion), S_3^{--} (trisulfide ion), NO_2^+ (nitronium ion), BH_4^- (borohydride ion), NH_4^+ (ammonium ion), $N(CH_3)_4^+$ (tetramethylammonium ion).

For each of these ions, what corresponding neutral molecule has the same electronic structure? (Example: HS^- has the same electronic structure as HCl.)

6-8. The difluorine cation F_2^+ is known to have dissociation energy 318 kJ mole^{-1} (to $F + F^+$). What electronic structure would you assign to the molecule? How would you describe the bond in this molecule? Discuss the bond energy in relation to the values for O—O, N—N, O=O, and N≡N in Tables VIII-1 and VIII-2.

6-9. The substance pentamethylene tetrazole, $C_6H_{10}N_4$, which is used in the treatment of mental illness, contains a ring of six carbon atoms and one nitrogen atom, fused with a five-membered ring. The large ring involves a sequence of five CH_2 groups. Draw a structural formula for the substance on the basis of this information, with each atom completing the helium or neon shell. How many double bonds are there in the molecule?

6-10. Write electronic structures for the molecules NH_3 and BF_3. These molecules combine to form the addition compound H_3NBF_3. What is the electronic structure of this compound? What similarity is there in the electronic rearrangements in the following chemical reactions?

$$NH_3 + H^+ \longrightarrow NH_4^+$$
$$NH_3 + BF_3 \longrightarrow H_3NBF_3$$

6-11. Spiropentane, C_5H_8, contains two three-membered rings with a common corner. Make a drawing showing the arrangement of bonds in the molecule. How many isomers would you expect for dimethylspiropentane, $C_5H_6(CH_3)_2$? How many of them would be members of enantiomeric pairs?

6-12. White phosphorus and sulfur combine, when heated, to form the compound P_4S_3, which is used in making safety matches. The molecule has been shown by x-ray studies to have a threefold axis of symmetry, and the low value (near zero) of the heat of formation indicates that the atoms have their normal covalences. Assign an electronic structure to the molecule.

6-13. Potassium burns in oxygen to form the orange-yellow paramagnetic crystalline substance potassium superoxide, KO_2. The paramagnetism results from the spin of the unpaired electron of the superoxide ion, O_2^-, which has bond length 1.28 Å. Discuss the electronic structure of the superoxide ion in relation to this bond length. (The bond length in hydrogen peroxide is 1.46 Å and that in the normal state of the oxygen molecule is 1.21 Å.)

6-14. The standard enthalpy of formation of $CO_2(g)$ is -394 kJ mole^{-1}, and that of $CS_2(g)$ is $+115$ kJ mole^{-1}. Calculate the enthalpies of formation of the molecules from the monatomic elements (Table VIII-3). By comparison with the C=O and C=S bond-energy values (Table VIII-2), calculate the resonance energy for each of these molecules. (Answer: 157, 204 kJ mole^{-1}.)

6-15. Make a similar calculation of the resonance energy for carbon oxysulfide, $OCS(g)$, standard enthalpy of formation -137 kJ mole^{-1}.

6-16. Assuming $x = 2.1$ for H, evaluate x for C, N, P, As, and S from values of the standard enthalpy of formation given in tables in Chapter 7, by using $100 (x - x_H)^2$ kJ mole^{-1} as the increase in bond energy resulting from partial ionic character.

6-17. Copper vapor contains some Cu_2 molecules, which have bond energy 190 kJ mole^{-1}. From the values of the standard enthalpy of formation of CuF(g), CuCl(g), CuBr(g), and CuI(g) given in Table 21-2 calculate the bond energies of these molecules (see also Table VIII-3), and from them the electronegativity of copper. (Answer: 1.93, 1.94, 1.79, 1.78, av. 1.86.)

6-18. The bond energies of Ag_2 and Au_2 are 157 and 215 kJ mole^{-1}, respectively, and the standard enthalpies of formation of AgCl(g) and AuCl(g) are 97 and 189 kJ mole^{-1}, respectively. Evaluate x for Ag and Au.

6-19. The bond energies of the gas molecules CuAg, CuAu, and AgAu are 175, 228, and 199 kJ mole^{-1}, respectively. Using $x = 1.86$ for copper (Exercise 6-17), evaluate x for Ag and Au. (Answer: 1.98, 2.36.)

6-20. From the standard enthalpies of formation of C_2H_6(g) and C_2H_5OH(g) given in Table 7-2 and of O(g) given in Table 7-1 evaluate $\Delta H°$ for the reaction

$$C_2H_6(g) + O(g) \longrightarrow C_2H_5OH(g)$$

From this value and the C—H and O—H bond energies in Table VIII-1 calculate a value for the C—O bond energy. (Answer: 353 kJ mole^{-1}.)

6-21. Use the method of the preceding Exercise to obtain $\Delta H°$ for the reaction

$$C_2H_6(g) + O(g) \longrightarrow (CH_3)_2O(g)$$

and another value for the C—O bond energy. Can you suggest a structural explanation for the difference in the two values of this bond energy? (Answer: 347 kJ mole^{-1}.)

6-22. Using $100(x_A - x_B)^2 - C(x_A - x_B)^4$ for the extra bond energy from partial ionic character and the values of x in Table 6-4, calculate the coefficient C by comparing the O—H bond energy with the average of O—O and H—H (Table VIII-1). Repeat the calculation for C—O, C—F, and H—F. (Answer: 5.9, 6.5, 6.9, 7.3; av. 6.7 kJ mole^{-1}.)

6-23. (a) Write electronic structural formulas for H_2O, H_2O_2, H_2S, H_2S_2, H_2S_3, H_2S_4, H_2S_5.
 (b) What is the structural explanation of the instability of hydrogen peroxide and the stability of the corresponding compound dihydrogen disulfide, as expressed by the enthalpies of decomposition:

$$H_2O_2(l) \longrightarrow H_2O(l) + \tfrac{1}{2} O_2(g) + 98 \text{ kJ mole}^{-1}$$
$$H_2S_2(l) \longrightarrow H_2S(g) + \tfrac{1}{8} S_8(c) - 3 \text{ kJ mole}^{-1}$$

Note that this explanation applies also to the nonexistence of H_2O_3, H_2O_4, H_2O_5; the analogous sulfur compounds are stable.

6-24. What is the electronic structure of S_6? If you could make the oxygen analog, O_6, would it, with liquid hydrogen, give better rocket propulsion than HF, or not so good? (See Tables 7-1 and 7-9 for enthalpy values.) (Answer: 44% better, per kg of fuel plus oxidant.)

6-25. The standard enthalpies of formation of CH_4, CH_3Cl, CH_2Cl_2, $CHCl_3$, and CCl_4 (all as gases) are -75, -86, -94, -105, and -109 kJ mole^{-1}, respectively. Use these values to calculate the four successive C—Cl bond-energy values. (Answer: 329, 326, 330, 323 kJ mole^{-1}.)

6-26. The observed value of the electric dipole moment of LiH(g), which has internuclear distance 1.60 Å, is 1.23 ϵÅ. To what amount of partial ionic character of the molecule does this correspond? (Answer: 77%.)

6-27. The observed value of the electric dipole moment of the OH radical (bond length 0.97 Å) is 0.33 ϵÅ. To what amount of partial ionic character and what electronegativity difference does this value correspond? (Compare with H_2O, Section 6-9.)

6-28. Discuss the electronic structure of hydrogen peroxide, H_2O_2, in relation to the observed value of the electric dipole moment, 0.43 ϵÅ. From the value for OH, what value for H_2O_2 would you expect for a cis planar structure? For a trans planar structure?

6-29. The electric dipole moment of H_2S (bond angle 92°) is 0.200 ϵÅ. To what value of the HS moment does this correspond? What values would you predict for H_2S_2 (observed 0.22 ϵÅ) for the cis planar structure, the trans planar structure, and the skew structure with dihedral angle 90°? (Answer: 0.144, 0.288, 0, 0.204.)

6-30. Assign electronic structures to the following gas molecules (*a*) from the extreme ionic point of view; (*b*) from the extreme covalent point of view; (*c*) from use of electronegativity values to determine partial ionic character of bonds. To what extent is there agreement with the electroneutrality principle? LiI, $MgCl_2$, $Pb(CH_3)_4$, H_3CF, HCN, H_2CO, P_2H_4.

6-31. Assign resonance structures, with each atom argononic, to the following molecules and ions: NO_3^- (nitrate ion); NO_2^+ (nitronium ion); H_3BO_3 (boric acid); O_3; H_3CNO_2 (nitromethane).

6-32. Assign transargononic structures to the following molecules: PCl_3F_2; $TeCl_4$; SF_6; S_2F_{10}; $HClO_4$; $Te(OH)_6$. What outer orbitals are used for bonds and unshared pairs by the transargononic atoms?

6-33. Briefly discuss the following facts in relation to the electronegativity values of the elements: (*a*) in 1787 Lavoisier did not include sodium, potassium, calcium, and aluminum in his list of known elementary substances; (*b*) fluorine combines with all other elements except the lighter argonons; (*c*) ethyl chloride, C_2H_5Cl, can be made by adding HCl(g) to ethanol, C_2H_5OH; (*d*) the most stable transargononic compounds are fluorides; (*e*) many metals can be made by reaction of their chlorides with sodium; (*f*) NCl_3 is explosive, whereas NF_3 is stable; (*g*) the polymer $(CF_2)_n$ is highly resistant to attack by corrosive chemicals; (*h*) the boron hydrides are used as fuel for rocket propulsion.

The Nonmetallic Elements and Some of Their Compounds

In the preceding chapters we have discussed some of the basic principles of chemistry, relating to the atomic and molecular structure of matter, the nature of the photon, the electron, and the nuclei of atoms, the electronic structure of atoms in relation to the periodic system of the elements, and the electronic structure of molecules and the nature of the chemical bond and of valence. This theoretical foundation permits us to discuss the properties of substances in a much more satisfying way than was possible fifty years ago. A correlation, even if only partial, of the observed physical and chemical properties of substances with their atomic and molecular structure makes descriptive chemistry more interesting and more easily remembered.

On the first page of this book there is the statement that chemistry is the science of *substances*—their structure, their properties, and the reactions that change them into other substances. Only a few substances have

been discussed in the preceding chapters. In this chapter and the following one many facts are given about many substances (the nonmetallic
elements and their compounds), with some effort to make a useful correlation of the facts with theoretical principles, especially the principles of
electronic and molecular structure. We shall then return to theoretical
chemistry in Chapter 9.

In the first section of this chapter some of the properties of the elements hydrogen, carbon, nitrogen, phosphorus, arsenic, antimony, bismuth, oxygen, sulfur, selenium, tellurium, fluorine, chlorine, bromine,
and iodine are described. The following sections are devoted to some of
their compounds with one another, especially the single-bonded normal-
valence compounds. Compounds of nonmetals with oxygen are discussed
in the following chapter.

7-1. The Elementary Substances

Hydrogen

Hydrogen is a very widely distributed element. It is found in most of
the substances that constitute living matter, and in many inorganic substances. There are more compounds of hydrogen known than of any
other element, carbon being a close second.

Free hydrogen, H_2, is a colorless, odorless, and tasteless gas. It is the
lightest of all gases, its density being about one-fourteenth that of air.
Its melting point ($-259°C$ or $14°K$) and boiling point ($-252.7°C$) are
very low, only those of helium being lower. Liquid hydrogen, with density
0.070 g cm^{-3}, is, as might be expected, the lightest of all liquids. Crystalline
hydrogen, with density 0.088 g cm^{-3}, is also the lightest of all crystalline
substances. Hydrogen is very slightly soluble in water; 1 liter of water at
$0°C$ dissolves only 21.5 ml of hydrogen gas under 1 atm pressure. The
solubility decreases with increasing temperature, and increases with increase in the pressure of the gas.

The electronic structure of the hydrogen molecule has been discussed in
Section 6-1.

In the laboratory hydrogen may be easily made by the reaction of an
acid such as sulfuric acid, H_2SO_4, with a metal such as zinc. The equation
for the reaction is

$$H_2SO_4(aq) + Zn(c) \longrightarrow ZnSO_4(aq) + H_2(g)$$

Hydrogen can also be prepared by the reaction of some metals with
water or steam. Sodium and the other alkali metals react very vigorously

TABLE 7-1
Standard Enthalpy of Compounds of Hydrogen and Oxygen at 25°C, in kJ mole^{-1}

$e^-(g)$	0*	$O^+(g)$	1567	$OH^-(g)$	-133
$H_2(g)$	0	$O^-(g)$	110	$OH^-(aq)$	-230
$O_2(g)$	0	$O_2^+(g)$	1184	$H_2O(g)$	-242
$H^+(aq)$	0	$O_2(aq)$	-16	$H_2O(l)$	-286
$H(g)$	218	$O_3(g)$	142	$H_2O_2(g)$	-133
$O(g)$	249	$O_4(g)$	-0.7	$H_2O_2(l)$	-188
$H^+(g)$	1536	$OH(g)$	42	$H_2O_2(aq)$	-191
$H^-(g)$	149				

*Values of the enthalpy of formation of elements in a standard state, of electron gas, and of hydrogen ion in aqueous solution are arbitrarily taken to be zero.

with water, so vigorously as to generate enough heat to ignite the liberated hydrogen. An alloy of lead and sodium, which reacts less vigorously, is sometimes used for the preparation of hydrogen.

Much of the hydrogen that is used in industry is produced by the reaction of iron with steam. The steam from a boiler is passed over iron filings heated to a temperature of about 600°C. The reaction that occurs is

$$3Fe(c) + 4H_2O(g) \longrightarrow Fe_3O_4(c) + 4H_2(g)$$

After a mass of iron has been used in this way for some time, it is largely converted into iron oxide, Fe_3O_4. The iron can then be regenerated by passing carbon monoxide, CO, over the heated oxide:

$$Fe_3O_4(c) + 4CO(g) \longrightarrow 3Fe(c) + 4CO_2(g)$$

There is, of course, nothing special about sodium and iron (except their low cost and availability) that causes them, rather than other metals, to be used for the preparation of hydrogen. Other metals with electronegativity about the same as that of sodium ($x = 0.9$) react with water as vigorously as sodium, and metals with electronegativity about the same as that of iron ($x = 1.8$) react with steam in about the same way as iron. Although other structural features, which we shall discuss later, are involved in chemical reactivity, bond energy, which is determined by electronegativity differences, is the most important one.

Values of the standard enthalpy of some compounds of hydrogen and oxygen are given in Table 7-1. In using this table and later tables you must remember that the standard enthalpy of a substance is the value of ΔH for the reaction of formation of the compound from the elements in their standard states; it is the negative of the heat of formation, Q_f, which we have defined as the heat evolved in the formation reaction ($Q_f = -\Delta H_f$).

Elementary Carbon

Carbon occurs in nature in its elementary state in two allotropic forms: *diamond*, one of the hardest substances known,* which often forms beautiful transparent and highly refractive crystals, used as gems; and *graphite*, a soft, black crystalline substance, used as a lubricant and in the "lead" of lead pencils. *Bort* and *black diamond* (*carbonado*) are imperfectly crystalline forms of diamond, which do not show the cleavage characteristic of diamond crystals. Their density is slightly less than that of crystalline diamond, and they are tougher and somewhat harder. They are used in diamond drills and saws and other grinding and cutting devices. Other uses of diamonds also depend upon their great hardness. For example, diamonds with a tapering hole drilled through them are used for drawing wires. Charcoal, coke, and carbon black (lampblack) are microcrystalline or amorphous forms of carbon. The density of diamond is 3.51 g cm^{-3} and that of graphite is 2.26 g cm^{-3}.

The great hardness of diamond is explained by the structure of the diamond crystal, as determined by the x-ray diffraction method. In the diamond crystal (Figure 6-16) each carbon atom is surrounded by four other carbon atoms, which lie at the corners of a regular tetrahedron about it. A structural formula can be written for a small part of a diamond crystal:

Valence bonds connect each carbon atom with four others. Each of these four is bonded to three others (plus the original one), and so on throughout the crystal. The entire crystal is a giant molecule, held together by covalent bonds (Section 6-4). To break the crystal, many of these bonds must be broken; this requires a large amount of energy, and hence the substance is very hard. The bond length in diamond has the single-bond value, 1.54 Å.

The commercial manufacture of diamonds was begun in the period around 1950, after techniques for obtaining very high pressures (over 70,000 atm) at high temperatures (2000°C) had been developed. The crystallization of the artificial diamonds is favored by the addition of a small amount of a metal such as nickel. It is significant that the length of

---

*Some of the boron carbides are harder than diamond, as are also bort and black diamond.

TABLE 7-2
Standard Enthalpy of Carbon Compounds at 25°C (kJ mole^{-1})

C(graphite)	0	C$_2$H$_2$(g)	227	CH$_2$O(g)	formaldehyde	-116
C(diamond)	1.9	C$_2$H$_4$(g)	52	CH$_3$CHO(g)	acetaldehyde	-166
C(g)	718	C$_2$H$_6$(g)	-85	(CH$_3$)$_2$CO(g)	acetone	-216
C$^+$(g)	1806	CF$_4$(g)	-912	HCOOH(g)	formic acid	-363
C$_2$(g)	982	CCl$_4$(g)	-107	CH$_3$COOH(g)	acetic acid	-435
CO(g)	-111	CHCl$_3$(g)	-100	CH$_3$OH(g)	methanol	-201
CO$^+$(g)	1224	CH$_2$Cl$_2$(g)	-88	C$_2$H$_5$OH(g)	ethanol	-237
CO$_2$(g)	-394	CH$_3$Cl(g)	-82	(CH$_3$)$_2$O(g)	dimethyl ether	-185
CH(g)	595	CBr$_4$(g)	50	C$_3$H$_6$(g)	cyclopropane	38
CH$_2$(g)	397	CS$_2$(g)	115	C$_6$H$_{12}$(g)	cyclohexane	-126
CH$_3$(g)	134	COS(g)	-137	C$_6$H$_{10}$(g)	cyclohexene	-6
CH$_4$(g)	-75	(CH$_3$)$_2$S(g)	-38	C$_6$H$_6$(g)	benzene	83

the edge of the unit cube of the nickel crystal, containing four nickel atoms in cubic closest packing, is 3.52 Å, nearly equal to that, 3.56 Å, of the unit cube of the diamond crystal, which contains eight carbon atoms in the arrangement shown in Figure 6-16. Artificial diamonds contain some nickel atoms replacing pairs of carbon atoms.

The structure of graphite is shown in Figure 6-17. It is a layer structure. Each atom forms two single bonds and one double bond with its three nearest neighbors, as shown in the lowest part of the drawing. The bonds resonate among the positions in each layer in such a way as to give each bond two-thirds single-bond character and one-third double-bond character. The interatomic distances in the layer are 1.42 Å, which is intermediate between the single-bond value, 1.54 Å, and the double-bond value, 1.33 Å. The distance between layers is 3.4 Å, over twice the bond length in a layer. The crystal of graphite can be described as built of giant flat molecules, loosely held together in a pile. The layers can be easily separated; hence graphite is a soft substance, which is even used as a lubricant. The lubricating property is determined to some extent by the presence of water; the mechanism of this dependence is not well understood.

From Table 7-2 we see that the heat of sublimation of diamond to carbon atoms is 716 kJ mole^{-1}. Sublimation of diamond involves breaking two C—C bonds per atom. The difference between the heat of sublimation and twice the C—C bond energy ($2 \times 348 = 696$ kJ mole^{-1}, Table VIII-1 of the Appendix), 20 kJ mole^{-1}, is the energy of van der Waals attraction between atoms other than nearest neighbors in the crystal (the energy of van der Waals attraction between bonded atoms is included in the bond energy).

The valence-bond structure for a graphite layer drawn in Figure 6-17 shows each carbon atom forming two single bonds and a double bond. The $C{=}C$ bond energy, 615 kJ mole^{-1} (Table VIII-2), is 81 kJ mole^{-1} less than twice the $C{-}C$ bond energy; hence, assuming the same van der Waals interaction for graphite as for diamond, this structure would make graphite less stable than diamond by $\frac{1}{2} \times 81 = 40.5$ kJ mole^{-1}. In fact, it is more stable, by 1.9 kJ mole^{-1} (Table 7-2). The extra stability of graphite, 42.4 kJ mole^{-1}, is attributed to the resonance energy of the double bond among the carbon-carbon positions. A layer of graphite is represented not by a single valence-bond structure, but by many, so that each carbon-carbon bond has about one-third double-bond character.

Hardness

The property of hardness is not a simple one to define. It has probably evaded precise definition because the concept of hardness represents a composite of several properties (tensile strength, resistance to cleavage, and others). Various scales of hardness and instruments for testing hardness have been proposed. One test consists of dropping a diamond-tipped weight on the specimen and measuring the height of rebound. In another test (the Brinell test) a hardened steel ball is pressed into the surface of the specimen, and the size of the produced indentation is measured.

A very simple test of hardness is the scratch test—a specimen that scratches another specimen and is not scratched by it is said to be harder than the second specimen. The scratch-test scale used by mineralogists is the *Mohs scale* (Friedrich Mohs, 1773–1839, German mineralogist), with reference points (the *Mohs hardness*) from 1 to 10, defined by the following ten minerals:

1. Talc, $Mg_3Si_4O_{10}(OH)_2$
2. Gypsum, $CaSO_4 \cdot 2H_2O$
3. Calcite, $CaCO_3$
4. Fluorite, CaF_2
5. Apatite, $Ca_5(PO_4)_3F$
6. Orthoclase, $KAlSi_3O_8$
7. Quartz, SiO_2
8. Topaz, $Al_2SiO_4F_2$
9. Corundum, Al_2O_3
10. Diamond, C

Diamond is far harder than corundum, and modifications of the Mohs scale have been suggested that assign a much larger hardness number, such as 15, to diamond. The hardness of graphite is between 1 and 2.

TABLE 7-3
Properties of the Elements of Group V

	Atomic Number	Atomic Weight	Melting Point	Boiling Point	Density of Solid	Color	Co-valent Radius	Van der Waals Radius
N	7	14.0067	−209.8°C	−195.8°C	1.026 g cm⁻³	White	0.70 Å	1.5 Å
P	15	30.9738	44.1°	280°	1.81	White	1.10	1.9
As	33	74.9216	814°*	715°†	5.73	Gray	1.21	2.0
Sb	51	121.75	630°	1380°	6.68	Silvery white	1.41	2.2
Bi	83	208.980	271°	1470°	9.80	Reddish white	1.51	2.3

*At 36 atm
†It sublimes.

Nitrogen and Its Congeners

Elementary nitrogen occurs in nature in the atmosphere, of which it constitutes 78% by volume. It is a colorless, odorless, and tasteless gas, composed of diatomic molecules, N_2. At 0°C and 1 atm pressure a liter of nitrogen weighs 1.2506 g. The gas condenses to a colorless liquid at −195.8°C, and to a white solid at −209.96°C. Nitrogen is slightly soluble in water, 1 liter of which dissolves 23.5 ml of the gas at 0°C and 1 atm. Some of the properties of nitrogen and other group-V elements are given in Table 7-3. Enthalpy values are given in Tables 7-4 and 7-5.

Phosphorus was discovered in 1669 by a German alchemist, Henning Brand, in the course of his search for the Philosopher's Stone. Brand heated the residue left on evaporation of urine, and collected the distilled phosphorus in a receiver. The name given the element (from Greek *phosphoros*, giving light) refers to its property of glowing in the dark.

Elementary phosphorus is made by heating calcium phosphate with silica and carbon in an electric furnace. The silica forms calcium silicate, displacing tetraphosphorus decoxide, P_4O_{10}, which is then reduced by the carbon. The phosphorus leaves the furnace as vapor, and is condensed under water to *white phosphorus*.

Phosphorus vapor is tetratomic: in the P_4 molecule each atom has one unshared electron pair and forms a single bond with each of its three neighbors (Figure 6-14).

At high temperatures the vapor is dissociated slightly, forming some diatomic molecules, P_2, with the structure :P≡P:, analogous to that of the nitrogen molecule.

TABLE 7-4
Standard Enthalpy of Some Nitrogen Compounds at 25°C (kJ mole⁻¹)

$N_2(g)$	0	$NH(g)$	331	$NO_2^-(aq)$	-106
$N(g)$	473	$NH_3(g)$	-46	$NO_3^-(aq)$	-207
$N^+(g)$	1883	$NH_3(aq)$	-81	$NH_2OH(c)$	-107
$NO(g)$	90	$NH_4^+(g)$	628	$NH_4OH(aq)$	-367
$NO_2(g)$	34	$NH_4^+(aq)$	-133	$H_2N_2O_2(aq)$	-57
$NO_3(g)$	54	$N_2H_4(l)$	50	$NH_4NO_3(c)$	-365
$N_2O(g)$	82	$HN_3(g)$	294	$NF_3(g)$	-114
$N_2O_3(g)$	84	$N_3^-(aq)$	245	$NCl_3(in\ CCl_4)$	229
$N_2O_4(g)$	44	$HNO_2(aq)$	-119	$NH_4F(c)$	-467
$N_2O_5(c)$	-42	$HNO_3(l)$	-173	$NH_4Cl(c)$	-315

TABLE 7-5
Standard Enthalpy of Some Compounds of Phosphorus, Arsenic, Antimony, and Bismuth at 25°C (kJ mole⁻¹)

	X = P	As	Sb	Bi
$X(c)*$	0	0	0	0
$X(g)$	315	254	254	208
$X^+(g)$	1380	1273	1094	917
$X_2(g)$	142	124	218	249
$X_4(g)$	55	149	204	
$XO(g)$	-41	20	188	67
$X_4O_6(c)$	-1640	-1314	-1409	-1154
$X_4O_{10}(c)$	-2984	-1829	-1961	
$XH_3(g)$	9	171		
$HXO_3(c)$	-955			
$H_3XO_2(aq)$	-609			
$H_3XO_3(aq)$	-972	-742		
$H_3XO_4(aq)$	-1289	-899	-902	
$XO_4^{---}(aq)$	-1279	-870		
$X_2O_7^{----}(aq)$	-2276			
$XCl_3(g)$	-255	-299	-315	-271
$XCl_5(g)$	-343		-393	
$XCl_3O(g)$	-592			
XBr_3	$-150(g)$	$-195(c)$	$-260(c)$	
$XBr_5(c)$	-276			
$XBr_3O(c)$	-479			
$XI_3(c)$	-46	-57	-96	
$XN(g)$	-85	29	311	

*The standard states are white phosphorus (cubic, P_4) and hexagonal arsenic, antimony, and bismuth. The enthalpy of black phosphorus is -43 kJ mole⁻¹.

Phosphorus vapor condenses to liquid white phosphorus, which freezes to solid white phosphorus, a soft, waxy, colorless material, soluble in carbon disulfide, benzene, and other nonpolar solvents. Both solid and liquid white phosphorus contain the same P_4 molecules as the vapor.

White phosphorus is metastable, and it slowly changes to a stable form, *red phosphorus*, in the presence of light or on heating. White phosphorus usually has a yellow color because of partial conversion to the red form. The reaction takes several hours even at 250°C; it can be accelerated by the addition of a small amount of iodine, which serves as a catalyst.* Red phosphorus is far more stable than the white form—it does not catch fire in air at temperatures below 240°C, whereas white phosphorus ignites at about 40°C, and oxidizes slowly at room temperature, giving off a white light ("phosphorescence"). Red phosphorus is not poisonous, but white phosphorus is very poisonous, the lethal dose being about 0.15 g; it causes necrosis of the bones, especially those of the jaw. White phosphorus burns are painful and slow to heal. Red phosphorus cannot be converted into white phosphorus except by vaporizing it. It is not appreciably soluble in any solvent. When heated to 500° or 600°C red phosphorus slowly melts (if under pressure) or vaporizes, forming P_4 vapor.

Several other allotropic forms of the element are known. One of these, *black phosphorus*, is formed from white phosphorus under high pressure. It is still less reactive than red phosphorus, and is the stable form of the element (standard enthalpy, relative to white phosphorus, -43 kJ mole^{-1}; that of red phosphorus is -18 kJ mole^{-1}).

The explanation of the properties of red and black phosphorus lies in their structure. These substances are high polymers, consisting of giant molecules extending throughout the crystal. In order for such a crystal to melt or to dissolve in a solvent, a chemical reaction must take place. This chemical reaction is the rupture of some P—P bonds and formation of new ones. Such processes are very slow. The structure of red phosphorus is not known in detail; that of black phosphorus involves puckered layers, somewhat similar to those of arsenic shown in Figure 6-15, but with a different sort of puckering.

Elementary arsenic exists in several forms. Ordinary *gray arsenic* is a semimetallic substance, steel-gray in color, with density 5.73 and melting point (under pressure) 814°C. It sublimes rapidly at about 450°C, forming gas molecules As_4, similar in structure to P_4. An unstable yellow crystalline allotropic form containing As_4 molecules, and soluble in carbon disulfide,

---

*A substance with the property of accelerating a chemical reaction without itself undergoing significant change is called a *catalyst*, and is said to *catalyze* the reaction; see Section 16-5.

TABLE 7-6
Properties of Oxygen, Sulfur, Selenium, and Tellurium

	Atomic Number	Atomic Weight	Melting Point	Boiling Point	Density	Co-valent Radius	Ionic Radius, X^{--}
Oxygen (gas)	8	15.9994	$-218.4°C$	$-183.0°C$	$1.429 \text{ g liter}^{-1}$	0.66 Å	1.40 Å
Sulfur (orthorhombic)	16	32.064	119.25° 112.8°*	444.6°	2.07 g cm^{-3}	1.04	1.84
Selenium (gray)	34	78.96	217°	685°	4.79	1.17	1.98
Tellurium (gray)	52	127.60	450°	1087°	6.25	1.37	2.21

*For monoclinic sulfur and (rapidly heated) orthorhombic sulfur, respectively.

TABLE 7-7
Standard Enthalpy of Compounds of Sulfur, Selenium, and Tellurium at 25°C ($kJ \, mole^{-1}$)

	X = S	Se	Te
X(c)*	0	0	0
X(g)	279	202	199
$X^+(g)$	1284	1149	1074
$X^{--}(g)$	524		
$X^{--}(aq)$	42	132	
$X_2(g)$	125	139	172
$X_6(g)$	106		
$X_8(g)$	101		
XO(g)	6	40	180
XO_2	$-297(g)$	$-230(c)$	$-325(c)$
$XO_3(g)$	-395		
$H_2X(g)$	-20	86	154
$H_2XO_3(aq)$	-633	-512	-605
H_2XO_4	$-811(l)$	$-538(c)$	
$H_2XO_4(aq)$	-908	-608	-697
$XCl_2(g)$		-41	
$X_2Cl_2(l)$	-60		-84
$XF_6(g)$	-1209	-1029	-1318

*The standard states are orthorhombic sulfur (S_8 molecules) and hexagonal selenium and tellurium (long chains of atoms).

also exists. The gray form has the layer structure shown in Figure 6-15, in which each atom forms three covalent bonds with neighboring atoms in the same layer. Elementary antimony and bismuth crystallize with the same layer structure.

Oxygen and Its Congeners

Oxygen is the most abundant element in the earth's mantle. It constitutes by weight 89% of water, 23% of air (21% by volume), and nearly 50% of the common minerals (silicates). The element consists of diatomic molecules, with the structure described in Section 6-6. It is a colorless, odorless gas, which is slightly soluble in water: 1 liter of water at 0° dissolves 48.9 ml of oxygen gas at 1 atm pressure. Its density at 0°C and 1 atm is 1.429 g liter^{-1}. Oxygen condenses to a pale blue liquid at its boiling point, $-183.0°C$, and on further cooling freezes, at $-218.4°C$, to a pale blue crystalline solid.

Oxygen may be easily prepared in the laboratory by heating potassium chlorate, $KClO_3$:

$$2KClO_3 \longrightarrow 2KCl + 3O_2(g)$$

The reaction proceeds readily at a temperature just above the melting point of potassium chlorate if a small amount of manganese dioxide, MnO_2, is mixed with it. Although the manganese dioxide accelerates the rate of evolution of oxygen from the potassium chlorate, it itself is not changed.

Oxygen is made commercially mainly by the distillation of liquid air. Nitrogen is more volatile than oxygen, and tends to evaporate first from liquid air. Nearly pure oxygen can be obtained by properly controlling the conditions of the evaporation. The oxygen is stored and shipped in steel cylinders, at pressures of 100 atm or more. Oxygen is also made commercially, together with hydrogen, by the electrolysis of water.

Some of the properties of oxygen and its congeners are listed in Table 7-6. Values of the enthalpy of compounds of oxygen are given in Table 7-1 and other tables, and those of compounds of the congeners of oxygen are given in Table 7-7.

Elementary sulfur exists in several allotropic forms. Ordinary sulfur is a yellow solid substance, which forms crystals with orthorhombic symmetry; it is called *orthorhombic sulfur* or, usually, *rhombic sulfur*. It is insoluble in water, but soluble in carbon disulfide (CS_2), carbon tetrachloride, and similar nonpolar solvents, giving solutions from which well-formed crystals of sulfur can be obtained.

At 112.8°C orthorhombic sulfur melts to form a straw-colored liquid. This liquid crystallizes in a monoclinic crystalline form, called *monoclinic sulfur*. The sulfur molecules in both orthorhombic sulfur and monoclinic sulfur, as well as in the straw-colored liquid, are S_8 molecules, with a staggered-ring configuration (Figure 6-13). The formation of this large molecule (and of the similar molecules Se_8 and Te_8) is the result of the tendency of the sixth-group elements to form two single covalent bonds, instead of one double bond. Diatomic molecules S_2 are formed by heating sulfur vapor (S_6 and S_8 at lower temperatures) to a high temperature, but these molecules are less stable than the large molecules containing single bonds. This fact is not isolated, but is an example of the generalization that stable double bonds and triple bonds are formed readily by the light elements carbon, nitrogen, and oxygen, but only rarely by the heavier elements (Section 6-6). Carbon disulfide, CS_2, and other compounds containing a carbon-sulfur double bond are the main exceptions to this rule.

Monoclinic sulfur is the stable form above 95.5°C, which is the *equilibrium temperature* (*transition temperature* or *transition point*) between it and the orthorhombic form. Monoclinic sulfur melts at 119.25°C. Sulfur that has just been melted is a mobile, straw-colored liquid. The viscosity of this liquid is low because the S_8 molecules that compose it are nearly spherical in shape (Figure 6-13) and roll easily over one another. When molten sulfur is heated to a higher temperature, however, it gradually darkens in color and becomes more viscous, finally becoming so thick (at about 200°C) that it cannot be poured out of its container. Most substances decrease in viscosity with increasing temperature because the increased thermal agitation causes the molecules to move around one another more easily. The abnormal behavior of liquid sulfur results from the production of molecules of a different kind—long chains, containing scores of atoms. These very long molecules get entangled with one another, causing the liquid to be very viscous. The dark red color is due to the ends of the chains, which consist of sulfur atoms with only one valence bond instead of the normal two.

The straw-colored liquid, S_8, is called λ-sulfur, and the dark red liquid consisting of very long chains is called μ-sulfur. When this liquid is rapidly cooled by being poured into water, it forms a rubbery *supercooled liquid*, insoluble in carbon disulfide. On standing at room temperature the long chains slowly rearrange into S_8 molecules, and the rubbery mass changes into an aggregate of crystals of orthorhombic sulfur.

A form of crystalline sulfur with rhombohedral symmetry can be made by extracting an acidified solution of sodium thiosulfate with chloroform and evaporating the chloroform solution. These crystals, which are orange in color, consist of S_6 molecules; they are unstable, and change into long chains and then into orthorhombic sulfur (S_8) in a few hours. A form of sulfur containing S_{12} molecules can also be made.

Sulfur boils at 444.6°C, forming S_8 vapor, which on a cold surface condenses directly to orthorhombic sulfur.

TABLE 7-8
Properties of the Halogens

	Atomic Number	Atomic Weight	Color and Form	Melting Point	Boiling Point	Ionic Radius*	Co-valent Radius	Heat of Disso-ciation
F_2	9	18.9984	Pale yellow gas	$-223°C$	$-187°C$	1.36 Å	0.64 Å	153 kJ mole^{-1}
Cl_2	17	35.453	Greenish yellow gas	$-101.6°$	$-34.6°$	1.81	.99	243
Br_2	35	79.909	Reddish brown liquid	$-7.3°$	$58.7°$	1.95	1.14	193
I_2	53	126.9044	Grayish black lustrous solid	$113.5°$	$184°$	2.16	1.27	151

*For negatively charged ion with ligancy 6, such as Cl^- in the NaCl crystal.

The elementary substances selenium and tellurium differ from sulfur in their physical properties in ways expected from their relative positions in the periodic table. Their melting points, boiling points, and densities are higher, as shown in Table 7-6. The stable forms of selenium and tellurium (gray) involve a hexagonal packing of infinitely long chains, each chain having a three-fold screw axis of symmetry. The red allotropic forms of selenium consist of Se_8 molecules.

The increase in metallic character with increase in atomic number is striking. Sulfur is a nonconductor of electricity, as is the red allotropic form of selenium. The gray form of selenium has a small but measurable electronic conductivity, and tellurium is a semiconductor, with conductivity a fraction of 1% of that of metals. An interesting property of the gray form of selenium is that its electric conductivity is greatly increased during exposure to visible light. This property is used in "selenium cells" for the measurement of light intensity, and is the basis of the xerographic method of copying documents.

The Halogens

The halogens consist of diatomic molecules, F_2, Cl_2, Br_2, and I_2. Some physical properties of the halogens are given in Table 7-8.

Fluorine, the lightest of the halogens, is the most reactive of all the elements, and it forms compounds with all the elements except the lighter inert gases. This great reactivity may be attributed to the large value of its electronegativity. Substances such as wood and rubber burst into flame when held in a stream of fluorine, and even asbestos (a silicate of mag-

TABLE 7-9
Standard Enthalpy of Halogen Compounds at 25°C (kJ mole⁻¹)

	X = F	Cl	Br	I
$X_2(g)$	0	0	31	62
X_2			0(l)	0(c)
$X_2(aq)$		−25	−5	21
$X(g)$	77	121	112	107
$X^+(g)$	1764	1378	1261	1120
$X^-(g)$	−256	−229	−218	−193
$X^-(aq)$	−329	−167	−121	−56
$HX(g)$	−269	−92	−36	26
$KX(c)$	−563	−436	−392	−328
$X_2O(g)$	23	76		
$HXO(aq)$		−118		−159
$HXO_2(aq)$		−52		
$HXO_3(aq)$		−98	−40	−230
$HXO_4(aq)$		−131		
$H_5XO_6(aq)$				−766

nesium and aluminum) reacts vigorously with it and becomes incandescent. Platinum is attacked only slowly by fluorine. Copper and steel can be used as containers for the gas; they are attacked by it, but become coated with a thin layer of copper fluoride or iron fluoride, which then protects them against further attack.

Because its electronegativity is greater than that of any other element, we cannot expect that fluorine could be prepared by reaction of any other element with a fluoride. It can, however, be made by electrolysis of fluorides, since the oxidizing power (electron affinity) of an electrode can be increased without limit by increasing the applied voltage. (We shall discuss this matter in Chapter 15.) It was by the electrolysis of a solution of KF in liquid HF that fluorine was first obtained, by the French chemist Henri Moissan (1852–1907), in 1886.

Chlorine (from Greek *chloros*, greenish yellow), the most common of the halogens, is a greenish-yellow gas, with a sharp irritating odor. It was first made by the Swedish chemist K. W. Scheele in 1774, by the action of manganese dioxide on hydrochloric acid. It is now manufactured on a large scale by the electrolysis of a strong solution of sodium chloride.

The element bromine (from Greek *bromos*, stench) occurs in the form of compounds in small quantities in seawater and in natural salt deposits. It is an easily volatile, dark reddish-brown liquid with a strong, disagreeable odor and an irritating effect on the eyes and throat. It produces painful sores when spilled on the skin. The free element can be made by treating a bromide with an oxidizing agent, such as chlorine.

The element iodine (from Greek *iodes*, violet) occurs as iodide ion, I^-, in very small quantities in seawater, and, as sodium iodate, $NaIO_3$, in deposits of Chile saltpeter. It is made commercially from sodium iodate obtained from saltpeter, from kelp, which concentrates it from the seawater, and from oil-well brines.

The free element is an almost black crystalline solid with a slightly metallic luster. On gentle warming it gives a beautiful blue-violet vapor. Its solutions in chloroform, carbon tetrachloride, and carbon disulfide are also blue-violet in color, indicating that the molecules I_2 in these solutions closely resemble the gas molecules. The solutions of iodine in water containing potassium iodide and in alcohol (tincture of iodine) are brown; this change in color suggests that the iodine molecules have undergone chemical reaction in these solutions. The brown compound KI_3, potassium triiodide, is present in the first solution, and a compound with alcohol in the second.

Values of the standard enthalpy of some halogen compounds are given in Table 7-9.

7-2. Hydrides of Nonmetals. Hydrocarbons

The elements carbon, nitrogen, oxygen, and fluorine and their congeners form simple hydrides with composition corresponding to their normal covalences (CH_4, NH_3, H_2O, HF). Some properties of these hydrides are given in Table 7-10.

The tetrahydrides (CH_4 to SnH_4) have a regular tetrahedral structure, corresponding to the use of sp^3 tetrahedral bond orbitals by the central atom (bond angles 109.5°, Section 6-4). The other hydrides have smaller bond angles, approaching 90°, the value for p bond orbitals (Section 6-5).

The stability of these hydrides is determined mainly by the difference in electronegativity of hydrogen and the other element, as has been discussed in Section 6-11 and Appendix VIII. In Figure 7-1 values are shown of the heat of formation per bond (per hydrogen atom) of the hydrides in the gaseous state from the elements in their standard states, compared with the calculated function $100(x - 2.1)^2 - 6.5(x - 2.1)^4$ kJ mole^{-1} (here 2.1 is the electronegativity of hydrogen). The values for H_2S, H_2Se, and HI would be brought closer to the curve by correction for the van der Waals attraction in the crystalline (standard) state of sulfur, selenium, and iodine. The low values for NH_3 and H_2O, shown by the open circles, are corrected (as indicated by the arrow) to points near the curve by adding the multiple-bond correction for nitrogen and oxygen discussed in Section 6-11.

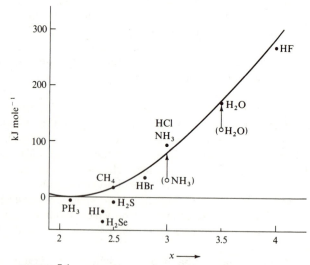

Standard heat of formation per bond of hydrides (g) of
nonmetallic elements, compared with values calculated by the
electronegativity equation.

TABLE 7-10
Some Properties of Hydrides of Nonmetallic Elements

Formula	Melting Point	Boiling Point	Density of Liquid	Standard Enthalpy (g)	Bond Length	Bond Angle
CH_4	−183°C	−161°C	0.54 g cm⁻³	−75 kJ mole⁻¹	1.09 Å	109.5°
SiH_4	−185°	−112°	0.68	−62	1.48	109.5°
GeH_4	−165°	−90°	1.52		1.53	109.5°
SnH_4	−150°	−52°			1.70	109.5°
NH_3	−78°	−33°	0.82	−46	1.01	107.3°
PH_3	−133°	−85°	0.75	9	1.42	93.1°
AsH_3	−114°	−55°		171	1.52	91.8°
SbH_3	−88°	−17°	2.26		1.71	91.3°
BiH_3		22°				
H_2O	0°	100°	1.00	−242	0.96	104.5°
H_2S	−86°	−61°		−20	1.33	92.2°
H_2Se	−64°	−42°	2.12	86	1.46	91.0°
H_2Te	33°	57°	2.57	154	1.7	90°
HF	−92°	19°	0.99	−269	0.92	
HCl	−112°	−84°	1.19	−92	1.27	
HBr	−89°	−67°	1.78	−36	1.41	
HI	−51°	−35°	2.85	26	1.61	

This correction is especially important for nitrogen. The nitrogen molecule, $:N{\equiv}N:$, is 469 kJ mole^{-1} more stable than a hypothetical single-bonded form of the element, and in consequence the heat of formation of $NH_3(g)$ from $\frac{1}{2}N_2(g)$ and $\frac{3}{2}H_2(g)$ is only 46 kJ mole^{-1}, rather than $46 + 235 = 281$ kJ mole^{-1}.

Methane and Other Alkanes

The *hydrocarbons* are compounds of hydrogen and carbon alone. *Methane*, CH_4, is the first member of a series of hydrocarbons called the *methane series* or *paraffin series*. Some of these compounds are listed in Table 7-11. They are called *alkanes*.

Natural gas, from oil wells or gas wells, is usually about 85% methane. The gas that rises from the bottom of a marsh is methane (plus some carbon dioxide and nitrogen), formed by the anaerobic (air-free) fermentation of vegetable matter.

Methane is used as a fuel. It is also used in large quantities for the manufacture of carbon black, by combustion with a limited supply of air:

$$CH_4 + O_2 \longrightarrow 2H_2O + C$$

The methane burns to form water, and the carbon is deposited as very finely divided carbon, which finds extensive use as a filler in rubber for automobile tires.

TABLE 7-11
Some Physical Properties of Normal Alkanes

Substance	Formula	Melting Point	Boiling Point	Density of Liquid
Methane	CH_4	$-183°C$	$-161°C$	0.54 g cm^{-3}
Ethane	C_2H_6	-172	-88	.55
Propane	C_3H_8	-190	-45	.58
Butane	C_4H_{10}	-135	-1	.60
Pentane	C_5H_{12}	-130	36	.63
Hexane	C_6H_{14}	-95	69	.66
Heptane	C_7H_{16}	-91	98	.68
Octane	C_8H_{18}	-57	126	.70
Nonane	C_9H_{20}	-54	151	.72
Decane	$C_{10}H_{22}$	-30	174	.73
Pentadecane	$C_{15}H_{32}$	10	271	.77
Eicosane	$C_{20}H_{42}$	38		.78
Triacontane	$C_{30}H_{62}$	70		.79

The name paraffin means "having little affinity." The compounds of this series are not very reactive chemically. They occur in petroleum. *Ethane* has the structure

It is a gas (Table 7-11), which occurs in large amounts in natural gas from some wells. *Propane*, the third member of the series, has the structure

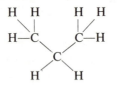

It is an easily liquefied gas, used as a fuel.

In the structural formula for propane there is a chain of three carbon atoms bonded together. The next larger alkane, *butane*, C_4H_{10}, can be obtained by replacing a hydrogen atom at one end of the chain by a methyl group,

Its formula is obtained by adding CH_2 to that of propane. These hydrocarbons, with longer and longer chains of carbon atoms, are called the normal alkanes (*n*-alkanes).

The lighter members of the paraffin series are gases, the intermediate members are liquids, and the heavier members are solid substances. The common name petroleum ether refers to the pentane-hexane-heptane mixture, used as a solvent and in dry cleaning. Gasoline is the heptane-to-nonane mixture (C_7H_{16} to C_9H_{20}), and kerosene the decane-to-hexadecane mixture ($C_{10}H_{22}$ to $C_{16}H_{34}$). Heavy fuel oil is a mixture of paraffins containing twenty or more carbon atoms per molecule. The lubricating oils, petroleum jelly ("Vaseline"), and solid paraffin are mixtures of still larger paraffin molecules.

The phenomenon of isomerism is shown first in the paraffin series by butane, C_4H_{10}. Isomerism (Section 6-2) is the existence of two or more compound substances having the same composition but different properties. The difference in properties is usually the result of difference in the way that the atoms are bonded together. There are two isomers of butane, called normal butane (*n*-butane) and isobutane. These substances have the

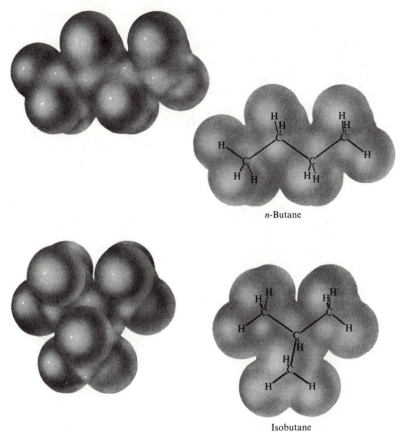

n-Butane

Isobutane

FIGURE 7-2
The structure of the isomers normal-butane and isobutane.

structures shown in Figure 7-2; normal butane has a "straight chain" (actually the carbon chain is a zigzag chain, because of the tetrahedral nature of the carbon atom), and the isobutane molecule has a branched chain. In general, the properties of these isomers are rather similar; for example, their melting points are $-135°C$ and $-145°C$, respectively. The branched-chain hydrocarbons are more stable than their straight-chain isomers [standard enthalpy -126 kJ mole^{-1} for n-$C_4H_{10}(g)$, -135 for iso-$C_4H_{10}(g)$; -146 for n-$C_5H_{12}(g)$, -154 for iso-$C_5H_{12}(g)$, -166 for neo-$C_5H_{12}(g)$; neopentane is tetramethylmethane, $C(CH_3)_4$]. The greater stability of the branched chains than the straight chains may be attributed to their more compact structure, which leads to greater van der Waals stabilization (attraction of pairs of nonbonded atoms).

The normal (straight-chain) hydrocarbons "knock" badly when burned in a high-compression gasoline engine, whereas the highly branched hydrocarbons, which burn more slowly, do not knock. The "octane number" (the antiknock rating) of a gasoline is measured by comparing it with varying mixtures of *n*-heptane and a highly branched octane, 2,2,4-trimethylpentane, with structural formula

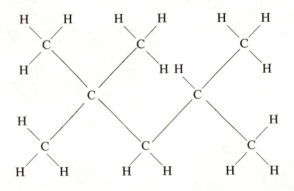

The octane number is the percentage of this octane in the mixture that has the same knocking properties as the gasoline being tested.

The substance *tetraethyl lead*, $Pb(C_2H_5)_4$, is widely used in gasoline as an antiknock agent. Gasoline containing it is called *ethyl gasoline.*

Names of Organic Compounds

Chemists have developed a rather complicated system of names for organic compounds. The student of general chemistry needs to know only a small part of this system.

The simpler substances usually have special names; for example, methane, ethane, propane, butane. From pentane on (Table 7-11), the names of the alkanes contain Greek prefixes that give. the number of carbon atoms.

The group obtained by removing a hydrogen atom from an alkane has the name of the alkane with the ending ane changed to yl. Thus the methyl group is —CH_3, the ethyl group is —C_2H_5, (as in lead tetraethyl, above), and so on. These groups are called *alkyl groups.*

A branched hydrocarbon is given a name that is based on the longest chain of carbon atoms in it. The carbon atoms are numbered from one end (1, 2, 3,...), and groups attached to them, in place of hydrogen atoms, are indicated. For example, the substance called isobutane above (in the discussion of isomerism) may be called 2-methylpropane. Another example is 2,2,4-trimethylpentane, the structural formula of which is given above.

Restricted Rotation about Single Bonds

Until about thirty-five years ago chemists assumed that the two ends of a molecule such as ethane, $H_3C—CH_3$, could rotate freely about the single bond connecting them. This assumption of free rotation about single bonds was made because all efforts to obtain isomers of substances such as 1,2-dichloroethane, $H_2ClC—CH_2Cl$, had failed. Then in 1937 the American chemists J. D. Kemp and K. S. Pitzer showed that the experimental value of the entropy of ethane requires that there be an energy barrier with height about 12.5 kJ mole^{-1}, restricting rotation of one methyl group in ethane relative to the other.

Many experimental values of the height of the potential barrier have been obtained, especially by E. B. Wilson, Jr., and his collaborators, through use of the techniques of microwave spectroscopy (study of absorption spectra of gas molecules in the wavelength region about 1 cm). The values 13.8 kJ mole^{-1} for $H_3C—CH_2F$ and 13.3 kJ mole^{-1} for $H_3C—CHF_2$ are nearly the same as for ethane; those for $H_3C—CH_2Cl$ and $H_3C—CH_2Br$ are somewhat higher, both 14.9 kJ mole^{-1}. In every case the stable configuration is the *staggered configuration* (bonds on opposite sides of the C—C axis, as illustrated in Figure 7-2). The unstable configuration, obtained by rotating a methyl group 60° about the C—C bond, is called the *eclipsed configuration*.

Spectroscopic studies have shown that 1,2-dichloroethane in the gas phase or in the solution is a mixture of three isomers. All three have staggered configurations; viewed along the C—C axis, they have the following aspects:

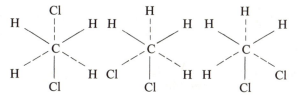

(The second and third constitute an enantiomeric pair, Section 6-3.) The energy of the barrier that restricts rotation is so low that the isomers are converted into one another too rapidly to permit their separation in the laboratory.

The cause of restricted rotation is still somewhat uncertain. The small dependence on the size of the atoms attached to the carbon atoms indicates that it is not mainly steric hindrance (hindrance resulting from contact of the atoms). The most likely explanation is that the barrier results from the repulsion of the H—C bond electrons of one methyl group and those of the other—that is, repulsion of the outer bonds. This explanation is supported by the values of the barrier height for $H_3C—NH_2$ and $H_3C—OH$, 7.9 and 4.5 kJ mole^{-1}, respectively.

Cyclic Hydrocarbons

It was mentioned in Section 6-4 that some hydrocarbon molecules involve rings of carbon atoms. The simplest cyclic hydrocarbon is *cyclopropane* (also called trimethylene), C_3H_6, with the structure shown in Figure 6-10. It is a colorless gas, with melting point $-126.6°C$, boiling point $-34.4°C$, and standard enthalpy of formation 20.4 kJ mole^{-1}. It is a good anesthetic agent, but it is dangerous: its mixtures with air may explode if ignited by an electrostatic spark.

Cyclohexane, C_6H_{12}, is a colorless liquid (boiling point 81°C, melting point 6.5°C) that is obtained in the distillation of petroleum and is used as a solvent. Its structure is that of a puckered hexagonal ring, with unstrained angles (tetrahedral, 109.5°), normal bond lengths (C—C = 1.54 Å C—H = 1.10 Å), and the stable (staggered) orientation around each carbon-carbon bond. Its standard enthalpy, for $C_6H_{12}(g)$, is -126 kJ mole^{-1} (Table 7-2). This value may be taken as the normal value for an unstrained $(CH_2)_n$ molecule; it can be written as $-21\ n$ kJ mole^{-1}. For a hypothetical unstrained cyclopropane molecule we thus expect the standard enthalpy -63 kJ mole^{-1}. The difference between this value and the observed standard enthalpy of cyclopropane, 38 kJ mole^{-1}, is 101 kJ mole^{-1}, the strain energy of the bent bonds in cyclopropane.

The strain energy of cyclopropane makes it more reactive than cyclohexane. Cyclopropane reacts with hydrogen at 80°C, in the presence of a catalyst (finely divided platinum):

$$C_3H_6(g) + H_2(g) \xrightarrow[\text{Pt, 80°C}]{} C_3H_8(g)$$

The symbols Pt and 80°C are written below the arrow in this equation to show that platinum and a temperature of 80°C are needed to cause the reaction to take place.

The ring molecule cyclobutane, C_4H_8, reacts similarly, to form butane, at a somewhat higher temperature:

$$C_4H_8(g) + H_2(g) \xrightarrow[\text{Pt, 120°C}]{} C_4H_{10}(g)$$

The larger rings, such as cyclohexane, are not opened by hydrogenation below 200°C.

Other 3-ring molecules, ethylene imine, $H_2C{-}{-}{-}CH_2$, and ethylene

$$\diagdown \diagup$$
$$N$$
$$H$$

oxide, $H_2C{-}{-}{-}CH_2$, have about the same strain as cyclopropane, and

$$\diagdown \diagup$$
$$O$$

have high chemical reactivity. The chemical reactivity of the epoxy group,

$$-\overset{|}{C}-\overset{|}{C}-,$$ permits the cross-linking of large molecules to take place

during the setting of epoxy glues.

Hydrazine, Hydrogen Peroxide, and Related Hydrides

Many other nonmetal hydrides are known in which there are two or more nonmetal atoms connected to one another by single bonds. For example, a solution of sodium sulfide, Na_2S, dissolves sulfur, S_8, to form a series of polysulfides, Na_2S_2, Na_2S_3, $Na_2S_4, \ldots$, and the hydrogen compounds H_2S_2, H_2S_3, $H_2S_4, \ldots$ are liberated on addition of hydrochloric acid. We would expect, in agreement with observation, that a reaction such as $8H_2S + S_8 \longrightarrow 8HS-SH$ would have only a very small value of ΔH, because the product molecules contain the same bonds (16 H—S, 8 S—S) as the reactant molecules. The various polysulfides are accordingly about as stable (in relation to the elementary nonmetal) as are the alkanes.

For nitrogen and oxygen, however, the corresponding reactions

$$H_2O(g) + \tfrac{1}{2}O_2(g) \longrightarrow H_2O_2(g)$$
$$4NH_3(g) + N_2(g) \longrightarrow 3N_2H_4(g)$$

are strongly endothermic, the values of ΔH being 109 kJ mole^{-1} for H_2O_2 and 155 kJ mole^{-1} for N_2H_4. Both hydrogen peroxide and hydrazine may be described as energy-rich molecules, because of the presence of the relatively unstable O—O and N—N bonds. Both substances are used as rocket propellants. Higher analogues, such as H_2O_3 and N_3H_5, have not been made.

When barium oxide, BaO, is heated to a dull red heat in a stream of air it adds oxygen to form a higher oxide, BaO_2, barium peroxide:

$$2BaO + O_2 \longrightarrow 2BaO_2$$

Hydrogen peroxide, H_2O_2, is made by treating barium peroxide with sulfuric acid or phosphoric acid, and distilling:*

$$BaO_2 + H_2SO_4 \longrightarrow BaSO_4 + H_2O_2$$

Pure hydrogen peroxide is a colorless, syrupy liquid, with density 1.47 g cm^{-3}, melting point $-1.7°C$, and boiling point 151°C. It is a very strong oxidizing agent, which spontaneously oxidizes organic substances. Its uses are in the main determined by its oxidizing power.

Commercial hydrogen peroxide is an aqueous solution, sometimes con-

---

*A method involving organic compounds is used in industry.

taining a small amount of a stabilizer, such as phosphate ion, to decrease its rate of decomposition to water and oxygen by the reaction

$$2H_2O_2 \longrightarrow 2H_2O + O_2$$

Drug-store hydrogen peroxide is a 3% solution (containing 3 g H_2O_2 per 100 g), for medical use as an antiseptic, or a 6% solution, for bleaching hair. A 30% solution and, in recent years, an 85% solution are used in chemical industries.

Oxygen in hydrogen peroxide is assigned the oxidation number -1. In the above reaction one oxygen atom of each molecule is oxidized to oxidation number 0 and the other is reduced to oxidation number -2. Such a reaction is called an *auto-oxidation-reduction reaction*. Hydrogen peroxide can act as an oxidizing agent and also as a reducing agent. It is its oxidizing power that permits it to be used for bleaching hair and is responsible for its effectiveness as a mild antiseptic. Oil paintings that have been discolored by the formation of lead sulfide, PbS, which is black, from the lead hydroxy-carbonate (called white lead) in the paint may be bleached by washing with hydrogen peroxide:

$$PbS(c) + 4H_2O_2(aq) \longrightarrow PbSO_4(c) + 4H_2O(l)$$

Its action as a reducing agent is shown, for example, by its decolorization of permanganate ion in acidic solution:

$$2MnO_4^- + 5H_2O_2 + 6H^+ \longrightarrow 2Mn^{++} + 5O_2(g) + 8H_2O$$

Hydrogen Sulfide and the Sulfides

Hydrogen sulfide, H_2S, is analogous to water. Its electronic structure is

It is far more volatile (m.p. $-85.5°C$, b.p. $-60.3°C$) than water. It is appreciably soluble in cold water (2.6 liters of gas dissolves in 1 liter of water at 20°C), forming a slightly acidic solution. The solution is slowly oxidized by atmospheric oxygen, giving a milky precipitate of sulfur.

Hydrogen sulfide has a powerful odor, resembling that of rotten eggs. It is very poisonous, and care must be taken not to breathe the gas while using it in the analytical chemistry laboratory.

Hydrogen sulfide is readily prepared by action of hydrochloric acid on ferrous sulfide:

$$2HCl(aq) + FeS(c) \longrightarrow FeCl_2(aq) + H_2S(g)$$

The *sulfides* of the alkali and alkaline-earth metals are colorless substances easily soluble in water. The sulfides of most other metals are in-soluble or only very slightly soluble in water, and their precipitation under varying conditions is an important part of the usual scheme of qualitative analysis for the metallic ions. Many metallic sulfides occur in nature; important sulfide ores include FeS, Cu_2S, CuS, ZnS, Ag_2S, HgS, and PbS.

7-3. Hydrocarbons Containing Double Bonds and Triple Bonds

The substance *ethylene*, C_2H_4, consists of molecules

in which there is a double bond between the two carbon atoms. This double bond confers upon the molecule the property of much greater chemical reactivity than is possessed by the alkanes. For example, whereas chlorine, bromine, and iodine do not readily attack the alkane hydro-carbons, they easily react with ethylene; a mixture of chlorine and ethylene reacts readily at room temperature in the dark, and with explosive violence in light, to form the substance *dichloroethane*, $C_2H_4Cl_2$:

$$C_2H_4 + Cl_2 \longrightarrow C_2H_4Cl_2$$

or

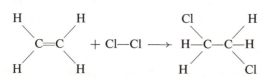

In the course of this reaction the double bond between the two carbon atoms has become a single bond, and the single bond between the two chlorine atoms has been broken. Two new bonds—single bonds between a chlorine atom and a carbon atom—have been formed. We may use the bond-energy values of Tables VIII-1 and VIII-2 to estimate the heat of the reaction:

Reactant bond energies		Product bond energies	
C=C	615	C—C	344
Cl—Cl	243	2C—Cl	656
	858 kJ mole^{-1}		1000 kJ mole^{-1}

We see that the bonds in the product molecules are more stable than those in the reactant molecules by 142 kJ mole^{-1}. Hence this reaction is exothermic, with a moderately large amount of heat evolved on reaction, 142 kJ mole^{-1}.

A reaction of this sort is called an *addition reaction*. *An addition reaction is a reaction in which a molecule adds to a molecule containing a double bond, converting the double bond into a single bond.*

Because of this property of readily combining with other substances such as the halogens, ethylene and related hydrocarbons are said to be *unsaturated*. Ethylene is the first member of a homologous series of hydrocarbons, called the *alkenes*.

Ethylene is a colorless gas (b.p. $-104°C$) with a sweetish odor. It can be made in the laboratory by heating ethyl alcohol, C_2H_5OH, with concentrated sulfuric acid, preferably in the presence of a catalyst (such as silica) to increase the rate of the reaction. Concentrated sulfuric acid is a strong dehydrating agent, which removes water from the alcohol molecule:

$$C_2H_5OH \xrightarrow[\text{H}_2\text{SO}_4]{} C_2H_4 + H_2O$$

Ethylene is made commercially by passing alcohol vapor over a catalyst (aluminum oxide) at about 400°C. The reaction is endothermic; a small amount of heat is absorbed when it takes place:

$$C_2H_5OH \longrightarrow C_2H_4 + H_2O - 47 \text{ kJ mole}^{-1}$$

Endothermic chemical reactions are in general favored by heating the reactants.

Ethylene has the interesting property of causing green fruit to ripen, and it is used commercially for this purpose. It is also used as an anesthetic.

Cis and Trans Isomers

The ethylene molecule is planar (Section 6-4), with greatly restricted rotation around the double bond. In consequence, a disubstituted ethylene such as 1,2-dichloroethylene, CHCl=CHCl, exists as two isomers, called *cis*-1,2-dichloroethylene and *trans*-1,2-dichloroethylene:

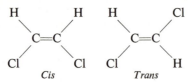

These two substances have different properties: the *cis* isomer has m.p. $-80.5°C$, b.p. 59.8°C, density of liquid 1.291 g cm^{-3}, and electric dipole moment 0.39 $_\epsilon$Å; and the *trans* isomer has m.p. $-50°C$, b.p. 48.5°C, density of liquid 1.265 g cm^{-3}, and electric dipole moment 0.

The potential energy barrier restricting rotation about the double bond has been found by experiment to be about 200 kJ mole^{-1}.*

Acetylene, H—C≡C—H, is the first member of a homologous series of hydrocarbons containing triple bonds. Aside from acetylene, these substances (called *alkynes*) have not found wide use, except for the manufacture of other chemicals.

Acetylene is a colorless gas (b.p. $-84°C$), with a characteristic garliclike odor. It is liable to explode when compressed in the pure state, and is usually kept in solution under pressure in acetone. It is used as a fuel, in the oxyacetylene torch and the acetylene lamp, and is also used as the starting material for making other chemicals.

Acetylene is most easily made from *calcium carbide* (calcium acetylide, CaC_2). Calcium carbide is made by heating lime (calcium oxide, CaO) and coke in an electric furnace:

$$CaO + 3C \longrightarrow CaC_2 + CO(g)$$

Calcium carbide is a gray solid that reacts vigorously with water to produce calcium hydroxide and acetylene:

$$CaC_2 + 2H_2O \longrightarrow Ca(OH)_2 + C_2H_2(g)$$

The existence of calcium carbide and other carbides with similar formulas and properties shows that acetylene is an acid, with two replaceable hydrogen atoms. It is an extremely weak acid, however, and its solution in water does not taste acidic.

Acetylene and other substances containing a carbon-carbon triple bond are very reactive. They readily undergo addition reactions with chlorine and other reagents, and they are classed as unsaturated substances.

7-4. Aromatic Hydrocarbons. Benzene

An important hydrocarbon is *benzene*, which has the formula C_6H_6. It is a volatile liquid (m.p. 5.5°C, b.p. 80.1°C, density 0.88 g cm^{-3}). Benzene and other hydrocarbons similar to it in structure are called the *aromatic hydrocarbons*. Their derivatives are called aromatic substances—many of them have a characteristic aroma (agreeable odor). Benzene itself was discovered in 1825 by Faraday, who found it in the illuminating gas made by heating oils and fats.

For many years there was discussion about the structure of the benzene

---

*The height of the barrier can be estimated from the spectroscopic value of the frequency of the torsional vibration of the molecule (twisting vibration of one end relative to the other end), and from the activation energy (Chapter 16) of the *cis-trans* isomerization reaction.

molecule. The German chemist August Kekulé (1829–1896) in 1865 proposed that the six carbon atoms form a regular hexagon in space, the six hydrogen atoms being bonded to the carbon atoms, and forming a larger hexagon. Kekulé suggested that, in order for a carbon atom to show its normal quadrivalence, the ring contains three single bonds and three double bonds in alternate positions, as shown below. A structure of this sort is called a Kekulé structure.

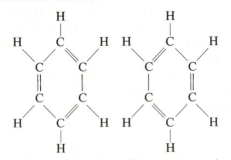

Other hydrocarbons, derivatives of benzene, can be obtained by replacing the hydrogen atoms by methyl groups or similar groups. Coal tar and petroleum contain substances of this sort, such as *toluene*, C_7H_8, and the three *xylenes*, C_8H_{10}. These formulas are usually written $C_6H_5CH_3$ and $C_6H_4(CH_3)_2$, to indicate the structural formulas:

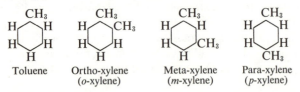

| Toluene | Ortho-xylene (*o*-xylene) | Meta-xylene (*m*-xylene) | Para-xylene (*p*-xylene) |

In these formulas the benzene ring of six carbon atoms is shown simply as a hexagon. This convention is used by organic chemists, who often also do not show the hydrogen atoms, but only other groups attached to the ring.

It is to be noted that we can draw two Kekulé structures for benzene and its derivatives. For example, for ortho-xylene the two Kekulé structures are

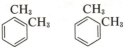

In the first structure there is a double bond between the carbon atoms to which methyl groups are attached, and in the second there is a single bond in this position. The organic chemists of a century ago found it impossible, however, to separate two isomers corresponding to these formulas. To

explain this impossibility of separation Kekulé suggested that the molecule does not retain one Kekulé structure, but rather slips easily from one to the other. In the modern theory of molecular structure the ortho-xylene molecule is described as a hybrid of these two structures, with each bond between two carbon atoms in the ring intermediate in character between a single bond and a double bond. Even though this resonance structure is accepted for benzene and related compounds, it is often convenient simply to draw one of the Kekulé structures, or just a hexagon, to represent a benzene molecule.

The structure of the benzene molecule was determined by the electron diffraction method in 1929 and the following years. It is a planar hexagon with carbon-carbon bond length 1.40 Å (C—H bond length 1.06 Å). This value for a bond with 50% double-bond character is reasonable in comparison with the values 1.54 Å for C—C, 1.33 Å for C=C, and 1.42 Å for $33\frac{1}{3}\%$ double-bond character (graphite). The planar configuration is required by the properties of the double bond (Section 6-4).

Benzene and its derivatives are extremely important substances. They are used in the manufacture of drugs, explosives, photographic developers, plastics, synthetic dyes, and many other substances. For example, the substance *trinitrotoluene*, $C_6H_2(CH_3)(NO_2)_3$, is an important explosive (TNT). The structure of this substance is

In addition to benzene and its derivatives, there exist many other aromatic hydrocarbons, containing two or more rings of carbon atoms. *Naphthalene*, $C_{10}H_8$, is a solid substance with a characteristic odor; it is used as a constituent of moth balls and in the manufacture of dyes and other organic compounds. *Anthracene*, $C_{14}H_{10}$, and *phenanthrene*, $C_{14}H_{10}$, are isomeric substances containing three rings fused together. These substances are also used in making dyes, and derivatives of them are important biological substances (cholesterol, sex hormones; see Chapter 24). For naphthalene, anthracene, and phenanthrene we may write the following structural formulas:

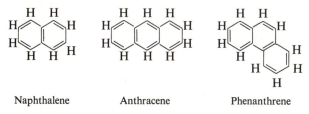

Naphthalene   Anthracene   Phenanthrene

These molecules also have hybrid structures: the structural formulas shown above do not represent the molecules completely, but are analogous to one Kekulé structure for benzene.

Resonance Energy

The heat evolved when a molecule of hydrogen is added to a double bond is about 120 kJ mole^{-1}. For cyclohexene, for example, the value determined by experiment is 119.6 kJ mole^{-1}:

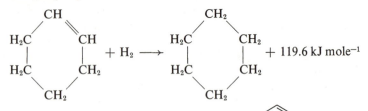

If the benzene molecule had one Kekulé structure, , we might well expect that the heat of hydrogenation of its three double bonds would be approximately three times the heat of hydrogenation of the one double bond in cyclohexene, $3 \times 119.6 = 358.8$ kJ mole^{-1}:

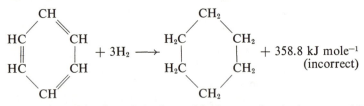

The experimental value of the heat of hydrogenation is, however, 150 kJ mole^{-1} smaller:

$$C_6H_6(g) + 3H_2(g) \longrightarrow C_6H_{12}(g) + 208.4 \ \text{kJ} \ \text{mole}^{-1}$$

We conclude that *the benzene molecule is 150 kJ mole^{-1} more stable than it would be if it were represented by a single Kekulé structure, with each of the three double bonds similar to the double bond in cyclohexene.* This extra stabilizing energy of 150 kJ mole^{-1} is called the *resonance energy* of benzene. It is attributed to the fact that the benzene molecule is not satisfactorily represented by a single Kekulé structure, but instead can be reasonably well described as a hybrid of the two Kekulé structures.*

---

*It would, of course, be surprising if the benzene molecule in its normal state were actually *less* stable than the hypothetical molecule with a single Kekulé structure; we would then ask why the molecule was prevented from assuming this more stable structure. The theory of resonance is based upon a theorem in quantum mechanics that the normal state of an atom or molecule is the most stable of all possible states.

The resonance energy of benzene makes the substance far less reactive chemically than alkenes or other unsaturated substances. For example, the reaction of addition of one hydrogen molecule to benzene to form cyclohexadiene,

is endothermic, not exothermic. The properties of benzene and other aromatic substances reflect the stability conferred upon them by the resonance energy.

7-5.   Ammonia and Its Compounds

Ammonia, NH_3, is an easily condensable gas (b.p. $-33.4°C$; m.p. $-77.7°C$), readily soluble in water. The gas is colorless and has a pungent odor, often detected around stables and manure piles, where ammonia is produced by decomposition of organic matter. The solution of ammonia in water, called ammonium hydroxide solution (or sometimes *aqua ammonia*), contains the molecular species NH_3, NH_4OH (ammonium hydroxide), NH_4^+, and OH^-. Ammonium hydroxide is a weak base, and is only slightly ionized to ammonium ion, NH_4^+, and hydroxide ion OH^-:

$$NH_3 + H_2O \rightleftarrows NH_4OH \rightleftarrows NH_4^+ + OH^-$$

The ammonium ion has the configuration of a regular tetrahedron. The NH_4^+ ion can be described as having electron pairs in four tetrahedral sp^3 orbitals. In the ammonium hydroxide molecule the ammonium ion and the hydroxide ion are held together by a hydrogen bond (Section 12-4).

The Preparation of Ammonia

Ammonia is easily made in the laboratory by heating an ammonium salt, such as ammonium chloride, NH_4Cl, with a strong alkali, such as sodium hydroxide or calcium hydroxide:

$$2NH_4Cl + Ca(OH)_2 \longrightarrow CaCl_2 + 2H_2O + 2NH_3(g)$$

The gas may also be made by warming concentrated ammonium hydroxide.

The principal commercial method of production of ammonia is the *Haber process*, the direct combination of nitrogen and hydrogen under high pressure (several hundred atmospheres) in the presence of a catalyst (usually iron, containing molybdenum or other substances to increase the catalytic activity). The gases used must be specially purified, to prevent "poisoning" the catalyst. The yield of ammonia at equilibrium is less at a high temperature than at a lower temperature. The gases react very slowly at low temperatures, however, and the reaction became practical as a commercial process only when a catalyst was found which speeded up the rate satisfactorily at 500°C. Even at this relatively low temperature the equilibrium is unfavorable if the gas mixture is under low pressure, less than 0.1% of the mixture at 1 atm being converted to ammonia. Increase in the total pressure favors the formation of ammonia; at 500 atm pressure the equilibrium mixture is over one-third ammonia (see Example 11-7).

Smaller amounts of ammonia are obtained as a by-product in the manufacture of coke and illuminating gas by the distillation of coal, and are made by the cyanamide process. In the *cyanamide process* a mixture of lime and coke is heated in an electric furnace, forming calcium acetylide (calcium carbide), CaC_2:

$$CaO + 3C \longrightarrow CO + CaC_2$$

Nitrogen, obtained by fractionation of liquid air, is passed over the hot calcium acetylide, forming calcium cyanamide, $CaCN_2$:

$$CaC_2 + N_2 \longrightarrow CaCN_2 + C$$

Calcium cyanamide may be used directly as a fertilizer, or may be converted into ammonia by treatment with steam under pressure:

$$CaCN_2 + 3H_2O \longrightarrow CaCO_3 + 2NH_3$$

Ammonium Salts

The ammonium salts are similar to the potassium salts and rubidium salts in crystal form, molar volume, color, and other properties. This similarity is due to the close approximation in size of the ammonium ion (radius 1.48 Å) to these alkali ions (radius of K^+, 1.33 Å, and of Rb^+, 1.48 Å). The ammonium salts are all soluble in water, and are completely ionized in aqueous solution.

Ammonium chloride, NH_4Cl, is a white salt, with a bitter salty taste. It is used in dry batteries and as a flux in soldering and welding. Ammonium sulfate, $(NH_4)_2SO_4$, is an important fertilizer; and ammonium nitrate, NH_4NO_3, mixed with other substances, is used as an explosive and as a fertilizer.

Liquid Ammonia as a Solvent

Liquid ammonia (b.p. $-33.4°C$) has a high dielectric constant, and is a good solvent for salts, forming ionic solutions. It also has the unusual power of dissolving the alkali metals and alkaline-earth metals without chemical reaction, to form blue solutions that have an extraordinarily high electric conductivity and a metallic luster. These metallic solutions slowly decompose, with evolution of hydrogen, forming amides, such as sodium amide, $NaNH_2$:

$$2Na + 2NH_3 \longrightarrow 2Na^+ + 2NH_2^- + H_2$$

The amides are ionized in the solution into sodium ion and the amide ion,

which is analogous to the hydroxide ion in aqueous systems. The ammonium ion in liquid ammonia is analogous to the hydronium ion in aqueous systems.

Derivatives of Ammonia

The alkylamines, such as methylamine, $H_3C\!-\!NH_2$, are gases or liquids that resemble ammonia in being able to add a proton to the unshared pair of the nitrogen atom, to form alkylammonium ions, such as the methylammonium ion, $CH_3NH_3^+$.

Hydrazine, N_2H_4, is also basic (that is, it can add protons). It forms two series of salts, those containing the ion $N_2H_5^+$ (such as N_2H_5Cl), and those containing the ion $N_2H_6^{++}$ (such as $N_2H_6Cl_2$). Hydroxylamine, $H_2N\!-\!OH$, forms salts containing the ion H_3NOH^+.

Hydronium Salts

Crystalline perchloric acid monohydrate, $HClO_4 \cdot H_2O$, has been found to have the same structure as ammonium perchlorate, NH_4ClO_4, with the hydronium ion, OH_3^+, replacing the ammonium ion, NH_4^+. The substance might be called hydronium perchlorate. Similar structures are found for the hydrates of other strong acids.

Phosphonium Salts

The proton affinity of the unshared electron pairs of other Vth-, VIth-, and VIIth-group atoms is small. Only phosphine, PH_3, forms a series of salts. Phosphonium bromide, PH_4Br, can be made by addition of phosphine to a cold aqueous solution of hydrobromic acid. The only other phosphonium salts known are the iodide, chloride, and sulfate.

TABLE 7-12
Properties of Some Chlorides of Nonmetals

	CCl_4	NCl_3	Cl_2O	ClF
m.p.	$-23°$	$-40°$	$-20°$	$-154°C$
b.p.	$77°$	$70°$	$4°$	$-100°$
Bond length	1.77 Å	1.73 Å	1.69 Å	1.63 Å
Bond angle	109.5°	110°	110°	

	$SiCl_4$	PCl_3	SCl_2	Cl_2
m.p.	$-70°$	$-112°$	$-78°$	$-102°$
b.p.	$60°$	$74°$	$59°$	$-34°$
Bond length	2.01 Å	2.04 Å	2.00 Å	1.99 Å
Bond angle	109.5°	100.0°	102°	

	$GeCl_4$	$AsCl_3$		$BrCl$
m.p.	$-50°$	$-18°$		
b.p.	$83°$	$130°$		
Bond length	2.09 Å	2.16 Å		2.14 Å
Bond angle	109.5°	99°		

	$SnCl_4$	$SbCl_3$	$TeCl_2$	ICl
m.p.	$-33°$	$73°$	$209°$	$27°$
b.p.	$114°$	$223°$	$327°$	$97°$
Bond length	2.32 Å	2.38 Å	2.34 Å	2.30 Å
Bond angle	109.5°	99°	99°	

7-6.   Other Normal-valence Compounds of the Nonmetals

The halogens form covalent compounds with most of the nonmetallic elements (including each other) and the metalloids. These compounds are usually molecular substances, with the relatively low melting points and boiling points characteristic of substances with small forces of inter-molecular attraction.

An example of a compound involving a covalent bond between a halogen and a nonmetal is chloroform, $CHCl_3$. In this molecule the carbon atom is attached by single covalent bonds to one hydrogen atom and three chlorine atoms. Chloroform is a colorless liquid, with a characteristic sweetish odor, b.p. 61°C, density 1.498 g ml⁻¹. Chloroform is only slightly soluble in water, but it dissolves readily in alcohol, ether, and carbon tetrachloride.

The halogenides of carbon and its congeners are tetrahedral (sp^3 bond orbitals). Those of nitrogen and oxygen and their congeners have bond angles near 100°, corresponding to p bond orbitals with a small amount of s character.

The melting points, boiling points, bond length, and bond angles of some chlorides are given in Table 7-12. Some values of the standard enthalpy are given in earlier tables in this chapter. The stability of the substances is determined by the electronegativity difference of the bonded atoms, with correction for compounds of nitrogen and oxygen.

Many of these substances react readily with water, to form a hydride of one element and a hydroxide of the other:

$$ClF + H_2O \longrightarrow HClO + HF$$
$$PCl_3 + 3H_2O \longrightarrow P(OH)_3 + 3HCl$$

In general, in a reaction of this sort, called *hydrolysis*, the more electronegative element combines with hydrogen, and the less electronegative element combines with the hydroxide group. This rule is seen to be followed in the above examples.

Resonance in Fluorinated Hydrocarbons

An interesting difference in properties is observed between methyl fluoride and methylene difluoride, trifluoromethane, and tetrafluoromethane. It is found that successive steps in chlorination of methane are accompanied by nearly the same enthalpy change (see Exercise 6-24), to within ± 4 kJ mole⁻¹. This constancy for reactions in which C—H is converted to C—Cl is expected from the bond-energy principle. For the fluoromethanes, however, the values differ greatly in the sequence of reactions:

	$CH_4(g)$	$CH_3F(g)$	$CH_2F_2(g)$	$CHF_3(g)$	$CF_4(g)$
$\Delta H°_f(298°K)$	−75	−234	−449	−691	−923 kJ mole⁻¹
Difference		−159	−215	−242	−232

The enthalpy value for CH_3F leads to the value 443 kJ mole⁻¹ for the C—F bond energy. This value is given in Table VIII-1. The other fluoromethanes are stabilized by resonance with structures other than the normal valence-bond structure.

The observed C—F bond length in methyl fluoride and ethyl fluoride is 1.385 Å, which is close to the value expected for a single bond. Significant shortening is found, however, for the other molecules, to 1.358 Å for

CH_2F_2, 1.334 Å for CHF_3, and 1.320 Å for CF_4. These values suggest an increasing amount of double-bond character, such as would result from resonance among the following structures for methylene fluoride:

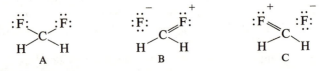

The electronegativity difference of carbon and fluorine corresponds to 43% of partial ionic character for the C—F bond, freeing a carbon orbital for double-bond formation with the other fluorine atom. With use of 443 kJ mole^{-1} for the C—F bond energy for the normal-valence structures (A for CH_2F_2) we obtain the following values for the resonance energy with structures such as B and C: 56 kJ mole^{-1} for CH_2F_2, 139 kJ mole^{-1} for CHF_3, and 212 kJ mole^{-1} for CF_4.

A significant amount of resonance energy is also found for CH_2ClF (25 kJ mole^{-1}), $CHClF_2$ (69), $CClF_3$ (132), $CHCl_2F$ (36), CCl_2F_2 (69), and CCl_3F (21). In these molecules some decrease in bond length is observed for C—Cl, but not so great as for C—F.

The decreased chemical reactivity resulting from this resonance energy has great practical importance. The chloromethanes, such as carbon tetrachloride, are toxic, because of the ease with which they hydrolyze. The fluorochloromethanes do not hydrolyze in this way, and can be used safely in the home and in industrial plants. There was rapid development of the home refrigerator industry after the discovery of these substances, especially the freons CCl_3F (b.p. 23.8°C) and CCl_2F_2 (b.p. −30°C). They are used also as the vehicle for aerosols.

The bond lengths in $SiCl_4$ and other halides of the heavier elements are usually less than the sum of the single-bond radii; for $SiCl_4$ the difference

TABLE 7-13
Some Physical Properties of the Chloromethanes

Substance	Formula	Melting Point	Boiling Point	Density of Liquid
Methyl chloride	CH_3Cl	−98°C	−24°C	0.92 g ml^{-1}
Dichloromethane	CH_2Cl_2	−97	40	1.34
Chloroform	$CHCl_3$	−64	61	1.50
Carbon tetrachloride	CCl_4	−23	77	1.60

is 0.16 Å. It is likely that this difference is to be interpreted as showing that the bonds have some double-bond character, resulting from resonance of the sort discussed above.

Substitution Reactions

Methane and other paraffins react with chlorine and bromine when exposed to sunlight or when heated to a high temperature. When a mixture of methane and chlorine is passed through a tube containing a catalyst (aluminum chloride, $AlCl_3$, mixed with clay) heated to about 300°C, the following reactions occur:*

$$CH_4 + Cl_2 \longrightarrow CH_3Cl + HCl$$
$$CH_3Cl + Cl_2 \longrightarrow CH_2Cl_2 + HCl$$
$$CH_2Cl_2 + Cl_2 \longrightarrow CHCl_3 + HCl$$
$$CHCl_3 + Cl_2 \longrightarrow CCl_4 + HCl$$

In each of these reactions a chlorine molecule, with structural formula Cl—Cl, is split into two chlorine atoms; one chlorine atom takes the place of a hydrogen atom bonded to carbon, and the other combines with the displaced hydrogen atom to form a molecule of hydrogen chloride, H—Cl. By use of values of bond energies from Table VIII-1, we calculate for the heat of each of these reactions the value $328 + 432 - 243 - 415 = 102$ kJ mole^{-1}. The reactions are not as strongly exothermic as the reaction of addition of chlorine to a double bond (142 kJ mole^{-1}).

Chemical reactions such as these four are called *substitution reactions*. *A substitution reaction* is the replacement of one atom or group of atoms in a molecule by another atom or group of atoms. The four chloromethanes are substitution products of methane. Substitution reactions and addition reactions (Section 7-4) are extensively used in practical organic chemistry.

Some physical properties of the chloromethanes are given in Table 7-13; their enthalpies of formation from the elements are given in Table 7-2. All four are colorless, with characteristic odors and with low boiling points, increasing with the number of chlorine atoms in the molecule. The chloromethanes do not ionize in water.

Chloroform and carbon tetrachloride are used as solvents; carbon tetrachloride is an important dry-cleaning agent. Chloroform was formerly used as a general anesthetic.

Care must be taken in the use of carbon tetrachloride that no large amount of its vapor is inhaled, because it damages the liver.

---

*The relative amounts of the four products may be varied somewhat by changing the ratio of methane and chlorine in the gas mixture used.

Methyl Alcohol and Ethyl Alcohol

An *alcohol* is obtained from a hydrocarbon by replacing one hydrogen atom by a hydroxyl group, —OH. Thus methane, CH_4, gives *methyl alcohol*, CH_3OH, and ethane, C_2H_6, gives *ethyl alcohol*, C_2H_5OH. The names of the alcohols are often written by using the ending *ol*; methyl alcohol is called *methanol*, and ethyl alcohol *ethanol*. They have the following structural formulas:

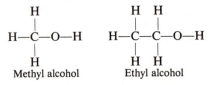

Methyl alcohol   Ethyl alcohol

To make methyl alcohol from methane, the methane may be converted to methyl chloride by treatment with chlorine, as described above, and the methyl chloride then converted to methyl alcohol by treatment with sodium hydroxide:

$$CH_3Cl + NaOH \longrightarrow CH_3OH + NaCl$$

Methyl alcohol is made by the destructive distillation of wood; it is sometimes called wood alcohol. It is a poisonous substance, which on ingestion causes blindness and death. It is used as a solvent and for the preparation of other organic compounds.

The most important method of making ethyl alcohol is by the fermentation of sugars with yeast. Grains and molasses are the usual raw materials for this purpose. Yeast produces an enzyme that catalyzes the fermentation reaction. In the following equation the formula $C_6H_{12}O_6$ is that of a sugar, glucose (also called dextrose and grape sugar; Chapter 23):

$$C_6H_{12}O_6 \longrightarrow 2CO_2 + 2C_2H_5OH$$

Ethyl alcohol is a colorless liquid (m.p. $-117°C$, b.p. 79°C) with a characteristic odor. It is used as a fuel, as a solvent, and as the starting material for preparing other compounds. Beer contains 3 to 5% alcohol, wine usually 10 to 12%, and distilled liquors such as whiskey, brandy, and gin 40 to 50%.

The *ethers* are compounds obtained by reaction of two alcohol molecules, with elimination of water. The most important ether is *diethyl ether* (ordinary ether), $(C_2H_5)_2O$. It is made by treating ethyl alcohol with concentrated sulfuric acid, which serves as a dehydrating agent:

$$2C_2H_5OH \underset{H_2SO_4}{\longrightarrow} C_2H_5OC_2H_5 + H_2O$$

It is used as a general anesthetic and as a solvent.

The Organic Acids

Ethyl alcohol can be oxidized by oxygen of the air to *acetic acid,* $HC_2H_3O_2$ or CH_3COOH:

$$C_2H_5OH + O_2 \longrightarrow CH_3COOH + H_2O$$

This reaction occurs easily in nature. If wine, containing ethyl alcohol, is allowed to stand in an open container, it undergoes acetic-acid fermentation and changes into vinegar by the above reaction. The change is brought about by microorganisms ("mother of vinegar"), which produce enzymes that catalyze the reaction.

Acetic acid has the following structural formula:

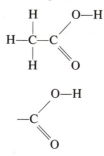

It contains the group

which is called the *carboxyl group.* It is this group that gives acidic properties to the organic acids.

Acetic acid melts at 17°C and boils at 118°C. It is soluble in water and alcohol. The molecule contains one hydrogen atom that ionizes from it in water, producing the *acetate ion,* $C_2H_3O_2^-$. The acid reacts with bases to form salts. An example is sodium acetate, $NaC_2H_3O_2$, a white solid:

$$HC_2H_3O_2 + NaOH \longrightarrow NaC_2H_3O_2 + H_2O$$

Formic acid, HCOOH, is the simplest of the carboxylic acids. Some others are discussed in Chapter 23.

The structure shown above for acetic acid is not completely satisfactory. The experimental value for the C—OH bond length is 1.36 Å, 0.07 Å less than that for a C—O single bond. The same value, 1.36 Å, is found in methyl formate, $HCOOCH_3$. The carboxylic acids are accordingly assigned the following resonance structure:

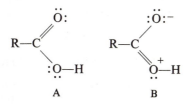

A   B

The bond lengths 1.22 Å and 1.36 Å indicate that structure A contributes about 80% and structure B about 20% for the acids and esters. For the carboxylate ion the two structures A′ and B′ are equivalent:

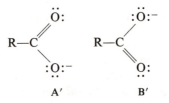

and each contributes 50% to the normal state of the ion. The resonance energy, relative to structure A or A′, is found to be about 65 kJ mole^{-1} for acids and esters and 130 kJ mole^{-1} for carboxylate ions. It is mentioned in Section 23-5 that the extra resonance energy of the carboxylate ions provides an explanation of the fact that the OH group is far more acidic in the carboxylic acids than in the alcohols.

Chemical Reactions of Organic Substances

In the above paragraphs we have discussed derivatives of methane and ethane in which a hydrogen atom is replaced by a chlorine atom, —Cl, a hydroxyl group, —OH, or a carboxyl group, —COOH. There are many other groups that can replace a hydrogen atom, to form other substances.

In general, the chemical reactions that can be used to convert methane into its derivatives can be applied also to the other hydrocarbons. By chemical analysis and the study of the chemical reactions of a new substance the chemist can determine its formula. For example, if a substance contains only carbon, hydrogen, and oxygen, and has acidic properties like those of acetic acid, the chemist assumes that it contains a carboxyl group, —COOH. An important part of organic chemistry is the use of special reactions that identify different groups in a molecule.

7-7.   Some Transargononic Single-bonded Compounds

When chlorine is passed over phosphorus it reacts with it to form phosphorus trichloride:

$$\tfrac{1}{4}P_4(c) + \tfrac{3}{2}Cl_2(g) \longrightarrow PCl_3(g) + 279 \text{ kJ mole}^{-1}$$

Another product, phosphorus pentachloride, is also formed:

$$PCl_3(g) + Cl_2(g) \longrightarrow PCl_5(g) + 92 \text{ kJ mole}^{-1}$$

In the PCl$_5$ molecule the phosphorus atom has a transargononic structure, with five shared electron pairs in the outer shell. It forms five covalent

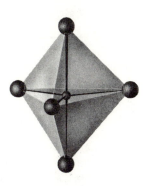

FIGURE 7-3
The structure of the molecule PCl₅, showing the arrangement of the five chlorine atoms at the corners of a trigonal bipyramid about the phosphorus atom.

bonds, with the bond orbitals formed by hybridization of a $3d$ orbital with the $3s$ orbital and the three $3p$ orbitals. The valence-bond structure for the molecule is

The molecule has the trigonal bipyramidal structure, with three equally-spaced chlorine atoms around the equator and one at each pole (Figure 7-3).

The P—Cl bond energy in PCl_3 is 317 kJ mole^{-1}:

$$P(g) + 3Cl(g) \longrightarrow PCl_3(g) + 3 \times 317 \text{ kJ mole}^{-1}$$

The effective P—Cl bond energy for each of the two added bonds in PCl_5 is 165 kJ mole^{-1}:

$$PCl_3(g) + 2Cl(g) \longrightarrow PCl_5(g) + 2 \times 165 \text{ kJ mole}^{-1}$$

A transargononic P—Cl bond is 152 kJ mole^{-1} less stable than an argononic P—Cl bond. This smaller stability results from the smaller stability of a $3d$ electron in phosphorus than a $3p$ electron, partially neutralized by the greater bond-forming power of *spd* hybrid bond orbitals (greater overlap) than of p bond orbitals. In addition, a significant contribution to the normal state of PCl_5 is made by ionic structures such as

which do not involve any $3p \longrightarrow 3d$ promotion energy.

TABLE 7-14
Formulas of Single-bonded Transargononic Molecules and Ions

SiF_6^{--}					
GeF_6^{--}					
SnF_6^{--}	$SnCl_6^{--}$	$SnBr_6^{--}$	SnI_6^{--}	$Sn(OH)_6^{--}$	
PF_5	PCl_5	PF_6^-	PCl_6^-		
PF_3Cl_2	PBr_3Cl_2	PCl_3F_2	PCl_3Br_2		
AsF_6^-					
SbF_6^-	$SbCl_5$	$SbCl_6^-$	$SbBr_6^-$	$Sb(OH)_6^-$	
SF_4	SF_6	S_2F_{10}	SCl_4		
SeF_4	SeF_6	$SeCl_4$	$SeBr_4$	$SeCl_6^{--}$	$SeBr_6^{--}$
TeF_4	TeF_6	$TeCl_4$	$TeBr_4$	$TeCl_6^{--}$	$Te(OH)_6$
ClF_3	ClF_5				
BrF_3	BrF_5	BrF_4^-			
IF_5	IF_7	ICl_3	ICl_2^-	ICl_4^-	

The importance of ionic structures in stabilizing transargononic compounds is shown by the great stability of transargononic fluorides. The bond energy for each of the two added fluorine atoms in PF_5 is 425 kJ mole^{-1}:

$$PF_3(g) + 2F(g) \longrightarrow PF_5(g) + 2 \times 425 \text{ kJ mole}^{-1}$$

This value is only 61 kJ mole^{-1} less than the normal (argononic) P—F bond energy, 486 kJ mole^{-1} (in PF_3), a far smaller difference than for P—Cl (152 kJ mole^{-1}). It is not surprising that the fluorides predominate among the transargononic molecules and ions, as listed in Table 7-14.

Most of the known single-bonded transargononic molecules and ions are listed in Table 7-14. Those with small molecular weight, especially the fluorides, are gases at room temperature (SF_6 has boiling point $-62°C$; $PCl_5(c)$ sublimes at 160°C and melts, under pressure, at 168°C).

SF_6 and other molecules and ions with ligancy six and no unshared electron pairs in the outer shell of the central atom have the six bonds directed towards the corners of a regular octahedron. In BrF_5 the five bonds are directed toward five corners of a regular octahedron, the sixth corner presumably being occupied by an unshared pair. The unshared pair, in an orbital that is largely *s* in character, occupies more space than a bonding pair; the angle between the bond to the polar fluorine atom and that to any of the four equatorial fluorine atoms is less than 90° (about 86°). In BrF_4^- the four bonds lie in a plane, with the two unshared pairs occupying the other two octahedral directions.

An interesting and not fully understood fact is that phosphorus and antimony form more stable transargononic compounds than their intermediate congener, arsenic.

The stability of transargononic compounds increases with increase in electronegativity difference of the ligated atom and the central atom, as shown by the following comparison:

$$ClF(g) + 2F_2(g) \longrightarrow ClF_5(g) + 188 \text{ kJ mole}^{-1}$$
$$BrF(g) + 2F_2(g) \longrightarrow BrF_5(g) + 370 \text{ kJ mole}^{-1}$$
$$IF(g) + 2F_2(g) \longrightarrow IF_5(g) + 727 \text{ kJ mole}^{-1}$$

The elements of the first period (C, N, O) do not form stable transargononic compounds. This general difference in their chemical properties from their heavier congeners is clearly explained by the nonexistence of $2d$ orbitals.

Transargononic compounds involving multiple bonds are discussed in the following chapter.

7-8. The Argonons

The members of group 0 of the periodic table, helium, neon, argon, krypton, xenon, and radon, which we call the argonons (Section 5-3), have in the past usually been called the inert gases or noble gases, because of their lack of chemical reactivity. For many years it was thought that the argonons could not form covalent compounds; the only compounds reported were the clathrate hydrates, such as $Xe_8(H_2O)_{46}$ (Section 12-6). In recent years a number of transargononic compounds of krypton and xenon have been made.

The electronic structures of the argonons have been discussed in Section 5-3. Their names, except radon, are from Greek roots: *helios*, sun: *neos*, new; *argos*, inert; *kryptos*, hidden; *xenos*, stranger.* Radon is named after radium, from which it is formed by radioactive decomposition. Some of their properties are given in Table 7-15. Note the regular dependence of melting point and boiling point on atomic number.

Helium

Helium is present in very small quantities in the atmosphere. Its presence in the sun is shown by the occurrence of its spectral lines in sunlight. These lines were observed in 1868, long before the element was discovered on earth, and the lines were ascribed to a new element, which was named helium† by Sir Norman Lockyer (1836–1920).

---

*The helium on earth, despite its name, did not originate in the sun. All of the original helium escaped from the earth's atmosphere, in consequence of its low molecular weight. The helium now on earth has been formed by alpha decomposition of the radioactive elements (Chapter 26).

†The ending "ium," which is otherwise used only for metallic elements, is from Lockyer's incorrect surmise that the new element was a metal. "Helion" would be a better name, as its ending is consistent with those of the names of the other argonons.

TABLE 7-15
Properties of the Argonons

	Symbol	Atomic Number	Atomic Weight	Melting Point	Boiling Point
Helium	He	2	4.0026	$-272.2°C*$	$-268.9°C$
Neon	Ne	10	20.183	$-248.67°$	$-245.9°$
Argon	Ar	18	39.948	$-189.2°$	$-185.7°$
Krypton	Kr	36	83.80	$-157°$	$-152.9°$
Xenon	Xe	54	131.30	$-112°$	$-107.1°$
Radon	Rn	86	222	$-77°$	$-61.8°$

*At 26 atm pressure. At smaller pressures helium remains liquid at still lower temperatures.

Helium occurs as a gas entrapped in some uranium minerals, from which it can be liberated by heating. It is also present in natural gas from some wells, especially in Texas and Canada; this is the principal source of the element.

Helium is used for filling balloons and dirigibles and for mixing with oxygen (in place of the nitrogen of the air) for breathing by divers, in order to avoid the "bends," caused by gas bubbles formed by release of the nitrogen of the atmosphere that had dissolved in the blood under increased pressure, and to avoid the narcotic action (anesthetizing action) of nitrogen under pressure.

Neon

The second argonon, neon, occurs in the atmosphere to the extent of 0.002%. It is obtained, along with the other argonons (except helium), by the distillation of liquid air (air that has been liquefied by cooling).

When an electric current is passed through a tube containing neon gas at low pressure, the atoms of neon are caused to emit light with their characteristic spectral lines. This produces a brilliant red light, used in advertising signs (neon signs). Other colors for signs are obtained by the use of helium, argon, and mercury, sometimes in mixtures with neon or with one another.

Argon

Argon composes about 1% of the atmosphere. It is used in incandescent light bulbs to permit the filament to be heated to a higher temperature, and thus to produce a whiter light than would be practical in a vacuum. The argon decreases the rate at which the metallic filament evaporates, by keeping vaporized metal atoms from diffusing away from the filament and permitting them to reattach themselves to it. Argon is also extensively used in industry to provide a chemically inert atmosphere, especially in welding and in making pure metals and alloys. The total production of argon for these purposes in the year 1963 was about 10^9 cu ft.

Krypton, Xenon, and Radon

Krypton and xenon, which occur in very small quantities in the air, have not found any significant use. Xenon is a good anesthetic agent, but it is too expensive for general use (it has been used in two major operations on human beings).

Radon, which is produced steadily by radium, is used in the treatment of cancer. It has been found that the rays given off by radioactive substances are often effective in controlling this disease. A convenient way of administering this radiation is to pump the radon that has been produced by a sample of radium into a small gold tube, which is then placed in proximity to the tissues to be treated.

The Discovery of the Argonons

The story of the discovery of argon provides an interesting illustration of the importance of attention to minor discrepancies in the results of scientific investigations.

For over a hundred years it was thought that atmospheric air consisted, aside from small variable amounts of water vapor and carbon dioxide, solely of oxygen (21% by volume) and nitrogen (79%). In 1785 the English scientist Henry Cavendish (1731–1810) investigated the composition of the atmosphere. He mixed oxygen with air and then passed an electric spark through the mixture, to form a compound of nitrogen and oxygen, which was dissolved in a solution in contact with the gas. The sparking was continued until there was no further decrease in volume, and the oxygen was then removed from the residual gas by treatment with another solution. He found that after this treatment only a small bubble of air remained unabsorbed, not more than $\frac{1}{120}$ of the original air. Although Cavendish did not commit himself on the point, it seems to have been assumed by chemists that if the sparking had been continued for a longer time there would have been no residue, and Cavendish's experiment was accordingly interpreted as showing that only oxygen and nitrogen were present in the atmosphere.

Then in 1894, more than 100 years later, Lord Rayleigh began an investigation involving the careful determination of the densities of the gases hydrogen, oxygen, and nitrogen. To prepare nitrogen he mixed dried air with an excess of ammonia, NH_3, and passed the mixture over red-hot copper. Under these conditions the oxygen reacts with ammonia, according to the equation

$$4NH_3 + 3O_2 \longrightarrow 6H_2O + 2N_2$$

The excess ammonia is then removed by bubbling the gas through sulfuric acid. The remaining gas, after drying, should have been pure nitrogen, derived in part from the ammonia and in part from air. The density of this gas was determined. Another sample of nitrogen was made simply by passing air over red-hot copper, which removed the oxygen by combining with it to form copper oxide:

$$O_2 + 2Cu \longrightarrow 2CuO$$

When the density of this gas was determined it was found to be about 0.1% greater than that from the sample of ammonia and air. In order to investigate this discrepancy, a third sample of nitrogen was made by the reaction of ammonia and pure oxygen. It was found that this sample of nitrogen had a density 0.5% less than that of the second sample.

Further investigations showed that nitrogen prepared entirely from air has a density 0.5% greater than nitrogen prepared from ammonia or in any other chemical way. Nitrogen obtained from air was found to have density 1.2572 g liter^{-1} at 0°C and 1 atm, whereas nitrogen made by chemical methods has density 1.2505 g liter^{-1}. Rayleigh and Ramsay then repeated Cavendish's experiment, and showed by spectroscopic analysis that the residual gas was indeed not nitrogen but a new element. They then searched for the other argonons and discovered them.

Transargononic Compounds of the Argonons

In 1933, on the basis of arguments about the electronic structure of molecules, it was pointed out that transargononic compounds of krypton, xenon, and radon with fluorine and oxygen should be stable. The known acids H_8SnO_6, H_7SbO_6, H_6TeO_6, and H_5IO_6, for example, strongly suggest that H_4XeO_6, perxenic acid, should also exist. An effort was made to synthesize XeF_6 by reaction of xenon and fluorine, but without success. Then in 1962 and 1963 several compounds of xenon were synthesized. The first one to be reported (by the English chemist Neil Bartlett (born 1933)) was xenon hexafluoroplatinate, $XePtF_6$, a yellow crystalline substance. Later (1963) he reported the synthesis of the corresponding rhodium compound, $XeRhF_6$. Scientists in the Argonne National Laboratory, and later other investigators, prepared several xenon fluorides, including XeF_2, XeF_4, and XeF_6. Similar compounds of krypton and radon have also been synthesized, including KrF_2, KrF_4, and RnF_4.

The molecule XeF_2 is linear, with bond length 2.00 Å; XeF_4 has the square planar configuration, with bond length 1.95 Å; and XeF_6, with bond length 1.90 Å, is known not to be a regular octahedron (it has one unshared electron pair, as well as six shared pairs, about the xenon atom). The decrease in Xe—F bond length in this sequence reflects the increasing amount of d character of the xenon bond orbitals.

The xenon fluorides react vigorously with water; for example,

$$XeF_6 + H_2O \longrightarrow XeOF_4 + 2HF$$
$$XeOF_4 + 2H_2O \longrightarrow XeO_3 + 4HF$$
$$XeO_3 + H_2O \longrightarrow H_2XeO_4$$

Xenon also forms the tetroxide, XeO_4, and the corresponding acid, H_4XeO_6. X-ray study of the crystal $Na_4XeO_6 \cdot 6H_2O$ has shown that the perxenate ion has the structure of a regular octahedron with Xe—O bond length 1.84 Å. Perxenic acid is a powerful oxidizing agent, which can oxidize manganous ion, Mn^{++}, to permanganate ion, MnO_4^-.

From measurement of the heat of reaction of xenon tetrafluoride with a solution of potassium iodide the heat of formation has been determined:

$$Xe(g) + 2F_2(g) \longrightarrow XeF_4(g) + 188 \text{ kJ mole}^{-1}$$

The bond energy of the Xe—F bond accordingly is equal to 126 kJ mole^{-1}. This value, compared with the values for transargononic I—F, Br—F, and Cl—F bonds, which are 727, 370, and 188 kJ mole^{-1}, respectively, suggests that the electronegativity of xenon is about 3.1.

The xenon oxides are unstable: XeO_3 explodes about as violently as TNT. The heat of explosion has been measured:

$$XeO_3(c) \longrightarrow Xe(g) + \tfrac{3}{2}O_2(g) + 402 \text{ kJ mole}^{-1}$$

The enthalpy of sublimation of XeO_3 is 80 kJ mole^{-1}; hence the Xe=O bond energy is 88 kJ mole^{-1}.

Exercises

7-1. Write the symbols for the electronic structures of atoms of C, N, O, and F in their normal states. How many electron pairs does each atom have in its outer shell (the *L* shell)? How many unpaired electrons? To what valence would you assign each atom in its normal state? Give the formula of one compound of each element corresponding to this value of the valence.

7-2. What electron configuration would you assign to the carbon atom in methane, CH_4? To the silicon atom in silane, SiH_4?

7-3. The angle between two bonds of the tetrahedral carbon atom (see Figure 6-9) is usually given as 109.5°. Calculate the ideal value of this angle to 0.001° by use of the geometric relation of the regular tetrahedron to the cube. Also evaluate the ideal angle between a single bond and a double bond formed by a carbon atom (Figure 6-11).

7-4. The molar enthalpy of formation of $CBr_4(g)$ from diamond and $2Br_2(g)$ is -14 kJ mole^{-1}. Derive this value from the standard molar enthalpy of formation 50 kJ mole^{-1} given in Table 7-2.

7-5. Values of the molar enthalpy of formation at 25°C of $CX_4(g)$ from diamond and $2X_2(g)$, X = H, Br, Cl, F, are -77, -14, -109, and -914 kJ mole^{-1}, respectively. Discuss the relation of these values to the structure of the reactants and products and the electronegativity of the elements.

7-6. From the value of the standard enthalpy of formation of $C_2H_6(g)$ given in Table 7-2 and the values for H(g) and C(g) calculate the value of $\Delta H°$ at 25°C for the reaction $2C(g) + 6H(g) \longrightarrow C_2H_6(g)$.

7-7. Compare the bonds in cyclohexane, C_6H_{12}, with those in methane and ethane. From the molar enthalpy of formation of $CH_4(g)$ from C(g) and 4H(g), -1662 kJ mole^{-1}, and of $C_2H_6(g)$ from 2C(g) and 6H(g), -2833 kJ mole^{-1}, calculate a value of that for $C_6H_{12}(g)$, on the assumption of constancy of bond energy. The observed value is -7032 kJ mole^{-1}. (Answer: -7026 kJ mole^{-1}.)

7-8. Derive the experimental value of $\Delta H° = -66$ kJ mole^{-1} at 25° for the reaction

$$6C_2H_6(g) \longrightarrow C_6H_{12}(g) + 6CH_4(g)$$

from the values of the standard enthalpy of the compounds given in Table 7-2. Why is the value close to zero?

7-9. From the standard enthalpy values in Table 7-2 calculate the value of $\Delta H°$ at 25°C for the reaction

$$3C_2H_6(g) \longrightarrow C_3H_6(g) + 3CH_4(g)$$

What structural interpretation do you give to the numerical value? (Answer: 68 kJ mole^{-1}.)

7-10. Can you give a structural explanation of the fact that the standard enthalpy of gaseous ethanol (-237 kJ mole^{-1}) lies below that for gaseous dimethyl ether (-185 kJ mole^{-1})?

7-11. Only two of the normal-valence compounds of chlorine (HCl, SCl$_2$, PCl$_3$, and so on) have positive values of the standard enthalpy of formation, NCl$_3$ and OCl$_2$ (Tables 7-4 and 7-9). Discuss this fact in relation to the structure of the compounds and the elements.

7-12. Hydrazine, N$_2$H$_4$, combines with hydrochloric acid to form two crystalline salts, with formulas N$_2$H$_5$Cl and N$_2$H$_6$Cl$_2$. Discuss the electronic structures of the ions N$_2$H$_5^+$ and N$_2$H$_6^{++}$, in comparison with the ammonium ion.

7-13. It is stated in Section 7-2 that the highly reactive substance ethylene oxide,
H$_2$C————CH$_2$, owes its high reactivity to the strain of the bent bonds in
$\diagdown$ $\diagup$
O
the three-membered ring. The experimental value of the standard enthalpy $\Delta H°$ of C$_2$H$_4$O(g) is -51 kJ mole^{-1}. With this value and Table VIII-3 calculate its enthalpy of formation from atoms, and with the bond-energy values of Table VIII-1 calculate the strain energy. (Answer: -2602, 102 kJ mole^{-1}.)

7-14. From values given in Table 7-2 calculate the value of $\Delta H°$ at 25°C for the reaction

$$H_2CO(g) + (CH_3)_2O(g) \longrightarrow HCOOCH_3(g) + CH_4(g)$$

Discuss the value in relation to the electronic structure of the ester methyl formate, HCOOCH$_3$. (Answer: -124 kJ mole^{-1}.)

7-15. Discuss similarly the values of the enthalpy change for the two following reactions, which produce formic acid and acetic acid, respectively:

$$H_2CO(g) + CH_3OH(g) \longrightarrow HCOOH(g) + CH_4(g)$$
$$CH_3CHO(g) + CH_3CH_2OH(g) \longrightarrow CH_3COOH(g) + C_2H_6(g)$$

(Answer: -121, -117 kJ mole^{-1}.)

7-16. It is stated in Section 7-3 that the ability of the halogens to attack unsaturated substances such as ethylene is related to the bond-energy values, especially for C=C and C—C. Discuss this relation, using bromine and ethylene as an example.

7-17. Although the benzene molecule is usually discussed by use of the Kekulé
structures ⬡ and ⬡, showing double bonds, the halogens do not
easily add to benzene, as they do to ethylene. Explain this difference in properties of benzene and ethylene in relation to their electronic structures and the electronic structures of the products of addition of X$_2$.

7-18. Assign electronic structures to the following substances: ClF_3, IF_5, IF_7, XeF_2, XeF_4, XeF_6. In each case state which orbitals are used for unshared electron pairs (the more stable orbitals) and which for bonding electrons.

7-19. The American scientist R. S. Mulliken has suggested that for univalent elements the electronegativity is proportional to the sum of the first ionization energy and the electron affinity. Using the enthalpy values of Table 7-9, calculate the divisor that gives the same sum for the four halogens as the sum of the values in Table 6-4. To what values of the electronegativity does this relation lead? (Answer: 531 kJ mole^{-1}; 3.94, 3.05, 2.81, 2.50.)

7-20. Why is no value for the enthalpy of formation of $AsCl_5$ given in Table 7-5? (See Section 7-7.)

7-21. From enthalpy values given in Table 7-7, evaluate the enthalpy of the reaction $8S(g) \longrightarrow S_8(g)$. To what value of the S—S bond energy does this correspond?

7-22. The standard enthalpy of $H_2S_2(g)$ is 4 kJ mole^{-1}. Evaluate the enthalpy of formation of H_2S_2 from atoms. Assuming the H—S bond energy to be the same as in H_2S, obtain a value for the S—S bond energy and compare with the value for the S_8 molecule.

7-23. From consideration of the bonds involved, estimate the enthalpy of the reaction $H_2S(g) + \frac{1}{8}S_8(g) \longrightarrow H_2S_2(g)$. Compare with the experimental value given by the standard enthalpies in Table 7-7 and the preceding exercise.

7-24. The minerals pyrite and marcasite are rather similar in structure. Each has composition FeS_2, with the sulfur atoms in pairs connected by a covalent bond. Their standard enthalpies of formation are −178 and −154 kJ mole^{-1}, respectively. One reacts more vigorously with acids and oxygen than the other. Which would you expect to be the more reactive? (Answer: Marcasite.)

7-25. The standard enthalpy of formation of $Cl_2O(g)$ is 76 kJ mole^{-1}. What is the enthalpy of formation of the molecule from atoms? To what value of the Cl—O bond energy does this value lead? (Answer: −415, 207.5 kJ mole^{-1}.)

7-26. No value has been reported for the standard enthalpy of formation of $Br_2O(g)$. What electronic structure would you assign to this molecule? Can you predict a value for the Br—O bond energy and the standard enthalpy of $Br_2O(g)$? (Answer: 216, 41 kJ mole^{-1}.)

7-27. It is stated in Section 7-6 that in hydrolysis of a binary compound the more electronegative element combines with hydrogen, and the less electronegative element combines with the OH group. Can you explain why this is to be expected? What would be the products of hydrolysis of ICl?

7-28. What is the electronic structure of sulfur dichloride, SCl_2? What value do you predict for the bond length? For the amount of ionic character of the S—Cl bond? For the electric dipole moment of the S—Cl bond? For the electric dipole moment of the SCl_2 molecule, assuming 102° for the bond angle?

7-29. What electronic structure would you assign to SF_6? To S_2F_{10}? Note that sulfur has 3d orbitals available, which can be hybridized with the 3s and 3p orbitals to form bond orbitals.

7-30. The enthalpy of formation of ClF(g) is -56 kJ mole^{-1}. What is the value of the Cl—F bond energy? (Answer: 256 kJ mole^{-1}.)

7-31. What is the structure of the ClF_3 molecule? What orbitals are used by the chlorine atom for outer unshared pairs of electrons, and what are used for bond orbitals?

7-32. The enthalpy of formation of ClF_3(g) is -162 kJ mole^{-1}.
    (*a*) What is the enthalpy of formation of ClF_3 from ClF and F_2?
    (*b*) What is the effective bond energy of the two additional Cl—F bonds?
    (*c*) What is the average bond energy of the three bonds? Why do you think the bonds are weaker than the bond in ClF?
    (Answer: -106; 132, 173 kJ mole $^{-1}$; use of 3d orbital.)

7-33. From the values of single-bond covalent radii of Sb, Te, and I given in Table 6-6, estimate a value for Xe. What is the predicted value for the Xe—F bond length? (By x-ray diffraction of XeF_4 crystals the experimental value 1.92 ± 0.03 Å has been determined.)

7-34. Predict values for the single-bond covalent radius of krypton and for the Kr—F bond length. (No experimental value has yet been reported.)

7-35. The values of the enthalpy of sublimation of hydrogen peroxide and hydrazine are 65 and 52 kJ mole^{-1}, respectively.
    (*a*) What do you estimate the enthalpy of sublimation of hydroxylamine to be? (No experimental value has been reported.)
    (*b*) From the standard enthalpy of formation of NH_2OH(c) given in Table 7-4 obtain that of NH_2OH(g). (Answer: -49 kJ mole^{-1}.)

7-36. (*a*) Using the standard enthalpy of formation -49 kJ mole^{-1} for NH_2OH(g) and the values for NH_3 and H_2O (Tables 7-4 and 7-1), compute the enthalpy of the reaction $NH_2OH(g) + H_2(g) \longrightarrow NH_3(g) + H_2O(g)$.
    (*b*) What bonds are broken and what are formed in the reaction? With use of the known bond-energy values evaluate the bond energy for N—O.
    (Answer: 179 kJ mole^{-1}.)

7-37. With use of the equation between bond energy and electronegativity, estimate a value for the N—O bond, and compare it with the value found in Exercise 7-36.

7-38. The engines in some large rockets are fueled with 1,1-dimethylhydrazine, $(CH_3)_2N$—NH_2, with liquid oxygen as oxidant. The products of combustion are $H_2O(g)$, $CO_2(g)$, and $N_2(g)$. Using bond-energy values, estimate a value of the enthalpy of formation of the fuel and from this obtain the heat of combustion. On a weight basis (fuel plus oxidant), is this combination better than hydrogen and oxygen? (Ignore heats of vaporization.)

7-39. Nitrogen and chlorine have the same electronegativity, 3.0, so that one might expect the heat evolved on formation of ammonia to be three times that of hydrogen chloride. But it is just half as great (46 vs. 92 kJ mole^{-1}). Why?

7-40. How do you account for the fact that although the atmosphere is mainly nitrogen, ocean salts contain almost no nitrogen compounds?

7-41. The compound $P_3N_3Cl_6$ (m.p. 114°C, b.p. 257°C) has been shown by x-ray diffraction to have a six-membered ring with P and N alternating, and two chlorine atoms attached to each phosphorus atom. Assign an electronic structure to the molecule.

8

Oxygen Compounds of Nonmetallic Elements

Of the 2468 inorganic compounds considered important enough to be listed in a table of physical properties in a reference book, over 49% (1220 compounds) involve nonmetallic elements bonded to oxygen atoms. Most of the oxygen-containing compounds of nonmetallic elements have transargononic structures, and it is the variety of these structures that makes a greater contribution than any other structural feature to the richness of inorganic chemistry.

In Section 8-1 of this chapter we shall examine the oxycompounds of chlorine, as an example. The following sections present a survey of the oxycompounds of other nonmetallic elements.

8-1.   The Oxycompounds of the Halogens

Chlorine forms a normal-valence oxide, Cl_2O $\left(\begin{array}{c} :\overset{\cdot\cdot}{\underset{|}{Cl}}: \\ :\overset{\cdot\cdot}{\underset{\cdot\cdot}{O}}-\overset{\cdot\cdot}{\underset{\cdot\cdot}{Cl}}: \end{array}\right)$, a normal-valence oxygen acid, $HClO$ $\left(\begin{array}{c} H \\ | \\ :\overset{}{\underset{\cdot\cdot}{O}}-\overset{\cdot\cdot}{\underset{\cdot\cdot}{Cl}}: \end{array}\right)$, and a rather large number of transargononic oxycompounds.

Dichlorine monoxide, Cl_2O, is a yellow gas obtained by passing chlorine over mercuric oxide:

$$2Cl_2(g) + HgO(c) \longrightarrow HgCl_2(c) + Cl_2O(g)$$

The gas condenses to a liquid at about 4°C. It is the anhydride of hypochlorous acid: that is, it reacts with water to give hypochlorous acid:

$$Cl_2O(g) + H_2O(l) \longrightarrow 2HClO(aq)$$

The standard enthalpy of $Cl_2O(g)$ is 24 kJ mole^{-1}.

An interesting transargononic chlorine oxide is *dichlorine heptoxide*, Cl_2O_7, a colorless liquid with melting point $-91°C$ and boiling point 82°C, which can be made by mixing tetraphosphorus decoxide, P_4O_{10}, and perchloric acid, $HClO_4$. We might write an argononic structural formula for Cl_2O_7:

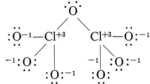

This formula is, however, unsatisfactory in that it places a large electric charge, $+3$, on each chlorine atom. Moreover, the electron-diffraction determination of the structure of the molecule has given the value 1.42 Å for the length of the six outer Cl—O bonds. This value is 0.28 Å less than the single-bond value 1.70 Å found in Cl_2O and the two central bonds in Cl_2O_7, and supports the assignment to the molecule of the transargononic structure

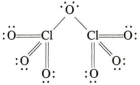

For this structure each chlorine atom has covalence 7, corresponding to its group in the periodic table. In the formation of seven covalent bonds the chlorine atom may make use of three $3d$ orbitals, together with its $3s$ and $3p$ orbitals.

The standard enthalpy of $Cl_2O_7(g)$ is 213 kJ mole^{-1}. The liquid is not a very sensitive explosive, but it explodes on percussion or ignition.

From the enthalpy values of the substances and of $O(g)$ (Table VIII-3) we write the equation

$$Cl_2O(g) + 6O(g) \longrightarrow Cl_2O_7(g) + 6 \times 216 \text{ kJ mole}^{-1}$$

The bond energy for a transargononic $Cl\!\!=\!\!O$ double bond is accordingly 216 kJ mole^{-1}.

Other experimental values of heats of reaction show that the value of the transargononic $Cl\!\!=\!\!O$ bond energy is nearly the same in other molecules as in Cl_2O_7. An example is the oxidation of ClF to ClO_3F:

$$ClF(g) + 3O(g) \longrightarrow ClO_3F(g) + 3 \times 239 \text{ kJ mole}^{-1}$$

The increase in value over Cl_2O_7 may be attributed to the greater ionic character of the Cl—F than of the Cl—O single bond, thus releasing an added part of an sp orbital and decreasing the promotion energy.

Oxidation Numbers of the Halogens

The halogens other than fluorine form stable compounds corresponding to nearly all values of the oxidation number from -1 to $+7$, as shown in the accompanying chart.

+7		$HClO_4$, Cl_2O_7		H_5IO_6
+6		Cl_2O_6		
+5		$HClO_3$	$HBrO_3$	HIO_3, I_2O_5
+4		ClO_2	BrO_2	IO_2
+3		$HClO_2$		
+2				
+1		$HClO$, Cl_2O	$HBrO$, Br_2O	HIO
0	F_2	Cl_2	Br_2	I_2
−1	HF, F$^-$	HCl, Cl$^-$	HBr, Br$^-$	HI, I$^-$

(Highly reactive molecules known only in the dilute gas phase or trapped in a crystal or supercooled liquid, such as OF and ClO, are not listed.)

The Oxygen Acids of Chlorine

From the preceding discussion it is not surprising that the transargononic oxygen acids $HClO_2$, $HClO_3$, and $HClO_4$ exist, as well as the normal-valence oxygen acid HClO (correctly written HOCl).

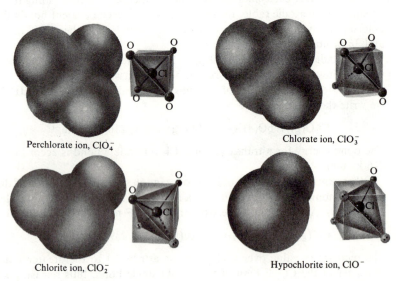

Perchlorate ion, ClO_4^- Chlorate ion, ClO_3^-

Chlorite ion, ClO_2^- Hypochlorite ion, ClO^-

FIGURE 8-1
The structure of ions of the four oxygen acids of chlorine.

These oxygen acids and their anions have the following names:

         $HClO_4$, perchloric acid ClO_4^-, perchlorate ion
         $HClO_3$, chloric acid ClO_3^-, chlorate ion
         $HClO_2$, chlorous acid ClO_2^-, chlorite ion
         $HClO$, hypochlorous acid ClO^-, hypochlorite ion

The structures of the four anions are shown in Figure 8-1.

The electronic structures shown below, which are in agreement with the electroneutrality principle but involve making use of the $3d$ orbitals for the chlorine atom (except for hypochlorous acid), may be assigned to the four acids:

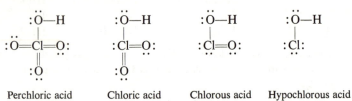

     Perchloric acid   Chloric acid   Chlorous acid   Hypochlorous acid

In the following sections these acids, their salts, and the oxides of chlorine are discussed in the order of increasing oxidation number of the halogen.

Hypochlorous Acid and the Hypochlorites

Hypochlorous acid, HClO, and most of its salts are known only in aqueous solution; they decompose when the solution is concentrated. A mixture of chloride ion and hypochlorite ion is formed when chlorine is bubbled through a solution of sodium hydroxide:

$$Cl_2 + 2OH^- \longrightarrow Cl^- + ClO^- + H_2O$$

A solution of *sodium hypochlorite*, NaClO, made in this way or by electrolysis of sodium chloride solution is a popular household sterilizing and bleaching agent. The hypochlorite ion is an active oxidizing agent, and its oxidizing power is the basis of its sterilizing and bleaching action.

Bleaching powder is a compound obtained by passing chlorine over calcium hydroxide:

$$Ca(OH)_2(c) + Cl_2(g) \longrightarrow CaCl(ClO)(c) + H_2O(l)$$

The formula CaCl(ClO), which approximates the composition of commercial bleaching powder, indicates it to be a calcium chloride-hypochlorite, containing the two anions Cl^- and ClO^-. Bleaching powder is a white, finely-powdered substance that usually smells of chlorine, because of its decomposition by water vapor in the air. It is often called by the incorrect name "chloride of lime." It is used as a household bleaching and sterilizing agent; in its former industrial use, for bleaching paper pulp and textile fabrics, it has been largely displaced by liquid chlorine. Pure *calcium hypochlorite*, Ca(ClO)$_2$, is also manufactured and used as a bleaching agent.

Hypochlorous acid is a weak acid. The solution obtained by adding another acid, such as sulfuric acid, to a solution of a hypochlorite contains molecules HClO, and very few hypochlorite ions ClO^-:

$$ClO^- + H^+ \longrightarrow HClO$$

Chlorous Acid and the Chlorites

When chlorine dioxide, ClO_2 (discussed below), is passed into a solution of sodium hydroxide or other alkali a chlorite ion and a chlorate ion are formed:

$$2ClO_2 + 2OH^- \longrightarrow ClO_2^- + ClO_3^- + H_2O$$

This is an auto-oxidation-reduction reaction, the chlorine with oxidation number $+4$ in chlorine dioxide being reduced and oxidized simultaneously to oxidation numbers $+3$ and $+5$. Pure sodium chlorite, NaClO$_2$, can be made by passing chlorine dioxide into a solution of sodium peroxide:

$$2ClO_2 + Na_2O_2 \longrightarrow 2Na^+ + 2ClO_2^- + O_2$$

In this reaction the peroxide oxygen serves as a reducing agent, decreasing the oxidation number of chlorine from $+4$ to $+3$.

Sodium chlorite is an active bleaching agent, used in the manufacture of textile fabrics.

Chloric Acid and Its Salts

Chloric acid, $HClO_3$, is an unstable acid that, like its salts, is a strong oxidizing agent. The most important salt of chloric acid is *potassium chlorate*, $KClO_3$, which is made by passing an excess of chlorine through a hot solution of potassium hydróxide or by heating a solution containing hypochlorite ion and potassium ion:

$$3ClO^- \longrightarrow ClO_3^- + 2Cl^-$$

The potassium chlorate can be separated from the potassium chloride formed in this reaction by crystallization, its solubility at low temperatures being much less than that of the chloride (3 g and 28 g, respectively, per 100 g of water at $0°C$). A cheaper way of making potassium chlorate is to electrolyze a solution of potassium chloride, using inert electrodes and keeping the solution mixed. The electrode reactions are

$$\text{Cathode reaction: } 2e^- + 2H_2O \longrightarrow 2OH^- + H_2$$
$$\text{Anode reaction: } Cl^- + 3H_2O \longrightarrow ClO_3^- + 6H^+ + 6e^-$$

In the stirred solution the hydroxide ions and the hydrogen ions are brought into contact with one another, and combine to form water. The over-all reaction is

$$Cl^- + 3H_2O \longrightarrow ClO_3^- + 3H_2$$

Potassium chlorate is a white crystalline substance, which is used as the oxidizing agent in matches and fireworks and in the manufacture of dyes.

A solution of the similar salt *sodium chlorate*, $NaClO_3$, is used as a weed killer. Potassium chlorate would be as good as sodium chlorate for this purpose; however, sodium salts are cheaper than potassium salts, and for this reason they are often used when only the anion is important. Sometimes the sodium salts have unsatisfactory properties, such as deliquescence (attraction of water from the air to form a solution), which make the potassium salts preferable for some uses, even though more expensive.

All the chlorates form sensitive explosive mixtures when mixed with reducing agents; *great care must be taken in handling them.* The use of sodium chlorate as a weed killer is attended with danger, because combustible material such as wood or clothing that has become saturated with the chlorate solution will ignite by friction after it has dried. Also, *it is very dangerous to grind a chlorate with sulfur, charcoal, or other reducing agent.*

Perchloric Acid and the Perchlorates

Potassium perchlorate, $KClO_4$, is made by heating potassium chlorate just to its melting point:

$$4KClO_3 \longrightarrow 3KClO_4 + KCl$$

At this temperature very little decomposition with evolution of oxygen occurs in the absence of a catalyst. Potassium perchlorate may also be made by long-continued electrolysis of a solution of potassium chloride, potassium hypochlorite, or potassium chlorate.

Potassium perchlorate and other perchlorates are oxidizing agents, somewhat less vigorous and less dangerous than the chlorates. Potassium perchlorate is used in explosives, such as the propellent powder of the bazooka and other rockets. This powder is a mixture of potassium perchlorate and carbon together with a binder; the equation for the principal reaction accompanying its burning is

$$KClO_4 + 4C \longrightarrow KCl + 4CO$$

Anhydrous magnesium perchlorate, $Mg(ClO_4)_2$, and barium perchlorate, $Ba(ClO_4)_2$, are used as drying agents (*desiccants*). These salts have a very strong attraction for water. Nearly all the perchlorates are highly soluble in water; potassium perchlorate is exceptional for its low solubility, 0.75 g in 100 g of water at 0°C. Sodium perchlorate, $NaClO_4$, made by the electrolytic method, is used as a weed killer; it is safer than sodium chlorate. In general the mixtures of perchlorates with oxidizable materials are less dangerous than the corresponding mixtures of chlorates.

Perchloric acid, $HClO_4 \cdot H_2O$, is a colorless liquid made by distilling, under reduced pressure, a solution of a perchlorate to which sulfuric acid has been added. The perchloric acid distills as the monohydrate, and it cools to form crystals of the monohydrate. These crystals are isomorphous with ammonium perchlorate, NH_4ClO_4, and the substance is presumably hydronium perchlorate, $(H_3O)^+(ClO_4)^-$.

Chlorine Oxides

The oxides ClO, ClO_2, ClO_3 (or Cl_2O_6), Cl_2O_7, and ClO_4 (possibly Cl_2O_8) are known, in addition to the normal-valence oxide Cl_2O, which has been discussed above.

Chlorine monoxide, ClO, has been characterized by analysis of its band spectrum. Its bond length, 1.55 Å, is intermediate between that for a single bond, 1.69 Å, and that for a double bond, about 1.42 Å (as in Cl_2O_7). Its bond energy, 269 kJ mole^{-1}, is 59 kJ mole^{-1} greater than that for the Cl—O single bond. This additional bond energy is ascribed to the formation of a three-electron bond, in addition to a single bond, and the electronic structure is written as $:\overset{..}{\underset{..}{Cl}}\overset{..}{\cdots}\overset{..}{\underset{..}{O}}:$, corresponding to resonance between the two structures $:\overset{..}{\underset{..}{Cl}}\!-\!\overset{..}{\underset{..}{O}}:$ and $:\overset{.}{\underset{..}{Cl}}\!-\!\overset{..}{\underset{..}{O}}:$.

Chlorine dioxide, ClO_2, is the only known compound of quadripositive chlorine. It is a reddish-yellow gas, which is very explosive, decomposing readily to chlorine and oxygen. The violence of this decomposition makes it very dangerous to add sulfuric acid or any other strong acid to a chlorate or to any dry mixture containing a chlorate.

Chlorine dioxide can be made by carefully adding sulfuric acid to potassium chlorate, $KClO_3$. It would be expected that this mixture would react to produce chloric acid, $HClO_3$, and then, because of the dehydrating power of sulfuric acid, to produce the anhydride of chloric acid, Cl_2O_5:

$$KClO_3 + H_2SO_4 \longrightarrow KHSO_4 + HClO_3$$
$$2HClO_3 \longrightarrow H_2O + Cl_2O_5$$

Dichlorine pentoxide, Cl_2O_5, however, is very unstable—its existence has never been verified. If it is formed at all, it decomposes at once to give chlorine dioxide and oxygen:

$$2Cl_2O_5 \longrightarrow 4ClO_2 + O_2$$

The molecule has a triangular structure, with O—Cl—O angle 118° and

bond lengths 1.49 Å. We assign it the structure $\overset{\displaystyle :\overset{..}{O}}{\underset{..}{\overset{\|}{Cl}}\cdots O:}$, with inter-

change of the two kinds of bonds (resonance).

From the value 269 kJ mole^{-1} for the $Cl\overset{\cdots}{}O$ bond energy of ClO and the value 216 kJ mole^{-1} for the transargononic Cl=O in Cl_2O_7 we would expect about -485 kJ mole^{-1} for the enthalpy of $ClO_2(g)$ relative to $Cl(g)$ and 2 O(g); the experimental value is -497 kJ mole^{-1}.

The Oxygen Acids and Oxides of Bromine

Bromine forms only two stable oxygen acids—hypobromous acid and bromic acid—and their salts:

HBrO, hypobromous acid   KBrO, potassium hypobromite
HBrO₃, bromic acid   KBrO₃, potassium bromate

Their preparation and properties are similar to those of the corresponding compounds of chlorine. They are somewhat weaker oxidizing agents than their chlorine analogues.

The bromite ion, BrO_2^-, has been reported to exist in solution. During many years no effort to prepare perbromic acid or any perbromate had succeeded; the preparation of perbromic acid has recently been reported.

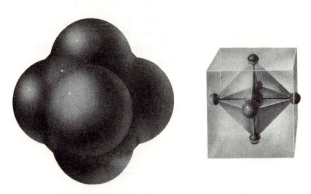

FIGURE 8-2
The periodate ion, IO_6^{5-}.

Three very unstable oxides of bromine, Br_2O, BrO_2, and Br_3O_8, have been described. The structure of Br_3O_8 is not known.

None of the oxygen compounds of bromine has found important practical use.

The Oxygen Acids and Oxides of Iodine

Iodine reacts with hydroxide ion in cold alkaline solution to form the *hypoiodite ion*, IO^-, and iodide ion:

$$I_2 + 2OH^- \longrightarrow IO^- + I^- + H_2O$$

The solution when warmed reacts further to form iodate ion, IO_3^-:

$$3IO^- \longrightarrow IO_3^- + 2I^-$$

The salts of hypoiodous acid and iodic acid may be made in these ways. *Iodic acid* itself, HIO_3, is usually made by oxidizing iodine with concentrated nitric acid:

$$I_2 + 10HNO_3 \longrightarrow 2HIO_3 + 10NO_2 + 4H_2O$$

Iodic acid is a white solid, which is only very slightly soluble in concentrated nitric acid; it accordingly separates out during the course of the reaction. Its principal salts, potassium iodate, KIO_3, and sodium iodate, $NaIO_3$, are white crystalline solids.

Periodic acid has the normal formula H_5IO_6, with an octahedral arrangement of the oxygen atoms around the iodine atom, as shown in Figure 8-2. This difference in composition from its analogue perchloric acid, $HClO_4$, results from the large size of the iodine atom, which permits this atom to coordinate six oxygen atoms about itself, instead of four. The ligancy of iodine in periodic acid is hence 6.

There exists a series of periodates corresponding to the formula H_5IO_6 for periodic acid, and also a series corresponding to HIO_4. Salts of the first series are dipotassium trihydrogen periodate, $K_2H_3IO_6$, silver periodate, Ag_5IO_6, and so on. Sodium periodate, $NaIO_4$, a salt of the second series, occurs in small amounts in crude Chile saltpeter.

The two forms of periodic acid, H_5IO_6 and HIO_4 (the latter being unstable, but forming stable salts), represent the same oxidation state of iodine, $+7$. The equilibrium between the two forms is a hydration reaction:

$$HIO_4 + 2H_2O \rightleftharpoons H_5IO_6$$

Iodine pentoxide, I_2O_5, is obtained as a white powder by gently heating either iodic acid or periodic acid:

$$2HIO_3 \longrightarrow I_2O_5 + H_2O$$
$$2H_5IO_6 \longrightarrow I_2O_5 + 5H_2O + O_2$$

The anhydride of periodic acid, I_2O_7, seems not to be stable; its preparation has never been reported.

The lower oxide of iodine, IO_2, can be made by treating an iodate with concentrated sulfuric acid and then adding water. This oxide is a paramagnetic yellow solid.

The Oxidizing Strength of the Oxygen Compounds of the Halogens

Elementary fluorine, F_2, is able to oxidize the halogen ions of its congeners to the free halogens, by reactions such as

$$F_2 + 2Cl^- \longrightarrow 2F^- + Cl_2$$

Fluorine is more electronegative than the other elements, and it accordingly is able to take electrons away from the anions of these elements. Similarly, chlorine is able to oxidize both bromide ion and iodide ion, and bromine is able to oxidize iodide ion:

$$Cl_2 + 2Br^- \longrightarrow 2Cl^- + Br_2$$
$$Cl_2 + 2I^- \longrightarrow 2Cl^- + I_2$$
$$Br_2 + 2I^- \longrightarrow 2Br^- + I_2$$

The order of strength as an oxidizing agent for the elementary halogens is accordingly $F_2 > Cl_2 > Br_2 > I_2$.

At first sight there seems to be an anomaly in the reactions involving the free halogens and their oxygen compounds. Thus, although chlorine is able to liberate iodine from iodide ion, iodine is able to liberate chlorine from chlorate ion, according to the reaction

$$I_2 + 2ClO_3^- \longrightarrow 2IO_3^- + Cl_2$$

In this reaction, however, it is to be noted that elementary iodine is acting as a reducing agent, rather than as an oxidizing agent. During the course of the reaction the oxidation number of iodine is increased, from 0 to +5, and that of chlorine is decreased, from +5 to 0. The direction in which the reaction takes place predominantly is accordingly that which would be predicted by the electronegativity scale; iodine, the heavier halogen and less electronegative element, tends to have a high positive oxidation number, and chlorine tends to have a low oxidation number. (Remember that in this case, as in nearly all chemical reactions, we may be dealing with chemical equilibrium. The foregoing statement is to be interpreted as meaning that at equilibrium there are present in the system larger amounts of iodate ion and free chlorine than of chlorate ion and free iodine.)

Similarly, hypochlorite ion, ClO^-, can oxidize bromine to hypobromite ion, and hypobromite ion can oxidize iodine to hypoiodite ion. These regularities fail, however, for the higher oxidation states of bromine: $HBrO_2$, $HBrO_3$, and $HBrO_4$ are very much less stable than their chlorine and iodine analogs. No satisfactory explanation of this property of bromine has been advanced. Selenium and arsenic in their high oxidation states show somewhat similar deviations in properties from their lighter and heavier congeners.

8-2. Oxycompounds of Sulfur, Selenium, and Tellurium

The transargononic oxycompounds of sulfur are more stable than those of chlorine, and those of phosphorus are still more stable. Perchloric acid and the perchlorates are strong oxidizing agents, whereas sulfuric acid and the sulfates are weak oxidizing agents, and phosphoric acid and the phosphates are still weaker. This difference in properties corresponds to the electronegativity values $x = 3$ for Cl, 2.5 for S, 2.1 for P, with Δx (relative to oxygen) = 0.5 for Cl, 1.0 for S, 1.4 for P. The following representative heats of reaction reflect the increasing values of Δx:

$$HCl(g) + 2O_2(g) \longrightarrow HClO_4(l) + 8 \text{ kJ mole}^{-1}$$
$$H_2S(g) + 2O_2(g) \longrightarrow H_2SO_4(l) + 790 \text{ kJ mole}^{-1}$$
$$H_3P(g) + 2O_2(g) \longrightarrow H_3PO_4(l) + 1250 \text{ kJ mole}^{-1}$$

The stable compounds of sulfur, selenium, and tellurium correspond to several values of oxidation number, from -2 to $+6$, as shown in the chart:

$+6$	SO_3, H_2SO_4, SF_6	H_2SeO_4, SeF_6	TeO_3, $Te(OH)_6$, TeF_6
$+4$	SO_2, H_2SO_3	SeO_2, H_2SeO_3	TeO_2
$+2$			
0	S_8, S_6, S_2	Se	Te
-2	H_2S, S^{--}	H_2Se	H_2Te

Oxides of Sulfur

The normal-valence oxide of sulfur, SO, is much less stable than the transargononic oxides SO_2 and SO_3. The heats of formation have the following values:

$$\tfrac{1}{8}S_8(c) + \tfrac{1}{2}O_2(g) \longrightarrow SO(g) - 7 \text{ kJ mole}^{-1}$$
$$\tfrac{1}{8}S_8(c) + O_2(g) \longrightarrow SO_2(g) + 297 \text{ kJ mole}^{-1}$$
$$\tfrac{1}{8}S_8(c) + \tfrac{3}{2}O_2(g) \longrightarrow SO_3(g) + 396 \text{ kJ mole}^{-1}$$

From the first two equations we see that the decomposition of sulfur monoxide to sulfur dioxide and sulfur is strongly exothermic:

$$2SO(g) \longrightarrow \tfrac{1}{8}S_8(c) + SO_2(g) + 311 \text{ kJ mole}^{-1}$$

It is accordingly not surprising that sulfur monoxide is not known as a stable substance, but only as a highly reactive molecule in a very dilute gas or frozen in a matrix. It has the structure $:S\!\cdots\!O:$, with two electrons with parallel spins, thus resembling the molecules O_2 and S_2.

Sulfur dioxide, SO_2, is formed by burning sulfur or a sulfide, such as pyrite (FeS_2):

$$S + O_2 \longrightarrow SO_2$$
$$4FeS_2 + 11O_2 \longrightarrow 2Fe_2O_3 + 8SO_2$$

It is colorless, and has a characteristic choking odor. Its melting point is $-75°C$ and its boiling point $-10°C$.

Sulfur dioxide is conveniently made in the laboratory by adding a strong acid to solid sodium hydrogen sulfite:

$$H_2SO_4 + NaHSO_3 \longrightarrow NaHSO_4 + H_2O + SO_2$$

It may be purified and dried by bubbling it through concentrated sulfuric acid.

The electronic structure of sulfur dioxide is

In this structure use is made of one $3d$ orbital, as well as the $3s$ orbital and the three $3p$ orbitals. The observed sulfur-oxygen bond length, 1.43 Å, is a little less than the value 1.49 Å expected for a double bond. The angle O—S—O has the value 119.5°.

Sulfur dioxide is used in great quantities in the manufacture of sulfuric acid, sulfurous acid, and sulfites. It destroys fungi and bacteria, and is used as a preservative in the preparation of dried prunes, apricots, and other fruits. A solution of calcium hydrogen sulfite, $Ca(HSO_3)_2$, made by reaction of sulfur dioxide and calcium hydroxide, is used in the manufacture of paper pulp from wood. The solution dissolves lignin, a substance that cements the cellulose fibers together, and liberates these fibers, which are then processed into paper.

Sulfur trioxide, SO_3, is formed in very small quantities when sulfur is burned in air. It is usually made by oxidation of sulfur dioxide by air, in the presence of a catalyst. The reaction of its formation from the elements is exothermic, but less exothermic, per oxygen atom, than that of sulfur dioxide. The equilibrium

$$SO_2(g) + \tfrac{1}{2}O_2(g) \rightleftharpoons SO_3(g)$$

is such that at low temperatures a satisfactory yield of SO_3 can be obtained; the reaction proceeds nearly to completion. The rate of the reaction, however, is so small at low temperatures as to make the direct combination of the substances unsuitable as a commercial process. At higher temperatures, at which the rate is satisfactory, the yield is low because of the unfavorable equilibrium.

The solution to this problem was the discovery of certain catalysts (platinum, vanadium pentoxide) that speed up the reaction without affecting the equilibrium. The catalyzed reaction proceeds not in the gaseous mixture, but on the surface of the catalyst, as the gas molecules strike it. In practice, sulfur dioxide, made by burning sulfur or pyrite, is mixed with air and passed over the catalyst at a temperature of 400° to 450°C. About 99% of the sulfur dioxide is converted into sulfur trioxide under these conditions. It is used mainly in the manufacture of sulfuric acid.

Sulfur trioxide is a corrosive gas, which combines vigorously with water to form sulfuric acid:

$$SO_3(g) + H_2O(l) \longrightarrow H_2SO_4(l) + 130 \text{ kJ mole}^{-1}$$

It also dissolves readily in sulfuric acid, to form *oleum* or *fuming sulfuric acid*, which consists mainly of disulfuric acid, $H_2S_2O_7$ (also called pyrosulfuric acid):

$$SO_3 + H_2SO_4 \rightleftharpoons H_2S_2O_7$$

Sulfur trioxide condenses at 44.5°C to a colorless liquid, which freezes at

16.8°C to transparent crystals. The substance is polymorphous, these crystals being the unstable form (the α-form). The stable form consists of silky asbestos-like crystals, which are produced when the α-crystals or the liquid stands for some time, especially in the presence of a trace of moisture. There exist also one or more other forms of this substance, which are hard to investigate because the changes from one form to another are very slow. The asbestos-like crystals slowly evaporate to SO_3 vapor at temperatures above 50°C.

The sulfur trioxide molecule, in the gas phase, the liquid, and the α-crystals, has the electronic structure

The molecule is planar, and the bonds have the same length, 1.43 Å, as in the sulfur dioxide molecule.

The properties of sulfur trioxide may be in large part explained as resulting from the instability of the sulfur-oxygen double bond, relative to two single bonds. Thus by reaction with water one double bond of sulfur trioxide can be replaced by two single bonds, in sulfuric acid:

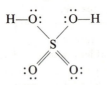

The increased stability of the product is reflected in the large amount of heat evolved in the reaction. A second sulfur trioxide molecule can eliminate one double bond by combining with a molecule of sulfuric acid to form a molecule of disulfuric acid:

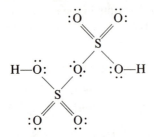

Similarly, molecules of trisulfuric acid, $H_2S_3O_{10}$, tetrasulfuric acid, $H_2S_4O_{13}$, and others can be formed (see Figure 8-3), culminating in a chain $HO_3SO(SO_3)_\infty SO_3H$ of nearly infinite length—essentially a high polymer

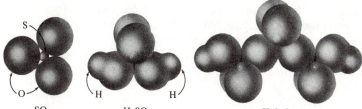

FIGURE 8-3
Sulfur trioxide and some oxygen acids of sulfur.

of sulfur trioxide, $(SO_3)_x$, with x large. It is these very long molecules that constitute the asbestos-like crystalline form of sulfur trioxide. We can understand why the crystals are fibrous, like asbestos—they consist of extremely long chain molecules, arranged together side by side, but easily separated into fibers, because, although the chains themselves are strong, the forces between them are relatively weak.

The molecular structures explain why the formation of the fibrous crystals and their decomposition to SO_3 vapor are slow processes, whereas crystallization and evaporation are usually rapid. In this case these processes are really *chemical reactions*, involving the formation of new chemical bonds. The role of a trace of water in catalyzing the formation of the fibrous crystals can also be understood; the molecules of water serve to start the chains, which can then grow to great length.

In the long-chain polymer of sulfur trioxide (Figure 8-3) the sulfur-oxygen distance for the oxygen atoms bonded to two sulfur atoms in the chain is about 1.62 Å, and that for the oxygen atoms bonded to one sulfur atom is about 1.43 Å. These two bond lengths are a little less than the single-bond and double-bond values given by the covalent radii. We are accordingly led to assign to the sulfur trioxide polymer the structure

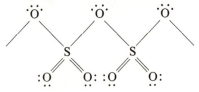

In this structure each sulfur atom is indicated as forming six bonds; the bond orbitals are hybrids of one 3*s* orbital, three 3*p* orbitals, and two 3*d* orbitals.

Sulfurous Acid

A solution of sulfurous acid, H_2SO_3, is obtained by dissolving sulfur dioxide in water. Both sulfurous acid and its salts, the sulfites, are active reducing agents. They form sulfuric acid, H_2SO_4, and sulfates on oxidation by oxygen, the halogens, hydrogen peroxide, and similar oxidizing agents.

The structure of sulfurous acid is

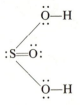

Sulfuric Acid and the Sulfates

Sulfuric acid, H_2SO_4, is one of the most important of all chemicals, finding use throughout the chemical industry and related industries. It is a heavy, oily liquid (density 1.838 g cm^{-3}), which fumes slightly in air, as the result of the liberation of traces of sulfur trioxide, which then combine with water vapor to form droplets of sulfuric acid. When heated, pure sulfuric acid yields a vapor rich in sulfur trioxide, and then boils, at 338°C, with the constant composition 98% H_2SO_4, 2% water. This is the ordinary "concentrated sulfuric acid" of commerce.

Concentrated sulfuric acid is very corrosive. It has a strong affinity for water, and a large amount of heat is liberated when it is mixed with water, as the result of the formation of hydronium ion:

$$H_2SO_4 + 2H_2O \rightleftharpoons 2H_3O^+ + SO_4^{--}$$

In diluting it, the concentrated acid should be poured into water in a thin stream, with stirring; *water should never be poured into the acid,* because it is apt to spatter and throw drops of acid out of the container. The diluted acid occupies a smaller volume than its constituents, the effect being a maximum at $H_2SO_4 + 2H_2O$ [$(H_3O)_2^+(SO_4)^{--}$].

The crystalline phases that form on cooling sulfuric acid containing varying amounts of sulfur trioxide or water are $H_2S_2O_7$, H_2SO_4, $H_2SO_4 \cdot H_2O$ [presumably $(H_3O)^+(HSO_4)^-$], $H_2SO_4 \cdot 2H_2O$ [$(H_3O)_2^+(SO_4)^{--}$], and $H_2SO_4 \cdot 4H_2O$.

The Manufacture of Sulfuric Acid

Sulfuric acid is made by two processes, the *contact process* and the *lead-chamber process*, which are now about equally important.

In the contact process sulfur trioxide is made by the catalytic oxidation of sulfur dioxide (the name of the process refers to the fact that reaction occurs on contact of the gases with the solid catalyst). The catalyst formerly used was finely divided platinum; it has now been largely replaced by vanadium pentoxide, V_2O_5. The gas containing sulfur trioxide is then bubbled through sulfuric acid, which absorbs the sulfur trioxide. Water is added at the proper rate, and 98% acid is drawn off.

In the lead-chamber process oxygen, sulfur dioxide, nitric oxide, and a small amount of water vapor are introduced into a large lead-lined chamber. White crystals of nitrosulfuric acid, $NOHSO_4$ (sulfuric acid in which one hydrogen ion is replaced by the nitronium ion, $:N{\equiv}O:^+$), are formed. When steam is introduced the crystals react to form drops of sulfuric acid, liberating oxides of nitrogen. In effect, the oxides of nitrogen serve to catalyze the oxidation of sulfur dioxide by oxygen. The complex reactions that occur may be summarized as

$$2SO_2 + NO + NO_2 + O_2 + H_2O \longrightarrow 2NOHSO_4$$
$$2NOHSO_4 + H_2O \longrightarrow 2H_2SO_4 + NO + NO_2$$

The oxides of nitrogen, NO and NO_2, that take part in the first reaction are released by the second reaction, and can serve over and over again.

The Chemical Properties and Uses of Sulfuric Acid

The uses of sulfuric acid are determined by its chemical properties, as an acid, a dehydrating agent, and an oxidizing agent.

Sulfuric acid has a high boiling point, 330°C, which permits it to be used with salts of more volatile acids in the preparation of these acids. Nitric acid, for example, can be made by heating sodium nitrate with sulfuric acid:

$$NaNO_3 + H_2SO_4 \longrightarrow NaHSO_4 + HNO_3$$

The nitric acid distills off at 86°C. Sulfuric acid is also used for the manufacture of soluble phosphate fertilizers, of ammonium sulfate for use as a fertilizer, of other sulfates, and in the manufacture of many chemicals and drugs. Steel is usually cleaned of iron rust (is "pickled") by immersion in a bath of sulfuric acid before it is coated with zinc, tin, or enamel. Sulfuric acid is used as the electrolyte in ordinary lead sulfate electric storage cells.

Sulfuric acid has such a strong affinity for water as to make it an effective dehydrating agent. Gases that do not react with the substance may be dried by being bubbled through sulfuric acid. The dehydrating power of the concentrated acid is great enough to cause it to remove hydrogen and oxygen as water from organic compounds, such as sugar:

$$C_{12}H_{22}O_{11} \longrightarrow 12C + 11H_2O$$
$$\text{Sugar (sucrose) } H_2SO_4$$

Many explosives, such as glyceryl trinitrate (nitroglycerine), are made by reaction of organic substances with nitric acid, producing the explosive substance and water:

$$C_3H_5(OH)_3 + 3HNO_3 \longrightarrow C_3H_5(NO_3)_3 + 3H_2O$$
$$\text{Glycerine} \qquad H_2SO_4 \text{ Glyceryl trinitrate}$$

These reversible reactions are made to proceed to the right by mixing the nitric acid with sulfuric acid, which by its dehydrating action favors the products.

Hot concentrated sulfuric acid is an effective oxidizing agent, the product of its reduction being sulfur dioxide. It will dissolve copper, and will even oxidize carbon:

$$Cu + 2H_2SO_4 \longrightarrow CuSO_4 + 2H_2O + SO_2$$
$$C + 2H_2SO_4 \longrightarrow CO_2 + 2H_2O + 2SO_2$$

The solution of copper by hot concentrated sulfuric acid illustrates a general reaction—*the solution of an unreactive metal in an acid under the influence of an oxidizing agent*. The reactive metals are oxidized to their cations by hydrogen ion, which is itself reduced to elementary hydrogen; for example,

$$Zn + 2H^+ \longrightarrow Zn^{++} + H_2(g)$$

Copper does not undergo this reaction. It can be oxidized to cupric ion, however, by a stronger oxidizing agent, such as chlorine or nitric acid or, as illustrated above, hot concentrated sulfuric acid.

Sulfates

Sulfuric acid combines with bases to form normal sulfates, such as K_2SO_4, potassium sulfate, and hydrogen sulfates or acid sulfates, such as $KHSO_4$, potassium hydrogen sulfate.

The less soluble sulfates occur as minerals: these include $CaSO_4 \cdot 2H_2O$ (gypsum), $SrSO_4$, $BaSO_4$ (barite), and $PbSO_4$. Barium sulfate is the least soluble of the sulfates, and its formation as a white precipitate is used as a test for sulfate ion.

Common soluble sulfates include $Na_2SO_4 \cdot 10H_2O$, $(NH_4)_2SO_4$, $MgSO_4 \cdot 7H_2O$ (Epsom salt), $CuSO_4 \cdot 5H_2O$ (blue vitriol), $FeSO_4 \cdot 7H_2O$, $(NH_4)_2Fe(SO_4)_2 \cdot 6H_2O$ (a well-crystallized, easily purified salt used in analytical chemistry in making standard solutions of ferrous ion), $ZnSO_4 \cdot 7H_2O$, $KAl(SO_4)_2 \cdot 12H_2O$ (alum), $NH_4Al(SO_4)_2 \cdot 12H_2O$ (ammonium alum), and $KCr(SO_4)_2 \cdot 12H_2O$ (chrome alum).

The Peroxysulfuric Acids

Sulfuric acid contains sulfur in its highest oxidation state. When a strong oxidizing agent (hydrogen peroxide or an anode at suitable electric potential) acts on sulfuric acid, the only oxidation that can occur is that of the oxygen atoms, from -2 to -1. The products of this oxidation, peroxysulfuric acid, H_2SO_5, and peroxydisulfuric acid, $H_2S_2O_8$, have the following structures:

Peroxysulfuric acid, H_2SO_5

Peroxydisulfuric acid, $H_2S_2O_8$

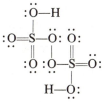

These acids and their salts are used as bleaching agents.

The Thio or Sulfo Acids

Sodium thiosulfate, $Na_2S_2O_3 \cdot 5H_2O$ (incorrectly called "hypo," from an old name, sodium hyposulfite), is a substance used in photography. It is made by boiling a solution of sodium sulfite with free sulfur:

$$SO_3^{--} + S \longrightarrow S_2O_3^{--}$$

Sulfite ion   Thiosulfate ion

Thiosulfuric acid, $H_2S_2O_3$, is unstable, and sulfur dioxide and sulfur are formed when a thiosulfate is treated with acid.

The structure of the thiosulfate ion, $S_2O_3^{--}$, is interesting in that the two sulfur atoms are not equivalent. This ion is a sulfate ion, SO_4^{--}, in which one of the oxygen atoms has been replaced by a sulfur atom. The central sulfur atom may be assigned oxidation number $+6$, and the attached sulfur atom oxidation number -2.

Thiosulfate ion is easily oxidized, especially by iodine, to tetrathionate ion, $S_4O_6^{--}$:

$$2S_2O_3^{--} \longrightarrow S_4O_6^{--} + 2e^-$$

or

$$2S_2O_3^{--} + I_2 \longrightarrow S_4O_6 + 2I^-$$

This reaction, between thiosulfate ion and iodine, is very useful in the quantitative analysis of oxidizing and reducing agents. The structure of tetrathionate ion is shown in Figure 8-4; it contains a disulfide group —S—S— in place of the peroxide group of the peroxydisulfate ion. The oxidation of thiosulfate ion to tetrathionate ion is analogous to the oxidation of sulfide ion to disulfide ion:

$$2S^{--} \longrightarrow S_2^{--} + 2e^-$$

Thiosulfuric acid is representative of a general class of acids, called *thio acids* or *sulfo acids*, in which one or more oxygen atoms of an oxygen acid are replaced by sulfur atoms. For example, diarsenic pentasulfide dissolves in a sodium sulfide solution to form the thioarsenate ion, AsS_4^{---}, completely analogous to the arsenate ion, AsO_4^{---}:

$$As_2S_5 + 3S^{--} \longrightarrow 2AsS_4^{---}$$

Diarsenic trisulfide also dissolves, to form the thioarsenite ion:

$$As_2S_3 + 3S^{--} \longrightarrow 2AsS_3^{---}$$

If disulfide ion, S_2^{--}, is present in the solution, the thioarsenite ion is oxidized to thioarsenate ion:

$$AsS_3^{---} + S_2^{--} \longrightarrow AsS_4^{---} + S^{--}$$

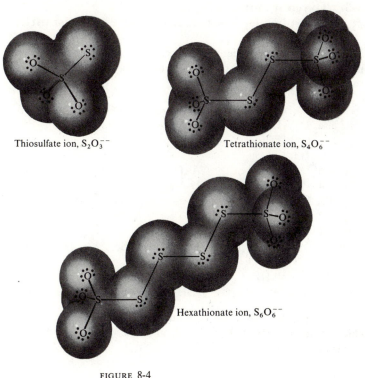

Thiosulfate ion, $S_2O_3^{--}$

Tetrathionate ion, $S_4O_6^{--}$

Hexathionate ion, $S_6O_6^{--}$

FIGURE 8-4
The thiosulfate ion and related ions.

An alkaline solution of sodium sulfide and sodium disulfide (or of the ammonium sulfides) is used in the usual systems of qualitative analysis as a means of separating the precipitated sulfides of certain metals and metalloids. This separation depends upon the capability of certain sulfides (HgS, As_2S_3, As_2S_5, Sb_2S_3, Sb_2S_5, SnS, SnS_2) to form thio anions (HgS_2^{--}, AsS_4^{---}, SbS_4^{---}, SnS_4^{----}), whereas others (Ag_2S, PbS, Bi_2S_3, CuS, CdS) remain undissolved.

Selenium and Tellurium

The transargononic compounds of selenium closely resemble those of sulfur. The selenates, salts of selenic acid, H_2SeO_4, are much like the sulfates. Telluric acid, however, has the formula $Te(OH)_6$, in which the large central atom has increased its ligancy from 4 to 6, as the iodine atom does in H_5IO_6.

8-3.   Oxycompounds of Phosphorus, Arsenic, Antimony, and Bismuth

Phosphorus and its heavier congeners form stable compounds correspond-ing to various values of the oxidation number between -3 and $+5$, as shown in the following chart:

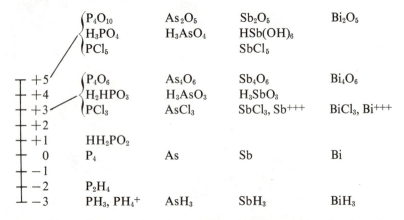

In addition, the properties of a number of highly reactive simple mole-cules, including PH, PH_2, PO, PS, and PN, have been determined by spectroscopic studies.

Oxides of Phosphorus

Tetraphosphorus hexoxide, P_4O_6, is made by burning phosphorus in a limited supply of oxygen or air. It has melting point 22.5°C and boiling point 173.1°C. Its molecular structure, shown in Figure 8-5, is that of a normal-valence compound. Its standard enthalpy, -2145 kJ mole^{-1} for $P_4O_6(g)$, leads to 415 kJ mole^{-1} for the P—O bond energy. The observed bond length is 1.66 Å, with bond angles 99° (O–P–O) and 128° (P–O–P).

Tetraphosphorus decoxide, P_4O_{10}, is formed when phosphorus is burned with a free supply of air. It reacts with water with great violence, to form phosphoric acid, and it is used in the laboratory as a drying agent for gases; it is the most efficient drying agent known. Its standard enthalpy, -2834 kJ mole^{-1} for $P_4O_{10}(g)$, corresponds to the value 584 kJ mole^{-1} for the energy of the transargononic P=O double bond:

$$P_4O_6(g) + 4O(g) \longrightarrow P_4O_{10}(g) + 4 \times 584 \text{ kJ mole}^{-1}$$

The molecule has the structure shown in Figure 8-5, with bond lengths P—O = 1.60 Å, P=O = 1.40 Å.

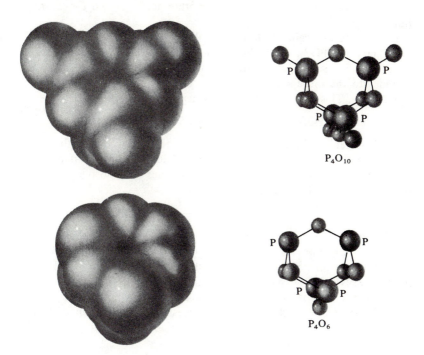

FIGURE 8-5
Molecules of the oxides of phosphorus.

Phosphoric Acid

The most important acid of phosphorus is phosphoric acid, H_3PO_4 (also called *orthophosphoric acid*). Pure phosphoric acid is a deliquescent crystalline substance with melting point 42°C. It is made by dissolving tetraphosphorus decoxide in water. Commercial phosphoric acid (86% H_3PO_4) is a viscous liquid.

Phosphoric acid is a weak acid. It is a stable substance, without effective oxidizing power.

Orthophosphoric acid forms three series of salts, with one, two, and three of its hydrogen atoms replaced by metal atoms. The salts are usually made by mixing phosphoric acid and the metal hydroxide or carbonate, in proper proportion. Sodium dihydrogen phosphate, NaH_2PO_4, is slightly acidic in reaction. It is used (mixed with sodium hydrogen carbonate) in baking powder, and also for treating boiler water to prevent formation of scale. Disodium hydrogen phosphate, Na_2HPO_4, is slightly basic in reaction. Trisodium phosphate, Na_3PO_4, is strongly basic. It is used as a detergent (for cleaning woodwork and other surfaces) and for treating boiler water.

Phosphates are valuable fertilizers. Phosphate rock itself (tricalcium phosphate, $Ca_3(PO_4)_2$, and hydroxy-apatite) is too slightly soluble to serve as an effective source of phosphorus for plants. It is accordingly converted into the more soluble substance calcium dihydrogen phosphate, $Ca(H_2PO_4)_2$. This may be done by treatment with sulfuric acid:

$$Ca_3(PO_4)_2 + 2H_2SO_4 \longrightarrow 2CaSO_4 + Ca(H_2PO_4)_2$$

Enough water is added to convert the calcium sulfate to its dihydrate, gypsum, and the mixture of gypsum and calcium dihydrogen phosphate is sold as "superphosphate of lime." Sometimes the phosphate rock is treated with phosphoric acid:

$$Ca_3(PO_4)_2 + 4H_3PO_4 \longrightarrow 3Ca(H_2PO_4)_2$$

This product is much richer in phosphorus than the "superphosphate"; it is called "triple phosphate." Over ten million tons of phosphate rock is converted into phosphate fertilizer each year. Ammonium dihydrogen phosphate, $NH_4H_2PO_4$, has recently come into use.

Phosphoric acid is present in the nucleic acids, which are of fundamental importance in the biological process of reproduction of living organisms.

The Condensed Phosphoric Acids

Phosphoric acid easily undergoes the process of *condensation*. Condensation is the reaction of two or more molecules to form larger molecules, either without any other products (in which case the condensation is also called *polymerization*), or with the elimination of small molecules, such as water. Condensation of two phosphoric acid molecules occurs by the reaction of two hydroxyl groups to form water and an oxygen atom held by single bonds to two phosphorus atoms.

When orthophosphoric acid is heated it loses water and condenses to *diphosphoric acid* or *pyrophosphoric acid*, $H_4P_2O_7$:

$$2H_3PO_4 \rightleftharpoons H_4P_2O_7 + H_2O$$

(The name pyrophosphoric acid is the one customarily used.) This acid is a white crystalline substance, with melting point 61°C. Its salts may be made by neutralization of the acid or by strongly heating the hydrogen orthophosphates or ammonium orthophosphates of the metals. Magnesium pyrophosphate, $Mg_2P_2O_7$, is obtained in a useful method for quantitative analysis for either magnesium or orthophosphate. A solution containing orthophosphate ion may be mixed with a solution of magnesium chloride (or sulfate), ammonium chloride, and ammonium hydroxide. The very slightly soluble substance magnesium ammonium phosphate, $MgNH_4PO_4 \cdot 6H_2O$, then slowly precipitates. The precipitate is washed with dilute ammonium hydroxide, dried, and heated to a dull red heat, causing it to form magnesium pyrophosphate, which is then weighed:

$$2MgNH_4PO_4 \cdot 6H_2O \longrightarrow Mg_2P_2O_7 + 2NH_3 + 13H_2O$$

Larger condensed phosphoric acids also occur, such as *triphosphoric acid*, $H_5P_3O_{10}$. The interconversion of triphosphates, pyrophosphates, and phosphates is important in many bodily processes, including the absorption and metabolism of sugar. These reactions occur at body temperature under the influence of special enzymes.

An important class of condensed phosphoric acids is that in which each phosphate tetrahedron is bonded by oxygen atoms to two other tetrahedra. These acids have the composition $(HPO_3)_x$, with $x = 3, 4, 5, 6, \ldots$. They are called the *metaphosphoric acids*. Among these acids are tetrametaphosphoric acid and hexametaphosphoric acid. The structures of the condensed phosphoric acids are similar to those of the polymers of sulfur trioxide, Figure 8-3.

Metaphosphoric acid is made by heating orthophosphoric acid or pyrophosphoric acid or by adding water to phosphorus pentoxide. It is a viscous sticky mass, which contains, in addition to ring molecules such as $H_4P_4O_{12}$, long chains approaching $(HPO_3)_\infty$ in composition. It is the long chains, which may also be condensed together to form branched chains, that, by becoming entangled, make the acid viscous and sticky.

The metaphosphates are used as water softeners. Sodium hexametaphosphate, $Na_6P_6O_{18}$, is especially effective for this purpose.

Phosphorous Acid

Phosphorous acid, H_2HPO_3, is a white substance, m.p. 74°C, which is made by dissolving diphosphorus trioxide in cold water:

$$P_4O_6 + 6H_2O \longrightarrow 4H_2HPO_3$$

It may also be conveniently made by the action of water on phosphorus trichloride:

$$PCl_3 + 3H_2O \longrightarrow H_2HPO_3 + 3HCl$$

Phosphorous acid is an unstable substance. When heated it undergoes auto-oxidation-reduction to phosphine and phosphoric acid:

$$4H_2HPO_3 \longrightarrow 3H_3PO_4 + PH_3$$

The acid and its salts, the *phosphites*, are powerful reducing agents. Its reaction with silver ion is used as a test for phosphite ion; a black precipi-

tate is formed, which consists of silver phosphate, Ag_3PO_4, colored black
by metallic silver formed by reduction of silver ion. Phosphite ion also reduces iodate ion to free iodine, which can be detected by the starch test
(blue color) or by its coloration of a small volume of carbon tetrachloride
shaken with the aqueous phase.

Phosphorous acid is a weak acid, which forms two series of salts.
Ordinary sodium phosphate is $Na_2HPO_3 \cdot 5H_2O$. Sodium hydrogen phosphite, $NaHHPO_3 \cdot 5H_2O$, also exists, but the third hydrogen atom cannot
be replaced by a cation. The nonacidic character of this third hydrogen
atom is due to its attachment directly to the phosphorus atom, rather than
to an oxygen atom:

The phosphite ion is HPO_3^{--}, not PO_3^{---}.

Hypophosphorous Acid

The solution remaining from the preparation of phosphite from phosphorus and alkali contains the *hypophosphite ion*, $H_2PO_2^-$. The corresponding acid, hypophosphorous acid, HH_2PO_2, can be prepared by using
barium hydroxide as the alkali, thus forming barium hypophosphite,
$Ba(H_2PO_2)_2$, and then adding to the solution the calculated amount of
sulfuric acid, which precipitates barium sulfate and leaves the hypophosphorous acid in solution.

Hypophosphorous acid is a weak monoprotic acid, forming only one
series of salts. The two nonacidic hydrogen atoms are bonded to the
phosphorus atom:

The acid and the hypophosphite ion are powerful reducing agents, able to
reduce the cations of copper and the more noble metals.

Oxidation-reduction Properties of Phosphorus Compounds

The compounds of phosphorus differ in a striking way from the analogous compounds of chlorine in their oxidation-reduction properties
(those of sulfur are intermediate). Thus the highest oxygen acid of phos-

phorus, H_3PO_4, is stable, and not an oxidizing agent, whereas that of chlorine, $HClO_4$, is a very strong oxidizing agent. The lower oxygen acids of phosphorus are strong reducing agents, and those of chlorine are strong oxidizing agents. The phosphide ion, P^{---}, is such a strong reducing agent that it cannot be obtained; sulfide ion, S^{--}, is a strong reducing agent; but chloride ion is stable.

These differences in properties may be accounted for in terms of the different electronegativities of the elements—3.0 for Cl, 2.5 for S, and 2.1 for P—which lead to increasing stability of the bonds with oxygen in the sequence Cl, S, P. The energies of addition of four oxygen atoms in the corresponding reactions of oxidation from the lowest to the highest oxidation states are the following:*

$$Cl^-(aq) + 4O(g) \longrightarrow ClO_4^-(aq) + 4 \times 240 \text{ kJ mole}^{-1}$$
$$S^{--}(aq) + 4O(g) \longrightarrow SO_4^{--}(aq) + 4 \times 485 \text{ kJ mole}^{-1}$$
$$P^{---}(aq) + 4O(g) \longrightarrow PO_4^{---}(aq) + 4 \times 620 \text{ kJ mole}^{-1}$$

These values correspond to the average energy 240 kJ mole^{-1} for the $Cl{=}\ddot{O}$: bond, 485 kJ mole^{-1} for the $S{=}\ddot{O}$: bond, and 620 kJ mole^{-1} for the $P{=}\ddot{O}$: bond; the increase from chlorine to phosphorus is roughly as expected from the increasing amount of ionic character of the bonds.

8-4. Oxycompounds of Nitrogen

Compounds of nitrogen are known representing all oxidation levels from -3 to $+5$. Some of these compounds are shown in the following chart:

$+5$	N_2O_5, dinitrogen pentoxide		HNO_3, nitric acid
$+4$	NO_2, nitrogen dioxide N_2O_4, dinitrogen tetroxide		
$+3$	N_2O_3, dinitrogen trioxide		HNO_2, nitrous acid
$+2$	NO, nitric oxide		
$+1$	N_2O, nitrous oxide		$H_2N_2O_2$, hyponitrous acid
0	N_2, free nitrogen		
-1	NH_2OH, hydroxylamine		
-2	N_2H_4, hydrazine		
-3	NH_3, ammonia		NH_4^+, ammonium ion

---

*The values are obtained from the standard enthalpy values given in Chapter 7, with use of the value 200 kJ mole $^{-1}$ for $P^{---}(aq)$ estimated from the values for PH_3, H_2S, S^{--}, HCl, and Cl$^-$.

The Oxides of Nitrogen

Nitrous oxide, N_2O, is made by heating ammonium nitrate:

$$NH_4NO_3 \longrightarrow 2H_2O + N_2O$$

It is a colorless, odorless gas, which has the power of supporting combustion, by giving up its atom of oxygen, leaving molecular nitrogen. When breathed for a short time the gas causes hysteria; this effect, discovered in 1799 by Humphry Davy, led to the use of the name *laughing gas* for the substance. Longer inhalation causes unconsciousness, and the gas, mixed with air or oxygen, is used as a general anesthetic for minor operations. The gas also finds use in making whipped cream; under pressure it dissolves in the cream, and when the pressure is released it fills the cream with many small bubbles, simulating ordinary whipped cream.

The electronic structure of nitrous oxide is

$$\left\{ : \ddot{N}\!=\!N\!=\!\ddot{O} : \qquad : N\!\equiv\!N\!-\!\ddot{\underset{..}{O}} : \right\}$$

The position of the oxygen atom at the end of the linear molecule explains the ease with which nitrous oxide acts as an oxidizing agent, with liberation of $: N\!\equiv\!N :$.

Nitric oxide, NO, can be made by reduction of dilute nitric acid with copper or mercury:

$$3Cu + 8H^+ + 2NO_3^- \longrightarrow 3Cu^{++} + 4H_2O + 2NO$$

When made in this way the gas usually contains impurities such as nitrogen and nitrogen dioxide. If the gas is collected over water, in which it is only slightly soluble, the nitrogen dioxide is removed by solution in the water.

A metal or other reducing agent may reduce nitric acid to any lower stage of oxidation, producing nitrogen dioxide, nitrous acid, nitric oxide, nitrous oxide, nitrogen, hydroxylamine, hydrazine, or ammonia (ammonium ion), depending upon the conditions of the reduction. Conditions may be found that strongly favor one product, but usually appreciable amounts of other products are also formed. Nitric oxide is produced preferentially under the conditions mentioned above.

Nitric oxide is a colorless, difficultly condensable gas (b.p. $-151.7°C$, m.p. $-163.6°C$). It combines readily with oxygen to form the red gas nitrogen dioxide, NO_2.

Dinitrogen trioxide, N_2O_3, can be obtained as a blue liquid by cooling an equimolal mixture of nitric oxide and nitrogen dioxide. It is the anhydride of nitrous acid, and produces this acid on solution in water:

$$N_2O_3 + H_2O \longrightarrow 2HNO_2$$

Nitrogen dioxide, NO_2, a red gas, and its dimer *dinitrogen tetroxide*,

N_2O_4, a colorless, easily condensable gas, exist in equilibrium with one another:

$$2NO_2 \rightleftarrows N_2O_4$$
$$\text{Red} \qquad \text{Colorless}$$

The mixture of these gases may be made by adding nitric oxide to oxygen, or by reducing concentrated nitric acid with copper:

$$Cu + 4H^+ + 2NO_3^- \longrightarrow Cu^{++} + 2H_2O + 2NO_2$$

It is also easily obtained by decomposing lead nitrate by heat:

$$2Pb(NO_3)_2 \longrightarrow 2PbO + 4NO_2 + O_2$$

The gas dissolves readily in water or alkali, forming a mixture of nitrate ion and nitrite ion.

Dinitrogen pentoxide, N_2O_5, the anhydride of nitric acid, can be made, as white crystals, by carefully dehydrating nitric acid with diphosphorus pentoxide or by oxidizing nitrogen dioxide with ozone. It is unstable, decomposing spontaneously at room temperature into nitrogen dioxide and oxygen.

The electronic structures of the oxides of nitrogen are shown below and on the next page. Most of these molecules are resonance hybrids, and the contributing structures are not all shown; for dinitrogen pentoxide, for example, the various single and double bonds may change places.

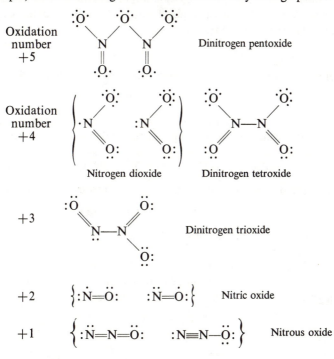

We may well ask why it is that two of the most stable of these substances, NO and NO_2, are odd molecules, representing oxidation levels for nitrogen not occurring in other compounds, and also why N_2O_3 and N_2O_5, the anhydrides of the important substances HNO_2 and HNO_3, are so unstable that they decompose at room temperature. The answer to these questions probably is that the resonance of the odd electron between the two or three atoms of the molecule stabilizes the substances NO and NO_2 enough to make them somewhat more stable than the two anhydrides.

Nitric Acid and the Nitrates

Nitric acid, HNO_3, is a colorless liquid with melting point $-42°C$, boiling point $86°C$, and density 1.52 g cm^{-3}. It is a strong acid, completely ionized to hydrogen ion and nitrate ion (NO_3^-) in aqueous solution; and it is a strong oxidizing agent. It attacks the skin, and gives it a yellow color.

Nitric acid can be made in the laboratory by heating sodium nitrate with sulfuric acid in an all-glass apparatus:

$$NaNO_3 + H_2SO_4 \longrightarrow NaHSO_4 + HNO_3$$

The substance is also made commercially in this way, from natural sodium nitrate (Chile saltpeter).

Much nitric acid is made by the oxidation of ammonia. This oxidation occurs in several steps. Ammonia mixed with air burns on the surface of a platinum catalyst to form nitric oxide:

$$4NH_3 + 5O_2 \longrightarrow 4NO + 6H_2O$$

On cooling, the nitric oxide is further oxidized to nitrogen dioxide:

$$2NO + O_2 \longrightarrow 2NO_2$$

The gas is passed through a tower packed with pieces of broken quartz through which water is percolating. Nitric acid and nitrous acid are formed:

$$2NO_2 + H_2O \longrightarrow HNO_3 + HNO_2$$

As the strength of the acid solution increases, the nitrous acid decomposes:

$$3HNO_2 \rightleftarrows HNO_3 + 2NO + H_2O$$

The nitric oxide is reoxidized by the excess oxygen present and again enters the reaction.

A method (the arc process) formerly used for fixation of atmospheric nitrogen is the direct combination of nitrogen and oxygen to nitric oxide at the high temperature of the electric arc. The reaction

$$N_2 + O_2 \longrightarrow 2NO$$

is slightly endothermic, and the equilibrium yield of nitric oxide increases with increasing temperature, from 0.4% at 1500°C to 5% at 3000°C. The reaction was carried out by passing air through an electric arc in such a way that the hot gas mixture was cooled very rapidly, thus "freezing" the high-temperature equilibrium mixture. The nitric oxide was then converted into nitric acid in the way described above.

Sodium nitrate, $NaNO_3$, forms colorless crystals closely resembling crystals of calcite, $CaCO_3$. This resemblance is not accidental. The crystals have the same structure, with Na^+ replacing Ca^{++} and NO_3^- replacing CO_3^{--}. The crystals of sodium nitrate have the same property of birefringence (double refraction) as calcite. Sodium nitrate is used as a fertilizer and for conversion into nitric acid and other nitrates. *Potassium nitrate*, KNO_3 (*saltpeter*), is used in pickling meat (ham, corned beef), in medicine, and in the manufacture of *gunpowder*, which is an intimate mixture of potassium nitrate, charcoal, and sulfur, which explodes when ignited in a closed space.

The nitrate ion has a planar structure, with each bond a hybrid of a single bond and a double bond:

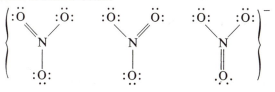

The nitrates of all metals are soluble in water.

Nitrous Acid and the Nitrites

Nitrous acid, HNO_2, forms in small quantity together with nitric acid when nitrogen dioxide is dissolved in water. Nitrite ion can be made together with nitrate ion by solution of nitrogen dioxide in alkali:

$$2NO_2 + 2OH^- \longrightarrow NO_2^- + NO_3^- + H_2O$$

Sodium nitrite, $NaNO_2$, and *potassium nitrite*, KNO_2, can be made also by decomposing the nitrates by heat:

$$2NaNO_3 \longrightarrow 2NaNO_2 + O_2$$

or by reduction with lead:

$$NaNO_3 + Pb \longrightarrow NaNO_2 + PbO$$

These nitrites are slightly yellow crystalline substances, and their solutions are yellow. They are used in the manufacture of dyes and in the chemical laboratory.

The nitrite ion is a reducing agent, being oxidized to nitrate ion by bromine, permanganate ion, chromate ion, and similar oxidizing agents. It is also itself an oxidizing agent, able to oxidize iodide ion to iodine. This property may be used, with the starch test (blue color) for iodine, to distinguish nitrite from nitrate ion, which does not oxidize iodide ion readily.

The electronic structure of the nitrite ion is

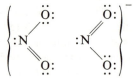

Other Compounds of Nitrogen

Hyponitrous acid, $H_2N_2O_2$, is formed in small quantity by reaction of nitrous acid and hydroxylamine:

$$H_2NOH + HNO_2 \longrightarrow H_2N_2O_2 + H_2O$$

It is a very weak acid, with structure

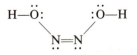

The acid decomposes to form nitrous oxide, N_2O; it is not itself formed in appreciable concentration by reaction of nitrous oxide and water. Its salts have no important uses.

Hydrogen cyanide, HCN (structural formula $H\!-\!C\!\equiv\!N:$), is a gas which dissolves in water and acts as a very weak acid. It is made by treating a cyanide, such as *potassium cyanide*, KCN, with sulfuric acid, and is used as a fumigant and rat poison. It smells like bitter almonds and crushed fruit kernels, which in fact owe their odor to it. Hydrogen cyanide and its salts are very poisonous.

Cyanides are made by action of carbon and nitrogen on metallic oxides. For example, barium cyanide is made by heating a mixture of barium oxide and carbon to a red heat in a stream of nitrogen:

$$BaO + 3C + N_2 \longrightarrow Ba(CN)_2 + CO$$

The cyanide ion, $[:C\!\equiv\!N:]^-$, is closely similar to a halide ion in its properties. By oxidation it can be converted into *cyanogen*, C_2N_2 ($:N\!\equiv\!C\!-\!C\!\equiv\!N:$), which is analogous to the halogen molecules F_2, Cl_2, etc. By suitable procedures three anions can be made that are similar in structure to the carbon dioxide molecule $:\!\overset{\cdot\cdot}{O}\!=\!C\!=\!\overset{\cdot\cdot}{O}\!:$ and the nitrous oxide molecule $:\overset{\cdot\cdot}{N}\!=\!N\!=\!\overset{\cdot\cdot}{O}:$ (these structures are hybridized with other structures, such as $:O\!\equiv\!C\!-\!\overset{\cdot\cdot}{O}:$ and its analogues). These anions are

$$\left[\;:\overset{\cdot\cdot}{N}\!=\!C\!=\!\overset{\cdot\cdot}{O}:\;\right]^{-}\quad\text{Cyanate}$$

$$\left[\;:\overset{\cdot\cdot}{C}\!=\!N\!=\!\overset{\cdot\cdot}{O}:\;\right]^{-}\quad\text{Fulminate ion}$$

$$\left[\;:\overset{\cdot\cdot}{N}\!=\!N\!=\!\overset{\cdot\cdot}{N}:\;\right]^{-}\quad\text{Azide ion}$$

A related ion is the thiocyanate ion, $[:\overset{\cdot\cdot}{N}\!=\!C\!=\!\overset{\cdot\cdot}{S}:]^{-}$, which forms a deep red complex with ferric ion, used as a test for iron. The azide ion also forms a deep red complex with ferric ion.

The fulminates and azides of the heavy metals are very sensitive explosives. *Mercuric fulminate*, $Hg(CNO)_2$, and *lead azide*, $Pb(N_3)_2$, are used as detonators.

8-5.   Oxycompounds of Carbon

Carbon burns to form the gases *carbon monoxide*, CO, and *carbon dioxide*, CO_2, the former being produced when there is a deficiency of oxygen or when the flame temperature is very high.

Carbon Monoxide

Carbon monoxide is a colorless, odorless gas with small solubility in water (35.4 ml per liter of water at 0°C and 1 atm). It is poisonous, because of its ability to combine with the hemoglobin in the blood in the same way that oxygen does; thus the carbon monoxide prevents the hemoglobin from combining with oxygen in the lungs and carrying it to the tissues. It causes death when about one-half of the hemoglobin in the blood has been converted into carbonmonoxyhemoglobin. The exhaust gas from automobile engines contains some carbon monoxide, and it is accordingly dangerous to be in a closed garage with an automobile whose engine is running. Carbon monoxide is a valuable industrial gas, for use as a fuel and as a reducing agent.

The heat of formation of carbon monoxide is 110.5 kJ mole^{-1} and its heat of combustion is 283.0 kJ mole^{-1} (Section 6-11). The blue lambent flame seen over a charcoal fire involves the combustion of the carbon monoxide that has been formed by the surface combustion of the charcoal.

In Example 6-8 it was suggested from consideration of its electric dipole moment that the carbon monoxide molecule has an electronic structure that is a hybrid of $:C\equiv O:$ and $:C\overset{..}{=}O:$. Support for this structure is provided by the heats of formation from atoms of the three molecules C_2, O_2, and CO

$$2C(g) \longrightarrow C_2(g) + 455 \text{ kJ mole}^{-1}$$
$$2O(g) \longrightarrow O_2(g) + 495 \text{ kJ mole}^{-1}$$
$$C(g) + O(g) \longrightarrow CO(g) + 1077 \text{ kJ mole}^{-1}$$

The bond energy of carbon monoxide, 1077 kJ mole^{-1}, is larger than that for any other diatomic molecule. We may first estimate the energy of the double bond, $:C\overset{..}{=}O:$. Its value is approximately the average* of the values for C_2 and O_2, with the addition of $2 \times 100 = 200$ kJ mole^{-1} for the partial ionic character of the two single C—O bent bonds constituting the double bond: $\frac{1}{2}(455 + 495) + 200 = 675$ kJ mole^{-1}. This value is much less than the observed bond energy, and the structure $:C\overset{..}{=}O:$ alone for the CO molecule is accordingly not acceptable. (The observed bond length 1.13 Å also rules out the double-bond structure, which would give bond length 1.22 Å.) For the triple-bond structure $:C\equiv O:$, in which all four orbitals of the carbon atom as well as of the oxygen atom are used, we estimate that the bond energy is about 1000 kJ mole^{-1}, 50% greater than for the double-bond structure. The experimental value, 1077 kJ mole^{-1}, is a little larger still; the difference is the resonance energy between the two structures $:C\equiv O:$ and $:C\overset{..}{=}O:$.

Carbon Dioxide

Carbon dioxide is a colorless, odorless gas with a weakly acid taste, owing to the formation of some carbonic acid when it is dissolved in water. It is about 50% heavier than air. It is easily soluble in water, one liter of water at 0°C dissolving 1713 ml of the gas under 1 atm pressure. Its melting point (freezing point) is higher than the point of vaporization at 1 atm of the crystalline form. When crystalline carbon dioxide is heated from a very low temperature, its vapor pressure reaches 1 atm at −79°C, at which temperature it vaporizes (sublimes) without melting. If the pressure is increased to 5.2 atm, the crystalline substance melts to a liquid at −56.6°C. Under ordinary pressure, then, the solid substance is changed directly to a gas. This property has made solid carbon dioxide (dry ice) popular as a refrigerating agent.

---

*The oxygen molecule contains a single bond and two three-electron bonds (Section 6-6). Its bond energy is approximately that of a double bond, since a three-electron bond (as well as a one-electron bond) is about half as strong as an electron-pair bond. The C_2 molecule also has a triplet structure, with two one-electron bonds and a single bond, $:C \div C:$.

The enthalpy of formation of carbon dioxide is -394 kJ mole^{-1}. Its heat of sublimation at $-78.48°C$ (1 atm pressure) is 25 kJ mole^{-1}. The molecule is linear, with carbon-oxygen bond length 1.159 Å.

Carbon dioxide is used for the manufacture of *sodium carbonate*, $Na_2CO_3 \cdot 10H_2O$ (washing soda); *sodium hydrogen carbonate*, $NaHCO_3$ (baking soda); and carbonated water, for use as a beverage (soda water). Carbonated water is charged with carbon dioxide under a pressure of 3 or 4 atm.

Carbon dioxide can be used to extinguish fires by smothering them. One form of portable fire extinguisher is a cylinder of liquid carbon dioxide—the gas can be liquefied at ordinary temperatures under pressures of about 70 atm. Some commercial carbon dioxide (mainly solid carbon dioxide) is made from the gas emitted in nearly pure state from gas wells in the western United States. Most of the carbon dioxide used commercially is a by-product (subsidiary substance produced in the process) of cement mills, limekilns, iron blast furnaces, and breweries.

Carbonic Acid and Carbonates

When carbon dioxide dissolves in water, some of it reacts to form carbonic acid:

$$CO_2 + H_2O \longrightarrow H_2CO_3$$

The structural formula of carbonic acid is

The acid is diprotic; with a base such as sodium hydroxide it may form both a normal salt, Na_2CO_3, and an acid salt, $NaHCO_3$. The normal salt contains the carbonate ion, CO_3^{--}, and the acid salt contains the hydrogen carbonate ion, HCO_3^-.

For many years chemists assigned the structural formula

to the carbonate ion. With this formula, one of the oxygen atoms is attached to the carbon atom by a double bond, and the other two are attached by single bonds. Then in 1914 W. L. Bragg carried out an x-ray diffraction study of calcite, $CaCO_3$, and found that the three bonds from the carbon atom to the three oxygen atoms in the carbonate ion in this crystal are identical. This new experimental fact required a change in the structural formula. The new structural formula was proposed in 1931, when the chemical resonance theory was developed (Section 6-7). For the carbonate ion the structure is a hybrid of three structures:

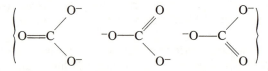

Each oxygen atom is attached to the carbon atom by a bond that is a hybrid of a double bond (one-third) and a single bond (two-thirds). The three carbon-oxygen bonds are thus identical.

Calcium Carbonate

The most important carbonate mineral is calcium carbonate, $CaCO_3$. This substance occurs in beautiful colorless hexagonal crystals as the mineral *calcite*. *Marble* is a microcrystalline form of calcium carbonate, and *limestone* is a rock composed mainly of this substance. Calcium carbonate is the principal constituent also of pearls, coral, and most sea shells. It also occurs in a second crystalline form, as the orthorhombic mineral *aragonite*.

When calcium carbonate is heated (as in a limekiln, where limestone is mixed with fuel, which is burned), it decomposes, forming calcium oxide (quicklime):

$$CaCO_3 \longrightarrow CaO + CO_2(g)$$

Quicklime is slaked by adding water, to form calcium hydroxide:

$$CaO + H_2O \longrightarrow Ca(OH)_2$$

Slaked lime prepared in this way is a white powder that can be mixed with water and sand to form *mortar*. The mortar hardens by first forming crystals of calcium hydroxide, which cement the grains of sand together; then on exposure to air the mortar continues to get harder by taking up carbon dioxide and forming calcium carbonate.

Large amounts of limestone are used also in the manufacture of Portland cement, described in Chapter 18.

Sodium carbonate (washing soda), $Na_2CO_3 \cdot 10H_2O$, is a white, crystalline substance used as a household alkali, for washing and cleaning, and as an industrial chemical. The crystals of the decahydrate lose water readily, forming the monohydrate, $Na_2CO_3 \cdot H_2O$. The monohydrate when heated to 100° changes to anhydrous sodium carbonate (soda ash), Na_2CO_3.

Sodium hydrogen carbonate (baking soda, bicarbonate of soda), $NaHCO_3$, is a white substance usually available as a powder. It is used in cooking, in medicine, and in the manufacture of baking powder. Baking powder is a leavening agent used in making biscuits, cakes, and other food. Its purpose is to provide bubbles of gas, to make the dough "rise." The same foods can be made by use of sodium hydrogen carbonate and sour milk, instead of baking powder. In each case the reaction that occurs involves the action of an acid on sodium hydrogen carbonate, to form carbon dioxide. When sour milk is used, the acid that reacts with the sodium hydrogen carbonate is lactic acid, $HC_3H_5O_3$, the equation for the reaction being

$$NaHCO_3 + HC_3H_5O_3 \longrightarrow NaC_3H_5O_3 + H_2O + CO_2(g)$$

The product $NaC_3H_5O_3$ is sodium lactate, the sodium salt of lactic acid (Section 23-5). Cream-of-tartar baking powder consists of sodium hydrogen carbonate, potassium hydrogen tartrate ($KHC_4H_4O_6$, commonly known as cream of tartar), and starch, the starch being added to keep water vapor in the air from causing the powder to form a solid cake. The reaction that occurs when water is added to a cream-of-tartar baking powder is

$$NaHCO_3 + KHC_4H_4O_6 \longrightarrow NaKC_4H_4O_6 + H_2O + CO_2(g)$$

The product sodium potassium tartrate, $NaKC_4H_4O_6$, has the common name "Rochelle salt." Baking powders are also made with calcium dihydrogen phosphate, $Ca(H_2PO_4)_2$, sodium dihydrogen phosphate, NaH_2PO_4, or sodium aluminum sulfate, $NaAl(SO_4)_2$, as the acidic constituent. The last of these substances is acidic because of the hydrolysis of the aluminum salt (Section 14-6).

The leavening agent in ordinary bread dough is yeast (Section 7-6).

8-6. Molecules Containing Bivalent Carbon. Free Radicals

Carbon monoxide may be considered to be a compound of bivalent carbon. The valence state of the carbon atom in this molecule is based upon the electron configuration $2s^2 2p^2$, involving an unshared pair of electrons in the $2s$ orbital and two valence electrons occupying separate $2p$ orbitals.

There are only a few other substances containing bivalent carbon. The isocyanides constitute a class of substances of this sort. The isocyanides are substances that are less stable than the corresponding isomeric substances called the cyanides. For example, methyl cyanide, which is assigned the electronic structure $CH_3-C\equiv N:$, has enthalpy of formation 88 kJ mole^{-1}, whereas its isomer methyl isocyanide, $CH_3-N\equiv C:$, has enthalpy of formation 150 kJ mole^{-1}. Methyl isocyanide is a volatile liquid (b.p. 59.6°C) with an evil odor, far more pungent than that of methyl cyanide. The alkyl isocyanides with small side chains (methyl, ethyl, isopropyl, *tert*-butyl) have the property of combining with hemoglobin. There are only a few substances with this property, including molecular oxygen and carbon monoxide; this property is important in the functioning of hemoglobin, in combining with oxygen in the lungs and releasing it in the tissues. The alkyl isocyanides, like carbon monoxide, are poisonous because of the property of combining with the hemoglobin in the blood and inhibiting its use as an oxygen carrier.

The cyanide ion, $[:C{\equiv}N:]^-$, and the fulminate ion, $[:C{\equiv}N\!-\!\overset{..}{\underset{..}{O}}:]^-$, may also be considered to involve bivalent carbon.

The molecules C_2 and C_3 are present in the vapor of graphite at very high temperature. C_2 in its normal state has two unpaired electrons; its electronic structure can be written $:C\dot{\div}C:$, with two one-electron bonds plus a single bond. A possible structure for C_3 is $:C{=}C{=}C:$.

Carbene or *methylene*, CH_2, can be made as a dilute gas by the photolysis of diazomethane, CH_2N_2:

$$CH_2N_2 + h\nu \longrightarrow CH_2 + N_2$$

The symbol $h\nu$ is used to represent a photon; for this reaction ultraviolet light, wavelength about 1415 Å, is used. Substituted carbenes can be made by a similar reaction; for example, diphenylcarbene, $(C_6H_5)_2C:$, by the photolysis of diazodiphenylmethane, $(C_6H_5)_2CN_2$.

Dichlorocarbene, CCl_2, is a useful chemical reagent. It cannot be prepared and stored in concentrated form, but it is easily generated in solution and it shows consistently characteristic properties of a highly reactive transient molecule (a molecule with a short lifetime). A convenient way of generating it is by the reaction of sodium ethoxide, C_2H_5ONa (the sodium derivative of ethanol), and chloroform:

$$C_2H_5ONa + CHCl_3 \longrightarrow C_2H_5OH + NaCl + CCl_2$$

Dichlorocarbene can be used to carry out many organic reactions that are hard to carry out in other ways. For example, it adds easily to a double bond, to form cyclopropane derivatives:

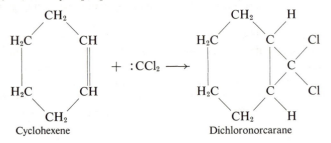

Cyclohexene   Dichloronorcarane

Another highly reactive carbene is *carbonyl carbene*, C_2O. It can be made by irradiating carbon suboxide, C_3O_2, with ultraviolet light.

Carbon suboxide (tricarbon dioxide) is an evil-smelling gas with boiling point 7°C and melting point -111°C. It is prepared by dehydrating malonic acid, $CH_2(COOH)_2$, with tetraphosphorus decoxide at about 150°C:

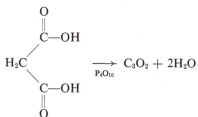

The structure of the molecule has been determined by the electron-diffraction method: it is linear, with bond lengths 1.16 Å for carbon-oxygen and 1.28 Å for carbon-carbon. These values correspond to the valence-bond structure $:\ddot{O}{=}C{=}C{=}C{=}\ddot{O}:$ with some contribution by $^+:O{\equiv}C{-}C{\equiv}C{-}\ddot{O}:^-$ and $^-:\ddot{O}{-}C{\equiv}C{-}C{\equiv}O:^+$.

Carbonyl carbene reacts with many molecules as a donor of a carbon atom. An example is its reaction with ethylene to form allene, $H_2C{=}C{=}CH_2$, and carbon monoxide. This reaction probably takes place by the formation of a cyclopropane ring, followed by immediate decomposition:

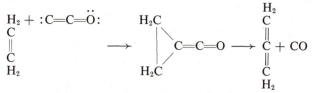

It has been shown by use of ^{14}C as a tracer (Section 26-5) that the carbene carbon atom (end carbon atom) of C_2O becomes the central carbon of the allene molecule. Malonic acid is synthesized with ^{14}C in the central position. On dehydration it gives $^{14}C{=}C{=}O$. Reaction with ethylene produces carbon monoxide that is free of ^{14}C.

Similarly, $^{14}C{=}C{=}O$ reacts with molecular oxygen to form ^{14}CO, CO_2, and O. The fact that the carbon dioxide is free of ^{14}C shows that one proposed mechanism is incorrect:

$$O_2 + :C{=}C{=}O \longrightarrow \begin{array}{c} O \\ \diagdown \\ \diagup \\ O \end{array} C{=}C{=}O \longrightarrow CO_2 + CO$$

The probably correct mechanism involves an intermediate with a four-membered ring:

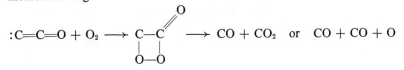

The Structure of Carbene

There are two reasonable electronic structures that might be suggested for the normal state of the carbene molecule. One is the structure

based upon the configuration $2s^2 2p^2$ for the carbon atom. The bonds from the carbon atom to the hydrogen atoms for this structure would involve carbon bond orbitals $2p$ in character. The other structure that can be suggested is the triplet structure,

With this structure the carbon atom is in the electron configuration $2s2p^3$, and the unshared electrons occupy two tetrahedral orbitals, the other two being occupied by the bond electrons. The carbon-hydrogen bonds are expected to be stronger for this structure than for the singlet structure, but this triplet structure requires promotion energy from the bivalent state of carbon to the quadrivalent state. Spectroscopic studies indicate that the normal state of the molecule is the triplet state, and that the singlet state is somewhat less stable; the relative energy of the two states has not yet been determined by experiment.

We can estimate the energy of the singlet state by use of bond-energy values. From the standard enthalpy of CH(g) given in Table 7-2 we may calculate that the heat of formation of CH from atoms is 338 kJ mole⁻¹. We might expect the energy of the second C—H bond in the singlet state of CH_2 to be the same; the heat of formation of $:CH_2$ from atoms would then be predicted to be about 676 kJ mole⁻¹. The observed heat of formation of the normal state ($\uparrow CH_2$, triplet state) from atoms is 754 kJ mole⁻¹ (calculated from the value of the enthalpy of formation given in Table 7-2). This value is greater than the value predicted for the singlet state, and accordingly we estimate that the singlet state lies about 78 kJ mole⁻¹ above the triplet state.

The C—H bond in CH has length 1.12 Å, essentially the same as in methane (1.11 Å). In the normal (triplet) state CH_2 has bond length 1.07 Å and bond angle 140°; in the singlet state the values are 1.12 Å and 103°. These values have been obtained by analysis of the band spectrum of CH_2.

Free Radicals

An atom or group of atoms with one or more unshared electrons, which may enter into chemical-bond formation, is called a *free radical*. (The same group in a molecule is called a radical; for example, the methyl radical in methyl cyanide or other molecules.) Free radicals are usually highly reactive and difficult to prepare in any except low concentration.

One way of making the methyl radical as a dilute gas is by heating mercury dimethyl, $Hg(CH_3)_2$, which decomposes to give metallic mercury and methyl radical. Methyl radical can also be made conveniently by the decomposition of diacetyl, $(CH_3CO_2)_2$, by either heat or ultraviolet light. The diacetyl molecule liberates two molecules of carbon dioxide and two methyl radicals.

The American chemist Moses Gomberg (1866–1947) discovered in 1900 that some hydrocarbon free radicals are stable. He attempted to synthesize the substance hexaphenylethane, $(C_6H_5)_3C$—$C(C_6H_5)_3$, which he expected to be a stable, white crystalline substance. Instead, he obtained a strongly colored solution, with the property of combining readily with oxygen. He concluded correctly that the solution did not contain the hexaphenyl derivative of ethane, but instead the free radical triphenylmethyl, with the formula $(C_6H_5)_3C\cdot$. Many similar hydrocarbon free radicals have been made, and it has been shown that they are paramagnetic, and accordingly contain unpaired electrons (the paramagnetism is due to the magnetic moment of the electron spin of the unpaired electron). The stability of the triphenylmethyl radical, which is responsible for the low bond energy of the carbon-carbon bond in the substituted ethane, is attributed to the resonance energy of the unpaired electron among the various carbon atoms of the molecule. Note that you can write eight valence-bond structures of type A, each involving nine double bonds with the unpaired electron on the central carbon atom in the triphenylmethyl radical, and 36 similar structures of type B, with the unpaired electron on one of the ring carbon atoms.

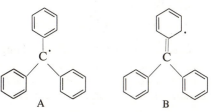

A   B

8-7.   Unstable and Highly Reactive Molecules

During the period of development of classical chemistry, from 1780 to 1920, the attention of chemists was limited almost entirely to the study of the properties of stable substances. It has now become possible, however, to investigate unstable and highly reactive molecules, especially by the use of spectroscopic and magnetic resonance techniques.

The analysis of molecular spectra has provided information about the electronic structure, internuclear distance, and vibrational frequency for several hundred diatomic molecules in their normal and excited states. Values of the bond length (internuclear distance) for the normal states of some molecules are given in Table 8-1. The electronic structure of a diatomic molecule is described by use of symbols, such as $^2\Sigma$, analogous to the Russell-Saunders symbols for atoms. The left superscript has the same significance as for atoms; it is equal to $2S + 1$, with S the resultant electron-spin quantum number, and hence is 1 greater than the number of unpaired electrons. The symbols Σ, Π, Δ, ... correspond to the values 0, 1, 2, ... for the component of the resultant orbital angular momentum of the electrons along the internuclear axis of the molecule.

The symbol $^1\Sigma$ for BH, for example, shows that there is zero electron spin ($S = 0$, electrons all paired) and that there is no component of orbital angular momentum along the B—H axis. We would, from the arguments given in Chapter 6, assign the structural formula : B—H to the molecule (the $1s$ electrons of B are not shown). The unshared pair occupies essentially the $2s$ orbital of boron (with a small amount of $2p_\sigma$ character; the symbol $2p_\sigma$ means the $2p$ orbital extending along the B—H axis). The bond is formed by the $1s$ orbital of H and the $2p_\sigma$ orbital of B, with a small amount of $2s$ character. The bond length, 1.233 Å, is close to the value expected for a single bond.

The structures with a single bond to the hydrogen atom correspond to the observed normal states for the other hydride molecules. For CH, for example, we write the structural formula : C̈—H. The structure is the same as for BH, with the addition of an unpaired electron. This electron occupies one of the remaining $2p$ orbitals of the carbon atom (one, the $2p_\sigma$ orbital, is used in bond formation). These two orbitals are called the $2p_\pi$ orbitals. One has component $+1$ and the other has component -1 of orbital angular momentum along the C—H axis. The state with one occupied is accordingly a $^2\Pi$ state.

The structure of NH is : N̈—H. The two odd electrons occupy the two $2p_\pi$ orbitals, with components $+1$ and -1 of orbital angular momentum. Their resultant is zero, so that the state is a Σ state. Hund's first rule (Section 5-3) leads to the conclusion that for the normal state the spins of the two electrons are parallel; hence the state is a $^3\Sigma$ state.

TABLE 8-1
Values of the Internuclear Distance for Some Diatomic Molecules in their Normal States

Molecule	Symbol	Bond length	Moleule	Symbol	Bond Length
BH	$^1\Sigma$	1.233 Å	Cl_2	$^1\Sigma$	1.988 Å
BH^+	$^2\Sigma$	1.215	Cl_2^+	$^2\Pi$	1.891
CH	$^2\Pi$	1.120	N_2	$^1\Sigma$	1.094
CH^+	$^1\Sigma$	1.131	N_2^+	$^2\Sigma$	1.116
NH	$^3\Sigma$	1.038	O_2	$^3\Sigma$	1.207
OH	$^2\Pi$	0.971	O_2^+	$^2\Pi$	1.123
OH^+	$^3\Sigma$	1.029	CF	$^2\Pi$	1.271
SiH	$^2\Pi$	1.520	SiN	$^2\Sigma$	1.572
PH	$^3\Sigma$	1.433	SiO	$^1\Sigma$	1.510
SH	$^2\Pi$	1.35	SiF	$^2\Pi$	1.603

TABLE 8-2
Normal State and Excited States of NO

	Energy	Symbol	Bond Length
D*	6.58 eV	$^2\Sigma$	1.09Å
C	6.47	$^2\Sigma$	1.09
B	5.65	$^2\Pi$	1.33
A	5.47	$^2\Sigma$	1.09
X	0	$^2\Pi$	1.15

*The normal state is usually designated by the letter X, and the excited states by A, B, ...

In molecules other than the hydrides ionization is usually accompanied by a change in the nature of the bond between the two atoms. The chlorine molecule-ion has the structure :Cl∵Cl:$^+$, with a single bond plus a three-electron bond; correspondingly the bond length is 0.10 Å less than in :Cl—Cl:. There is, however, little change from N_2 to N_2^+; in each molecule there is a triple bond.

The normal state of NO (Table 8-2) has the symbol $^2\Pi$, showing that an odd electron occupies a p_π orbital. We may combine the two electronic structures :N⚌O: and :N⚌O: in either the symmetric or the antisymmetric way. The symmetric combination stabilizes the molecule, which is said then to have a three-electron bond in addition to the double bond. The bond length 1.15 Å observed for the normal state corresponds to this structure, $:N\overset{...}{=}O:$. The antisymmetric $^2\Pi$ state would be expected to have a double bond weakened by a three-electron antibond, and to have bond length between that for a double bond and that for a single bond, as observed for state B (1.33 Å). The other three excited states, A, C, and D, all have bond length 1.09 Å, which corresponds to a triple bond. These three states may be described as involving an NO^+ core with an electron in an outer orbital with total quantum number 3 or greater. The structural formula $(:N\equiv O:)\cdot$ may be used for each of these states.

Many highly reactive polyatomic molecules have been subjected to spectroscopic study. An example is silicon dihydride, SiH_2, which is obtained by subjecting $SiH_4(g)$ to flash photolysis (decomposition by a burst of ultraviolet photons). The band spectrum of SiH_2 shows that the H—Si—H angle is 92.1°, with Si—H = 1.521 Å. These structural parameters correspond to the structure :Si⟨ (with H above and H below). The silicon bond orbitals have large p character, and the unshared electron pair is largely $3s$ in character.

Reactive molecules can be preserved by being trapped in an unreactive matrix, such as solidified argon, or in a crystal in which they are formed. An example is the CO_2^- anion, which can be made by x-irradiation of sodium formate. The CO_2^- ions occupy some of the HCO_2^- positions. Electron-spin resonance spectroscopic study of the CO_2^- ion, which contains an unpaired electron, has shown it to be similar in structure to the NO_2 molecule, with O—C—O angle about 134°.

Tautomerism

Some apparently pure substances have the chemical and physical properties expected for two different structures. For example, the substance usually called acetaldehyde reacts with other substances in the way characteristic of a carbon-carbon double bond as well as of that characteristic of a carbon-oxygen double bond. The properties suggest that the substance, as gas, liquid, or solute, consists largely of molecules with structure A (the keto structure), with some molecules present with structure B (the enol structure).

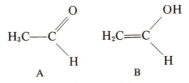

A   B

The rate of interconversion is so great that the two forms cannot be separated under ordinary conditions.

The liquid substance called acetoacetic ester (or ethyl acetoacetate) at room temperature reacts with bromine to the extent of 7% within a second, and continues to 100% bromination at a lower rate. This fact indicates that the equilibrium mixture in the liquid is 93% keto (A) and 7% enol (B).

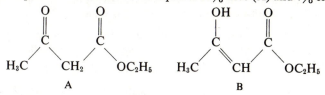

During recent decades it has become possible to study tautomeric mixtures by physical methods. The hydroxyl group, for example, has a stretching vibrational frequency about 3500 cm^{-1}, somewhat greater than that for C—H, about 3000 cm^{-1}. By measuring the intensity of the OH absorption band in the infrared the amount of the substance in the enol form can be determined. The nuclear magnetic resonance technique is especially useful, in providing information not only about the relative amounts of tautomers but also about their rates of interconversion.

Exercises

8-1. What are the formulas and structures of the oxygen acids of the $+5$ oxidation states of the fifth-group elements?

8-2. What are the formulas and structures of the oxygen acids of the $+3$ oxidation states of the fifth-group elements (include $Bi(OH)_3$ in this tabulation)? How do the properties of these compounds vary with atomic number?

8-3. Write the equations for the hydrolysis of phosphorus tribromide and of phosphorus pentachloride.

8-4. To make a soluble phosphate for use in fertilizers, $Ca_3(PO_4)_2$ is allowed to react with sulfuric acid to produce a mixture of $CaSO_4 \cdot 2H_2O$ and $Ca(H_2PO_4)_2$ or with phosphoric acid to produce $Ca(H_2PO_4)_2$ alone. Write equations for the reactions. What percentage of phosphorus is there in the first product (called "superphosphate")? In the second product ("triple phosphate")?

8-5. What are the electronic structures of phosphorous acid (H_3PO_3) and hypophosphorous acid (H_3PO_2)?

8-6. If phosphorous acid is heated to 200°C it forms phosphine and phosphoric acid. Write a balanced equation for the reaction.

8-7. Hydrogen bromide can be prepared by the reaction between phosphorus tribromide and water. Write the equation. What other phosphorus compound might be produced, if the mixture is heated?

8-8. Arsenic reacts with hot concentrated nitric acid to form arsenic acid, nitrogen dioxide, and water. Write the equation for the reaction.

8-9. What is the structure of trimetaphosphoric acid, $H_3P_3O_9$?

8-10. Write a chemical equation for the reduction of Ag^+ by a solution of sodium phosphite.

8-11. Write chemical equations illustrating the acidic and the basic properties of the $+3$ oxidation state of antimony.

8-12. Discuss the enthalpies of formation of trichlorides, tribromides, and triiodides of phosphorus, arsenic, and antimony in relation to electronegativity values.

8-13. Write chemical equations for the preparation of each of the substances H_2S, SO_2, and SO_3 by
    (*a*) a chemical reaction in which there is an oxidation or reduction of the sulfur atom;
    (*b*) a chemical reaction in which there is no change in the oxidation number of sulfur.

8-14. List some examples of the use of concentrated sulfuric acid for the preparation of more volatile acids. Why cannot this method be applied to the preparation of hydrogen iodide gas?

8-15. Write chemical reactions illustrating the three important kinds of uses of sulfuric acid.

8-16. What are the electronic structures of disulfuric acid, peroxysulfuric acid, and peroxydisulfuric acid?

8-17. (*a*) Carbon disulfide, which has boiling point 46.3°C, is made by passing sulfur vapor over red-hot carbon. The carbon burns in the sulfur vapor, forming carbon disulfide. Write the equation for this reaction.
    (*b*) The standard enthalpy of formation of $CS_2(g)$ is 115 kJ mole^{-1}. Using other values from Table 7-7, calculate the enthalpy of the reaction

$$S_2(g) + C(\text{graphite}) \longrightarrow CS_2(g)$$

Is the reaction exothermic or endothermic?

8-18. In the presence of iodine as catalyst, carbon disulfide reacts with chlorine to form carbon tetrachloride and disulfur dichloride, S_2Cl_2. Write the equation for this reaction.

8-19. By microwave spectroscopy it has been found that in the sulfur dioxide molecule the bonds have length 1.432 Å, the bond angle has the value 119.5°, and the electric dipole moment is 0.331 eÅ. What electric charges on the atoms (assumed to be at nuclear positions) would account for the dipole moment? (Answer: +0.46 on S, −0.23 on each O.)

8-20. In the discussion of sulfur dioxide, Section 8-2, it is mentioned that the bond length 1.49 Å is expected for the S=O bond. How is this value obtained?

8-21. Assuming that a double bond is two bent single bonds with the same properties as straight single bonds and that each of the two oxygen atoms is attached to the sulfur atom by a double bond, calculate from the electronegativity values for sulfur and oxygen the ionic character of each of the four bent single bonds in the sulfur dioxide molecule, and the corresponding value of the electric charge on the sulfur atom and on each oxygen atom. (Answer: 22%, +0.88, −0.44.)

8-22. The American chemists G. N. Lewis and Irving Langmuir assumed that the sulfur atom in its compounds would have the electronic structure of argon. What two resonance structures for SO_2 correspond to this postulate? To what electric charges on sulfur and oxygen do they correspond? (Answer: +1.66, −0.83.)

8-23. The observed sulfur-oxygen bond length in sulfur dioxide suggests that the bonds have some triple-bond character (Exercises 8-19, 8-20). To what bond length and what electric charge on the sulfur atom would the pair of resonance structures

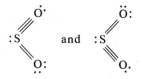

correspond? (Answer: 1.36 Å, −0.10 on sulfur.)

8-24. Using the results of Exercises 8-21 and 8-23, calculate the relative contributions of the structure

and the two structures of Exercise 8-23 that lead to the observed electric dipole moment of sulfur dioxide (charge +0.46 on S; Exercise 8-19). (Answer: 57% for the structure with two double bonds.)

8-25. Discuss the values of the enthalpy of formation of the dioxides, hexa-fluorides, and oxygen acids of sulfur, selenium, and tellurium in relation to position in the periodic table (see discussion of the stability of the higher oxidation states of bromine at the end of Section 8-1.)

8-26. The standard enthalpy of formation of liquid dimethylsulfone, $(CH_3)_2S{=}O$, is -196 kJ mole^{-1}, and its enthalpy of vaporization is 53 kJ mole^{-1}. What is the enthalpy of formation of $(CH_3)_2SO(g)$ from atoms? Subtract the sum of the bond energies of the single bonds in the molecule to obtain a value of the S$=$O bond energy. (Answer: 3409, 401 kJ mole^{-1}.)

8-27. What do you suggest as the electronic structure of H_5IO_6? What orbitals are used in forming the hybrid bond orbitals of the iodine atom?

8-28. The reactive molecule ClO, which has one unpaired electron, has been studied as a gas by spectroscopic methods. What electronic structure do you suggest for the molecule? Its standard enthalpy of formation is 138 kJ mole^{-1}. What is its enthalpy of formation from atoms? The three-electron bond (like the one-electron bond) is usually estimated to have about 50% or 60% of the bond energy of the corresponding electron-pair bond (single bond). What conclusion do you draw from the enthalpy of formation from atoms about the amount of three-electron-bond character in this molecule? (Answer: 231 kJ mole^{-1}.)

8-29. What electronic structure would you suggest for the odd molecule ClO_2 (one unpaired electron)? Note that the chlorine-oxygen bond length is 1.49 Å in this molecule, as compared with 1.69 Å in Cl_2O. The O—Cl—O bond angle has the value 116°.

8-30. The standard enthalpy of formation of $ClO_2(g)$ is 103 kJ mole^{-1}, and that of ClO(g) is 138 kJ mole^{-1}. What is the enthalpy of the reaction ClO(g) $+$ O(g) $\longrightarrow$ $ClO_2(g)$? How does this value compare with the average value for acid anions given in Section 8-3? (Answer: -2.84 kJ mole^{-1}.)

8-31. The crystals $NOClO_4$ and $(NO)_2SnCl_6$ closely resemble NH_4ClO_4 and $(NH_4)_2SnCl_6$, respectively, and are assumed to contain the nitronium ion NO^+. What electronic structure would you assign to this ion? What bond length?

8-32. Hydrazoic acid, HN_3, is a dangerously explosive gas. Its standard enthalpy of formation is 294 kJ mole^{-1}. How much heat would be liberated if the products of explosion were molecular hydrogen and nitrogen? Ammonia and nitrogen? NH(g) and nitrogen?

8-33. Discuss the electronic structure of the azide ion, N_3^-, in comparison with nitrous oxide and carbon dioxide.

8-34. From the tabulated enthalpy values calculate the heats of formation of SO(g) and PO(g) from atoms, and compare with the values given in the text for the S$=$O bond energy and the P$=$O bond energy in the ions of the oxygen acids.

8-35. Assign electronic structures to P_4O_6 and P_4O_{10}. What bond orbitals are used by the phosphorus atom in these molecules?

8-36. The enthalpy of formation of $P_4O_{10}(c)$ is given in Table 7-5 as -2984 kJ mole^{-1}, and that of $P_4O_6(c)$ as -1640 kJ mole^{-1}. Using 100 and 67 kJ mole^{-1}, respectively, for enthalpy of sublimation, evaluate the P=O bond energy.

8-37. (a) Using information in the preceding Exercise, obtain a value for the P—O bond energy.
(b) For comparison, calculate a value for the P—O bond energy from the values for P—P and O—O, with the correction for electronegativity difference.

8-38. The observed bond lengths in P_4O_{10} are 1.65 Å and 1.39 Å (the latter to oxygen atoms attached to only one phosphorus atom). The first corresponds to a single bond with a small amount of double-bond character, and the second to a triple bond. Draw a valence-bond structure involving single bonds and triple bonds. What orbitals are used as bond orbitals by the bridging oxygen atoms? By the outer oxygen atoms? How many unshared pairs do the oxygen atoms have? What orbitals are used as bond orbitals by the phosphorus atoms?

8-39. Using the structure with triple bonds to the outer oxygen atoms and assigning each bond the amount of ionic character corresponding to the electronegativity difference, calculate the electric charges on the phosphorus and oxygen atoms. Does this structure give better or worse agreement with the electroneutrality principle (atomic charges 0 ± 1) than the structure with double bonds to the outer oxygen atoms?

8-40. What do you estimate the value of the electric dipole moment of the P_4O_6 molecule to be? Of the P_4O_{10} molecule?

8-41. The heat of solution of hyponitrous acid in water can be estimated to be about 75 kJ mole^{-1} (compare 58 for H_2O_2, 16 for O_2). With this value, the standard enthalpy of formation of $H_2N_2O_2(aq)$ given in Table 7-4, and bond-energy values for the other bonds, evaluate the N=N bond energy. Compare with the values for N—N, N≡N, C—C, C=C, and C≡C. (Answer: 586 kJ mole^{-1}.)

8-42. Ammonium dihydrogen phosphate, $NH_4H_2PO_4$, can be made by passing ammonia into a water-phosphoric acid solution and spray-drying the product. It is said that this fertilizer can be delivered from England to countries such as India and Malaya at 20% less cost (per P_2O_5 unit) than other phosphates (see Exercise 8-4). Explain this statement.

9

Gases: Quantum Mechanics and Statistical Mechanics

A major part of chemical (and physical) theory was developed in connection with the experimental investigation of the properties of gases. In this chapter we shall discuss some of these properties, especially in their relation to the general theories of quantum mechanics and statistical mechanics.

It is interesting that it was not until the early years of the seventeenth century that the word "gas" was used. This word was invented by a Belgian physician, J. B. van Helmont (1577–1644), to fill the need caused by the new idea that different kinds of "airs" exist. Van Helmont discovered that a gas (the gas that we now call carbon dioxide) is formed when limestone is treated with acid, and that this gas differs from air in that when respired it does not support life and that it is heavier than air. He also found that the same gas is produced by fermentation, and that it is present in the Grotto del Cane, a cave in Italy in which dogs were observed to become unconscious (carbon dioxide escaping from fissures in the floor displaces the air in the lower part of the cave).

During the seventeenth and eighteenth centuries other gases were discovered, including hydrogen, oxygen, and nitrogen, and many of their properties were investigated. It was not until nearly the end of the eighteenth century, however, that these three gases were recognized as elements. When Lavoisier recognized that oxygen is an element, and that combustion is the process of combining with oxygen, the foundation of modern chemistry was laid.

Gases differ remarkably from liquids and solids in that the volume of a sample of gas depends in a striking way on the temperature of the gas and the applied pressure. The volume of a sample of liquid water, say 1 kg of water, remains essentially the same when the temperature and pressure are changed somewhat. Increasing the pressure from 1 atm to 2 atm causes the volume of a sample of liquid water to decrease by less than 0.01% and increasing the temperature from 0°C to 100°C causes the volume to increase by only 2%. On the other hand, the volume of a sample of air is cut in half when the pressure is increased from 1 atm to 2 atm, and it increases by 36.6% when the temperature is changed from 0°C to 100°C.

We can understand why these interesting phenomena attracted the attention of scientists during the early years of development of modern chemistry through the application of quantitative experimental methods of investigation of nature, and why many physicists and chemists during the past century have devoted themselves to the problem of developing a sound theory to explain the behavior of gases. A part of this theory is presented in the second part of this chapter.

In addition to the desire to understand this part of the physical world, there is another reason, a practical one, for studying the gas laws. This reason is concerned with the measurement of gases. The most convenient way to determine the amount of material in a sample of a solid is to weigh it on a balance. This can also be done conveniently for liquids; or we may measure the volume of a sample of a liquid, and, if we want to know its weight, multiply the volume by its density, as found by a previous experiment. The method of weighing is, however, not conveniently used for gases, because their densities are very small; volume measurements can be made much more accurately and easily by the use of containers of known volume. It is partly for this reason that study of the pressure-volume-temperature properties of gases is a part of chemistry.

Another important reason for studying the gas laws is that the density of a dilute gas is related in a simple way to its molecular weight, whereas there is no similar simple relation for liquids and solids. This relation for gases was of great value in the original decision as to the correct atomic weights of the elements, and it is still of great practical significance, permitting the direct calculation of the approximate density of a gas of known molecular composition, or the experimental determination of the effective (average) molecular weight of a gas of unknown molecular composition by the measurement of its density. These uses are discussed in the following sections.

It has been found by experiment that *at low density all ordinary gases behave in nearly the same way.* The nature of this behavior is described by the *perfect-gas laws* (often referred to briefly as the *gas laws*).

It is found by experiment that—to within the reliability of the gas laws—the volume of a sample of any gas is determined by only three quantities: the *pressure* of the gas, the *temperature* of the gas, and the *number of molecules* in the sample of the gas. The law describing the dependence of the volume of the gas on the pressure is called *Boyle's law*; that describing the dependence of the volume on the temperature is called the *law of Charles and Gay-Lussac*; and that describing the dependence of the volume on the number of molecules in the sample of gas is called *Avogadro's law*.

It is also found by experiment that the internal energy (Section 6-11) of a dilute gas changes very little with change in volume, at constant temperature. For a perfect gas the internal energy is independent of the volume. We shall see later, in our discussion of statistical mechanics (kinetic theory of gases), that the assumption that the molecules of a gas (a perfect gas) do not interact with one another (attract or repel one another) leads to this property as well as to the perfect-gas equation.

9-1. The Perfect-gas Equation

The Dependence of Gas Volume on Pressure

Experiments on the volume of a gas as a function of the pressure have shown that, for nearly all gases, the volume of a sample of gas at constant temperature is inversely proportional to the pressure; that is, the product of pressure and volume under these conditions is constant:

$$PV = \text{constant (temperature constant, moles of gas constant)} \quad (9\text{-}1)$$

This equation expresses Boyle's law. The law was inferred from experimental data by the English natural scientist Robert Boyle (1627–1691) in 1662.

Whereas all of the common gases, such as oxygen, hydrogen, nitrogen, carbon monoxide, carbon dioxide, and the rest, behave in the way described by Boyle's law, there are some gases that do not. One of these is the gas nitrogen dioxide, NO_2, the molecules of which can combine to form double molecules, of dinitrogen tetroxide, N_2O_4. A sample of this gas under ordinary conditions contains some molecules NO_2 and some molecules N_2O_4. When the pressure on the sample of the gas is changed the number of molecules of each of these kinds changes, causing the volume to depend on the pressure in a complicated way, rather than in the simple way described by Boyle's law. This effect will be discussed in Chapter 11.

The Partial Pressures of Components of a Gas Mixture

It is found by experiment (Dalton, 1801) that when two samples of gas at the same pressure are mixed there is no change in volume. If the two samples of gas were originally present in containers of the same size, at a pressure of 1 atm, each container after the mixing was completed would contain a mixture of gas molecules, half of them of one kind and half the other. It is reasonable to assume that each gas in this mixture exerts the pressure of $\frac{1}{2}$ atm, as it would if the other gas were not present. Dalton's *law of partial pressures* states that in a gas mixture the molecules of gas of each kind exert the same pressure as they would if present alone, and that the total pressure is the sum of the partial pressures exerted by the different gases in the mixture.

Correction for the Vapor Pressure of Water

When a sample of gas is collected over water, the pressure of the gas is due in part to the water vapor in it. The pressure due to the water vapor in the gas in equilibrium with liquid water is equal to the vapor pressure of water. Values of the vapor pressure of water at different temperatures are given in Appendix IX.

The Dependence of Gas Volume on Temperature

After the discovery of Boyle's law, it was more than one hundred years before the dependence of the volume of a gas on the temperature was investigated. Then in 1787 the French physicist Jacques Alexandre Charles (1746–1823) reported that different gases expand by the same fractional amount for the same rise in temperature. Dalton in England continued these studies in 1801, and in 1802 Joseph Louis Gay-Lussac (1778–1850) extended the work, and determined the amount of expansion per degree Centigrade. He found that all gases expand by $\frac{1}{273}$ of their volume at 0°C for each degree Centigrade that they are heated above this temperature. Thus a sample of gas with volume 273 ml at 0°C has the volume 274 ml at 1°C and the same pressure, 275 ml at 2°C, 373 ml at 100°C, and so on.

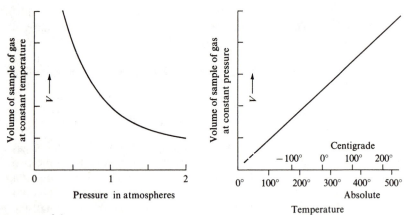

FIGURE 9-1
Curves showing, at the left, the dependence of the volume of a sample of gas at constant temperature and containing a constant number of molecules on the pressure, and, at the right, the dependence of the volume of a sample of gas at constant pressure and containing a constant number of molecules on the temperature.

We now state the law of the dependence of the volume of a gas on temperature, the *law of Charles and Gay-Lussac*, in the following way: if the pressure and the number of moles of a sample of gas remain constant, the volume of the sample of gas is proportional to the absolute temperature:

$$V = \text{constant} \times T \text{ (pressure constant, number of moles constant)} \quad (9\text{-}2)$$

The dependence of volume on the absolute temperature is a direct proportionality, whereas the volume is inversely proportional to the pressure. The nature of these two relations is illustrated in Figure 9-1.

Standard Conditions

It is customary to refer the volumes of gases to conditions of temperature 0°C and pressure 1 atm. This temperature and pressure are called *standard conditions*. A sample of gas is said to be reduced to standard conditions when its volume is calculated at this temperature and pressure.

Avogadro's Law

In 1805 Gay-Lussac began a series of experiments to find the volume percentage of oxygen in air. The experiments were carried out by mixing a certain volume of hydrogen with air and exploding the mixture, and then testing the remaining gas to see whether oxygen or hydrogen had been present in excess. He was surprised to find a very simple relation: 1000 ml of oxygen required just 2000 ml of hydrogen, to form water. Continuing the study of the volumes of gases that react with one another, he found that 1000 ml of hydrogen chloride combined exactly with 1000 ml of ammonia, and that 1000 ml of carbon monoxide combined with 500 ml of oxygen to form 1000 ml of carbon dioxide. On the basis of these observations he formulated the law of combining volumes: the volumes of gases that react with one another or are produced in a chemical reaction are in the ratios of small integers.

Such a simple empirical law as this called for a simple theoretical interpretation, and in 1811 Amedeo Avogadro (1776–1856), professor of physics in the University of Turin, proposed a hypothesis to explain the law. Avogadro's hypothesis was that *equal numbers of molecules are contained in equal volumes of all dilute gases under the same conditions*. This hypothesis has been thoroughly verified to within the accuracy of approximation of real gases to ideal behavior, and it is now called a law— *Avogadro's law*.*

During the last century Avogadro's law provided the most satisfactory and the only reliable way of determining which multiples of the equivalent weights of the elements should be accepted as their atomic weights; the arguments involved are discussed in the following sections. But the value of this law remained unrecognized by chemists from 1811 to 1858. In this year Stanislao Cannizzaro (1826–1910), an Italian chemist working in Geneva, showed how to apply the law systematically, and immediately the uncertainty about the correct atomic weights of the elements and the correct formulas of compounds disappeared. Before 1858 many chemists used the formula HO for water and accepted 8 as the atomic weight of oxygen; since that year H_2O has been accepted as the formula for water by everyone.†

Avogadro's Law and the Law of Combining Volumes

Avogadro's law requires that the volumes of gaseous reactants and products (under the same conditions) be approximately in the ratios of small integers; the numbers of molecules of reactants and products in a chemical reaction are in integral ratios, and the same ratios represent the relative gas volumes. Some simple diagrams illustrating this for several reactions are given in Figure 9-2. Each cube in these diagrams represents the volume occupied by four gas molecules.

---

*Dalton had considered and rejected the hypothesis that equal volumes of gases contain equal numbers of atoms; the idea that elementary substances might exist as polyatomic molecules (H_2, O_2) did not occur to him.

†The failure of chemists to accept Avogadro's law during the period from 1811 to 1858 seems to have been due to a feeling that molecules were too "theoretical" to deserve serious consideration.

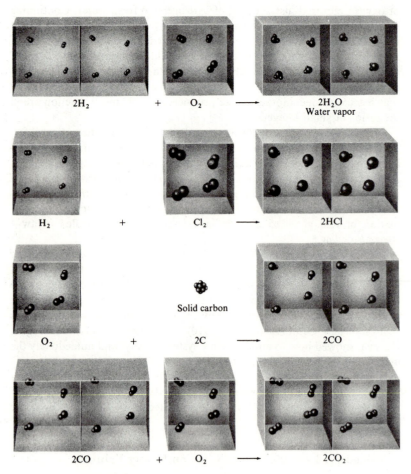

FIGURE 9-2
The relative volumes of gases involved in chemical reactions.

The Use of Avogadro's Law in the Determination of the Correct Atomic Weights of Elements

The way in which Avogadro's law was applied by Cannizzaro in 1858 for the selection of the correct approximate atomic weights of elements was essentially the following. Let us accept as the molecular weight of a substance the weight in grams of 22.4 liters of the gaseous substance reduced to standard conditions. (Any other volume could be used—this would correspond to the selection of a different base for the atomic weight scale.) Then it is probable that, of a large number of compounds of a particular element, at least one compound will have only one atom of the element per molecule; the weight of the element in the standard gas volume of this compound is its atomic weight.

For gaseous compounds of hydrogen the weight per standard volume and the weight of the contained hydrogen per standard volume are given in the table below.

	Weight of gas (in grams)	Weight of contained hydrogen (in grams)
Hydrogen (H_2)	2	2
Methane (CH_4)	16	4
Ethane (C_2H_6)	30	6
Water (H_2O)	18	2
Hydrogen sulfide (H_2S)	34	2
Hydrogen cyanide (HCN)	27	1
Hydrogen chloride (HCl)	36	1
Ammonia (NH_3)	17	3
Pyridine (C_5H_5N)	79	5

In these and all other compounds of hydrogen the minimum weight of hydrogen in the standard gas volume is found to be 1 g, and the weight is always an integral multiple of the minimum weight; hence 1 can be accepted as the atomic weight of hydrogen. The elementary substance hydrogen then is seen to consist of diatomic molecules H_2, and water is seen to have the formula H_2O_x, with x still to be determined.

For oxygen compounds the following similar table of experimental data can be set up.

	Weight of gas (in grams)	Weight of contained oxygen (in grams)
Oxygen (O_2)	32	32
Water (H_2O)	18	16
Carbon monoxide (CO)	28	16
Carbon dioxide (CO_2)	44	32
Nitrous oxide (N_2O)	44	16
Nitric oxide (NO)	30	16
Sulfur dioxide (SO_2)	64	32
Sulfur trioxide (SO_3)	80	48

From the comparison of oxygen and water in this table it can be concluded rigorously that the oxygen molecule contains two atoms or a multiple of two atoms; we see that the standard volume of oxygen contains twice as much oxygen (32 g) as is contained by the standard volume of water vapor (16 g of oxygen). The data for the other compounds provide no evidence that the atomic weight of oxygen is less than 16; hence this value may be adopted. Water thus is given the formula H_2O.

Note that this application of Avogadro's law provided rigorously only a maximum value of the atomic weight of an element. The possibility was not eliminated that the true atomic weight was a submultiple of this value.

The Complete Perfect-gas Equation

Boyle's law, the law of Charles and Gay-Lussac, and Avogadro's law can be combined into a single equation,

$$PV = nRT \qquad (9\text{-}3)$$

In this equation P is the pressure acting on a given sample of gas, V is the volume occupied by the sample of gas, n is the number of moles of gas in the sample, R is a quantity called the *gas constant*, and T is the absolute temperature.

The gas constant R has a numerical value depending on the units in which it is measured (that is, the units used for P, V, n, and T). If P is measured in atmospheres, V in liters, n in moles, and T in degrees Kelvin, the value of R is 0.0820 liter atmospheres per degree mole (more accurately 0.08206 l atm $\deg^{-1}$ mole^{-1}); R also has the value 8.3146 J $\deg^{-1}$ mole^{-1}. The value of R/N (N is Avogadro's number) is called *Boltzmann's constant, k;* its value is 13.805×10^{-24} J $\deg^{-1}$.

The perfect-gas equation can also be written in the form

$$n = \frac{PV}{RT} \qquad (9\text{-}4)$$

This equation expresses Avogadro's law.

The volume of one mole of a perfect gas at standard conditions is 22.415 liters (22.415×10^{-3} m^3).

Some of the ways in which the perfect-gas equation can be used in the solution of chemical problems are illustrated in the following examples.

Example 9-1.   What is the density of oxygen at standard conditions?

Solution.   The volume of one mole of O_2, with mass 31.9988 g, is 22.415 l. Hence the density is 31.9988/22.415 = 1.428 g l^{-1}.

Example 9-2.   An experiment is made to find out how much oxygen is liberated from a given amount of potassium chlorate, $KClO_3$. The quantity 2.00 g of this salt is weighed out, mixed with some manganese dioxide (to serve as a catalyst), and introduced into a test tube, which is provided with a cork and delivery tube leading to a bottle filled with water and inverted in a trough. The test tube is heated, and the heating is continued until the evolution of gas ceases. The volume of the liberated gas was determined to be 591 ml. The temperature was 18°C, and the pressure was 0.9851 atm. What was the weight of oxygen liberated? How does this compare with the theoretical yield?

Solution. The atmospheric pressure, 0.9851 atm, is balanced in part by the pressure of the oxygen collected in the bottle, and in part by the pressure of the water vapor dissolved by the oxygen as it bubbles through the water. By reference to Appendix IX we see that the equilibrium vapor pressure of liquid water at 18°C is 0.0204 atm. Accordingly, the partial pressure of the oxygen in the bottle is 0.9851 − 0.0204 = 0.9647 atm. The volume at this pressure is 591 ml. We know that increasing the pressure (to 1 atm) will make the volume smaller; hence we multiply 591 ml by 0.9647/1, to obtain 570 ml as the volume at 1 atm and 18°C (291.15°K). This is corrected to the smaller volume at standard conditions by multiplying by the ratio 273.15/291.15, with the result 535 ml. (The same result, to three significant figures, is obtained by multiplying by 273/291.) From Example 9-1 we know that the density of oxygen at standard conditions is 1.428 g l^{-1}. Hence the weight of oxygen liberated is 0.535 $\times$ 1.428 = 0.764 g.

To answer the second question, let us calculate the theoretical yield of oxygen from 2.00 g of potassium chlorate. The equation for the decomposition of potassium chlorate is

$$KClO_3 \longrightarrow KCl + \tfrac{3}{2}O_2$$

We see that 1 gram formula weight of $KClO_3$, 122.6 g, should liberate 3 gram-atoms of oxygen, 48.0 g. Hence the amount of oxygen that should be liberated from 2.00 g of potassium chlorate is 48.0/122.6 $\times$ 2.00 g = 0.783 g.

The observed amount of oxygen liberated is seen to be less than the theoretical amount by 0.019 g, which is 2.4%.

Example 9-3. Determination of the Molecular Weight of a Substance by the Mercury-column Method. A chemist isolated a substance in the form of a yellow oil. He found on analysis that the oil contained only hydrogen and sulfur, and the amount of water obtained when a sample of the substance was burned showed that it consisted of about 3% hydrogen and 97% sulfur. To determine the molecular weight he prepared a very small glass bulb, weighed the glass bulb, filled it with the oil, and weighed it again; the difference in the two weighings, which is the weight of the oil, was 0.0302 g. He then introduced the filled bulb into the evacuated space above the mercury column in a tube, as shown in Figure 9-3. The level of the mercury dropped to 118 mm below its original level, after the oil had been completely vaporized. The temperature of the tube was 30°C. The volume of the gas phase above the mercury at the end of the experiment was 73.2 ml. Find the molecular weight and formula of the substance.

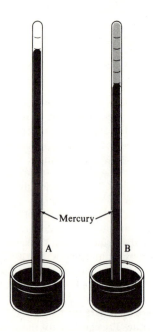

Mercury

A   B

FIGURE 9-3
The Hofmann method for determining
the density of a vapor.

Solution. The vapor of the substance is stated to occupy the volume 73.2 ml at temperature 30°C and pressure 118 mm Hg. Its volume corrected to standard conditions is seen to be

$$73.2 \text{ ml} \times \frac{273}{303} \times \frac{118}{760} = 10.24 \text{ ml}$$

One mole of gas at standard conditions occupies 22,415 ml; hence the number of moles in the sample of the substance is $10.24/22{,}415 = 0.000457$. The weight of this fraction of a mole is stated to be 0.0302 g; hence the weight of one mole is this weight divided by the number of moles:

$$\text{Molar weight of substance} = \frac{0.0302 \text{ g}}{0.000457 \text{ mole}} = 66.1 \text{ g mole}^{-1}$$

The substance was found by analysis to contain 3% hydrogen and 97% sulfur. If we had 100 g of the oil, it would contain 3 g of hydrogen, which is 3 gram-atoms, and 97 g of sulfur, which is also 3 gram-atoms (the atomic weight of sulfur is 32). Hence the molecule contains equal numbers of hydrogen atoms and sulfur atoms. If its formula were HS, its molecular weight would be the sum of the atomic weights of hydrogen and sulfur, 33. It is evident from the observed molecular weight that the formula is H_2S_2, the molecular weight of which is 66.15.

9-2.   Quantum Mechanics of a Monatomic Gas

We shall begin our theoretical treatment of gases by discussing the quantized states of a monatomic molecule in a cubical box.

Let us discuss first the one-dimensional problem, that of a particle (a molecule) restricted to motion along a line, the x-axis, from $x = 0$ to $x = a$; that is, it is in a one-dimensional box with length a. The potential energy is taken as 0 in the region $0 < x < a$. In this region the particle moves as a free particle, with constant velocity v and constant momentum mv, which change sign, but not magnitude, when the particle bounces back from the wall at $x = a$.

It has been pointed out in Section 3-8 that an electron moving with the momentum mv has the wavelength $\lambda = h/mv$. It can be represented by the wave function for a pure sinusoidal wave with this wavelength, which is sin $2\pi x/\lambda$ or cos $2\pi x/\lambda$, through the range $0 \leq x \leq a$. The barriers at $x = 0$ and $x = a$ prevent the particle from getting outside the box. Hence the probability function for the particle, which is the square of the wave function, must fall to zero at $x = 0$ and also at $x = a$. The cosine functions are thus ruled out (they have value 1 at $x = 0$). Moreover, all of the sine functions are ruled out except those with sin $2\pi a/\lambda$ equal to zero, which requires that $2a/\lambda$ be equal to an integer, $n\,(n = 1, 2, 3, \ldots)$.

We are thus led to the conclusion that the quantized states for the particle in a one-dimensional box with length a are those with

$$\lambda = \frac{h}{mv} = \frac{2a}{n} \tag{9-5}$$

with $n = 1, 2, 3, \ldots$.

The allowed wave functions are

$$\Psi_n(x) = \left(\frac{2}{a}\right)^{1/2} \sin \frac{\pi n x}{a} \tag{9-6}$$

Several of these functions and their squares (probability distribution functions for the particle) are represented in Figure 9-4. The factor $(2/a)^{1/2}$ is a normalization factor, which leads to total probability unity that the particle is in the box.

We note that the length of the "orbit" of the particle (motion from $x = 0$ to $x = a$ and back to $x = 0$) is $2a$, which is required to be equal to $n\lambda$, that is, to be an integral number of wavelengths. This relation is more meaningful than the corresponding relation for circular Bohr orbits in the hydrogen atom, pointed out by de Broglie (Section 3-8), because the wavelength is constant for a free particle (even in a box, when not just bouncing from a wall), whereas that of the electron in a hydrogen atom is not constant.

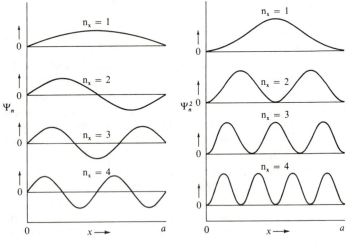

FIGURE 9-4

The wave function $\Psi_n(x)$ and the probability distribution function $\Psi_n^2(x)$ for the particle in a box for the first four values of n.

The energy of the particle is the kinetic energy $\frac{1}{2}mv^2$, which is found, with use of the relation $mv = nh/2a$ from Equation 9-5, to have the value $n^2h^2/8ma^2$:

$$E_n = \tfrac{1}{2}mv^2 = \frac{(mv)^2}{2m} = \frac{n^2h^2}{8ma^2} \tag{9-7}$$

The treatment of a particle in a three-dimensional cubical box ($0 \leq x \leq a$, $0 \leq y \leq a$, $0 \leq z \leq a$) is closely similar. The velocity vector has three components, v_x v_y, and v_z, parallel to the three edges of the box. Each must have such value that the corresponding wavelength is an integral multiple of $2a$; that is,

$$\left.\begin{array}{l} mv_x = n_xh/2a \\ mv_y = n_yh/2a \\ mv_z = n_zh/2a \end{array}\right\} \tag{9-8}$$

with $n_x, n_y, n_z = 1, 2, 3, \ldots$. The corresponding wave functions are

$$\Psi_{n_x n_y n_z} = \left(\frac{2}{a}\right)^{3/2} \sin\frac{\pi n_x x}{a} \sin\frac{\pi n_y y}{a} \sin\frac{\pi n_z z}{a} \tag{9-9}$$

The energy of the quantized states, $\frac{1}{2}m(v_x^2 + v_y^2 + v_z^2) = \frac{1}{2}mv^2$, is given by the following equation:

$$E(n_x, n_y, n_z) = \frac{(n_x^2 + n_y^2 + n_z^2)h^2}{8ma^2} \tag{9-10}$$

The allowed sets of values of the three quantum numbers n_x, n_y, n_z are 1, 1, 1; 2, 1, 1; 1, 2, 1; 1, 1, 2; 2, 2, 1; and so on. They are illustrated in Figure 9-5, and the energy levels are shown in Figure 9-6. The lowest level is a single state; it is called a nondegenerate level. The second level is three states; it is called a three-fold degenerate level.

If the particle were an electron, which can orient its spin in either one of two ways, then the lowest level would correspond to two states, 1, 1, 1 with $m_s = \frac{1}{2}$ and 1, 1, 1 with $m_s = -\frac{1}{2}$; and the degree of degeneracy of each of the other levels would be doubled.

Let us consider two states, A and B (for example, Ψ_{111} and Ψ_{211}) and two identical particles. We may write four wave functions for the system:

$$A(1)\ A(2)$$
$$B(1)\ B(2)$$
$$A(1)\ B(2)$$
$$B(1)\ A(2)$$

The function $A(1)\ A(2)$ means that particle 1 is in state A and particle 2 is in state A. The probability distribution function is $\{A(1)\ A(2)\}^2 = \{A(1)\}^2\{A(2)\}^2$; that is, each particle has a time-average distribution in the box shown by the lowest curve in Figure 9-5. A complication arises for the third function, $A(1)\ B(2)$, for which the probability distribution function is $\{A(1)\}^2\{B(2)\}^2$. The functions A^2 and B^2 are different; hence the "identical" particles have different properties in this state, and presumably could be distinguished from one another by a suitable experiment. But this is contrary to observation; if they could be so distinguished from one another we would not describe the two particles as identical.

The resolution of this problem is that the two functions $A(1)\ B(2)$ and $B(1)\ A(2)$ are not acceptable wave functions for the system of two identical particles, but their sum and difference are. We are thus led to the following four wave functions:

Symmetric in 1 and 2

$$\left.\begin{array}{c} A(1)\ A(2) \\ B(1)\ B(2) \\ A(1)\ B(2) + B(1)\ A(2) \end{array}\right\} \qquad (9\text{-}11)$$

Antisymmetric in 1 and 2

$$A(1)\ B(2) - B(1)\ A(2) \qquad (9\text{-}12)$$

It is easy to verify that the probability distribution function for each of these four functions (the square of the function) is the same for particle 1 as for particle 2. These functions are acceptable for two identical particles.

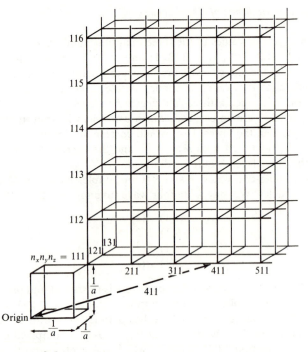

FIGURE 9-5
A geometrical representation of the energy levels for a particle in a rectangular box.

$\dfrac{8ma^2W}{h^2}$		$n_x n_y n_z$
27	$p = 4$	511 etc., 333
26	$p = 6$	431 etc.
24	$p = 3$	422 etc.
22	$p = 3$	332 etc.
21	$p = 6$	421 etc.
19	$p = 3$	331 etc.
18	$p = 3$	411 etc.
17	$p = 3$	322 etc.
14	$p = 6$	321, 132, 213, 312, 231, 123
12	$p = 1$	222
11	$p = 3$	311, 131, 113
9	$p = 3$	221, 122, 212
6	$p = 3$	211, 121, 112
3	$p = 1$	111
0		

FIGURE 9-6
Energy levels, degrees of degeneracy, and quantum numbers for a particle in a cubic box.

Bosons and Fermions

It is observed that some particles occupy only symmetric wave functions. They are called *bosons*; they are named after the Indian physicist S. N. Bose, who in 1924 discovered that photons are bosons. (Albert Einstein made the same discovery, at the same time.) Other particles, including the electron, the proton, and the neutron, are *fermions*; they are named after Enrico Fermi (1901–1954), who, together with W. Pauli and P. A. M. Dirac, was responsible for their understanding. Bosons have integral spin (0, 1, . . .), and fermions have half-integral spin ($\frac{1}{2}$, $\frac{3}{2}$, . . .).

Compound particles are fermions if they contain an odd number of fermions; otherwise they are bosons. For example, the $^3_2He_1$ nucleus (two protons and one neutron) is a fermion, and the $^4_2He_2$ nucleus is a boson. The helium-4 atom (nucleus and two electrons) is a boson.

Two identical bosons in a system can occupy the three symmetric states, Equations 9-11, but not the antisymmetric state, Equation 9-12; two identical fermions can occupy the antisymmetric state, Equation 9-12, but not the other three; two nonidentical particles can occupy all four.

The number of quantized states for a system of many identical particles is decreased tremendously by this symmetry requirement. Let us consider 26 nonidentical particles, 1 to 26, and 26 different one-particle states, A to Z. The first particle might occupy any of the 26 functions A to Z; the second has 25 choices for each choice by the first, and so on, giving a total of 26! different states for the system.* But there is only one completely symmetric combination of these 26! functions (such as $A(1) B(2) \ldots Z(26)$), and only one completely antisymmetric combination. For identical particles, either bosons or fermions, the number of states of the system (not involving double occupancy, such as $A(1) A(2) C(3) \ldots$) is reduced by division by $N!$, where N is the number of particles.

9-3. The Wave Equation

A particle in field-free one-dimensional space has wavelength $\lambda = h/mv$ and can be represented by the sinusoidal wave functions $\sin 2\pi xmv/h$ and $\cos 2\pi xmv/h$. These wave functions are solutions of the differential equation

$$\frac{d^2\Psi(x)}{dx^2} = -\frac{8\pi^2 m}{h^2} E\Psi(x) \qquad (9\text{-}13)$$

In this equation E is the energy of the quantized state, equal to the kinetic energy, $\frac{1}{2}mv^2$, for the free particle, with potential energy zero.

---

*Factorial N, $N!$, is equal to $1 \times 2 \times 3 \times \cdots \times N$.

This equation is the Schrödinger wave equation, discovered by Erwin Schrödinger in 1925 (published in 1926). For a particle moving in one dimension in a force field described by the potential energy function $V(x)$ the Schrödinger equation is

$$\frac{d^2\Psi}{dx^2} = \frac{8\pi^2 m}{h^2} \{V(x) - E\}\Psi(x) \tag{9-14}$$

The particle in a one-dimensional box has $V(x) = 0$ for $0 < x < a$, and $V(x)$ very large for $x \leq 0$ and $x \geq a$. The allowed solutions of the wave equation 9-14 are the functions 9-6, which correspond to the values of the energy E given in Equation 9-7.

The Schrödinger wave equation for a particle in three-dimensional space is

$$\frac{\partial^2\Psi}{\partial x^2} + \frac{\partial^2\Psi}{\partial y^2} + \frac{\partial^2\Psi}{\partial z^2} = \frac{8\pi^2 m}{h^2}(V - E)\Psi \tag{9-15}$$

Here Ψ is $\Psi(x, y, z)$, a function of the three coordinates of the particle.

The Number of Quantum States

In the discussion, later in this chapter, of the kinetic theory of gases we shall want to know the number of quantum states of a particle in the range of energy E to $E + dE$. Let us consider all states with translational energy from 0 to E_{max}. From Equation 9-10 we see* that these are the states with

$$r^2 = n_x^2 + n_y^2 + n_z^2 \leq \frac{8m \, V^{2/3} \, E_{max}}{h^2} \tag{9-16}$$

Here V, the volume of the box, is equal to a^3. The symbol r represents the radius (distance from the origin to the point n_x, n_y, n_z) in Figure 9-5. The volume of the octant with radius r is $\frac{1}{8} \times \frac{4\pi r^3}{3}$, and there is one quantum state per unit volume; hence this volume is the number of states with energy between 0 and E_{max}:

$$N(E_{max}) = \frac{\pi}{6} r^3 = \frac{\pi}{6}\left(\frac{8mV^{2/3} E_{max}}{h^2}\right)^{3/2}$$

Hence we obtain

$$N(E) = \frac{8\pi 2^{1/2} m^{3/2} V \, E^{3/2}}{3h^3} \tag{9-17}$$

(The subscript max on E has been dropped.) By differentiation we obtain

$$dN = \frac{4\pi 2^{1/2} m^{3/2} V \, E^{1/2}}{h^3} \, dE \tag{9-18}$$

---

*Note that Equation 9-10 states that $n_x^2 + n_y^2 + n_z^2 = \frac{8ma^2 E}{h^2}$.

9-4. The Kinetic Theory of Gases

During the nineteenth century the concepts that atoms and molecules are in continual motion and that the temperature of a body is a measure of the intensity of this motion were developed. The idea that the behavior of gases could be accounted for by considering the motion of the gas molecules had occurred to several people (Daniel Bernoulli in 1738, J. P. Joule in 1851, A. Kronig in 1856), and in the years following 1858 this idea was developed into a detailed kinetic theory of gases by Clausius, Maxwell, Boltzmann, and many later investigators. The subject is discussed in courses in physics and physical chemistry, and it forms an important part of the branch of theoretical science called statistical mechanics.

In a gas at temperature T the molecules are moving about, different molecules having at a given time different speeds v and different kinetic energies of translational motion $\frac{1}{2}mv^2$ (m being the mass of a molecule). It has been found that the average kinetic energy per molecule, $\frac{1}{2}m[v^2]_{average}$, is the same for all gases at the same temperature, and that its value increases with the temperature, being directly proportional to T.

The average (root-mean-square*) velocity of hydrogen molecules at 0°C is 1.84×10^3 m s^{-1}. At higher temperatures the average velocity is greater; it reaches twice as great a value, 3.68×10^3 m s^{-1}, for hydrogen molecules at 820°C, corresponding to an increase by 4 in the absolute temperature.

Since the average kinetic energy is equal for different molecules, the average value of the square of the velocity is seen to be inversely proportional to the mass of the molecule, and hence the average velocity (root-mean-square average) is inversely proportional to the square root of the molecular weight. The molecular weight of oxygen is just 16 times that of hydrogen; accordingly, molecules of oxygen move with a speed just one quarter as great as molecules of hydrogen at the same temperature. The average speed of oxygen molecules at 0°C is 0.46×10^3 m s^{-1}.

The explanation of Boyle's law given by the kinetic theory is simple. A molecule on striking the wall of the container of the gas rebounds, and contributes momentum to the wall; in this way the collisions of the molecules of the gas with the wall produce the gas pressure, which balances the external pressure applied to the gas. If the volume is decreased by 50%, each of the molecules strikes the walls twice as often, and hence the pressure is doubled. The explanation of the law of Charles and Gay-Lussac is

---

*The root-mean-square average of a quantity x is the square root of the average value of the square of the quantity, $\left\{ \left(\sum\limits_{i=1}^{n} x_i^2 \right) \Big/ n \right\}^{1/2}$.

equally simple. If the absolute temperature is doubled, the speed of the molecules is increased by the factor $\sqrt{2}$. This causes the molecules to make $\sqrt{2}$ times as many collisions as before, and each collision is increased in force by $\sqrt{2}$, so that the pressure itself is doubled ($\sqrt{2} \times \sqrt{2} = 2$) by doubling the absolute temperature. Avogadro's law is also explained by the fact that the average kinetic energy is the same at a given temperature for all gases.

The Effusion and Diffusion of Gases; the Mean Free Paths of Molecules

There is an interesting dependence of the rate of effusion of a gas through a small hole on the molecular weight of the gas. The speeds of motion of different molecules are inversely proportional to the square roots of their molecular weights. If a small hole is made in the wall of a gas container, the gas molecules will pass through the hole into an evacuated region outside at a rate determined by the speed at which they are moving (these speeds determine the probability that a molecule will strike the hole). Accordingly, the kinetic theory requires that the rate of *effusion* of a gas through a small hole be inversely proportional to the square root of its molecular weight. This law was discovered experimentally before the development of the kinetic theory—it was observed that hydrogen effuses through a porous plate four times as rapidly as oxygen.

In the foregoing discussions we have ignored the appreciable sizes of gas molecules, which cause the molecules to collide often with one another. In an ordinary gas, such as air at standard conditions, a molecule moves only about 500 Å, on average, between collisions; that is, its *mean free path* under these conditions is only about two hundred times its own diameter.

The value of the mean free path is significant for phenomena that depend on molecular collisions, such as the viscosity and the thermal conductivity of gases. Another such phenomenon is the *diffusion* of one gas through another or through itself (such as of radioactive molecules of a gas through the nonradioactive gas). In the early days of kinetic theory it was pointed out by skeptics that it takes minutes or hours for a gas to diffuse from one side of a quiet room to the other, even though the molecules are attributed velocities of about a mile per second. The explanation of the slow diffusion rate is that a molecule diffusing through a gas is not able to move directly from one point to another a long distance away, but instead is forced by collisions with other molecules to follow a tortuous path, making only slow progress in its resultant motion. Only when diffusing into a high vacuum can the gas diffuse with the speed of molecular motion.

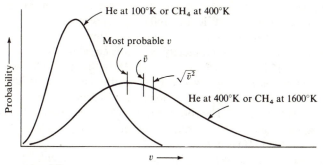

FIGURE 9-7
The velocity distribution function for helium atoms at 100°K and for
helium atoms at 400°K. These two curves also apply to methane
at temperatures 400°K and 1600°K, respectively.

9-5.   The Distribution Law for Molecular Velocities

In 1860 the English physicist James Clerk Maxwell (1831–1879) derived
an equation that correctly gives the fraction of gas molecules with velocities in the range v to $v + dv$. This equation is called the *Maxwell
distribution law* (or *Maxwell-Boltzmann distribution law*) for molecular
velocities. In a perfect gas at temperature T, containing N molecules, each
with mass m, we ask how many molecules dN have velocities lying between v and $v + dv$. The velocity v may be described as a vector with
components v_x, v_y, and v_z in velocity space. The volume of the spherical
shell bounded by the surfaces v and $v + dv$ is $4\pi v^2 dv$. It was found by
Maxwell through analysis of the transfer of momentum from one molecule
to another during a molecular collision that this volume element must be
multiplied by the exponential factor $\exp(-\frac{1}{2}mv^2/kT)$. (This factor, called
the Boltzmann factor, is discussed in the following section.) The normalizing factor $(m/2\pi kT)^{3/2}$ is also needed in order that the integral of dN over
all velocities ($v = 0$ to $v = \infty$) should be equal to N. The distribution law
for molecular velocities is

$$dN = 4\pi N \left(\frac{m}{2\pi kT}\right)^{3/2} \exp\left(-\frac{\frac{1}{2}mv^2}{kT}\right) v^2 \, dv \qquad (9\text{-}19)$$

The distribution function calculated for helium atoms at 100°K and also
for helium atoms at 400°K is shown as Figure 9-7. We see that in Equation 9-19 the mass and the absolute temperature occur only in the ratio
m/T. Accordingly, the two curves that are shown apply also to methane,

CH_4, with molecular weight four times that of helium, at temperatures four times as great, 400°K and 1600°K, respectively.

The maximum for the distribution function occurs at the value of v called the most probable velocity, v_{mp}; it is equal to $(2kT/m)^{1/2}$, which has the value $128.95(T/M)^{1/2}$ m s^{-1} (here M is the molecular weight). This value is represented by a vertical line on the curve for helium at 400°K.

The average value of the velocity is $(8kT/\pi m)^{1/2}$, which is equal to $145.51(T/M)^{1/2}$ m s^{-1}. The root-mean-square value of the velocity, which is the square root of the average value of v^2, is equal to $(3kT/m)^{1/2}$, with value $157.94(T/M)^{1/2}$ m s^{-1}.

We see that *the average kinetic energy per molecule*, $\frac{1}{2}m(v^2)_{average}$, *is equal to $\frac{3}{2}kT$*. It accordingly has the same value for all gases at the same temperature, as stated in the second paragraph of Section 9-4.

This result, called the *equipartition of energy*, is one of the most important consequences of the kinetic theory. Some of the ways in which it can be used in the discussion of the properties of gases have been mentioned above.

The law of equipartition of energy is sometimes expressed in the following words: *the average kinetic energy of molecules or other particles in the region in which classical theory applies is $\frac{1}{2}kT$ per degree of freedom*. A molecule is considered to have three degrees of freedom of translational motion, corresponding to the three components of velocity v_x, v_y, and v_z, and the average kinetic energy of the molecule at temperature T is accordingly $\frac{3}{2}kT$. This value for the average kinetic energy applies also to liquids and crystals, if the temperature is high enough for the classical theory to be valid. At lower temperatures, at which the quantum effect of decrease in heat capacity below the equipartition value occurs, the average energy per degree of freedom is less than the equipartition value.

Example 9-4.   The amount of heat required to raise the temperature of unit quantity (1 mole or 1 g) of a substance by 1°C without change in phase is called the *heat capacity* (sometimes called *specific heat*) of the substance. What is the heat capacity of helium gas at constant volume?

Solution.   In helium gas the molecules (atoms of helium) interact with one another only very weakly, and we need to consider only the kinetic energy of the molecules. The average value of the kinetic energy is $\frac{3}{2}kT$ per molecule, which is equal to $\frac{3}{2}RT$ per mole. The increase in energy accompanying an increase by dT in temperature is

$$\frac{d}{dT}(\tfrac{3}{2}RT) = \tfrac{3}{2}R$$

The value of R is 8.315 J deg^{-1} mole^{-1}; accordingly the heat capacity at constant volume, C_V, is $\frac{3}{2} \times 8.315 = 12.47$ J deg^{-1} mole^{-1}. This value is found by experiment for all monatomic gases.

Pressure-Volume Work

When the volume of a system changes by amount dV at constant pressure P an amount of work PdV is done. The perfect-gas equation is $PV = nRT$. Let us consider an increase in temperature dT, with pressure held constant, of a sample of gas in a cylinder with a moveable piston. The amount of work done (assuming no friction) is PdV. By differentiation we find that $PdV = nRdT$ (for P constant). The amount of work done by expansion of the gas is accordingly $nRdT$; that is, the amount R per mole and per degree increase in temperature. This amount of thermal energy is extracted from the environment, in addition to the thermal energy that increases the internal energy of the gas, when a gas is heated at constant pressure. Accordingly the heat capacity at constant pressure, C_P, of a mole of gas is larger than the heat capacity at constant volume, C_V, by the amount R, equal to 8.3146 J deg^{-1} mole^{-1}.

This conclusion agrees with experiment. The heat capacity of a monatomic gas is 12.47 J deg^{-1} mole^{-1} at constant volume and 20.76 J deg^{-1} mole^{-1} at constant pressure.

Experimental Verification of the Distribution Law

Many deductions from the distribution law were found to be in agreement with experiment, and none in disagreement. For example, Maxwell showed that according to kinetic theory the viscosity of a gas should be independent of pressure (except at very small and very large pressures), and should increase with increasing temperature, rather than decrease. These surprising properties were verified by experiment, and the kinetic theory of gases, including the distribution law for molecular velocities, was accepted long before a direct experimental determination of the velocity distribution function could be carried out. By 1920 the experimental techniques of physics, especially the ability to obtain a high vacuum, had developed enough to permit direct determinations to be made. Otto Stern (born 1888) carried out the first experiment of this sort. He studied a beam of silver atoms emitted from the silver coating on a tungsten wire heated to about 1200°C. The beam was defined by a system of slits, and it then impinged on the surface of a rotating drum. One of the slits was also rotating in such a way that the atoms of silver could pass through the slit only during a small fraction of the time of revolution of the drum. The fast atoms struck the drum very quickly, before it had rotated far, whereas the slow atoms were delayed in striking the drum. This experiment gave a rough verification of the distribution function.

Later experiments have led to essentially complete experimental verification of the distribution function. One of these experiments, carried out by I. Estermann, O. C. Simpson, and O. Stern in 1947, is illustrated in

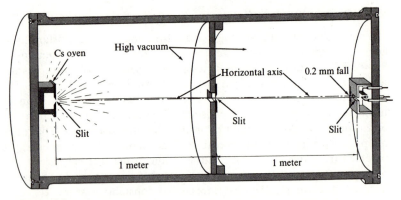

FIGURE 9-8
Apparatus used in experimental test of the distribution law for molecular velocities.

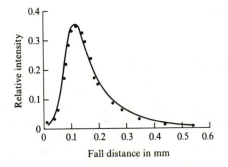

FIGURE 9-9
Velocity distribution of cesium
atoms, determined experimentally
by use of the apparatus of
Figure 9-8.

Figure 9-8. The entire apparatus is evacuated. At the left there is shown an
oven containing some cesium metal. The oven is heated to a temperature
of about 450°K, at which the vapor pressure of cesium is sufficiently great
to cause many cesium atoms to be present in the gas phase. A beam of
cesium atoms coming from a small horizontal slit in the oven is further
defined by another slit, 1 m away. Then after traveling another meter, the
beam passes through another slit and impinges on the detector, which
consists of a fine horizontal tungsten wire, electrically heated, and an
adjacent plate with a negative electric charge. Whenever a cesium atom
strikes the wire it is ionized; it leaves the wire as a cesium ion, Cs^+, and is
attracted to the negatively charged plate. The number of cesium ions
reaching the plate can be measured by measuring the electric current between the wire and the plate.

The atoms of cesium constituting the beam defined by the slit in the
oven and the slit in the center of the tube move along parabolic paths
under the influence of the earth's gravitational attraction. The average

amount of deflection in this apparatus is about 0.2 mm for cesium atoms at 450°K.

The distribution observed—the number of cesium ions reaching the plate as a function of the vertical coordinate of the tungsten wire—is shown in Figure 9-9. From this distribution the distribution of velocities of the molecules (atoms) of cesium in the oven can be calculated. It is found to be that given by the Maxwell-Boltzmann distribution law for molecular velocities, Equation 9-19. The experiments provide verification of the law to within about 1%.

9-6.   The Boltzmann Distribution Law

In this section we shall derive the Boltzmann distribution law, of which the distribution law for molecular velocities, given above without derivation (Equation 9-19), is a special case.

We consider a system composed of a very large number, N, of identical parts, which might be the molecules in a sample of gas. (The equation that we shall derive applies also to more complex systems, such as a mixed gas.) We assume that molecules (that is, the parts of the system) interact with one another only weakly, and that the total energy of the system, E_{total}, can be expressed as the sum of the energies of the molecules; that is, we write

$$E_{total} = E_1 + E_2 + \cdots + E_N = \sum_{i=1}^{N} E_i \qquad (9\text{-}20)$$

The possible values of E_i are those corresponding to the quantum states of a molecule. For a monatomic molecule in a container with volume V they are given by Equation 9-9.

For example, one molecule might have all the energy, $E_i = E_{total}$, and all the others then would have zero energy. Or the energy might be divided equally among the molecules, each having exactly the energy E_{total}/N. Our task is to find the most probable way of distributing the total energy of the system among the molecules. Our basic assumption is that *all quantum states are equally important*. Hence we want to find the way of distributing the total energy that corresponds to the greatest number of quantum states of the system.

We can find this distribution by making use of the theory of combinations and permutations to calculate the number of quantum states for an arbitrary distribution of the total energy among the molecules, and then finding the maximum value of this number.

Let us group the states for one molecule into groups with essentially the same energy and an equal number of states, the n states of the jth group all having the energy E_j. We consider the assignment of different numbers, N_j, of molecules to the groups. Let us first make the calculation for N nonidentical molecules, labeled 1 to N, and then later introduce a factor to correct for identity of the particles.

We arrange the molecules in a row, in a random way. The molecule labeled 1 may be in any one of the N places in this row. For each position for molecule 1 there are $N - 1$ positions for molecule 2; and so on. The total number of ways of arranging the N labeled molecules in a row is $N!$.

Now let the first N_1 molecules in the row be in the first group. There are n states accessible to each molecule in the group, so that the number of states for a particular set of N_1 labeled molecules is n^{N_1}. But the distribution discussed in the preceding paragraph counts $N_1!$ ways of placing the same set of N_1 labeled molecules in the first group. (For example, with $N_1 = 3$ the first group might consist of molecules 4, 8, and 9 in the six sequences 489, 948, 894, 498, 849, and 984.) Thus the first group contributes the factor $\dfrac{n^{N_1}}{N_1!}$, the second contributes the factor $\dfrac{n^{N_2}}{N_2!}$, and so on. For all groups the factor is

$$\frac{n^N}{N_1! N_2! N_3! \cdots} = \frac{n^N}{\displaystyle\prod_{j=1}^{\infty} N_j!}$$

Here the numerator has been simplified by use of the relation $N_1 + N_2 + N_3 + \cdots = N$. The symbol Π means the product of the terms.

The number of states for a system of N labeled molecules with N_1, $N_2, \ldots$ molecules in each of the groups of n molecular states is obtained by multiplying this quantity by the number of ways of ordering the N labeled molecules in a row, which is given above as $N!$:

$$\text{Number of states of labeled molecules} = \frac{N! n^N}{\displaystyle\prod_{j=1}^{\infty} N_j!} \qquad (9\text{-}21)$$

It was pointed out in Section 9-3, however, that N identical particles, either fermions or bosons, occupying a set of N single-particle states correspond to one allowed wave function (either totally antisymmetric or totally symmetric) for the system, and not to $N!$ wave functions. (In this calculation we ignore the states with two or more identical particles in the same one-particle state, which do not occur for fermions (the Pauli exclusion principle) and have extra weight for bosons.) Hence we divide by $N!$, obtaining the result

$$\text{Number of states of identical molecules} = W = \frac{n^N}{N_1! N_2! N_3! \cdots} \qquad (9\text{-}22)$$

We want to find the values of N_1, N_2, ... that make this quantity W a maximum; this maximum corresponds to the most probable distribution of the molecules over the states.*

Let us assume that N_1, N_2, N_3, ... have the values that give W its maximum value, and ask how W is changed by moving some molecules from one molecular state to another. Let us take two molecules from the lth group, with energy E_l, and add one to the kth group, with energy $E_l - \Delta E$, and the other to the mth group, with energy $E_l + \Delta E$. Neither the number of molecules nor the energy of the system has been changed by this small redistribution of the molecules. The new value of W, which we may call W', is seen to be

$$W' = \frac{n^N}{N_1! N_2! \cdots (N_k + 1)! \cdots (N_l - 2)! \cdots (N_m + 1)! \cdots} \quad (9\text{-}23)$$

Let us divide W' by W:

$$\frac{W'}{W} = \frac{(N_l - 1)N_l}{(N_k + 1)(N_m + 1)}$$

The fact that W is a maximum requires that W' be smaller than W, and hence that W'/W be less than 1; hence we may write

$$\frac{(N_l - 1)N_l}{(N_k + 1)(N_m + 1)} < 1 \quad (9\text{-}24)$$

If we transfer a molecule from group k to group l and one from group m to group l we obtain, in the same way, the result

$$\frac{N_k N_m}{(N_l + 1)(N_l + 2)} < 1$$

which may be inverted to read

$$\frac{(N_l + 1)(N_l + 2)}{N_k N_m} > 1 \quad (9\text{-}25)$$

Because the number of molecules in an ordinary sample of gas is so large, we are allowed to ignore the numbers 1 and 2 in comparison with N_k, N_l, and N_m in the inequalities 9-24 and 9-25, which would then become

$$\frac{N_l^2}{N_k N_m} - \text{a small term} < 1$$

and

$$\frac{N_l^2}{N_k N_m} + \text{a small term} > 1$$

---

*We might argue that all states of the system (with the same total energy) should be counted, not just those that correspond to the most probable distribution of the molecules over the molecular states. It is found, however, that the same answer is obtained as with the above treatment, because the less probable distributions contribute a number of states of the system far smaller than that contributed by the most probable distribution.

We see that the quantity $N_l^2/N_k N_m$ has a value very close to 1; hence we write

$$\frac{N_l^2}{N_k N_m} = 1$$

or, by rearrangement,

$$\frac{N_l}{N_k} = \frac{N_m}{N_l} \tag{9-26}$$

This equation holds for all pairs of groups of states differing in energy by ΔE; hence we may write

$$\frac{N_l(E_l)}{N_k(E_k)} = \text{constant, for a fixed value of } E_l - E_k \tag{9-27}$$

The ratio of the occupancy numbers of two groups of molecular states, with n molecular states in each group, is a function only of the difference in the energy of a molecule in a state of the second group and that of a molecule in a state of the first group.

You probably remember that this property is characteristic of exponential functions. The most general function of x with this property is $a \exp(bx)$, in which a and b are constants. We see that $a \exp(bx_2)$ divided by $a \exp(bx_1)$ is equal to $a \exp\{b(x_2 - x_1)\}$, which is determined solely by the difference $x_2 - x_1$.

We are thus led to write for N_i, the occupancy number of a group of molecular states with energy E_i, the equation

$$N_i = CN \exp(-\beta E_i) \tag{9-28}$$

Here we have introduced CN in place of a and $-\beta$ in place of b. The value of C must be such that the sum of all the values of N_i is equal to N, the total number of molecules:

$$C \sum_{i=1}^{\infty} \exp(-\beta E_i) = 1$$

$$C = \frac{1}{\displaystyle\sum_{i=1}^{\infty} \exp(-\beta E_i)} \tag{9-29}$$

Equation 9-28 is the Boltzmann distribution law in the form that applies to quantized systems (see Appendix XI for the classical form). When this equation is applied to the quantum states of a molecule in a box, Equation 9-10, it is found that the average energy per molecule is $\frac{3}{2\beta}$. The scale of absolute temperature, T, is defined in such a way that the average energy of a monatomic molecule in a gas at temperature T is $\frac{3}{2}kT$, where k is the Boltzmann constant, with value 13.805×10^{-24} J deg^{-1}. Hence we replace β by $1/kT$ in Equation 9-28, and obtain the Boltzmann distribution law in its usual form:

$$N_i = NQ^{-1} \exp\left(-\frac{E_i}{kT}\right) \tag{9-30}$$

with the normalizing constant Q given by

$$Q = \sum_{i=1}^{\infty} \exp\left(-\frac{E_i}{kT}\right) \tag{9-31}$$

Derivation of the Maxwell Distribution Law for Molecular Velocities

We now can derive the distribution law for molecular velocities. From Equation 9-18 we see that the number of one-molecule quantized states in the energy range E to $E + dE$ is proportional to $E^{\frac{1}{2}}dE$. The Boltzmann distribution law, Equation 9-30, shows that the occupancy of a state is proportional to the Boltzmann factor $\exp(-E/kT)$. Hence the number of molecules in the energy range E to $E + dE$ is

$$dN = \text{constant} \times \exp(-E/kT)\, E^{\frac{1}{2}}dE \tag{9-32}$$

Since $E = \frac{1}{2}mv^2$, we have $dE = mvdv$ and $E^{\frac{1}{2}}dE = 2^{-\frac{1}{2}}m^{\frac{3}{2}}v^2dv$. The constant can be evaluated by integration from 0 to ∞, with $\int dN = N$. This leads to Equation 9-19.

The Distribution Laws for Bosons and Fermions

The Boltzmann distribution law applies to systems of bosons or fermions at high temperatures. At lower temperatures, such that the number of translational states with energy less than $\frac{3}{2}kT$ is nearly the same as the number of molecules, there occurs deviation from this distribution, because of the assignment of different weights to states with two or more molecules in the same level. The correct law for bosons is called the Bose-Einstein distribution law, and that for fermions is called the Fermi-Dirac distribution law. Some of the properties of liquid ^{4}He and of liquid ^{3}He, which consist of bosons and fermions, respectively, give evidence of the corresponding deviations from the Boltzmann distribution. The Bose-Einstein distribution law, which is given here without derivation, is

$$N_i = \frac{1}{C' \exp\left(\dfrac{E_i}{kT}\right) - N} \tag{9-33a}$$

and the Fermi-Dirac distribution law is

$$N_i = \frac{1}{C'' \exp\left(\dfrac{E_i}{kT}\right) + N} \tag{9-33b}$$

In these equations the constants C' and C'' have values such that ΣN_i is equal to N, the total number of particles.

Example 9-5. At what height above sea level is the atmospheric pressure 0.5 atm? Assume the temperature to be constant, 20°C.

Solution. This problem illustrates a way of using the Boltzmann distribution law. Let us think of a box of air at sea level and a similar box at height z, with the two boxes connected by a small tube. The translational states for a molecule in the upper box are the same as those for a molecule in the lower box, but the value of the energy for a state in the upper box is greater than that for a state in the lower box by the difference in the potential energy of the molecule in the earth's gravitational field. This difference is mgz, where m is the weight of the molecule and g is the gravitational acceleration. We note that the Boltzmann factor can be written as the product of two exponentials: $\exp(-E/kT) = \exp(-\text{kinetic energy}/kT - \text{potential energy}/kT) = \exp(-\text{kinetic energy}/kT) \times \exp(-\text{potential energy}/kT)$. Hence the occupancy of each state in the upper box is $\exp(-mgz/kT)$ times that of the corresponding state (with the same kinetic energy) in the lower box. The problem states that this ratio of occupancies is 0.5. Hence we write

$$\exp(-mgz/kT) = 0.5$$
$$-mgz/kT = 2.303 \log 0.5$$
$$mgz = 2.303 \, kT \log 2$$
$$z = \frac{2.303 \, kT \log 2}{mg}$$

For convenience we multiply numerator and denominator by Avogadro's number:

$$z = \frac{2.303 \, RT \log 2}{Mg}$$

With $R = 8.315$ J mole^{-1}, $T = 293°$, $M = 0.0288$ kg mole^{-1} (average for N_2 and O_2), and $g = 9.807$ m s^{-2}, we obtain the result $z = 5980$ m. We accordingly have calculated that the atmospheric pressure is 0.5 atm at a height of about 6 km above sea level.

9-7. Deviations of Real Gases from Ideal Behavior

Real gases differ in their behavior from that represented by the perfect-gas equation for two reasons. First, the molecules have a definite size, so that each molecule prevents others from making use of a part of the volume of the gas container. This causes the volume of a gas to be larger than that calculated for ideal behavior. Second, the molecules even when some distance apart do not move independently of one another, but attract one another slightly. This tends to cause the pressure of a gas to be smaller than the calculated pressure.

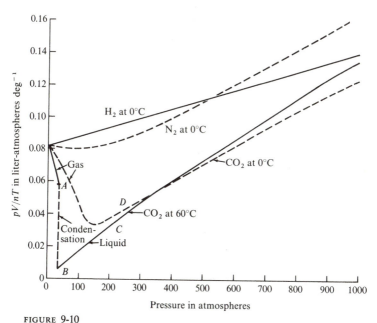

FIGURE 9-10
The value of PV/nT for some gases, showing deviation from the perfect-gas law at high pressures.

The amounts of the deviation for some gases are shown in Figure 9-10. It is seen that for hydrogen at 0°C the deviation is positive at all pressures—it is due essentially to the volume of the molecules, as the effect of their attraction at this high temperature (relative to the boiling point, $-252.8°C$) is extremely small.

At pressures below 120 atm, nitrogen (at 0°C) shows small negative deviations from ideal behavior, intermolecular attraction having a greater effect than the finite size of the molecules.

The deviation of hydrogen and nitrogen at 0°C from ideal behavior is seen to be less than 10% at pressures less than 300 atm. Oxygen, helium, and other gases with low boiling points also show only small deviations from the perfect-gas law. For these gases the perfect-gas law holds to within 1% at room temperature or higher temperatures and at pressures below 10 atm.

Larger deviations are shown by gases with higher boiling points: in general, the deviations from ideal behavior become large as the gas approaches condensation. It is seen from the figure that for carbon dioxide at 60°C the volume of the gas is only about 30% as great at 120 atm pressure as the volume calculated by the perfect-gas equation.

If the temperature is low the deviations are shown in a pronounced way by the condensation of the gas to a liquid (see the curve for carbon dioxide at 0°C). After carbon dioxide has been compressed to about 40 atm at 0°C, the effect of the attraction of the molecules for one another becomes so great that they cling together, forming a liquid, the system then consisting of two phases, the gaseous phase and the liquid phase. On further compression the volume decreases without change in pressure (region A of the figure) until all the gas is condensed (point B). The volume of the liquid decreases much less rapidly with increase in pressure from point B on than would that of a gas, because the molecules of the liquid are effectively in contact; hence the curve rises (region C).

An extraordinary phenomenon, the *continuity of the liquid and gaseous states*, was discovered about a century ago by Thomas Andrews (1813–1885). He found that above a temperature characteristic of the gas, called the *critical temperature*, the transition from the gaseous state to the liquid state occurs without a sharp change in volume on increasing the pressure.

The critical temperature of carbon dioxide is 31.1°C. Above this temperature (at 60°C, for example, corresponding to the curve shown in the figure), all the properties of the substance change continuously, showing no signs that the gas had condensed to a liquid. Nevertheless, when the pressure becomes greater than about 200 atm the substance behaves like carbon dioxide liquid, rather than like a gas (region D of Figure 9-10). It is, indeed, possible to change from the gas at 0°C and 1 atm pressure to the liquid at 0°C and 50 atm either by the ordinary process of condensing the gas to the liquid, passing through the two-phase stage, or, without condensation or any discontinuity, by heating to 60°, increasing the pressure to about 200 atm, cooling to 0°, and then reducing the pressure to 50 atm. The liquid could then be made to boil, simply by reducing the pressure and keeping the temperature at 0°C; and then, by repeating the cycle, it could be brought back to 0°C and 50 atm pressure without condensation, and be made to boil again.

Values of the critical temperature, critical pressure, and critical density of some substances are given in Table 9-1.

The possibility of continuous transition from the gaseous to the liquid state is understandable in view of the common characteristic of randomness of structure of these phases, as discussed in Chapter 2. It is, on the other hand, difficult to imagine the possibility of a gradual transition from a disordered state (liquid) to a completely ordered state (crystal); and correspondingly it has not been found possible to crystallize substances or to melt crystals without passing through a discontinuity at the melting point —there is no critical temperature for melting a crystal.

TABLE 9-1
Van der Waals Constants and Critical Constants of Some Substances

Sub-stance	a	b	Critical Temperature	Critical Pressure	Critical Volume
He	$0.0341\ l^2\,atm\,mole^{-2}$	$0.0237\ l\,mole^{-1}$	$5.3°K$	2.26 atm	58 ml $mole^{-1}$
Ne	.211	.0171	44.5	25.9	42
Ar	1.35	.0322	151	48	75
Kr	2.32	.0398	210	54	107
Xe	4.19	.0550	300	58	112
H_2	0.244	.0266	73.3	12.8	65
N_2	1.39	.0391	126.1	33.5	90
O_2	1.36	.0318	154.4	49.7	74
Cl_2	6.49	.0562	129	76	125
CO	1.49	.0399	134	35	90
CO_2	3.59	.0427	304	73	96
N_2O	3.78	.0441	309.7	72.6	98
CH_4	2.25	.0428	91	46	99
C_2H_6	5.49	.0638	305	49	143
SO_2	6.71	.0564	430	78	123
CCl_4	20.39	.138	556	45	276
$SnCl_4$	26.91	.164	592	37	352
H_2O	5.46	.0305	647.2	217.7	45
NH_3	4.17	.0371	406	112	72
Hg	8.09	.0170	1823	200	45

The van der Waals Equation of State

An equation relating the volume of a sample of substance to the temperature and pressure is called an *equation of state* for the substance.

The perfect-gas equation has only limited applicability to real gases, and many other equations, involving constants characteristic of the substance in addition to the general constant R, have been proposed.

The most useful simple equation of this sort was discovered in 1873 by the Dutch physicist Johannes Diderik van der Waals (1837–1923). It is

$$\left(P + \frac{n^2a}{V^2}\right)(V - nb) = nRT \qquad (9\text{-}34)$$

This equation contains two constants, characteristic of the substance, a and b. (They are usually called van der Waals' a and van der Waals' b.) Values of a and b found by experiment (measurement of the deviation from the perfect-gas law) for several substances are given in Table 9-1.

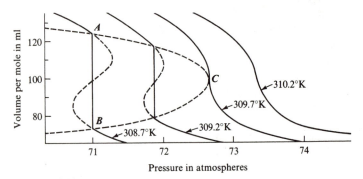

FIGURE 9-11
Curves calculated by use of the van der Waals equation, giving the molar volume of nitrous oxide for four values of the temperature.

Van der Waals' b has the dimensions of molar volume. It is usually about 25% greater than the molar volume of the liquid substance. Representative values, in ml mole^{-1}, are the following: Ar, $b = 32$, molar volume of liquid $= 28$; Cl_2, 56, 46; SO_2, 56, 45; CCl_4, 138, 98.

The value of a/V is the effective energy of attraction of the molecules. For V equal to the molar volume of the liquid, a/V is found usually to be about two-thirds of the heat of vaporization of the liquid. Representative values of a/V (converted to kJ mole^{-1}) and the heat of vaporization (in kJ mole^{-1}) are 4.78, 6.53 for Ar; 14.5, 20.4 for Cl_2; 15.2, 25.4 for SO_2; 21.4, 30.0 for CCl_4.

Both van der Waals' b, 0.0237 l mole^{-1}, and the molar volume of the liquid, 0.0317 l mole^{-1}, for helium are larger than the values for neon, 0.0171 and 0.017 l mole^{-1}, respectively, although other evidence, especially the theoretical treatment of the electron distribution for these atoms (Section 6-13), leads us to believe that the helium atom is smaller than the neon atom. This anomaly is thought to be a quantum effect.

The van der Waals equation provides an interesting explanation of the transition between the gaseous state and the liquid state and of the critical point. In Figure 9-11 there are shown four curves of the volume per mole as a function of the pressure, calculated by Equation 9-34, with a and b given the values for N_2O, for four values of T in the neighborhood of the critical temperature. The calculated curve for 308.7°K, one degree less than the critical temperature, shows a region in which the pressure would decrease with decrease in volume (center of dashed portion). Instead, the fluid separates into two phases, and with decreasing volume the pressure remains constant as the phase point moves along the straight line from A to B, until only one phase remains (at B, liquid phase only).

The point C is the critical point, at which the properties of the gaseous phase and the liquid phase become identical. This is a point of inflection for the curve. By evaluating the point of inflection of Equation 9-34, the following expressions are found for the critical constants:

$$\text{Critical temperature } = \frac{8a}{27bR} \tag{9-35}$$

$$\text{Critical pressure } = \frac{a}{27b^2} \tag{9-36}$$

$$\text{Critical molar volume } = 3b \tag{9-37}$$

It can be seen by reference to the values in Table 9-1 that these relations are satisfied approximately. The major deviation is shown by the critical molar volume, which is usually close to $2.25b$.

Exercises

9-1. Calculate the volume occupied at 25°C and 1 atm pressure by the gas evolved from 1 cm³ of solid carbon dioxide (density 1.56 g cm⁻³).

9-2. The density of helium at 0°C and 1 atm is 0.1786 g l⁻¹. Calculate its density at 100°C and 200 atm.

9-3. Avogadro was the first person to describe water as consisting of triatomic molecules (H_2O in modern symbols). What experimental facts led him to this conclusion? Imagining yourself in Avogadro's place, in the year 1811, write a brief communication to a scientific journal in which your argument leading to this conclusion is presented as rigorously as possible.

9-4. In 1811 Avogadro assigned the formula $C_{10}H_{16}O$ to camphor, on the basis of its vapor density and chemical analysis. The observed density of camphor vapor at 210°C and 1 atm is 3.84 g l⁻¹. To what value of the molecular weight does this density correspond?

9-5. The observed density of hydrogen cyanide at 26°C and 1 atm is 1.210 g l^{-1}. Calculate the apparent molecular weight of the gas. Assuming that the gas contains the dimer, $(HCN)_2$, as well as the monomer, calculate the fraction that is dimerized. (Answer: 29.70, 18.2%.)

9-6. At 3000°K, hydrogen gas at a total pressure of 1 atm is 9.03% dissociated into hydrogen atoms. What is the density of the gas? What would the density of the gas at 1 atm and 3000°K be if the diatomic molecules did not dissociate into atoms?

9-7. At 2500°K and 1 atm pressure the observed density of CO_2 gas is 0.1991 g l^{-1}. What is the average molecular weight of the molecules in the gas? Assuming that the molecular weight is decreased because the carbon dioxide has partially decomposed into carbon monoxide and oxygen, calculate what fraction of the CO_2 molecules have decomposed in this way.

9-8. A gas was observed to have a density of 1.63 g l^{-1} at 25°C and 1 atm. What is the molecular weight of the gas? Its heat capacity at constant volume was found on measurement to be 0.31 J deg^{-1} g^{-1}. How many atoms are there in the molecule of the gas? Can you identify this gas?

9-9. The density of ethylene at very low pressure corresponds to the ideal density 1.251223 g l^{-1} at standard conditions. The formula of ethylene is C_2H_4. Calculate from this information a precise value of the molecular weight of ethylene. Assuming the atomic weight of carbon to be 12.01115, calculate the atomic weight of hydrogen.

9-10. Would deuterium (atomic weight 2.0147) effuse through a porous plate more rapidly or less rapidly than hydrogen? Calculate the relative rates of effusion of the two molecules. What would be the relative rate of effusion of a molecule made of one light hydrogen atom and one deuterium atom?

9-11. In the so-called gaseous diffusion method of separating ^{235}U from natural uranium use is made of gaseous UF_6 (normal boiling point 56°C) effusing through small holes (about 10 nm diameter) in a metal barrier. What is the maximum increase in the $^{235}U/^{238}U$ ratio that would be expected in a single stage of the separation? (Answer: 1.00429; in practice a value about one-third as great is achieved.)

9-12. It is shown in textbooks of physics that the speed of sound waves in a gas is equal to $(\gamma P/\rho)^{1/2}$, in which γ is the ratio of C_P to C_V for the gas, P is the pressure (in N m^{-2}), and ρ is the density (in kg m^{-3}).
   (a) With use of the perfect-gas equation, show that this quantity is equal to $(1000 \gamma RT/MW)^{1/2}$, in which MW is the molecular weight.
   (b) What is the speed of sound in helium at 0°C and 1 atm?
   (c) What is the ratio of the speed of sound to the most probable molecular speed?

9-13. Calculate the speed of sound in neon, argon, krypton, and xenon at 0°C. (Answer: 169.8 m s^{-1} for xenon.)

9-14. The observed speed of sound in oxygen at 0°C is 316 m s⁻¹. What is the corresponding value of γ? To what values of C_V and C_P does this value of γ lead? (Answer: 1.406; 2.5 R, 3.5 R; see Section 10-10.)

9-15. What is the calculated speed of sound in mercury vapor (monatomic molecules) at 25°C? The value observed for liquid mercury at this temperature is 1450 m s⁻¹. Can you suggest a simple qualitative interpretation of the difference between the two values? (See the discussion near the end of Section 11-3.)

9-16. By measuring areas under a curve in Figure 9-7 calculate the approximate fraction of molecules in a gas that have velocities greater than Mach 1 (the speed of sound). Than Mach 2.

9-17. The root-mean-square velocity of molecules in a gas at temperature T is $(3\ kT/m)^{1/2}$, corresponding to the value $\frac{3}{2}kT$ for the average kinetic energy of a molecule. At what fraction of the root-mean-square velocity is the Boltzmann factor of Equation 9-19 equal to 0.99? What fraction of the molecules in a gas have velocities less than this velocity?

9-18. Crystalline xenon has the cubic close-packed structure (Figure 2-3), with cube edge $a = 6.24$ Å at 88°K. What is the value of the molar volume of Xe(c) at this temperature? What are the ratios of van der Waals' b and of the critical volume to this molar volume? (36.6 ml mole⁻¹, 1.50, 3.06.)

9-19. A sample of a perfect gas is subjected to adiabatic expansion from volume V to volume $V + dV$ (adiabatic means without transfer of thermal energy to or from the environment).
   (*a*) How much work has been done by the system?
   (*b*) The change in internal energy of the gas is $C_V dT$, in which dT is the change in temperature resulting from the expansion. How is this energy quantity related to the preceding one?
   (*c*) From this relation and the perfect gas equation, $PV = nRT$, derive a differential equation in V and T.
   (*d*) Solve this equation and introduce the initial conditions (V_0, P_0, T_0) to obtain the relation between V and T during adiabatic expansion or compression of a perfect gas.
   (*e*) With use of the perfect-gas equation convert this equation into an equation between P and T, and also into an equation between P and V. (Last answer: $P/P_0 = (V_0/V)^\gamma$, with $\gamma = C_P/C_V$.)

9-20. A sample of xenon initially at 0°C and 1 atm is compressed adiabatically to half its volume. What are the final values of P and T?

9-21. A primitive method of making fire is to quickly compress the air in a tube containing a combustible powder by rapidly pushing down on a piston. Assuming that the temperature 300°C must be reached, calculate how far down the tube the piston must be pushed. The value of C_V for air is $\frac{5}{2}R$.

9-22. Make a graph of the function $\sin 2\pi xmv/h$ for values of x from 0 to $2h/mv$. Use this graph in a brief discussion of the statement in the first sentence of Section 9-3 that this function and the corresponding cosine function represent a particle in field-free one-dimensional space with the de Broglie wavelength $\lambda = h/mv$.

9-23. Carry out the differentiation twice with respect to x of the sine and cosine functions of the preceding Exercise, and show that the functions are solutions of Equation 9-13.

9-24. A tank contains $CO_2(g)$ at the critical temperature and pressure, 31°C and 73 atm. How much more does the tank hold than the amount calculated by the perfect-gas equation? (Answer: 256% more.)

9-25. It is seen from Figure 9-11 that at the critical point both dP/dV and d^2P/dV^2 are equal to zero. Rewrite Equation 9-34 to obtain P as a function of V, calculate the two derivatives, equate them to zero, and solve for the critical constants (Equations 9-35, 9-36, 9-37).

10

Chemical Thermodynamics

In this chapter we shall develop and discuss some of the principles of chemical thermodynamics. Our development of these principles will be based on the discussions of quantum states and the Boltzmann distribution law presented in the preceding chapter. It is my opinion that this approach gives the student a better understanding of chemical thermodynamics than any other.

10-1. Heat and Work. Energy and Enthalpy

We shall find it convenient to divide the universe into two parts: the *system* under discussion, and its *environment*, which is the rest of the universe. The system may be an isolated system, with no interaction with the environment. We shall for the most part discuss systems containing a fixed amount of matter (that is, with the boundaries of the system impermeable to atoms), but capable of transferring heat to the environment or receiving heat from the environment, and of doing work on the environment or having work done on the system by the environment.

The relation between heat and work is treated in courses in physics, and may be briefly reviewed here. Work is done by a directed force acting through a distance; the amount of work done by a force of one newton acting through a distance of one meter is called one joule. If this amount of work is done in putting an object initially at rest into motion, we say that the moving object has a kinetic energy of 1 J. All of this kinetic energy may be used to do work, as the moving object is slowed down to rest; for example, a string attached to the moving object might serve to lift a small weight to a certain height above its original position.

Another way in which the moving object can be slowed down to rest is through *friction*. The process that then occurs is that the kinetic energy of the directed motion of the moving body is converted into energy of randomly directed motion of the molecules of the bodies between which friction occurs.

This increase in vigor of molecular motion corresponds to an increase in temperature of the bodies. We say that heat has been added to the bodies, causing their temperatures to rise. Thus if one of the bodies was 1 g of water, and if its temperature rose by 1 deg, we would say that 1 calorie of heat had entered it. The calorie, which in the past has been generally used as the unit of quantity of heat, used to be defined as the amount of energy required to cause 1 g of water to increase in temperature by 1 deg; that is, the heat capacity of water was taken to be 1 cal deg^{-1} g^{-1}.

The question at once arises as to how much work must be done to produce this much heat. This question was answered by experiments carried out in Manchester, England, between 1840 and 1878 by James Prescott Joule (1818–1889), after Count Rumford (Benjamin Thompson, 1753–1814, an American Tory) had shown in 1798 that the friction of a blunt borer in a cannon caused an increase in temperature of the cannon. Joule's work led to nearly the same value as that now accepted for the mechanical equivalent of heat; that is, for the relation between heat and work:

$$1 \text{ cal} = 4.1840 \text{ joule}$$

One joule is also equal to the work done by the flow of one coulomb of electricity through a potential difference of one volt:

$$1 \text{ joule} = 1 \text{ volt coulomb} = 1 \text{ watt second}$$

It was pointed out in Section 6-11 that the word "heat" is used for a quantity of energy that is transferred across the boundaries of a system in a certain way—namely, by the thermal processes of conduction and radiation.

10-2. The First Law of Thermodynamics

The energy E of a system is determined by the state of the system (composition, pressure, temperature; sometimes other factors, such as strength of gravitational field, electric field, magnetic field). *The change in energy ΔE accompanying change of a system from an initial state to a final state is determined exactly by the initial state and the final state, and is independent of the path by way of which this change is effected.*

The foregoing statement is a statement of the law of conservation of energy. It is also a statement of the *first law of thermodynamics*.

In a change from a specified initial state to a specified final state of a system the sum of the heat transferred to the system and the work done on the system constitutes the change in energy of the system. This sum is independent of the path between the initial state and the final state. The heat transferred depends on the path, and the work done depends on the path.

To illustrate, let us consider the reaction of liquid nitrogen trichloride at 25°C and 1 atm pressure to form nitrogen and chlorine at the same temperature and pressure:

$$2NCl_3(l) \longrightarrow N_2(g) + 3Cl_2(g)$$

The value of ΔE for this change in state is found by experiment to be -451.9 kJ for 2 moles of NCl_3.

At 25°C and 1 atm pressure two moles of liquid nitrogen trichloride occupy the volume 0.148 l. Under the same conditions the products, which are gases, occupy the volume 97.900 l. We may carry out the change from the initial state to the final state in such a way that no work is done on the system by the environment or on the environment by the system. Two moles (240.8 g) of nitrogen trichloride at 25°C and 1 atm is sealed in a thin glass bulb, which is placed in a steel vessel with volume 97.900 l. The vessel is then evacuated, and allowed to come into thermal equilibrium with the immediate environment, a thermostat maintained at 25°C. The system is then in its initial state. Then the nitrogen trichloride is exploded (caused to react) by passing a spark between two wires sealed into the bulb. When thermal equilibrium has been reached with the environment (at 25°C), the pressure inside the steel vessel is 1 atm. No work has been done. The amount of heat transferred to the system by the environment is exactly ΔE, which is -451.9 kJ; that is, 451.9 kJ of heat has been transferred from the system to the environment.

Another path from the same initial state of the system to the same final state is the following. Let us place a small internal-explosion engine inside the steel vessel, with 240.8 g of nitrogen trichloride (at 25°C) in its fuel tank, and with a shaft passing through the wall of the vessel to permit work to be done on the environment. The vessel, with volume 97.900 l (not including the solid parts of the engine), is evacuated. The engine is then allowed to operate until the fuel is exhausted, and the system is allowed to stand until thermal equilibrium with the environment (at 25°C) is reached.

Let us assume that the engine resembles good Diesel engines in having 43% efficiency. The work done by the system on the environment would then be 43% × 451.9 kJ = 194.3 kJ.

The heat transferred from the system to the environment would amount to 451.9 − 194.3 = 257.6 kJ.

The change from the same initial state to the same final state of the system by other paths would in general involve different values of the heat exchanged with the environment and of the work done.

Most changes in state that are studied in the laboratory are carried out at 1 atm pressure, under conditions such that the volume of the system can change (rather than in an evacuated vessel, as described above). The system may increase in volume, by the amount ΔV; it then does the work $P\Delta V$ on the environment, in which P is the pressure. For example, if $2NCl_3(l)$ were to decompose slowly at 25°C and 1 atm pressure, in a cylinder with movable piston, the work 1 atm × (97.900 − 0.148) l = 97.852 atm l would be done by the system on the environment. (This work consists in pushing back the atmosphere.) Using the relation 1 atm l = 101.3 J we find that the pressure-volume work done by the system is 9.91 kJ. The heat transferred to the environment during the change in state by this path is hence 451.9 − 9.9 = 442.0 kJ.

It has been mentioned in Section 6-11 that the enthalpy H of a system is defined as the energy E plus the term PV:

$$H = E + PV \qquad (10\text{-}1)$$

For any change of a system from an initial state to a final state carried out at constant pressure P in such a way that the only work done is the pressure-volume work, the value of ΔH is exactly equal to the heat transferred from the environment to the system. To obtain ΔE, it is necessary to measure also the change in volume and to apply the corresponding correction for the pressure-volume work.

The simplification resulting from the use of enthalpy (with the PV term) rather than energy was first recognized by the great American scientist J. Willard Gibbs (1839–1903) in the course of his theoretical studies, published in 1876 and 1878 under the title *On the Equilibrium of Heterogeneous Substances*, in which he laid the foundations of the whole field of chemical thermodynamics.

Gibbs and other workers in this field also attacked and solved the question of the maximum amount of work that can be done by a system on the environment during a change in state of the system. We shall return to this question later in this chapter.

Example 10-1. What is the value of the change in internal energy, ΔE, for the reaction

$$H_2(g) + \tfrac{1}{2}O_2(g) \longrightarrow H_2O(l)$$

at 25°C and 1 atm?

Solution. The value of ΔH for this reaction is given in Table 7-1 as -286 kJ mole^{-1}. A more accurate experimental value is -285.840 kJ mole^{-1}. To obtain ΔE we must add a correction term $-\Delta(PV)$, as shown by Equation 10-1. The value of PV for $H_2(g)$ is equal to RT, which is $8.3146 \times 298.15 = 2479$ J mole$^{-1} = 2.479$ kJ mole^{-1}. The value for $\tfrac{1}{2}O_2(g)$ is one half as great, 1.240 kJ mole^{-1}. The value for liquid water, with molar volume 18 ml mole^{-1}, is 0.002 kJ mole^{-1}, giving $\Delta(PV) = -3.717$ kJ mole^{-1}. Hence ΔE for the reaction is $-285.840 + 3.717 = -282.123$ kJ mole^{-1}.

This change in internal energy accompanying the reaction is largely the change in electronic energy of the molecules; the difference in rotational and vibrational energy is small.

10-3. Heat Capacity. Heats of Fusion, Vaporization, and Transition

The value of the standard enthalpy of a substance (its enthalpy at 1 atm and 273.15°K) can be combined with the experimentally determined heat capacity to obtain the value of its enthalpy at another temperature. Also, the enthalpy of the substance in another state of aggregation can be calculated with use of the heat of fusion, vaporization, or transition associated with the change in phase at constant pressure.

Heat Capacity

It has been mentioned earlier that the amount of heat required to raise the temperature of unit quantity (1 mole or 1 g) of a substance by 1°C without change in phase is called the *heat capacity* (or *specific heat*) of the substance. Values of the heat capacity of substances are given in tables that may be found in reference books. The heat capacity is usually measured at constant pressure. The heat capacity per mole is equal to the heat capacity per gram multiplied by the molecular weight.

It was pointed out in 1819 by Dulong and Petit in France that for the heavier solid elementary substances (with atomic weights above 35) the product of the heat capacity per gram and the atomic weight (molar heat capacity) is approximately constant, with value about 26 J deg^{-1} mole^{-1}. This is called the *rule of Dulong and Petit*. The extent of validity of the rule is indicated by the following examples:

Element	Experimental values of molar heat capacity at 20°C
Al	24.3 J deg^{-1} mole^{-1}
Fe	25.2
Ni	26.0
Ag	25.5
Au	25.2
Pb	26.8

The low value for aluminum (atomic weight 27) illustrates the deviation from the rule that occurs for elements with low atomic weight (see Section 10-16).

During the first half of the nineteenth century the rule was used to get rough values of the atomic weight of some elements, by dividing 26 by the measured heat capacity per gram of the solid elementary substance. For example, the heat capacity of bismuth is 0.123 J. By dividing this into 26 we obtain 211 as the rough value of the atomic weight of bismuth given by the rule of Dulong and Petit; the actual atomic weight of bismuth is 209.

Kopp's Rule

A rough estimate of the molar heat capacity of a solid substance can be obtained by adding the following values for the various atoms in the formula of the substance (this procedure is called *Kopp's rule*):

H	Li	Be	B	C	N	O	F	All others	
10	21	15	13	8	13	18	20	26	J deg^{-1} mole^{-1}

These values are for room temperature. Liquids have somewhat larger heat capacity than the crystalline substances, usually about 15% larger. The heat capacity of liquid water is unusually large; it is 75.4 J deg^{-1} mole^{-1}, whereas that of ice is 38 J deg^{-1} mole^{-1}, agreeing with the rule. The interpretation of the large value for water is given in Chapter 12.

Example 10-2. What is the value of ΔH for the reaction of Example 10-1,

$$H_2(g) + \tfrac{1}{2}O_2(g) \longrightarrow H_2O(l)$$

at 100°C and 1 atm? The experimental values of the average molar heat capacity at constant pressure, C_P, over the range 25°C to 100°C are 28.9, 29.4, and 75.5 J deg^{-1} mole^{-1}, for $H_2(g)$, $O_2(g)$, and $H_2O(l)$, respectively.

Solution. To clarify the problem we draw a diagram representing the system in several states. It is helpful to have the vertical coordinate represent qualitatively the enthalpy of the system.

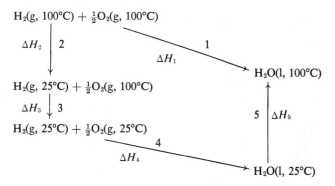

In the diagram there are shown two ways for the system to change from the initial state, $H_2(g, 100°C) + \tfrac{1}{2}O_2(g, 100°C)$, to the final state, $H_2O(l, 100°C)$. The enthalpy of a system is determined by the state of the system, and is independent of the way in which the system has reached that state; hence the value of the enthalpy change for the first path, 1, is equal to that for the other path, 2, 3, 4, 5:

$$\Delta H_1 = \Delta H_2 + \Delta H_3 + \Delta H_4 + \Delta H_5$$

The value of ΔH_4 is given as -285.840 kJ mole^{-1} in Example 10-1. The value of ΔH_5 is the integral of the heat capacity at constant pressure for $H_2O(l)$, $\int_{25°}^{100°} C_P dT = C_P \int_{25°}^{100°} dT = 75.5 \times 75° = 5663$ J mole^{-1} = 5.663 kJ mole^{-1}. The values of ΔH_2 and ΔH_3 are similarly found to be -2.168 kJ mole^{-1} and -1.103 kJ mole^{-1}, respectively. The sum $\Delta H_2 + \Delta H_3 + \Delta H_4 + \Delta H_5$, -283.448 kJ mole^{-1}, is the answer to the problem.

Heat of Vaporization

When heat is added to a liquid at its boiling point, the temperature does not change. Instead a certain amount of liquid becomes a gas. The heat absorbed on vaporization at the boiling point is the *heat of vaporization*; for water its value is 40.6 kJ mole^{-1}.

For many substances a rough value of the heat of vaporization can be predicted from *Trouton's rule*, which states that the quotient of the molar heat of vaporization by the absolute boiling point has a constant value, about 84 J deg^{-1} mole^{-1}. For example, this rule predicts that the molar heat of vaporization of carbon disulfide, b.p. 319°K, is $84 \times 319 = 26.8$ kJ mole^{-1}; the experimental value is 26.7 kJ mole^{-1}. The heat of vaporization of water is larger than expected from Trouton's rule because of the strong intermolecular forces in the liquids, which result from the action of hydrogen bonds (see Section 11-3 for a more detailed discussion of the relation between the heat of vaporization and the boiling point).

Heat of Transition

The transition of a substance from one crystalline modification to another crystalline modification stable in a higher temperature range is accompanied by the absorption of the *heat of transition*. The value of this quantity for the transition of red phosphorus to white phosphorus, for example, is 15.5 J mole^{-1}, and that for red mercury(II) iodide to yellow mercury(II) iodide is 12.6 J mole^{-1}.

10-4. Entropy. The Probable State of an Isolated System

Let us consider an isolated system. We define the system first by stating what atoms are present within the boundaries of the system (or what molecules or chemical substances). The macrostate of the system may be further defined by stating what the volume is and what the value of the energy of the system is. (Instead of the two macroscopic parameters volume and energy, another pair, such as volume and temperature or pressure and temperature, may be used to define the macrostate of the system. Additional macroscopic parameters, such as intensity of gravitational field, electrostatic field, or magnetic field, may also be introduced in order to discuss the associated properties of the system.)

Our desire is to develop a theory of the system that will permit us to discuss its behavior (that is, its macroscopic properties) in terms of its composition and structure (that is, in terms of the atoms or molecules of which it is composed, and their quantum states). We restrict our discussion to macroscopic systems, containing a very large number of identical atoms or molecules, or perhaps a mixture of several kinds.

We have developed a theory of this sort in Section 9-5. In that discussion it was pointed out, for the special case of a pure monatomic gas (particles of one kind), that the macrostate of the system, defined by the

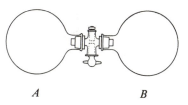

FIGURE 10-1
An isolated system with volume 20
liters, containing 1 mole of helium,
with total energy corresponding to
temperature 0°C.

A   *B*

volume and energy, corresponds to a very great number of quantum states (microstates). The assumption was made that each of the quantum states (with the assumed amount of energy for the system) has the same importance as any other, and the conclusion was drawn that the most probable macrostate of the system is the one that represents the maximum number of quantum states of the system.

We shall call the number of quantum states associated with the macrostate of a system its *multiplicity*, and use the symbol *W* for it. (Here we are following Boltzmann: *W* is the initial letter of the German word *Wahrscheinlichkeit*, which means probability.) The probability of a macrostate of an isolated system is proportional to its multiplicity. Unless there is a restraint on the system, it will assume the macrostate with the greatest multiplicity, the greatest value of *W*. If there is a restraint on the isolated system that prevents it from achieving the state with the greatest multiplicity, and this restraint is removed, the system will change spontaneously toward the macrostate with the maximum multiplicity.

The foregoing sentence is one way of stating the *second law of thermodynamics*.

Let us consider, for example, the system shown in Figure 10-1. The system consists of two bulbs, each with volume 10 liters, connected by a tube and stopcock, and containing 1 mole of helium. We introduce a constraint, by closing the stopcock. The initial macrostate of the system is that in which all the helium molecules are in bulb *A* and the total energy is $\frac{3}{2}R \times 273°$; that is, it corresponds to 273°K (Section 9-4).

We now ask what macrostate the isolated system will assume after the stopcock has been opened. Our common sense and experience, of course, tell us that the gas will distribute itself equally between the two bulbs. The temperature does not change (the total energy, which is all kinetic energy of the molecules, remains the same), and the pressure drops to half the original pressure in bulb *A*.

The probability of the final macrostate is 2^N times that of the initial (restrained) macrostate, with *N* the number of molecules. It need not surprise us that spontaneous fluctuations in pressure in bulb *A* (or bulb *B*) after the system has been allowed time to reach equilibrium are very hard to detect.

The ratio $W_{\text{final}}/W_{\text{initial}}$ can be derived in the following way. We see from Equation 9-18 that the number of quantum states in the energy range E to $E + dE$ for a particle in a box is directly proportional to the volume. The number of quantum states for a system of N identical molecules is given in Equation 9-22 as

$$W = \frac{n^N}{\displaystyle\prod_{j=1}^{\infty} N_j!} \tag{10-2}$$

The value of N_j, the number of molecules in a set of n molecular quantum states with energy very close to E_j, is n times the probability of occupancy of one quantum state with energy E_j, as given by the Boltzmann distribution law (Equation 9-30, with $E_i = E_j$). We see that if we place $n = n_0 V/V_0$, that is, take n proportional to V, the values of N_j are the same for volume V as for volume V_0, and hence the denominator in Equation 10-2 is the same. W is accordingly proportional to n^N, which is equal to $(n_0 V/V_0)^N$. In this way we obtain the equation

$$\frac{W}{W_0} = \left(\frac{V}{V_0}\right)^N \tag{10-3}$$

For $V/V_0 = 2$ the ratio is 2^N. The ratio $(V/V_0)^N$ is also given by a very simple probability argument: the relative chances of finding a molecule in the volumes V and V_0 is V/V_0; the total chance for N molecules behaving independently of one another is $(V/V_0)^N$.

Entropy

The concept of entropy (from Greek *trope*, a turning or change) was introduced about 1850 as a thermodynamic quantity in the analysis of the efficiency of heat engines. Thirty years later Boltzmann suggested that this thermodynamic quantity (which is given the symbol S) could be identified with $k \ln W$, where k is Boltzmann's constant and W is the multiplicity of the system:

$$S = k \ln W \tag{10-4}$$

(His work was done in the late nineteenth century, before the development of quantum theory; he was able to discuss only ratios of multiplicities, W_2/W_1, and the corresponding differences in entropy values, $S_2 - S_1$.)

From Equation 10-4 we see that an increase in W results in an increase in S. We concluded in the preceding paragraphs that a spontaneous change in macrostate of an *isolated* system is always accompanied by an increase in W. We are thus led to conclude that it is always accompanied by an increase in entropy. This is an alternative statement of the second law of thermodynamics.

The universe may be looked on as an isolated system. Both the first law and the second law of thermodynamics are presented in the following sentence, written by the German physicist Rudolf Clausius (1822–1888): *Die Energie der Welt bleibt konstant*; *die Entropie strebt einem Maximum zu.*

We have discussed one example of a spontaneous process in an isolated system, the opening of a stopcock (Figure 10-1) and the expansion of a gas into a larger volume. For a mole of perfect gas expanding from V_1 to V_2 without doing work (that is, in an isolated system) the value of the increase in entropy is $\Delta S = S_2 - S_1 = R \ln (V_2/V_1)$, with $R = 8.315$ J deg^{-1} mole^{-1}. For a doubling of the volume we have $\Delta S = R \ln 2 = 8.315 \times 2.303 \times 0.301 = 5.764$ J deg^{-1} mole^{-1}.

Entropy of Mixing

We consider one mole of gas, consisting of x_1 mole of gas G_1 and x_2 mole of gas G_2, with $x_1 + x_2 = 1$. (x_1 and x_2 are called the mole-fractions of the two components.) Let the initial state be that in which gas G_1 occupies the bulb A and G_2 the bulb B of Figure 10-1, with A and B proportional in volume to x_1 and x_2; there is then no change in pressure when the stopcock is opened. The quantum states for the gas G_1 in flask A and in the total volume are the same as for the preceding case of expansion of a gas. Hence the increase in entropy for the x_1 moles of gas G_1 is equal to $x_1 R \ln (V_2/V_1) = x_1 R \ln (1/x_1)$, and that for gas G_2 is $x_2 R \ln (1/x_2)$. Hence the entropy of mixing of one mole of the mixed gas, at constant pressure, is

$$\Delta S = -R\{x_1 \ln x_1 + x_2 \ln x_2\} \tag{10-5}$$

with $x_1 + x_2 = 1$. Mixing of the two gases would take place spontaneously in an isolated system when the stopcock was opened. For $x_1 = x_2 = \frac{1}{2}$ the increase in entropy is $R \ln 2 = 5.764$ J deg^{-1} per mole of the mixed gas.

Conversion of Potential Energy to Heat

Let our system be two connected bulbs, one above the other, with the upper bulb filled with liquid mercury (Figure 10-1 rotated through 90°). When the stopcock is opened there occurs a spontaneous change in the macrostate of the isolated system: the mercury falls from the upper bulb to the lower bulb, and remains there. This change in macrostate of the isolated system takes place with such vigor (splashing and vibration of the liquid mercury) that there must be a considerable increase in entropy associated with it. What is the nature of the increase in entropy?

The answer to this question is that the potential energy Mgz of the mass of mercury in the upper bulb, at height z above its final elevation, is ultimately converted into thermal energy. The liquid mercury in the lower bulb of the isolated system is at higher temperature than when it was in the upper bulb. In the discussion of the Boltzmann distribution law we have seen that the number of quantum states of a system with specified internal energy (not including the Mgz term) increases with increase in the internal energy, that is, it increases with temperature. We shall see later that the increase in entropy for this spontaneous process is Mgz/T, where T is the temperature of the mercury.

10-5.   The Absolute Entropy of a Perfect Gas

We have accepted $k \ln W$ as the entropy of a system (Equation 10-5). The quantity $\ln W$ for a system of N identical particles has been evaluated (Equation 9-22), as a function of the distribution of the particles among groups of single-particle quantum states. The distribution of the particles among their quantum states, and hence among groups of quantum states, has also been evaluated; it is given in general form by the Boltzmann distribution law, Equation 9-30, and for the special case of a perfect gas by Equation 9-32. By substituting this distribution in Equation 9-22 we can evaluate the entropy of a perfect gas. This evaluation is carried out in Appendix XII.

It is found that the entropy of a monatomic perfect gas is a function of the molecular weight, the volume, and the temperature (or, alternatively, the molecular weight, the pressure, and the temperature). The two equations derived in Appendix XII for the molar entropy are the following:

$$S_{\text{molar}} = \tfrac{3}{2}R \ln M + \tfrac{3}{2}R \ln T + R \ln V + 11.11 \text{ J deg}^{-1} \text{ mole}^{-1} \quad (10\text{-}6)$$

and

$$S_{\text{molar}} = \tfrac{3}{2}R \ln M + \tfrac{5}{2}R \ln T - R \ln P - 9.69 \text{ J deg}^{-1} \text{ mole}^{-1} \quad (10\text{-}7)$$

In these equations M is the molecular weight in the ordinary units (12 for ^{12}C), V the molar volume in liter mole^{-1}, and P the pressure in atmospheres.

These equations are valid only when the multiplicity of the system is large enough to justify the use of Stirling's approximation for $\ln N!$. At extremely low temperatures only the quantum states with the smallest energy would be occupied. In the limit $T \longrightarrow 0°K$ only one quantum state would represent the system, W would have the value 1, and S the value 0.

Let us put $S_{\text{molar}} = 0$ for a gas at 1 atm pressure, with $M = 4.003$ (helium), and use Equation 10-7 to solve for T. The value obtained is $T = 1.44°K$. In fact, at 1 atm pressure helium condenses to the liquid state at $4.3°K$, before it reaches the temperature range in which serious deviation from ideal-gas behavior would be expected to result from the small number of accessible quantum states. All other gases, too, condense to liquids before reaching the low-temperature quantum degeneration region.

Change in Entropy with Temperature

Let the volume of the system be held constant while enough heat is transmitted to the system (one mole of perfect monatomic gas) to increase its temperature by the small amount δT. By what amount does the entropy increase? Equation 10-6 provides the answer: the value of dS_{molar}/dT is seen to be $\frac{3}{2}R/T$, and hence δS_{molar} has the value $\frac{3}{2}R\delta T/T$.

We know that C_V, the molar heat capacity at constant volume, is equal to $\frac{3}{2}R$ (Example 9-4). Hence the heat q transferred to the system to raise the temperature by δT amounts to $\frac{3}{2}R\delta T$. The increase in entropy of the system is seen to be q/T.

Now let us carry out the process of heating the system with the pressure kept constant. By differentiating the right side of Equation 10-7 we find that $\delta S_{molar} = \frac{5}{2}R\delta T/T$. Also, we know (Section 9-4) that the molar heat capacity at constant pressure is $\frac{5}{2}R$, and hence that the value of q is $\frac{5}{2}R\delta T$. In this case, too, the increase in entropy of the system is equal to q/T.

We now consider a larger change in temperature, from T_1 to T_2. From Equation 10-6 we obtain the following expression for ΔS_{molar} for the case of constant volume:

$$\Delta S_{molar} = \tfrac{3}{2}R \ln \frac{T_2}{T_1} \qquad \text{at constant volume} \qquad (10\text{-}8)$$

and from Equation 10-7 a similar expression for the case of constant pressure:

$$\Delta S_{molar} = \tfrac{5}{2}R \ln \frac{T_2}{T_1} \qquad \text{at constant pressure} \qquad (10\text{-}9)$$

10-6. Reversible and Irreversible Changes in State

Let us consider a system and its environment. The environment contains machines that can do work on the system or have work done on them by the system, and also contains many heat reservoirs, at different temperatures. Since we know a great deal about the properties of perfect monatomic gases, we assume that each heat reservoir consists of a number of moles of a perfect monatomic gas.

Let the system be one mole of a perfect monatomic gas, with constant volume V and initial temperature T_1. We put the system in thermal contact with a heat reservoir at temperature T_2 (greater than T_1) containing a very large number of moles of a monatomic perfect gas. The quantity $q = C_V(T_2 - T_1) = \frac{3}{2}R(T_2 - T_1)$ of heat flows from the reservoir to the system, raising its temperature from T_1 to T_2.

The change in entropy of the reservoir (which, because it is very large, has remained at temperature very close to T_2) is $-q/T_2 = -\frac{3}{2}R(T_2 - T_1)/T_2$. The change in entropy of the system is $\frac{3}{2}R \ln (T_2/T_1)$ (Equation 10-8). The change in entropy of the universe associated with this spontaneous change, the transfer of heat from a warmer body to a cooler body, is seen to be positive, in agreement with the second law of thermodynamics, as stated in Section 10-4. The amount of increase, the sum of the two terms given above, can be written in the following way, by replacing $T_2 - T_1$ by ΔT and making use of the series expansion $\ln x = a + \frac{1}{2}a^2 + \frac{1}{3}a^3 + \dots$, with $a = (x - 1)/x$:

$$\text{Total } \Delta S = \frac{3}{2} R \left\{ \frac{1}{2}\left(\frac{\Delta T}{T_2}\right)^2 + \frac{1}{3}\left(\frac{\Delta T}{T_2}\right)^3 + \cdots \right\} \qquad (10\text{-}10)$$

If the system were now to be cooled to its original temperature T_1 by bringing it into thermal contact with a large heat reservoir at temperature T_1, there would again be an increase in total entropy of the universe. The system would then be in its initial state, with its initial entropy; but a quantity of heat $\frac{3}{2}R\Delta T$ would have been removed from the first large reservoir, at temperature T_2, and transferred to the second large reservoir, at temperature T_1, with the net increase in entropy for the two reservoirs given by the equation

$$\Delta S = \frac{3}{2} R\Delta T \left(\frac{1}{T_1} - \frac{1}{T_2}\right) = \frac{3}{2} R \frac{(\Delta T)^2}{T_1 T_2} \qquad (10\text{-}11)$$

The foregoing process, in which the system is first warmed by contact with a warm reservoir and then cooled to its original temperature (its initial state) is called an *irreversible cyclic process*. In each of the two steps there is an increase in entropy of the universe.

A *reversible cyclic process* is defined as a process in which the system is returned to its initial state (the system completes a cycle of changes in state) with no increase in entropy of the universe.

We can devise a process that approaches this ideal much more closely than the one described above. Let us bring the system, at temperature T_1, into thermal contact with a heat reservoir at temperature $T_1 + \delta T$, where $\delta T = (T_2 - T_1)/n$. The temperature of the system is thus raised to $T_1 + \delta T$. A second step, involving use of a heat reservoir with temperature $T_1 + 2\delta T$, will raise the temperature of the system to $T_1 + 2\delta T$, and after n such steps it will reach T_2. Similarly, it can be returned to the initial temperature T_1 by n steps.

From the form of Equation 10-11 we see that the increase in entropy of the environment is only $1/n$ times as great as for the original cyclic process. By making n infinite we would achieve a reversible cyclic process, with no change in entropy of the universe.

A complication is that the process of conduction of heat, to achieve thermal equilibrium between the system and the universe, takes time, and an infinite time would be needed for a process involving an infinite number of steps. *No actual process is reversible.* Actual processes can, however, approximate reversibility as closely as our patience permits.

At every step in a reversible process an infinitesimal change in the environment can reverse the direction of the process. Thus if, in the process of heating the system, the temperature of the heat reservoir with which the system is in contact is decreased (by $2\delta T$, say, in the immediately preceding discussion) then the system will drop back one step.

If the system is held at constant pressure, rather than constant volume, it expands when heated, doing the work $P\Delta V$ on the environment. For reversibility we assume that this work is stored up, by raising a weight, say, with no frictional losses. The potential energy stored up in the environment can then do the work of compressing the system to its original volume.

Since every step in a reversible process must be reversible, in order that the entire process be reversible, we see that for each step the entropy increase for the system is just equal to the entropy decrease of the environment.

10-7. The Efficiency of a Heat Engine

In 1824 a young French physicist, Sadi Carnot (1796–1832), published a paper entitled *Réflexions sur la puissance motrice du feu et sur les machines propres à développer cette puissance*. In this paper he discussed the nature of heat and work, with special reference to the efficiency of steam engines (heat engines). Although the first law of thermodynamics had not yet been recognized, Carnot was led to formulate a fundamental principle equivalent to the second law of thermodynamics.

Carnot discussed a cycle of changes in state of a gas interacting with its environment (the Carnot cycle). As shown in Figure 10-2, the cycle involves four steps: first, the isothermal expansion of the gas from volume V_1 to V_2, with work done on the environment and heat absorbed from the environment at temperature T_{hot}; second, an adiabatic (no heat transfer) expansion to reach the volume V_3 at the lower temperature T_{cold}; third, the isothermal compression at T_{cold} to the volume V_4; and fourth, the adiabatic compression to the original volume. All processes are carried out reversibly, so that there is no entropy change in the universe. The final state of the system (the gas) is identical with the initial state; hence the entropy change of the system is zero; and in consequence the entropy change of the environment is also zero.

The second and fourth steps in the cycle are adiabatic ($q = 0$), and hence involve no change in entropy of the environment. The first step involves the transfer from the environment to the system of the amount of heat q_{hot}, with consequent change $-q_{hot}/T_{hot}$ in the entropy of the environment. The third step involves the transfer to the environment of the amount of heat q_{cold}, with consequent change q_{cold}/T_{cold} in the entropy of the environment. The total entropy change of the environment is, however, zero; hence we write

$$-\frac{q_{hot}}{T_{hot}} + \frac{q_{cold}}{T_{cold}} = 0$$

or

$$\frac{q_{hot}}{q_{cold}} = \frac{T_{hot}}{T_{cold}} \tag{10-12}$$

From the law of conservation of energy we write

$$\text{Work done by system} = w = q_{hot} - q_{cold}$$

which with Equation 10-12 gives

$$w = q_{hot}\left(\frac{T_{hot} - T_{cold}}{T_{hot}}\right) \tag{10-13}$$

Thus we see that the efficiency of a perfect heat engine (a reversible one), the fraction of the heat withdrawn from the hot reservoir that is converted into work, is equal to the difference in temperature of the hot and cold heat reservoirs divided by the temperature of the hot reservoir. An actual heat engine operates in such a way that there is an increase in entropy of the universe with each cycle, and hence the efficiency is less.

A *heat pump* uses work to transport heat from a cold reservoir to a hot reservoir. The effectiveness of a perfect heat pump, q_{hot}/w, is equal to $T_{hot}/(T_{hot} - T_{cold})$. Thus a perfect heat pump using electric power in freezing weather (0°C) to warm a house to 25°C is $298°/25° = 11.9$ times more effective than ordinary electric heaters. Actual heat pumps have a somewhat smaller coefficient of performance.

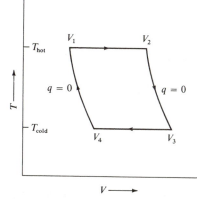

FIGURE 10-2
The Carnot cycle.

10-8.   Change in Entropy of Any System with Temperature

We now consider the change in entropy of any system when the temperature is increased from T_1 to T_2.

Let δq_{rev} be the increment of heat transferred to the system from the environment during one step, at temperature T, of the reversible process of heating the system. (The symbol δq, rather than dq, is conventionally used here.) From the preceding discussion we know that the change in entropy for each step is $\delta q_{rev}/T$; hence the total change in entropy is

$$\Delta S = S_2 - S_1 = \int_{T_1}^{T_2} \frac{\delta q_{rev}}{T} \tag{10-14}$$

We may replace δq_{rev} by $C\,dT$, where C is the heat capacity of the system (the reversible heat capacity, of course):

$$\Delta S = S_2 - S_1 = \int_{T_1}^{T_2} C\,\frac{dT}{T} \tag{10-15}$$

The heat capacity may be either C_V (when the volume is held constant) or C_P (when the pressure is held constant).

The enthalpy has been defined (Section 6-11) as $H = E + PV$. We see that the increase in enthalpy ΔH for a system heated at constant pressure from T_1 to T_2 under conditions such that the only work done is $P\Delta V$ is $\int_{T_1}^{T_2} \delta q$, which is equal to $\int_{T_1}^{T_2} C_P\,dT$ plus the enthalpy associated with any changes of state that may occur.

Example 10-4. The molar entropy of ice at 0°C is known to be 51.84 J deg^{-1} mole^{-1}. What is the molar entropy of water at 0°C? At 25°C? (Pressure 1 atm.)

Solution. During the melting of one mole of ice the heat of fusion, 6010 J mole^{-1}, is absorbed by the system without change in temperature (Section 10-3). Hence the entropy increases by 6010/273.15 = 22.00 J deg^{-1} mole^{-1}, to the value 73.84 J deg^{-1} mole^{-1}, which is the molar entropy of water at 0°C.

To obtain the molar entropy of water at 25°C we add

$$\int_{273°}^{298°} C_P \frac{dT}{T} =$$

$$C_P \int_{273°}^{298°} d\ln T = C_P \ln(298°/273°) = C_P \ln 1.092 = 0.0880\, C_P$$

The value of C_P (constant over this range) is 75.3 J deg^{-1} mole^{-1}, giving $\Delta S = 6.63$ and the value 80.47 J deg^{-1} mole^{-1} for the molar entropy of water at 25°C.

The molar enthalpy of water at 0°C is, of course, 6010 J mole^{-1} greater than that of ice at 0°C and 1 atm.

10-9.   The Third Law of Thermodynamics

The third law of thermodynamics is comprised in the statement that *the entropy of a pure substance in the form of a perfect crystal is zero at the absolute zero.*

The Boltzmann equation $S = k \ln W$ connecting entropy and multiplicity leads directly to the third law. Let us consider a crystal of copper, for example (Section 2-4). We have described the atomic arrangement in this crystal as that of cubic closest packing, with each atom carrying out vibrational motions in the neighborhood of one of the points of a face-centered cubic lattice. As the temperature becomes lower the amplitude of the vibrational motion decreases, and the number of vibrational quantum states accessible to each atom decreases. In the limit as T approaches 0°K only one quantum state, that with lowest energy, remains accessible; all others, with larger values of the energy, are ruled out by the Boltzmann factor $\exp(-E/kT)$, which approaches zero as T approaches zero.

The third law of thermodynamics was discovered by the German chemist Walther Nernst (1864–1941). In the period around 1906 he determined by experiment the difference in entropy of a number of crystalline compounds and the elements (crystalline) from which they could be

TABLE 10-1
Third-law Value of the Entropy of Lead (gas) at 2020°K

0° to 5°	0.11 J deg^{-1} mole^{-1}
5° to 298.15°	64.74
298.15° to 600.5° (m.p.)	19.3
Entropy of fusion	8.29
600.5° to 2020°K	38.8
Entropy of vaporization	85
Entropy of Pb(g)	216
Value from Equation 10-7	215.04

made.* He also measured their heat capacities from room temperature to low temperatures, which permitted him to evaluate the entropy difference at low temperatures. He found that in every case the entropy difference seemed to approach zero as the temperature approached 0°K. In 1911 he stated the law essentially as given in the first sentence of this section.

The Entropy of Lead Vapor

In Section 10-5 we derived expressions for the entropy of a monatomic perfect gas (Equations 10-6 and 10-7). We can use Equation 10-7 to calculate the entropy of Pb(g), pressure 1 atm, at the boiling point; 2017°K. With $M = 207.19$, the atomic weight, we obtain the value $S = 215.04$ J deg^{-1} mole^{-1}.

Another way of obtaining a value for the entropy of Pb(g) is to accept the third-law value $S = 0$ for Pb(c) at 0°K, add $\int C_P \, d \ln T$ for the temperature range 0°K to the melting point, add $\Delta H_{\text{fusion}}/T_{\text{M.P.}}$ for the transition to the liquid, then $\int C_P \, d \ln T$ from the melting point, and finally $\Delta H_{\text{vaporization}}/T_{\text{B.P.}}$ for the transition to the vapor. Comparison of this value of the entropy of Pb(g) with that given by Equation 10-7 provides a test both of the third law and the quantum statistical theory used to derive Equation 10-7.

The heat capacity of Pb(c), as measured in the laboratory, is shown by one of the curves in Figure 10-10. The value decreases rapidly with decreasing temperature. Around 5°K it is given by the expression $C_P = 0.0026 \, T^3$ J deg^{-1} mole^{-1}. With use of this expression we evaluate S at 5°K as $\int_0^{5°} C_P \, d \ln T = 0.11$ J deg^{-1} mole^{-1}. The value is entered in the first line of Table 10-1.

---

*The experimental methods of determining this difference in entropy will be discussed in Chapter 11.

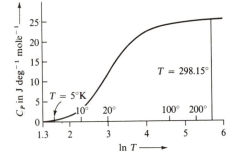

FIGURE 10-3
The molar heat capacity of lead (crystal) plotted against ln T. The area under the curve from 5°K to 298.15°K is the difference in molar entropy of Pb(c) at the two temperatures.

TABLE 10-2
Calculated and Observed Values of the Entropy of the Argonons at 1 atm and 0°C

	Entropy Calculated by Equation 10-7	Experimental Value ($S = 0$ at 0°K)
He	126.07 J deg^{-1} mole^{-1}	126.0 ± 0.5 J deg^{-1} mole^{-1}
Ne	146.24	146.5 ± 0.4
Ar	154.76	154.6 ± 0.8
Kr	164.01	163.9 ± 0.5
Xe	169.61	170.3 ± 1.0

The value of $\int_{5°}^{298.15°} C_P \, d \ln T$ has been obtained by plotting the curve of the experimental values of C_P against ln T, as shown in Figure 10-3, and measuring the area under the curve between the limits 5° and 298.15°K. The value found for the area is 64.74 J deg^{-1} mole^{-1}.

Between 298.15°K and the melting point, 600.5°K, C_P increases from 25 to 30 J deg^{-1} mole^{-1}. As an approximation we use the average value, 27.5, and 27.5 ln (600.5/298.15) = 19.3 J deg^{-1} mole^{-1} for ΔS over this range.

The heat of fusion, determined by direct experiment, is 4.98 kJ mole^{-1}, leading to 8.29 J deg^{-1} mole^{-1} for the entropy of fusion.

The value of C_P for the liquid is about 32 J deg^{-1} mole^{-1} in the range 600° to 900°K. If we accept this value up to the boiling point, 2017°K, we obtain ΔS = 32 ln (2017/600.5) = 38.8 J deg^{-1} mole^{-1}. The heat of vaporization has been determined only approximately, from the variation of the vapor pressure of Pb(l) with temperature; the reported value is 170 kJ mole^{-1}, which gives 84 J deg^{-1} mole^{-1} for the entropy of vaporization.

The sum of these values, 216 J deg^{-1} mole^{-1}, agrees with the much more accurate value from Equation 10-7 to within experimental error.

The same sort of comparison is made for the five argonons helium to xenon in Table 10-2. The experimental error in the third-law values is small, averaging ± 0.6, and the deviation of the values from the theoretical values is smaller, ± 0.3.

Exceptions to the Third Law

There are no known exceptions to the third law, as stated above: that the entropy of a pure substance in the form of a perfect crystal is zero at the absolute zero. There are some crystalline pure substances, however, that retain some entropy when they are cooled to very low temperatures under laboratory conditions. An example is nitrous oxide, N_2O, which has a residual entropy 4.77 J deg^{-1} mole^{-1} at very low temperature. (This value is the difference between the theoretical value for the gas, given by Equation 10-7 plus terms for rotation and vibration of the gas molecules, and the integrated values of $C_P \ln T$ plus entropy of fusion and vaporization.) The accepted explanation of the residual entropy is that the N_2O crystal is not a perfect crystal, but is disordered, under the conditions of the experiment. The molecule is linear, with the oxygen atom at one end. If the crystal were perfect, its structure might be represented as follows:

NNO   NNO   NNO   NNO   NNO   . . .

ONN   ONN   ONN   ONN   ONN   . . .

The molecule in each position in the crystal has only one orientation: the oxygen atom to the right in the first row, to the left in the second, and so on. But the oxygen atom and the nitrogen atom have nearly the same size, and have nearly zero electric charge (the electric dipole moment is only 0.0346 ϵÅ). The energy difference for the two orientations may be so small that they are nearly equally probable under the conditions of the measurement of C_P. The multiplicity of the crystal would be 2^N, if each molecule could assume either one of the two orientations, independently of the adjacent molecules. $R \ln 2$ has the value 5.76 J deg^{-1} mole^{-1}, only 21% larger than the experimental value.

We can understand why the crystal is disordered. At temperatures a little below the freezing point, 183°K, a molecule can turn over easily in the crystal, and it assumes the two orientations nearly equally (the energy difference for the two being much less than kT). At lower temperatures the rate of reversing direction is extremely small, and the molecules are "frozen in" to one of the two orientations, essentially at random.

Carbon monoxide is also found to have residual entropy, about 4.5 J deg^{-1} mole^{-1}. It is attributed to the same sort of disorder.

The residual entropy of ice will be discussed in Section 12-5.

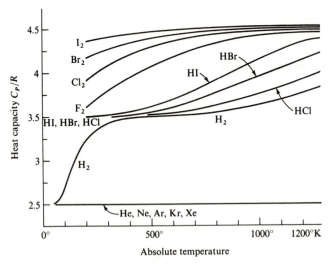

Experimental values of the heat capacity at constant pressure of some diatomic gases.

10-10. The Heat Capacity of Diatomic Gases

The heat capacity at constant pressure of monatomic gases is equal to $\frac{5}{2}R$. The value $\frac{5}{2}R$ is also found experimentally for diatomic hydrogen, H_2, in the temperature range from the boiling point, $20.4°K$, to about $50°K$. It then begins to increase, as shown in Figure 10-4, and approximates the value $\frac{7}{2}R$ from about $400°K$ to $700°K$, above which a further increase occurs.

For other diatomic gases the observed value of C_P increases from $\frac{7}{2}R$ at low temperatures to $\frac{9}{2}R$ at high temperatures.

Equipartition of Energy

In classical kinetic theory and also quantum thermodynamics (at high temperature) the average amount of kinetic energy $\frac{1}{2}kT$ is associated with each degree of freedom of a molecule. The average kinetic energy of a monatomic molecule is $\frac{3}{2}kT$, and correspondingly the value of C_V for a monatomic gas is $\frac{3}{2}R$ (Section 9-4). The value of C_P is larger by R because of the $P\Delta V$ work done in the thermal expansion.

A diatomic molecule has six degrees of freedom, three for each of the two atoms. The degrees of freedom (the independent coordinates required to specify the configuration of the system) may be taken to be the three Cartesian coordinates of the center of mass, the distance between the two atoms, and the two polar angles θ and ϕ determining the orientation of the internuclear axis in space. The first three degrees of freedom correspond to the translational energy $\frac{3}{2}kT$, the next one to the vibrational energy kT (of which $\frac{1}{2}kT$ is the equipartition value of the average kinetic energy and another $\frac{1}{2}kT$ is the value of the average potential energy), and the last two to kT, for the kinetic energy of the two degrees of freedom of rotational motion. The sum gives $C_V = \frac{7}{2}R$, and hence $C_P = \frac{9}{2}R$, in agreement with the high-temperature experimental values for the heavier molecules represented in Figure 10-4.

The decrease to $C_P = \frac{7}{2}R$ at lower temperatures is a quantum effect involving the "freezing-in" of the vibrational motion; that is, the restriction of the molecule to the lowest vibrational quantum state. Similarly, the further decrease shown by H_2 involves the freezing-in of the rotational motion. These quantum effects are discussed further in the following sections.

10-11.   Quantum States of the Rigid Rotator

As a first approximation a diatomic molecule can be considered to be two point masses with a fixed value of the interatomic distance. A classical dumbbell molecule can be described as rotating about an axis through the center of mass with angular velocity ω radians per second, angular momentum $I\omega$, and kinetic energy of rotation $\frac{1}{2}I\omega^2$. Here I, the moment of inertia, is $M_1r_1^2 + M_2r_2^2$, with $r_1/r_2 = M_2/M_1$ and $r_1 + r_2$ equal to the internuclear distance.

We have seen earlier (Section 5-1) that in his quantum theory of the hydrogen atom Niels Bohr assigned to the electron the orbital angular momentum $n\hbar$, with $n = 1, 2, 3, \ldots$. When quantum mechanics was developed it was found that in fact the orbital angular momentum of the electron has the values

$$\{l(l + 1)\}^{1/2}\hbar$$

with $l = 0, 1, 2, \ldots$. It is not surprising that the solution of the Schrödinger wave equation for the rigid rotator gives the result that the allowed states of rotation are those with angular momentum $I\omega = \{J(J + 1)\}^{1/2}\hbar$, with the rotational quantum number J having values $0, 1, 2, \ldots$. The rotational energy $\frac{1}{2}I\omega^2$ is easily evaluated:

$$E_{\text{rotation}} = \frac{J(J + 1)\hbar^2}{2I} \tag{10-16}$$

For each value of J there are $2J + 1$ quantum states, corresponding to the values $-J, -J + 1, \ldots, J$ for the magnetic quantum number M (the component of angular momentum in the direction of an applied magnetic field is $M\hbar$).

Pure Rotation Spectra

The energy levels, as given by Equation 10-16, are shown in Figure 10-5. In classical theory a rotating electric dipole emits and absorbs light with frequency equal to the frequency of rotation of the molecule. A quantum rotator such as the HCl molecule has a related property; the quantum number J changes by only ± 1 in the absorption or emission of a photon. The allowed transitions for molecules with an electric dipole moment, such as HCl, are indicated by arrows in Figure 10-5.

It is observed, for example, that $^1H^{35}Cl$ gas has a strong absorption line in the far infrared (microwave) region, at wavelength 0.474 mm. There are also fainter lines at $\frac{1}{2}, \frac{1}{3}, \ldots$, times this wavelength. These lines correspond to the transitions $J = 0 \longrightarrow 1, 1 \longrightarrow 2, 2 \longrightarrow 3, \ldots$, respectively. From the wavelength we calculate that ν for the photon has the value 6.33×10^{12} s^{-1}, and $h\nu$, the energy, has the value 4.19×10^{-22} J. This is the energy difference of the first rotational state, $J = 1$, and the normal state, $J = 0$, of HCl, and it is equal to $\hbar^2/I$ (Equation 10-16). Hence we find $I = 2.65 \times 10^{-47}$ kg m^2. The value of the moment of inertia for $^1H^{35}Cl$, the principal isotopic species, is $1.626\ r^2 \times 10^{-27}$ kg m^2, where r is the internuclear distance. Hence we find that r, the average internuclear distance for $^1H^{35}Cl$, has the value 1.277×10^{-10} m $= 1.277$ Å. The same value is found for the other isotopic species, such as $^2H^{35}Cl$.

10-12. The Rotational Entropy of Diatomic Gases

The equilibrium distribution of molecules among the vibrational states, with energy given by Equation 10-16, corresponds to the Boltzmann distribution law. The contribution of rotation to the entropy of a diatomic gas is found by the method of Appendix XII to be

$$S_{\text{rot}} = R + R \ln T - R \ln \frac{\hbar^2}{2Ik} - R \ln \sigma \qquad (10\text{-}17)$$

In this equation, which is valid only in the high-temperature range (in which the rotational heat capacity has the value R), the quantity σ, called the *symmetry number*, has the value 1 if the two atoms in the molecule are different and 2 if they are identical. With two identical atoms, either bosons or fermions, the symmetry requirements of the wave function allow only half the states, either those with J even (for which the rotational wave

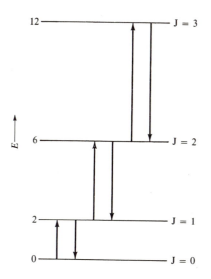

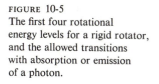

FIGURE 10-5
The first four rotational
energy levels for a rigid rotator,
and the allowed transitions
with absorption or emission
of a photon.

functions are symmetric in the two atoms) or those with J odd (anti-symmetric). (The spins of the nuclei and the symmetry character of the electronic state are also involved; this matter is too complex to be discussed in detail here.)

As an example, let us evaluate the entropy of hydrogen, $H_2(g)$, at 500°K and 1 atm. The translational entropy is found from Equation 10-7 with $M = 2.016$ to have the value 125.31 J deg^{-1} mole^{-1}. The rotational entropy is given by Equation 10-17, with $\sigma = 2$. Analysis of the spectrum of molecular hydrogen has given the value 4.61×10^{-48} kg m^2 for I (corresponding to internuclear distance 0.742 Å), which leads to $S_{rot} = 17.06$ J deg^{-1} mole^{-1}. The total entropy of $H_2(g)$ at 500°K and 1 atm is the sum of these two terms, 142.37 J deg^{-1} mole^{-1}. Values at other temperatures can be calculated from this value by the $\int C_P \, d \ln T$ method.

Equation 10-17 gives low values for S_{rot} at low temperatures, where the rotational heat capacity falls below the value R; that is, for H_2 below room temperature. For HCl(g) the equation is still accurate at the normal boiling point of the liquid, 188°K. It gives $S_{rot} = 0$ at 5.6°K, showing that the rotational quantum effects are important at this temperature.

A Simplified Discussion

We may carry out an approximate treatment at low temperature by considering only the normal state, $J = 0$, and the three first excited states, with $J = 1$ (and $M = -1$, 0, and $+1$). We assume $E (J = 1) = 4.20 \times 10^{-22}$ J, as for HCl. The value of E/k is then 30.4°, and the relative proba-

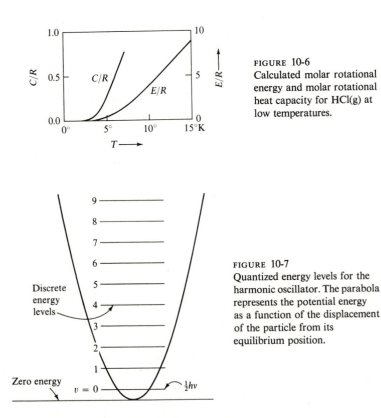

FIGURE 10-6
Calculated molar rotational
energy and molar rotational
heat capacity for HCl(g) at
low temperatures.

FIGURE 10-7
Quantized energy levels for the
harmonic oscillator. The parabola
represents the potential energy
as a function of the displacement
of the particle from its
equilibrium position.

bilities of the normal state and each of the three states with $J = 1$ are 1:
$\exp(-30.4/T)$. The average rotational energy per molecule at temperature
T is

$$E_{av}/k = \frac{3 \times 30.4 \exp(-30.4/T)}{1 + 3 \exp(-30.4/T)}$$

Values obtained by substituting values of T are shown plotted in Figure
10-6. Moreover, we can differentiate this expression with respect to T to
obtain the heat capacity, also shown in Figure 10-6. Graphical integration
gives the value 0.06 J deg^{-1} $mole^{-1}$ for the rotational entropy of HCl(g) at
5.6°K. This is the correct value; the low value (0) given by Equation 10-17
is the result of applying this equation in the temperature range in which it
is not valid (because of neglect of terms in $1/T^2$, $1/T^3$, . . . in evaluating the
distribution series).

10-13. Quantum States of the Harmonic Oscillator

We have so far discussed the quantum states of motion corresponding to five of the six degrees of freedom of the hydrogen chloride molecule: the three translational and two rotational degrees of freedom. The motion associated with the sixth degree of freedom, the distance r between the two nuclei, is the vibrational motion of the two nuclei relative to one another, or, more accurately, their synchronous vibrational motion relative to the center of mass. Newton's equations of motion show that this vibration is equivalent to the vibration of a particle with mass $(M_1^{-1} + M_2^{-1})^{-1}$ vibrating about a fixed point, with the same potential energy function as for the two nuclei.

Let us consider a particle free to move in one dimension (the x-axis), and bound to the point $x = 0$ by a force proportional to its displacement. Let k be the force constant, such that the restoring force for displacement x is $-kx$ and the potential energy is $\frac{1}{2}kx^2$. It is shown in courses in physics that the motion of the particle is simple harmonic, described by the equation

$$x = a \sin 2\pi\nu t \tag{10-18}$$

Here a is the amplitude and ν is the frequency of the oscillation. The frequency ν is related to the force constant k and the mass m of the particle by the relation

$$k = 4\pi^2 m\nu^2 \tag{10-19}$$

and the energy is given by the equation

$$E = 2\pi^2 m\nu^2 a^2 \tag{10-20}$$

It is easily shown that for the harmonic oscillator in any state of motion the time-average potential energy (time average of $\frac{1}{2}kx^2$) and the time-average kinetic energy (time average of $\frac{1}{2}mv^2$, with v the velocity) are equal; each is equal to one-half of the total energy, E. (This is called the *virial theorem* for the harmonic oscillator; it holds for both classical mechanics and quantum mechanics.)

Solution of the Schrödinger wave equation for the harmonic oscillator gives the result that the allowed energy levels are equally spaced: they have the values (Figure 10-7)

$$E_v = (v + \tfrac{1}{2})h\nu \tag{10-21}$$

Here the vibrational quantum number, v, has the allowed values 0, 1, 2, ... It is interesting that the vibration does not cease even in the lowest quantum state: the energy $\frac{1}{2}h\nu$ for the lowest state is called the *zero-point energy* of the oscillator.

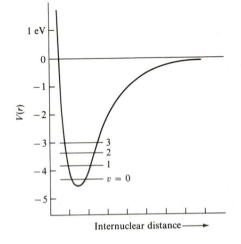

FIGURE 10-8
A curve representing the electronic energy of a diatomic molecule (HCl) as a function of the distance between the nuclei. The zero for energy is the energy for the separated atoms. The minimum of the curve corresponds to the equilibrium value of the internuclear distance.

10-14.   Vibrational States of Diatomic Molecules

In a molecule the vibrational motion of the atoms is determined by a potential energy function which is equal to the electronic energy (calculated by solving the Schrödinger wave equation for fixed positions of the nuclei).* For hydrogen chloride, for example, the potential energy function has the form shown in Figure 10-8. Here the value $V(r) = 0$ corresponds to separate atoms H and Cl in their normal states. The lower part of the curve can be approximated by a parabola. The observed energy levels are not quite equally separated; the $v = 2$, $v = 1$ energy difference is 5% less than for $v = 1$, $v = 0$. The transition $v = 0$ to $v = 1$ gives the first vibrational absorption band of HCl, in the infrared, with center at 3.40 μm. (It is a band, a collection of closely-spaced lines, because the rotational quantum number J also changes.) The transition $v = 0$ to $v = 2$ is observed as a weak line at 1.73 μm. This line can be called the first overtone of the fundamental line at 3.40 μm.

The wavelength 3.40 μm for the transition $v = 0$ to $v = 1$ shows that the first vibrational level lies $E_1 - E_0 = h\nu = hc/\lambda = 5.85 \times 10^{-20}$ J above the normal state. This value provides an explanation of the temperature range over which vibration begins to contribute to the heat capacity of hydrogen chloride, as shown by the following calculation.

---

*This is the Born-Oppenheimer principle, formulated by Max Born and Robert Oppenheimer in 1927.

Example 10-5. From Figure 10-4 it is seen that the contribution of vibration to the heat capacity of HCl reaches 0.1 R at about 750°K. What value for this temperature would be expected from the spectroscopic value 5.85×10^{-20} J for $E_2 - E_1$?

Solution.   We need consider only the two lowest vibrational states. Their relative weights are 1 to $\exp\{-(E_1 - E_0)/kT\}$. The energy, relative to the normal state, is just the energy of the excited state multiplied by the weight of this state; that is, $(E_1 - E_0) \exp\{-(E_1 - E_0)/kT\}$. (Note that the denominator $1 + \exp\{-(E_1 - E_0)/kT\}$ has been approximated by 1; we shall see later that this is a justified approximation.) The molar heat capacity is N times the derivative of the energy with respect to T:

$$C_{\text{vibration}} = N\frac{(E_1 - E_0)^2}{kT^2} \exp\{-(E_1 - E_0)/kT\}$$

$$= Rx^2 \exp(-x)$$

with $x = (E_1 - E_0)/kT$. For $C_{vib} = 0.1\ R$ we have $x^2 \exp(-x) = 0.1$, which can be written as $10x^2 = \exp x$. This equation can be solved by inspection of a table of values of $\exp x$; one finds $x = 5.83$, which leads to $T = (E_1 - E_0)/k \times 5.83 = 5.85 \times 10^{-20}/13.805 \times 10^{-24} \times 5.83 = 727°K$. Hence we calculate that the vibrational heat capacity of HCl(g) should reach 0.1 R at 727°K.

The mole fraction of molecules in the excited state, $v = 1$, at this temperature is $e^{-x} = 0.003$; we were accordingly justified in ignoring this term in the denominator.

10-15.   Energy, Heat Capacity, and Entropy of a Harmonic Oscillator

Let us now apply the Boltzmann distribution law to obtain the probability p_v of the state v of the harmonic oscillator. From Equation 9-30 we write

$$p_v = Cx^v$$

for the probability. Here C is a normalizing factor and x has the value

$$x = \exp\left(-\frac{h\nu}{kT}\right)$$

To evaluate C we note that the sum of p_v for all values of v is

$$\sum_{v=0}^{\infty} p_v = C(1 + x + x^2 + x^3 + \cdots)$$

We recognize that the series $1 + x + x^2 + x^3 + \ldots$ is $(1 - x)^{-1}$. [This can be verified by expanding $(1 - x)^{-1}$ by the binomial theorem.] The sum of probabilities $\sum\limits_{v=0}^{\infty} p_v$ is unity; hence we obtain

$$C(1-x)^{-1} = 1$$

or

$$C = 1 - x = 1 - \exp\left(-\frac{h\nu}{kT}\right)$$

The probability p_v hence has the value

$$p_v = (1-x)x^v = \left[1 - \exp\left(-\frac{h\nu}{kT}\right)\right]\exp\left(-\frac{vh\nu}{kT}\right) \qquad (10\text{-}22)$$

The energy E_v of the vth state is $(v + \frac{1}{2})h\nu$ (Equation 10-21). The average energy $\bar{E}$ is the sum of the products p_vE_v:

$$\bar{E} = \sum_{v=0}^{\infty} p_v E_v$$

$$= h\nu \sum_{v=0}^{\infty} p_v v + \tfrac{1}{2}h\nu \sum_{v=0}^{\infty} p_v$$

Introducing $(1-x)x^v$ for p_v (Equation 10-22) and noting that $\Sigma p_v = 1$ (in the second term), we obtain

$$\bar{E} = h\nu(1-x)\left(\sum_{v=0}^{\infty} vx^v\right) + \tfrac{1}{2}h\nu$$

The sum is easily evaluated by use of the binomial theorem:

$$\sum_{v=0}^{\infty} vx^v = x + 2x^2 + 3x^3 + \cdots$$

$$= x(1 + 2x + 3x^2 + \cdots)$$

$$= x(1-x)^{-2}$$

Hence we obtain the result

$$E_{\text{vib}} = \frac{Nh\nu \exp(-h\nu/kT)}{1 - \exp(-h\nu/kT)} + \tfrac{1}{2}Nh\nu \qquad (10\text{-}23)$$

The heat capacity is $\dfrac{dE}{dT}$. It is found to have the value

$$C_{\text{vib}} = R\left(\frac{h\nu}{kT}\right)^2 \frac{\exp(h\nu/kT)}{\{\exp(h\nu/kT) - 1\}^2} \qquad (10\text{-}24)$$

The entropy of vibration can be evaluated as $\int_0^T C_{\text{vib}}d\ln T$; it is

$$S_{\text{vib}} = \frac{R(h\nu/kT)}{\exp(h\nu/kT) - 1} - R\ln\left\{1 - \exp\left(-\frac{h\nu}{kT}\right)\right\} \qquad (10\text{-}25)$$

The three functions E_{vib}, C_{vib}, and S_{vib} are shown in Figure 10-9. At high temperature the energy and the heat capacity approach the equipartition

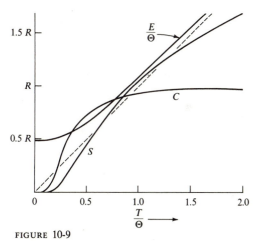

FIGURE 10-9
Values of vibrational heat capacity, energy (divided
by the characteristic temperature Θ), and entropy,
as functions of T/Θ, calculated by the Planck-
Einstein equations (harmonic oscillator equations).

values RT and R, respectively, and the entropy approaches the value
$R\{\ln (kT/h\nu) + 1\}$. The equipartition values of the heat capacity of a
diatomic gas are $C_V = \frac{7}{2}R$, $C_P = \frac{9}{2}R$.

A nonlinear polyatomic molecule has $3n$ degrees of freedom (n = number of atoms), of which 3 are translational, 3 rotational, $3n - 6$ vibrational. Water vapor, for example, has $C_P = 4.04\ R$ at 298°K. Of this
total R is the PV term, $\frac{3}{2}R$ the equipartition value for the three translational degrees of freedom, and $\frac{3}{2}R$ that for the three rotational degrees of
freedom. Two of the remaining three degrees of freedom involve high
vibrational frequencies, corresponding to stretching the two O—H bonds.
The other one involves a lower vibrational frequency, corresponding to
the bending of bonds (wagging motion of the hydrogen atoms). We expect the latter to come into operation at lower temperatures than the
former. The stretching vibrations have about 20% greater frequency than
for HCl, for which $C_{\text{vib}} = \frac{1}{2}R$ at about 1400°K, and the bending vibration has about 55% of this frequency.* We expect C_P to reach $5R$ at
about 1000°K and $7R$ at over twice this temperature. The observed value
at 1000°K is 4.96 R.

---

*Molecular vibrational frequencies are usually given in tables in units cm⁻¹ (the re-
ciprocal of the wavelength in cm). The value for HCl (0 to 1) is 2940 cm⁻¹. The two
stretching vibrations of $H_2O(g)$ are at 3660 cm⁻¹ and 3760 cm⁻¹, and the bending vibration
is at 1600 cm⁻¹.

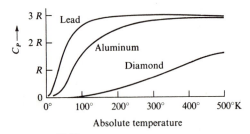

FIGURE 10-10
Experimental values of heat capacity of diamond,
aluminum, and lead.

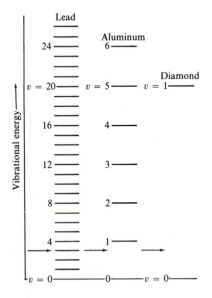

FIGURE 10-11
Diagram showing relative vibrational
energy levels of atoms in crystals of
diamond, aluminum, and lead,
according to the Einstein theory.

10-16. The Quantum Theory of
Low-temperature Heat Capacity of Crystals

One of the early successes of the quantum theory was the explanation, by
Einstein, of the decrease in heat capacity toward zero observed for solids
at low temperature, as shown in Figure 10-10.

In the preceding section we have shown that the average energy of a
harmonic oscillator at temperature T has the value kT if the temperature
is high, so that kT is greater than $h\nu$, the energy difference of adjacent
energy levels.

For a classical harmonic oscillator this energy difference of adjacent energy levels is zero, and the expression $\bar{E} = kT$ should be valid at all values of the temperature.

A crystal containing N atoms (one mole) may be described as equivalent to $3N$ harmonic oscillators. The molar energy of vibration is then

$$E_{\text{molar}} = 3NkT$$

The heat capacity at constant volume can be obtained by differentiating with respect to T:

$$C_V = \frac{dE_{\text{molar}}}{dT} = 3R \text{ (classical)}$$

Thus the classical theory of statistical mechanics leads directly to the conclusion that the molar heat capacity at constant volume for elements should have the value $3R$; the value at constant pressure should be only a little larger. As mentioned in Section 10-3, this agrees with the experimental values at room temperature for elements with large atomic weight, but the light elements at room temperature and all elements at sufficiently low·temperature are observed to have much smaller values, approaching zero as the temperature approaches 0°K.

The force constant for the bonds between each atom and its neighboring atoms in a crystal can be calculated from the measured value of the compressibility of the crystal. From the value of the force constant and the mass of the atom the frequency of the vibration of the atom relative to its neighbors can be calculated. The value of ν found in this way for aluminum is 4 times that for lead, and the value for diamond is 20 times that for lead. The corresponding energy levels (relative to the state with $v = 0$ for each element) are shown in Figure 10-11. We see that if a piece of lead, a piece of aluminum, and a diamond were in contact with a substance at such a temperature that the average collision between atoms could provide the amount of energy indicated by the horizontal arrow, the lead atoms could accept the energy. On the other hand, this amount of energy would be far less than needed to raise a carbon atom of diamond from the lowest vibrational state ($v = 0$) to the first excited level ($v = 1$). Hence increase in the temperature would not cause the diamond to accept energy from the surroundings, and its heat capacity would be close to zero.

The curves of Figure 10-9, with the factor 3 for the three degrees of vibrational freedom of each atom (along the x, y, and z axes), correspond to the Einstein theory of crystals. We see that the heat capacity is calculated to be $\frac{3}{2}R$, half the normal value, when T has approximately the value $\frac{1}{3}h\nu/k$.

Einstein calculated the heat-capacity function in this way in 1907. The calculated curves were found to fall off with decreasing temperature more rapidly than the experimental values. P. Debye then developed a more refined theory. He took into consideration the vibrational motion of a group of two or more atoms relative to the surrounding atoms. His theoretical heat-capacity function was found to be in complete agreement with experiment, and scientists then recognized (in 1912) that it is necessary to use the quantum theory to understand the properties of substances in terms of their atomic and molecular structures. In the Debye theory the heat capacity is $\frac{3}{2}R$ when T has approximately the value $\frac{1}{4}h\nu/k$ (more accurately $0.249\ h\nu_{max}/k$).

Debye pointed out that the low frequencies of vibration correspond to long wavelengths, many times the atomic diameter, and can be discussed by use of the ordinary theory of deformation of elastic solids. There are two kinds of vibrations: *compressional waves* (ordinary *sound waves, phonons*) and *transverse waves*, in which the displacement of atoms is perpendicular to the direction of propagation. A simple analysis (similar to that of a particle in a box, Figures 9-4 and 9-5) leads to the following distribution function of frequencies:

$$N(\nu)d\nu = 9N\ \frac{\nu^2}{\nu_{max}^3}\ d\nu \tag{10-26}$$

We see that $\int_0^{\nu_{max}} N(\nu)\ d\nu = 3N$. The maximum frequency, ν_{max}, corresponds to the Einstein frequency.

The energy, heat capacity, and entropy can be calculated by integrating the corresponding Einstein functions (Equations 10-23, 10-24, 10-25) over the range of frequencies from 0 to ν_{max}. We shall not give the Debye equations, which can be found in reference books, together with tabulated numerical values of the Debye functions. The Debye theory is found to agree extremely well with the measured heat capacities of nonmetallic crystals with simple structure (all atoms equivalent). For other crystals, especially molecular crystals, a more complex treatment is needed.

The Debye heat-capacity function and the Einstein function are compared in Figure 10-12. An important difference is observed at low T; the Debye function gives C proportional to T^3 in this region, whereas the Einstein function decreases much more rapidly with decreasing T.

The Debye low-temperature heat capacity is found to be given by the equation

$$C_V = \frac{12\pi^4}{5}\left(\frac{k}{h\nu_{max}}\right)^3 RT^3 = 233.78\left(\frac{k}{h\nu_{max}}\right)^3 RT^3 \tag{10-27}$$

The error is less than 1% for T less than $0.08\ h\nu_{max}/k$.

For metals the observed heat capacity at low temperatures has the form $\gamma T + \alpha T^3$; there is a linear term as well as the Debye T^3 term. The linear term results from excitation of the valence electrons (conduction electrons), as will be described in Chapter 18.

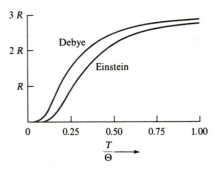

FIGURE 10-12
Curves representing the
Einstein and Debye functions
for the heat capacity of a crystal.

Approximate expressions for the Debye E, C_V, and S at high temperatures are the following ($\Theta = h\nu_{\max}/k$):

$$E = 3RT \left\{ 1 + \frac{1}{20} \left(\frac{\Theta}{T} \right)^2 - \cdots \right\} \tag{10-28}$$

$$C_V = 3R \left\{ 1 - \frac{1}{20} \left(\frac{\Theta}{T} \right)^2 + \cdots \right\} \tag{10-29}$$

$$S = 3R \left\{ \ln \left(\frac{T}{\Theta} \right) + 1.333 + \cdots \right\} \tag{10-30}$$

We shall use these equations in the following chapter.

Exercises

10-1. The molar enthalpy of vaporization of methyl alcohol at 25°C is 37.4 kJ mole^{-1}. From this value and information given in Tables 7-1 and 7-2 calculate the enthalpy change for the reaction

$$CH_3OH(l) + \tfrac{3}{2}O_2(g) \longrightarrow CO_2(g) + 2H_2O(l)$$

at 25°C and 1 atm.

10-2. The heat of combustion of liquid benzene, $C_6H_6(l)$, to form $CO_2(g)$ and $H_2O(l)$ at 25°C is 3273 kJ mole^{-1}. What is the standard enthalpy of formation of $C_6H_6(l)$?

10-3. The enthalpy of vaporization of chlorine, Cl_2, at the standard boiling point, 284°K (vapor pressure 1 atm), is 20.41 kJ mole^{-1}. What is the difference ΔE in internal energy of the gas and the liquid? (Answer: 18.05 kJ mole^{-1}.)

10-4. One mole of liquid chlorine at 284°K is injected into an evacuated 25-l bulb in a thermostat at 284°K. How much thermal energy would need to flow into the bulb to return it to 284°K?

10-5. (a) With use of values of C_P from Figure 10-4 (reasonably extrapolated) calculate the amount of energy required to heat one mole of HCl(g) from 25°C to 1200°C.

(b) The standard enthalpy of formation of HCl(g) is -92 kJ mole^{-1} (Table 7-9). Would you think that a hydrogen-chlorine flame might reach a flame temperature of 1200°C? (Ignore the possibilities of incomplete combustion and of dissociation into H(g) and Cl(g).)

10-6. By integration under the curve of C_P for aluminum in Figure 10-10, calculate the value of $H(500°K) - H(298°K)$, the difference in molar enthalpy of Al(c) at 500°K and 298°K.

10-7. Under certain conditions water can be cooled considerably below 0°C without freezing (can be supercooled). Assuming C_P to be constant for water and ice, with the values of 0°C (79.95 and 37.72 J deg^{-1} mole^{-1}, respectively) formulate an expression for the enthalpy of fusion of ice as a function of the temperature (the value at 0°C is 6.008 kJ mole^{-1}). What amount of supercooling would be needed in order that the water could freeze entirely without transfer of heat to the environment?

10-8. In the perfect-gas approximation, what is the change in internal energy at 25°C and 1 atm on mixing 1 mole of $O_2(g)$ and 4 moles of $N_2(g)$ (to obtain a gas approximating air in composition)? What is the change in entropy?

10-9. What is the difference in entropy, in J deg^{-1} mole^{-1}, of helium at 1 atm and at 5 atm pressure, at 25°C? At 100°C? Of helium at 1 atm pressure and at 0.2 atm, at each of these temperatures?

10-10. A system consists of two bulbs, one containing 1 mole of helium at 5 atm and the other containing 1 mole of helium at 0.2 atm, in a thermostat at 25°C. The stopcock in the tube connecting the bulbs is opened, and the system reaches thermal equilibrium with the thermostat. What change in entropy of the helium has occurred?

10-11. The third-law value of the absolute entropy of S_8 (orthorhombic) at 25°C is 255.1 J deg^{-1} mole^{-1}, and the heat capacity is 181 J deg^{-1} mole^{-1}.

(a) Assuming that the heat capacity is constant, evaluate the entropy of the crystal at the transition temperature to the monoclinic form, 95.4°C.

(b) The enthalpy of transition S_8 (orthorhombic) $\longrightarrow$ S_8 (monoclinic) at 95.4°C is observed to be 3.01 kJ mole^{-1}. What is the value of the entropy of transition?

(c) Of the absolute entropy of S_8 (monoclinic)?

10-12. The third-law entropy of S_8 (monoclinic), obtained from measurement of the heat capacity of the crystals (which are metastable of low temperatures; that is, change only slowly to the orthorhombic form) is 260.4 J deg^{-1} mole^{-1} at 25°C, and C_P at this temperature is 189 J deg^{-1} mole^{-1}. Discuss the test of the third law of thermodynamics provided by this information and that given in the preceding Exercise.

10-13. Use Equation 10-7 to calculate a value for the entropy of Li(g) at the normal boiling point, 1336°C. (The third-law value is larger than this value by $R \ln 2$: see the following Exercise.)

10-14. Consider the lowest energy level of a particle in a box, Equation 9-10, quantum numbers $n_x = n_y = n_z = 1$.

(a) How much kinetic energy would a lithium atom in a cubical box 10 cm on edge have?

(b) The normal state of the lithium atom is $^2S_{1/2}$. How many values can the quantum number M_J have? How many quantum states of the atom have the energy calculated in (a), assuming zero magnetic field?

(c) How many more quantum states does a mole of Li(g) have than a mole of monatomic gas with the same molecular weight and state $^1S_0$?

(d) To what additional entropy does the odd electron in Li(g) lead?

10-15. The third-law value of the entropy of HI(g) at 25°C and 1 atm is 206.3 J deg^{-1} mole^{-1}. The value of the heat capacity (Figure 10-4) shows that the entropy is essentially all due to translation and rotation.

(a) Calculate the translational entropy.

(b) Calculate the rotational entropy, and compare the sum with the third-law value. The interatomic distance H—I is found from the band spectrum to have the value $r_0 = 1.61$ Å.

10-16. The vibrational level $v = 1$ of the hydrogen iodide molecule lies 4.59×10^{-20} J above the level with $v = 0$. What is the contribution of vibration to the entropy of hydrogen iodide at 25°C (see Equation 10-25)?

10-17. The Debye characteristic temperature Θ is 4.02 times the temperature at which C_V equals $\frac{3}{2} R$. Estimate Θ for Al(c) from the curve in Figure 10-10. Use Equation 10-27 to evaluate C_V for Al(c) at 10°K and at 20°K.

10-18. Integrate C_V to obtain an equation for $H(T) - H(0°K)$ valid for small T. Also integrate $C_V \ln T$ to obtain an equation for $S(T)$. (Note that the PdV term is negligible for crystals at very low temperatures.)

10-19. Use the results of Exercises 10-17 and 10-18 to evaluate the enthalpy and entropy of aluminum at 10°K and 20°K.

10-20. Make a graph of the heat capacity of aluminum from 20°K to 298°K against ln T by reading values from Figure 10-10, and evaluate S (298°K) − S (20°K) by measuring the area. Add S (20°K), from Exercise 10-19, to obtain S (298°K). The reference-book value of S (25°C) is 28.32 J deg^{-1} mole^{-1}.

10-21. Using the above value for S (25°C) and the following information, calculate the entropy of Al(g) at 2767°K: C_P for Al(c), 29.3 J deg^{-1} mole^{-1} at 700°K, 33.1 at 900°K; melting point, 933°K; enthalpy of fusion, 10.71 kJ mole^{-1}; C_P of Al(l), 31.8 J deg^{-1} mole^{-1}; boiling point at 1 atm, 2767°K; enthalpy of vaporization, 290.8 kJ mole^{-1}. (Answer: about 211.0 J deg^{-1} mole^{-1}.)

10-22. The normal state of the aluminum atom, $^2P_{1/2}$, has quantum weight 2 (two quantum states, $M_J = +\frac{1}{2}$ and $-\frac{1}{2}$, with the same energy, in the absence of a magnetic field). The first excited state, $^2P_{3/2}$, has quantum weight 4. It lies 112 cm^{-1} (= 2.22×10^{-21} J) above the normal state. What is the value of kT at 2767°K? What fractions of the atoms are in the two states at this temperature? (The next excited state, at 25350 cm^{-1}, and the still higher ones can be neglected.)

10-23. Evaluate the translational entropy of Al(g) at 1 atm and 2767°K by use of Equation 10-7. Discuss the answer in relation to Exercises 10-21 and 10-22. (Answer: 196.14 J deg^{-1} mole^{-1}.)

10-24. It was found by J. K. Koehler and W. F. Giauque in 1958, by use of the theoretical value of the entropy of $ClO_3F(g)$, calculated with the known values of the moments of inertial and vibrational frequencies of the molecule, and the experimental values of the heat capacity and heat of sublimation of the crystal, that at very low temperatures the crystal has residual entropy 10.13 J deg^{-1} mole^{-1}. What structure would you assign to the molecule? To what sort of disorder do you attribute the residual entropy? What is the calculated value for complete disorder of this sort?

11

Chemical Equilibrium

In this chapter we shall discuss chemical equilibrium and the "driving force" of chemical reactions.

11-1. The Thermodynamic Condition for Chemical Equilibrium

Let us consider a system that might exist in either of two different states, *A* and *B*, at a certain pressure and temperature. The system might, for example, be one mole of sulfur, which is observed to be stable in the form of orthorhombic crystals at temperatures below 368.54°K and in the form of monoclinic crystals above this temperature. At this temperature the two forms coexist, with no tendency for one to go over to the other; they are in equilibrium with one another at this temperature (1 atm pressure).

We now ask what special attribute is associated with this temperature; in general, under what conditions could the two states *A* and *B* coexist, be in equilibrium.

Let us consider the system at first in state A. We now carry out the reversible conversion of the system to state B, with the final pressure and temperature equal to the initial pressure and temperature. This reversible conversion involves our transferring the heat q_{rev} to the system from the environment (at the fixed temperature T) and doing the work w_{rev} on the system (not including the work $P\Delta V$). The first law of thermodynamics permits us to write

$$\Delta H = H_B - H_A = q_{rev} + w_{rev} \tag{11-1}$$

The entropy change of the environment is $-q_{rev}/T$, and that of the system is $S_B - S_A$. The second law of thermodynamics requires that the sum of these two terms be zero:

$$-\frac{q_{rev}}{T} + S_B - S_A = 0$$

and hence

$$q_{rev} = T(S_B - S_A) \tag{11-2}$$

From Equations 11-1 and 11-2 we derive the equation

$$w_{rev} = H_B - H_A - T(S_B - S_A) \tag{11-3}$$

which we rewrite as

$$w_{rev} = G_B - G_A \tag{11-4}$$

with $G_B = H_B - TS_B$ and $G_A = H_A - TS_A$.

The new thermodynamic function, $G = H - TS$, is called the *Gibbs free energy* of the system,* named in honor of the great American theoretical physicist J. Willard Gibbs (1839–1903), who made very important contributions to chemical thermodynamics and statistical mechanics.

Let us now discuss the system under the condition that no work is done on the system. If $\Delta G = G_B - G_A$ happens to be zero the change of the system from A to B (or from B to A) is reversible, and not associated with any change in entropy of the universe. We conclude that the *condition for equilibrium at P and T constant, with no work other than $P\Delta V$ work done on the system, is that the free energy* (Gibbs free energy) *of the system be constant*:

$$\Delta G = \Delta H - T\Delta S = 0 \tag{11-5}$$

Spontaneous changes in state are those leading to an increase in entropy of the universe. Analysis of our problem shows that an increase in entropy of the universe occurs for ΔG negative: hence *spontaneous changes in a system at constant pressure and temperature are accompanied by a decrease in the free energy of the system.*

---

*In the older literature the symbol F, rather than G, is usually used. The Helmholtz free energy, $A = E - TS$, is used to some extent, mainly in discussing systems held at constant volume. In some books use is made of special symbols, such as $\bar{H}$, $\bar{G}$, $\bar{S}$, for thermodynamic properties referred to one mole of substance.

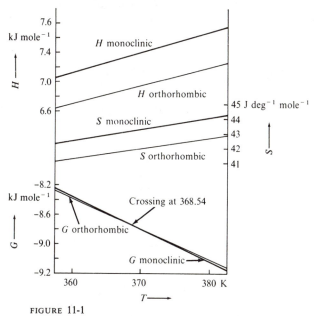

FIGURE 11-1
Thermodynamic functions of two crystalline forms
of sulfur (per sulfur atom).

Some thermodynamic functions of orthorhombic sulfur and monoclinic sulfur are shown in Figure 11-1, over the temperature range 360° to 380°K. The experimental value of the enthalpy for orthorhombic sulfur has been obtained by integrating the heat capacity C_P from 0°K to T°K (using the Debye T^3 law in the region below 15°K, for which measurements have not been made). The value $H = 0$ at $T = 0$°K is assumed for this modification of the element (the stable one). The value of the enthalpy of monoclinic sulfur is obtained by use of the heat of transition at the equilibrium temperature (402 J mole^{-1}) and the measured heat capacity. The values of the entropy of both modifications are the third-law experimental values, $\int_0^T C_P \, d\ln T$; the transition of monoclinic sulfur to orthorhombic sulfur is slow enough to allow the heat-capacity measurements to be made. The values of G are calculated from the equation $G = H - TS$.

We see that the difference in enthalpy favors the orthorhombic form and that in entropy favors the monoclinic form. There is a delicate balance between the terms ΔH and $-T\Delta S$ that contribute to ΔG. With increase in temperature the second term becomes increasingly important; hence the monoclinic form, with the larger entropy, is the high-temperature form.

Because of the small value of ΔG the transformation is a sluggish one.

It has been mentioned in Section 7-1 that orthorhombic sulfur can be superheated; it melts at 386°K, about 6° lower than monoclinic sulfur. The lower melting point is, of course, required by its larger free energy.

Change of Free Energy with Temperature and Pressure

Let us consider a reversible increase by δT in the temperature of a system with pressure constant. From the definition of G as $H - TS$ we see that

$$\delta G = \delta H - T\delta S - S\delta T$$

However, δS is equal to $\delta H/T$; hence the first and second terms on the right side of this equation cancel, and we obtain the result

$$\delta G = -S\delta T \tag{11-6}$$

It is customary to write this equation with use of the partial derivative of G with respect to T:*

$$\left(\frac{\partial G}{\partial T}\right)_P = -S \tag{11-7}$$

The subscript P means that the pressure is held constant.

To evaluate $\left(\dfrac{\partial G}{\partial P}\right)_T$ we write $G = E + PV - TS$, and obtain for a reversible increase δP in pressure at constant temperature the expression

$$\delta G = \delta E + V\delta P + P\delta V - T\delta S$$

The sum of δE, the change in internal energy of the system, and $P\delta V$, the work done by the system on the environment, is equal to q_{rev}, which divided by T is δS, the change in entropy. Thus these two terms cancel the last term, and we obtain the result

$$\left(\frac{\partial G}{\partial P}\right)_T = V \tag{11-8}$$

11-2. The Vapor Pressure of a Liquid or Crystal

Let us consider the reaction of evaporation of a liquid:

$$X(\text{l}) \rightleftarrows X(\text{g}) \tag{11-9}$$

X is the formula of the substance. Let G_1° and G_g° be the molar free

---

*The partial derivative $\left(\dfrac{\partial G}{\partial T}\right)_P$ is the ratio of the change in G to the change in T with P held constant.

energies of the liquid and the gas, respectively, at standard pressure (1 atm) and temperature T. The question that we pose is the following: At what pressure will the two phases be in equilibrium with one another? In other words, what is the vapor pressure of the liquid?

We answer this question by making use of Equation 11-8. For the liquid the volume changes only slightly with small changes (a few atmospheres) in pressure, and we can use for V the molar volume at 1 atm, V_1:

$$dG_1 = V_1 dP$$
$$G_1 - G_1^\circ = V_1(P - P_0) \tag{11-10}$$

For the gas we use the perfect gas equation, which for one mole is $PV = RT$, or $V_g = RT/P$, and write

$$dG_g = RT \frac{dP}{P}$$
$$G_g - G_g^0 = RT \ln \frac{P}{P_0} \tag{11-11}$$

By subtracting Equation 11-10 from Equation 11-11 we obtain

$$G_g - G_1 - (G_g^0 - G_1^0) = RT \ln \frac{P}{P_0} - V_1(P - P_0) \tag{11-12}$$

We have seen in Section 11-1 that the condition for equilibrium between the two phases is that their free energies be equal—that is, that $G_g - G_1 = 0$; moreover, the term $-V_1(P - P_0)$ is usually so small relative to $RT \ln (P/P_0)$ that it can be neglected. Hence Equation 11-12 leads to the vapor-pressure equation

$$-\Delta G^0 = RT \ln \frac{P}{P_0} \tag{11-13}$$

with $\Delta G^\circ = G_g^\circ - G_1^\circ$ and $P_0 = 1$ atm.

We can determine ΔG° as a function of T by use of Equation 11-7, which we rewrite as

$$\frac{d(-\Delta G^0)}{dT} = \Delta S^0 = S_g - S_1$$

which becomes, when $-\Delta G^\circ$ is replaced by its value as given in Equation 11-13, the following:

$$\frac{d \ln (P/P_0)}{dT} = \frac{\Delta S^0}{RT} \tag{11-14}$$

Now, ΔS°, the entropy of vaporization, is equal to $\Delta H^\circ/T$, the enthalpy of vaporization divided by the temperature:

$$\frac{d \ln (P/P_0)}{dT} = \frac{\Delta H_{\text{vap}}^0}{RT^2} \tag{11-15}$$

This equation, which is called the *Clausius-Clapeyron equation*, can be integrated for the case ΔH_{vap}^0 constant (a good approximation over a small range of temperature) to give

$$\ln (P/P_0) = -\frac{\Delta H_{vap}}{RT} + \text{constant} \qquad (11\text{-}16)$$

(Comparison with Equation 11-13 shows that the constant is $\Delta S°/R$.)

We see from Equation 11-16 that a graph of $\ln(P/P_0)$ or $\log (P/P_0)$ against $1/T$ should be a straight line over the range of temperature for which ΔH_{vap}^0 is constant. Thus measurement of the vapor pressure of a liquid or crystal over a range of temperature provides a value of the enthalpy of vaporization or sublimation, as well as of $\Delta G°$ and $\Delta S°$.

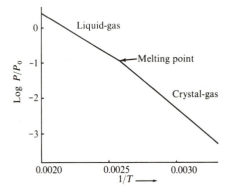

FIGURE 11-2
Plot of the logarithm of the vapor pressure of $I_2(c)$ and $I_2(l)$ against $1/T$.

TABLE 11-1
The Vapor Pressure of Iodine as a Function of Temperature

Temperature	Vapor Pressure of Iodine Crystals	Temperature	Vapor Pressure of Liquid Iodine
30°C	0.00062 atm	114.2°C (m.p.)	0.118 atm
40	.00136	120	.146
50	.00284	130	.207
60	.00567	140	.286
70	.0108	150	.387
80	.0199	160	.518
90	.0353	170	.686
100	.0599	180	.893
110	.0986	184.35 (b.p.)	1.000
114.2 (m.p.)	.118	190	1.143

Example 11-1. Experimental values of the vapor pressure of iodine (both $I_2(c)$ and $I_2(l)$) are given in Table 11-1 (see also Figure 2-16). What are the values of some thermodynamic quantities that can be derived from them?

Solution. The values of log (P/P_0) are plotted against $1/T$ in Figure 11-2. (The points do not deviate appreciably from the two straight lines in the figure.) The measured slopes are -3.19×10^3 deg and -2.35×10^3 deg, respectively. We multiply by $-2.303\ R = -2.303 \times 8.315$ J deg^{-1} to obtain the value 61.1 kJ mole^{-1} for the enthalpy of sublimation and 45.0 kJ mole^{-1} for the enthalpy of vaporization. Their difference, 16.1 kJ mole^{-1}, is the enthalpy of fusion of iodine. The entropy of fusion is 41.5 J deg^{-1} mole^{-1}, obtained by dividing the enthalpy of fusion by the crystal-liquid equilibrium temperature, 387.4°K. The standard entropy of vaporization is 98.4 J deg^{-1} mole^{-1}, obtained by dividing the enthalpy of vaporization by the standard boiling point, 457.5°K.

11-3. Entropy of Transition, Fusion, and Vaporization

The structural basis of the difference in thermodynamic properties of orthorhombic sulfur and monoclinic sulfur is the difference in the ways in which the molecules are packed together in the two crystals. The molecules are the same: puckered 8-atom rings (Figure 6-13), with S—S distance 2.06 Å and bond angle 105°. The density is 2.07 g cm^{-3} for the orthorhombic crystal and 1.96 g cm^{-3} for the monoclinic crystal; hence the molecules fit together better in the former than in the latter. The enthalpy of sublimation of S_8 (orthorhombic) to $S_8(g)$ at 298°K is 101.3 kJ mole^{-1} and that of S_8 (monoclinic) to $S_8(g)$ is 98.0 kJ mole^{-1}. We assume that there are about $98/101 = 97\%$ as many intermolecular contacts in the monoclinic crystal as in the better-packed, more stable orthorhombic form. The molecules vibrate at a lower frequency in the less tightly packed form; hence the Debye characteristic temperature Θ is lower and the entropy is larger (Equation 10-30). Monoclinic sulfur at the transition temperature has entropy 3% greater than orthorhombic sulfur.

Entropy of Fusion

Some experimental values of the entropy of fusion are given in Table 11-2. For the substances in the first column, which are monatomic in the gas phase and have simple crystal structures, values from 0.8 R to 1.7 R are observed. The increase in volume on melting is about 3% for the alkali metals, 5% for Cu, Ag, and Au, and 15% for the four argonons; there is a clear correlation with the entropy of fusion.

TABLE 11-2
Entropy of Fusion of Some Substances

	m.p.	$\Delta S/R$		m.p.	$\Delta S/R$
Li	453°K	0.77	HCl	158.9°K	1.51
Na	371	0.86	HBr	186.3	1.56
K	337	0.85	H_2S	187.6	1.53
Cu	1357	1.15	H_2Se	206.2	1.46
Ag	1234	1.11	CH_4	90.7	1.25
Au	1336	1.15	CF_4	89.5	0.94
Ne	24.6	1.64	SiF_4	182.9	4.64
Ar	83.9	1.69	C_2H_4	104.0	3.88
Kr	116.0	1.70	CO_2	217.0	4.63
Xe	161.3	1.71	Cl_2	172.2	4.48
			Br_2	265.9	4.78

X-ray diffraction studies of monatomic liquids have shown that the distance between neighboring atoms is nearly the same as for the crystal, but that the average ligancy (number of neighbors) is less, about 10 instead of 12. We might describe each atom in the liquid as having ten atoms and two holes (or nine and three, or eleven and one) as its neighbors, with the holes distributed at random, and the long-range order of the crystal destroyed. The entropy of mixing one mole of atoms and 0.2 mole of holes is 0.54 R (Section 10-4). Moreover, the vibrational frequencies of the liquid are lower than those of the crystal, because of the smaller ligancy. Hence the Debye Θ is smaller, which contributes the amount 3 R $\ln(\Theta_{crystal}/\Theta_{liquid})$ to the entropy of fusion, amounting to 0.55 R for a value of $\Theta_{crystal}$ 20% larger than that of Θ_{liquid}. These rough considerations make the observed range 0.8 R to 1.7 R seem reasonable.

The values around 4.5 R for CO_2, Cl_2, and Br_2 (and many other diatomic and triatomic molecules not listed) suggest that an additional entropy of orientational disorder, about 3 R, exists in these liquids. If we write the Boltzmann relation $S = k \ln W = k \ln w^N = R \ln w = 3 R$, we find that w is equal to about 20. This value is somewhat larger than we might expect for the number of orientational quantum states of a molecule in the liquid; there may well be some additional decrease in the Debye Θ of the crystal because of the orientational disorder, accounting for part of the large entropy of fusion.

Several substances listed in Table 11-2 have such small values of ΔS_{fusion}, about R to 1.5 R, as to suggest that fusion does not increase orientational disorder. All of these substances have one or more crystal-crystal transitions at temperatures below the melting point, with entropy of transition 2 R to 3 R. These transitions are interpreted as involving a great increase in rotational disorder, leading to as much freedom of rotation in the high-temperature crystals as in the liquid. The process of fusion of these polyatomic crystals differs little from that for monatomic crystals.

Entropy of Vaporization

For many sustances the entropy of vaporization at the standard boiling point (1 atm pressure) is approximately 85 J deg^{-1} mole^{-1}. This statement, called *Trouton's rule*, is useful for estimating values of the heat of vaporization when the boiling point is known.

Values of the standard boiling point and the heat of vaporization for some liquids whose gases are monatomic are given in Table 11-3 (the first 10 entries). The boiling points cover a wide range, from 27.2°K for neon to 2933°K for gold. Over the range of liquids from neon to gold the heat of vaporization also varies widely, but the ratio of heat of vaporization to temperature, which is the entropy of vaporization, ranges only from 66 J deg^{-1} mole^{-1} to 106 J deg^{-1} mole^{-1}. The values of the entropy of vaporization for many liquids with polyatomic molecules also lie within this range.

An improvement on Trouton's rule was made in 1915 by the American chemist J. H. Hildebrand (born 1881). He found that the entropy of vaporization at the equilibrium temperature between the liquid and its vapor at standard molar volume (22.4 liters) is nearly the same for many substances. Values of this temperature for molar gas volume 22.4 liters are given in the third column of Table 11-3. The last column of the table is the entropy of vaporization to the gas at this standard molar volume.

The calculation of boiling point at standard volume is made by use of Equation 11-16, as illustrated in the following example.

Example 11-2.   The standard boiling point of hydrogen is 20.39°K, with enthaply of vaporization 904 J mole^{-1}. To what extent does hydrogen conform to Hildebrand's rule?

Solution.   The standard entropy of vaporization, 904/20.39 = 44 J deg^{-1} mole^{-1}, is very low. To correct to the Hildebrand boiling point we use Equation 11-16 to convert from T_1, at which the vapor pressure is 1 atm, to T_2, at which the vapor pressure P_2 is such as to give molar volume $V_2 = 22.4$ l. From Equation 11-16 we write

$$\ln(P_1/P_0) = -\frac{\Delta H^{\circ}_{\text{vap}}}{RT_1} + \frac{\Delta S^{\circ}}{R}$$

$$\ln(P_2/P_0) = -\frac{\Delta H^{\circ}_{\text{vap}}}{RT_2} + \frac{\Delta S^{\circ}}{R}$$

We have $P_1 = P_0 = 1$ atm; by subtracting the first equation from the second we obtain

$$\ln P_2 = \frac{\Delta H^\circ_{vap}}{R} \left(\frac{1}{T_1} - \frac{1}{T_2} \right)$$

The gas-law equation $P_2 V_2 = RT_2$ permits us to write

$$\ln \left(\frac{RT_2}{V_2} \right) = \frac{\Delta H^\circ_{vap}}{R} \left(\frac{1}{T_1} - \frac{1}{T_2} \right)$$

With $V_2 = 22.4$ liter we have $R/V_2 = 0.08206/22.4 = 0.00366$. Replacing ln by 2.303 log and reordering the terms, we obtain the equation

$$\frac{1}{T_2} = \frac{1}{T_1} - \frac{2.303\,R}{\Delta H^\circ_{vap}} \log(0.00366\,T_2)$$

or, with the value of ΔH°_{vap} introduced,

$$\frac{1}{T_2} = \frac{1}{T_1} - 0.0212 \log(0.00366\,T_2)$$
$$= 0.0490 - 0.0212 \log(0.00366\,T_2)$$

This is the sort of equation that one solves by successive approximations. We substitute a trial value for T_2 in the slowly changing log term on the right, and solve. A safe trial value is that of T_1, 20.390, which gives $T_2 = 13.72°$. Repetition with this value substituted in the right side gives 13.07°, then 12.99° and 12.98°. The value of the entropy of vaporization at the Hildebrand boiling point 12.98°K is $904/12.98 = 69.6$ J deg^{-1} mole^{-1}, which is 18% lower than the normal value 85 J deg^{-1} mole^{-1}.

Example 11-3. The substance POFClBr has standard boiling point 352°K. Estimate its enthalpy of vaporization.

Solution. From Table 11-3 we see that at this temperature the Hildebrand boiling point is about 9° higher than the standard boiling point. Hence we write $\Delta H^\circ_{vap} = (352° + 9°) \times 84.9 = 30650$ J mole^{-1}. Hence our estimated value of the enthalpy of vaporization is 30.65 kJ mole^{-1}.

We may give a simple interpretation to Hildebrand's rule by considering only the translational entropy of the gas and the liquid—that is, by assuming that the rotational and intramolecular vibrational entropy terms are the same for the two phases. The translational entropy for the gas contains a term $R \ln V_g$. We assume that a similar term $R \ln V_1$ applies to the liquid. The rule then permits us to write

$$R \ln \left(\frac{V_g}{V_1} \right) = 84.9 \text{ J deg}^{-1} \text{mole}^{-1}$$

Solution of this equation gives $V_1 = 0.91$ cm^3 mole^{-1}, which is 1.37 Å^3 per molecule.

TABLE 11-3
Values of Entropy of Vaporization of Some Liquids

	Boiling Point			Molar Entropy of Vaporization (gas at standard volume)
Substance	At 1 atm	At Standard Volume	Molar Heat of Vaporization	
Ne	27.2°K	20.6°K	1.80 kJ mole⁻¹	87.4 J deg⁻¹ mole⁻¹
Ar	87.3	76.6	6.53	85.2
Kr	119.9	108.9	9.04	83.0
Xe	165	155	12.64	81.5
Cu	2855	3560	305	85.7
Ag	2466	3030	254	83.8
Au	2933	3680	310	84.2
Zn	1180	1367	115	84.1
Cd	1040	1190	100	84.0
Hg	630	688	58.1	84.4
HCl	188.1	181	16.2	89.2
HBr	206.4	200	17.6	88.0
HI	238	234	19.7	84.2
H_2S	213	208	18.7	89.9
H_2Se	232	228	19.3	84.6
PH_3	185	177	14.6	82.5
AsH_3	211	205	17.4	84.9
CH_4	111.7	100	8.20	82.0
SnH_4	221	216	18.5	85.6
CCl_4	350	359	30.0	83.6
O_2	90.2	79.5	6.82	85.8
N_2	77.3	66.5	5.56	83.6
CO	81.7	71.1	6.07	85.4
F_2	85.2	74.4	6.32	84.9
Cl_2	239	236	20.4	86.6
P_4	553	596	49.8	83.6
				Average 84.9 ± 1.5

We have thus calculated that for the substances in Table 11-3 the accessible volume per molecule at the Hildebrand boiling point is very nearly constant, equal to 1.37 Å³ per molecule. This volume may seem to be surprisingly small. We know, for example (Chapter 2), that the distance between copper atoms in the copper crystal is 2.55 Å, and the volume of a sphere with this diameter is 8.6 Å³. In the liquid, which has a smaller density, the volume per atom of copper is somewhat larger. However, the center of the atom cannot move throughout the entire volume. It is instead restrained by the surrounding molecules to a rather small region. The

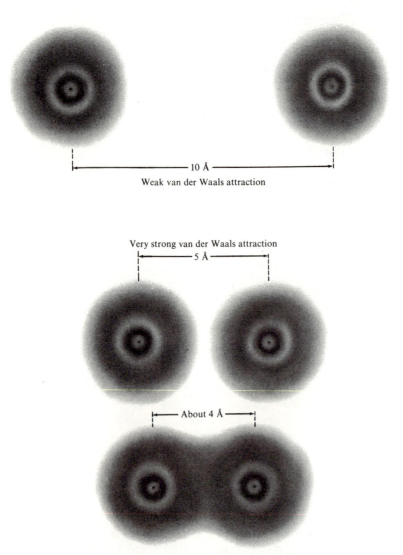

Weak van der Waals attraction

Very strong van der Waals attraction

Van der Waals attraction just balanced by repulsive
forces due to interpenetration of outer electron shells

FIGURE 11-3
Diagram illustrating van der Waals attraction and repulsion in relation to electron
distribution of monatomic molecules of argon.

volume $1.37\,Å^3$ accessible to the center of the atom corresponds to motion of the center within a sphere of radius $0.69\,Å$. Hence we conclude that in liquids such as those in Table 11-3 each molecule is free to move about $0.69\,Å$ from its average position.

No one knows why liquids so different as the noble gases and the metals have the same free volume at their boiling points, $1.37\,Å^3$ per molecule. The theory of liquids is still in a rather primitive state. We may hope that the explanation of this interesting fact will be provided before many years.

11-4. Van der Waals Forces. Melting Points and Boiling Points

All molecules exert a weak attraction upon one another. This attraction, the *electronic van der Waals attraction*, is the result of the mutual interaction of the electrons and nuclei of the molecules; it has its origin in the electrostatic attraction of the nuclei of one molecule for the electrons of another, which is largely but not completely compensated by the repulsion of electrons by electrons and nuclei by nuclei. The van der Waals attraction is significant only when the molecules are very close together—almost in contact with one another. At small distances (about $4\,Å$ for argon, for example) the force of attraction is balanced by a force of repulsion due to interpenetration of the outer electron shells of the molecules (Figure 11-3).

It is these intermolecular forces of electronic van der Waals attraction that cause substances such as the argonons, the halogens, and others to condense to liquids and to freeze into solids at sufficiently low temperatures. The boiling point is a measure of the amount of molecular agitation necessary to overcome the forces of van der Waals attraction, and hence is an indication of the magnitude of these forces. In general, the electronic van der Waals attraction between molecules increases with increase in the number of electrons per molecule. Since the molecular weight is roughly proportional to the number of electrons in the molecule, usually about twice the number of electrons, the van der Waals attraction usually increases with increase in the molecular weight. Large molecules (containing many electrons) attract one another more strongly than small molecules (containing few electrons); hence normal molecular substances with large molecular weight have high boiling points, and those with small molecular weight have low boiling points.

This generalization is indicated in Figure 11-4, in which the boiling points of some molecular substances are shown. The steady increase in boiling point for sequences such as He, Ne, Ar, Kr, Xe, Rn and H_2, F_2, Cl_2, Br_2, I_2 is striking.

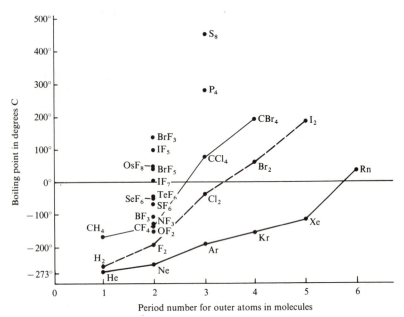

FIGURE 11-4
Diagram showing increase in boiling point with increase in molecular complexity.

The similar effect of increase in the number of atoms (with nearly the same atomic number) in the molecule is shown by the following sequences:

	Ar	Cl_2	P_4	S_8		
Boiling point	$-185.7°$	$-34.6°$	280°	444.6°C		

	Ne	F_2	CF_4	SF_6	IF_7	OsF_8
Boiling point	$-245.9°$	$-188°$	$-128°$	$-62°$	4.5°	47.5°C

The theory of the van der Waals force of attraction between molecules was developed by the physicist F. London in 1929. It had been suggested that the van der Waals attraction between two HCl molecules (or other molecules with a permanent electric dipole moment; Section 6-8) was the result of the interaction of the permanent dipole moments. Careful calculations of this energy of attraction for two HCl molecules, however, gave a result only 10% of the observed interaction energy. Moreover, the interaction energy of molecules of xenon (boiling point $-107°C$) is nearly as great as that of molecules of hydrogen chloride (boiling point $-84°C$), although the molecules of xenon, which are single atoms, have no permanent electric dipole moment.

London showed that between any two molecules there is a van der Waals force of attraction that can be described as resulting from the interaction of the instantaneous electric dipole moments of the molecules. For example, the 54 electrons in an atom of xenon might happen at one instant of time to be arranged in such a way that their center of charge would coincide with the nucleus, but at other instants it would lie to one side or the other of the nucleus, and the atom would have an instantaneous electric dipole moment. London found by quantum-mechanical calculation that the stable relative orientations of the instantaneous dipole moments of two atoms, such as $+\!\!\rightarrow$ $+\!\!\rightarrow$, occur more often than the unstable ones, such as $+\!\!\rightarrow$ $\leftarrow\!\!+$, and that the resulting effect is attraction. The London equation for the interaction energy of molecules A and B is

$$\text{Energy of attraction} = -\frac{3}{2}\frac{\alpha_A \alpha_B}{r_{AB}^6}\frac{I_A I_B}{(I_A + I_B)} \tag{11-17}$$

This corresponds to an inverse-seventh-power force of attraction (found by differentiating with respect to r_{AB}):

$$\text{Force of attraction} = 9\frac{\alpha_A \alpha_B}{r_{AB}^7}\frac{I_A I_B}{(I_A + I_B)} \tag{11-18}$$

In these equations r_{AB} is the distance between the centers of the molecules, I_A and I_B are excitation energies of molecules A and B (somewhat larger than their first ionization energies), and α_A and α_B are the *electronic polarizabilities* of the molecules.

The electronic polarizability of a molecule expresses the extent to which the electrons are shifted by an electric field E, to give the molecule an induced electric dipole moment, $\mu = \alpha E$. The dimensions of α are those of volume. The electric polarizability of a conducting sphere (a metallic sphere) with radius r is equal to r^3. The polarizability of the xenon atom is 4.16 Å^3 (4.16 $\times$ 10^{-30} m^3), which is equal to (1.61 Å)^3. The value 1.61 Å, which we might call the polarizability radius of the xenon atom, agrees roughly with other radii; it is, for example, 16% less than the average of the crystal radii of its isoelectronic ionic neighbors Cs$^+$ and I$^-$.

Light traversing a substance has a velocity different from light traversing a vacuum. The ratio of the velocity of light in a vacuum to that in a substance is the *index of refraction* of the substance. The interaction between the light and the substance that causes the index of refraction of the substance to differ from unity is the polarization of the atoms or molecules of the substance by the electric vector of the light. The relation between the index of refraction n and the molecular polarizability α is given by the *Lorenz-Lorentz equation*, which was derived in 1880 by the Danish physicist Ludwig Valentin Lorenz (1829–1891) and the Dutch physicist

Hendrik Antoon Lorentz (1853–1928). This equation, in which M is the molecular weight and d the density of the substance, is

$$\alpha = \frac{3}{4\pi N} \times \frac{n^2 - 1}{n^2 + 2} \times \frac{M}{d} \tag{11-19}$$

The following example shows how the molecular polarizability can be calculated from the observed value of the index of refraction.

Example 11-4. The value of the index of refraction n for xenon at standard conditions is 1.000703. (This is the value for the sodium D lines, wavelength* 5894 Å; values of n are dependent on the wavelength of the light; for example, n for xenon varies from 1.000713 at 4800 Å to 1.000697 at 6700 Å.) What is the polarizability of the xenon atom?

Solution. The quantity M/d for a gas at standard conditions is 22,414 cm³, and n^2 for xenon is equal to 1.001406. Hence from the Lorenz-Lorentz equation we write

$$\begin{aligned}
\alpha &= \frac{3}{4\pi N} \times \frac{n^2 - 1}{n^2 + 2} \times \frac{M}{d} \\
&= \frac{3}{4\pi \times 0.602 \times 10^{24}} \times \frac{1.406 \times 10^{-3}}{3.001} \times 22.414 \times 10^3 \\
&= 4.16 \times 10^{-24} \, \text{cm}^3
\end{aligned}$$

Accordingly the polarizability of the xenon atom has the value $\alpha = 4.16$ Å³.

Values of α for some molecules and ions are given in Table 11-4. A set of values of atomic polarizabilities, applicable to atoms in molecules, is also given; the polarizability of a molecule is obtained by summing the values for the atoms in the molecule (Table 11-5).

It is found that the London equation, which involves an approximation, provides moderately reliable values of the energy and force of van der Waals attraction when the excitation energy I is taken to be about 60% greater than the first ionization energy of the molecule. A rough average for the first ionization energy is 12 eV = 1160 kJ mole⁻¹. We may accordingly rewrite Equation 11-17 as

$$\text{Energy of attraction} = -\frac{\alpha_A \alpha_B}{r_{AB}^6} \times 1400 \text{ kJ mole}^{-1} \tag{11-20}$$

Example 11-5. The xenon crystal has the cubic closest-packed structure, with each atom surrounded by twelve others at the distance 4.41 Å. What is the value of the van der Waals energy of attraction of the atoms? Compare with the observed heat of sublimation, 14.9 kJ mole⁻¹.

*Average of 5890.12 Å and 5896.16 Å, intensity ratio 1:2.

TABLE 11-4
Electric Polarizability of Some Atoms, Ions, and Molecules (Sodium D lines)

		H	0.67 Å³				
		He	.20	Li⁺	0.03 Å³	Be⁺⁺	0.008 Å³
F⁻	1.05 Å³	Ne	.40	Na⁺	.18	Mg⁺⁺	.09
Cl⁻	3.69	Ar	1.64	K⁺	.84	Ca⁺⁺	.47
Br⁻	4.81	Kr	2.48	Rb⁺	1.42	Sr⁺⁺	.86
I⁻	7.16	Xe	4.16	Cs⁺	2.44	Ba⁺⁺	1.56
H₂	0.80	HCN	2.58	F₂	1.16	P₄	14.71
HCl	2.64	H₂O	1.49	Cl₂	4.60	BCl₃	8.31
HBr	3.62	NH₃	2.23	Br₂	6.90	BBr₃	11.87
HI	5.45	H₂S	3.80	N₂	1.76	CCl₄	10.53

TABLE 11-5
Electric Polarizability Terms for Atoms and Structural Features in Organic Molecules

H—	0.40 Å³	F—	0.58 Å³	
C in CH₄	1.03	Cl—	2.30	Three-membered ring 0.24 Å³
C in diamond	0.83	Br—	3.45	
$\diagdown\!\!\!\overset{}{\underset{\diagup}{C}}\!\!=$	1.34	I—	5.50	Four-membered ring 0.13 Å³
—C≡	1.42	$\overset{\diagup}{\underset{\diagdown}{O}}$	0.69	
$\overset{\diagup}{\underset{\diagdown}{N}}\!\!-$	0.97	O=	.85	
$\diagdown\!\!\!\underset{}{N}\!\!=$	.90	$\overset{\diagup}{\underset{\diagdown}{S}}$	3.00	
N≡	.83	S=	3.14	

Solution. We substitute $\alpha = 4.16$ Å³ and $r_{AB} = 4.41$ Å in Equation 11-20 to obtain the energy of van der Waals attraction per pair of xenon atoms: $1400 \times 4.16^2/4.41^6 = 3.29$ kJ mole⁻¹. There are six nearest-neighbor interactions per atom;* hence the calculated energy of van der Waals attraction of nearest neighbors is $6 \times 3.29 = 19.8$ kJ mole⁻¹, somewhat larger than the heat of sublimation. To obtain a better calculated value we would need to take into consideration the energy of van der Waals attraction between more distant atoms and also the energy of the van der Waals repulsion. The repulsion energy for the xenon crystal is about half as great in magnitude as the attraction energy. If we take the repulsion energy as B/r^{12}, we find that with energy of attraction $-A/r^6$ the equilibrium value of r (obtained from the force equation $-6A/r^7 + 12B/r^{13} = 0$) leads to this result, that the repulsion term is one-half the attraction term in the energy expression, with changed sign.

*Remember that each Xe—Xe interaction involves two xenon atoms.

Bond Type and Atomic Arrangement

Fifty years ago, before modern structural chemistry had been developed, it was thought that an abrupt change in melting point or boiling point in a series of related compounds could be accepted as proof of a change in type of bond. The fluorides of the elements of the second period, for example, have the following melting points and boiling points.

	NaF	MgF_2	AlF_3	SiF_4	PF_5	SF_6
Melting point	995°	1263°	>1257°	−90°	−94°	−51°C
Boiling point	1704°	2227°	1257°*	−95°*	−85°	−64°C*

*Note that aluminum trifluoride, silicon tetrafluoride, and sulfur hexafluoride have the interesting property of subliming at 1 atm pressure without melting. The temperatures given in the table as the boiling points of these substances are in fact the subliming points, when the vapor pressure of the crystals becomes equal to 1 atm.

The great change between aluminum trifluoride and silicon tetrafluoride is, however, not due to any great change in bond type (the bonds are in all cases intermediate in character between extreme ionic bonds $M^+ F^-$ and normal covalent bonds $M:\ddot{F}:$), but rather to a change in atomic arrangement. The three easily volatile substances exist as discrete molecules SiF_4, PF_5, and SF_6 (with no dipole moments) in the liquid and crystalline states as well as the gaseous state (Figure 11-5), and the thermal agitation necessary for fusion or vaporization is only that needed to overcome the weak intermolecular forces, and is essentially independent of the strength or nature of the interatomic bonds within a molecule. But the other three substances in the crystalline state are giant molecules, with strong bonds between neighboring ions holding the whole crystal together. To melt such a crystal some of these strong bonds must be broken, and to boil the liquid more must be broken; hence the melting point and boiling point are high. A detailed discussion of these substances and their properties in terms of the relative sizes of the atoms (ionic radius ratio) is given in Section 18-2.

The extreme case is that in which the entire crystal is held together by very strong covalent bonds; this occurs for diamond, with sublimation point 4600°K at 1 atm and melting point at some higher value.

The Dependence of Melting Point on Molecular Symmetry

The foregoing discussion has indicated that the melting points and the boiling points of substances are determined by several factors. One of these is the symmetry of the molecules, which has a pronounced effect on the melting point, but not on the boiling point: the greater the symmetry

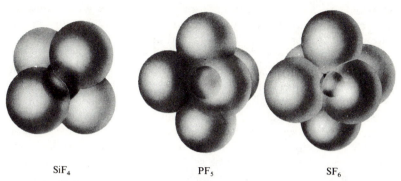

SiF₄   PF₅   SF₆

FIGURE 11-5

Molecules of silicon tetrafluoride, phosphorus pentafluoride, and sulfur hexafluoride, three very volatile substances.

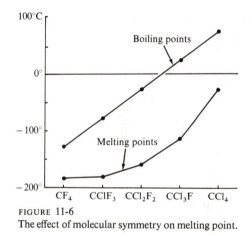

FIGURE 11-6

The effect of molecular symmetry on melting point.

of the molecule, the higher the melting point of the substance. This effect is shown by many organic compounds, and it is strikingly evident for the series of tetrahedral molecules CF_4, CF_3Cl, CF_2Cl_2, $CFCl_3$, CCl_4 (Figure 11-6). The boiling points of these substances are very nearly a linear function of the number of chlorine atoms in the molecule, but the melting points show a pronounced deviation from linearity.

This deviation can be explained in the following way. In a liquid, which has a structure characterized by some randomness of molecular arrangement, the molecules are piled together in various ways, not determined by their symmetry. In the crystal, however, there is a striking difference: a molecule of CF_2Cl_2 can fit into its place in the crystal in only two ways (differing in rotation of the molecule through 180° around a certain axis), whereas a molecule of CF_3Cl can fit into its place in a crystal of this substance in three ways (differing by a rotation through 120° around an axis through the carbon atom and the chlorine atom), and a molecule of CF_4 can fit into its place in the crystal of the substance in twelve ways. At the melting point, where there is equilibrium between the crystal and the liquid, equal numbers of molecules must be leaving the crystal and attaching themselves to it. The chance of leaving the crystal, under the influence of thermal agitation, is the same for a molecule of high symmetry as for one of low symmetry, whereas the chance of striking the crystal in suitable orientation to stick to it is greater for the molecule of high symmetry than for that of low symmetry. Thus a molecule of CF_4 can strike the crystal and adhere to it in any one of twelve orientations, whereas the molecule of CF_2Cl_2 can strike its crystal and adhere to it in only two orientations. Accordingly, substances with molecules of high symmetry crystallize more readily than those of low symmetry; that is, they have higher melting point. Relative to CF_2Cl_2 (symmetry number 2; that is, the molecule can fit into the crystal in two ways), this effect causes an increase in melting point of about 14° for CF_3Cl and $CFCl_3$ (symmetry number 3) and of 57° for CF_4 and CCl_4 (symmetry number 12).

For some sequences of substances the effect of the symmetry number on the melting point is overcome by an opposite effect of the electric dipole moment. This effect can be recognized because it affects the boiling point as well as the melting point. One example of the many that might be given is methylene chloride, CH_2Cl_2, which has a large value of the electric dipole moment, 0.34 ϵÅ. Its melting point, 176°K, is not about 57° lower than the average of those of the more symmetric molecules CH_4 (m.p. 90.6°K) and CCl_4 (m.p. 250°K), but is 6° higher. That the electric dipole is effective in this case is indicated by the boiling point of CH_2Cl_2, 313°K, which is 82° higher than the average of those of CH_4 (111.6°K) and CCl_4 (350.0°K).

11-5. Chemical Equilibrium in Gases

We have seen in Section 11-1 that spontaneous changes in state of a system at constant pressure and temperature are those associated with a decrease of the free energy G; that is, ΔG is negative.

Sometimes a chemical reaction begins, continues for a while, and then appears to stop before any one of the reactants is used up: the reaction is said to have reached equilibrium. The reaction between nitrogen dioxide, NO_2, and dinitrogen tetroxide, N_2O_4, provides an interesting example. The gas that is obtained by heating concentrated nitric acid with copper is found to have a density at high temperatures corresponding to the formula NO_2, and a density at low temperatures and high pressures approximating the formula N_2O_4. At high temperatures the gas is deep red in color, and at low temperatures it becomes lighter in color; the crystals formed when the gas is solidified are colorless. The change in the color of the gas and in its other properties with change in temperature and change in pressure can be accounted for by assuming that the gas is a mixture of the two molecular species NO_2 and N_2O_4, in equilibrium with one another according to the equation

$$\underset{\text{Colorless}}{N_2O_4} \rightleftarrows \underset{\text{Red}}{2NO_2} \tag{11-21}$$

It has been found by experiment that the amounts of nitrogen dioxide and dinitrogen tetroxide in the gas mixture are determined by a simple equation:

$$\frac{[NO_2]^2}{[N_2O_4]} = K \tag{11-22}$$

This equation, which is called the *equilibrium equation* for the reaction, is seen to involve in the numerator the concentration of the substance on the right side of the chemical equation (Equation 11-21), with the exponent 2, which is the coefficient shown in the chemical equation. The denominator contains the concentration of the substance on the left side. Its exponent is 1, because in the equation as written the coefficient of N_2O_4 is 1. The symbol [A] is used for the concentration of A, in moles per liter.

The quantity K is called the *equilibrium constant* of the reaction of dissociation of dinitrogen tetroxide to nitrogen dioxide. The equilibrium constant is independent of the pressure of the system, or of the concentration of the reacting substances. It is, however, dependent on the temperature.

The General Equation for the Equilibrium Constant

The chemical equation for a general reaction can be written in the following form:

$$aA + bB + \cdots \rightleftarrows dD + eE + \cdots \tag{11-23}$$

Here the capital letters A, B, D, E are used to represent different molecular species, the reactants and the products, and the small letters *a, b, d, e* are the numerical coefficients that tell how many molecules of the different sorts are involved in the reaction.

The equilibrium equation for this reaction is

$$\frac{[D]^d[E]^e \cdots}{[A]^a[B]^b \cdots} = K \tag{11-24}$$

Here K is the equilibrium constant for the reaction.

It is customary to write the concentration ratio in the way given in Equation 11-24 for a chemical equation such as Equation 11-23; that is, *the concentrations of the products, to the appropriate powers, are written in the numerator and the concentrations of the reactants in the denominator.* This is a convention that has been accepted by all chemists.

The equilibrium constant for gases is often written with partial pressure replacing concentration:

$$\frac{P_D^d P_E^e \cdots}{P_A^a P_B^b \cdots} = K_P \tag{11-25}$$

Unless the reaction involves no change in number of molecules, the numerical value of K_P differs from that of K.

The validity of the equilibrium equation, with K a constant at constant temperature, is a consequence of the laws of thermodynamics, if the reactants and the products are gases obeying the perfect gas laws or are solutes in dilute solution. In gases under high pressure and in concentrated solutions there occur some deviations from this equation, similar in magnitude to the deviations from the perfect gas laws. Sometimes these deviations are taken into account by introducing *activity coefficients*, as discussed for ions in solution in Chapter 13.

Examples of the use of the general equilibrium equation will be given in the following chapters of this book. This simple equation permits the chemist to answer many important questions that arise in his work.

Thermodynamic Derivation of the Equilibrium Equation

Let us assume that each of the reactants and products in the reaction 11-23 is at 1 atm partial pressure. The free energy change accompanying the reaction is then the difference in the standard free energies of the products and the reactants:

$$\Delta G^\circ = dG^\circ(D) + eG^\circ(E) + \cdots - aG^\circ(A) - bG^\circ(B) - \cdots \tag{11-26}$$

Now we evaluate ΔG for partial pressures $P_A, P_B, \ldots$. From the argument leading to Equation 11-11 we know that the difference in free energy of a mole of perfect gas at partial pressure P_A (in atm) and at partial pressure 1 atm is

$$G(P_A) - G^\circ(A \text{ at } 1 \text{ atm}) = RT \ln P_A$$

and the difference for a moles of A is

$$a\{G(P_A) - G^\circ(A \text{ at } 1 \text{ atm})\} = RT \ln P_A^a$$

With use of this equation and similar equations we obtain the result

$$\Delta G - \Delta G^\circ = RT \ln \frac{P_D^d P_E^e \cdots}{P_A^a P_B^b \cdots} \tag{11-27}$$

At equilibrium ΔG is equal to zero; there is then no driving force for the reaction in either direction. Thus we obtain as the condition for the equilibrium the following equation:

$$-\Delta G° = RT \ln \frac{P_D^d P_E^e \cdots}{P_A^a P_B^b \cdots} \qquad (11\text{-}28)$$

This equation can be rewritten as

$$\frac{P_D^d P_E^e \cdots}{P_A^a P_B^b \cdots} = \exp \frac{-\Delta G°}{RT} = K_P \qquad (11\text{-}29)$$

Equation 11-29 expresses a relation between the equilibrium constant for a reaction and the change in standard free energy for the reaction. The use of this equation is illustrated later in this Chapter.

Relation to the Boltzmann Distribution Law

We may use the relation $\Delta G° = \Delta H° - T\Delta S°$ to rewrite Equation 11-29:

$$K_P = \exp \frac{\Delta S°}{R} \times \exp \frac{-\Delta H°}{RT} \qquad (11\text{-}30)$$

Comparison with the Boltzmann distribution law suggests that the factor $\exp(\Delta S°/R)$ may be interpreted as the ratio of the number of quantum states accessible to the products under standard conditions to the number accessible to the reactants, and the factor $\exp(-\Delta H°/RT)$ as the ratio of the Boltzmann exponential factors for products and reactants.

Example 11-6. Hydrogen iodide, HI, is not a very stable substance. The pure gas is colorless, but when it is made in the laboratory the gas may have a violet color, indicating the presence of free iodine. In fact, hydrogen iodide decomposes to an appreciable amount according to the equation

$$2HI(g) \underset{\longleftarrow}{\overset{\longrightarrow}{\rightleftharpoons}} H_2(g) + I_2(g)$$

The equilibrium constant for this decomposition reaction has been found by experiment to have the value 0.00271 at 100°C. To what extent does hydrogen iodide decompose at this temperature?

Solution.   In this example the value of the equilibrium constant is given without a statement as to its dimensions. Let us write the expression for the equilibrium constant:

$$K = \frac{[H_2][I_2]}{[HI]^2} = 0.00271$$

Each of the concentrations H_2, I_2, and HI has the dimensions moles/liter.

Hence we see that for this reaction the dimensions of K are those of a pure number:

$$\text{Dimensions of } K = \frac{(\text{moles/liter})(\text{moles/liter})}{(\text{moles/liter})^2} = 1$$

When hydrogen iodide decomposes, equal numbers of molecules of hydrogen and iodine are formed. Accordingly the concentrations of hydrogen and iodine are equal in the gas formed when hydrogen iodide undergoes some decomposition.

Let us use the symbol x to represent the concentration of hydrogen and also the concentration of iodine:

$$[H_2] = [I_2] = x$$

Then we have

$$\frac{x^2}{[HI]^2} = 0.00271$$

or

$$x^2 = 0.00271[HI]^2$$
$$x = (0.00271)^{1/2}[HI] = 0.0521[HI]$$

By solving this equation we have found that after the hydrogen iodide has decomposed enough to produce the equilibrium state at 100°C the molar concentration of hydrogen is equal to 0.052 times the molar concentration of HI. The molar concentration of iodine is also equal to 0.052 times the molar concentration of HI. The question "To what extent does hydrogen iodide decompose at this temperature?" is to be interpreted as meaning "What percentage of pure hydrogen iodide originally produced decomposes to give hydrogen and iodine?" The equation for the chemical reaction shows that two molecules of HI on reaction form only one molecule of H_2 and one of I_2. Accordingly, there must have been 10.4% more moles of HI present initially than when equilibrium is reached. Hence the extent of decomposition of the hydrogen iodide is 0.104/1.104 = 0.094; that is, 9.4%.

Example 11-7. At 500°C the equilibrium constant for the formation of ammonia by the reaction

$$N_2 + 3H_2 \rightleftarrows 2NH_3 \tag{11-31}$$

has the value 1.50×10^{-5}, expressed in terms of partial pressures in atmospheres. What fraction of a stoichiometric mixture of nitrogen and hydrogen could be converted to ammonia with the total pressure kept at 1 atm? At 500 atm?

Solution. The equilibrium equation for this reaction is

$$\frac{P_{NH_3}^2}{P_{N_2}P_{H_2}^3} = 1.50 \times 10^{-5} \text{ atm}^{-2}$$

The problem states that

$$P_{H_2} = 3P_{N_2}$$

because the stoichiometric ratio N_2 to $3H_2$ is specified for these gases. The total pressure is the sum of the partial pressures of the three gases:

$$P_{N_2} + P_{H_2} + P_{NH_3} = P_{total}$$

Let

$$x = P_{NH_3}$$

Then

$$P_{N_2} + P_{H_2} = P_{total} - x$$

and

$$P_{N_2} = \tfrac{1}{4}(P_{total} - x)$$
$$P_{H_2} = \tfrac{3}{4}(P_{total} - x)$$

Therefore

$$\frac{x^2}{\dfrac{27}{256}(P_{total} - x)^4} = 1.50 \times 10^{-5}$$

$$\frac{x^2}{(P_{total} - x)^4} = 1.58 \times 10^{-6}$$

$$x = 1.26 \times 10^{-3} \times (P_{total} - x)^2$$

This quadratic equation might be solved directly. It is seen, however, that when P_{total} is small x is small compared with P_{total}. Hence as a first approximation we neglect x in comparison with P_{total} in the term on the right where these quantities are subtracted, obtaining

$$x = 0.00126 \times P^2_{total}$$

For total pressure 1 atm this gives $x = P_{NH_3} = 0.00126$ atm. Remembering that two molecules of ammonia are formed from four molecules of the reactants, we see that only 0.25% of the gas mixture is converted into ammonia.

For $P_{total} = 500$ atm this approximate method gives $x = P_{NH_3} = 0.00126 \times 500^2 = 315$ atm, which is not negligible compared with $P_{total} = 500$ atm. Direct solution of the quadratic equation gives $P_{NH_3} = 152$ atm, which leads to $P_{N_2} = 87$ atm and $P_{H_2} = 261$ atm. From these values it is calculated that at this total pressure 46.6% of the original mixture is converted into ammonia.

The principal commercial method of production of ammonia is the *Haber process*, the direct combination of nitrogen and hydrogen under high pressure (several hundred atmospheres) in the presence of a catalyst (usually iron containing molybdenum or other substances to increase the catalytic activity). The gases used must be specially purified, to prevent "poisoning" the catalyst. The reaction 11-31 is exothermic, and accordingly the yield of ammonia at equilibrium is less at a high temperature than at lower temperature (Section 11-6). The gases react very slowly at low temperatures, however, and the reaction could be used as a commercial process only when a catalyst was found which speeded up the rate satisfactorily at 500°C. As shown by the calculation above, the yield is unsatisfactory at this temperature at 1 atm total pressure, but satisfactory at 500 atm.

11-6.   Change of Equilibrium with Temperature

To determine the temperature dependence of the equilibrium constant we write Equation 11-28 in the form

$$\ln K_P = - \frac{\Delta G^\circ}{RT} \qquad (11\text{-}32)$$

Differentiation with respect to T (P constant) gives the result

$$\frac{d \ln K_P}{dT} = \frac{1}{R} \left\{ \frac{\Delta G^\circ}{T^2} - \frac{1}{T} \left(\frac{\partial \Delta G^\circ}{\partial T} \right)_P \right\} \qquad (11\text{-}33)$$

We have found (Equation 11-7) that the second law of thermodynamics requires that

$$\left(\frac{\partial \Delta G^\circ}{\partial T} \right)_P = -\Delta S^\circ = \frac{\Delta G^\circ - \Delta H^\circ}{T} \qquad (11\text{-}34)$$

Substitution of this value in Equation 11-33 gives the *van't Hoff equation:*[*]

$$\frac{d \ln K_P}{dT} = \frac{\Delta H^\circ}{RT^2} \qquad (11\text{-}35)$$

Over a small range of temperature ΔH° can be considered to be constant. For example, for the formation of $2NH_3(g)$ from the elements (Equation 11-31), ΔH° changes from -107.3 kJ mole^{-1} at 800°K to -108.9 kJ mole^{-1} at 900°K. Integration of Equation 11-35 with ΔH° constant gives the equation

$$\ln \left\{ \frac{K_P(T_2)}{K_P(T_1)} \right\} = \frac{\Delta H^\circ}{R} \left(\frac{1}{T_1} - \frac{1}{T_2} \right) \qquad (11\text{-}36)$$

The use of this equation is illustrated in the following example.

Example 11-8.   How much increase in the yield of ammonia by the Haber process could be achieved by a 10° decrease in the operating temperature, to 490°C?

Solution.   By introducing the values $\Delta H^\circ = -107.3$ kJ mole^{-1}, $T_1 = 773$°K, and $T_2 = 763$°K in Equation 11-36 we obtain the result

$$\ln \left\{ \frac{K_P(T_2)}{K_P(T_1)} \right\} = 0.219$$

$$\frac{K_P(490°C)}{K_P(500°C)} = 3.2$$

---

[*]This equation was first derived by J. H. van't Hoff in 1884, with use of Equation 11-34 (called the Gibbs-Helmholtz equation), which had been derived independently a few years earlier by J. Willard Gibbs and the German scientist H. L. F. Helmholtz (1821-1894).

With $K_P = 1.50 \times 10^{-5}$ atm^{-2} at 500°C, this equation gives $K_P = 4.80 \times 10^{-5}$ atm^{-2} at 490°C. By the method of Example 11-7 we find that the yield of ammonia is 57.3%, which is 23% greater than at the 10° higher temperature.

Determination of Standard Thermodynamic Properties of Substances

The standard free energy of a substance at temperature T can be determined by evaluating the equilibrium constant for a reaction involving this substance and other substances with known values of the standard free energies, and making use of Equation 11-29. A common convention is that $G°$, as well as $H°$, is taken as zero at every temperature for every element in its standard state, which usually is the stable crystalline modification at low temperatures, the liquid above the melting point, and the gas at 1 atm pressure above the standard boiling point.

Measurement of the ratio of equilibrium constants at two temperatures permits the evaluation of $\Delta H°$, by Equation 11-35 or 11-36, and from $\Delta G°$ and $\Delta H°$ the value of $\Delta S°$ can be calculated. These methods, together with those discussed in Chapter 15, have been used to obtain tables of values of the thermodynamic properties of thousands of substances.

11-7.   Equilibrium in Heterogeneous Systems

For heterogeneous systems, with more than one phase present, the equilibrium conditions are closely related to those for homogeneous systems. There are two important principles that apply.

1. *The activity of a substance in a system at equilibrium has the same value for each phase of the system of which the substance is a component.* For example, in a system containing ice, water, and water-vapor at 0°C the activity of water is the same for all three phases. Activity is usually expressed as partial pressure or as concentration; in this case the vapor pressure of ice and the vapor pressure of liquid water are both equal to the partial pressure of water vapor in the gas phase. If some ether were also present, forming another liquid phase, the amount of water dissolved in the ether would be such as to give the same partial pressure of water for the ether-water phase as for the other phases.

2. *The activity of a substance present in a system as a pure liquid or crystalline phase is constant at constant temperature.* In any system in which there is ice, for example, at −10°C, the activity of water has the same value, equal to the vapor pressure of ice at this temperature.

Since the activity of a pure liquid or crystalline phase is constant at constant temperature, the equilibrium constant for a reaction may be written without including the substance composing this phase.

Let us consider, for example, the decomposition of silver oxide:

$$2Ag_2O(c) \rightleftarrows 4Ag(c) + O_2(g)$$

We might write as the equilibrium constant

$$\frac{P_{O_2}P_{Ag}^4}{P_{Ag_2O}^2} = K$$

Since, however, P_{Ag} and P_{Ag_2O} are constants, the simplified equation $P_{O_2} = K$ may be written. This states that at equilibrium at constant temperature the partial pressure of oxygen produced by partial decomposition of silver oxide is constant. At 400°C, for example, the decomposition constant of Ag_2O has the value 0.145 atm. Hence silver oxide held at this temperature would decompose so long as the partial pressure of oxygen in the gas in contact with the solid were less than 0.145 atm. Since the partial pressure of oxygen in air is 0.21 atm, finely divided silver at 400°C combines with atmospheric oxygen to form the oxide. At 426°C the dissociation pressure of the oxide reaches 0.21 atm; hence at temperatures above this silver remains unoxidized in air.

The vapor-pressure equation discussed in Section 11-2 represents a simple case of equilibrium in heterogeneous systems.

11-8. Le Chatelier's Principle

An important and interesting qualitative principle about equilibrium is the *principle of Le Chatelier*. This principle, which is named after the French chemist Henry Louis Le Chatelier (1850–1936), may be expressed in the following way: *if the conditions of a system, initially at equilibrium, are changed, the equilibrium will shift in such a direction as to tend to restore the original conditions, if such a shift is possible.*

From the equilibrium equation 11-29 at constant temperature, we see that increasing the partial pressure of a reactant A or B causes the equilibrium to shift in such a way as to decrease the partial pressure of that reactant toward its initial value (with changes also in the other partial pressures). Increasing the partial pressure of a product, D or E, causes the reaction to go in the opposite direction, decreasing the partial pressure of that product toward its original value.

Increasing the total pressure causes the equilibrium to shift in such a way as to decrease the total pressure, if the equilibrium is dependent on the total pressure. We have seen that increase in total pressure of the system shifts the ammonia equilibrium in the direction of the product, NH_3; this shift decreases the total pressure, because the volume of $2NH_3(g)$ is only half that of $N_2(g) + 3H_2(g)$. The reaction $H_2(g) + I_2(g) \rightleftarrows 2HI(g)$ is an example of one in which change in total pressure causes no shift in the equilibrium; it is the existence of systems of this sort that requires the final qualifying clause in our statement of Le Chatelier's principle.

The van't Hoff equation, Equation 11-35, shows that increasing the temperature of a system causes the equilibrium constant to change in the direction that corresponds to the sign of $\Delta H°$. If the system were rapidly heated and then isolated, the equilibrium would shift in the direction that uses up thermal energy, and the temperature would decrease somewhat toward its initial value. We see that increase in temperature will drive a reaction further toward completion (by increasing the equilibrium constant) if the reaction is endothermic and will drive it back (by decreasing the equilibrium constant) if the reaction is exothermic.

For example, let us consider the NO_2—N_2O_4 equilibrium mixture at room temperature. Heat is absorbed when an N_2O_4 molecule dissociates into two NO_2 molecules:

$$N_2O_4(g) \longrightarrow 2NO_2(g) \qquad \Delta H° = 63 \text{ J mole}^{-1}$$

If the reaction mixture were to be increased in temperature by a few degrees the equilibrium would be changed, according to the principle of Le Chatelier, in such a way as to tend to restore the original temperature— that is, in such a way as to lower the temperature of the system, by using up some of the heat energy. This would be achieved by the decomposition of some additional molecules of dinitrogen tetroxide. Accordingly, in agreement with the statement above, the equilibrium constant would change in such a way as to correspond to the dissociation of more of the N_2O_4 molecules.

This principle is of great practical importance. For example, as mentioned above, the synthesis of ammonia from nitrogen and hydrogen is exothermic; hence the yield of ammonia is made a maximum by keeping the temperature as low as possible. The commercial process of manufacturing ammonia from the elements became practicable when catalysts were found that cause the reaction to proceed rapidly at low temperatures.

The reaction

$$N_2 + O_2 \rightleftarrows 2NO$$

is, on the other hand, endothermic (90 kJ is absorbed per mole of NO formed); hence in order to fix atmospheric nitrogen by forming nitric oxide a very high temperature is needed. The temperature used in practice, that of an electric arc, is about 2000°C. The gas mixture is rapidly chilled from this temperature, so rapidly that the reverse reaction does not have time to occur in response to the changed conditions.

11-9.   The Phase Rule—a Method of Classifying All Systems in Equilibrium

We have so far discussed a number of examples of systems in equilibrium. These examples include, among others, a crystal or a liquid in equilibrium with its vapor, a crystal and its liquid in equilibrium with its vapor at its melting point, and silver oxide and metallic silver in equilibrium with oxygen.

These systems appear to be quite different from one another. However, it was discovered by J. Willard Gibbs in the course of his early work on chemical thermodynamics that *a simple, unifying principle holds for all systems in equilibrium*. This principle is called the *phase rule*.

The phase rule is a relation among the number of independent components, the number of phases, and the variance of a system in equilibrium. The independent components (or, briefly, the components) of a system are the substances that must be added to realize the system. The phases have been defined earlier (Section 1-5). Thus a system containing ice, water, and water vapor consists of three phases but only one component (water-substance), since any two of the phases can be formed from the third. The *variance* of the system is the number of independent ways in which the system can be varied; these ways may include varying the temperature and the pressure, and also varying the composition of any solutions (gaseous, liquid, or crystalline) that exist as phases in the system.

The nature of the phase rule can be induced from some simple examples. Consider the system represented in Figure 11-7. It is made of water-substance (water in its various forms), in a cylinder with movable piston (to permit the pressure to be changed), placed in a thermostat with changeable temperature. If only one phase is present, both the pressure and the temperature can be arbitrarily varied over wide ranges: the variance is 2. For example, liquid water can be held at any temperature from its freezing point to its boiling point under any applied pressure. But if two phases are present, the pressure is automatically determined by the temperature: the variance is reduced to 1. For example, pure water vapor in equilibrium with water at a given temperature has a definite pressure, the vapor pressure of water at that temperature. And if three phases are present in equilibrium, ice, water, and water vapor, both the temperature and the pressure are exactly fixed: the variance is 0. This condition is called the triple point of ice, water, and water vapor. It occurs at temperature $+0.0099°C$ and pressure 0.0060 atm.

We see that for this simple system, with one component, the sum of the number of phases and the variance is equal to 3. It was discovered by

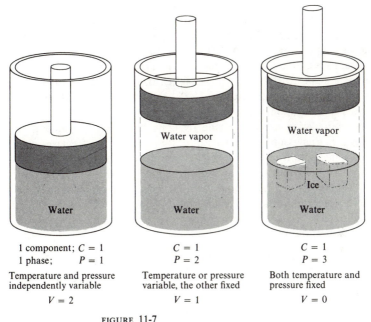

1 component; $C = 1$	$C = 1$	$C = 1$
1 phase; $P = 1$	$P = 2$	$P = 3$
Temperature and pressure independently variable	Temperature or pressure variable, the other fixed	Both temperature and pressure fixed
$V = 2$	$V = 1$	$V = 0$

FIGURE 11-7
A simple system illustrating the phase rule.

Gibbs that *for every system in equilibrium the sum of the number of phases and the variance is 2 greater than the number of components:*

Number of phases + variance = number of components + 2

or, using the abbreviations P, V, and C,

$$P + V = C + 2 \tag{11-37}$$

This is the phase rule.

Let us now consider its application. We ask: Is it ever possible for four phases to exist together in equilibrium? The answer is seen to be that it is, provided that there are at least two components. If there are only two components the four phases can coexist only at exactly fixed temperature and pressure. For example, we might add sodium chloride to the water in our system. The components would then be two in number ($C = 2$). When ice, liquid solution, and water vapor were present ($P = 3$) the temperature could still be varied somewhat, by varying the concentration of sodium chloride in the liquid solution. The variance would then be 1, with three phases. On lowering the temperature there would ultimately be formed crystals of sodium chloride dihydrate, $NaCl \cdot 2H_2O$. The system would then be at fixed temperature ($-21.2°C$), fixed composition of the liquid phase (22.42 g NaCl per 100 g of solution), and fixed pressure (the vapor pressure of ice at that temperature, 0.00091 atm). The variance is zero, as required for $C = 2$ and $P = 4$. There is another invariant point for the system, with the four phases $H_2O(g)$, salt solution, $NaCl \cdot 2H_2O(c)$, and $NaCl(c)$, at temperature $0.15°C$, concentration of solution 26.3 g NaCl per 100 g, and pressure 0.0055 atm.

We can derive the phase rule by considering the free energy of the system. Let us assume, for generality, that each of the P phases is a solution (solid, liquid, or gas) of all C components. The composition of each phase can be specified by the mole fractions $x_1, x_2, \ldots, x_C$ of the components, with their sum equal to 1. Hence $C - 1$ values are required to specify the composition of each phase. The state of the system can be specified by $P(C - 1)$ compositional parameters, plus two more, the temperature and pressure:

$$\text{Total number of variables} = PC - P + 2$$

The equilibrium state of the system at fixed temperature and pressure is the state of minimum free energy. Let us consider a variation from the stable state, consisting in removing a small amount of the first component from the first phase and adding it to the second phase. Because G is at its minimum, this variation leaves G unchanged. Hence there is a restraint on the compositions of the first and second phases: the change in free energy of the first phase on removing a small amount of the first component is the negative of that of the second phase on adding this small amount of the first component. (This is equivalent to saying that it equals that of the second phase on removal of the same small amount.) Similarly, the third phase and other phases are restrained in composition, with respect to the first component; $P - 1$ restraints for each component, and a total of $C(P - 1)$:

$$\text{Number of restraints} = CP - C$$

The variance of the system, V, is the total number of variables minus the number of restraints:

$$V = PC - P + 2 - PC + C$$

or

$$P + V = C + 2$$

The foregoing argument is essentially identical with that used by Gibbs in his discovery of the phase rule in 1876.

11-10. The Conditions under Which a Reaction Proceeds to Completion

A reaction such as the synthesis of hydrogen iodide

$$H_2(g) + I_2(g) \rightleftarrows 2HI(g)$$

does not proceed to completion; the relative amounts of the three substances must correspond to the equilibrium expression, and even under the most favorable conditions the conversion to the product is not complete.

Some reactions, in particular heterogeneous reactions, may proceed to completion, one of the reacting substances being used up completely. If, for example, in the system of Figure 11-7 the pressure is greater than the vapor pressure of water the reaction of condensation of water vapor will continue to completion, until no more of the vapor phase remains. Similarly, for the reaction

$$Ag_2O(c) \rightleftarrows 2Ag(c) + \tfrac{1}{2}O_2(g)$$

the reaction will go to completion toward the right if the partial pressure of oxygen is less than the equilibrium value, and toward the left (conversion of all the silver to oxide) if it is greater. In general, a reaction proceeds to completion if the initial state and variance are not in accord with the phase rule.

The conditions under which a given reaction may proceed to completion or sufficiently near to completion for the purpose at hand can usually be found by considering the equilibrium problem which is involved. Under ordinary conditions (T and P constant) the equilibrium is determined by the value of $\Delta G°$. If $\Delta H°$ has a large negative value (a strongly exothermic reaction), larger than $T\Delta S°$, the reaction will go nearly to completion. For some reactions, with $\Delta H°$ small in magnitude and $T\Delta S°$ large, the equilibrium is determined largely by the entropy change.

Values of $\Delta H°$, $\Delta G°$, $-T\Delta S°$, and $\Delta S°$ at 298°K for some reactions are given in Table 11-6. We see that the value of $\Delta S°$ is small for those reactions involving only solid (crystalline) reactants and products. For these reactions the equilibrium is essentially determined by the value of $\Delta H°$, which can often be estimated from the electronegativity differences of the atoms bonded to one another or by the use of bond energies.

The effect of having one of the substances in the liquid state is illustrated in the third reaction in the table. The entropy of fusion of $Br_2(c)$ is 40 J deg^{-1} mole^{-1} at the melting point, 266°K, and somewhat larger at 298°K. Hence the extra entropy of $\tfrac{1}{2}Br_2(l)$ is responsible for over half the value of $\Delta S°$ for the reaction.

TABLE 11-6

Values of $\Delta H°$, $\Delta G°$, $-T\Delta S°$, and $\Delta S°$ for Some Reactions (298°K) ($\Delta G° = \Delta H° - T\Delta S°$)

Reaction	$\Delta H°$ kJ mole^{-1}	$\Delta G°$ kJ mole^{-1}	$-T\Delta S°$ kJ mole^{-1}	$\Delta S°$ J deg^{-1} mole^{-1}
$Si(c) + C(c) \rightleftarrows SiC(c)$	-86.4	-84.0	2.4	-8
$4B(c) + C(c) \rightleftarrows B_4C(c)$	-38.9	-38.3	0.6	-2
$Na(c) + \frac{1}{2}Br_2(l) \rightleftarrows NaBr(c)$	-361.4	-349.3	12.1	-41
$\frac{1}{2}Cl_2(g) + \frac{1}{2}F_2(g) \rightleftarrows ClF(g)$	-50.8	-52.3	-1.5	5
$\frac{1}{2}H_2(g) + \frac{1}{2}F_2(g) \rightleftarrows HF(g)$	-271.1	-273.2	-2.1	7
$C(c) + O_2(g) \rightleftarrows CO_2(g)$	-393.5	-394.4	-0.9	3
$CH_4(g) \rightleftarrows C(c) + 2H_2(g)$	74.9	50.8	-24.1	81
$NH_3(g) \rightleftarrows \frac{1}{2}N_2(g) + \frac{3}{2}H_2(g)$	45.9	16.4	-29.5	99
$2H_2O(g) \rightleftarrows 2H_2(g) + O_2(g)$	483.7	457.2	-26.5	89
$2OF_2(g) \rightleftarrows O_2(g) + 2F_2(g)$	29.4	2.2	-27.2	91
$\frac{2}{3}H_2O(l) \rightleftarrows \frac{2}{3}H_2(g) + \frac{1}{3}O_2(g)$	190.6	158.1	-32.4	109
$2ClO_2(g) \rightleftarrows Cl_2(g) + 2O_2(g)$	-209.2	-244.6	-35.4	119
$\frac{2}{3}HNO_3(g) \rightleftarrows \frac{1}{3}H_2(g) + \frac{1}{3}N_2(g) + O_2(g)$	89.5	49.3	-40.2	135
$CF_4(g) \rightleftarrows C(c) + 2F_2(g)$	922.6	877.9	-44.7	150
$SiF_4(g) \rightleftarrows Si(c) + 2F_2(g)$	1614.9	1572.6	-42.4	142
$\frac{1}{2}SF_6(g) \rightleftarrows \frac{1}{2}S(c) + \frac{3}{2}F_2(g)$	610.4	558.5	-51.9	174
$2Ag_2O(c) \rightleftarrows 4Ag(c) + O_2(g)$	61.2	21.6	-39.6	133

The next three reactions involve the same number of reactant and product gas molecules. The values of $\Delta S°$ are small, and the equilibrium is largely determined by the value of $\Delta H°$.

The remaining eleven reactions are written with numerical coefficients such that there is one more product gas molecule than reactant gas molecule. The sublimation of a crystal is a reaction of this sort:

$$A(c) \rightleftarrows A(g)$$

Representative values for the standard entropy of sublimation at 298°K are 103 J deg^{-1} mole^{-1} for P_4 and 117 J deg^{-1} mole^{-1} for I_2. The corresponding values for vaporization are in the neighborhood of 85 J deg^{-1}

mole^{-1}, the Trouton constant. The difference is the entropy of fusion, which ranges from about 10 J deg^{-1} mole^{-1} for monatomic molecules (and molecules rotating in the crystal) to 50 to 100 J deg^{-1} mole^{-1} for complex molecules.

The standard entropy increase associated with the formation of a gas molecule is seen to range from 81 to 174 J deg^{-1} mole^{-1} for the eleven reactions listed in Table 11-6. For $\Delta H°$ small the direction in which a reaction of this sort takes place spontaneously is determined by $\Delta S°$, and this direction is that in which the number of gas molecules increases. Even with an unfavorable value of $\Delta H°$ a reaction may be made to go in this direction by increasing the temperature, thus making the term $T\Delta S°$ more important. We see from Table 11-6 that at 298°K $\Delta G°$ for the reaction of disilver oxide is positive, but at twice this temperature the entropy term causes it to be negative.

Equilibrium involving ions in aqueous solutions will be discussed in Chapter 13.

11-11.   Tabulated Values of the Thermodynamic Properties of Substances

During the last eighty years many chemists have worked to determine the values of the thermodynamic properties of substances. Many values of the enthalpy of formation and of the heat capacity were obtained in the period before 1900, and during recent years these properties have been measured with increased accuracy over a wide range of temperatures. Studies of chemical equilibria have been made to determine $\Delta G°$ and (from its temperature coefficient) $\Delta H°$, with $T\Delta S°$ given by their difference. The third law of thermodynamics has been applied to measure heat-capacity values to give values of the absolute entropy of many substances. For the smaller molecules molecular properties obtained by spectroscopic and diffraction methods have been used in the calculation of the entropy and heat capacity. An extensive table of thermodynamic properties is available: *Selected Values of Chemical Thermodynamic Properties*, Circular No. 500 of the U.S. National Bureau of Standards, 1952. Smaller tables can be found in the standard chemical handbooks and other reference books.

A small table of values of $\Delta \tilde{H}_f°$, $\Delta \tilde{G}_f°$, $\tilde{S}°$, and $\tilde{C}_P$ is given in Appendix XV of this book. Values $\Delta \tilde{H}_f°$ are given in several tables in the text. The uses of the tables are illustrated in the Exercises.

Exercises

11-1. The value of the enthalpy of Hg(c) at the standard melting point, 234.29°K, is found to be 5.228 kJ mole^{-1}, relative to $H(0°K) = 0$, by integrating C_P. The third-law value of the entropy at this temperature is 59.5 J deg^{-1} mole^{-1}, and the enthalpy of fusion is 2.295 kJ mole^{-1}. What are the values of H, S, and G of liquid mercury at this temperature?

11-2. Is ΔG for the reaction S_8 (orthorhombic) $\longrightarrow$ S_8 (monoclinic) positive or negative at 90°C? At 95.4°C (the equilibrium temperature)? At 100°C?

11-3. A substance exists in two crystalline forms, α (stable for $T < T_{tr}$) and β (stable for $T > T_{tr}$). At $T = T_{tr}$ is the molar enthalpy of α less than or greater than that of β? Is the molar entropy of α less than or greater than that of β at this temperature?

11-4. Gray tin is stable below 18°C and white tin is stable above this temperature. Their heat-capacity curves are similar in shape to the Debye functions. Is the Debye characteristic temperature Θ for gray tin less than or greater than for white tin? What argument leads you to your conclusion?

11-5. The third-law value of the entropy of Sn(gray) at 18°C is 44.2 J deg^{-1} mole^{-1} and that of Sn(white) is 53.7 J deg^{-1} mole^{-1}. What is the value of ΔH for the reaction Sn(gray) $\longrightarrow$ Sn(white) at this temperature?

11-6. The freezing point of water decreases nearly linearly with increasing pressure. The amount of decrease can be calculated from thermodynamic equations with use of other properties of water and ice.

　(*a*) Use Equation 11-8, with ΔG in place of G and ΔV in place of V, to calculate ΔG for the reaction $H_2O(c) \longrightarrow H_2O(l)$ at 0°C and 100 atm (equal to 10.1325 MN m^{-2}). The value of ΔV is 1.45 cm^3 mole^{-1}.

　(*b*) Assuming that ΔS is the same at this pressure as at 1 atm (22.0 J deg^{-1} mole^{-1}), use Equation 11-7 to calculate the temperature at which ΔG has returned to zero. (Answer: 14.6 J mole^{-1}; -0.66°C.)

11-7. What is the ratio of the small change in temperature δT and small change in pressure δP of a system that leave the free energy unchanged, in terms of the entropy and volume? Note that this ratio can be written as

$$\left(\frac{\partial T}{\partial P} \right)_G. \quad \text{(Answer: } V/S\text{)}$$

11-8. Le Chatelier's principle requires that the melting point of a substance increase with increase in pressure if the volume of the liquid is greater than that of the crystal, and decrease if it is less. The result of Exercise 11-7 shows that ΔT (melting) has the same sign as $\Delta V / \Delta S$. Explain the apparent discrepancy.

11-9. At what temperature would the transition S_8 (orthorhombic) $\longrightarrow$ S_8 (monoclinic) occur at pressure 100 atm? The value of ΔH is 0.39 kJ mole^{-1} at the normal transition temperature, 95.4°C. The density of orthorhombic sulfur at this temperature is 2.04 g cm^{-3} and that of monoclinic sulfur is 1.93 g cm^{-3}.

11-10. It is observed that for most molecular crystals with two crystalline forms the form stable at high temperatures has a smaller density than the form stable at low temperatures. Can you suggest an explanation of this generalization? (Consider the number of close van der Waals contacts in relation to the Debye heat-capacity function.)

11-11. Carbon tetrachloride (like carbon tetrafluoride) has two crystalline forms. The low-temperature form changes to the high-temperature form at 225.5°K, with entropy of transition 20 J deg^{-1} mole^{-1}. The entropy of fusion is 10 J deg^{-1} mole^{-1} (melting point 250.3°K). The intermediate substance CF_2Cl_2 has only one crystalline form. It has melting point 118°C and entropy of fusion 35 J deg^{-1} mole^{-1}. Discuss these facts in relation to the probable structure of the crystals.

11-12. The substance $GeCl_4(l)$ has standard enthalpy of formation -544 kJ mole^{-1}.
   (a) What is the standard enthalpy of formation of $GeCl_4(g)$? The boiling point at standard volume of the vapor is 74°C.
   (b) To what value of the Ge—Cl bond energy does this lead? (Answer (b): 332 kJ mole^{-1}.)

11-13. The substance $CHCl_2F(l)$ has density 1.426 g cm^{-3} and index of refraction (sodium D lines) 1.372. What value of the electric polarizability of the molecule does this information give (Equation 11-19)? Compare with the sum of atomic polarizabilities (Table 11-5). (Answer: 6.51 Å^3.)

11-14. The index of refraction of water (sodium D lines) at 25°C is 1.333. Calculate the electronic polarizability of the water molecule and compare with the value in Table 11-4. Calculate the index of refraction of ice at 0°C and of water vapor at 100°C and 1 atm.

11-15. Let the interaction energy of two molecules be $V = -A\,r^{-6} + B\,r^{-n}$, with the ratio B/A determined by the value r_0 of the equilibrium distance, at which V has its minimum value. Determine this ratio by differentiating V with respect to r and equating to zero. By what factor should $-A\,r_0^{-6}$, the energy of van der Waals attraction at $r = r_0$, be multiplied to obtain $V(r_0)$, the total energy of interaction of the two molecules? (Answer: $6\,r_0^{n-6}/n$; $1 - 6/n$.)

11-16. Tin tetraiodide has a crystal structure such that each iodine atom is in contact with nine other iodine atoms, 4.21 Å away, in addition to the three in the same molecule, 4.29 Å away. The density of SnI_4(c) is 4.473 g cm^{-3} and its index of refraction is 2.106.

(*a*) Evaluate the electric polarizability of the molecule.

(*b*) Assigning one quarter of α to each iodine atom and using the exponent $n = 12$ in the energy of repulsion (Exercise 11-15), calculate the contribution of the van der Waals attraction and repulsion to the enthalpy of sublimation. (The experimental value of the enthalpy of sublimation is 137 kJ mole^{-1}; it includes not only the van der Waals energy but also the $P\Delta V$ energy and the difference in rotational and vibrational energy in the liquid and the gas.) (Answer: 29.6 Å^3; 124 kJ mole^{-1}.)

11-17. The substance HI(g) at temperatures somewhat above room temperature has a violet color, because of partial dissociation to H_2(g) and I_2(g). The color becomes more intense as the temperature of the gas is increased, at constant pressure. What is the sign of ΔH for the reaction 2HI(g) $\longrightarrow$ H_2(g) + I_2(g)?

11-18. The value of K for the above reaction is 0.00271 at 100°C (Example 11-6). To what extent would HI decompose (a) if it were mixed with an equimolar amount of N_2 at room temperature and then heated to 100°C? (b) If it were mixed with an equimolar amount of H_2 at room temperature and then heated to 100°C?

11-19. The values of $\Delta H°_f$, $\Delta G°_f$, and $S°$ of I_2(g) at 25°C are 62.2 kJ mole^{-1}, 19.4 kJ mole^{-1}, and 260.6 J deg^{-1} mole^{-1}, respectively. With use of information given in Appendix XV, calculate the value of K for the reaction

$$2HI(g) \rightleftharpoons H_2(g) + I_2(g)$$

at this temperature. To what extent would HI(g) dissociate to these gaseous products at this temperature? (Answer: 0.00114.)

11-20. With use of information given in the preceding Exercise and in Appendix XV calculate the vapor pressure of I_2(c) at 25°C.

11-21. By what fraction of its value does the equilibrium constant K for the reaction

$$2HI(g) \rightleftharpoons H_2(g) + I_2(g)$$

increase on increase of the temperature from 25°C to 26°C? (See Equation 11-34; note that the value of $\Delta H°$ at 25°C can be obtained from information given in Exercise 11-18 and Appendix XV.) (Answer: 0.0141.)

11-22. With the assumption that $\Delta H°$ is constant over the range of temperatures 25°C to 100°C, use the experimental values of K given in Exercises 11-18 and 11-19 for these temperatures to evaluate $\Delta H°$ (see Equation 11-35). (Note that this value is close to the value for the average temperature 62.5°C, as given by the value for 25°C and the correction by ΔC_P.)

11-23. (*a*) A bulb with volume 1 liter contains 17.55 g sulfur. At 750°K the observed pressure of the sulfur vapor (no condensed phase present) is 0.237 atm, showing that essentially all of the sulfur is $S_8(g)$. At 885°K the observed total pressure is 0.73 atm. Evaluate the partial pressures of S_8 and S_2, and calculate K for the equilibrium $S_8(g) \rightleftarrows 4S_2(g)$ at 885°K from this information. (Ignore the correction for the presence of $S_6(g)$ that should be made.) (Answer: 0.99 atm³.)

11-24. The experiment of the preceding Exercise gave total pressure 0.824 atm at 900°K. What is the value of K at this temperature? What value of $\Delta H°$ for the reaction does the change in K from 885°K to 900°K correspond to?

11-25. The equilibrium $H_2S(g) \rightleftarrows H_2(g) + \frac{1}{2} S_2(g)$ has been studied at high temperatures by methods similar to that mentioned in the two preceding Exercises. The following values of K have been reported:

$$T = 1000°K \qquad\qquad K = 0.00752 \text{ atm}^{1/2}$$
$$1200 \qquad\qquad\qquad .0452$$
$$1400 \qquad\qquad\qquad .164$$
$$1600 \qquad\qquad\qquad .423$$

Calculate $R \ln K$ and T^{-1}, and make a plot with these quantities along the y and x axes, respectively. Determine the slope dy/dx of a straight line drawn through the points, and from it obtain a value for the enthalpy of the reaction.

11-26. The vapor pressure of Al(l) has been found by experiment to have the value 0.132 atm at 2353°K and 0.526 atm at 2593°K. From this information calculate an approximate value of ΔH for the reaction Al(l) $\longrightarrow$ Al(g).

11-27. A normal liquid with standard boiling point 350°K (such as CCl_4, Table 11-3) has 85.7 J deg⁻¹ mole⁻¹ entropy of vaporization at this boiling point. From this information calculate approximate values of the vapor pressure of normal liquids with standard boiling point 350°K at 250°K, 300°K, 400°K, and 450°K, and make a graph of the results. Note that a family of curves could be drawn in this way, from which the approximate dependence of vapor pressure on temperature could be obtained for a normal liquid from knowledge of its vapor pressure at one temperature.

11-28. From measurements of the electromotive force of an electric cell (Chapter 15) it is found that $\Delta G°$ at 25°C for the reaction

$$Hg_2Cl_2(c) \longrightarrow 2Hg(l) + Cl_2(g)$$

has the value 210.7 kJ mole⁻¹. What is the expression for the equilibrium constant K_P for this reaction? What is its value at 25°C? What is the partial pressure of Cl_2 in equilibrium with a mixture of calomel crystals (Hg_2Cl_2) and liquid mercury at this temperature?

11-29. From information given in Appendix XV calculate an approximate value of the partial pressure of $Cl_2(g)$ in equilibrium with calomel crystals and liquid mercury at 100°C.

12

Water

Water is one of the most important of all chemical substances. It is a major constituent of our bodies and of the environment in which we live. Its physical properties are strikingly different from those of other substances, in ways that determine the nature of the physical and biological world.

12-1. The Composition of Water

Water was thought by the ancients to be an element. Henry Cavendish in 1781 showed that water is formed when hydrogen is burned in air, and Lavoisier first recognized that water is a compound of the two elements hydrogen and oxygen.

The formula of water is H_2O. The relative weights of hydrogen and oxygen in the substance have been very carefully determined as 2.016: 16.000. This determination has been made both by weighing the amounts of hydrogen and oxygen liberated from water by electrolysis and by determining the weights of hydrogen and oxygen that combine to form water.

Purification of Water by Distillation

Ordinary water is impure; it usually contains dissolved salts and dissolved gases, and sometimes organic matter. For chemical work water is purified by distillation. Pure tin vessels and pipes are often used for storing and transporting distilled water. Glass vessels are not satisfactory, because the alkaline constituents of glass slowly dissolve in water. Distilling apparatus and vessels made of fused silica are used in making very pure water.

The impurity that is hardest to keep out of distilled water is carbon dioxide, which dissolves readily from the air.

Removal of Ionic Impurities from Water

Ionic impurities can be effectively and cheaply removed from water by an interesting process that involves the use of *giant molecules*—molecular structures that are so big as to constitute visible particles. A crystal of diamond is an example of such a giant molecule (Chapter 6). Some complex inorganic crystals, such as the minerals called *zeolites*, are of this nature. These minerals are used to "soften" hard water. Hard water is water containing cations of calcium, magnesium, and iron, which are undesirable because they form a precipitate with ordinary soap. The zeolite is able to remove these ions from the water, replacing them by sodium ion.

A zeolite is an aluminosilicate, with formula such as $Na_2Al_2Si_4O_{12}$ (Chapter 18). It consists of a rigid framework formed by the aluminum, silicon, and oxygen atoms, honeycombed by corridors in which sodium ions are located. These ions have some freedom of motion, and, when hard water flows over zeolite grains, some of the sodium ions run out of the corridors into the solution and are replaced by ions of calcium, magnesium, and iron. In this way the hardness of the water is removed. After most of the sodium ion has been replaced, the zeolite is regenerated by allowing it to stand in contact with a saturated brine; the reaction is then reversed, Na^+ replacing Ca^{++} and the other cations in the corridors of the zeolite.

The reactions that occur may be written with symbols. If Z^- is used to represent a small portion of the zeolite framework, carrying one negative charge, the replacement of calcium ion in the water by sodium ion may be written*

$$\underline{2Na^+Z^-} + Ca^{++} \longrightarrow \underline{Ca^{++}(Z^-)_2} + 2Na^+$$

*A line is drawn under the formula for a substance to indicate that it is a solid.

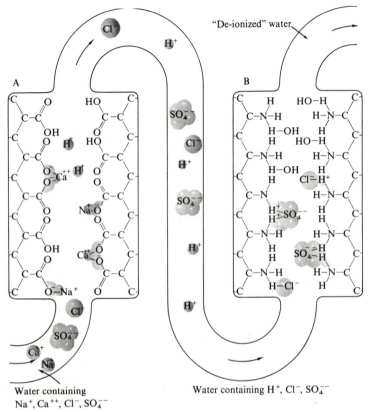

Water containing
$Na^+, Ca^{++}, Cl^-, SO_4^{--}$

Water containing H^+, Cl^-, SO_4^{--}

FIGURE 12-1
The removal of ions from water by use of giant molecules with attached acidic
and basic groups.

When concentrated salt solution (brine) is run through the zeolite, the
reverse reaction occurs:

$$2Na^+ + Ca^{++}(Z^-)_2 \longrightarrow 2Na^+Z^- + Ca^{++}$$

The reason that giant molecules—the aluminosilicate framework—are
important here is that these molecules, which look like large grains of
sand, are not carried along in the water, but remain in the water-softening tank.

Both the positive ions and the negative ions can be removed from water
by a similar method, illustrated in Figure 12-1. The first tank, A, contains
grains that consist of giant organic molecules in the form of a porous

framework to which acidic groups are attached. These groups are represented in the figure as *carboxyl groups*, —COOH:*

The reactions that occur when a solution containing salts passes through
tank A may be written as

$$\underline{RCOOH + Na^+} \longrightarrow \underline{(RCOO)^-Na^+} + H^+$$

$$\underline{2RCOOH + Ca^{++}} \longrightarrow \underline{(RCOO)^-_2Ca^{++}} + 2H^+$$

That is, sodium ions and calcium ions are removed from the solution by
the acidic framework, and hydrogen ions are added to the solution. The
solution is changed from a salt solution (Na^+, Cl^-) to an acid solution
(H^+, Cl^-).

This acid solution then runs through tank B, which contains grains of
giant organic molecules with basic groups attached. These groups are
shown in Figure 12-1 as *substituted ammonium hydroxide* groups,
$(RNH_3)^+(OH)^-$:

$$\left[\begin{array}{c} H \\ | \\ R{-}N{-}H \\ | \\ H \end{array} \right]^+ \qquad \left[\; \ddot{\underset{..}{O}}{-}H \; \right]^-$$

The hydroxide ion of these groups combines with the hydrogen ion in
the water:

$$OH^- + H^+ \longrightarrow H_2O$$

The negative ions then remain, held by the ammonium ions of the framework. The reactions are

$$\underline{(RNH_3)^+(OH)^-} + Cl^- + H^+ \longrightarrow \underline{(RNH_3)^+Cl^-} + H_2O$$

$$\underline{2(RNH)_3^+(OH)^-} + SO_4^{--} + 2H^+ \longrightarrow \underline{(RNH_3)^+_2(SO_4)^{--}} + 2H_2O$$

The water that passes out of the second tank contains practically no
ions, and may be used in the laboratory and in industrial processes in
place of distilled water.

The giant molecules in tank A may be regenerated after use by passing

---

*R represents a part of the framework, shown as a carbon atom in Figure 12-1.

a moderately concentrated solution of sulfuric acid through the tank, to restore the acid groups:

$$2(RCOO)^-Na^+ + 2H^+ \longrightarrow 2RCOOH + 2Na^+$$

Those in tank B may be regenerated by use of a moderately concentrated solution of sodium hydroxide:

$$(RNH_3)^+Cl^- + OH^- \longrightarrow (RNH_3)^+OH^- + Cl^-$$

Other Ways of Softening Water

Hard water may also be softened by chemical treatment. In practice the use of giant organic molecules (synthetic resins) for deionizing water, described above, is restricted to industries requiring very pure water, as in making medicinal products. The zeolite method is sometimes used on a large scale, to treat the water for an entire city, but it is more often applied only for an individual house or building. Water for a city is usually treated by the addition of chemicals, followed by sedimentation when the water is allowed to stand in large reservoirs, and then by filtration through beds of sand. The settling process removes suspended matter in the water together with precipitated substances that might be produced by the added chemicals, and some living microorganisms. After filtration, the remaining living organisms may be destroyed by treatment with chlorine, bleaching powder, sodium hypochlorite or calcium hypochlorite, or ozone.

The hardness of water is due mainly to calcium ion, ferrous ion (Fe^{++}), and magnesium ion; it is these ions that form insoluble compounds with ordinary soap. Hardness is usually reported in parts per million (ppm), calculated as calcium carbonate. Domestic water with hardness less than 100 ppm is good, and that with hardness between 100 and 200 ppm is fair.

Ground water in limestone regions may contain a large amount of calcium ion and hydrogen carbonate ion, HCO_3^-. Although calcium carbonate itself is insoluble, calcium hydrogen carbonate, $Ca(HCO_3)_2$, is a soluble substance. A water of this sort (which is said to have *temporary hardness*) can be softened simply by boiling, which causes the excess carbon dioxide to be driven off, and the calcium carbonate to precipitate:

$$Ca^{++} + 2HCO_3^- \longrightarrow CaCO_3(c) + H_2O + CO_2(g)$$

This method of softening water cannot be applied economically in the treatment of the water supply of a city, however, because of the large fuel cost. Instead, the water is softened by the addition of calcium hydroxide, slaked lime:

$$Ca^{++} + 2HCO_3^- + Ca(OH)_2 \longrightarrow 2CaCO_3(c) + 2H_2O$$

If sulfate ion or chloride ion is present in solution instead of hydrogen carbonate ion, the hardness of the water is not affected by boiling—the water is said to have *permanent hardness*. Permanently hard water can be softened by treatment with sodium carbonate:

$$Ca^{++} + CO_3^{--} \longrightarrow CaCO_3(c)$$

The sodium ions of the sodium carbonate are left in solution in the water, together with the sulfate or chloride ions that were already there.

In softening water by use of calcium hydroxide or sodium carbonate, enough of the substance is used to cause magnesium ion to be precipitated as magnesium hydroxide and iron as ferrous hydroxide or ferric hydroxide. Sometimes, in addition to the softening agent, a small amount of aluminum sulfate, alum, or ferric sulfate is added as a coagulant. These substances, with the alkaline reagents, form a flocculent, gelatinous precipitate of aluminum hydroxide, $Al(OH)_3$, or ferric hydroxide, $Fe(OH)_3$, which entraps the precipitate produced in the softening reaction, and helps it to settle out. The gelatinous precipitate also tends to adsorb coloring matter and other impurities in the water.*

A water that is used in a steam boiler often deposits a scale of calcium sulfate, which is left as the water is boiled away. In order to prevent this, boiler water is sometimes treated with sodium carbonate, causing the precipitation of calcium carbonate as a sludge, and preventing the formation of the calcium sulfate scale. Sometimes trisodium phosphate, Na_3PO_4, is used, leading to the precipitation of calcium as hydroxyapatite, $Ca_5(PO_4)_3OH$, as a sludge. In either case the sludge is removed from the boiler by draining at intervals.

12-2.   The Water Molecule

Many properties of the water molecule have been determined by the analysis of the band spectrum emitted or absorbed by the molecules in the gas phase. The O—H internuclear distance is 0.9584 Å (this value, called r_e, corresponds to the minimum value of the electronic and internuclear energy), and the bond angle H—O—H has the value 104.54°. Its vibrational frequencies are 3657 cm⁻¹, 3756 cm⁻¹, and 1595 cm⁻¹ (values in

---

Adsorption is the adhesion of molecules of a gas, liquid, or dissolved substance or of particles to the surface of a solid substance. *Absorption* is the assimilation of molecules into a solid or liquid substance, with the formation of a solution or a compound. Sometimes the word *sorption* is used to include both of these phenomena. We say that a heated glass vessel *adsorbs* water vapor from the air on cooling, and becomes coated with a very thin layer of water: a dehydrating agent such as concentrated sulfuric acid *absorbs* water, forming hydrates.

Hz, cycles per second, are obtained by multiplying by c in cm s^{-1}). The low frequency corresponds to the bending vibration, in which the bond angle increases and decreases, with little change in the bond length. The other two frequencies are those of the unsymmetric and symmetric bond-stretching vibrations.

The electric dipole moment of the molecule has the value 0.387 ϵÅ. As was mentioned in Section 6-9, this corresponds to 33% ionic character of each O—H bond.

12-3. The Properties of Water

The Ionic Dissociation of Water

An acidic solution contains hydrogen ions, H^+ (actually hydronium ions, H_3O^+). A basic solution contains hydroxide ions, OH^-. A number of years ago chemists asked, and answered, the question, "Are these ions present in pure neutral water?" The answer is that they are present, in equal but very small concentrations.

Pure water contains hydrogen ions in concentration 1×10^{-7} mole per liter, and hydroxide ions in the same concentration. These ions are formed by the dissociation of water:

$$H_2O \rightleftharpoons H^+ + OH^-$$

When a small amount of acid is added to pure water the concentration of hydrogen ion is increased. The concentration of hydroxide ion then decreases, *but not to zero*. Acidic solutions contain hydrogen ion in large concentration and hydroxide ion in very small concentration. The equation connecting these concentrations is discussed in Chapter 14.

The Physical Properties of Water

Water is a clear, transparent liquid, colorless in thin layers. Thick layers of water have a bluish-green color.

The physical properties of water are used to define many physical constants and units. The freezing point of water (saturated with air at 1 atm pressure) is 0°C, and the boiling point of water at 1 atm is 100°C. The unit of mass in the metric system was chosen so that 1 cm^3 of water at 4°C (the temperature of its maximum density) weighs approximately 1 gram. A similar relation holds in the English system: 1 cu. ft. of water weighs approximately 1,000 ounces.

Most substances diminish in volume, and hence increase in density, with decrease in temperature. Water has the very unusual property of

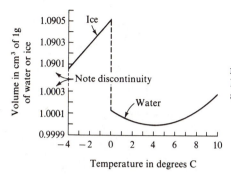

FIGURE 12-2
Dependence of the volume of ice
and water on temperature.

having a temperature at which its density is a maximum. This temperature is 4°C. With further cooling below this temperature the volume of a sample of water increases somewhat (Figure 12-2).

A related phenomenon is the increase in volume that water undergoes on freezing.

The Abnormal Melting and Boiling Points of Hydrogen Fluoride, Water, and Ammonia

The melting points and boiling points of the hydrides of some nonmetallic elements are shown in Figure 12-3. The variation for a series of congeners is normal for the sequence CH_4, SiH_4, GeH_4, and SnH_4, but is abnormal for the other sequences. The curves through the points for H_2Te, H_2Se, and H_2S show the expected trend, but when extrapolated they indicate values of about $-100°C$ and $-80°C$, respectively, for the melting point and boiling point of water. The observed value of the melting point is 100° greater, and that of the boiling point is 180° greater, than would be expected for water if it were a normal substance; and hydrogen fluoride and ammonia show similar, but smaller, deviations.

Dielectric Constant

Water and some other liquids have very large dielectric constants, that of water being 78.5 at 298°K, and that of liquid hydrocyanic acid being 110. For many liquids the value of the dielectric constant is approximately that indicated by the line near the bottom of Figure 12-4, which reaches the value 10 when the dipole moment of the molecule (measured in the gas phase) reaches 0.4 $_\epsilon$Å. Water, HF(l), H_2O_2(l), and HCN(l) have very much higher values, and NH_3(l) and CH_3OH(l), as well as other alcohols not shown in the figure, also lie somewhat above the line.

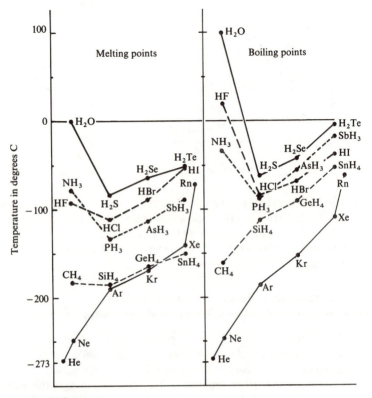

FIGURE 12-3
Melting points and boiling points of hydrides of nonmetallic elements, showing
abnormally high values for hydrogen fluoride, water, and ammonia,
caused by hydrogen-bond formation.

12-4. The Hydrogen Bond—the Cause of the Unusual Properties of Water

The unusual properties of water mentioned above are due to the power of
its molecules to attract one another especially strongly. This power is
associated with a structural feature which is called the *hydrogen bond*.

The Hydrogen Bond

The hydrogen ion is a bare nucleus, with charge $+1$. If hydrogen
fluoride, HF, had an extreme ionic structure, it could be represented as in

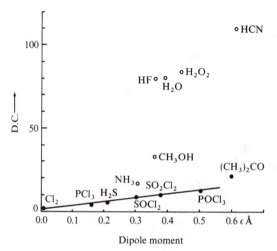

FIGURE 12-4

Values of the dielectric constant of some liquids at 298°K, plotted against values of the electric dipole moment of their molecules.

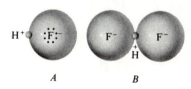

FIGURE 12-5

The hydrogen fluoride molecule (A) and the hydrogen difluoride ion, containing a hydrogen bond (B).

A of Figure 12-5. The positive charge of the hydrogen ion could then strongly attract a negative ion, such as a fluoride ion, forming an $[F^-H^+F^-]^-$ or HF_2^- ion, as shown in B. This does indeed occur, and the stable ion HF_2^-, called the *hydrogen difluoride ion*, exists in considerable concentration in acidic fluoride solutions, and in salts such as KHF_2, potassium hydrogen difluoride. The bond holding this complex ion together, called the *hydrogen bond*, is weaker than ordinary ionic or covalent bonds, but stronger than ordinary van der Waals forces of intermolecular attraction.

Hydrogen bonds are also formed between hydrogen fluoride molecules, causing the gaseous substance to be largely polymerized into the molecular species H_2F_2 H_3F_3, H_4F_4, H_5F_5, and H_6F_6. The last of these seems to be especially stable, probably because of its ability to form an extra hydrogen bond by assuming a ring structure (Figure 12-6).

In a hydrogen bond the hydrogen ion is usually attached more strongly to one of the two electronegative atoms that it holds together than to the

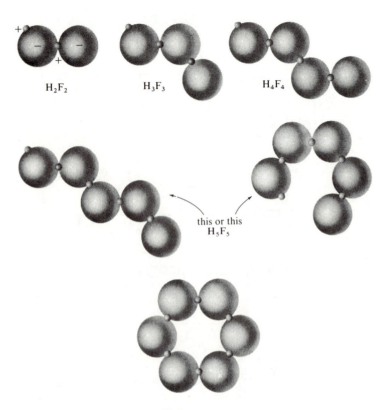

FIGURE 12-6
Some polymers of hydrogen fluoride.

other. The structure of the dimer of hydrogen fluoride may be represented by the formula

$$F^- \!\!-\!\! H^+ \text{-} \text{-} \text{-} F^- \!\!-\!\! H^+$$

in which the dashed line represents the hydrogen bonding.

Because of the electrostatic origin of the hydrogen bond, only the most electronegative atoms—fluorine, oxygen, nitrogen—form these bonds. Usually an unshared electron pair of the attracted atom approaches closely to the attracting hydrogen ion. Water is an especially suitable substance for hydrogen-bond formation, because each molecule has two attached hydrogen atoms and two unshared electron pairs, and hence can form four hydrogen bonds. The tetrahedral arrangement of the shared and unshared electron pairs causes these four bonds to extend in the four tetrahedral directions in space, and leads to the characteristic crystal structure of ice (Figure 12-7). This structure, in which each molecule is surrounded

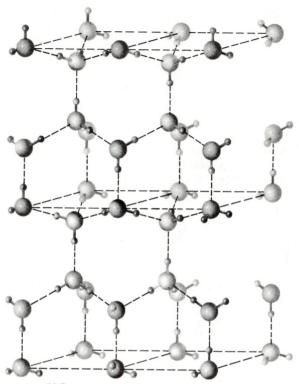

FIGURE 12-7

A small part of a crystal of ice. The molecules above are shown
with approximately their correct size (relative to the interatomic
distances). Note hydrogen bonds, and the open structure that
gives ice its low density. The molecules are indicated
diagrammatically as small spheres for oxygen atoms and still
smaller spheres for hydrogen atoms.

by only four immediate neighbors, is a very open structure, and accordingly ice is a substance with abnormally low density. When ice melts, this
tetrahedral structure is partially destroyed, and the water molecules are
packed more closely together, causing water to have greater density than
ice. Many of the hydrogen bonds remain, however, and aggregates of
molecules with the open tetrahedral structure persist in water at the freezing point. With increase in temperature some of these aggregates break up,
causing a further increase in density of the liquid; only at 4°C does the
normal expansion due to increase in molecular agitation overcome this
effect, and cause water to begin to show the usual decrease in density with
increasing temperature.

The abnormally large dielectric constant of water, which is responsible for the striking power of water to dissolve ionic substances, is due to its power to form hydrogen bonds. Two separate dipole molecules have much less power to neutralize an applied electric field than has the complex of the two, with dipole moment doubled. The only substances with dielectric constants greater than 40, and with great power to dissolve electrolytes, are water, liquid hydrogen fluoride, hydrogen peroxide, and liquid hydrogen cyanide (HCN). All of these substances polymerize through hydrogen-bond formation.

The O—H$\cdots$O distance in ice is 2.76 Å. Neutron diffraction studies of deuterium oxide have given the distances O—H = 1.00 Å and H$\cdots$O = 1.76 Å. In some substances with stronger hydrogen bonds the O—H$\cdots$O distance is as small as 2.40 Å and the hydrogen atom seems to be midway between the two oxygen atoms.

The heat of sublimation of ice is 51 kJ mole^{-1}. We can estimate by the methods described in Section 11-4 that the electronic van der Waals forces contribute only about 11 kJ mole^{-1}, leaving 40 kJ mole^{-1} as the energy of the two hydrogen bonds per molecule. We see that the energy of the O—H$\cdots$O hydrogen bond in ice, 20 kJ mole^{-1}, is only 4.3% of the energy of the O—H covalent bond. It is large enough, however, to permit this structural feature to have an important effect on the properties of water and of other substances.

In the vapor of methanol, CH$_3$OH, there occurs an equilibrium between the monomer and the tetramer, (CH$_3$OH)$_4$, with the structure

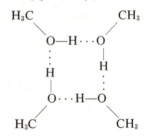

The enthalpy of formation corresponds to the value 26 kJ mole^{-1} for the energy of the hydrogen bond, somewhat larger than for ice.

In general, the energy of a hydrogen bond is (in kJ mole^{-1}) approximately 15 to 20 times the difference in electronegativity of the electronegative atom and hydrogen. This formula gives the range 28 to 38 kJ mole^{-1} for F—H$\cdots$F (observed 28 in (HF)$_6$), 21 to 28 kJ mole^{-1} for O—H$\cdots$O (observed 28 in the dimer of acetic acid), and 13 to 18 kJ mole^{-1} for N—H$\cdots$N (observed 12 to 20 in several substances). Averages of these values apply to hydrogen bonds between different atoms, such as N—H$\cdots$O.

12-5. The Entropy of Ice

The entropy of $H_2O(g)$ can be calculated from the structural constants (moments of inertia and vibrational frequencies) of the molecule, and also from measured values of $\Delta G°$ and $\Delta H°$ of formation from $H_2(g)$ and $O_2(g)$ and the known values of the entropy for these elements. The values of $S°_{273°}$ obtained in these two ways agree with one another; they are, however, larger by 3.40 J deg^{-1} mole^{-1} than the value of $\int_{0°}^{273°} C_p d \ln T$ for ice, calculated from the experimental heat capacity of ice. This comparison shows that ice near 0°K has a residual entropy of about 3.40 J deg^{-1} mole^{-1}. This residual entropy is attributed to an orientational disorder of the molecules in ice, somewhat similar to that of the NNO molecules in a nitrous oxide crystal (Section 10-9).

It is possible to develop a simple theory of this residual entropy. We assume that each water molecule is so oriented that its two hydrogen atoms are directed approximately toward two of the four surrounding oxygen atoms, that only one hydrogen atom lies along each oxygen-oxygen line, and that under ordinary conditions the interaction of non-adjacent molecules is such as not to stabilize appreciably any one of the many configurations satisfying these conditions with reference to the others. Thus we assume that an ice crystal can exist in any one of a large number of configurations, each corresponding to certain orientations of the water molecules. It can change from one configuration to another by rotation of some of the molecules or by motion of some of the hydrogen nuclei, each moving 0.76 Å from a position 1.00 Å from one oxygen atom to the similar position near the other bonded atom. The protons will tend to jump in this way in groups, so as to leave each oxygen atom with two protons attached; ice is so similar to water that we are assured that the concentrations of $(OH)^-$ and $(H_3O)^+$ ions present in ice are very small. It is probable that both processes occur. The fact that at temperatures above about 200°K the dielectric constant of ice is of the order of magnitude of that of water shows that the molecules can reorient themselves with considerable freedom, the crystal changing in the stabilizing presence of the electric field from unpolarized to polarized configurations satisfying the above conditions.

When a crystal of ice is cooled to very low temperatures it is caught in some one of the many possible configurations, but it does not assume (in a reasonable period of time) a uniquely determined configuration with no randomness of molecular orientation. It accordingly retains the residual entropy $k \ln W$, in which k is the Boltzmann constant and W is the number of configurations accessible to the crystal.

Let us now calculate W. In a mole of ice there are $2N$ hydrogen nuclei. If each nucleus had the choice of two positions along its O—O axis, one closer to one oxygen atom and the other closer to the second oxygen atom, there would be 2^{2N} configurations. Many of these configurations are ruled out, however, by the condition that each oxygen atom have two attached hydrogen atoms. Let us consider a particular oxygen atom and the four surrounding hydrogen nuclei. There are 16 arrangements of this OH_4 group; one with all four hydrogen nuclei close to the oxygen atom, corresponding to the ion $(H_4O)^{++}$, four corresponding to $(H_3O)^+$, six to H_2O, four to $(OH)^-$, and one to O^{--}. The acceptable arrangements assigning two strongly bonded hydrogen nuclei to this oxygen atom accordingly comprise six-sixteenths or three-eighths of the total possible arrangements. Of these arrangements, only three-eighths are suitable with respect to the second oxygen atom, and so on; the number of configurations W is hence $2^{2N}(\frac{3}{8})^N = (\frac{3}{2})^N$.

This calculation thus gives the value $k \ln (\frac{3}{2})^N = R \ln \frac{3}{2} = 3.37$ J deg^{-1} mole^{-1} for the residual entropy of ice, in agreement with the experimental value.*

12-6. The Importance of Water as an Electrolytic Solvent

Salts are insoluble in most solvents. Gasoline, benzene, carbon disulfide, carbon tetrachloride, alcohol, ether—these substances are "good solvents" for grease, rubber, organic materials generally; but they do not dissolve salts.

The reasons that water is so effective in dissolving salts are that *it has a very high dielectric constant* (about 80 at room temperature) and *its molecules tend to combine with ions, to form hydrated ions*. Both of these properties are related to the large electric dipole moment of the water molecule.

The force of attraction or repulsion of electric charges is inversely proportional to the dielectric constant of the medium surrounding the charges. This means that two opposite electric charges in water attract each other with a force only $\frac{1}{80}$ as strong as in air (or a vacuum). It is clear that the ions of a crystal of sodium chloride placed in water could dissociate away from the crystal far more easily than if the crystal were in air, since the electrostatic force bringing an ion back to the surface of the crystal from the aqueous solution is only $\frac{1}{80}$ as strong as from air. It is accordingly not surprising that the thermal agitation of the ions in a salt crystal at room temperature is not great enough to cause the ions to dissociate away into the air, but that it is great enough to overcome the

---

*The calculation given in the text is only approximately correct; a more careful counting of the acceptable arrangements leads to $R \ln 1.5068$ for the residual entropy of ice.

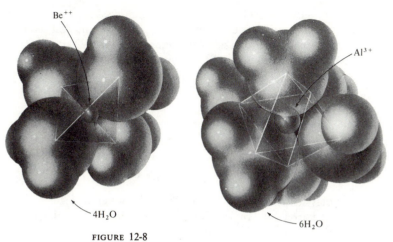

FIGURE 12-8
Diagrams showing the structure of hydrated ions.

relatively weak attraction when the crystal is surrounded by water, thus allowing large numbers of the ions to dissociate into aqueous solution.

The Hydration of an Ion

A related effect that stabilizes the dissolved ions is the formation of *hydrates* of the ions. Each negative ion attracts the positive ends of the adjacent water molecules, and tends to hold several water molecules attached to itself. The positive ions, which are usually smaller than the negative ions, show this effect still more strongly; each positive ion attracts the negative ends of the water molecules, and binds several molecules tightly about itself, forming a hydrate, which may have considerable stability, especially for the bipositive and terpositive cations.

The number of water molecules attached to a cation, its *ligancy*,* is determined by the size of the cation. The small cation Be^{++} forms the tetrahydrate† $Be(OH_2)_4^{++}$. A somewhat larger ion, such as Mg^{++} or Al^{+++}, forms a hexahydrate $Mg(OH_2)_6^{++}$ or $Al(OH_2)_6^{+++}$ (Figure 12-8).

The forces between cations and water molecules are so strong that the ions often retain a layer of water molecules in crystals. This water is called

---

*The ligancy of an atom is the number of atoms bonded to it or in contact with it. The ligancy was formerly called the *coordination number*.

†In these formulas water is written OH_2 instead of H_2O, to indicate that the oxygen atom of the water molecule is near the metal ion, the hydrogen atoms being on the outside. Usually the formulas are written $Be(H_2O)_4^{++}$, and so on.

water of crystallization. This effect is more pronounced for bipositive and terpositive ions than for unipositive ions. The tetrahedral complex $Be(H_2O)_4^{++}$ occurs in various salts, including $BeCO_3 \cdot 4H_2O$, $BeCl_2 \cdot 4H_2O$, and $BeSO_4 \cdot 4H_2O$, and is no doubt present also in solution. The following salts contain larger ions, with six water molecules in octahedral coordination:

$$MgCl_2 \cdot 6H_2O \qquad\qquad AlCl_3 \cdot 6H_2O$$
$$Mg(ClO_3)_2 \cdot 6H_2O \qquad KAl(SO_4)_2 \cdot 12H_2O$$
$$Mg(ClO_4)_2 \cdot 6H_2O \qquad Fe(NH_4)_2(SO_4)_2 \cdot 6H_2O$$
$$MgSiF_6 \cdot 6H_2O \qquad\quad Fe(NO_3)_2 \cdot 6H_2O$$
$$NiSnCl_6 \cdot 6H_2O \qquad\quad FeCl_3 \cdot 6H_2O$$

In a crystal such as $FeSO_4 \cdot 7H_2O$, six of the water molecules are attached to the iron ion, in the complex $Fe(OH_2)_6^{++}$, and the seventh occupies another position, being packed near a sulfate ion of the crystal. In alum, $KAl(SO_4)_2 \cdot 12H_2O$, six of the twelve water molecules are coordinated about the aluminum ion and the other six about the potassium ion.

Crystals also exist in which some or all of the water molecules have been removed from the cations. For example, magnesium sulfate forms the three crystalline compounds $MgSO_4 \cdot 7H_2O$, $MgSO_4 \cdot H_2O$, and $MgSO_4$.

Clathrate Compounds

The argonons, simple hydrocarbons, and many other substances form crystalline hydrates; for example, xenon forms the hydrate $Xe \cdot 5\frac{3}{4}H_2O$, stable at about 2°C and partial pressure of xenon 1 atm, and methane forms a similar hydrate, $CH_4 \cdot 5\frac{3}{4}H_2O$. X-ray investigation has shown that these crystals have a structure in which the water molecules form a hydrogen-bonded framework, resembling that of ice in that each water molecule is tetrahedrally surrounded by four others, at 2.76 Å, but with a more open arrangement, such as to provide cavities (pentagonal dodecahedra and other polyhedra with pentagonal or hexagonal faces) that are big enough to permit occupancy by the noble gas atoms or other molecules. Crystals of this sort are called clathrate crystals.

The structure of xenon hydrate and the hydrates of argon, krypton, methane, chlorine, bromine, hydrogen sulfide, and some other substances is shown in Figure 12-9. The cubic unit of structure has edge about 12 Å and contains 46 water molecules. Chloroform hydrate, $CHCl_3 \cdot 17H_2O$, has a somewhat more complicated structure, in which the chloroform molecule is surrounded by a 16-sided polyhedron formed by 28 water molecules.

Clathrate compounds also can be made in which the hydrogen-bonded framework is formed by organic molecules such as urea, $(H_2N)_2CO$.

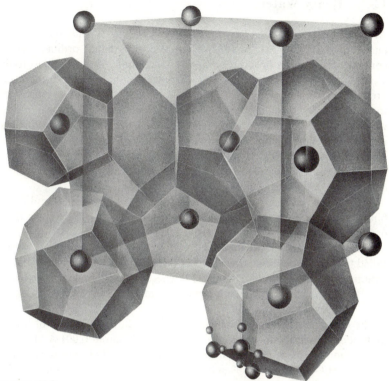

FIGURE 12-9

The structure of a clathrate crystal, xenon hydrate. The xenon atoms occupy cavities (eight per unit cube) in a hydrogen-bonded three-dimensional network formed by the water molecules (46 per unit cube). The $O—H \cdots O$ distance is 2.76 Å, as in ice. Two xenon atoms, at 0, 0, 0, and $\frac{1}{2}$, $\frac{1}{2}$, $\frac{1}{2}$, are at the centers of nearly regular pentagonal dodecahedra. The other six, at 0, $\frac{1}{4}$, $\frac{1}{2}$; 0, $\frac{3}{4}$, $\frac{1}{2}$; $\frac{1}{2}$, 0, $\frac{1}{4}$; $\frac{1}{2}$, 0, $\frac{3}{4}$; $\frac{1}{4}$, $\frac{1}{2}$, 0; and $\frac{3}{4}$, $\frac{1}{2}$, 0, are at the centers of tetrakaidecahedra. Each tetrakaidecahedron (one is outlined, right center) has 24 corners (water molecules), two hexagonal faces, and 12 pentagonal faces.

Other Electrolytic Solvents

Some liquids other than water can serve as ionizing solvents, with the power of dissolving electrolytes to give electrically conducting solutions. These liquids include hydrogen peroxide, hydrogen fluoride, liquid ammonia, and hydrogen cyanide. All of these liquids, like water, have large dielectric constants. Liquids with low dielectric constants, such as benzene and carbon disulfide, do not act as ionizing solvents.

Liquids with large dielectric constants are sometimes called *dipolar liquids* (or simply *polar liquids*).

12-7. Heavy Water

After the discovery of the heavy isotopes of oxygen, ^{17}O and ^{18}O, in 1929, and deuterium, $D = {}^2H$, in 1932, it was recognized that ordinary water consists of molecules of several different kinds, built out of these isotopic atoms in various ways. Since these molecules have almost identical properties except for mass, the density of a sample of water is proportional to the average molecular weight of the molecules in it. If the sample of water consisted of ordinary oxygen combined only with deuterium, its molecular weight would be 20 instead of 18, and its density would accordingly be over 10% greater than that of ordinary water. The term *heavy water* is used to refer to this form of water, which may also be called *deuterium oxide*.

Still heavier water can be made, by isolating the isotope ^{18}O and combining it with deuterium. This water has density about 20% greater than ordinary water.

There is, in fact, a still heavier form of water. The isotope $T = {}^3H$, called tritium, is a radioactive substance with half-life 12.4 years. Ordinary tritium oxide has molecular weight 22, whereas water made from tritium and ^{18}O would have molecular weight 24, and would be over 30% denser than ordinary water.

The bond lengths and bond angles of the different kinds of water molecules are nearly the same (the same to within about 0.001 Å in r_e and 0.1° in bond angle). The vibrational frequencies of D_2O are lower than those of H_2O by about $2^{-1/2} = 0.707$. Values of the translational, rotational, and vibrational entropy are all larger for D_2O than for H_2O. At ordinary temperatures an important contribution to the difference in ΔH of reactions of deuterium compounds and those of protium compounds is the zero-point vibration energy, $\frac{1}{2}h\nu$.

The density of D_2O at 20°C is 1.1059 g cm^{-3}, its freezing point is 3.82°C, its boiling point is 101.42°C, and its temperature of maximum density is 11.6°C.

12-8. Deviation of Water and Some Other Liquids from Hildebrand's Rule

Many substances with polyatomic molecules have normal values of the entropy of vaporization (Table 11-6). This fact is interpreted as showing that the molecules are about as free to assume various orientations in the liquid as in the gas phase, where they are completely free. Substances whose molecules differ from a roughly spherical shape are, however, found to have values of the entropy of vaporization larger than the Hildebrand

TABLE 12-1

Values of Entropy of Vaporization Deviating from Hildebrand's Rule

Substance	Boiling Point		Molar Heat of Vaporization	Molar Entropy of Vaporization (gas at standard volume)
	At 1 atm	At Standard Volume		
H_2O	373.2°K	383°K	40.7 kJ mole^{-1}	106 J deg^{-1} mole^{-1}
H_2O_2	423	435	54.4	125
CH_3OH	338	344	35.3	103
C_2H_5OH	352	358	38.6	108
$(CH_2OH)_2$	470	489	56.9	116
HNO_3	353	360	39.5	110
NH_3	240	237	23.4	99
C_2N_2	252	250	23.4	94
C_2H_2	189	184	17.6	96

value 85 J deg^{-1} mole^{-1}. Two examples, cyanogen ($:N\equiv C—C\equiv N:$) and acetylene ($H—C\equiv C—H$), are given in Table 12-1. Acetylene has entropy of vaporization 10.5 J deg^{-1} mole^{-1} greater than the normal value. If this excess is attributed to the restriction in freedom of orientation of the rodlike molecules by their neighbors in the liquid, the solid angle accessible to the axis of the average molecule is calculated to be 30% of the value 4π corresponding to complete freedom of orientation.

Substances whose molecules form hydrogen bonds have large values of the entropy of vaporization. For water, hydrogen peroxide, methanol, ethanol, ethylene glycol, and nitric acid the excess over the Hildebrand value ranges from 17 to 40 J deg^{-1} mole^{-1} (Table 12-1). These values correspond to restriction in orientation by the factors 0.08 to 0.01; the accessible solid angles to which the molecules are restrained by the hydrogen bonds in the liquids are only 8% to 1% of the free-orientation values.

12-9.   The Dense Forms of Ice

Under pressure, the open structure of ordinary ice (ice I) becomes unstable, and several denser structures take its place, such as ice II, shown in Figure 12-10. Phases of ice that occur at high pressure are identified in Figure 12-11. In the dense ice phases each water molecule forms hydrogen bonds with four near neighbors, but (except in ice VII and ice VIII) the neighbors lie at the corners of a tetrahedron that is considerably distorted from the ideal configuration seen in ice I. This distortion allows one or

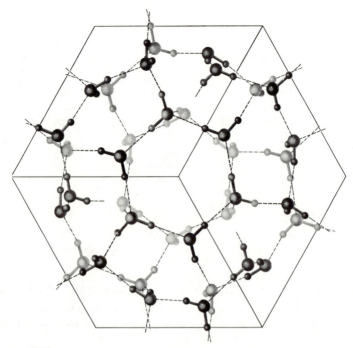

FIGURE 12-10

The structure of ice II, viewed along a hexagonal axis. The structure is more compact than that of ice I. Each water molecule has four nearest neighbors, at about 2.8 Å, to which it is attached by bent hydrogen bonds, and one other near neighbor, at 3.24 Å.

more additional neighbors to approach almost as closely as the four bonded neighbors, thus increasing the density of molecular packing. Distortion from the ideal tetrahedral geometry occurs by bending of the hydrogen bonds, which requires additional energy, and in consequence these dense forms of ice are unstable relative to ice I at low pressures. Under high pressure the extra energy is compensated by the work done in the compression from ice I to the denser phases, thus stabilizing them. The needed energy is offset to some extent by increased van der Waals attraction in the denser phases, resulting from the generally shorter interatomic distances that occur (Equation 11-17).

An important difference between the ice structures in Figures 12-7 and 12-10 is that in ice II each water molecule is constrained to one definite orientation in the crystal, whereas in ice I the molecule can assume any of the six orientations that allow hydrogen bonds to be formed with its four neighbors. This type of randomness or disorder in the water-molecule

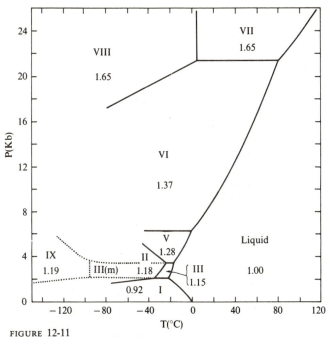

FIGURE 12-11

Phase diagram of water, showing stability fields of the various forms of
ice (roman numerals) and of liquid water. For each phase, the approximate density (in g cm⁻³) at the low-pressure limit of stability is given.
III(m) represents ice III cooled metastably into the stability field of ice II.
The pressure is in kilobars.

orientations in ice I contributes 3.37 J deg⁻¹ mole⁻¹ to the entropy of the
crystal (Section 12-5).

The entropy of disorder of ice I is observed directly in the ice II ⟶
ice I transition, for which there is an observed entropy increase of 3.2 J
deg⁻¹ mole⁻¹. Ices I, III, V, VI, and VII have disordered water-molecule
orientations, whereas ices II, VIII, and IX are ordered. The state of order
or disorder has a great effect on many properties of the ice phases.

Ices III, V, and VI have structures resembling that of ice II in that each
water molecule forms four bent hydrogen bonds and has other water
molecules closer than the next-nearest neighbors in ice I, which are at
4.51 Å. Ice Ic, which is slightly less stable than ice I at all temperatures and
pressures, is the cubic analogue of ice I; the oxygen atoms are in the
positions occupied by carbon atoms in diamond. Ice VII (with protons
disordered) and ice VIII (with protons ordered) have a structure of two
hydrogen-bonded ice-Ic frameworks interpenetrating one another. The

TABLE 12-2
Properties of Forms of Ice

Form	Density 1 atm, 110°K	$H° - H°(I)$ 1 atm, 0°C	$S° - S°(I)$ 1 atm, 0°C	Average* H-bond length 1 atm, 110°K	$\Delta\theta$† r.m.s.
I	0.94 g cm⁻³	0.00 kJ mole⁻¹	0.00 J deg⁻¹ mole⁻¹	2.75 Å	0°
Ic	0.94	0.08	<0.1	2.75	0°
II	1.18	0.04	−3.22	2.80	17°
III	1.15	1.00	1.09	2.78	16°
IX	1.16	0.40	−2.29	2.77	16°
V	1.23	1.30	0.84	2.80	18°
VI	1.31	1.72	0.79	2.81	23°
VII	1.50	3.85	0.46	2.95	0°
VIII	1.50	2.76	−3.43	2.95	0°
$H_2O(l)$	1.00 (0°C)	6.02	22.1	2.84 (0°C)	

*The values at 0°C are 0.01 Å greater.
†Root-mean-square deviation from the tetrahedral value, 109.5°, of the angles between pairs of hydrogen bonds.

two frameworks fit tightly, and van der Waals repulsion of the atoms increases the hydrogen-bond lengths to 2.95 Å.

Ice IX is the low-temperature form of ice III, with protons ordered.* Some of the properties of the various forms of ice are given in Table 12-2.

The Structure of Liquid Water

The enthalpy change $\Delta H° = 6.0$ kJ mole⁻¹ for the transition ice I $\longrightarrow$ $H_2O(l)$ can be ascribed to the breaking of some hydrogen bonds or to their bending (as in ice II, for example). It is likely that both structural features occur in liquid water. The x-ray diffraction pattern of liquid water at 4°C agrees closely with that calculated for a mixture of microcrystals of ice I, ice II, and ice III in the ratios 50%:33%:17%. A possible description of water is that the environment of each molecule shifts from that for ice I (for tetrahedrally-directed unstrained hydrogen bonds) to those for ice II and ice III (four bent hydrogen bonds, with non-hydrogen-

---

*The name ice IV was assigned by P. W. Bridgman, who investigated the phase diagram of H_2O in 1912 and that of D_2O in 1935, to a phase in the ice-V field of Figure 12-11. Later investigators have had difficulty in obtaining this phase for x-ray investigation.

bonded neighbors at about 3.5 Å and 4.5 Å, as in ice I and ice II), as well as occasionally to structures with only three hydrogen bonds. The density of liquid water is intermediate between that of ice I and those of ice II and ice III.

12-10.   The Phase Diagram of Water

Figure 12-11 is the phase diagram for water substance, as determined by experiment. It shows the range of conditions of pressure and temperature under which liquid water and the various solid phases of ice are stable. The phase rule (Section 11-9) indicates that when only a single phase is present in a one-component system, such as water substance, the variance of the system is 2; hence the pressure and temperature can be varied independently. This is the case within the single-phase fields outlined in Figure 12-11. When a second phase appears the variance is reduced to 1, which holds along the two-phase equilibrium curves bounding the phase fields in the figure. In the presence of a third phase the variance is reduced to 0, and both the pressure and temperature are fixed. In the phase diagram this occurs where three equilibrium curves meet at a point, called a triple point.

The equilibrium curves in Figure 12-11 provide thermodynamic information on the phases, by means of the general Clausius-Clapeyron equation. This equation states that on the equilibrium curve between phases A and B the slope of the curve, dP/dT, is given by

$$\frac{dP}{dT} = \frac{\Delta S^\circ}{\Delta V} \tag{12-1}$$

where ΔS is the molar entropy difference ($\Delta S = S_B - S_A$) and ΔV is the molar volume difference ($\Delta V = V_B - V_A$) between the two phases. (Compare with Equation 11-15, which applies to an equilibrium between a perfect gas and a condensed phase, $\Delta S = \Delta H/T$, $\Delta V \cong V(\text{gas}) = RT/P$.) When the equilibrium curve is known, a measurement of the change in density between two phases thus gives their entropy difference. This value in turn fixes the enthalpy difference ΔH of the phase change, because $\Delta H = T\Delta S$ in the case of a reversible transformation at fixed pressure. The determination of ΔH can accordingly be made without any calorimetric measurements. Finally, the energy difference between the phases can be calculated from the relation $\Delta E = \Delta H - P\Delta V$. For the ice phases this energy difference is related to the hydrogen-bond bending energy discussed in Section 12-9.

When the enthalpy of a phase transformation is known, as in the melting of ice I to water, Equation 12-1 gives information on the relative densities of the phases. Thus ice I melts with an increase in density ($\Delta V < 0$), and the slope dP/dT of the melting curve is negative; that is, the melting point decreases with pressure (see Figure 12-11). The melting curves of the dense ice phases all have positive slopes ($dP/dT > 0$), showing that all of these phases melt with decrease in density, as is normal.

The above interpretation of the slope of the melting curve is valid for equilibrium curves in general. The phase on the higher-temperature side of an equilibrium curve always has a higher entropy than the phase on the lower temperature side, and the phase on the higher-pressure side of an equilibrium curve always has a smaller molar volume. These statements are connected by Equation 12-1, and the validity of either one is a consequence of the second law of thermodynamics. Both statements are examples of Le Chatelier's principle: heating a system promotes reactions that absorb heat (with increase in entropy), and compressing a system promotes reactions that tend to reduce the pressure (by decreasing the volume). Examples of phase changes with a variety of different thermodynamic relationships can be found in Figure 12-11: for example, the phase change ice VI $\longrightarrow$ ice VIII has large ΔV and small ΔS, and the change VIII $\longrightarrow$ VII has large ΔS and practically zero ΔV. These relations are reflected in the slopes of the respective equilibrium lines.

Exercises

12-1. Write the fundamental chemical equations for the softening of water by a zeolite, and the regeneration of the zeolite.

12-2. Write the fundamental chemical equations for the removal of most of the ionic impurities in water by the "ion-exchange" process. Why do you suppose this process is sometimes preferred to distillation by factories for the preparation of moderately pure water? What do you think is the simplest method of determining when the absorbers in tanks A and B of Figure 12-1 are saturated with ions and should be regenerated?

12-3. By reference to Figure 12-3, estimate the melting points and boiling points that hydrogen fluoride, water, and ammonia would be expected to have if these substances did not form hydrogen bonds. What would you expect the relative density of ice and water to be if hydrogen bonds were not formed?

12-4. The value of r_e for OH(g), determined by analysis of its band spectrum, is 0.9706 Å, and the value of the vibrational frequency is 3735 cm^{-1}. What is the value of the zero-point vibrational energy (the energy of the molecule in the state $v = 0$, relative to the bottom of the electronic energy curve; see Sections 10-13 and 10-14)? What values of the vibrational frequency and zero-point energy would you calculate for OD?

12-5. The mutual electrostatic energy of two electric charges ϵ_1 and ϵ_2 at the points x_1 and x_2 ($x_2 > x_1$) along the x axis is $V = \epsilon_1\epsilon_2/(x_2 - x_1)$. (With ϵ_1 and ϵ_2 in stoneys and $x_2 - x_1$ in meters, the energy is in joules.) Note that the partial differential of this quantity with respect to x_2 is the mutual energy of the charge ϵ_1 at x_1 and a charge ϵ_2 at $x_2 + \delta x_2$ and of ϵ_1 at x_1 and a charge $-\epsilon_2$ at x_2. It is accordingly equal to the mutual energy of ϵ_1 at x_1 and an electric dipole with moment $\mu = \epsilon_2 \, \delta x_2$ lying along the x axis with its negative end toward ϵ_1. Evaluate this mutual energy. (Answer:

$$V(\epsilon_1, \mu = \epsilon_2\delta x_2) = \frac{\partial V(\epsilon_1, \epsilon_2)}{\partial x_2} \, \delta x_2 = -\frac{\epsilon_1\epsilon_2\delta x_2}{(x_2 - x_1)^2} = -\epsilon_1\mu_2/r^2,$$

with $r = x_2 - x_1$.)

12-6. By the method of the preceding Exercise evaluate the mutual electrostatic energy of two dipoles μ_1 and μ_2 the distance r apart, pointing along an axis with the positive end of one directed toward the negative end of the other. (Answer: $V(\mu_1, \mu_2) = -2\mu_1\mu_2/r^3$.)

12-7. By use of the expression found in Exercise 12-6 calculate the mutual electrostatic energy of the two HF dipoles in the molecule HF$\cdots$HF. The fluorine-fluorine distance is 2.55 Å and the dipole moment of HF is 0.398 ϵ Å. (Answer: -26.5 kJ mole^{-1}.)

12-8. Corresponding potassium and ammonium salts are often isomorphous. An exception is ammonium fluoride, which forms hexagonal crystals with a structure like that of ice, with ammonium ions and fluoride ions alternating in the oxygen positions (Figure 12-7), whereas KF(c) has the sodium chloride structure (Figure 6-18). What is the explanation of this deviation from isomorphism?

12-9. The density of ammonium fluoride is 1.009 g cm^{-3}. What is the length of the N—H$\cdots$F hydrogen bond in this crystal? (Note that the hexagonal crystal, resembling ice, has the same molar volume as a cubic crystal resembling diamond, with NH_4^+ and F^- alternating in the C positions and with the same bond length.) (Answer: 2.71 Å.)

12-10. Ammonium fluoride is one of the very few substances with an appreciable solubility in ice (that is, with the power of forming a crystalline solution with ice). Can you explain this unusual property?

13

The Properties of Solutions

One of the most striking properties of water is its ability to dissolve many substances, forming *aqueous solutions*. Solutions are very important kinds of matter—important for industry and for life. The ocean is an aqueous solution that contains thousands of components: ions of the metals and nonmetals, complex inorganic ions, many different organic substances. It was in this solution that the first living organisms developed, and from it that they obtained the ions and molecules needed for their growth and life. In the course of time organisms were evolved that could leave this aqueous environment, and move out onto the land and into the air. They achieved this ability by carrying the aqueous solution with them, as tissue fluid, blood plasma, and intracellular fluids containing the necessary supply of ions and molecules.

The properties of solutions have been extensively studied, and it has been found that they can be correlated in large part by some simple laws. These laws and some descriptive information about solutions are discussed in the following sections.

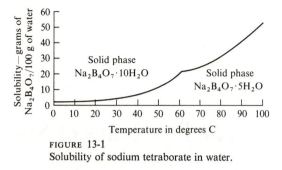

FIGURE 13-1
Solubility of sodium tetraborate in water.

13-1. Types of Solutions. Nomenclature

In Chapter 1 a solution was defined as a homogeneous material that does not have a definite composition.

The most common solutions are liquids. Carbonated water, for example, is a *liquid solution* of carbon dioxide in water. Air is a *gaseous solution* of nitrogen, oxygen, carbon dioxide, water vapor, and the argonons. Coinage silver is a *solid solution* or *crystalline solution* of silver and copper. The structure of this crystalline solution is like that of crystalline copper, described in Chapter 2. The atoms are arranged in the same regular way, cubic closest packing, but atoms of silver and atoms of copper follow one another in a largely random sequence.

If one component of a solution is present in larger amount than the others, it may be called the *solvent*; the others are called *solutes*.

The concentration of a solute is often expressed as the number of grams per 100 g of solvent or the number of grams per liter of solution. It is often convenient to give the number of gram formula weights per liter of solution (the *formality*), the number of gram molecular weights per liter of solution (the *molarity*), or the number of equivalent weights per liter of solution (the *normality*). Sometimes they are referred to 1000 g of solvent; they are then called the *weight-formality*, *weight-molarity*, and *weight-normality*, respectively.

Example 13-1.   A solution is made by dissolving 64.11 g of $Mg(NO_3)_2 \cdot 6H_2O$ in water enough to bring the volume to 1 liter. Describe the solution.

Answer.   The formula weight of $Mg(NO_3)_2 \cdot 6H_2O$ is 256.43; hence the solution is 0.25 F (0.25 formal) in this substance. The salt is, however, completely ionized in solution, to give magnesium ions Mg^{++} and nitrate ions NO_3^-. The solution is 0.25 M (0.25 molar) in Mg^{++} and 0.50 M in NO_3^-. It is also 0.50 N (0.50 normal) in Mg^{++} and 0.50 N in NO_3^-.

It is worth noting that a 1 M aqueous solution cannot be made up accurately by dissolving one mole of solute in 1 liter of water, because the volume of the solution is in general different from that of the solvent. Nor is it equal to the sum of the volumes of the components; for example, 1 liter of water and 1 liter of alcohol on mixing give 1.93 liters of solution; there occurs a volume contraction of 3.5%. There is no reliable way of predicting the density of a solution; tables of experimental values for important solutions are given in reference books.

13-2. Solubility

An isolated system is in *equilibrium* when its properties, in particular the distribution of components among phases, remain constant with the passage of time.

If the system in equilibrium contains a solution and another phase that is one of the components of the solution in the form of a pure substance, the concentration of that substance in the solution is called the *solubility* of the substance. The solution is called a *saturated solution*.

For example, at 0°C a solution of borax containing 1.3 g of anhydrous sodium tetraborate, $Na_2B_4O_7$, in 100 g of water is in equilibrium with the solid phase $Na_2B_4O_7 \cdot 10H_2O$, sodium tetraborate decahydrate; on standing the system does not change, the composition of the solution remaining constant. The solubility of $Na_2B_4O_7 \cdot 10H_2O$ in water is hence 1.3 g of $Na_2B_4O_7$ per 100 g, or, correcting for the water of hydration, 2.5 g of $Na_2B_4O_7 \cdot 10H_2O$ per 100 g.

Change in the Solid Phase

The solubility of $Na_2B_4O_7 \cdot 10H_2O$ increases rapidly with increasing temperature; at 60°C it is 20.3 g of $Na_2B_4O_7$ per 100 g (Figure 13-1). If the system is heated to 70° and held there for some time, a new phenomenon occurs. A third phase appears, a crystalline phase with composition $Na_2B_4O_7 \cdot 5H_2O$, and the other solid phase disappears. At this temperature the solubility of the decahydrate is greater than that of the pentahydrate; a solution saturated with the decahydrate is supersaturated with respect to the pentahydrate, and will deposit crystals of the pentahydrate.* The process of solution of the unstable phase and crystallization of the stable phase will then continue until none of the unstable phase remains.†

In this case the decahydrate is less soluble than the pentahydrate below

---

*The addition of "seeds" (small crystals of the substance) is sometimes necessary to cause the process of crystallization to begin.

†The third hydrate of sodium tetraborate, kernite, $Na_2B_4O_7 \cdot 4H_2O$, is more soluble than the other phases.

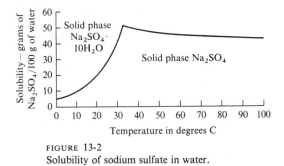

FIGURE 13-2
Solubility of sodium sulfate in water.

61°C, and is hence the stable phase below this temperature. The solubility curves of the two hydrates cross at 61°, the pentahydrate being stable in contact with solution above this temperature.

Change other than solvation may occur in the stable solid phase. Thus rhombic sulfur is less soluble in suitable solvents than is monoclinic sulfur at temperatures below 95.5°C, the transition temperature between the two forms; above this temperature the monoclinic form is the less soluble. The principles of thermodynamics require that the temperature at which the solubility curves of the two forms cross be the same for all solvents, and be also the temperature at which the vapor-pressure curves intersect.

The Dependence of Solubility on Temperature

The solubility of a substance may either increase or decrease with increasing temperature. An interesting case is provided by sodium sulfate. The solubility of $Na_2SO_4 \cdot 10H_2O$ (the stable solid phase below 32.4°C) increases very rapidly with increasing temperature, from 5 g of Na_2SO_4 per 100 g of water at 0° to 55 g at 32.4°C. Above 32.4° the stable solid phase is Na_2SO_4; the solubility of this phase decreases rapidly with increasing temperature, from 55 g at 32.4° to 42 g at 100° (Figure 13-2).

Most salts show increased solubility with increase in temperature; a good number (NaCl, K_2CrO_4) change only slightly in solubility with increase in temperature; and a few, such as Na_2SO_4, $FeSO_4 \cdot H_2O$, and $Na_2CO_3 \cdot H_2O$, show decreased solubility (Figures 13-3 and 13-4).

The thermodynamic discussion of equilibrium in Chapter 11 applies also to solubility. If the partial molar enthalpy of solution of the substance (the change in enthalpy when one mole of the crystalline substance is dissolved in a very large amount of the nearly saturated solution at constant temperature and pressure) is positive, the solubility increases with increasing temperature, and if it is negative the solubility decreases. This is, of course, in agreement with Le Chatelier's principle.

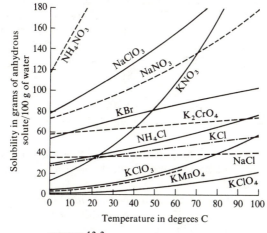

FIGURE 13-3
Solubility curves for some salts in water.

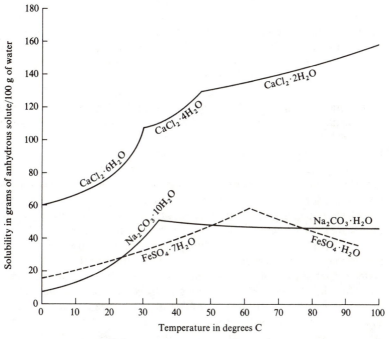

FIGURE 13-4
Solubility curves for salts forming two or three hydrates.

Most salts, corresponding to their positive temperature coefficients of solubility, have positive enthalpies of solution in water. For example, the enthalpy of solution of $Na_2SO_4 \cdot 10H_2O$ in water is 79 kJ per gram formula weight. The formal enthalpy of solution of sodium chloride is 5.4 kJ and that of Na_2SO_4 is -23 kJ.

13-3.   The Dependence of Solubility on the Nature of Solute and Solvent

Substances vary greatly in their solubilities in various solvents. There are a few general rules about solubility, which, however, apply in the main to organic compounds.

One of these rules is that *a substance tends to dissolve in solvents that are chemically similar to it.* For example, the hydrocarbon naphthalene, $C_{10}H_8$, has a high solubility in gasoline, which is a mixture of hydrocarbons; it has a somewhat smaller solubility in ethyl alcohol, C_2H_5OH, whose molecules consist of short hydrocarbon chains with hydroxide groups attached, and a very small solubility in water, which is much different from a hydrocarbon. On the other hand, boric acid, $B(OH)_3$, a hydroxide compound, is moderately soluble in both water and alcohol, which themselves contain hydroxide groups, and is insoluble in gasoline. In fact, the three solvents themselves show the same phenomenon—both gasoline and water are soluble in alcohol, whereas gasoline and water dissolve in each other only in very small amounts.

The explanation of these facts is the following. Hydrocarbon groups (involving only carbon and hydrogen atoms) attract hydrocarbon groups only weakly, as is shown by the low melting and boiling points of hydrocarbons, relative to other substances with similar molecular weights. But hydroxide groups and water molecules show very strong intermolecular attraction; the melting point and boiling point of water are higher than those of any other substance with low molecular weight. This strong attraction is due to the partial ionic character of the O—H bonds, which places electric charges on the atoms. The positively charged hydrogen atoms are then attracted to the negative oxygen atoms of other molecules, forming hydrogen bonds and holding the molecules firmly together. The reason that substances such as gasoline and naphthalene do not dissolve in water is that their molecules in solution would prevent water molecules from forming as many of these strong hydrogen bonds as in pure water; on the other hand, boric acid is soluble in water because the decrease in the number of water-water bonds is compensated by the formation of strong hydrogen bonds between the water molecules and the hydroxide groups of the boric acid molecules.

13-4. Solubility of Salts and Hydroxides

In the study of inorganic chemistry, especially qualitative analysis, it is useful to know the approximate solubility of common substances. The simple rules of solubility are given below. These rules apply to compounds of the common cations Na^+, K^+, NH_4^+, Mg^{++}, Ca^{++}, Sr^{++}, Ba^{++}, Al^{+++}, Cr^{+++}, Mn^{++}, Fe^{++}, Fe^{+++}, Co^{++}, Ni^{++}, Cu^{++}, Zn^{++}, Ag^+, Cd^{++}, Sn^{++}, Hg_2^{++}, Hg^{++}, and Pb^{++}. By "soluble" it is meant that the solubility is more than about 1 g per 100 ml (roughly 0.1 M in the cation), and by "insoluble" that the solubility is less than about 0.1 g per 100 ml (roughly 0.01 M); substances with solubilities within or close to these limits are described as *sparingly soluble*.

Class of mainly soluble substances:

All *nitrates* are soluble.

All *acetates* are soluble.

All *chlorides*, *bromides*, and *iodides* are soluble except those of silver, mercurous mercury (mercury with oxidation number $+1$), and lead. $PbCl_2$ and $PbBr_2$ are sparingly soluble in cold water (1 g per 100 ml at 20°C) and more soluble in hot water (3 g, 5 g, respectively, per 100 ml at 100°C).

All *sulfates* are soluble except those of barium, strontium, and lead. $CaSO_4$, Ag_2SO_4, and Hg_2SO_4 (mercurous sulfate) are sparingly soluble.

All salts of *sodium*, *potassium*, and *ammonium* are soluble except $NaSb(OH)_6$ (sodium antimonate), K_2PtCl_6 (potassium hexachloroplatinate), $(NH_4)_2PtCl_6$, $K_3Co(NO_2)_6$ (potassium cobaltinitrite), and $(NH_4)_3Co(NO_2)_6$.

Class of mainly insoluble substances:

All *hydroxides* are insoluble except those of the alkali metals, ammonium, and barium. $Ca(OH)_2$ and $Sr(OH)_2$ are sparingly soluble.

All normal *carbonates* and *phosphates* are insoluble except those of the alkali metals and ammonium. Many hydrogen carbonates and phosphates, such as $Ca(HCO_3)_2$, $Ca(H_2PO_4)_2$, etc., are soluble.

All *sulfides* except those of the alkali metals, ammonium, and the alkaline-earth metals are insoluble.*

---

*The sulfides of aluminum and chromium are hydrolyzed by water, precipitating $Al(OH)_3$ and $Cr(OH)_3$.

13-5.   The Solubility-Product Principle

In many cases the solubility of a substance is not changed very much by the addition of other substances (in small concentrations) to the solution. For example, ordinarily a nonelectrolytic (nonionizing) solute, such as sugar or iodine, has little effect on the solubility of a salt in water, and conversely a salt such as sodium nitrate has little effect on the solubility of iodine in water. Also the presence of a salt that has no ion in common with another salt whose solubility is under consideration produces only a rather small effect on the solubility of the second salt, usually a small increase; this small increase is due to the electrostatic interaction of the ions in the solution, which decreases their activity somewhat, as discussed in Section 13-11.

Sometimes, however, the solubility of a substance is greatly changed by the presence of other solutes. For example, iodine is very much more soluble in a solution containing iodide ion than in pure water, and potassium perchlorate is much less soluble in a solution containing either another potassium salt or another perchlorate than in pure water. The *increased* solubility of iodine is due to the *formation of a complex*, the triiodide ion I_3^-; this phenomenon is discussed in Chapter 19. The other effect, the *decrease* in solubility of a salt by another salt with a *common ion*, is the subject of the following discussion.

In a saturated solution of potassium perchlorate the ion-concentration product is constant (at a given temperature):

$$[K^+][ClO_4^-] = K_{SP} \text{ at saturation}$$

The constant K_{SP} is called the *solubility-product constant* of the salt, or simply its *solubility product*.

The constancy of this product follows from the equilibrium expression

$$\frac{[K^+][ClO_4^-]}{[KClO_4]} = K \tag{13-1}$$

for the ionization reaction

$$KClO_4 = K^+ + ClO_4^-$$

We replace $[KClO_4]$ by a constant, which is justified whenever the solution is in equilibrium with pure crystalline $KClO_4$, *and only then. The solubility-product principle applies only to saturated solutions of the salt.* The product $[K^+][ClO_4^-]$ can, of course, have any value less than K_{SP} for an unsaturated solution.

Potassium perchlorate has at 0°C the solubility 7.5 g/l in water. The ion concentrations $[K^+]$ and $[ClO_4^-]$ are hence both equal to 7.5/138.56 = 0.054 M, and K_{SP} is equal to $0.054^2 = 29.0 \times 10^{-4}$ mole^2/l^2. This value can now be used to calculate the solubility of potassium perchlorate in a solution of another potassium salt or another perchlorate. Its solubility in such a solution is less than in pure water, because of the action of the ion (K^+ or ClO_4^-) already present in the solution. This decrease in solubility is said to result from the *common-ion effect*.

Let us consider, for example, a solution initially 0.2 *F* in KCl. The initial ion concentrations are

$$[K^+] = [Cl^-] = 0.200$$

Now let this solution be saturated with $KClO_4$. If *x* moles of $KClO_4$ dissolve per liter, the final ion concentrations are

$$[K^+] = 0.200 + x$$
$$[Cl^-] = 0.200$$
$$[ClO_4^-] = x$$

The solubility-product principle is then expressed by the equation

$$[K^+][ClO_4^-] = (0.200 + x)x = K_{SP} = 29.0 \times 10^{-4} \text{ mole}^2 \text{ l}^{-2}$$

On solution for *x* this gives *x* = 0.014. Hence $[K^+]$ = 0.214 and $[ClO_4^-]$ = 0.014. The solubility of potassium perchlorate is 0.2 *F* KCl solution is accordingly 0.014 formula weights per liter, which is only about one-fourth of its solubility in pure water.

Values of the solubility product are given for many salts in Table 13-1.

The reliability of calculations made with use of the principle is determined mainly by the total concentration of ions in the solution. In very dilute solutions (0.001 *F*) the calculations are reliable to within about 4%. In more concentrated solutions, in which the activity coefficients of the ions are considerably less than unity, the actual solubilities of salts are usually somewhat larger than calculated. This increased solubility amounts for salts consisting of singly charged ions (such as K^+, Cl^-) to about 10% for 0.01 *F* solutions and about 20% for 0.1 *F* solutions.

Example 13-2.   The solubility of mercurous chloride (calomel), Hg_2Cl_2, is 3.0×10^{-5} g per 100 ml. What is its solubility product? What volume of 0.01 *F* NaCl solution would be needed to dissolve the amount of mercurous chloride which would be dissolved by 1 l of pure water?

Answer.   The solubility is 3.0×10^{-4} g/l or $3.0 \times 10^{-4}/472.1$ = 0.64 $\times$ 10^{-6} formula weight per liter (472.1 is the formula weight of Hg_2Cl_2). The mercurous ion is Hg_2^{++} (Chapter 21), and the molecule Hg_2Cl_2 dissociates into one Hg_2^{++} ion and two Cl^- ions. The ion concentrations are hence

$$[Hg_2^{++}] = 0.64 \times 10^{-6} \text{ mole l}^{-1}$$
$$[Cl^-] = 1.28 \times 10^{-6} \text{ mole l}^{-1}$$

TABLE 13-1
Solubility-Product Constants at Room Temperature ($18°–25°C$)

Halides	K_{SP}	Halides	K_{SP}
AgCl	1.6×10^{-10}	Hg_2I_2*	1×10^{-28}
AgBr	5×10^{-13}	MgF_2	6×10^{-9}
AgI	1×10^{-16}	PbF_2	3.2×10^{-8}
BaF_2	1.7×10^{-6}	$PbCl_2$	1.7×10^{-5}
CaF_2	3.4×10^{-11}	$PbBr_2$	6.3×10^{-6}
CuCl	1×10^{-7}	PbI_2	9×10^{-9}
CuBr	1×10^{-8}	SrF_2	3×10^{-9}
CuI	1×10^{-12}	TlCl	2.0×10^{-4}
Hg_2Cl_2*	1×10^{-18}	TlBr	4×10^{-6}
Hg_2Br_2*	5×10^{-23}	TlI	6×10^{-8}

Carbonates	K_{SP}	Carbonates	K_{SP}
Ag_2CO_3	8×10^{-12}	$FeCO_3$	2×10^{-11}
$BaCO_3$	5×10^{-9}	$MnCO_3$	9×10^{-11}
$CaCO_3$	4.8×10^{-9}	$PbCO_3$	1×10^{-13}
$CuCO_3$	1×10^{-10}	$SrCO_3$	1×10^{-9}

Chromates	K_{SP}	Chromates	K_{SP}
Ag_2CrO_4	1×10^{-12}	$PbCrO_4$	2×10^{-14}
$BaCrO_4$	2×10^{-10}	$SrCrO_4$	3.6×10^{-5}

Hydroxides	K_{SP}	Hydroxides	K_{SP}
$Al(OH)_3$	1×10^{-33}	$Fe(OH)_3$	1×10^{-38}
$Ca(OH)_2$	8×10^{-6}	$Mg(OH)_2$	6×10^{-12}
$Cd(OH)_2$	1×10^{-14}	$Mn(OH)_2$	1×10^{-14}
$Co(OH)_2$	2×10^{-16}	$Ni(OH)_2$	1×10^{-14}
$Cr(OH)_3$	1×10^{-30}	$Pb(OH)_2$	1×10^{-16}
$Cu(OH)_2$	6×10^{-20}	$Sn(OH)_2$	1×10^{-26}
$Fe(OH)_2$	1×10^{-15}	$Zn(OH)_2$	1×10^{-15}

Sulfates	K_{SP}	Sulfates	K_{SP}
Ag_2SO_4	1.2×10^{-5}	Hg_2SO_4*	6×10^{-7}
$BaSO_4$	1×10^{-10}	$PbSO_4$	2×10^{-8}
$CaSO_4 \cdot 2H_2O$	2.4×10^{-5}	$SrSO_4$	2.8×10^{-7}

Sulfides	K_{SP}	Sulfides	K_{SP}
HgS	10^{-54}	ZnS	10^{-24}
CuS	10^{-40}	FeS	10^{-22}
CdS	10^{-28}	CoS†	10^{-21}
PbS	10^{-28}	NiS†	10^{-21}
SnS	10^{-28}	MnS†	10^{-16}

*The solubility-product expressions for mercurous salts involve the concentration $[Hg_2^{++}]$.
†CoS and NiS are probably dimorphous; the less soluble forms, with K_{SP} about 10^{-27}, are not easily precipitated from acid solutions. MnS is dimorphous; the value given is for the usual flesh-colored form, the green form having $K_{SP} = 10^{-22}$.

and the solubility product is

$$K_{SP} = [Hg_2^{++}][Cl^-]^2 = 1.0 \times 10^{-18} \text{ mole}^3 \text{ l}^{-3}$$

In 1 liter of pure water 0.64×10^{-6} formula weight of calomel can be dissolved. In $0.01 F$ NaCl solution $[Cl^-]$ equals 0.01; hence if this solution is saturated with calomel, the mercurous-ion concentration is given by the equation

$$[Hg_2^{++}](0.01)^2 = K_{SP} = 1.0 \times 10^{-18}$$

or

$$[Hg_2^{++}] = 1.0 \times 10^{-14} \text{ mole l}^{-1}$$

The solubility of calomel in this salt solution is hence $1.0 \times 10^{-14}/0.64 \times 10^{-6} = 1.6 \times 10^{-8}$ times smaller than in pure water, and a volume of 0.64×10^8 l (that is, 64 million liters) would be required to dissolve the 0.003 g of calomel that will dissolve in 1 l of pure water.

13-6. The Solubility of Gases in Liquids: Henry's Law

Air is somewhat soluble in water: at room temperature (20°C) one liter of water dissolves 19.0 ml of air at 1 atm pressure. (The amount of dissolved air decreases with increasing temperature.) If the pressure is doubled, the solubility of the air is doubled. This proportionality of the solubility of air to its pressure illustrates *Henry's law,** which may be stated in the following way: *At constant temperature, the partial pressure in the gas phase of one component of a solution is, at equilibrium, proportional to the concentration of the component in the solution, in the region of low concentration.*

This is equivalent to saying that *the solubility of a gas in a liquid is proportional to the partial pressure of the gas.*

Example 13-3.   The solubility of atmospheric nitrogen† in water at 0°C is 23.54 ml l^{-1}, and that of oxygen is 48.89 ml l^{-1}. Air contains 79% N_2 and 21% O_2 by volume. What is the composition of the dissolved air?

Answer.   The solubilities of nitrogen and oxygen at partial pressures 0.79 and 0.21 atm respectively are $0.79 \times 23.54 = 18.60$ ml l^{-1} and $0.21 \times 48.89 = 10.27$ ml l^{-1} respectively. The composition of the dissolved air is hence $\dfrac{18.60}{18.60 + 10.27} = 64.4\%$ nitrogen and $\dfrac{10.27}{18.60 + 10.27} = 35.6\%$ oxygen.

The solubilities of most gases in water are of the order of magnitude of that of air. Exceptions are those gases which combine chemically with water or which dissociate largely into ions, including CO_2 (solubility 1,713 ml l^{-1} at 0°C), H_2S (4,670), and SO_2 and NH_3, which are extremely soluble.

---

*Discovered by the English chemist William Henry (1775–1836).
†98.8% N_2 and 1.2% A.

The Partition of a Solute Between Two Solvents

If a solution of iodine in water is shaken with chloroform, most of the iodine is transferred to the chloroform. The ratio of concentrations of iodine in the two phases, called the *distribution ratio*, is a constant in the range of small concentrations of the solute in each solution. For iodine in chloroform and water the value of this ratio at room temperature is 250; hence whenever a solution of iodine in chloroform is shaken with water or a solution in water is shaken with chloroform until equilibrium is reached, the iodine concentration in the chloroform phase is 250 times that in the water phase.

It is seen on consideration of the various equilibria that the distribution ratio of a solute between two solvents is equal to the ratio of the solubilities of the solute (as a crystalline or liquid phase) in the two solvents, provided that the solubilities are small. Moreover, the distribution ratio of a gaseous solute between two solvents is proportional to the ratio of its solubilities in the two solvents.

The method of shaking a solution with an immiscible solvent is of great use in organic chemistry, especially the chemistry of natural products, for removing one of several solutes from a solution. In inorganic chemistry it is useful in another way—for determining concentrations of particular molecular species. Thus iodine combines with iodide ion to form the triiodide ion: $I_2 + I^- \longrightarrow I_3^-$. The concentration of molecular iodine, I_2, in a solution containing both I_2 and I_3^- can be determined by shaking with chloroform, analyzing the chloroform solution, and dividing by the distribution ratio. (Triiodide ion is not soluble in chloroform.)

13-7. The Freezing Point and Boiling Point of Solutions

It is well known that the freezing point of a solution is lower than that of the pure solvent; for example, in cold climates it is customary to add a solute such as alcohol or glycerol or ethylene glycol to the radiator water of automobiles to keep it from freezing. Freezing-point lowering by the solute also underlies the use of a salt-ice mixture for cooling, as in freezing ice cream; the salt dissolves in the water, making a solution, which is in equilibrium with ice at a temperature below the freezing point of water.

Let us consider what happens as a salt solution (concentration 1 formal, say) is cooled, with solid carbon dioxide, for example. The temperature falls to a little below the freezing point of the solution, $-3.4°C$, and then, as ice begins to form, it rises to this value and remains constant. As ice continues to form, however, the salt concentration of the solution slowly increases, and its freezing point drops. When half of the water has frozen to ice the solution is $2\,F$ in NaCl, and the temperature is $-6.9°C$. Ice

TABLE 13-2

Solvent	Freezing Point	Molar* Freezing-Point Constant
Benzene	5.6°C	4.90°C
Acetic acid	17	3.90
Phenol	40	7.27
Camphor	180	40

*Molar = weight-molar, moles per 1000 g of solvent.

forms, the solution increases in concentration, and the temperature drops until it reaches the value $-21.1°C$. At this temperature the solution becomes saturated with respect to the solute, which begins to crystallize out as the solid phase $NaCl \cdot 2H_2O$ (sodium chloride dihydrate). The system then remains at this temperature, called the *eutectic temperature*, as the solution freezes completely, without change in composition, to form a fine-grained mixture of two solid phases, ice and $NaCl \cdot 2H_2O$. This mixture is called the *eutectic mixture* or *eutectic*.

It is found by experiment that the freezing-point lowering of a dilute solution is proportional to the concentration of the solute. In 1883 the French chemist F. M. Raoult made the very interesting discovery that *the weight-molar freezing-point lowering produced by different solutes is the same for a given solvent*. Thus the following freezing points are observed for 0.1 M solutions of the following solutes in water:

Hydrogen peroxide	H_2O_2	$-0.186°C$
Methanol	CH_3OH	-0.181
Ethanol	C_2H_5OH	-0.183
Dextrose	$C_6H_{12}O_6$	-0.186
Sucrose	$C_{12}H_{22}O_{11}$	-0.188

The *molar freezing-point constant* for water has the value 1.86°C, the freezing point of a solution containing c moles of solute per 1000 g of water being $-1.86\,c$ in degrees C. For other solvents the values of this constant are shown in Table 13-2.

The Determination of Molecular Weight by the Freezing-Point Method

The freezing-point method is a very useful way of determining the molecular weights of substances in solution. Camphor, with its very large constant, is of particular value for the study of organic substances.

TABLE 13-3

Solvent	Boiling Point	Molar Boiling-Point Constant
Water	100°C	0.52°C
Ethyl alcohol	78.5	1.19
Ethyl ether	34.5	2.11
Benzene	79.6	2.65

Example 13-4. The freezing point of a solution of 0.244 g of benzoic acid in 20 g of benzene was observed to be 5.232°C, and that of pure benzene to be 5.478°. What is the molecular weight of benzoic acid in this solution?

Answer. The solution contains $\dfrac{0.244 \times 1,000}{20} = 12.2$ g of benzoic acid per 1,000 g of solvent. The number of moles of solute per 1,000 g of solvent is found from the observed freezing-point lowering 0.246° to be $\dfrac{0.246}{4.90} = 0.0502$. Hence the molecular weight is $\dfrac{12.2}{0.0502} = 243$. The explanation of this high value (the formula weight for benzoic acid, C_6H_5COOH, being 122.05) is that in this solvent the substance forms double molecules, $(C_6H_5COOH)_2$.

Evidence for Electrolytic Dissociation

One of the strongest arguments advanced by the Swedish chemist Svante Arrhenius in 1887 in support of the theory of electrolytic dissociation was the fact that the freezing-point lowering of salt solutions is much larger than that calculated for undissociated molecules, the observed lowering for a salt such as NaCl or $MgSO_4$ in very dilute solution being just twice as great and for a salt such as Na_2SO_4 or $CaCl_2$ just three times as great as expected.

Elevation of Boiling Point

The boiling point of a solution is higher than that of the pure solvent by an amount proportional to the molar concentration of the solute. Values of the proportionality factor, the *molar boiling-point constant*, are given in Table 13-3 for some important solvents.

Boiling-point data for a solution can be used to obtain the molecular weight of the solute in the same way as freezing-point data.

13-8.   The Vapor Pressure of Solutions: Raoult's Law

It was found experimentally by Raoult in 1887 that the partial pressure of solvent vapor in equilibrium with a dilute solution is directly proportional to the mole fraction of solvent in the solution. It can be expressed by the equation

$$P = P_0 x \qquad (13\text{-}2)$$

in which P is the partial pressure of the solvent above the solution, P_0 is the vapor pressure of the pure solvent, and x is the mole fraction of solvent in the solution.

Determination of Molecular Weight from Vapor Pressure

Raoult's law permits the direct calculation of the effective molecular weight of a solute from data on the solvent vapor pressure of the solution, as shown by the following example.

> **Example 13-5.**   A 10-g sample of an unknown nonvolatile substance is dissolved in 100 g of benzene, C_6H_6. A stream of air is then bubbled through the solution, and the loss in weight of the solution (through saturation of the air with benzene vapor) is determined as 1.205 g. The same volume of air passed through pure benzene at the same temperature caused a loss of 1.273 g. What is the molecular weight of the solute?
>
> **Answer.**   The loss in weight by evaporation of benzene is proportional to the vapor pressure. Hence the mole fraction of benzene in the solution is $1.205/1.273 = 0.947$, and the mole fraction of solute is 0.053. The molecular weight of benzene, C_6H_6, is 78; and the number of moles of benzene in 100 g is $100/78$. Let x be the molecular weight of the solute; the number of moles of solute in the solution (containing 10 g of solute) is $10/x$. Hence
>
> $$\frac{10/x}{100/78} = \frac{0.053}{0.947}$$
>
> or
>
> $$x = \frac{78}{100} \times \frac{0.947}{0.053} \times 10 = 139$$
>
> This is the molecular weight of the substance.

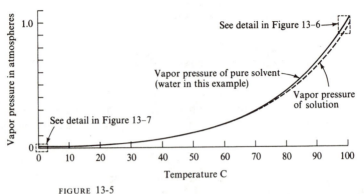

FIGURE 13-5
Vapor-pressure curves of water in the range 0° to 100°C.

The Derivation of Freezing-Point Depression and Boiling-Point Elevation from Raoult's Law

The laws of freezing-point lowering and boiling-point raising can be derived from Raoult's law in the following way. We first consider boiling-point raising. In Figure 13-5 the upper curve represents the vapor pressure of pure solvent as a function of the temperature. The temperature at which this becomes 1 atm is the boiling point of the pure solvent. The lower curve represents the vapor pressure of a solution of a nonvolatile solute; Raoult's law requires that it lie below the curve for the pure solvent by an amount proportional to the molar concentration of solute, and that the same curve apply for all solutes, the molar concentration being the only significant quantity. This curve intersects the 1-atm line at a temperature higher than the boiling point of the solvent by an amount proportional to the molar concentration of the solute (for dilute solutions), as expressed in the boiling-point law (Figure 13-6).

The argument for freezing-point lowering is similar. In Figure 13-7 the vapor-pressure curves of the pure solvent in the crystalline state and the liquid state are shown intersecting at the freezing point of the pure solvent. At higher temperatures the crystal has higher vapor pressure than the liquid, and is hence unstable relative to it, and at lower temperatures the stability relation is reversed. The solution vapor-pressure curve, lying below that of the liquid pure solvent, intersects the crystal curve at a temperature below the melting point of the pure solvent. This is the melting point of the solution.

Note that the assumption is made that the solid phase obtained on freezing the solution is pure solvent; if a crystalline solution is formed, as sometimes occurs, the freezing-point law does not hold.

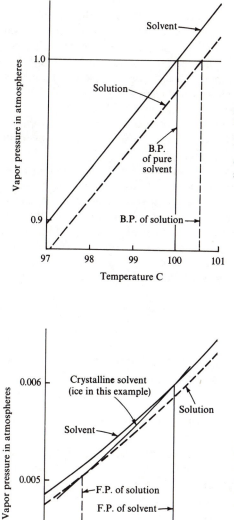

FIGURE 13-6
Vapor-pressure curves of water
and an aqueous solution near the
boiling point, showing elevation
of the boiling point of the
solution.

FIGURE 13-7
Vapor-pressure curves of water,
ice, and an aqueous solution
near the freezing point, showing
depression of freezing point
of the solution.

13-9.   The Osmotic Pressure of Solutions

If red blood corpuscles are placed in pure water they swell, become round, and finally burst. This is the result of the fact that the cell wall is permeable to water but not to some of the solutes of the cell solution (hemoglobin, other proteins); in the effort to reach a condition of equilibrium (equality of water vapor pressure) between the two liquids water enters the cell. If the cell wall were sufficiently strong, equilibrium would be reached when the hydrostatic pressure in the cell had reached a certain value, at which the water vapor pressure of the solution equals the vapor pressure of the pure water outside the cell. This equilibrium hydrostatic pressure is called the *osmotic pressure* of the solution.

A semipermeable membrane is a membrane with very small holes in it, of such a size that molecules of the solvent are able to pass through but molecules of the solute are not. A useful semipermeable membrane for measurement of osmotic pressure is made by precipitating cupric ferrocyanide, $Cu_2Fe(CN)_6$, in the pores of an unglazed porcelain cup, which gives the membrane mechanical support to enable it to withstand high pressures. Accurate measurements have been made in this way to over 250 atm. Cellophane membranes may also be used, if the osmotic pressure is not large.

It is found experimentally that the osmotic pressure of a dilute solution satisfies the equation

$$\pi V = n_1 R T \qquad (13\text{-}3)$$

with n_1 the number of moles of solute (to which the membrane is impermeable) in volume V, π the osmotic pressure, R the gas constant, and T the absolute temperature. This relation was discovered by van't Hoff in 1887. It is striking that the equation is identical in form with the perfect-gas equation; van't Hoff emphasized the similarity of a dissolved substance and a gas.

For inorganic substances and simple organic substances the osmotic-pressure method of determining molecular weight offers no advantages over other methods, such as the measurement of freezing-point lowering. It has, however, been found useful for substances of very high molecular weight; the molecular weight of hemoglobin was first reliably determined in this way by Adair in 1925. The value found by Adair, 68,000, has been verified by measurements made with the ultracentrifuge.

> **Example 13-6.**   (*a*) Horse hemoglobin is a protein in the red cells of the blood that is found on analysis of the dehydrated substance to contain 0.328% iron. What is the minimum molecular weight of horse hemoglobin?

(b) Adair in one experiment found that a solution with 80 g of hemoglobin per liter gave the osmotic pressure $\pi = 0.026$ atm at 4°C. What is the correct molecular weight?

Answer. (a) The smallest possible molecule would contain one iron atom, atomic weight 55.85, and would have molecular weight 55.85/0.00328 = 17,027.

(b) From Equation 13-3, with $\pi = 0.026$ atm, $V = 1$ l, $R = 0.082$ l atm deg^{-1} mole^{-1}, and $T = 277°K$, we obtain

$$n_1 = \frac{0.026}{0.082 \times 277} = 0.001145$$

Hence the molecular weight is 80/0.001145 $\cong$ 70,000. The number of significant figures suggests that the iron analysis is about ten times as accurate as the osmotic-pressure measurement. We conclude from the latter that the molecule contains four atoms of iron, and that the molecular weight is $4 \times 17,027 = 68,100$.

13-10.   The Escaping Tendency and the Chemical Potential

The various laws of solutions described in the foregoing sections were discovered independently. They are, however, closely related to one another and to the second law of thermodynamics, as shown by the following discussion.

A system is in thermal equilibrium with its environment when the transfer of an infinitesimal amount of heat from the system to the environment is accompanied by no change in total entropy. We might say that the *escaping tendency* of heat from the system is then equal to the escaping tendency of heat from its environment (that part of the universe with which it is in thermal contact). The temperature is a sort of potential function for the escaping tendency of heat. We say that two systems have the same temperature when they have equal escaping tendencies for heat.

We may similarly discuss the escaping tendency of a component of a phase from that phase. If the escaping tendencies for a component are equal for two phases, there is equilibrium between the phases with respect to that component. If they are unequal, and the phases are in contact, the component moves from the phase in which it has the greater escaping tendency to that in which it has the smaller.

The escaping tendency of the ith component of a phase can be related to the symbol μ_i, called the *chemical potential* of the ith component in the phase. The chemical potential is defined in such a way (as discussed later in this section) that the total free energy of the phase, G_{total}, is the sum of the chemical potentials of its components, multiplied by their amounts:

$$G_{total} = \sum_i n_i \mu_i \qquad (13\text{-}4)$$

The molar free energy of the phase is then

$$G = \sum_i x_i \mu_i \qquad (13\text{-}5)$$

For a pure perfect gas the chemical potential is seen from Equation 11-11 to be

$$\mu(gas) = \mu^\circ(T) + RT \ln P \qquad (13\text{-}6)$$

(The argument of the logarithmic term, which should have the dimensions of number, may be considered to be $P/1$ atm. $\mu^\circ(T)$ is then the chemical potential at 1 atm.) For a perfect gaseous solution the chemical potential of the ith component is

$$\mu_i(gas) = \mu_i^\circ(T) + RT \ln P_i \qquad (13\text{-}7)$$

The condition for equilibrium between the different states of aggregation of a pure substance, equality of the molar free energy, is equivalent to the condition of equality of the chemical potential. For example, at the triple point water vapor, liquid water, and ice have the same chemical potential, which is related to the vapor pressure at the triple point, 0.00603 atm, by Equation 13-6. For a multicomponent system at equilibrium the value of μ_i for each component must be the same in all the phases.

Let us consider a binary system, with two components, A and B. For a sufficiently dilute solution of A in B the chemical potential of A is given by the equation

$$\mu_A(x_A) = \mu_A^\circ(T) + RT \ln x_A \qquad (13\text{-}8)$$

or

$$\mu_A(c_A) = \mu_A^\circ(T) + RT \ln c_A \qquad (13\text{-}9)$$

Comparison of 13-9 with 13-7, with $\mu_i(gas) = \mu_A(c_A)$, shows that Henry's law, proportionality of the partial pressure of A to its solubility, applies to such a solution. For some solutions (called ideal solutions) Equation 13-8 is closely approximated for all values of x_A, but for others, especially ionic solutions (Section 13-11), large deviations from ideality are found even at very small concentrations.

It is possible to derive Raoult's law, the osmotic-pressure law, and other laws of dilute (or ideal) solutions from Equation 13-8. All of these laws are thermodynamically related to Henry's law, and are valid over the concentration ranges in which Henry's law is valid.

The Chemical Potential and the Partial Molar Free Energy

Let us consider a pure substance, A, in equilibrium with a solution of A and B. Let the solution be composed of n_A moles of A and n_B moles of B, with both n_A and n_B very large. We transfer one mole of A from its pure phase to the solution. The free energy of the pure phase A is decreased by G_A, the molar free energy of A in that phase. The free energy of the solution (which has undergone only a differential change in composition, because the number of moles of solution is very large) is increased by the amount

$$\bar{G}_A = \left(\frac{\partial G_{total}}{\partial n_A} \right)_{P,\, T,\, n_B} \qquad [13\text{-}10]$$

The quantity G_{total} is the free energy of n moles of solution, equal to nG, with G the molar free energy of the solution; $\bar{G}_A$ is called the *partial molar free energy* of the component A in the solution. We see that it satisfies the definition of the chemical potential given earlier in this Section.

The Gibbs-Duhem Equation

An important thermodynamic equation, called the Gibbs equation or the Gibbs-Duhem equation (after J. Willard Gibbs, who derived it in 1878, and the French chemist Pierre Duhem, who rediscovered it in 1887), can be obtained in the following way. Let us consider a system, at constant pressure and temperature, consisting of a large amount of a solution of A (mole fraction x_A) and B (mole fraction $x_B = 1 - x_A$), with molar free energy G. To this solution we add x_A moles of A. The increase in free energy of the system is $x_A \bar{G}_A$. Then we add $1 - x_A$ moles of B, with increase $(1 - x_A)\bar{G}_B$ in the free energy of the system. The total change in nature of the system is that it now has one more mole of solution than initially, and accordingly its free energy has increased by the amount G. Hence we see that, for a binary solution, the molar free energy of the solution is related to the partial molar free energies of the components in the following way:

$$G = x_A \bar{G}_A + (1 - x_A)\bar{G}_B \qquad (13\text{-}11)$$

Let us now consider n moles of solution, with total free energy $G_{total} = nG$. From Equation 13-8 and the relation $n_A = nx_A$ we obtain the equation

$$G_{total} = n_A \bar{G}_A + n_B \bar{G}_B$$

We now differentiate both sides with respect to n_A, with P, T, and n_B constant. The left side of the equation is seen from Equation 13-10 to be just $\bar{G}_A$, and we obtain

$$\bar{G}_A = \bar{G}_A + n_A \frac{\partial \bar{G}_A}{\partial n_A} + n_B \frac{\partial \bar{G}_B}{\partial n_A}$$

and hence, replacing n_A and n_B by x_A and x_B,

$$x_A \frac{\partial \bar{G}_A}{\partial x_A} + (1 - x_A) \frac{\partial \bar{G}_B}{\partial x_A} = 0 \tag{13-12}$$

In this equation $\bar{G}_A$ is μ_A and $\bar{G}_B$ is μ_B. From Equation 13-7 we see that the derivatives with respect to x_A are equal to the derivatives of $\ln P_A$ and $\ln P_B$ (with the common factor RT, which we can cancel from the equation). We are thus led to the Gibbs-Duhem equation

$$x_A \left(\frac{\partial \ln P_A}{\partial x_A} \right)_{P, T} = -x_B \left(\frac{\partial \ln P_B}{\partial x_A} \right)_{P, T}$$

or

$$\frac{x_A}{P_A} \left(\frac{\partial P_A}{\partial x_A} \right)_{P, T} = -\frac{x_B}{P_B} \left(\frac{\partial P_B}{\partial x_A} \right)_{P, T} \tag{13-13}$$

In Figure 13-8 there are shown the experimental curves for the partial pressures of the two components of the solution of 1,2-dibromoethane and 1,2-dibromopropane. Each curve is a straight line: P_A is proportional to x_A and P_B to x_B. Equation 13-13 shows that if this relation holds for A over the whole range $0 \leq x_A \leq 1$ it must also hold for B over the whole range. This liquid is an example of an ideal solution.

A nonideal solution, acetic acid and benzene, is represented in Figure 13-9. Equation 13-13 requires that the slopes of the two curves be in the ratio x_B/x_A when they cross ($P_A = P_B$); measurement of the slopes in the figure shows that this relation is verified. For any value of x_A the slopes are in the ratio $-x_A P_B / x_B P_A$. This means that if one curve is known over the whole range the other curve can be constructed over the whole range, except for a scale factor (which is determined if the vapor pressure of the pure liquid is known).

If P_A is proportional to x_A for small values of x_A (Henry's law), then we may write $P_A = C x_A$ and $P_A/x_A = C$, and we see that the left side of Equation 13-13 is equal to 1. Hence we write (using ordinary differentials)

$$\frac{dP_B}{P_B} = d \ln P_B = \frac{-dx_A}{1 - x_A} = d \ln (1 - x_A)$$

which on integration gives

$$P_B = P_B^\circ (1 - x_A) \tag{13-14}$$

This is Raoult's law. We have thus verified that Raoult's law follows from Henry's law, in consequence of the second law of thermodynamics. The osmotic-pressure law can also be derived in a similar way.

Equation 13-14 requires that a straight line with the slope of P_B in the region near $x_B = 1$ pass through the origin $P_B = 0$, $x_B = 0$. We see from the dashed lines in Figure 13-9 that this holds for acetic acid, but not for

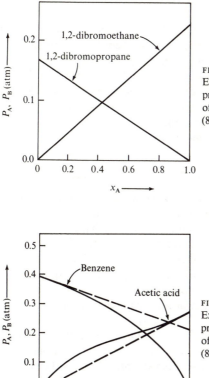

FIGURE 13-8
Experimental curves of partial
pressures of the two components
of a nearly ideal liquid solution
(85°C).

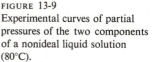

FIGURE 13-9
Experimental curves of partial
pressures of the two components
of a nonideal liquid solution
(80°C).

benzene. This failure of experiment to agree with theory might cause us to doubt the validity of the laws of thermodynamics (and set out on an effort to design a perpetual-motion machine); but there is in fact another explanation. The experimental values for this figure were obtained from an old Landolt-Börnstein reference book. In this reference book the mole fractions are incorrectly calculated. We now know that acetic acid is present in solution in benzene as double molecules, $(CH_3COOH)_2$, and that the correct value of x_A should be calculated with use of molecular weight twice that of CH_3COOH. It was this discrepancy and similar discrepancies in other properties of solutions (boiling point elevation, freezing point depression) that led to the discovery of the dimers of the carboxylic acids.

13-11. The Properties of Ionic Solutions

During the two or three decades after the discovery by Arrhenius of the dissociation of electrolytes into ions the theory was verified by measurement of vapor pressure and other properties of the solutions. It was found that for rather dilute solutions of salts such as NaCl the changes in aqueous vapor pressure, boiling-point elevation, and freezing-point depression were intermediate between those calculated for undissociated NaCl molecules as solute and the twice as large values calculated for $Na^+(aq) + Cl^-(aq)$. The attempt to explain these results by assuming that the molecules and ions are in equilibrium

$$NaCl(aq) \rightleftarrows Na^+(aq) + Cl^-(aq)$$

was not successful; no value of the equilibrium constant that would account for the observed properties could be found. Moreover, it was observed that the absorption spectra of colored salts, such as copper sulfate, remain the same over a great range of concentrations, from those at which other properties indicated essentially complete ionization to those for which very little was indicated. The idea was developed, about 1904, that many salts are completely ionized even in concentrated solution, but that the thermodynamic properties of the ions may be greatly affected by their electrostatic interactions.

In 1913 the British scientist S. R. Milner applied the Boltzmann distribution law to this problem, with some success. A general and comparatively simple method of treating the problem was then discovered by P. Debye and E. Hückel in 1923.

For a salt such as NaCl in solution the chemical potentials μ^+ and μ^- can be related to the concentrations by the equations

$$\mu^+ = \mu^{+\circ} + RT \ln \gamma^+ c^+ \qquad (13\text{-}15a)$$

and

$$\mu^- = \mu^{-\circ} + RT \ln \gamma^- c^- \qquad (13\text{-}15b)$$

In these equations c^+ and c^- are the concentrations of the ions, and γ^+ and γ^- are the *activity coefficients* required to make the equations valid. Since the effects of the cations and anions are not separately determined, the geometric average $\gamma^\pm$ is used:

$$\gamma^\pm = (\gamma^+ \gamma^-)^{1/2}$$

The attraction of oppositely charged ions causes each ion in the solution to have a predominance of oppositely charged ions about it (called its ion-atmosphere). The sodium chloride crystal shows how such an arrangement is possible. We see that the electrical forces decrease the free energy of the solution by an amount that increases with increasing con-

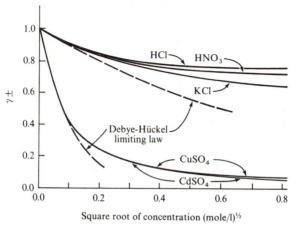

FIGURE 13-10

Values of the activity coefficient $\gamma\pm$ for some uniunivalent and bibivalent electrolytes (25°C).

centration, with the result that $\gamma^\pm$ decreases as c increases. Some experimental values of $\gamma^\pm$ are shown in Figure 13-10. It is seen that uniunivalent electrolytes (HCl, HNO_3, KCl, others not shown) have closely similar activity coefficients, and that bibivalent electrolytes ($CuSO_4$, $CdSO_4$, others not shown) deviate from ideality ($\gamma^\pm = 1$) by much larger amounts, as would be expected from the four-times greater interionic forces.

The Debye-Hückel theory is interesting, but too complicated to be presented in this book. The theory in its simplest form (called the limiting law) gives the following equations:

$$\ln \gamma^\pm = -1.172 \ c^{1/2} \quad \text{(uniunivalent)} \tag{13-16}$$

$$\ln \gamma^\pm = -9.378 \ c^{1/2} \quad \text{(bibivalent)} \tag{13-17}$$

The calculated curves are seen in Figure 13-10 to agree well with experiment at low concentrations.

The general equation for the Debye-Hückel limiting law, with the numerical coefficient for aqueous solutions at 25°C, is

$$\ln \gamma^\pm = -1.172 \ |z_+z_-| \ I^{1/2} \tag{13-18}$$

in which

$$I = \tfrac{1}{2} \sum_i c_i z_i^2 \tag{13-19}$$

Here z_i is the electric charge of the ith ion (for Na_2SO_4, for example, $z_+ = +1$ and $z_- = -2$) and $|z_+z_-|$ is the absolute value of the product of z_+ and z_-. The quantity I, called the *ionic strength* of the solution, is one half the sum of $c_i z_i^2$ over all the ions in solution.

Example 13-7. The solubility product of AgCl is 1.6×10^{-10} mole^2 l^{-2}. Calculate the solubility of AgCl in 0.25 F MgSO$_4$ solution.

Answer. An increase in solubility is expected because the activities of the ions, γ^+ (Ag$^+$) and γ^- (Cl$^-$), are less than the concentrations. The solubility-product equation is

$$\gamma^+[\text{Ag}^+] \, \gamma^-[\text{Cl}^-] = (\gamma^\pm)^2 \, [\text{Ag}^+][\text{Cl}^-] = 1.8 \times 10^{-10}$$

To evaluate $\gamma^\pm$ we first note that the ionic strength I of 0.25 F MgSO$_4$ is equal to 1. Equation 13-18 then gives for Ag$^+$Cl$^-$ (with $z_+z_- = 1$) the value $\ln \gamma^\pm = -1.172$; hence $\log \gamma^\pm = -1.172/2.303 = -0.509$, and $\gamma^\pm = 0.31$. With $x = [\text{Ag}^+] = [\text{Cl}^-]$ the above equation becomes

$$0.31^2x^2 = 1.6 \times 10^{-10}$$
$$x^2 = 16.65 \times 10^{-10}$$
$$x = 4.1 \times 10^{-5} \text{ mole l}^{-1}$$

Hence the solubility of silver chloride is about three times as great in 0.25 F MgSO$_4$ solution as in pure water, because of the decreased activity of the silver ion and chloride ion that results from their electrostatic attraction to the oppositely charged sulfate and magnesium ions in the solution.

The Effective Volume of Ions in Aqueous Solution

When 12.04 g (which is 0.1 gfw) of MgSO$_4$(c) is dissolved in 1 l of water the resulting solution has the volume 0.99982 l. The solution occupies 0.18 ml less volume than the pure water contained in it, and 4.73 ml less volume than the water plus the crystalline magnesium sulfate (the molar volume of MgSO$_4$(c) is 45.3 ml). The sum of the effective molar volumes of Mg^{++}(aq) and SO$_4^{--}$(aq) in very dilute aqueous solutions is -6.4 ml mole^{-1}, which is 51.7 ml mole^{-1} less than the molar volume of MgSO$_4$(c).

It is reasonable for us to expect the ions themselves, Mg^{++} and SO$_4^{--}$, to have about the same space-filling property (the same size) in aqueous solution as in the crystal. We attribute the decrease of 51.7 ml mole^{-1} in molar volume to the assumption by the water molecules of a more close-packed structure in the neighborhood of the ions than in ordinary water, as a response to the attraction of the ions for the water molecules; this decrease in volume is called electrostriction.

The sum of the molar volumes of Na$^+$(aq) and Cl$^-$(aq) in dilute solution is 16.6 ml mole^{-1}, which is 10.4 ml mole^{-1} less than the molar volume for NaCl(c). We see that the decrease in volume is about five times as great for the bibivalent salt as for the uniunivalent salt.

The density of ice II is 1.18 g cm^{-3}, and the conversion of liquid water to ice II is accompanied by a decrease in volume of 2.75 ml mole^{-1}. The observed properties of the solutions of sodium chloride and magnesium sulfate could be explained by assuming that each univalent ion causes two

TABLE 13-4
Molar Volume of Aqueous Ions (in ml mole $^{-1}$)*

H_3O^+	8.4	Li^+	−10.6	Be^{++}	−38	F^-	7.5
OH^-	4.4	Na^+	−8.3	Mg^{++}	−36	Cl^-	27.7
NH_4^+	10.3	K^+	−0.9	Ca^{++}	−35	Br^-	34.6
		Rb^+	5.1	Sr^{++}	−32	I^-	46.3
		Cs^+	11.5	Ba^{++}	−28		
NO_3^-	38	CO_3^{--}	18	Mn^{++}	−40	Al^{+++}	−70
ClO_3^-	48	SO_3^{--}	28	Fe^{++}	−42	Fe^{+++}	−64
ClO_4^-	54	SO_4^{--}	34	Co^{++}	−42		
				Ni^{++}	−42		
				Cu^{++}	−38		
				Zn^{++}	−38		

*Apparent molar volume in very dilute solution at 20°C.

water molecules in its neighborhood and each bivalent ion causes nine to undergo such a change in their relations to other water molecules.

Measurements of the density of dilute salt solutions provide the sums of the molar volumes of the ions present in the solutions. The values of individual ions in Table 13-4 have been obtained by giving $H_3O^+(aq)$ a value 4 ml mole^{-1} greater than OH^-, as estimated from the van der Waals radii of the atoms.

The molar volumes for $H_3O^+(aq)$ and $OH^-(aq)$ may be expected to average 18 ml mole^{-1} (the value for H_2O) plus the effect of electrostriction. The electrostriction term is seen to be −11.6 ml mole^{-1} for these ions.

Thermodynamic Aspects of Ionic Equilibria

We have seen that the entropy term $-T\Delta S$ in the free energy ΔG favors the dissociation of one gas molecule into two (Table 11-6). We might expect a similar effect for ionic solutions. Thus for the reaction

$$H_2O(aq) \rightleftarrows H^+(aq) + OH^-(aq)$$

a positive value of ΔS would be indicated. In fact, the equilibrium constant for the reaction increases tenfold, from 1×10^{-14} to 1×10^{-13} mole^2 l^{-2}, when the temperature is increased from 25°C to 61°C. The rate of increase of K leads to $\Delta H = 55.9$ kJ mole^{-1}, and the value of K at 25°C corresponds to $\Delta G° = 79.9$ kJ mole^{-1}. The difference gives $T\Delta S° = -24$ kJ mole^{-1} and $\Delta S° = -80$ J deg^{-1} mole^{-1}, rather than the similar value with positive sign that would be anticipated from Table 11-6 (somewhat smaller, because the standard state 1 mole l^{-1} for solutes is more condensed than 1 mole in 24 l for gases at 1 atm pressure).

TABLE 13-5

Enthalpy of Hydration of Ions (ΔH in kJ mole^{-1} for $A^{\pm}(g) \longrightarrow A^{\pm}(aq)$

	ΔH	r	$r - R$
NH_4^+	−336	2.04 Å	
H_3O^+	−335	2.03	
OH^-	−335	2.05	
Li^+	−534	1.28	0.68 Å
Na^+	−426	1.61	.66
K^+	−341	2.01	.68
Rb^+	−316	2.17	.69
Cs^+	−284	2.42	.73
F^-	−498	1.38	.02
Cl^-	−363	1.89	.08
Br^-	−328	2.09	.14
I^-	−288	2.38	.22
Be^{++}	−2512	1.09	.78
Mg^{++}	−1963	1.40	.75
Ca^{++}	−1633	1.68	.69
Sr^{++}	−1486	1.85	.72
Ba^{++}	−1344	2.04	.69
Al^{+++}	−4718	1.31	.71

The explanation is that the electrostatic attraction of water molecules by ions that causes the decrease in volume discussed above also imposes on the bound water a rigidity that decreases its entropy. At the melting point the entropy of water is 22 J deg^{-1} mole^{-1} greater than that of ice. If the two ions bound six water molecules effectively enough to impose the same rigidity on them as in ice, the observed negative value of $\Delta S°$ would be accounted for.

The Enthalpy of Hydration of Ions

Experimental values are available for the sums of the enthalpies of hydration of neutral sets of ions ($Na^+ + Cl^-$, $Mg^{++} + 2Cl^-$, and so on). For example, the standard enthalpy of formation of $Na^+(aq) + Cl^-(aq)$ is the sum of the standard enthalpy of formation of NaCl(c) and the enthalpy of solution of the crystal in a large amount of water. The enthalpy of $Na^+(g)$ is the sum of the enthalpy of sublimation of Na(c) and the enthalpy of ionization of Na(g), and the enthalpy of $Cl^-(g)$ is the sum of half the enthalpy of dissociation of $Cl_2(g)$ and the enthalpy of adding an electron. By taking the difference the sum of the enthalpies of hydration for $Na^+(g) + Cl^-(g) \longrightarrow Na^+(aq) + Cl^-(aq)$ is obtained.

Values for individual ions are often tabulated by taking the standard enthalpy of formation of H^+(aq) to be zero. The values in Table 13-5 are based on a different assumption: namely, that the enthalpies of hydration of H_3O^+ and OH^- are equal.

The values of r given in the table are the effective radii of the aqueous ions calculated by a method devised by Max Born in 1920. It is the radius of a charged sphere such that the hydration enthalpy is equal to the difference in electric field energy for a vacuum (dielectric constant $D = 1$) and a homogeneous medium with dielectric constant 80 (the macroscopic value for water) surrounding the sphere. Let us put the charge q on the sphere. The energy required to bring a differential charge dq to the surface of the sphere is qdq/rD. The total energy required to build the charge up from 0 to $z\epsilon$ ($\epsilon = $ the electronic charge in stoneys) is

$$\int_0^{z\epsilon} \frac{qdq}{rD} = \frac{z^2\epsilon^2}{2rD}$$

equal to $z^2\epsilon^2/160r$ for water and $z^2\epsilon^2/2r$ for a vacuum. (The energy of rearranging the charges within the sphere is independent of the dielectric constant outside the sphere, in the case of spherical symmetry.) The enthalpy of hydration thus is equal to $-z^2\epsilon^2(\frac{1}{2} - \frac{1}{160})/r$. For r in Å and ΔH in kJ mole^{-1} the effective radius is $r = 686\, z^2/\Delta H$.

It is interesting that for the five alkali ions, the five alkaline-earth ions, and the aluminum ion the values of r are within ± 0.04 Å of $R + 0.69$ Å, with R the crystal radius, given in Table 6-2. For the halide ions r lies between $R + 0.02$ Å and $R + 0.22$ Å.

13-12.   Colloidal Solutions

It was found by the British chemist Thomas Graham (1805–1869) in the years around 1860 that substances such as glue, gelatin, albumin, and starch diffuse very slowly in solution, their diffusion rates being as small as one-hundredth of those for ordinary solutes (such as salt or sugar). Graham also found that substances of these two types differ markedly in their ability to pass through a membrane such as parchment paper or collodion; if a solution of sugar and glue is put into a collodion bag and the bag is placed in a stream of running water, the sugar soon dialyzes through the bag into the water, and the glue remains behind. This process of dialysis gives a useful method of separating substances of these two kinds.

We now recognize that these differences in ability to pass through the pores of a membrane and in rates of diffusion are due to differences in size of the solute molecules. Graham thought that there was a deeper difference between ordinary, easily crystallizable substances and the slowly diffusing, nondialyzing substances, which he was unable to crystallize. He named the substances of the latter class *colloids* (Greek *kolla*, glue), in contradistinction to ordinary *crystalloids*. It is now known that there is not a sharp line of demarcation between the two classes (many substances of large molecular weight have been crystallized), but it has been found convenient to retain the name "colloid" for substances of large molecular weight.

Some colloids consist of well-defined molecules, with constant molecular weight and definite molecular shape, permitting them to be ordered in a crystalline array. Proteins have molecular weights ranging from about 10,000 to several hundred thousand.

Graham introduced the words *sol* for a colloidal solution (a dispersion of a solid substance in a fluid medium) and *gel* for a dispersion that has developed a structure that prevents it from being mobile. A solution of gelatin in water at high temperatures is a sol and at low temperatures a gel. A *hydrosol* is a dispersion in water, and an *aerosol* is a dispersion of a solid substance in air.

Inorganic sols may be made by dispersing a solid substance that is normally insoluble, such as gold, ferric oxide, and arsenious sulfide, in water. Gold sols, made by adding a reducing agent to a dilute solution of gold chloride, were known to the alchemists of the seventeenth century, and were studied by Michael Faraday. They often have striking colors— ruby red, blue, green, and others—that are the result of diffraction of light by the gold sol particles with dimensions approaching the wavelengths of light. The sols are stabilized by the presence of electric charge on the surface of the particles, a negative charge in the case of gold sols. Faraday found that on addition of a small amount of a salt the ruby gold sols turned blue. This is the result of the aggregation of smaller particles to form larger ones, which scatter light of longer wavelengths. With more salt the particles coagulate. The coagulation results from the action of small ions with opposite charge (Na^+, Mg^{++}, Al^{+++}) attaching themselves to the negative charges on the surface of the gold particles and neutralizing the charge, so as to permit the particles to approach closely enough to form the coagulum. The coagulating power of the cations is approximately proportional to the sixth power of their charge: an aluminum salt is effective as a coagulant at a concentration 700 times less than a sodium salt.

An *emulsion* is a colloidal dispersion of one liquid in another. The most common emulsions are those of oil in water or of water in oil. Emulsions are stabilized by the presence of emulsifying agents, such as soap, proteins, gums, or carbohydrates. A molecule of an effective emulsifying agent may usually be described as having one end soluble in oil and the other end soluble in water. The oil-soluble end may be an alkyl chain, and the water-soluble end an ionic group (carboxylate ion, ammonium ion) or a hydro-gen-bond-forming group, such as hydroxyl. Emulsifying agents such as soap are used for dispersing solid fatty substances in water, as well as liquid oils.

Exercises

13-1. A solution contains 10.00 g of anhydrous cupric sulfate, $CuSO_4$, in 1000 ml of solution. What is the formality of this solution in $CuSO_4$?

13-2. Saturated salt solution (20°C) contains 35.1 g NaCl per 100 g of water. What is its weight-formality? The density of the solution is 1.197 g ml^{-1}. What is its formality?

13-3. A 2 wt F solution of HCl is neutralized with 2 wt F NaOH. What is the weight formality of NaCl in the resulting solution?

13-4. Calculate the mole fraction of each component in the following solutions:
(a) 2.000 g of chloroform, $CHCl_3$, in 10.00 g of carbon tetrachloride, CCl_4.
(b) 1.000 g of acetic acid, $C_2H_4O_2$, in 20.00 g of benzene, recognizing that acetic acid actually exists in benzene solution as the dimer, $(C_2H_4O_2)_2$.

13-5. The density of constant-boiling hydrochloric acid is 1.10 g ml^{-1}. It contains 20.24% HCl. Calculate the weight molarity, the volume molarity, and the mole fraction of HCl in the solution.

13-6. Sodium perchlorate is very soluble in water. What would happen if a solution of about 70 g of $NaClO_4$ in 100 ml of water were to be mixed with a solution of about 40 g of KCl in 100 ml of water, at 20°C? (See Figure 13-3.)

13-7. Make qualitative predictions about the solubility of the following:
(a) Ethyl ether, $C_2H_5OC_2H_5$, in water, alcohol, and benzene.
(b) Hydrogen chloride in water and gasoline.
(c) Ice in liquid hydrogen fluoride and in cooled gasoline.
(d) Sodium tetraborate in water, in ether, and in carbon tetrachloride.
(e) Iodoform, HCI_3, in water and in carbon tetrachloride.
(f) Decane, $C_{10}H_{22}$, in water and in gasoline.

13-8. The density of sodium chloride is 2.16 g ml^{-1}, that of its saturated aqueous solution, containing 366.0 g NaCl per kg of water, is 1.197 g ml^{-1}, and that of a solution containing 333.0 g NaCl per kg of water is 1.180 g ml^{-1}. Would the solubility be increased or decreased by increasing the pressure? Give your calculations.

13-9. By referring to Figures 13-2, 13-3, 13-4, find three salts that on dissolving in a nearly saturated solution give out heat, and three that absorb heat.

13-10. Would heat be evolved or absorbed if some $Na_2CO_3 \cdot 10H_2O$ were dissolved in its nearly saturated aqueous solution at 30°C? If some $Na_2CO_3 \cdot H_2O$ were dissolved in this solution at 30°C?

13-11. What can you say about the heat of solution of common salt? (See Figure 13-3.)

13-12. The solubility of potassium hydrogen sulfate is 51.4 g per 100 g of water at 20°C, and 67.3 g per 100 g at 40°C. If you add some of the salt to a partially saturated solution and stir, will the system become colder or warmer?

13-13. Calculate approximately how much ethanol (C_2H_5OH) would be needed per gallon of radiator water to keep it from freezing at temperatures down to 10°C below the freezing point.

13-14. A solution containing 1 g of aluminum bromide in 100 g of benzene has a freezing point 0.099°C below that of pure benzene. What are the apparent molecular weight and the correct formula of the solute?

13-15. The solubility of nitrogen at 1 atm partial pressure in water at 0°C is 23.54 ml l^{-1}, and that of oxygen is 48.89 ml l^{-1}. Calculate the amount by which the freezing points of air-saturated water and air-free water differ.

13-16. An aqueous solution of amygdalin (a sugarlike substance obtained from almonds) containing 96 g of solute per liter was found to have osmotic pressure 4.74 atm at 0°C. What is the molecular weight of the solute?

13-17. A 3% aqueous solution of gum arabic (simplest formula $C_{12}H_{22}O_{11}$) was found to have an osmotic pressure of 0.0272 atm at 25°C. What are the average molecular weight and degree of polymerization of the solute?

13-18. A solution containing 2.30 g of glycerol in 100 ml of water was found to freeze at −0.465°C. What is the approximate molecular weight of glycerol dissolved in water? The formula of glycerol is $C_3H_5(OH)_3$. What would you predict as to the miscibility of this substance with water (its solubility)?

13-19. When 0.412 g of naphthalene ($C_{10}H_8$) was dissolved in 10.0 g of camphor, the freezing point was found to be 13.0° below that of pure camphor. What is the weight molar freezing-point constant for camphor, calculated from this observation? Can you explain why camphor is frequently used in molecular weight determinations? (Answer: 40.4°.)

13-20. A sample of a substance weighing 1.00 g was dissolved in 8.55 g of camphor, and was found to produce a depression of 9.5°C in the freezing point of the camphor. Using the value of the molar freezing-point constant found in the preceding problem, calculate the molecular weight of the substance.

13-21. The freezing point of ethylene glycol is $-17°C$, yet when mixed with water in a car radiator it can prevent freezing at even lower temperatures. How is this explained?

13-22. Adding alcohol to water lowers the freezing point of water, but also lowers the boiling point. Explain.

13-23. Find the osmotic pressure at $17°C$ of a solution containing 17.5 g of sucrose ($C_{12}H_{22}O_{11}$) in 150 ml of solution. (Answer: 8.1 atm.)

13-24. Considering that the cell wall is an osmotic membrane, explain why a lettuce salad containing salt and vinegar becomes limp in a few hours.

13-25. Why do red blood cells burst when placed in distilled water?

13-26. The value of K_{SP} for $PbCl_2$ is 1.7×10^{-5} mole^3 l^{-3}. What is its solubility in water? in 0.05 F NaCl?

13-27. From the solubility of Ag_2CrO_4, 0.0030 g per 100 ml, calculate its solubility product, and from this its solubility in 0.01 F AgNO$_3$ and in 0.01 F K$_2$CrO$_4$.

13-28. The silver halides have the following solubility products:

$$
\begin{array}{ll}
\text{AgCl} & 1.6 \times 10^{-10} \text{ mole}^2 \text{ l}^{-2} \\
\text{AgBr} & 5.0 \times 10^{-13} \text{ mole}^2 \text{ l}^{-2} \\
\text{AgI} & 1.0 \times 10^{-16} \text{ mole}^2 \text{ l}^{-2}
\end{array}
$$

What would happen if a silver chloride precipitate were allowed to stand in a solution of sodium iodide? If a silver iodide precipitate were allowed to stand in a solution of sodium chloride?

13-29. The values of K_{SP} for AgCl and Ag_2CrO_4 are 1.6×10^{-10} mole^2 l^{-2} and 1.0×10^{-12} mole^3 l^{-3}, respectively. Which will precipitate first when a drop of Ag$^+$ solution is added to a solution which contains 0.1 F NaCl and 0.1 F Na$_2$CrO$_4$? If additional Ag$^+$ is added (say from a 0.1 F solution), when will the other precipitate begin to form? Look up the color of Ag_2CrO_4 in a reference book; suggest a method for the determination of Ag$^+$ or Cl$^-$ on the basis of these calculations.

13-30. Calculate the activity coefficient of Ag$^+$ and I$^-$ ions in 0.1 F MgSO$_4$ solution from the Debye-Hückel limiting law. Compare the calculated solubility of silver iodide in this solution with that in pure water; use $K_{SP} = 1.0 \times 10^{-16}$ mole^2 l^{-2}.

13-31. The solubility of nonpolar gases in a solvent is determined in part by the enthalpy of solution, which is itself roughly proportional to the electric polarizability of the solute molecule. Some values of the solubility in water at $20°C$ are the following, all in cm^3 of gas per liter of solvent: He, 7; Ne, 13; Ar, 46; Kr, 70; Xe, 119. Make a graph of these values against the polarizability (Table 11-4). Use this relation to predict values of the solubility of H_2(g) and CH_4(g) in water.

13-32. The density of CsI(c), which has the structure shown in Figure 6-19, is 4.510 g cm^{-3}. What is its molar volume? Compare with the sum of the apparent molar volumes of Cs$^+$(aq) and I$^-$(aq), Table 13-4. Note that electrostriction is expected to be less important for large ions than for small ions.

13-33. Make a similar calculation for LiF(c), density 1.392 g cm^{-3}. What is the average decrease by electrostriction for Li$^+$(aq) and F$^-$(aq), assuming that the ions themselves in solution occupy the same volumes as in the crystal?

13-34. (*a*) With use of Equation 10-7, calculate the absolute entropy of Li$^+$(g) and that of F$^-$(g) at 25°C and 1 atm, assuming perfect-gas behavior.
(*b*) Make the calculation also for the pressure corresponding to the concentration 1 mole l^{-1}. (Answer, (*b*): 106.33, 118.90 J deg^{-1} mole^{-1}.)

13-35. By integrating the heat capacity of LiF(c) (divided by T) the absolute entropy of the substance at 25°C and 1 atm is found to be 35.86 J deg^{-1} mole^{-1}. The enthalpy of solution of LiF(c) in a large amount of nearly saturated solution, concentration 0.10 mole l^{-1}, is 4.56 kJ mole^{-1}.
(*a*) What is the change in free energy, ΔG, for this change in state of the system?
(*b*) What is the value of ΔS for this change in state?
(*c*) The sum of this quantity and the absolute entropy of LiF(c) may be taken as the entropy of Li$^+$(aq) and F$^-$(aq) at 0.10 M. Correct to the standard concentration 1 M, assuming perfect-solute behavior. (Answer, (*c*): 12.9 J deg^{-1} mole^{-1}.)

13-36. The theoretical difference in translational entropy of Li$^+$(g) and Li$^+$(aq), both at 1 mole l^{-1}, and that of F$^-$(g) and F$^-$(aq) are zero, assuming perfect-gas and perfect-solute behavior. The value -212 J deg^{-1} mole^{-1} for the sum for these ions may be ascribed in small part to interionic interactions in the solution (Section 13-11), and mainly to the decreased entropy of the water molecules that are tightly bound to the ions forming an "iceberg" about each ion. Using the entropy of fusion of ice, 22.0 J deg^{-1} mole^{-1}, calculate the approximate number of water molecules in the icebergs.

13-37. Carry out calculations for cesium iodide similar to those of Exercises 13-34, 13-35, and 13-36 for lithium fluoride. The standard absolute entropy of CsI(c), from the third-law calculation, is 130 J deg^{-1} mole^{-1}, its enthalpy of solution is -63 kJ mole^{-1}, and its solubility in water at 25°C is 0.35 mole l^{-1}. The amount of iceberg water for the large ions Cs$^+$ and I$^-$ should be much less than for the small ions Li$^+$ and F$^-$.

14

Acids and Bases

An acid may be defined as a hydrogen-containing substance that dissociates on solution in water to produce hydrogen ions, and a base may be defined as a substance containing the hydroxide ion or the hydroxyl group that dissociates in aqueous solution as the hydroxide ion. Acidic solutions have a characteristic sharp taste, due to the hydronium ion, H_3O^+, and basic solutions have a characteristic brackish taste, due to the hydroxide ion, OH^-. An alternative definition is that an acid is a proton donor and a base is a proton acceptor. The ordinary mineral acids (hydrochloric acid, nitric acid, sulfuric acid) are completely ionized (dissociated) in solution, producing one hydrogen ion for every acidic hydrogen atom in the formula of the acid, whereas other acids, such as acetic acid, produce only a smaller number of hydrogen ions. Acids such as acetic acid are called weak acids. A 1 F solution of acetic acid does not have nearly so sharp a taste and does not react nearly so vigorously with an active metal, such as zinc, as does a 1 F solution of hydrochloric acid, because the 1 F solution of acetic acid contains a great number of undissociated molecules $HC_2H_3O_2$, and only a relatively small number of ions H_3O^+ and $C_2H_3O_2^-$. There exists in a solution of acetic acid a steady state, corresponding to the equation

$$HC_2H_3O_2 + H_2O \rightleftarrows H_3O^+ + C_2H_3O_2^-$$

In order to understand the properties of acetic acid it is necessary to formulate the equilibrium expression for this steady state; by use of this equilibrium expression the properties of acetic acid solutions of different concentrations can be predicted.

The general principles of chemical equilibrium can similarly be used in the discussion of a weak base, such as ammonium hydroxide, and also of salts formed by weak acids and weak bases. In addition, these principles are important in providing an understanding of the behavior of indicators, which are useful for determining whether a solution is acidic, neutral, or basic. These principles are of further importance in permitting a discussion of the relation between the concentrations of hydronium ion and hydroxide ion in the same solution.

14-1. Hydronium-ion (Hydrogen-ion) Concentration

It was mentioned in the chapter on water (Chapter 12) that pure water does not consist simply of H_2O molecules, but also contains hydronium ions in concentration about 1×10^{-7} moles per liter (at 25°C), and hydroxide ions in the same concentration. These ions are formed by the *autoprotolysis* of water (reaction of one molecule of water acting as an acid with another molecule of water acting as a base; this reaction is also called the ionization of water):

$$2H_2O \rightleftarrows H_3O^+ + OH^-$$

A molecule, such as the water molecule, that can both lose a proton and add a proton is called an *amphiprotic* molecule (Greek *amphi*, both). Only amphiprotic molecules or ions can undergo autoprotolysis.

The means by which it has been found that pure water contains hydronium ions and hydroxide ions is the measurement of the electric conductivity of water. The mechanism of the electric conductivity of a solution was discussed in Section 6-8. According to this discussion, electric charge is transferred through the body of the solution by the motion of cations from the region around the anode to the region around the cathode, and anions from the region around the cathode to the region around the anode. If pure water contained no ions its electric conductivity would be zero. When investigators made water as pure as possible, by distilling it over and over again, it was found that the electric conductivity approached a certain small value, about one ten-millionth that of a 1 F solution of hydrochloric acid or sodium hydroxide. This indicates that the autoprotolysis of water occurs to such an extent as to give hydronium ions and hydroxide ions in concentration about one ten-millionth mole per liter. Refined measurements have provided the value $1.00 \times 10^{-7} M$ for $[H_3O^+]$ and $[OH^-]$ in pure water at 25°C.*

Although the hydronium ion, H_3O^+, is present in water and confers acidic properties upon aqueous solutions, it is customary to use the symbol H^+ in place of H_3O^+ and to speak of hydrogen ion in place of hydronium. In the later sections of this chapter we shall follow this custom except when the discussion is based upon the proton donor-acceptor theory (the Brønsted-Lowry theory). Accordingly, we shall use the symbol H^+ with two meanings: to represent the unhydrated proton when the Brønsted-Lowry theory is being applied, and to represent the hydronium ion, H_3O^+, when this theory is not being applied. We shall refer to H^+ in the Brønsted-Lowry applications as the proton and to H^+ (representing H_3O^+) in other discussions as the hydrogen ion (representing the hydronium ion).

The *p*H

Instead of saying that the concentration of hydrogen ion in pure water is $1.00 \times 10^{-7} M$, it is customary to say that the *p*H of pure water is 7. This symbol, *p*H, is defined in the following way: *the pH is the negative common logarithm of the hydrogen-ion activity* (approximately the hydrogen-ion concentration):

$$pH = -\log [H^+]$$

or

$$[H^+] = 10^{-pH} = \text{antilog}\,(-pH)$$

We see from this definition of *p*H that a solution containing 1 mole of hydrogen ions per liter, that is, with a concentration 10^{-0} in H^+, has *p*H zero. A solution only one-tenth as strong in hydrogen ion, containing 0.1 mole of hydrogen ions per liter, has $[H^+] = 10^{-1}$, and hence has *p*H 1. The relation between the hydrogen-ion concentration and the *p*H is shown for simple concentrations along the left side of Figure 14-1.

---

*The extent of autoprotolysis depends somewhat on the temperature. At 0°C $[H_3O^+]$ and $[OH^-]$ are $0.83 \times 10^{-7} M$, and at 100°C they are $6.9 \times 10^{-7} M$. When a solution of a strong acid and a solution of a strong base are mixed, a large amount of heat is given off. This shows that the reaction gives off heat, and accordingly that the reaction of dissociation of water absorbs heat. In accordance with Le Chatelier's principle, increase in the temperature would shift the equilibrium of dissociation of water in such a way as to tend to restore the original temperature; that is, the reaction would take place in the direction that absorbs heat. This direction is the dissociation of water to hydrogen ions and hydroxide ions, and accordingly the principle requires that increase in temperature cause an increased amount of dissociation of water, as is found experimentally.

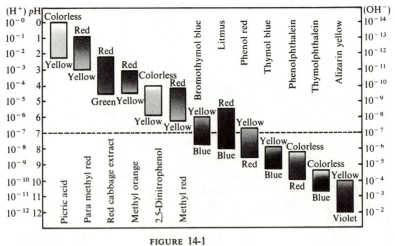

FIGURE 14-1
Color changes of indicators.

In science and medicine it is customary to describe the acidity of a solution by saying "The pH of the solution is 3," for example, instead of saying "The hydrogen-ion concentration of the solution is 10^{-3}." It is evident that the quantity pH is useful, in permitting the exponential expression to be avoided.

The chemical reactions involved in biological processes are often very sensitive to the hydrogen-ion concentration of the medium. In industries such as the fermentation industry the control of the pH of the materials being processed is very important. It is not surprising that the symbol pH was introduced by a Danish biochemist, S. P. L. Sørensen, while he was working on problems connected with the brewing of beer.

Example 14-1. What is the pH of a solution with $[H^+] = 0.0200$?

Solution. The log of 0.0200 is equal to the log of 2×10^{-2}, which is $0.301 - 2 = -1.699$. The pH of the solution is the negative of the logarithm of the hydrogen-ion concentration. Hence the pH of this solution is 1.699.

Example 14-2. What is the hydrogen-ion concentration of a solution with pH 4.30?

Solution. A solution with pH 4.30 has $\log [H^+] = -4.30$, or $0.70 - 5$. The antilog of 0.70 is 5.0, and the antilog of -5 is 10^{-5}. Hence the hydrogen-ion concentration in this solution is 5.0×10^{-5}.

14-2. The Equilibrium between Hydrogen Ion and Hydroxide Ion in Aqueous Solution

The equation for the autoprotolysis of water is

$$2H_2O \rightleftarrows H_3O^+ + OH^-$$

The expression for the equilibrium constant, in accordance with the principle developed in the preceding chapter, is

$$\frac{[H_3O^+][OH^-]}{[H_2O]^2} = K_1$$

In this expression the symbol $[H_2O]$ represents the activity (concentration) of water in the solution (see Sections 11-7 and 13-11 for the discussion of activity). Since the activity of water in a dilute aqueous solution is nearly the same as that for pure water, it is customary to omit the activity of water in the equilibrium expression for dilute solutions. Accordingly, the product of K_1 and $[H_2O]^2$ may be taken as another constant K_w, and we may write

$$[H_3O^+] \times [OH^-] = K_w$$

This expression states that the product of the hydronium-ion concentration and the hydroxide-ion concentration in water and in dilute aqueous solutions is a constant, at given temperature. The value of K_w is 1.00×10^{-14} moles^2/liter^2 at 25°C. *Hence in pure water both H_3O^+ and OH^- have the concentration 1.00×10^{-7} moles per liter at 25°C, and in acidic or basic solutions the product of the concentrations of these ions equals 1.00×10^{-14}.*

Thus a neutral solution contains both hydrogen ions (hydronium ions) and hydroxide ions at the same concentration, 1.00×10^{-7}. A slightly acidic solution, containing 10 times as many hydrogen ions (concentration 10^{-6}, pH 6), also contains some hydroxide ions, one-tenth as many as a neutral solution. A solution containing 100 times as much hydrogen ion as a neutral solution (concentration 10^{-5}, pH 5) contains a smaller amount of hydroxide ion, one one-hundredth as much as a neutral solution; and so on. A solution containing 1 mole of strong acid per liter has hydrogen-ion concentration 1, and pH 0; such a strongly acidic solution also contains some hydroxide ion, the concentration of hydroxide ion being 1×10^{-14}. Although this is a very small number, it still represents a large number of actual ions in a macroscopic volume. Avogadro's number is 0.602×10^{24}, and accordingly a concentration of 10^{-14} moles per liter corresponds to 0.602×10^{10} ions per liter, or 0.602×10^7 ions per milliliter.

14-3. Indicators

Indicators such as litmus may be used to tell whether a solution is acidic, neutral, or basic. The change in color of an indicator as the pH of the solution changes is not sharp, but extends over a range of one or two pH units. This is the result of the existence of chemical equilibrium between the two differently colored forms of the indicator, and the dependence of the color on the hydrogen-ion concentration is due to the participation of hydrogen ion in the equilibrium.

Thus the red form of litmus may be represented by the formula HIn and the blue form by In$^-$, resulting from the dissociation reaction

$$HIn \rightleftharpoons H^+ + In^-$$

Red   Blue
acidic form   basic form

In alkaline solutions, with [H$^+$] very small, the equilibrium is shifted to the right, and the indicator is converted almost entirely into the basic form (blue for litmus). In acidic solutions, with [H$^+$] large, the equilibrium is shifted to the left, and the indicator assumes the acidic form.

Let us calculate the relative amount of the two forms as a function of [H$^+$]. The equilibrium expression for the indicator reaction written above is

$$\frac{[H^+][In^-]}{[HIn]} = K_{In}$$

in which K_{In} is the *equilibrium constant for the indicator*. We rewrite this as

$$\frac{[HIn]}{[In^-]} = \frac{[H^+]}{K_{In}}$$

This equation shows how the ratio of the two forms of the indicator depends on [H$^+$]. When the two forms are present in equal amounts, the ratio of acidic form to alkaline form, [HIn]/[In$^-$], has the value 1, and hence [H$^+$] = K_{In}. *The indicator constant K_{In} is thus the value of the hydrogen-ion concentration at which the change in color of the indicator is half completed.* The corresponding pH value is called the pK of the indicator.

When the pH is decreased by one unit the value of [H$^+$] becomes ten times K_{In} and the ratio [HIn]/[In$^-$] then equals 10. Thus at a pH value 1 less than the pK of the indicator (its midpoint) the acidic form of the indicator predominates over the basic form in the ratio 10:1. In this solution 91% of the indicator is in the acidic form, and 9% in the basic form. Over a range of 2 pH units the indicator accordingly changes from 91% acidic form to 91% basic form. For most indicators the color change detectable by the eye occurs over a range of about 1.2 to 1.8 units.

Indicators differ in their *pK* values; pure water, with *p*H 7, is neutral to litmus (which has *pK* equal to 6.8), acidic to phenolphthalein (*pK* 8.8), and basic to methyl orange (*pK* 3.7).

A chart showing the color changes and effective *p*H ranges of several indicators is given in Figure 14-1. The approximate *p*H of a solution can be determined by finding by test the indicator toward which the solution shows a neutral reaction. Test paper, made with a mixture of indicators and showing several color changes, is now available with which the *p*H of a solution can be estimated to within about 1 unit over the *p*H range 1 to 13.

In titrating a weak acid or a weak base the indicator must be chosen with care. The way of choosing the proper indicator is described in Section 14-6.

It is seen that an indicator behaves as a weak organic acid; the equilibrium expression for an indicator is the same as that for an ordinary weak acid.

By the use of color standards for the indicator, the *p*H of a solution may be estimated to about 0.1 unit by the indicator method. A more satisfactory general method of determining the *p*H of a solution is by use of an instrument that measures the hydrogen-ion concentration by measuring the electric potential of a cell with cell reaction involving hydrogen ions. Modern glass-electrode *p*H meters are now available that cover the *p*H range 0 to 14 with an accuracy approaching 0.01 (Figure 14-2).

The structure of the glass electrode is shown at the right in Figure 14-2. The Ag-AgCl electrode provides a reversible electric connection between the terminal wire and the HCl solution. The glass bulb at the bottom is made of a special glass that conducts the electric current by accepting protons, passing them from one oxygen atom to another, and liberating protons on the other side—this glass is not permeable to other ions. The mercury-calomel electrode shown at the left permits a second reversible electric contact, not dependent on the hydrogen-ion concentration, to be made with the solution. In measuring the *p*H of a solution the ends of the two electrodes are placed in the solution and the electromotive force that is developed is measured with a vacuum-tube voltmeter. Since the conduction of current through the cell involves the transfer of hydrogen ions from a solution with one hydrogen-ion activity to a solution with another hydrogen-ion activity (the solutions on the two sides of the glass membrane) and the hydrogen-ion activities are not significantly involved in the other conduction steps, the electromotive force depends upon the *p*H of the solution being tested. It is linear in the *p*H, changing by 0.059 volt per *p*H unit at 25°C (Section 15-7).

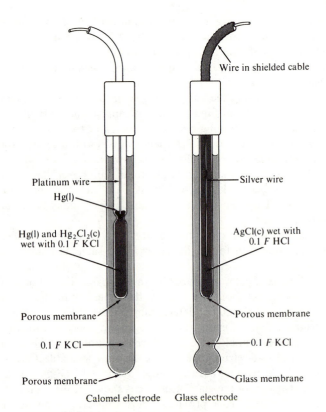

Platinum wire

Hg(l)

Hg(l) and Hg_2Cl_2(c) wet with 0.1 F KCl

Silver wire

AgCl(c) wet with 0.1 F HCl

Wire in shielded cable

Porous membrane

0.1 F KCl

Porous membrane

Porous membrane

0.1 F KCl

Glass membrane

Calomel electrode   Glass electrode

FIGURE 14-2

A representation of the two electrodes of a glass-electrode pH
meter. The glass membrane is made of a special glass that
has an effective permeability to hydrogen ions. The potential
of this electrode is determined by the concentration of hydrogen
ions in the medium surrounding the glass membrane.

14-4. Equivalent Weights of Acids and Bases

A solution containing one gram formula weight of hydrochloric acid,
HCl, per liter is 1 F in hydrogen ion. Similarly a solution containing 0.5
gram formula weight of sulfuric acid, H_2SO_4, per liter is 1 F in replaceable
hydrogen. Each of these solutions is neutralized* by an equal volume of a
solution containing one gram formula weight of sodium hydroxide,

---

*The meaning of "neutralized" in the case of weak acids or bases is discussed in a later
section of this chapter.

NaOH, per liter, and the weights of the acids are hence equivalent to one gram formula weight of the alkali.

The quotient of the gram formula weight of an acid by the number of hydrogen atoms which are replaceable for the reaction under consideration is called the *equivalent weight of the acid*. Likewise, the quotient of the gram formula weight of a base by the number of hydroxyl groups which are replaceable for the reaction under consideration is called the *equivalent weight of the base*.

One equivalent weight of an acid neutralizes one equivalent weight of a base. It is important to note that the equivalent weight of a polyprotic acid is not invariant; for H_3PO_4 it may be the gram formula weight, one half this, or one third, depending on whether one, two, or three hydrogens are effective in the reaction under consideration.

The *normality* of a solution of an acid or base is the number of equivalents of acid or base per liter; a 1 N solution contains 1 equivalent per liter of solution. By determining, with use of an indicator such as litmus, the relative volumes of acidic and alkaline solutions that are equivalent, the normality of one solution can be calculated from the known value of the other. This process of *acid-base titration* (the determination of the *titer* or strength of an unknown solution) with use of special apparatus such as graduated burets and pipets is an important method of volumetric quantitative analysis.

Example 14-3.   It is found by experiment that 25.0 ml of a solution of sodium hydroxide is neutralized by 20.0 ml of a 0.100 N acid solution. What are the normality of the alkaline solution and the weight of NaOH per liter?

Solution.   The unknown normality x of the alkaline solution is found by solving the equation that expresses the equivalence of the portions of the two solutions:

$$25.0x = 20.0 \times 0.100$$

$$x = \frac{20.0 \times 0.100}{25.0} = 0.080$$

The weight of NaOH per liter is 0.080 times the equivalent weight, 40.0, or 3.20 g.

You may find it useful to fix in your mind the following equation:

$$V_1N_1 = V_2N_2$$

Here V_1 is the volume of a solution with normality N_1 and V_2 is the equivalent volume (containing the same number of replaceable hydrogens or hydroxyls) of a solution with normality N_2. In solving the above exercise we began by writing this equation; $25.0x$ is V_1N_1, and 20.0×0.100 is V_2N_2, in this case.

14-5. Weak Acids and Bases

Ionization of a Weak Acid

A 0.1 N solution of a strong acid such as hydrochloric acid is 0.1 N in hydrogen ion, since this acid is very nearly completely dissociated into ions except in very concentrated solutions. But a 0.1 N solution of acetic acid contains hydrogen ions in much smaller concentration, as is seen by testing with indicators, observing the rate of attack of metals, or simply by tasting. Acetic acid is a weak acid; the acetic acid molecules hold their protons so firmly that not all of them are transferred to water molecules to form hydronium ions. Instead, there is an equilibrium reaction,

$$HC_2H_3O_2 + H_2O \rightleftharpoons H_3O^+ + C_2H_3O_2^-$$

or, ignoring the hydration of the proton,

$$HC_2H_3O_2 \rightleftharpoons H^+ + C_2H_3O_2^-$$

The equilibrium expression for this reaction is

$$\frac{[H^+][C_2H_3O_2^-]}{[HC_2H_3O_2]} = K$$

In general, for an acid HA in equilibrium with ions H^+ and A^- the equilibrium expression is

$$\frac{[H^+][A^-]}{[HA]} = K_a$$

The constant K_a, characteristic of the acid, is called its *acid constant* or *ionization constant*.

Values of acid constants are found experimentally by measuring the pH of solutions of the acids. A table of values is given later in this chapter.

Example 14-4. The pH of a 0.100 N solution of acetic acid is found by experiment to be 2.874. What is the acid constant, K_a, of this acid?

Solution. To calculate the acid constant we note that acetic acid added to pure water ionizes to produce hydrogen ions and acetate ions in equal quantities. Moreover, since the amount of hydrogen ion resulting from the dissociation of water is negligible compared with the total amount present, we have

$$[H^+] = [C_2H_3O_2^-] = \text{antilog}\,(-2.874) = 1.34 \times 10^{-3}$$

The concentration $[HC_2H_3O_2]$ is hence $0.100 - 0.001 = 0.099$, and the acid constant has the value

$$K_a = (1.34 \times 10^{-3})^2/0.099 = 1.80 \times 10^{-5}$$

The hydrogen-ion concentration of a weak acid (containing no other electrolytes that react with it or its ions) in 1 N concentration is approximately equal to the square root of its acid constant, as is seen from the following example.

Example 14-5.   What is [H⁺] of a 1 N solution of HCN, hydrocyanic acid, which has $K_a = 4 \times 10^{-10}$?

Solution.   Let $x = $ [H⁺]. Then we can write [CN⁻] $= x$ (neglecting the amount of hydrogen ion due to ionization of the water), and [HCN] $= 1 - x$. The equilibrium equation is

$$\frac{x^2}{1 - x} = K_a = 4 \times 10^{-10}$$

We know that x is going to be much smaller than 1, since this weak acid is only very slightly ionized. Hence we replace $1 - x$ by 1 (neglecting the small difference between unionized hydrocyanic acid and the total cyanide concentration), obtaining

$$x^2 = 4 \times 10^{-10}$$
$$x = 2 \times 10^{-5} = \text{[H⁺]}$$

The neglect of the ionization of water is also seen to be justified, since even in this very slightly acidic solution the value of [H⁺] is 200 times the value for pure water.

Successive Ionizations of a Polyprotic Acid

A polyprotic acid has several acid constants, corresponding to dissociation of successive hydrogen ions. For phosphoric acid, H_3PO_4, there are three equilibrium expressions:

$$H_3PO_4 \rightleftarrows H^+ + H_2PO_4^-$$

$$K_1 = \frac{[H^+][H_2PO_4^-]}{[H_3PO_4]} = 7.5 \times 10^{-3} = K_{H_3PO_4}$$

$$H_2PO_4^- \rightleftarrows H^+ + HPO_4^{--}$$

$$K_2 = \frac{[H^+][HPO_4^{--}]}{[H_2PO_4^-]} = 6.2 \times 10^{-8} = K_{H_2PO_4^-}$$

$$HPO_4^{--} \rightleftarrows H^+ + PO_4^{---}$$

$$K_3 = \frac{[H^+][PO_4^{---}]}{[HPO_4^{--}]} = 10^{-12} = K_{HPO_4^{--}}$$

Note that these constants have the dimensions of concentration, mole/liter.

The ratio of successive ionization constants for a polybasic acid is usually about 10^{-5}, as in this case. We see that with respect to its first hydrogen phosphoric acid is a moderately strong acid—considerably stronger than acetic acid. With respect to its second hydrogen it is weak, and to its third very weak.

Ionization of a Weak Base

A weak base dissociates in part to produce hydroxide ions:

$$\text{MOH} \rightleftharpoons \text{M}^+ + \text{OH}^-$$

The corresponding equilibrium expression is

$$\frac{[\text{M}^+][\text{OH}^-]}{[\text{MOH}]} = K_b$$

The constant K_b is called the *basic constant* of the base.

Ammonium hydroxide is the only common weak base. Its basic constant has the value 1.81×10^{-5} at 25°C. The hydroxides of the alkali metals and the alkaline-earth metals are strong bases.

Example 14-6. What is the *p*H of a 0.1 *F* solution of ammonium hydroxide?

Solution. Our fundamental equation is

$$\frac{[\text{NH}_4^+][\text{OH}^-]}{[\text{NH}_4\text{OH}]} = K_b = 1.81 \times 10^{-5}$$

Since the ions NH_4^+ and OH^- are produced in equal amounts by the dissociation of the base and the amount of OH^- from dissociation of water is negligible, we put

$$[\text{NH}_4^+] = [\text{OH}^-] = x$$

The concentration of NH_4OH is accordingly $0.1 - x$, and we obtain the equation

$$\frac{x^2}{0.1 - x} = 1.81 \times 10^{-5}$$

(Here we have made the calculation as though all the undissociated solute were NH_4OH. Actually there is dissolved NH_3 present; however, since the equilibrium $\text{NH}_3 + \text{H}_2\text{O} \rightleftharpoons \text{NH}_4\text{OH}$ is of such a nature that the ratio $[\text{NH}_4\text{OH}]/[\text{NH}_3]$ is constant, we are at liberty to write the equilibrium expression for the base as shown above, with the symbol $[\text{NH}_4\text{OH}]$ representing the total concentration of the undissociated solute, including the molecular species NH_3 as well as NH_4OH.)

Solving this equation, we obtain the result

$$x = [\text{OH}^-] = [\text{NH}_4^+] = 1.34 \times 10^{-3}$$

The solution is hence only slightly alkaline—its hydroxide-ion concentration is the same as that of a 0.00134 N solution of sodium hydroxide.

This value of [OH⁻] corresponds to $[H^+] = (1.00 \times 10^{-14})/(1.34 \times 10^{-3}) = 7.46 \times 10^{-12}$, as calculated from the water equilibrium equation

$$[H^+][OH^-] = 1.00 \times 10^{-14}$$

The corresponding pH is 11.13, which is the answer to the problem.

Very many problems in solution chemistry are solved with use of the acid and base equilibrium equations. The uses of these equations in discussing the titration of weak acids and bases, the hydrolysis of salts, and the properties of buffered solutions are illustrated in the following sections of this chapter.

The student while working a problem should not substitute numbers in the equations in a routine way, but should think carefully about the chemical reactions and equilibria involved and the magnitudes of the concentrations of the different molecular species. Every problem solved should add to his understanding of solution chemistry. The ultimate goal is such an understanding of the subject that the student can estimate the orders of magnitude of concentrations of the various ionic and molecular species in a solution without having to solve the equilibrium equations.

14-6. The Titration of Weak Acids and Bases

A liter of solution containing 0.2 mole of a strong acid such as hydrochloric acid has $[H^+] = 0.2$ and $pH = 0.7$. The addition of a strong base, such as 0.2 N NaOH, causes the hydrogen-ion concentration to diminish through neutralization by the added hydroxide ion. When 990 ml of strong base has been added, the excess of acid over base is $0.2 \times 10/1000 = 0.002$ mole, and since the total volume is very close to 2 liters the value of $[H^+]$ is 0.001, and the pH is 3. When 999 ml has been added, and the neutralization reaction is within 0.1% of completion, the values are $[H^+] = 0.0001$ and $pH = 4$. At $pH = 5$ the reaction is within 0.01% of completion, and at pH 6 within 0.001%. Finally pH 7, neutrality, is reached when an amount of strong base has been added exactly equivalent to the amount of strong acid present. A very small excess of strong base causes the pH to increase beyond 7.

We see that to obtain the most accurate results in titrating a strong acid and a strong base an indicator with indicator constant about 10^{-7} ($pK = 7$) should be chosen, such as litmus or bromthymol blue. The titration curve calculated above, and given in Figure 14-3, shows, however, that the choice of an indicator is in this case not crucial; any indicator with pK between 4

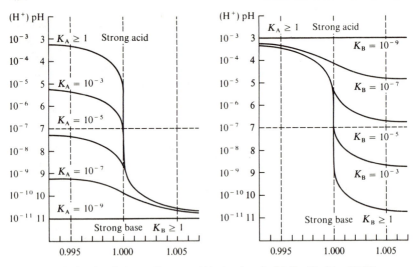

Ratio of equivalents of base to acid in titration of a 0.2 N acid with a 0.2 N base,
either acid or base being strong

FIGURE 14-3
Acid-base titration curves.

(methyl orange) and 10 (thymolphthalein) could be used with error less than 0.2%.

In titrating a weak acid (with a strong base) or a weak base (with a strong acid) greater care is needed in the selection of an indicator. Let us consider the titration of 0.2 N acetic acid, a moderately weak acid with $K_a = 1.80 \times 10^{-5}$, with 0.2 N sodium hydroxide. When an amount of the alkali equivalent to that of the acid has been added, the resultant solution is the same as would be obtained by dissolving 0.1 mole of the salt $NaC_2H_3O_2$ in a liter of water. The solution of this salt is not neutral, with *p*H 7, however, but is alkaline, as can be seen from the following argument, based on the Brønsted-Lowry theory.

The salt $NaC_2H_3O_2$ is completely dissociated into ions, Na^+ and $C_2H_3O_2^-$, when it is dissolved in water. The ion Na^+ has no protons, and hence is not an acid. The acetate anion, $C_2H_3O_2^-$, is, however, a base— it is the base conjugate to the acid $HC_2H_3O_2$, and it can accept a proton from an acid, such as H_2O:

$$H_2O + C_2H_3O_2^- \rightleftarrows HC_2H_3O_2 + OH^-$$

This reaction takes place to the extent determined by the value of its equilibrium constant:

$$\frac{[HC_2H_3O_2][OH^-]}{[H_2O][C_2H_3O_2^-]} = K_b$$

We may say that an aqueous solution of the neutral salt sodium acetate is alkaline, with pH greater than 7, because it contains the base (proton acceptor) acetate ion.

The value of K_b for a base is closely related to the value of the acid constant K_a of the homologous acid. Let us consider these constants for an acid HA and its homologous base A^-:

$$K_a = \frac{[H_3O^+][A^-]}{[HA][H_2O]}$$

$$K_b = \frac{[HA][OH^-]}{[H_2O][A^-]}$$

On multiplying the left sides and the right sides of these equations, we obtain

$$K_a K_b = \frac{[H_3O^+][A^-]}{[HA][H_2O]} \frac{[HA][OH^-]}{[H_2O][A^-]} = \frac{[H_3O^+][OH^-]}{[H_2O]^2}$$

The expression on the right side is seen to be the autoprotolysis constant for water, with value 1.00×10^{-14} (for $[H_2O]$ taken as 1; see Section 14-2); hence we have obtained the relation

$$K_b = \frac{K_w}{K_a} = \frac{1.00 \times 10^{-14}}{K_a}$$

between the acid constant and the base constant of a conjugate pair.

The acid constant of acetic acid is 1.80×10^{-5} (Section 14-5); the base constant of acetate ion accordingly has the value $1.00 \times 10^{-14}/1.80 \times 10^{-5} = 5.56 \times 10^{-10}$.

Let x be the number of acetate ions (mole/liter) that have undergone reaction with water to produce $HC_2H_3O_2$ and OH^-. The concentrations of solutes are seen to be

$$[HC_2H_3O_2] = [OH^-] = x$$
$$[C_2H_3O_2^-] = 0.1 - x$$

and the equilibrium equation is

$$\frac{x^2}{0.1 - x} = K_b = 5.56 \times 10^{-10}$$

Solution of this equation gives

$$x = 0.75 \times 10^{-5} \text{ mole/liter}$$

Hence $[OH^-] = 0.75 \times 10^{-5}$ and $[H^+] = 1.34 \times 10^{-9}$.

The pH of the solution of sodium acetate is hence 8.87. By reference to Figure 14-1 we see that *phenolphthalein, with $pK = 9$, is the best indicator to use for titrating a moderately weak acid such as acetic acid.*

The complete titration curve, showing the pH of the solution as a function of the amount of strong base added, can be calculated in essentially this way. Its course is shown in Figure 14-3 ($K_a = 10^{-5}$). We see that the solution has pH 7 when there is about 1% excess of acid; hence if litmus were used as the indicator an error of about 1% would be made in the titration.

The basic constant of ammonium hydroxide has about the same value as the acid constant of acetic acid. Hence to *titrate a weak base such as ammonium hydroxide with a strong acid methyl orange (pK 3.8) may be used as the indicator.*

It is possible by suitable selection of indicators to titrate separately a strong acid and a weak acid or a strong base and a weak base in a mixture of the two. Let us consider, for example, a solution of sodium hydroxide and ammonium hydroxide (ammonia). If strong acid is added until the pH is 11.1, which is that of 0.1 N ammonium hydroxide solution, the strong base will be within 1% of neutralization (Figure 14-3). Hence by using alizarine yellow (pK 11) as indicator the concentration of the strong base can be determined, and then by a second titration with methyl orange the concentration of ammonium hydroxide can be determined.

The Acidic Properties of Hydrated Ions of Metals other than the Alkalis and Alkaline Earths

Metal salts of strong acids, such as $FeCl_3$, $CuSO_4$, and $KAl(SO_4)_2 \cdot 12H_2O$ (alum), produce acidic solutions; the sour taste of these salts is characteristic.

It will be recalled from the discussion in Chapter 12 that the aluminum ion in aqueous solution is hydrated, having the formula $Al(H_2O)_6^{+++}$, with the six water molecules arranged octahedrally about the aluminum ion. The hydrolysis of aluminum salts may be represented by the equations

$$Al(H_2O)_6^{+++} + H_2O \rightleftharpoons H_3O^+ + Al(H_2O)_5OH^{++}$$
$$Al(H_2O)_5OH^{++} + H_2O \rightleftharpoons H_3O^+ + Al(H_2O)_4(OH)_2^+$$
$$Al(H_2O)_4(OH)_2^+ + H_2O \rightleftharpoons H_3O^+ + Al(H_2O)_3(OH)_3$$
$$\rightleftharpoons Al(OH)_3(c) + 3H_2O + H_3O^+$$

In these reactions the hydrated ions of aluminum lose protons, forming successive hydroxide complexes; the final neutral complex then loses water to form the insoluble hydroxide $Al(OH)_3$.

The complex ions $Al(H_2O)_5(OH)^{++}$ and $Al(H_2O)_4(OH)_2^+$ remain in solution, whereas the hydroxide $Al(OH)_3$ is very slightly soluble and precipitates if more than a very small amount is formed; precipitation occurs when the pH is greater than 3.

The protolysis of hydrated ferric ion occurs to such an extent that the color of ferric ion itself, $Fe(H_2O)_6^{+++}$, is usually masked by that of the hydroxide complexes. Ferric ion is nearly colorless; it seems to have a very pale violet color, seen in crystals of ferric alum, $KFe(SO_4)_2 \cdot 12H_2O$, and ferric nitrate, $Fe(NO_3)_3 \cdot 9H_2O$, and in ferric solutions strongly acidified with nitric or perchloric acid. Solutions of ferric salts ordinarily have the characteristic yellow to brown color of the hydroxide complexes $Fe(H_2O)_5OH^{++}$ and $Fe(H_2O)_4(OH)_2^+$, or even the red-brown color of colloidal particles of hydrated ferric hydroxide.

14-7. Buffered Solutions

Very small amounts of strong acid or base suffice to change the hydrogen-ion concentration of water in the slightly acidic to slightly basic region; one drop of strong concentrated acid added to a liter of water makes it appreciably acidic, increasing the hydrogen-ion concentration by a factor of 5000, and two drops of strong alkali would then make it basic, decreasing the hydrogen-ion concentration by a factor of over a million. Yet there are solutions to which large amounts of strong acid or base can be added with only very small resultant change in hydrogen-ion concentration. Such solutions are called *buffered solutions*.

Blood and other physiological solutions are buffered; the *p*H of blood changes only slowly from its normal value (about 7.4) on addition of acid or base. Important among the buffering substances in blood are the serum proteins (Chapter 24), which contain basic and acidic groups that can combine with the added acid or base.

A drop of concentrated acid, which when added to a liter of pure water increases $[H^+]$ 5000-fold (from 10^{-7} to 5×10^{-4}), produces an increase of $[H^+]$ of less than 1% (from 1.00×10^{-7} to 1.01×10^{-7}, for example) when added to a liter of buffered solution such as the phosphate buffer made by dissolving 0.2 gram formula weight of phosphoric acid in a liter of water and adding 0.3 gram formula weight of sodium hydroxide.

This is a half-neutralized phosphoric acid solution; its principal ionic constituents and their concentrations are Na^+, 0.3 M; HPO_4^{--}, 0.1 M; $H_2PO_4^-$, 0.1 M; H^+, about 10^{-7} M. From the titration curve of Figure 14-4 we see that this solution is a good buffer; to change its *p*H from 7 to 6.5 or 7.5 (tripling the hydrogen ion or hydroxide ion concentration) about one-twentieth of an equivalent of strong acid or base is needed per liter, whereas this amount of acid or base in water would cause a change of 5.7 *p*H units (an increase or decrease of $[H^+]$ by the factor 500,000). Such a solution, usually made by dissolving the two well-crystallized salts KH_2PO_4 and $Na_2HPO_4 \cdot 2H_2O$ in water, is widely used for buffering in the neutral region (*p*H 5.3 to 8.0).* Other useful buffers are sodium citrate-hydrochloric acid (*p*H 1 to 3.5), acetic acid-sodium acetate (*p*H 3.6 to

---

*A concentrated neutral buffer solution containing one-half gram formula weight of each salt per liter may be kept in the laboratory to neutralize either acid or base spilled on the body.

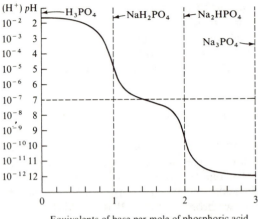

Equivalents of base per mole of phosphoric acid
Titration curve for H_3PO_4 (0.1 F) and strong base

FIGURE 14-4
Titration curve for phosphoric acid and a strong base.

5.6), boric acid-sodium hydroxide (*p*H 7.8 to 10,0), and glycine-sodium hydroxide (*p*H 8.5 to 13).

The behavior of a buffer can be understood from the equilibrium equation for the acid dissociation. Let us consider the case of acetic acid-sodium acetate. The solution contains $HC_2H_3O_2$ and $C_2H_3O_2^-$ in equal or comparable concentrations. The equilibrium expression

$$\frac{[H^+][C_2H_3O_2^-]}{[HC_2H_3O_2]} = K_a$$

may be written as

$$[H^+] = \frac{[HC_2H_3O_2]}{[C_2H_3O_2^-]} K_a$$

This shows that when $[C_2H_3O_2^-]$ and $[HC_2H_3O_2]$ are equal, as in an equimolal mixed solution of $HC_2H_3O_2$ and $NaC_2H_3O_2$, the value of $[H^+]$ is just that of K_a, 1.80×10^{-5}, and hence the *p*H is 4.7. A 1:5 mixture of $HC_2H_3O_2$ and $NaC_2H_3O_2$ has $[H^+] = \frac{1}{5}K_a$ and *p*H 5.4, and a 5:1 mixture has $[H^+] = 5K_a$ and *p*H 4.0. By choosing a suitable ratio of $HC_2H_3O_2$ to $NaC_2H_3O_2$, any desired hydrogen-ion concentration in this neighborhood can be obtained.

It is seen from the equilibrium expressions that *the effectiveness of a buffer depends on the concentrations of the buffering substances*; a tenfold dilution of the buffer decreases by the factor 10 the amount of acid or base per liter that can be added without causing the *p*H to change more than the desired amount.

For the phosphate buffer in the *p*H 7 region the equilibrium constant of interest is that for the reaction

$$H_2PO_4^- \rightleftharpoons HPO_4^{--} + H^+$$

The value of $K_{H_2PO_4^-}$ is 6.2×10^{-8}; this is accordingly the value of $[H^+]$ expected for a solution with $[H_2PO_4^-] = [HPO_4^{--}]$.

If the buffered solution is dilute, this is its hydrogen-ion concentration. Because the activities of ions are affected by other ions, however, there is appreciable deviation from the calculated values in salt solutions as concentrated as 0.1 *M*. This fact accounts for the small discrepancies between the *p*H values calculated from equilibrium constants and those given in the buffer tables.

14-8.   The Strengths of the Oxygen Acids

The oxygen acids, which consist of oxygen atoms O and hydroxide groups OH attached to a central atom ($HClO_4 = ClO_3(OH)$, $H_2SO_4 = SO_2(OH)_2$, and so on), vary widely in strength, from very strong acids such as perchloric acid, $HClO_4$, to very weak ones such as boric acid, H_3BO_3. It is often useful to know the approximate strengths of these acids. Fortunately there have been formulated some simple and easily remembered rules regarding these acid strengths.

The Rules Expressing the Strengths of the Oxygen Acids

The strengths of the oxygen acids are expressed approximately by the following two rules:

Rule 1. The successive acid constants K_1, K_2, K_3, ... are in the ratios $1:10^{-5}:10^{-10}: \cdots$.

We note the examples of phosphoric acid

$$K_{H_3PO_4} = 7.5 \times 10^{-3} \quad K_{H_2PO_4^-} = 6.2 + 10^{-8} \quad K_{HPO_4^{--}} = 1 \times 10^{-12}$$

and sulfurous acid

$$K_{H_2SO_3} = 1.2 \times 10^{-2} \quad K_{HSO_3^-} = 1 \times 10^{-7}$$

The rule holds well for all the acids of the class under consideration.

Rule 2. The value of the first ionization constant is determined by the value of m in the formula $XO_m(OH)_n$: if m is zero (no excess of oxygen atoms over hydrogen atoms, as in $B(OH)_3$) the acid is very weak, with $K_1 \leqq 10^{-7}$; for m = 1 the acid is weak, with $K_1 \cong 10^{-2}$; for m = 2 ($K_1 \cong 10^3$) or m = 3 ($K_1 \cong 10^8$) the acid is strong.

Note the occurrence of the factor 10^{-5} in both this rule and the first one. The applicability of this rule is shown by the tables beginning at the bottom of this page.

The second rule can be understood in the following way. The force attracting H^+ to ClO^- to form $ClOH$ (hypochlorous acid) is that of an O—H valence bond. But the force between H^+ and either one of the two oxygen atoms of the ion ClO_2^- to form $ClOOH$ (chlorous acid) may be smaller than that for an O—H valence bond because the total attraction for the proton is divided between the two oxygen atoms, and hence this acid (of the second class) may well be expected to be more highly dissociated than hypochlorous acid. An acid of the third class would be still more highly dissociated, since the total attraction for the proton would be divided among three oxygen atoms.

With use of these rules we can answer questions about the choice of indicators for titration without referring to tables of acid constants.

Example 14-7. What reaction to litmus would be expected of solutions of the following salts: $NaClO$, $NaClO_2$, $NaClO_3$, $NaClO_4$?

Solution. The corresponding acids are shown by the rule to be very weak, weak, strong, and very strong, respectively. Hence $NaClO$ and $NaClO_2$ would through hydrolysis give basic solutions, and the other two salts would give neutral solutions.

Example 14-8. What indicator could be used for titrating periodic acid, H_5IO_6?

Solution. This acid has one extra oxygen atom, and is hence of the second class, as is phosphoric acid. We accordingly refer to Figures 14-4 and 14-1, and see that methyl orange should be satisfactory for titrating the first hydrogen, or phenolphthalein for titrating the first two hydrogens.

Experimental Values of Acid Constants

Some values of acid constants that have been determined by experiment are given in the following tabulation.

First class; Very weak acids $X(OH)_n$ or H_nXO_n

First acid constant about 10^{-7} or less

	K_1
Hypochlorous acid, $HClO$	3.2×10^{-8}
Hypobromous acid, $HBrO$	2×10^{-9}
Hypoiodous acid, HIO	1×10^{-11}
Silicic acid, H_4SiO_4	1×10^{-10}

	K_1
Germanic acid, H_4GeO_4	3×10^{-9}
Boric acid, H_3BO_3	5.8×10^{-10}
Arsenious acid, H_3AsO_3	6×10^{-10}
Antimonous acid, H_3SbO_3	1×10^{-11}

Second Class; Weak acids $XO(OH)_n$ or H_nXO_{n+1}

First acid constant about 10^{-2}

	K_1
Chlorous acid, $HClO_2$	1.1×10^{-2}
Sulfurous acid, H_2SO_3	1.2×10^{-2}
Selenious acid, H_2SeO_3	0.3×10^{-2}
Phosphoric acid, H_3PO_4	0.75×10^{-2}
Phosphorous acid,* H_2HPO_3	1.6×10^{-2}
Hypophosphorous acid,* HH_2PO_2	1×10^{-2}
Arsenic acid, H_3AsO_4	0.5×10^{-2}
Periodic acid, H_5IO_6	1×10^{-3}
Nitrous acid, HNO_2	0.45×10^{-3}
Acetic acid, $HC_2H_3O_2$	1.80×10^{-5}
Carbonic acid,† H_2CO_3	0.45×10^{-6}

Third Class; Strong acids $XO_2(OH)_n$ or H_nXO_{n+2}

First acid constant about 10^3
Second acid constant about 10^{-2}

	K_1	K_2
Chloric acid, $HClO_3$	Large	
Sulfuric acid, H_2SO_4	Large	1.2×10^{-2}
Selenic acid, H_2SeO_4	Large	1×10^{-2}

Fourth Class; Very strong acids $XO_3(OH)_n$ or H_nXO_{n+3}

First acid constant about 10^8

Perchloric acid, $HClO_4$	Very strong
Permanganic acid, $HMnO_4$	Very strong

---

*It is known that phosphorous acid has the structure

$$H-\overset{\overset{\displaystyle O}{\|}}{\underset{\underset{\displaystyle OH}{|}}{P}}-OH$$

and hypophosphorous acid the structure

$$H-\overset{\overset{\displaystyle O}{\|}}{\underset{\underset{\displaystyle H}{|}}{P}}-OH;$$

the hydrogen atoms that are bonded to the phosphorus atom are not counted in applying the rule.

†The low value for carbonic acid is due in part to the existence of some of the un-ionized acid in the form of dissolved CO_2 molecules rather than H_2CO_3. The proton dissociation constant for the molecular species H_2CO_3 is about 2×10^{-4}.

Other Acids

There is no simple way of remembering the strengths of acids other than those discussed above. HCl, HBr, and HI are strong, but HF is weak, with $K_a = 7.2 \times 10^{-4}$. The homologues of water are weak acids, with the following reported acid constants:

	K_1	K_2
Hydrosulfuric acid, H_2S	1.1×10^{-7}	1.0×10^{-14}
Hydroselenic acid, H_2Se	1.7×10^{-4}	1×10^{-12}
Hydrotelluric acid, H_2Te	2.3×10^{-3}	1×10^{-11}

The hydrides NH_3 and PH_3 function as bases by adding protons rather than as acids by losing them.

Oxygen acids that do not contain a single central atom have strengths corresponding to reasonable extensions of our rules, as shown by the following examples.

Very weak acids: $K_1 = 10^{-7}$ or less

	K_1	K_2
Hydrogen peroxide, HO—OH	2.4×10^{-12}	
Hyponitrous acid, HON—NOH	9×10^{-8}	1×10^{-11}

Weak acids: $K_1 = 10^{-2}$

	K_1	K_2
Oxalic acid, HOOC—COOH	5.9×10^{-2}	6.4×10^{-5}

The following acids are not easily classified:

	K_1
Hydrocyanic acid, HCN	4×10^{-10}
Cyanic acid, HOCN	Strong
Thiocyanic acid, HSCN	Strong
Hydrazoic acid, HN_3	1.8×10^{-5}

Acid Strength and Condensation

It is observed that the tendency of oxygen acids to condense to larger molecules is correlated with their acid strengths. Very strong acids, such as $HClO_4$ and $HMnO_4$, condense only with difficulty, and the substances formed, Cl_2O_7 and Mn_2O_7, are very unstable. Less strong acids, such as

H_2SO_4, form condensation products such as $H_2S_2O_7$, disulfuric acid, on strong heating, but these products are not stable in aqueous solution. Phosphoric acid forms diphosphate ion and other condensed ions in aqueous solution, but these ions easily hydrolyze to the orthophosphate ion; other weak acids behave similarly. The very weak oxygen acids, including silicic acid (Chapter 18) and boric acid, condense very readily, and their condensation products are very stable substances.

This correlation is reasonable. The un-ionized acids contain oxygen atoms bonded to hydrogen atoms, and the condensed acids contain oxygen atoms bonded to two central atoms:

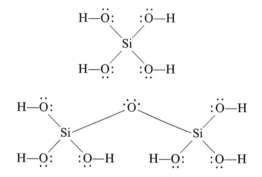

It is hence not surprising that stability of the un-ionized acid (low acid strength) should be correlated with stability of the condensed molecules.

14-9.   The Solution of Carbonates in Acid; Hard Water

The normal carbonates of metals other than the alkalies and ammonium are only slightly soluble in water. We all know, however, that carbonates dissolve easily in acid. This is due in part to the fact that carbon dioxide produced by the reaction escapes from the solution, thus diminishing the carbonate-ion concentration. It is also due in large measure to the existence of most of the carbonate ion in acid solution in the form of HCO_3^- or H_2CO_3 rather than CO_3^{--}; only the last enters into the solubility-product expression of a normal carbonate, and accordingly a larger amount of carbonate can be dissolved by an acidic solution than by a neutral or basic solution before the ion-concentration product equals the solubility product of the carbonate.

This is responsible for the solution of limestone by acid ground water; the quantitative discussion of this effect is given in the following paragraph.

Quantitative Treatment of the Solubility of Carbonates

The solubility product of $CaCO_3$ is 4.8×10^{-9} (Table 13-1).

The solubility of the substance in solutions sufficiently alkaline for all carbonate to exist as the carbonate ion CO_3^{--} is hence 7×10^{-5} mole/l or 0.001 g/l.

In water with pH 7 the ion species HCO_3^- predominates over CO_3^{--}; the equilibrium expression

$$\frac{[H^+][CO_3^{--}]}{[HCO_3^-]} = K_{HCO_3^-} = 4.7 \times 10^{-11}$$

leads to

$$\frac{[CO_3^{--}]}{[HCO_3^-]} = \frac{4.7 \times 10^{-11}}{[H^+]}$$

or

$$[CO_3^{--}]/[HCO_3^-] = 4.7 \times 10^{-4} \text{ for } [H^+] = 10^{-7}$$

Moreover, with $K_{H_2CO_3} = 4.3 \times 10^{-7}$, the ratio $[HCO_3^-]/[H_2CO_3]$ equals 4.3 at pH 7. Hence in neutral solution the total carbonate is divided in the ratios 19% H_2CO_3, 81% HCO_3^-, and only 0.038% CO_3^{--}, the total carbonate concentration being 2600 times the CO_3^{--} concentration. The equilibrium expression

$$[Ca^{++}][CO_3^{--}] = 4.8 \times 10^{-9}$$

can accordingly be rewritten as

$$[Ca^{++}][\text{total carbonate}] = 4.8 \times 10^{-9} \times 2600 = 1.25 \times 10^{-5}$$

If no calcium ion or carbonate were initially present, the two concentrations $[Ca^{++}]$ and [total carbonate] resulting from solution of $CaCO_3$ would be equal and each would equal $\sqrt{1.25 \times 10^{-5}} = 0.0035$ mole/l or 0.35 g/l, which is 51 times that in alkaline solutions. In acid solutions the solubility is much greater still.

Some natural waters are acid because of the presence of a large amount of dissolved carbon dioxide. On heating such water carbon dioxide is driven off, and the hydrogen-ion concentration of the water becomes much less. Even though the total carbonate concentration is decreased, the concentration of CO_3^{--} may be increased greatly because of the increased ionization of HCO_3^- resulting from the change in pH. This may cause the product $[Ca^{++}][CO_3^{--}]$ to reach the solubility product, resulting in the deposition of $CaCO_3$. Water of this kind, from which the dissolved calcium can be precipitated by boiling, is said to have *temporary hardness*. In practice this hardness is removed by adding lime, $Ca(OH)_2$, which neutralizes the acid, and causes all the calcium to precipitate as carbonate (Section 12-1).

The solubilities of salts of other weak acids, such as phosphates, acetates, and sulfides, are also dependent on hydrogen-ion concentration.

14-10. The Precipitation of Sulfides

In most systems of qualitative analysis for the metal ions a very important procedure is sulfide precipitation, with the aid of which about fifteen of the twenty-three or twenty-four metals commonly tested for are precipitated. The great usefulness of sulfides in qualitative analysis depends on two factors—the great range of the solubility products of the sulfides and the great range of concentrations of the sulfide ion S^{--} that can be obtained by varying the acidity of the solution.

Some solubility products of sulfides are given in Table 13-1.

The acid constants for H_2S are

$$K_{H_2S} = \frac{[H^+][HS^-]}{[H_2S]} = 1.1 \times 10^{-7}$$

$$K_{HS^-} = \frac{[H^+][S^{--}]}{[HS^-]} = 1.0 \times 10^{-14}$$

From these equations we obtain

$$\frac{[H^+]^2[S^{--}]}{[H_2S]} = 1.1 \times 10^{-21}$$

or

$$[S^{--}] = \frac{1.1 \times 10^{-21}[H_2S]}{[H^+]^2}$$

In the system of qualitative analysis part of the procedure consists in saturating a solution of appropriately adjusted hydrogen-ion concentration with hydrogen sulfide. In a saturated solution (with $P_{H_2S} = 1$ atm) the value of $[H_2S]$ is about 0.1 M. The foregoing equation hence for this case becomes

$$[S^{--}] = \frac{1.1 \times 10^{-22}}{[H^+]^2}$$

We see that by changing the pH from 0 ($[H^+] = 1$) to 12 ($[H^+] = 10^{-12}$) the sulfide-ion concentration can be varied throughout the great range from 10^{-22} to over 1.

In some analytical procedures the metals of the hydrogen sulfide group are precipitated from 0.3 N acid solution. Let us use the solubility products given above to predict which of these sulfides would be precipitated. With $[H^+] = 0.3$ we have $[S^{--}] = 10^{-21}$. About 0.5 or 1 mg of metal in 100 ml of solution is the smallest amount that should escape detection

in a system of qualitative analysis. The concentration of metal corresponding to this, assuming atomic weight 50 or 100, is $[M^{++}] = 10^{-4}$. Hence, if
the solubility product of a sulfide MS were 10^{-25} or smaller, any quantity
of the ion M^{++} greater than 0.5 or 1 mg per 100 ml would be precipitated
under these conditions. From the table we see that Hg^{++}, Cu^{++}, Cd^{++},
Pb^{++}, and Sn^{++} should precipitate with the H_2S group, and that Zn^{++},
Fe^{++}, Co^{++}, Ni^{++}, and Mn^{++} should not. This is correct—the hydrogen
sulfide group consists of the ions Hg^{++}, Cu^{++}, Cd^{++}, Pb^{++}, Sn^{++}, Sn^{4+},
As^{3+}, As^{++}, Sb^{+++}, Sb^{5+}, and Bi^{+++}. The solubility products for the sulfides
SnS_2, As_2S_3, As_2S_5, Sb_2S_3, Sb_2S_5, and Bi_2S_3 have values compatible with their
inclusion in this group.

After separation of the precipitate by filtration, the filtrate is made
neutral or basic by adding ammonium hydroxide. With $[H^+]$ less than
10^{-7}, $[S^{--}]$ becomes greater than 10^{-8}. Under these conditions any sulfide
MS with K_{SP} less than 10^{-12} would precipitate; this class includes the
sulfides of Zn^{++}, Fe^{++}, Co^{++}, Ni^{++}, and Mn^{++}.

14-11. Nonaqueous Amphiprotic Solvents

In Section 14-1 the autoprotolysis of the amphiprotic substance water was
discussed. Every solvent whose molecules contain one or more protons
and one or more unshared electron pairs in outer shells can act as an
amphiprotic solvent. One of its molecules can donate a proton to a sufficiently strong base or accept a proton from a sufficiently strong acid.
For example, perchloric acid dissolved in pure sulfuric acid undergoes the
following reaction, in which H_2SO_4 acts as a base:

$$HClO_4 + H_2SO_4 \rightleftharpoons H_3SO_4^+ + ClO_4^-$$

A prediction of the probable behavior of a solute in an amphiprotic
solvent can be made by comparing the acid constant of the solute when
dissolved in water with the acid constant of the solvent when dissolved in
water. For example, in water perchloric acid is about 10^5 times as strong
as sulfuric acid. Its great proton-donating power probably applies also
when sulfuric acid is the solvent.

Phosphoric acid, on the other hand, is in aqueous solution a weaker acid
than sulfuric acid; it reacts as a base when dissolved in pure sulfuric acid:

$$H_2SO_4 + H_3PO_4 \rightleftharpoons H_4PO_4^+ + HSO_4^-$$

Sulfuric acid also undergoes autoprotolysis:

$$2H_2SO_4 \rightleftharpoons H_3SO_4^+ + HSO_4^-$$

TABLE 14-1
Values of the Autoprotolysis Constant

Solvent	Autoprotolysis Constant*
H_2O	$[H_3O^+][OH^-] = 1.0 \times 10^{-14}$ (mole/liter)^2
NH_3	$[NH_4^+][NH_2^-] = 1 \times 10^{-33}$
H_2SO_4	$[H_3SO_4^+][HSO_4^-] = 2 \times 10^{-4}$
HCOOH (formic acid)	$[HC(OH)_2^+][HCOO^-] = 6 \times 10^{-7}$
CH_3COOH (acetic acid)	$[CH_3C(OH)_2^+][CH_3COO^-] = 1 \times 10^{-13}$
CH_3OH (methyl alcohol)	$[CH_3OH_2^+][CH_3O^-] = 2 \times 10^{-17}$
C_2H_5OH (ethyl alcohol)	$[C_2H_5OH_2^+][C_2H_5O^-] = 3 \times 10^{-20}$

*All values are for 25°C except that for ammonia, which is for −33°C. The value for water at 100°C is 0.5×10^{-12}.

Values of the autoprotolysis constants of sulfuric acid and some other amphiprotic solvents, as determined by measuring the electric conductivity of the solvent and some of its solutions, are given in Table 14-1.

The Lewis Theory of Acids and Bases

A still more general theory of acids and bases than the proton donor-acceptor theory was introduced by G. N. Lewis. He called a base anything that has available an unshared pair of electrons, such as NH_3, $:N—H$, with H above and H below,

and an acid anything that might attach itself to such a pair of electrons, such as H^+, to form NH_4^+, or BF_3, to form $F_3B—NH_3$.

This concept explains many phenomena. An example is the effect of certain substances other than hydrogen ion in changing the color of indicators. Another interesting application of the concept is its explanation of salt formation by reaction of acidic oxides and basic oxides.

Exercises

14-1. Which of these oxides are acid anhydrides and which basic anhydrides? Write an equation for each representing its reaction with water.

P_2O_3 N_2O_5 Na_2O Mn_2O_7

Cl_2O B_2O_3 I_2O_5 SO_2

Cl_2O_7 CO_2 SO_3

14-2. What is the normality of a solution of a strong acid 25.00 ml of which is rendered neutral by 28.75 ml of 0.1063 N NaOH solution?

14-3. The poisonous *botulinus* organism does not grow in canned vegetables if the *p*H is less than 4.5. Some investigators (*Journal of Chemical Education*, 22, 409, 1945) have recommended that in home canning of nonacid foods, such as beans, without a pressure canner a quantity of hydrochloric acid be added. The amount of hydrochloric acid recommended is 25 ml of 0.5 N hydrochloric acid per half-liter jar. Calculate the *p*H that this solution would have, assuming it originally to be neutral, and neglecting the buffering action of the organic material. Also calculate the amount of baking soda ($NaHCO_3$), measured in teaspoonfuls, that would be required to neutralize the acid after the jar is open. One teaspoon equals 4 grams of baking soda.

14-4. What is the *p*H, to the nearest *p*H unit, of 1 N HCl? of 0.1 N HCl? of 10 N HCl? of 0.1 N NaOH? of 10 N NaOH?

14-5. Calculate the hydrogen-ion concentration in the following solutions:

(*a*) 1 F $HC_2H_3O_2$, $K = 1.8 \times 10^{-5}$

(*b*) 0.06 F HNO_2, $K = 0.45 \times 10^{-3}$

(*c*) 0.004 F NH_4OH, $K_B = 1.8 \times 10^{-5}$

(*d*) 0.1 F HF, $K = 6.7 \times 10^{-4}$

What are the *p*H values of the solutions?

14-.6 Calculate the concentrations of the various ionic and molecular species in a solution prepared by mixing equal volumes of 1 N NaOH and 0.5 N NH_4OH.

14-7. Calculate the *p*H of a solution that is 0.1 F in HNO_2 and 0.1 F in HCl.

14-8. Calculate the concentrations of the various ionic and molecular species in the following solutions:
(a) 0.1 F H_2Se ($K_1 = 1.7 \times 10^{-4}$, $K_2 = 1 \times 10^{-11}$)
(b) 0.01 F H_2CO_3 ($K_1 = 4.5 \times 10^{-7}$, $K_2 = 6 \times 10^{-11}$)
(c) 1 F H_2CrO_4 ($K_1 = 0.18$, $K_2 = 3.2 \times 10^{-7}$)
(d) 0.5 F H_3PO_4 ($K_1 = 7.5 \times 10^{-3}$, $K_2 = 0.6 \times 10^{-7}$, $K_3 = 1 \times 10^{-12}$)
(e) 1 F H_2SO_4 ($K_2 = 1.20 \times 10^{-2}$)
(f) 0.01 F H_2SO_4

14-9. Calculate the pH of a solution that is
(a) 0.1 F in NH_4Cl, 0.1 F in NH_4OH
(b) 0.05 F in NH_4Cl, 0.15 F in NH_4OH
(c) 1.0 F in $HC_2H_3O_2$, 0.3 F in $NaC_2H_3O_2$
(d) prepared by mixing 10 ml 1 F $HC_2H_3O_2$ with 90 ml 0.05 F NaOH

14-10. Calculate the pH of a solution that is prepared from
(a) 10 ml 1 F HCN, 10 ml 1 F NaOH
(b) 10 ml 1 F NH_4OH, 10 ml 1 F HCl
(c) 10 ml 1 F NH_4OH, 10 ml 1 F NH_4Cl

14-11. Calculate the concentration of the various ionic and molecular species in
(a) 0.4 F NH_4Cl
(b) 0.1 F $NH_4C_2H_3O_2$
(c) 0.1 F $NaHCO_3$
(d) 0.1 F Na_2CO_3

14-12. Calculate the concentration of the various ionic and molecular species in a solution that is
(a) 0.3 F in HCl, and 0.1 F in H_2S
(b) buffered to a pH of 4, and 0.1 F in H_2S
(c) 0.2 F in KHS
(d) 0.2 F in K_2S

14-13. Boric acid loses only one hydrogen ion. In 0.1 M H_3BO_3, $[H^+] = 1.05 \times 10^{-5}$. Calculate the ionization constant for boric acid.

14-14. A patent medicine for stomach ulcers contains 2.1 g of $Al(OH)_3$ per 100 ml. How far wrong is the statement on the label that the preparation is "capable of combining with 16 times its volume of $N/10$ HCl"?

14-15. Which of these substances form acidic solutions, which neutral, and which basic? Write equations for the reactions which give H^+ or OH^-.

NaCl	$(NH_4)_2SO_4$	$CuSO_4 \cdot 5H_2O$
NaCN	$NaHSO_4$	$FeCl_2$
Na_3PO_4	NaH_2PO_4	$KAlSO_4 \cdot 12H_2O$
NH_4Cl	Na_2HPO_4	$Zn(ClO_4)_2$
NH_4CN	$KClO_4$	BaO

14-16. Approximately how much acetic acid must be added to a 0.1 N solution of sodium acetate to make the solution neutral (pH 7)?

14-17. What indicators should be used in titrating the following acids?

$$K_A$$

HNO₂	4.5×10^{-4}
H₂S (first hydrogen)	1.1×10^{-7}
HCN	$4 \ \times 10^{-10}$

HNO$_2$ 4.5×10^{-4}

With what indicators could you titrate separately for HCl and $HC_2H_3O_2$ in a solution containing both acids?

14-18. What relative weights of KH_2PO_4 and $Na_2HPO_4 \cdot 2H_2O$ should be taken to make a buffered solution with pH 6.0?

14-19. Carbon dioxide, produced by oxidation of substances in the tissues, is carried by the blood to the lungs. Part of it is in solution as carbonic acid, and part as hydrogen carbonate ion, HCO_3^-. If the pH of the blood is 7.4, what fraction is carried as the ion?

14-20. The value of K_1 for H_2S is 1.1×10^{-7}. What is the ratio $[H_2S]/[HS^-]$ at pH 8? If hydrogen sulfide at 1 atm pressure is 0.1 F soluble in acid solution, what would be its solubility at pH 8?

14-21. Would water act as an acid or a base when dissolved in liquid H_2S? Would H_2Se act as an acid or a base?

14-22. Hydrogen cyanide, H—C≡N:, is amphiprotic. What is its conjugate acid? Its conjugate base?

14-23. What reaction would you expect to take place when HCN is dissolved in pure sulfuric acid?

14-24. What is the concentration of the cation $H_3SO_4^+$ in pure sulfuric acid? Of the anion HSO_4^-? (See Table 14-1.) (Answer: 0.014 mole/liter, 0.014 mole/liter.)

14-25. What is the concentration of the cation $H_2C_2H_3O_2^+$ in pure acetic acid? Of the anion $C_2H_3O_2^-$?

14-26. A 0.0001 F solution of sodium acetate in acetic acid has electric conductivity about 300 times that of pure acetic acid. From this fact, calculate a rough value of the autoprotolysis constant of acetic acid. Why is this value only approximate?

14-27. An exact value of the autoprotolysis constant can be obtained by use of the measured values of the electric conductivities of pure acetic acid and three acetic acid solutions: 0.0001 F $NaC_2H_3O_2$, 0.0001 F $HClO_4$, and 0.0001 F $NaClO_4$. Can you explain how the calculation is made?

14-28. Discuss the reaction $F^- + BF_3 \longrightarrow BF_4^-$ in terms of the Lewis theory of acids and bases.

14-29. Explain why a metal hydroxide such as ferric hydroxide, $Fe(OH)_3$, is much more soluble in acidic solution than it is in a basic solution.

14-30. Using the solubility product of silver acetate, $[Ag^+][C_2H_3O_2^-] = 3.6 \times 10^{-3}$, and the ionization constant of acetic acid, calculate the solubility of silver acetate in a basic solution, a solution with pH 5.0, and a solution with pH 3.5.

14-31. What is the formula of the xenic acid formed by hydrolysis of xenon tetrafluoride? Write the equation for the reaction. It has been reported that this acid is weak. What value do you predict the acid constant to have? (Answer: Less than 10^{-7}.)

14-32. It was suggested long ago that xenon with oxidation number $+8$ should form the xenic acid H_4XeO_6, but this acid was first made in 1963, and its acid constants have not yet been reported. Would you predict it to be a strong acid or a weak acid? Estimate values of the four successive acid constants for this acid.

15

Oxidation-Reduction Reactions. Electrolysis

It was mentioned in Section 6-8 that in the period from 1884 to 1887 Svante Arrhenius developed the theory that electrolytes (salts, acids, bases) in aqueous solution are dissociated into electrically charged atoms or groups of atoms, called cations and anions, and some of the properties of these solutions were discussed in Section 13-11. The present chapter is devoted in part to the phenomena involved in the interaction of molten salts and ionic solutions with an electric current. It is found that the electron reactions that take place at electrodes can be described as involving oxidation or reduction of atoms or groups of atoms, and that the chemical reactions called oxidation-reduction reactions (sometimes shortened to redox reactions) can often be conveniently described in terms of two electrode reactions.

15-1. The Electrolytic Decomposition of Molten Salts

The discovery of ions resulted from the experimental investigations of the interaction of an electric current with chemical substances. These investigations were begun early in the nineteenth century, and were carried on effectively by Michael Faraday (1791–1867), in the period around 1830.

The Electrolysis of Molten Sodium Chloride

Molten sodium chloride (the salt melts at 801°C) conducts an electric current, as do other molten salts. During the process of conducting the current a chemical reaction occurs; the salt is decomposed. If two electrodes (carbon rods) are dipped into a crucible containing molten sodium chloride and an electric potential, from a battery or generator, is applied, metallic sodium is produced at the cathode and chlorine gas at the anode. Such electric decomposition of a substance is called electrolysis.

The Mechanism of Ionic Conduction

Molten sodium chloride, like the crystalline substance, consists of equal numbers of sodium ions and chloride ions. These ions are very stable, and do not gain electrons or lose electrons easily. Whereas the ions in the crystal are firmly held in place by their neighbors, those in the molten salt move about with considerable freedom.

An electric generator or battery forces electrons into the cathode and pumps them away from the anode—electrons move freely in a metal or a semimetallic conductor such as graphite. But electrons cannot ordinarily get into a substance such as salt; the crystalline substance is an insulator, and the electric conductivity shown by the molten salt is not electronic conductivity (metallic conductivity), but is conductivity of a different kind, called ionic conductivity or electrolytic conductivity. This sort of conductivity results from the motion of the ions in the liquid; the cations, Na^+, move toward the negatively charged cathode, and the anions, Cl^-, move toward the anode (Figure 15-1).

The Electrode Reactions

The preceding statement describes the mechanism of the conduction of the current through the liquid. We must now consider the way in which the current passes between the electrodes and the liquid; that is, we consider the electrode reactions.

The process that occurs at the cathode is this: sodium ions, attracted to the cathode, combine with the electrons carried by the cathode to form

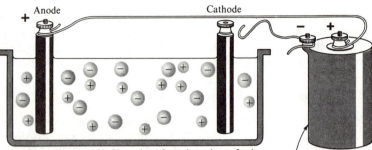

A molten salt such as NaCl consists of equal numbers of anions (Cl^-) and cations (Na^+).

Battery or other source of direct current.

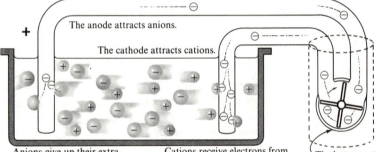

When the circuit is closed electrons flow as through a tube.

The anode attracts anions.

The cathode attracts cations.

Anions give up their extra electrons to the anode and become neutral atoms.

Cations receive electrons from the cathode and also become neutral atoms.

The battery acts as an electron pump.

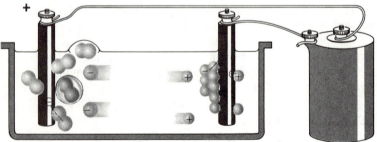

Neutral atoms of chlorine unite to form bubbles of chlorine gas (Cl_2).

Neutral sodium atoms form a layer of metallic sodium.

FIGURE 15-1
Electrolysis of molten sodium chloride.

sodium atoms; that is, to form sodium metal. The cathode reaction accordingly is

$$Na^+ + e^- \longrightarrow Na \qquad (15\text{-}1)$$

The symbol e^- represents an electron, which in this case comes from the cathode. Similarly, at the anode chloride ions give up their extra electrons to the anode, and become chlorine atoms, which are combined as the molecules of chlorine gas. The anode reaction is

$$2Cl^- \longrightarrow Cl_2 + 2e^- \qquad (15\text{-}2)$$

The Over-all Reaction

The whole process of electric conduction in this system thus occurs in the following steps.

1. An electron is pumped into the cathode.
2. The electron moves out of the cathode onto an adjacent sodium ion, converting it into an atom of sodium metal.
3. The charge of the electron is conducted across the liquid by the motion of the ions.
4. A chloride ion gives its extra electron to the anode, and becomes half of a molecule of chlorine gas.
5. The electron moves out of the anode toward the generator or battery.

The over-all reaction for the electrolytic decomposition is the sum of the two electrode reactions. Since two electrons are shown on their way around the circuit in Equation 15-2, we must double Equation 15-1:

$$2Na^+ + 2e^- \longrightarrow 2Na$$
$$\underline{2Cl^- \longrightarrow Cl_2 + 2e^-}$$
$$2Na^+ + 2Cl^- \underset{\text{electr.}}{\longrightarrow} 2Na + Cl_2 \qquad (15\text{-}3)$$

$$2NaCl \underset{\text{electr.}}{\longrightarrow} 2Na + Cl_2 \qquad (15\text{-}4)$$

The equations 15-3 and 15-4 are equivalent; they both represent the decomposition of sodium chloride into its elementary constituents. The abbreviation "electr." (for electrolysis) is written beneath the arrow to indicate that the reaction occurs on the passage of an electric current.

Ionic Conduction by Crystals

Metals conduct electricity by the motion of electrons from atom to atom within the crystal. The conductivity decreases with increasing temperature; the electrons are scattered by the disorder of the atoms associated with thermal agitation. Metalloids and other semiconductors, which also conduct electricity by motion of electrons, have smaller conductivity and a positive, rather than negative, temperature coefficient of conductivity. Some crystalline substances have large ionic conductivity, with positive temperature coefficient.

The relation between the current, I (in amperes), the applied voltage, E (in volts), and the resistance, R (in ohms), of a conductor is given by Ohm's law, $E = IR$. For a wire or other conductor with cross section A (in cm^2) and length l (in cm) the resistivity ρ (in ohm cm) is equal to RA/l. The reciprocal of the resistivity, $\sigma = \rho^{-1}$, is called the conductivity. It is usually reported in the units ohm^{-1} cm^{-1}, and is equal to the current in amperes that flows through a conductor with area 1 cm^2 under the potential 1 V per centimeter of length.

The conductivity of metals at 20°C ranges from about 1×10^4 ohm^{-1} cm^{-1} for the poorer conductors, such as barium ($\sigma = 1.7 \times 10^4$) and gadolinium ($\sigma = 0.7 \times 10^4$), to 0.7×10^6 for the best conductor, silver.

For ionic crystals, such as sodium chloride, the increase in conductivity with increasing temperature can be represented by an exponential factor:

$$\sigma(T) = \sigma_0 \exp(E^*/RT) \tag{15-5}$$

The value of E^*, which can be interpreted as the excitation energy required to move a sodium ion from its normal position in the crystal, is 190 kJ mole^{-1}. The conductivity remains very low, about 1×10^{-4} ohm^{-1} cm^{-1}, even at 800°C, only a degree below the melting point.

Silver iodide is an example of a crystal with large ionic conductivity, which reaches the value 2.5 ohm^{-1} cm^{-1} at 555°C, 3° below the melting point. At the melting point the conductivity of the crystal is greater than that of the liquid.

The very large ionic conductivity of silver iodide crystal is explained by its structure. The crystal is cubic, with the four iodide ions in the unit cell in the close-packed positions $0\ 0\ 0$, $0\ \frac{1}{2}\ \frac{1}{2}$, $\frac{1}{2}\ 0\ \frac{1}{2}$, $\frac{1}{2}\ \frac{1}{2}\ 0$. The silver ions might be in the octahedral positions $\frac{1}{2}\ \frac{1}{2}\ \frac{1}{2}$, and so on, which would give the sodium chloride structure, or the tetrahedral positions $\frac{1}{4}\ \frac{1}{4}\ \frac{1}{4}$, and so on, which would give the sphalerite structure, or in positions midway between adjacent iodide ions (ligancy 2 for silver, as found in the ion AgI_2^-). In fact, the x-ray diffraction pattern shows that the silver ions are distributed among all of these positions. They move with nearly complete freedom from one position to an adjacent (unoccupied) position. The potential barrier associated with this motion is small; the observed temperature coefficient of the conductivity corresponds to the value 5.1 kJ mole^{-1} for the excitation energy E^*.

15-2. The Electrolysis of an Aqueous Salt Solution

Although pure water does not conduct electricity very well (conductivity 4.4×10^{-4} ohm^{-1} cm^{-1} at 20°C), a solution of salt (or acid or base) is a good conductor. During electrolysis chemical reactions take place at the electrodes, as described below.

The phenomena that occur when a current of electricity is passed through such a solution are analogous to those described in the preceding section for molten salt. The five steps are the following.

1. Electrons are pumped into the cathode.
2. Electrons jump from the cathode to adjacent ions or molecules, producing the cathode reaction.
3. The current is conducted across the liquid by the motion of the dissolved ions.
4. Electrons jump from ions or molecules in the solution to the anode, producing the anode reaction.
5. The electrons move out of the anode toward the generator or battery.

Let us consider a dilute solution of sodium chloride (Figure 15-2). The process of conduction through this solution (step 3) is closely similar to that for molten sodium chloride. Here it is the dissolved sodium ions that move toward the cathode and the dissolved chloride ions that move toward the anode. By the motion of the ions in this way, negative electric charge is carried toward the anode and away from the cathode.

But the electrode reactions for dilute salt solutions are entirely different from those for molten salts. Electrolysis of dilute salt solution produces hydrogen at the cathode and oxygen at the anode, whereas electrolysis of molten salt produces sodium and chlorine.

The *cathode reaction* for a dilute salt solution is

$$2e^- + 2H_2O \longrightarrow H_2 + 2OH^- \qquad (15\text{-}6a)$$

Two electrons from the cathode react with two water molecules to produce a molecule of hydrogen and two hydroxide ions. The molecular hydrogen bubbles off as hydrogen gas (after the solution near the cathode has become saturated with hydrogen) and the hydroxide ions stay in the solution. The *anode reaction* is

$$2H_2O \longrightarrow O_2 + 4H^+ + 4e^- \qquad (15\text{-}6b)$$

Four electrons enter the anode from two water molecules, which decompose to form an oxygen molecule and four hydrogen ions.

These electrode reactions, like other chemical reactions, may occur in

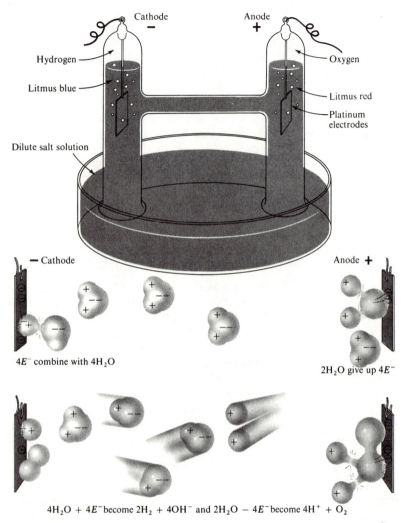

Cathode −

Anode +

Hydrogen

Oxygen

Litmus blue

Litmus red

Platinum electrodes

Dilute salt solution

− Cathode

Anode +

4E^- combine with 4H_2O

2H_2O give up 4E^-

4H_2O + 4E^- become 2H_2 + 4OH^- and 2H_2O − 4E^- become 4H^+ + O_2

Hydrogen gas evolved

4OH^- + 4H^+ form 4H_2O

Oxygen gas evolved

FIGURE 15-2
Electrolysis of a dilute aqueous salt solution.

steps; the description given in the preceding sentence of the course of the anode reaction is not to be interpreted as giving the necessary sequence of events.

The *over-all reaction* is obtained by multiplying Equation 15-6a by 2 and adding Equation 15-6b:

$$6H_2O \xrightarrow[\text{electr.}]{} \underset{\text{Cathode}}{2H_2} + \underset{\text{Anode}}{O_2} + \underset{\text{Anode}}{4H^+} + \underset{\text{Cathode}}{4OH^-} \qquad (15\text{-}7)$$

It is seen that in the electrolysis of the salt solution the solution around the anode becomes acidic, because of the production of hydrogen ions, and that around the cathode becomes basic, because of the production of hydroxide ions. This reaction could accordingly be used for the manufacture of acids such as hydrochloric acid and bases such as sodium hydroxide.

In the course of time, if the system were allowed to stand, the hydrogen ions produced near the anode and the hydroxide ions produced near the cathode would diffuse together and combine to form water:

$$H^+ + OH^- \longrightarrow H_2O$$

This reaction would, in particular, occur if the solution of electrolytes were to be stirred during the electrolysis. If this reaction of neutralization of hydrogen ion by hydroxide ion occurs completely, the over-all electrolysis reaction is

$$2H_2O \xrightarrow[\text{electr.}]{} \underset{\text{Cathode}}{2H_2} + \underset{\text{Anode}}{O_2} \qquad (15\text{-}8)$$

In discussing the electrode reactions we have made little use of the fact that the electrolyte is sodium chloride. Indeed, *the electrode reactions are the same for almost all dilute aqueous electrolytic solutions*, and even for pure water as well. When electrodes are placed in pure water and an electric potential is applied, the electrode reactions shown in Equations 15-6a and 15-6b begin to take place. Very soon, however, a large enough concentration of hydroxide ions is built up near the cathode and of hydrogen ions near the anode to produce a back electric potential that tends to stop the reactions. Even in pure water there are a few ions (hydrogen ions and hydroxide ions); these ions move slowly toward the electrodes, and neutralize the ions (OH^- and H^+, respectively) formed by the electrode reactions. It is the smallness of the current that the very few ions that are present in pure water can carry through the region between the electrodes that causes the electrolysis of pure water to proceed only very slowly.

Equations 15-6a and 15-6b show water molecules undergoing decomposition at the electrodes. These equations probably represent the usual molecular reactions in neutral salt solutions. In acidic solutions, however, in which there is a high concentration of hydrogen ions, the cathode reaction may well be simply the reaction

$$2H^+ + 2e^- \longrightarrow H_2$$

and in basic solutions, in which there is a high concentration of hydroxide ions, the anode reaction may be

$$4OH^- \longrightarrow O_2 + 2H_2O + 4e^-$$

The ions in an electrolytic solution can carry a larger current between the electrodes than can the very few ions in pure water. In a sodium chloride solution undergoing electrolysis, sodium ions move to the cathode region, where their positive electric charges compensate the negative charges of the hydroxide ions that have been formed by the cathode reaction. Similarly, the chloride ions that move toward the anode compensate electrically the hydrogen ions that have been formed by the anode reaction.

Production of hydroxide ions at the cathode and of hydrogen ions at the anode during the electrolysis can be demonstrated by means of litmus or a similar indicator.

The electrolysis of dilute aqueous solutions of other electrolytes is closely similar to that of sodium chloride, producing hydrogen and oxygen gases at the electrodes. Concentrated electrolytic solutions may behave differently; concentrated brine (sodium chloride solution) on electrolysis produces chlorine at the anode, as well as oxygen. We may understand this fact by remembering that in concentrated brine there are a great many chloride ions near the anode, and some of these give up electrons to the anode, and form chlorine molecules.

15-3. Oxidation-Reduction Reactions

From the rules for assigning oxidation numbers given in Section 6-14 we see that the anode reaction 15-2 involves an increase in oxidation number for chlorine, from -1 to 0, accompanied by loss of an electron to the electrode (the anode). Increase in oxidation number is described as oxidation. The anode reaction is an oxidation reaction. Similarly, the cathode reaction, Equation 15-1, in which an electron from the cathode causes the decrease in oxidation number of sodium from $+1$ to 0, is described as reduction: the cathode reaction is a reduction reaction.

We see that oxidation can be described as de-electronation, and reduction as electronation. In ordinary oxidation-reduction reactions the two processes take place simultaneously, sometimes by direct transfer of electrons from the atoms that are oxidized to those that are reduced.

It is often convenient in writing the equation for an oxidation-reduction reaction to write and balance equations for two electrode reactions (which may be hypothetical), and then to add these reactions in such a way that the electrons cancel.

The first step is to be sure that you know what the reactants are and what the products are. Then you identify the reducing agent and the oxidizing agent, write the equations for the de-electronation and electronation reactions, and combine these equations, as illustrated in the following example.

Example 15-1. The oxidizing agent permanganate ion, MnO_4^-, on reduction in acid solution forms manganous ion, Mn^{++}. Ferrous ion, Fe^{++}, can accomplish this reduction. Write the equation for the reaction between permanganate ion and ferrous ion in acid solution.

Solution. The oxidation number of manganese in permanganate ion is $+7$, $[Mn^{+7}(O^{-2})_4]^-$. That of manganous ion is $+2$. Hence five electrons are involved in the reduction of permanganate ion. The electron reaction is

$$[Mn^{+7}(O^{-2})_4]^- + 5e^- + \text{other reactants} \longrightarrow Mn^{++} + \text{other products} \quad (15\text{-}9)$$

In reactions in aqueous solutions water, hydrogen ion, and hydroxide ion may come into action as reactants or products. For example, in an acid solution hydrogen ion may be either a reactant or a product, and water may also be either a reactant or a product in the same reaction. In acid solutions hydroxide ion exists only in extremely low concentration, and would hardly be expected to enter into the reaction. Hence water and hydrogen ion may enter into the reaction now under consideration.

Reaction 15-9 is not balanced electrically; there are six negative charges on the left side and two positive charges on the right side. The only other ion that can enter into the reaction is hydrogen ion, and the number needed to give conservation of electric charge is 8. Thus we obtain

$$MnO_4^- + 5e^- + 8H^+ \longrightarrow Mn^{++} + \text{other products} \quad (15\text{-}10)$$

Oxygen and hydrogen occur here on the left side and not the right side of the reaction; conservation of atoms is satisfied if $4H_2O$ is written in as the "other products":

$$MnO_4^- + 5e^- + 8H^+ \longrightarrow Mn^{++} + 4H_2O \quad (15\text{-}11)$$

We check this equation on three points—*proper change in oxidation number* (5 electrons used, with change of -5 in oxidation number of manganese, from Mn^{+7} to Mn^{+2}), *conservation of electric charge* (from -1 -5 $+8$ to $+2$), and *conservation of atoms of each kind*—and convince ourselves that it is correct.

The electron reaction for the oxidation of ferrous ion is now written:

$$Fe^{++} \longrightarrow Fe^{+++} + e^- \tag{15-12}$$

This equation checks on all three points.

The equation for the oxidation-reduction reaction is obtained by combining the two electron reactions in such a way that the electrons liberated in one are used up in the other. We see that this is achieved by multiplying Equation 15-12 by 5 and adding it to Equation 15-11:

$$5Fe^{++} \longrightarrow 5Fe^{+++} + 5e^-$$
$$\underline{MnO_4^- + 5e^- + 8H^+ \longrightarrow Mn^{++} + 4H_2O}$$
$$MnO_4^- + 5Fe^{++} + 8H^+ \longrightarrow Mn^{++} + 5Fe^{+++} + 4H_2O$$

It is good practice to check this final equation also on all three points, to be sure that no mistake has been made.

It is not always necessary to carry through this entire procedure. Sometimes an equation is so simple that it can be written at once and verified by inspection. An example is the reduction of silver ion, Ag^+, by metallic zinc:

$$Zn(c) + 2Ag^+(aq) \longrightarrow 2Ag(c) + Zn^{++}(aq)$$

Sometimes, too, the conditions determine the reaction, as when a single substance decomposes. Thus ammonium nitrite decomposes to give water and nitrogen:

$$NH_4NO_2 \longrightarrow N_2 + 2H_2O$$

Here N^{+3} (of NO_2^-) oxidizes N^{-3} (of NH_4^+), both going to N^0 (of N_2).

Oxidation Equivalents and Reduction Equivalents

The *oxidizing capacity* or *reducing capacity* of an oxidizing agent or reducing agent is equal to the number of electrons involved in its reduction or oxidation. An oxidation-reduction equation is balanced when the amounts of oxidizing agent and reducing agent indicated as reacting have the same capacities.

An *oxidation equivalent* or *reduction equivalent* of a substance is the amount that takes up or gives up one electron (one mole of electrons). Thus the gram equivalent weight (equivalent weight expressed in grams) of potassium permanganate as an oxidizing agent in acid solution (see Equation 15-11) is one-fifth of the gram formula weight, whereas the gram equivalent weight of ferrous ion as a reducing agent is just the gram atomic weight.

Equivalent weights of oxidizing agents and reducing agents react exactly with one another, since they involve the taking up or giving up of the same number of electrons.

Normal Solutions of Oxidizing and Reducing Agents

A solution of an oxidizing agent or reducing agent containing 1 gram equivalent weight per liter of solution is called a 1 normal (1 N) solution. In general the *normality* of the solution is the number of gram equivalent weights of the oxidizing or reducing agent present per liter.

It is seen from the definition that equal volumes of an oxidizing solution and a reducing solution of the same normality react exactly with each other.

Example 15-2. What is the normality, for use as an oxidizing agent in acid solution, of a permanganate solution made by dissolving one-tenth of a gram formula weight of $KMnO_4$ ($\frac{1}{10} \times 158.03$ g) in water and diluting to a volume of 1 l?

Answer. The reduction of permanganate ion in acid solution involves 5 electrons (Equation 15-11). Hence 1 gram molecular weight is 5 equivalents. The solution is accordingly 0.5 *N*.

It is necessary to take care that the conditions of the use of a reagent are known in stating its normality. Thus permanganate ion is sometimes used as an oxidizing agent in neutral or basic solution, in which it is reduced by only three steps, to manganese dioxide, MnO_2, in which manganese has oxidation number 4. The above solution would have normality 0.3 for this use.

15-4. Quantitative Relations in Electrolysis

In 1832 and 1833 Michael Faraday reported his discovery by experiment of the fundamental laws of electrolysis. These are

1. *The weight of a substance produced by a cathode or anode reaction in electrolysis is directly proportional to the quantity of electricity passed through the cell.*

2. *The weights of different substances produced by the same quantity of electricity are proportional to the equivalent weights of the substances.*

These laws are now known to be the result of the fact that electricity is composed of individual particles, the electrons. Quantity of electricity can be expressed as number of electrons; the number of molecules of a substance produced by an electrode reaction is related in a simple way to the number of electrons involved; hence the amount of substance produced is proportional to the quantity of electricity, in a way that involves the molecular weight or chemical equivalent weight of the substance.

The amounts of substances produced by a given quantity of current can be calculated from knowledge of the electrode reactions. The magnitude of the charge of one mole of electrons is 96,490 C. This is called a *faraday*: 1 F = 1 faraday = 96,490 C.

It is customary to define the faraday as this quantity of positive electricity. The quantitative treatment of electrochemical reactions is made in the same way as the calculation of weight relations in ordinary chemical reactions, with use of the faraday to represent one mole of electrons.

Example 15-3. For how long a time would a current of 10 A have to be passed through a cell containing fused sodium chloride to produce 23 g of metallic sodium at the cathode? How much chlorine would be produced at the anode?

Solution. The cathode reaction is

$$Na^+ + e^- \longrightarrow Na$$

Hence 1 mole of electrons passing through the cell would produce 1 mole of sodium atoms. One mole of electrons is 1 faraday, and 1 mole of sodium atoms is a gram atom of sodium, 23.00 g. Hence the amount of electricity required is 96,490 coulombs. One coulomb is 1 ampere second. Hence 96,490 coulombs of electricity passes through the cell if 10 A flows for 9649 seconds. The answer is thus 9649 s.

The anode reaction is

$$2Cl^- \longrightarrow Cl_2 + 2e^-$$

To produce 1 mole of molecular chlorine, Cl_2, 2 F must pass through the cell. One faraday will produce $\frac{1}{2}$ mole of Cl_2, or 1 gram atom of chlorine, which is 35.46 g. Note that two moles of electrons are required to produce one mole of molecular chlorine.

Example 15-4. Two cells are set up in series, and a current is passed through them. Cell A contains an aqueous solution of silver sulfate, and has platinum electrodes (which are unreactive). Cell B contains a copper sulfate solution, and has copper electrodes. The current is passed through until 1.6 g of oxygen has been liberated at the anode of cell A. What has occurred at the other electrodes? (See Figure 15-3.)

Solution. At the anode of cell A the reaction is

$$2H_2O \longrightarrow O_2(g) + 4H^+ + 4e^-$$

Hence 4 F of electricity liberates 32 g of oxygen, and 0.2 F liberates 1.6 g.

At the cathode of cell A the reaction is

$$Ag^+ + e^- \longrightarrow Ag(c)$$

One gram atom of silver, 107.880 g, would be deposited by 1 F, and the passage of 0.2 F through the cell would hence deposit 21.576 g of silver on the platinum cathode.

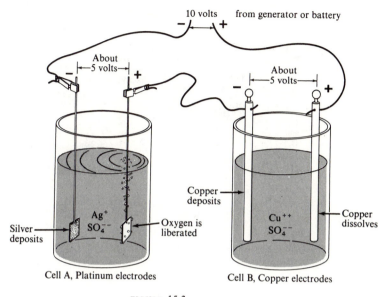

FIGURE 15-3
Two electrolytic cells in series.

At the cathode in cell B the reaction is

$$Cu^{++} + 2e^- \longrightarrow Cu(c)$$

One gram atom of copper, 63.57 g, would hence be deposited on the cathode by 2 F of electricity, and 6.357 g by 0.2 F.

The same amount of copper, 6.357 g, would be dissolved from the anode of cell B, since the same number of electrons flows through the anode as through the cathode. The anode reaction is

$$Cu(c) \longrightarrow Cu^{++} + 2e^-$$

Note that the total voltage difference supplied by the generator or battery (shown as 10 volts) is divided between the two cells in series. The division need not be equal, as indicated, but is determined by the properties of the two cells.

15-5.   The Electromotive-force Series of the Elements

It is found that if a piece of one metal is put into a solution containing ions of another metallic element the first metal may dissolve, with the deposition of the second metal from its ions. Thus a strip of zinc placed in a

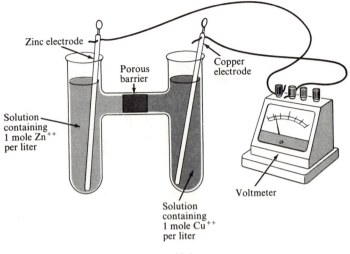

FIGURE 15-4
A copper-zinc cell.

solution of a copper salt causes a layer of metallic copper to deposit on the zinc, as the zinc goes into solution:

$$\frac{\begin{array}{l}Zn(c) \longrightarrow Zn^{++} + 2e^- \\ Cu^{++} + 2e^- \longrightarrow Cu(c)\end{array}}{Zn(c) + Cu^{++} \longrightarrow Zn^{++} + Cu(c)}$$

On the other hand, a strip of copper placed in a solution of a zinc salt does not cause metallic zinc to deposit.

It is not strictly correct to say that zinc can replace copper in solution and that copper cannot replace zinc. If a piece of metallic copper is placed in a solution containing zinc ion in appreciable concentration (1 mole per liter, say) and no cupric ion at all, the reaction

$$Cu(c) + Zn^{++} \longrightarrow Cu^{++} + Zn(c)$$

will occur to a very small extent, stopping when a certain very small concentration of copper ion has been produced. If metallic zinc is added to a 1-molar solution of cupric ion, the reaction

$$Zn(c) + Cu^{++} \longrightarrow Zn^{++} + Cu(c)$$

the reverse of the preceding reaction, will take place almost to completion, stopping when the concentration of copper ion has become very small. The principles of thermodynamics require that the ratio of concentrations of the two ions Cu^{++} and Zn^{++} in equilibrium with solid copper and solid zinc be the same whether the equilibrium is approached from the

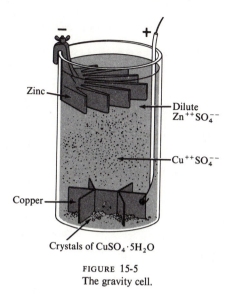

FIGURE 15-5
The gravity cell.

Cu, Zn^{++} side or the Zn, Cu^{++} side. The statement "zinc replaces copper from solution" means that at equilibrium the ratio of cupric-ion concentration to zinc-ion concentration is small.

By experiments of this sort, the elements can be arranged in a table showing their ability to reduce ions of other metals. This table is given as Table 15-1. The metal with the greatest reducing power is at the head of the list. This series is called the electromotive-force series because the tendency of one metal to reduce ions of another can be measured by setting up an electric cell and measuring the voltage that it produces. ("Electromotive force" is here a synonym for "voltage.") It is conventional to tabulate the voltage corresponding to unit activity for each ion; that is, the concentration of the ion multiplied by the activity coefficient (Section 13-11). Thus the cell shown in Figure 15-4 would be used to measure the voltage between the electrodes at which occur the electrode reactions

$$Zn(c) \rightleftarrows Zn^{++}(aq, a = 1) + 2e^-$$

and

$$Cu^{++}(aq, a = 1) + 2e^- \rightleftarrows Cu(c)$$

This cell produces a voltage* of about 1.107 V, the difference of the values of $E°$ shown in the table. The cell is used to some extent in practice; it is called the gravity cell when made as shown in Figure 15-5.

---

*Correction by a few millivolts must be made for the *liquid-junction potential*, which results from effects in the connecting arm where the two solutions meet.

TABLE 15-1
Standard Oxidation-Reduction Potentials and Equilibrium Constants
The values apply to temperature 25°C, with standard concentration
for aqueous solutions $1M$ and standard pressure of gases 1 atm.

	$E°$	K
$Li \rightleftarrows Li^+ + e^-$	3.05	4×10^{50}
$Cs \rightleftarrows Cs^+ + e^-$	2.92	1×10^{49}
$Rb \rightleftarrows Rb^+ + e^-$	2.92	1×10^{49}
$K \rightleftarrows K^+ + e^-$	2.92	1×10^{49}
$\frac{1}{2}Ba \rightleftarrows \frac{1}{2}Ba^{++} + e^-$	2.90	5×10^{48}
$\frac{1}{2}Sr \rightleftarrows \frac{1}{2}Sr^{++} + e^-$	2.89	4×10^{48}
$\frac{1}{2}Ca \rightleftarrows \frac{1}{2}Ca^{++} + e^-$	2.87	2×10^{48}
$Na \rightleftarrows Na^+ + e^-$	2.712	4.0×10^{45}
$\frac{1}{3}Al + \frac{4}{3}OH^- \rightleftarrows \frac{1}{3}Al(OH)_4^- + e^-$	2.35	3×10^{39}
$\frac{1}{2}Mg \rightleftarrows \frac{1}{2}Mg^{++} + e^-$	2.34	2×10^{39}
$\frac{1}{2}Be \rightleftarrows \frac{1}{2}Be^{++} + e^-$	1.85	1×10^{31}
$\frac{1}{3}Al \rightleftarrows \frac{1}{3}Al^{+++} + e^-$	1.67	1×10^{28}
$\frac{1}{2}Zn + 2OH^- \rightleftarrows \frac{1}{2}Zn(OH)_4^{--} + e^-$	1.216	2.7×10^{20}
$\frac{1}{2}Mn \rightleftarrows \frac{1}{2}Mn^{++} + e^-$	1.18	7×10^{19}
$\frac{1}{2}Zn + 2NH_3 \rightleftarrows \frac{1}{2}Zn(NH_3)_4^{++} + e^-$	1.03	2×10^{17}
$Co(CN)_6^{----} \rightleftarrows Co(CN)_6^{---} + e^-$	0.83	1×10^{14}
$\frac{1}{2}Zn \rightleftarrows \frac{1}{2}Zn^{++} + e^-$	.762	6.5×10^{12}
$\frac{1}{3}Cr \rightleftarrows \frac{1}{3}Cr^{+++} + e^-$	.74	3×10^{12}
$\frac{1}{2}H_2C_2O_4(aq) \rightleftarrows CO_2 + H^+ + e^-$	.49	2×10^8
$\frac{1}{2}Fe \rightleftarrows \frac{1}{2}Fe^{++} + e^-$	.440	2.5×10^7
$\frac{1}{2}Cd \rightleftarrows \frac{1}{2}Cd^{++} + e^-$	.402	5.7×10^6
$\frac{1}{2}Co \rightleftarrows \frac{1}{2}Co^{++} + e^-$	.277	4.5×10^4
$\frac{1}{2}Ni \rightleftarrows \frac{1}{2}Ni^{++} + e^-$	.250	1.6×10^4
$I^- + Cu \rightleftarrows CuI(s) + e^-$	.187	1.4×10^3
$\frac{1}{2}Sn \rightleftarrows \frac{1}{2}Sn^{++} + e^-$	.136	1.9×10^2
$\frac{1}{2}Pb \rightleftarrows \frac{1}{2}Pb^{++} + e^-$	.126	1.3×10^2
$\frac{1}{2}H_2 \rightleftarrows H^+ + e^-$	.000	1
$\frac{1}{2}H_2S \rightleftarrows \frac{1}{2}S + H^+ + e^-$	−0.141	4.3×10^{-3}

	E°	K
$Cu^+ \rightleftarrows Cu^{++} + e^-$	-0.153	2.7×10^{-3}
$\frac{1}{2}H_2O + \frac{1}{2}H_2SO_3 \rightleftarrows \frac{1}{2}SO_4^{--} + 2H^+ + e^-$	-0.17	1×10^{-3}
$\frac{1}{2}Cu \rightleftarrows \frac{1}{2}Cu^{++} + e^-$	-0.345	1.6×10^{-6}
$Fe(CN)_6^{----} \rightleftarrows Fe(CN)_6^{---} + e^-$	-0.36	9×10^{-7}
$I^- \rightleftarrows \frac{1}{2}I_2(s) + e^-$	-0.53	1×10^{-9}
$MnO_4^{--} \rightleftarrows MnO_4^- + e^-$	-0.54	1×10^{-9}
$\frac{4}{3}OH^- + \frac{1}{3}MnO_2 \rightleftarrows \frac{1}{3}MnO_4^- + \frac{2}{3}H_2O + e^-$	-0.57	3×10^{-10}
$\frac{1}{2}H_2O \rightleftarrows \frac{1}{4}O_2 + H^+ + e^-$	-0.682	3.5×10^{-12}
$Fe^{++} \rightleftarrows Fe^{+++} + e^-$	-0.771	1.1×10^{-13}
$Hg \rightleftarrows \frac{1}{2}Hg_2^{++} + e^-$	-0.799	3.7×10^{-14}
$Ag \rightleftarrows Ag^+ + e^-$	-0.800	3.5×10^{-14}
$H_2O + NO_2 \rightleftarrows NO_3^- + 2H^+ + e^-$	-0.81	3×10^{-14}
$\frac{1}{2}Hg \rightleftarrows \frac{1}{2}Hg^{++} + e^-$	-0.854	4.5×10^{-15}
$\frac{1}{2}Hg_2^{++} \rightleftarrows Hg^{++} + e^-$	-0.910	5.0×10^{-16}
$\frac{1}{2}HNO_2 + \frac{1}{2}H_2O \rightleftarrows \frac{1}{2}NO_3^- + \frac{3}{2}H^+ + e^-$	-0.94	2×10^{-16}
$NO + H_2O \rightleftarrows HNO_2 + H^+ + e^-$	-0.99	2×10^{-17}
$\frac{1}{2}ClO_3^- + \frac{1}{2}H_2O \rightleftarrows \frac{1}{2}ClO_4^- + H^+ + e^-$	-1.00	2×10^{-17}
$Br^- \rightleftarrows \frac{1}{2}Br_2(l) + e^-$	-1.065	1.3×10^{-18}
$H_2O + \frac{1}{2}Mn^{++} \rightleftarrows \frac{1}{2}MnO_2 + 2H^+ + e^-$	-1.23	2×10^{-21}
$Cl^- \rightleftarrows \frac{1}{2}Cl_2 + e^-$	-1.358	1.5×10^{-23}
$\frac{7}{6}H_2O + \frac{1}{3}Cr^{+++} \rightleftarrows \frac{1}{6}Cr_2O_7^{--} + \frac{7}{3}H^+ + e^-$	-1.36	1×10^{-23}
$\frac{1}{2}H_2O + \frac{1}{6}Cl^- \rightleftarrows \frac{1}{6}ClO_3^- + H^+ + e^-$	-1.45	4×10^{-25}
$\frac{1}{3}Au \rightleftarrows \frac{1}{3}Au^{+++} + e^-$	-1.50	6×10^{-26}
$\frac{4}{5}H_2O + \frac{1}{5}Mn^{++} \rightleftarrows \frac{1}{5}MnO_4^- + \frac{8}{5}H^+ + e^-$	-1.52	3×10^{-26}
$\frac{1}{2}Cl_2 + H_2O \rightleftarrows HClO + H^+ + e^-$	-1.63	4×10^{-28}
$H_2O \rightleftarrows \frac{1}{2}H_2O_2 + H^+ + e^-$	-1.77	2×10^{-30}
$Co^{++} \rightleftarrows Co^{+++} + e^-$	-1.84	1×10^{-31}
$F^- \rightleftarrows \frac{1}{2}F_2 + e^-$	-2.65	4×10^{-44}

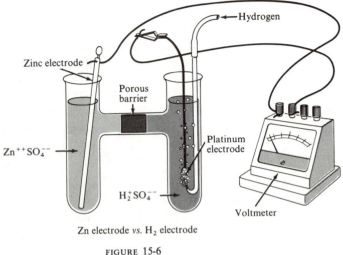

Zn electrode *vs.* H_2 electrode

FIGURE 15-6
A cell with a hydrogen electrode.

The standard reference point in the EMF series is the *hydrogen electrode*, which consists of gaseous hydrogen at 1 atm bubbling over a platinum electrode in an acidic solution with activity 1 for the hydrogen ion (Figure 15-6). Similar electrodes can be made for some other nonmetallic elements, and a few of these elements are included in the table, as well as some other oxidation-reduction pairs.

15-6. Equilibrium Constants for Oxidation-Reduction Couples

The electric cell shown in Figure 15-6 is conventionally written as

$$Zn(c)|Zn^{++}(a = 1)|H^+(a = 1)|H_2(1 \text{ atm})$$

Here the vertical lines denote contact between two phases. The electromotive force is taken as $E = E(\text{left}) - E(\text{right})$. For this cell $E(\text{right})$ is equal to zero, since the right half-cell in this example is taken as the standard, with $E° = 0$, and the value of E is $E(\text{left})$. From Table 15-1 we find that $E(\text{left})$ is 0.762 V, and hence that the EMF of the cell is 0.762 V. The positive value means that there is a greater electron pressure at the left electrode than at the right electrode.

If two moles of electrons (2 F) were to flow through the wire from the left electrode to the right electrode, with the corresponding electron reactions taking place at the electrodes, the change in the system would correspond to the equation

$$Zn(c) + 2H^+(a = 1) \longrightarrow Zn^{++}(a = 1) + H_2(1 \text{ atm})$$

The electric current could be used completely in doing work, the amount of work being $E \times 2F$ (or, in general, nEF, with n the number of electrons involved in the electrode reactions). We have seen in Chapter 11 that the amount of work done by a system when it undergoes a reversible change in state at constant pressure and temperature is $-\Delta G$, the decrease in the free energy of the system. Hence we write

$$nEF = -\Delta G \tag{15-13}$$

If the reactants and products in the cell are in their standard states the decrease in free energy is $-\Delta G°$:

$$nE°F = -\Delta G° \tag{15-14}$$

We have seen, however, that $-\Delta G°$ is related to the equilibrium constant K of the reaction (Equation 11-27):

$$-\Delta G° = RT \ln K \tag{15-15}$$

We may rewrite this equation as

$$K = \exp(-\Delta G°/RT) \tag{15-16}$$

or

$$K = \exp(nE°F/RT) \tag{15-17}$$

Many equilibrium constants (and free-energy values) have been determined experimentally by EMF measurements of electric cells. One difficulty that has not been overcome for many possible half-cells is that of finding an electrode surface that permits the half-cell (electron) reaction to take place in a reversible manner. A platinum electrode covered with finely divided platinum (platinum black) is effective for many half-cells.

The electron reactions are all written in Table 15-1 so as to produce one electron. This is done for convenience; with this convention the ratio of two values of K gives the equilibrium constant for the reaction obtained by subtracting the equation for one couple from that for another. It is sometimes desirable to clear the equation of fractions by multiplying by a suitable factor; this involves raising the equilibrium constant to the power equal to this factor.

Many questions about chemical reactions can be answered by reference to a table of standard oxidation-reduction potentials. In particular it can be determined whether or not a given oxidizing agent and a given reducing agent can possibly react to an appreciable extent, and the extent of possible reaction can be predicted. It cannot be said, however, that the reaction

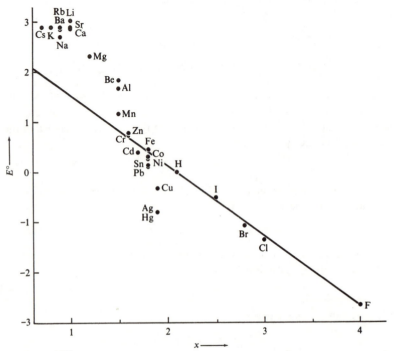

FIGURE 15-7
Diagram illustrating the relation between the standard electrode potentials of some elements and their electronegativity values.

will necessarily proceed at a significant rate under given conditions; the table gives information only about the state of chemical equilibrium and not about the rate at which equilibrium is approached. For this reason the most valuable use of the table is in connection with reactions that are known to take place, to answer questions as to the extent of reaction; but the table is also valuable in telling whether or not it is worth while to try to make a reaction go by changing conditions.

The great simplification introduced by this procedure can be seen by examining Table 15-1. This table contains only 56 entries, which correspond to 56 different electron reactions. By combining any two of these electron reactions the equation for an ordinary oxidation-reduction reaction can be written. There are 1540 (56 × $\frac{55}{2}$) of these oxidation-reduction reactions that can be formed from the 56 electron reactions. The 56 numbers in the table can be combined in such a way as to give the 1540 values of their equilibrium constants; accordingly, this small table permits a prediction to be made as to whether any one of these 1540 reactions will tend to go in a forward direction or the reverse direction.

A more extensive table in the book *The Oxidation States of the Elements and Their Potentials in Aqueous Solutions*, by W. M. Latimer, occupies eight pages; the information given on these eight pages permits one to calculate values of the equilibrium constants for about 85,000 reactions, which, if written out, would occupy 1750 pages of the same size as the pages in Latimer's book; and, moreover, it is evident that if the equilibrium constants were independent of one another, and had to be determined by separate experiments, 85,000 experiments would have had to be carried out, instead of only about 400.

Electrode Potentials and Electronegativity

There is a rough general correlation between standard electrode potentials and electronegativity values, illustrated in Figure 15-7. It is seen that to within the uncertainty in the electronegativity values, ± 0.05, the points for chlorine, bromine, and iodine lie on the straight line connecting the points for fluorine and hydrogen. The contribution of the entropy term to the standard free energy change (which determines $E°$) is different for the different kinds of oxidation-reduction couples, and a close correlation between $E°$ and x is not to be expected.

Some ways in which the values in Table 15-1 can be used are illustrated below.

Example 15-5. Would you expect reaction to occur on mixing solutions of ferrous sulfate and mercuric sulfate?

Answer. The ferrous-ferric couple has potential -0.771 V and the mercurous-mercuric couple -0.910 V; hence the latter couple is the stronger oxidizing of the two, and the reaction

$$2Fe^{++} + 2Hg^{++} \longrightarrow 2Fe^{+++} + Hg_2^{++}$$

would occur, and proceed well toward completion.

Example 15-6. In the manufacture of potassium permanganate a solution containing manganate ion is oxidized by chlorine. Would bromine or iodine be as good?

Answer. From the table we see that the values of $E°$ and K are the following:

		$E°$	K
$MnO_4^{--} \rightleftarrows MnO_4^- + e^-$		-0.54	1×10^{-9}
$Cl^- \rightleftarrows \frac{1}{2}Cl_2 + e^-$		-1.358	2×10^{-23}
$Br^- \rightleftarrows \frac{1}{2}Br_2(l) + e^-$		-1.065	1×10^{-18}
$I^- \rightleftarrows \frac{1}{2}I_2(s) + e^-$		-0.535	1×10^{-9}

The value for iodine is so close to that for manganate-permanganate that effective oxidation by iodine (approaching completion) would not occur; hence iodine would be unsatisfactory. Bromine would produce essentially complete reaction, and in this respect would be as good as chlorine; but it costs ten times as much, and so should not be used.

Example 15-7. What equilibrium composition would you expect to obtain by mixing equal volumes of a 0.2 F solution of $K_4Co(CN)_6$ and a 0.2 F solution of $K_3Fe(CN)_6$?

Answer. From Table 15-1 we obtain the following information:

	$E°$	K
$Co(CN)_6{}^{----} \rightleftarrows Co(CN)_6{}^{---} + e^-$	0.83	1×10^{14}
$Fe(CN)_6{}^{----} \rightleftarrows Fe(CN)_6{}^{---} + e^-$	-0.36	9×10^{-7}

The electron-reaction equilibrium equations are

$$\frac{[Co(CN)_6{}^{---}][e^-]}{[Co(CN)_6{}^{----}]} = 1 \times 10^{14}$$

$$\frac{[Fe(CN)_6{}^{---}][e^-]}{[Fe(CN)_6{}^{----}]} = 9 \times 10^{-7}$$

By dividing the first by the second we obtain the equilibrium equation for the over-all reaction:

$$\frac{[Co(CN)_6{}^{---}][Fe(CN)_6{}^{----}]}{[Co(CN)_6{}^{----}][Fe(CN)_6{}^{---}]} = \frac{1 \times 10^{14}}{9 \times 10^{-7}} = 1 \times 10^{20}$$

The stated conditions of the reaction are such that the equilibrium concentrations of the two reactants are equal to each other, and those of the two products are equal to each other. We write

$$x = [Co(CN)_6{}^{----}] = [Fe(CN)_6{}^{---}]$$
$$0.1 - x = [Co(CN)_6{}^{---}] = [Fe(CN)_6{}^{----}]$$

Hence

$$\frac{(0.1 - x)^2}{x^2} = 1 \times 10^{20}$$
$$x^2 = 1 \times 10^{-20} \times (0.1 - x)^2 \cong 1 \times 10^{-22}$$
$$x = 1 \times 10^{-11} \text{ mole } l^{-1}$$

Hence the reaction proceeds essentially to completion, with the concentration of each of the products 10^{10} times that of each of the reactants.

Example 15-8. What products would you expect on mixing equal volumes of a 0.2 F solution of $K_3Fe(CN)_6$ and a 0.2 F solution of $FeSO_4$?

Answer. We follow the method of the preceding example:

	E°	K
$Fe(CN)_6^{----} \rightleftharpoons Fe(CN)_6^{---} + e^-$	-0.36	9×10^{-7}
$Fe^{++} \rightleftharpoons Fe^{+++} + e^-$	-0.771	1.1×10^{-13}

$$\frac{[Fe(CN)_6^{---}][Fe^{++}]}{[Fe(CN)_6^{----}][Fe^{+++}]} = \frac{9 \times 10^{-7}}{1.1 \times 10^{-13}} = 8 \times 10^6$$

We see that the reaction should proceed almost to completion: ferro-cyanide ion is a stronger reducing agent than ferrous ion. Let $x = Fe^{+++} = [Fe(CN)_6^{----}]$, $0.1 - x = [Fe(CN)_6^{---}] = [Fe^{++}]$. The equilibrium equation can be rewritten as

$$x^2 = \frac{(0.1 - x)^2}{8 \times 10^6} = 12 \times 10^{-10}$$

$$x = 3 \times 10^{-5} \text{ mole } l^{-1}$$

Hence we calculate that the final concentration of each of the reactants, ferrocyanide ion and ferrous ion, is only 3×10^{-5} mole l^{-1}, and that of each of the products, ferrocyanide ion and ferric ion, is 0.1 mole l^{-1}. In fact, however, the products react to form a precipitate of ferric ferro-cyanide (in this case, with potassium ion present, of $KFeFe(CN)_6 \cdot H_2O$), and the concentration of the reactants becomes still smaller.

15-7.   The Dependence of the Electromotive Force of Cells on Concentration

The electromotive force E of a cell is related to the free energy change ΔG of the cell reaction by the equation $-\Delta G = nFE$ (Equation 15-13). E° corresponds to a cell with reactants and products at unit activity—for example, for the cell

$$Zn(c)|Zn^{++}|H^+|H_2$$

shown in Figure 15-6 with $a(Zn^{++}) = 1$ mole l^{-1}, $a(H^+) = 1$ mole l^{-1}, and $P(H_2) = 1$ atm. The change in free energy corresponding to a change in activity from a_1 to a_2 is $RT \ln(a_2/a_1)$. Consequently the value of ΔE, the difference in EMF of a cell with a reactant or product at activity a_1 and another with reactant or product at activity a_2, is

$$\Delta E = \pm \frac{RT}{nF} \ln \frac{a_2}{a_1} \tag{15-18}$$

At 25°C this equation corresponds to

$$\Delta E = \pm \frac{0.0592 \text{ V}}{n} \times \log \frac{a_2}{a_1} \tag{15-19}$$

The choice of sign depends on the way in which the substance with activity a_1 or a_2 enters into the cell reaction.

We see that for a cell involving a hydrogen electrode the value of E changes by 0.0592 V for each tenfold change in $a(H^+)$—that is, for each unit change in pH. This dependence of E on pH is the basis of the pH meter (Section 14-3).

> **Example 15-9.** By how much would the EMF of a cell involving a hydrogen electrode change if the pressure of $H_2(g)$ were changed from 1 atm to 0.1 atm?
>
> **Answer.** The one-electron electrode equation is $H^+ \rightleftarrows \frac{1}{2}H_2 + e^-$; hence the value of n for use with H_2 is 2, and the value of ΔE is
>
> $$\Delta E = \pm \frac{0.0592 \text{ V}}{2} \times \log \frac{P_2}{P_1}$$
>
> which gives 0.0296 V as the change in EMF for a tenfold change in $P(H_2)$. The sign of ΔE is the same for a decrease in $P(H_2)$ as for an increase in $a(H^+)$.

15-8. Primary Cells and Storage Cells

The production of an electric current through chemical reaction is achieved in *primary cells* and *storage cells*.

Primary cells are cells in which an oxidation-reduction reaction can be carried out in such a way that its driving force produces an electric potential. This is achieved by having the oxidizing agent and the reducing agent separated; the oxidizing agent then removes electrons from one electrode and the reducing agent gives electrons to another electrode, the flow of current through the cell itself being carried by ions.

Storage cells are similar cells, which, however, can be returned to their original state after current has been drawn from them (can be *charged*) by applying an impressed electric potential between the electrodes, and thus reversing the oxidation-reduction reaction.

The Common Dry Cell

One primary cell, the gravity cell, has been described in the preceding section. This cell is called a *wet cell*, because it contains a liquid electrolyte. A very useful primary cell is the *common dry cell*, shown in Figure 15-8. The common dry cell consists of a zinc cylinder that contains as electrolyte a paste of ammonium chloride (NH_4Cl), a little zinc chloride

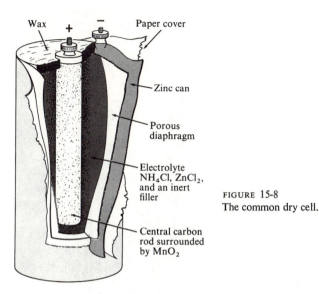

Wax + − Paper cover

Zinc can

Porous
diaphragm

Electrolyte
NH_4Cl, $ZnCl_2$,
and an inert
filler

Central carbon
rod surrounded
by MnO_2

FIGURE 15-8
The common dry cell.

($ZnCl_2$), water, and diatomaceous earth or other filler. The dry cell is not dry; water must be present in the paste that serves as electrolyte. The central electrode is a mixture of carbon and manganese dioxide, embedded in a paste of these substances. The electrode reactions are

$$Zn \longrightarrow Zn^{++} + 2e^-$$
$$2NH_4^+ + 2MnO_2 + 2e^- \longrightarrow 2MnO(OH) + 2NH_3$$

(The zinc ion combines to some extent with ammonia to form the zinc-ammonia complex ion, $Zn(NH_3)_4^{++}$.) This cell produces a potential of about 1.48 V.

The Lead Storage Battery

The most common storage cell is that in the *lead storage battery* (Figure 15-9). The electrolyte in this cell is a mixture of water and sulfuric acid with density about 1.290 g cm^{-3} in the charged cell (38% H_2SO_4 by weight). The plates are lattices made of a lead alloy, the pores of one plate being filled with spongy metallic lead, and those of the other with lead dioxide, PbO_2. The spongy lead is the reducing agent and the lead dioxide the oxidizing agent in the chemical reaction that takes place in the cell. The electrode reactions that occur as the cell is being discharged are

$$Pb + SO_4^{--} \longrightarrow PbSO_4 + 2e^-$$
$$PbO_2 + SO_4^{--} + 4H^+ + 2e^- \longrightarrow PbSO_4 + 2H_2O$$

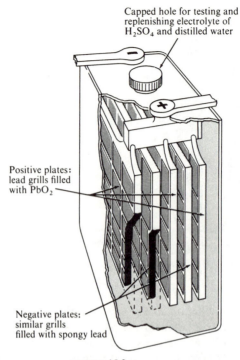

Capped hole for testing and
replenishing electrolyte of
H_2SO_4 and distilled water

Positive plates:
lead grills filled
with PbO_2

Negative plates:
similar grills
filled with spongy lead

FIGURE 15-9
The lead storage cell.

Each of these reactions produces the insoluble substance $PbSO_4$, lead sulfate, which adheres to the plates. As the cell is discharged, sulfuric acid is removed from the electrolyte, which decreases in density. The state of charge or discharge of the cell can accordingly be determined with use of a hydrometer, by measuring the density of the electrolyte.

The cell can be charged again by applying an electric potential across the terminals, and causing the above electrode reactions to take place in the opposite directions. The charged cell produces an electromotive force of slightly over 2 volts. A 6-V battery consists of three cells in series, and a 12-V battery of six cells.

It is interesting that in this cell the same element changes its oxidation state in the two plates: the oxidizing agent is PbO_2 (containing lead with oxidation number $+4$, which changes to $+2$ as the cell discharges), and the reducing agent is Pb (lead with oxidation number 0, which changes to $+2$).

15-9. Electrolytic Production of Elements

Many metals and some nonmetals are made by electrolytic methods. Hydrogen and oxygen are produced by the electrolysis of water containing an electrolyte. The alkali metals, alkaline-earth metals, magnesium, aluminum, and many other metals are manufactured either entirely or for special uses by electrochemical reduction of their compounds.

The Production of Sodium and Chlorine

Many electrochemical processes depend for their success on ingenious devices for securing the purity of the product. As an illustration we may consider a cell used for making metallic sodium and elementary chlorine from sodium chloride.

The molten sodium chloride (usually with some sodium carbonate added to reduce its melting point) is in a vessel containing a carbon anode and iron cathode, separated by an iron screen which leads to pipes, as indicated in Figure 15-10. The gaseous chlorine is led off through one pipe, and the molten sodium, which is lighter than the electrolyte, rises and is drawn off into a storage tank.

Only about 8% of the chlorine used in the United States is produced in this way. Most of it is produced in connection with the production of sodium hydroxide and hydrogen by electrolysis of brine.

If concentrated brine is electrolyzed, chlorine is formed at the anode and hydroxide ion at the cathode:

$$2Cl^- \longrightarrow Cl_2 + 2e^-$$
$$2H_2O + 2e^- \longrightarrow H_2 + 2OH^-$$

The alkali produced is then present in solution together with some undecomposed salt, from which it must be separated by crystallization. This troublesome step is avoided in the Castner-Kellner process. This process involves use of a slate tank divided into three compartments by slate partitions not quite touching the floor of the tank (Figure 15-11). A layer of mercury covers the floor. Dilute sodium hydroxide solution is put in the central compartment, and brine in the end compartments. The brine is electrolyzed between graphite anodes and the mercury, which serves as a cathode. The anode reaction is that given above, producing chlorine. The cathode reaction is

$$e^- + Na^+ \longrightarrow Na(amalgam)$$

["Na(amalgam)" means a solution of sodium in mercury; the alloys of mercury with other metals are called amalgams.] During the electrolysis the cell is rocked, causing the amalgam to flow back and forth.

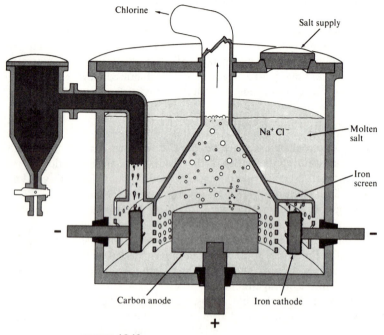

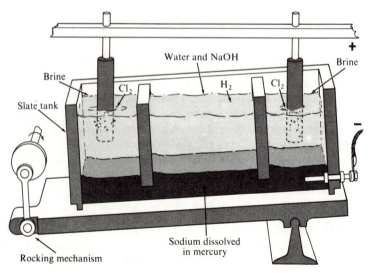

FIGURE 15-11
The electrolytic production of sodium hydroxide and chlorine.

The amalgam reacts with the water in the central compartment:

$$2Na(amalgam) + 2H_2O \longrightarrow H_2(g) + 2NaOH(aq)$$

In this way a pure solution of sodium hydroxide is produced. It is in practice found convenient to introduce iron electrodes in the central compartment, and to assist this reaction to occur by use of an applied electric potential, with the amalgam as anode and the iron electrode as cathode. The electrode reactions are the following:

Anode reaction: $Na(amalgam) \longrightarrow Na^+ + e^-$
Cathode reaction: $2H_2O + 2e^- \longrightarrow H_2(g) + 2OH^-$

The Electrolytic Production of Aluminum

All commercial aluminum is made electrolytically, by a process discovered in 1886 by a young American, Charles M. Hall (1863–1914), and independently, in the same year, by a young Frenchman, P. L. T. Héroult (1863–1914). A carbon-lined iron box, which serves as cathode, contains the electrolyte, which is the molten mineral cryolite, Na_3AlF_6 (or a mixture of AlF_3, NaF, and sometimes CaF_2, to lower the melting point), in which aluminum oxide, Al_2O_3, is dissolved (Figure 15-12). The aluminum oxide is obtained from the ore *bauxite* by a process of purification, which is described below. The anodes in the cell are made of carbon. The passage of the current provides heat enough to keep the electrolyte molten, at about 100°C. The aluminum metal that is produced by the process of electrolysis sinks to the bottom of the cell, and is tapped off. The cathode reaction is

$$Al^{+++} + 3e^- \longrightarrow Al$$

The anode reaction involves the carbon of the electrodes, which is converted into carbon dioxide:

$$C + 2O^{--} \longrightarrow CO_2 + 4e^-$$

The cells operate at about 5 V potential difference between the electrodes. Bauxite is a mixture of aluminum minerals ($AlHO_2$, $Al(OH)_3$), which contains some iron oxide. It is purified by treatment with sodium hydroxide solution, which dissolves hydrated aluminum oxide, as the aluminate ion, $Al(OH)_4^-$, but does not dissolve iron oxide:

$$Al(OH)_3 + OH^- \longrightarrow Al(OH)_4^-$$

The solution is filtered, and is then acidified with carbon dioxide, which reverses the above reaction by forming hydrogen carbonate ion, HCO_3^-:

$$Al(OH)_4^- + CO_2 \longrightarrow HCO_3^- + Al(OH)_3$$

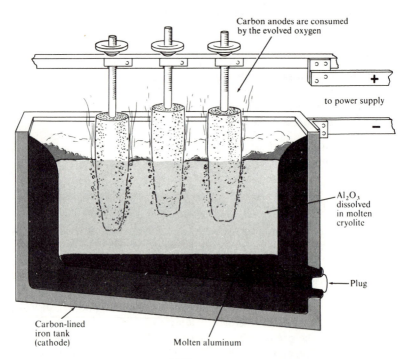

Carbon anodes are consumed
by the evolved oxygen

+

to power supply

—

Al$_2$O$_3$
dissolved
in molten
cryolite

Plug

Carbon-lined
iron tank
(cathode)

Molten aluminum

FIGURE 15-12
The electrolytic production of aluminum.

The precipitated aluminum hydroxide is then dehydrated by ignition (heating to a high temperature), and the purified aluminum oxide is ready for addition to the electrolyte.

The Electrolytic Refining of Metals

Several metals, won from their ores by either chemical or electro-chemical processes, are further refined by electrolytic methods.

Metallic copper is sometimes obtained by leaching a copper ore with sulfuric acid and then depositing the metal by electrolysis of the copper sulfate solution. Most copper ores, however, are converted into crude copper by chemical reduction. This crude copper is cast into anode plates about 2 cm thick, and is then refined electrolytically. In this process the anodes of crude copper alternate with cathodes of thin sheets of pure copper coated with graphite, which makes it possible to strip off the deposit. The electrolyte is copper sulfate. As the current passes through, the crude copper dissolves from the anodes and a purer copper deposits on the cathodes. Metals below copper in the EMF series, such as gold, silver, and platinum, remain undissolved, and fall to the bottom of the tank as sludge, from which they can be recovered. More active metals, such as iron, remain in the solution.

15-10.   The Reduction of Ores. Metallurgy

Metals are obtained from ores. An *ore* is a mineral or other natural material that may be profitably treated for the extraction of one or more metals.

The process of extracting a metal from the ore is called *winning* the metal. *Refining* is the purification of the metal that has been extracted from the ore. *Metallurgy* is the science and art of winning and refining metals, and preparing them for use.

The simplest processes for winning metals are those used to obtain the metals that occur in nature in the elementary state. Thus nuggets of gold and of the platinum metals may be picked up by hand, in some deposits, or may be separated by a hydraulic process (use of a stream of water) when the nuggets occur mixed with lighter materials in a placer deposit.* A quartz vein containing native gold may be treated by mining it, pulverizing the quartz in a stamp mill, and then mixing the rock powder with mercury. The gold dissolves in the mercury, which is easily separated from the rock powder because of its great density, and the gold can be recovered from the amalgam by distilling off the mercury.

The chemical processes involved in the winning of metals are mainly the reduction of a compound of the metal, usually oxide or sulfide. The principal reducing agent that is used is carbon, often in the form of coke. An example is the reduction of iron oxide with coke in a blast furnace (Chapter 20). Occasionally reducing agents other than carbon are used; thus antimony is won from stibnite, Sb_2S_3, by heating it with iron:

$$Sb_2S_3 + 3Fe \longrightarrow 3FeS + 2Sb$$

Whether or not a reaction may be used for winning a metal depends upon the free energies of the reactants and the products. For some reactions, especially those in which the products are similar to the reactants, the change in free energy for the reaction is nearly equal to the enthalpy change. An example is the reaction of stibnite and iron given above. Whether such a reaction is exothermic or endothermic can be predicted with reasonable confidence by use of electronegativity values. In the above

---

*A placer deposit is a glacial deposit or alluvial deposit (made by a river, lake, or arm of the sea), as of sand or gravel, containing gold or other valuable material.

reaction six Sb—S bonds are broken and six Fe—S bonds are formed. The electronegativity values are 1.9 for Sb, 2.5 for S, and 1.8 for Fe. The bond energy per metal-sulfur bond is $100(x_A - x_B)^2$ kJ mole^{-1}; its values are $100(2.5 - 1.9)^2 = 36$ kJ mole^{-1} for Sb—S and $100(2.5 - 1.8)^2 = 49$ kJ mole^{-1} for Fe—S. We conclude that the reaction as written is exothermic and probably also exergonic (accompanied by the evolution of free energy).

From the foregoing argument we would conclude that any element more electropositive than antimony could be used to prepare antimony from stibnite. Reference to Table 15-1 shows that many metals are in this class. Iron is used in practice rather than some other metal because it is the cheapest metal.

Carbon is more effective for reducing metal oxides than might be expected from its electronegativity, 2.5, for two reasons: first, the special stability of CO (low enthalpy) corresponding to the great strength of multiple bonds between first-row elements, and second, the large entropy of the gaseous product. It can be used to reduce oxides of metals with electronegativity as low as 1.6.

The Metallurgy of Copper

Copper occurs in nature as *native copper*; that is, in the free state. Other ores of copper include *cuprite*, Cu_2O; *chalcocite*, Cu_2S; *chalcopyrite*, $CuFeS_2$; *malachite*, $Cu_2CO_3(OH)_2$; and *azurite*, $Cu_3(CO_3)_2(OH)_2$. Malachite, a beautiful green mineral, is sometimes polished and used in jewelry.

An ore containing native copper may be treated by grinding it and then washing away the gangue (the associated rock or earthy material), and melting and casting the copper. Oxide or carbonate ores may be leached with dilute sulfuric acid, to produce a cupric solution from which the copper can be deposited by electrolysis. High-grade oxide and carbonate ores may be reduced by heating with coke mixed with a suitable flux. (A flux is a material, such as limestone, that combines with the silicate minerals of the gangue to form a slag that is liquid at the temperature of the furnace, and can be easily separated from the metal.)

Sulfide ores are smelted by a complex process. Low-grade ores are first concentrated, by a process such as *flotation*. The finely ground ore is treated with a mixture of water and a suitable oil. The oil wets the sulfide minerals, and the water wets the silicate minerals of the gangue. Air is then blown through to produce a froth, which contains the oil and the sulfide minerals; the silicate minerals sink to the bottom.

The concentrate or the rich sulfide ore is then roasted in a furnace through which air is passing. This removes some of the sulfur as sulfur dioxide, and leaves a mixture of Cu_2S, FeO, SiO_2, and other substances.

This roasted ore is then mixed with limestone to serve as a flux, and is heated in a furnace. The iron oxide and silica combine with the limestone to form a slag, and the cuprous sulfide melts and can be drawn off. This impure cuprous sulfide is called *matte*. It is then reduced by blowing air through the molten material:

$$Cu_2S + O_2 \longrightarrow SO_2 + 2Cu$$

Some copper oxide is also formed by the blast of air, and this is reduced by stirring the molten metal with poles of green wood. The copper obtained in this way has a characteristic appearance, and is called *blister copper*. It contains about 1% of iron, gold, silver, and other impurities, and is usually refined electrolytically, as described in the last part of Section 15-9.

The Metallurgy of Silver and Gold

The principal ores of silver are *native silver*, Ag; *argentite*, Ag_2S; and *cerargyrite* or horn-silver, AgCl. The *cyanide process* of winning the metal from these ores is widely used. This process involves treating the crushed ore with a solution of sodium cyanide, NaCN, for about two weeks, with thorough aeration to oxidize the native silver. The reactions producing the soluble complex ion $Ag(CN)_2^-$ may be written in the following way:

$$4Ag + 8CN^- + O_2 + 2H_2O \longrightarrow 4Ag(CN)_2^- + 4OH^-$$
$$AgCl + 2CN^- \longrightarrow Ag(CN)_2^- + Cl^-$$
$$Ag_2S + 4CN^- \longrightarrow 2Ag(CN)_2^- + S^{--}$$

The silver is then obtained from the solution by reduction with metallic zinc:

$$Zn + 2Ag(CN)_2^- \longrightarrow 2Ag + Zn(CN)_4^{--}$$

The *amalgamation process* is used for native silver. The ore is treated with mercury, which dissolves the silver. The liquid amalgam is then separated from the gangue and distilled, the mercury collecting in the receiver and the silver remaining in the retort.

Silver is obtained as a by-product in the refining of copper and lead. The sludge from the electrolytic refining of copper may be treated by simple chemical methods to obtain its content of silver and gold. The small amount of silver in lead is obtained by an ingenious method, the *Parkes process*. This involves stirring a small amount (about 1%) of zinc into the molten lead. Liquid zinc is insoluble in liquid lead, and the solubility of silver in liquid zinc is about 3000 times as great as in liquid lead. Hence most of the silver dissolves in the zinc. The zinc-silver phase comes to the top, solidifies as the crucible cools, and is lifted off. The zinc can then be distilled away, leaving the silver. Gold present in the lead is also obtained by this process.

Gold is obtained from its ores, such as gold-bearing quartz, by pulverizing the ore and washing it over plates of copper coated with a layer of amalgam. The gold dissolves in the amalgam, which is then scraped off and separated by distillation. The tailings may then be treated with cyanide solution, and the gold be won from the cyanide solution by electrolysis or treatment with zinc:

$$4Au + 8CN^- + O_2 + 2H_2O \longrightarrow 4Au(CN)_2^- + 4OH^-$$
$$2Au(CN)_2^- + Zn \longrightarrow 2Au + Zn(CN)_4^{--}$$

The Metallurgy of Zinc, Cadmium, and Mercury

The principal ore of zinc is *sphalerite* or *zinc blende*, ZnS. Less important ores include *zincite*, ZnO; *smithsonite*, $ZnCO_3$; *willemite*, Zn_2SiO_4; *calamine*, $Zn_2SiO_3(OH)_2$; and *franklinite*, Fe_2ZnO_4.

Many ores of zinc are concentrated by flotation before smelting. Sulfide ores and carbonate ores are then converted to oxide by roasting:

$$2ZnS + 3O_2 \longrightarrow 2ZnO + 2SO_2$$
$$ZnCO_3 \longrightarrow ZnO + CO_2$$

The zinc oxide is mixed with carbon and heated in a fire-clay retort to a temperature high enough to vaporize the zinc:

$$ZnO + C \longrightarrow Zn(g) + CO(g)$$

The zinc vapor is condensed in fire-clay receivers. At first the zinc is condensed in the cool condenser as a fine powder, called *zinc dust*, which contains some zinc oxide. After the receiver becomes hot the vapor condenses to a liquid, which is cast in ingots called *spelter*. Spelter contains small amounts of cadmium, iron, lead, and arsenic. It can be purified by careful redistillation.

The zinc oxide can also be reduced by electrolysis. It is dissolved in sulfuric acid, and electrolyzed with aluminum sheets as cathodes. The deposited zinc, which is about 99.95% pure, is stripped off the cathodes, melted, and cast into ingots, for use where pure zinc is needed, as in the production of brass. The sulfuric acid is regenerated in the process, as is seen from the reactions:

Solution of zinc oxide:	$ZnO + 2H^+ \longrightarrow Zn^{++} + H_2O$
Cathode reaction:	$Zn^{++} + 2e^- \longrightarrow Zn$
Anode reaction:	$H_2O \longrightarrow \frac{1}{2}O_2 + 2H^+ + 2e^-$
Over-all reaction:	$ZnO \longrightarrow Zn + \frac{1}{2}O_2$

Cadmium is obtained mainly as a by-product in the smelting and refining of zinc; it occurs to the amount of about 1% in many zinc ores. The sulfide of cadmium, CdS, is called *greenockite*. Cadmium is more volatile than zinc, and in the reduction of zinc oxide containing cadmium oxide it is concentrated in the first portions of dust collected in the receivers.

Mercury occurs as the native metal, in small globules of pure mercury, and as crystalline silver amalgam. Its most important ore is the red mineral *cinnabar*, HgS. Cinnabar is smelted simply by heating it in a retort in a stream of air, and condensing the mercury vapor in a receiver:

$$HgS + O_2 \longrightarrow Hg(g) + SO_2(g)$$

The Metallurgy of Tin and Lead

The principal ore of tin is *cassiterite*, SnO_2, the main deposits of which are in Bolivia and Malaya. The crude ore is ground and washed in a stream of water, which separates the lighter gangue from the heavy cassiterite. The ore is then roasted, to oxidize the sulfides of iron and copper to products that are removed by leaching with water. The purified ore is then mixed with carbon and reduced in a reverberatory furnance. The crude tin produced in this way is resmelted at a gentle heat, and the pure metal flows away from the higher-melting impurities, chiefly compounds of iron and arsenic. Some tin is purified by electrolysis.

The principal ore of lead is *galena*, PbS, which occurs, often in beautiful cubic crystals, in large deposits in the United States, Spain, and Mexico. The ore is first roasted until part of it has been converted into lead oxide, PbO, and lead sulfate, $PbSO_4$. The supply of air to the furnace is then cut off, and the temperature is raised. Metallic lead is then produced by the reactions

$$PbS + 2PbO \longrightarrow 3Pb + SO_2$$

and

$$PbS + PbSO_4 \longrightarrow 2Pb + 2SO_2$$

Some lead is also made by heating galena with scrap iron:

$$PbS + Fe \longrightarrow Pb + FeS$$

Silver is often removed from lead by the Parkes process, described above. Some pure lead is made by electrolytic refining.

Reduction of Metal Oxides or Halides by Strongly Electropositive Metals

Some metals, including titanium, zirconium, hafnium, lanthanum, and the lanthanons, are most conveniently obtained by reaction of their oxides or halides with a more electropositive metal. Sodium, potassium, calcium, and aluminum are often used for this purpose. Thus titanium may be made by reduction of titanium tetrachloride by calcium:

$$TiCl_4 + 2Ca \longrightarrow Ti + 2CaCl_2$$

Titanium, zirconium, and hafnium are purified by the decomposition of their tetraiodides on a hot wire. The impure metal is heated with iodine in an evacuated flask, to produce the tetraiodide as a gas:

$$Zr + 2I_2 \longrightarrow ZrI_4$$

The gas comes into contact with a hot filament, where it is decomposed, forming a wire of the purified metal:

$$ZrI_4 \longrightarrow Zr + 2I_2$$

The process of preparing a metal by reduction of its oxide by aluminum is called the *aluminothermic process*. For example, chromium can be prepared by igniting a mixture of powdered chromium(III) oxide and powdered aluminum:

$$Cr_2O_3 + 2Al \longrightarrow Al_2O_3 + 2Cr$$

The heat liberated by this reaction is so great as to produce molten chromium. The aluminothermic process is a convenient way of obtaining a small amount of liquid metal, such as iron for welding.

Exercises

15-1. A current of 10 A is drawn from a lead storage cell for 1 hour. How much $PbSO_4$ is formed from Pb at one electrode plate? How much is formed from PbO_2 at a plate of opposite polarity?

15-2. Assign oxidation numbers to elements in the following compounds:

Sodium peroxide, Na_2O_2 Sodium oxide, Na_2O
Permanganate ion, MnO_4^- Peroxysulfate ion, SO_5^{--}
Cuprous oxide, Cu_2O Cupric oxide, CuO
Ferrous oxide, FeO Ferric oxide, Fe_2O_3
Magnetite, Fe_3O_4 Borax, $Na_2B_4O_7 \cdot 10H_2O$
Garnet, $Ca_3Al_2Si_3O_{12}$ Topaz, $Al_2SiO_4F_2$
Sodium hydride, NaH Ammonia, NH_3
Nitric acid, HNO_3 Lead sulfide, PbS
Lead sulfate, $PbSO_4$ Phosphorus, P_4
Potassium chromate, K_2CrO_4 Nitrous acid, HNO_2
Silica, SiO_2 Ammonium nitrite, NH_4NO_2
Ammonium chloride, NH_4Cl

15-3. Complete and balance the following oxidation-reduction equations:

$$Cl_2 + I^- \longrightarrow I_2 + Cl^-$$
$$Sn + I_2 \longrightarrow SnI_4$$
$$KClO_3 \longrightarrow KClO_4 + KCl$$
$$MnO_2 + H^+ + Cl^- \longrightarrow Mn^{++} + Cl_2$$
$$ClO_4^- + Sn^{++} \longrightarrow Cl^- + Sn^{++++}$$

15-4. Write electrode equations for the electrolytic production of (a) magnesium metal from molten magnesium chloride; (b) perchlorate ion, ClO_4^-, from chlorate ion, ClO_3^-, in aqueous solution; (c) permanganate ion, MnO_4^-, from manganate ion, MnO_4^{--}, in aqueous solution; (d) fluorine from fluoride ion in a molten salt. State in each case whether the reaction occurs at the anode or at the cathode.

15-5. What weight of 3.00% hydrogen peroxide solution would be required to oxidize 2.00 g of lead sulfide, PbS, to lead sulfate, $PbSO_4$?

15-6. Would you expect zinc to reduce cadmium ion? (Refer to the electromotive-force series.) Iron to reduce mercuric ion? Zinc to reduce lead ion? Potassium to reduce magnesium ion?

15-7. Which metal ions would you expect gold to reduce?

15-8. What would you expect to happen if a large piece of lead were put in a beaker containing a solution of stannous salt? Note the values of the electromotive force in Table 15-1.

15-9. What would happen if chlorine gas were bubbled into a solution containing fluoride ion and bromide ion? If chlorine were bubbled into a solution containing both bromide ion and iodide ion?

15-10. The standard enthalpy of formation of $MgCl_2(c)$ is -642 kJ mole^{-1} (Table 18-4). The third-law values of the entropy at 25°C and 1 atm of $Mg(c)$, $Cl_2(g)$, and $MgCl_2(c)$ are 32.5, 33.9, and 89.5 J deg^{-1} mole^{-1}, respectively. The solubility of $MgCl_2(c)$ in water is about 0.6 mole l^{-1}. Calculate the EMF of an electric cell with magnesium metal at one electrode, a saturated aqueous solution of magnesium chloride as electrolyte, and chlorine gas bubbling over platinum at the other electrode.

15-11. The following electric cell is set up:

$$Hg(l), Hg_2Cl_2(c)|Cl^-(1\ F)|Cl_2(g, 1\ atm)$$

The aqueous phase (1 F KCl in water) is saturated with Hg_2Cl_2 in the neighborhood of the mixture of calomel crystals and liquid mercury. There are platinum electrodes at each end of the cell.

(*a*) Write an equation for the reaction accompanying the passage of one faraday through the cell.

(*b*) The EMF of the cell at 25°C is -1.091 V. What is the standard free energy change for the reaction at this temperature?

15-12. Explain how measurements made with the electric cell described in the preceding Exercise could be used to obtain a value for the standard enthalpy and entropy of formation of $Hg_2Cl_2(c)$ at 25°C.

16

The Rate of Chemical Reactions

16-1. Factors Influencing the Rate of Reactions

Two questions may be asked in the consideration of a proposed chemical process, such as the preparation of a useful substance. One question—"Are the stability relations such that it is possible for the reaction to occur?"—is answered by the use of chemical thermodynamics, which has been discussed in Chapters 10 and 11. The second question, discussed in this chapter, is equally important—"Under what conditions will the reaction proceed sufficiently rapidly for the method to be practicable?"

Every chemical reaction requires some time for its completion, but some reactions are very fast and some are very slow. Reactions between ions in solution without change in oxidation state are usually extremely fast. An example is the neutralization of an acid by a base, which proceeds as fast as the solutions can be mixed. Presumably nearly every time a hydronium ion collides with a hydroxide ion reaction occurs, and the number of collisions is very great, so that there is little delay in the reaction. The formation of a precipitate, such as that of silver chloride when a solution containing silver ion is mixed with a solution containing chloride ion, may require a few seconds, to permit the ions to diffuse together to form the crystalline grains of the precipitate:

$$Ag^+(aq) + Cl^-(aq) \longrightarrow AgCl(c)$$

On the other hand, ionic oxidation-reduction reactions are sometimes very slow. An example is the oxidation of stannous ion by ferric ion:

$$2Fe^{+++} + Sn^{++} \longrightarrow 2Fe^{++} + Sn^{++++}$$

This reaction does not occur every time a stannous ion collides with one or two ferric ions. In order for reaction to take place, the collision must be of such a nature that electrons can be transferred from one ion to another, and collisions that permit this electron transfer to occur may be rare.

An example of a reaction that is extremely slow at room temperature is that between hydrogen and oxygen:

$$2H_2 + O_2 \longrightarrow 2H_2O$$

A mixture of hydrogen and oxygen can be kept for years without appreciable reaction.

The factors that determine the rate of a reaction are manifold. The rate depends not only upon the composition of the reacting substances, but also upon their physical form, the intimacy of their mixture, the temperature and pressure, the concentrations of the reactants, special physical circumstances such as irradiation with visible light, ultraviolet light, x-rays, neutrons, or other waves or particles, and the presence of other substances that affect the reaction but are not changed by it.

Homogeneous and Heterogeneous Reactions

A reaction that takes place in a homogeneous system (consisting of a single phase) is called a *homogeneous reaction*. The most important of these reactions are those in gases (such as the formation of nitric oxide in the electric arc, $N_2 + O_2 \rightleftharpoons 2NO$) and those in liquid solutions. The effects of temperature, pressure, and the concentrations of the reactants on the rates of homogeneous reactions are discussed below.

A *heterogeneous reaction* is a reaction involving two or more phases. An example is the oxidation of carbon by potassium perchlorate:

$$KClO_4(c) + 2C(c) \longrightarrow KCl(c) + 2CO_2(g)$$

This reaction and similar reactions occur when perchlorate propellants are burned. (These propellants, used for assisted take-off of airplanes and for propulsion of rockets, consist of intimate mixtures of very fine grains of carbon black and potassium perchlorate held together by a plastic binder.) Another example is the solution of zinc in acid:

$$Zn(c) + 2H^+(aq) \longrightarrow Zn^{++}(aq) + H_2(g)$$

In this reaction three phases are involved: the solid zinc phase, the solution, and the gaseous phase formed by the evolved hydrogen.

The Rate of Heterogeneous Reactions

A heterogeneous reaction takes place at the surfaces (the *interfaces*) of the reacting phases, and it can be made to go faster by *increasing the extent of the surfaces.* Thus finely divided zinc reacts more rapidly with acid than does coarse zinc, and the rate of burning of a perchlorate propellant is increased by grinding the potassium perchlorate to a finer crystalline powder.

Sometimes a *reactant is exhausted* in the neighborhood of the interface, and the reaction is slowed down. Stirring the mixture then accelerates the reaction, by bringing fresh supplies of the reactant into the reaction region.

Catalysts may accelerate heterogeneous as well as homogeneous reactions. A small amount of cupric ion speeds up the solution of zinc in acid, and manganese dioxide increases the rate of evolution of oxygen from potassium chlorate.

The rates of nearly all chemical reactions depend greatly on the *temperature.* It is nearly universally true that reactions are speeded up by increase in temperature.

Special devices may be utilized to accelerate certain chemical reactions. The formation of a zinc amalgam on the surface of the grains of zinc by treatment with a small amount of mercury increases the speed of the reduction reactions.

The solution of zinc in acid is retarded somewhat by the bubbles of hydrogen, which prevent the acid from achieving contact with the zinc over its entire surface. This effect can be avoided by bringing a plate of unreactive metal, such as copper or platinum, into electrical contact with the zinc. The reaction then proceeds as two separate electron reactions. Hydrogen is liberated at the surface of the copper or platinum, and zinc dissolves at the surface of the zinc plate:

$$2H^+ + 2e^- \longrightarrow H_2(g) \qquad \text{at copper surface}$$
$$Zn \longrightarrow Zn^{++} + 2e^- \qquad \text{at zinc surface}$$

The electrons flow from the zinc plate to the copper plate through the electrical contact, and electric neutrality in the different regions of the solution is maintained by the migration of ions.

Homogeneous Reactions

Most actual chemical processes are very complicated, and the analysis of their rates is very difficult. As a reaction proceeds, the reacting substances are used up and new ones are formed; the temperature of the system is changed by the heat evolved or absorbed by the reaction; and other effects may occur that influence the reaction in a complex way. For example, when a drop of a solution of potassium permanganate is added to a solution containing hydrogen peroxide and sulfuric acid no detectable reaction may occur for several minutes. Then the reaction speeds up, and finally the rate may become so great as to decolorize a steady stream of permanganate solution as rapidly as it is poured into the reducing solution. This effect of the speeding-up of the reaction is due to the vigorous catalytic action of the products of reduction of permanganate ion: the reaction is extremely slow in the absence of the products, and is rapidly accelerated as soon as they are formed.

The *explosion* of a gaseous mixture, such as hydrogen and oxygen, and the *detonation* of a high explosive, such as glyceryl trinitrate (nitroglycerin), are interesting chemical reactions; but the analysis of the rates of these reactions is made very difficult because of the great changes in temperature and pressure that accompany them.

The great rate at which a chemical reaction may occur is illustrated by the rate at which a detonation surface moves through a specimen of glyceryl trinitrate or similar high explosive. This *detonation rate* is about 20,000 feet per second. A specimen of high explosive weighing several grams may accordingly be completely decomposed within a millionth of a second. Another reaction that can occur very rapidly is the fission of the nuclei of heavy atoms. The nuclear fission of several pounds of ^{235}U or ^{239}Pu may take place in a few millionths of a second in the explosion of an atomic bomb (see Chapter 26).

In order to obtain an understanding of the rates of reactions, chemists have attempted to simplify the problem as much as possible. A good understanding has been obtained of homogeneous reactions (in a gaseous or liquid solution) that take place at constant temperature. Experimental studies are made by placing the reaction vessel in a *thermostat*, which is held at a fixed temperature. The quantitative theory of reaction rate, discussed below, has special interest because of its relation to the theory of chemical equilibrium, which is treated in the following chapter.

16-2. The Rate of a First-order Reaction at Constant Temperature

If a molecule, which we represent by the general symbol A, has a tendency to decompose spontaneously into smaller molecules

$$A \longrightarrow B + C$$

at a rate that is not influenced by the presence of other molecules, we expect that *the number of molecules that decompose by such a unimolecular process in unit time will be proportional to the number present.* If the volume of the system remains constant, the concentration of A will decrease at a rate proportional to this concentration. The symbol [A] represents the concentration of A (in moles per liter). The rate of decrease in concentration with time is $-d[A]/dt$. For a unimolecular decomposition we accordingly may write the equation

$$-\frac{d[A]}{dt} = k[A] \tag{16-1}$$

as *the differential equation determining the rate of the reaction.* The factor k is called the *first-order rate constant.* A reaction of this kind is called a *first-order reaction*; the order of a reaction is the sum of the powers of the concentration factors in the rate expression (on the right side of the rate equation).

For example, the rate constant k may have the value 0.001, with the time t measured in seconds. The equation would then state that during each second 1/1000 of the molecules present would decompose. Suppose that at the time $t = 0$ there were 1,000,000,000 molecules per milliliter in the reaction vessel. During the first second 0.1% of these molecules would decompose, and there would remain at $t = 1$ second only 999,000,000 molecules undecomposed. During the next second 999,000 molecules would decompose, and there would remain 998,001,000 molecules.* After some time (about 693 seconds) half of the molecules would have decomposed, and there would remain only 500,000,000 undecomposed molecules per milliliter. Of these about 500,000 would decompose during the next second, and so on. After another 693 seconds there would remain about 250,000,000 undecomposed molecules, and after still another 693 seconds there would remain about 125,000,000, and so on. During each period of 693 seconds the concentration of reactant is reduced to one-half its value at the beginning of the period.

---

*These numbers are not precise. The molecules decompose at random, at the *average* rate given by Equation 16-1, and a small statistical fluctuation from this rate is to be expected.

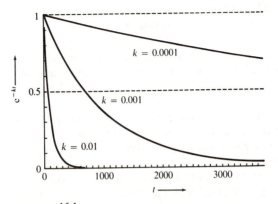

FIGURE 16-1
Curves showing decrease with time of the amount
remaining of a substance decomposing by a first-order
reaction, with indicated values of the reaction-rate
constant.

This relation between concentration of reactant and the time t is shown
by the curves in Figure 16-1. It is seen that each curve decreases by one-
half its value during successive constant time intervals. The negative slope
of each curve at any point is proportional to the value of the curve at that
point, as required by Equation 16-1.

The algebraic equation which expresses this relation is

$$[A] = [A]_0 \, e^{-kt} \qquad\qquad (16\text{-}2)$$

This is the equation that is obtained by integrating Equation 16-1; it is
equivalent to Equation 16-1. Equation 16-2 is called the *reaction-rate
expression for a first-order reaction in the integrated form*. It states that the
concentration of the reactant at time t is equal to the concentration at
time 0, $[A]_0$, multiplied by the exponential factor $\exp(-kt)$. The nature
of this function can be seen from Figure 16-1.

The first-order character of a reaction may be tested by observing the
rate of disappearance of reactant (the amount disappearing in a short
time) at various concentrations, and comparing with Equation 16-1, or
by observing the concentration itself for a particular sample for a long
time, and comparing with Equation 16-2.

Examples of First-order Reactions

Not many first-order reactions are known. One very important class of
reactions of this type is *radioactive decomposition*. Each nucleus of radium
or other radioactive element decomposes independently of the presence

of other atoms; and in consequence the rate of radioactive decay corresponds to Equation 16-1.

The time required for one-half of a first-order reactant to decompose is called its *half-life*. This time is 693 times the time required for 0.1% of the reactant to decompose. The half-life of radium itself is 1733 years. Hence every $1733/693 = 2\frac{1}{2}$ years 0.1% of the radium nuclei in a sample of radium undergo radioactive disintegration.

A representative first-order gas-phase chemical reaction is the decomposition of azomethane, CH_3—N=N—CH_3, into ethane and nitrogen:*

$$CH_3NNCH_3 \longrightarrow C_2H_6 + N_2$$

The molecular mechanism of this reaction is indicated in Figure 16-2. Most of the azomethane molecules have the configuration shown as *A*, with the methyl groups at opposite sides of the N=N axis. This is called the *trans* configuration. A few molecules have the *cis* configuration, *B*. If a molecule with the *cis* configuration collides very vigorously with another molecule, it may be set into very violent vibration, sufficient to bring the two carbon atoms close together, as shown in *C*. In this configuration the two N—C bonds tend to break, forming a C—C bond and another N—N bond. This tendency is indicated by the dashed valence-bond lines in *C*. The molecule *C* may either return to configuration *B*, or break the C—N bonds and separate into two molecules, *D*.

Two other first-order chemical reactions are the decomposition of dinitrogen pentoxide, N_2O_5, into nitrogen dioxide and oxygen, and the decomposition of dimethyl ether, CH_3—O—CH_3, into methane, carbon monoxide, and hydrogen:

$$2N_2O_5 \longrightarrow 4NO_2 + O_2$$
$$CH_3OCH_3 \longrightarrow CH_4 + CO + H_2$$

It is to be noted that *the order of a reaction cannot be predicted from the stoichiometric over-all equation*. The equation for the N_2O_5 decomposition makes use of two molecules of the reactant, but the reaction is in fact a first-order reaction. This shows that the reaction takes place in steps; first there occurs a first-order decomposition, probably

$$N_2O_5 \longrightarrow NO_3 + NO_2$$

This is followed by other reactions, such as

$$2NO_3 \longrightarrow 2NO_2 + O_2$$

A simple method of studying the rate of a gaseous reaction in which the number of molecules of product is either greater or smaller than the number of molecules of reactant is by measuring the change in pressure

---

*Some other products also are formed, by other reactions of azomethane.

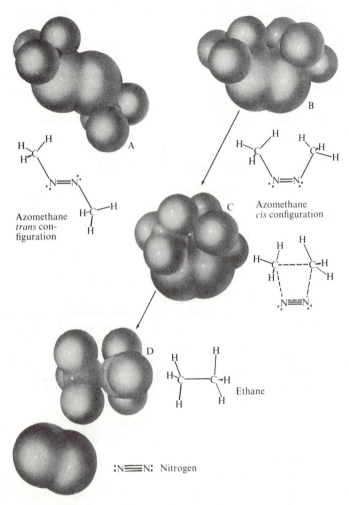

Azomethane *trans* configuration

Azomethane *cis* configuration

Ethane

:N≡N: Nitrogen

FIGURE 16-2
The unimolecular decomposition of azomethane into nitrogen and ethane.

during the reaction. The reactant is placed in the reaction vessel in the thermostat, and the gas pressure is then measured from time to time by a gauge. This method is illustrated by the following example.

Example 16-1. At a given temperature a sample of dimethyl ether is found to have initial pressure [of pure $(CH_3)_2O$] of 300 torr (mm Hg), and 10 seconds later the pressure has increased to 308.1 torr. How long would it take for the pressure to reach 600 torr, and how long would it then take for the pressure to change to 608.1 torr?

Solution.   Let us answer the second question first. The reaction is

$$(CH_3)_2O \longrightarrow CH_4 + CO + H_2$$

Hence the number of molecules is tripled by the decomposition; the final pressure will be 900 torr. When half the ether has decomposed, the pressure will be 600 torr ($P_{ether} = 150$, $P_{CH_4} = 150$, $P_{CO} = 150$, $P_{H_2} = 150$). Since there is then only half as much ether present as initially, the rate of decomposition will be only half as great, and it will accordingly take 20 seconds for the pressure to increase by 8.1 torr, which initially required 10 seconds.

The first question is answered in the following way. From Equation 16-1 we derive

$$-\frac{dP_{ether}}{dt} = kP_{ether}$$

since the partial pressure of a gas at constant temperature is proportional to concentration. The increase in total pressure of 8.1 torr in 10 sec means a decrease of 4.05 mm in P_{ether} in 10 s, or 0.405 torr s^{-1}:

$$-\frac{dP_{ether}}{dt} = 0.405 \text{ torr s}^{-1}$$

Hence

$$k = \frac{0.405}{300} = 0.00135 \text{ s}^{-1}$$

The total pressure will reach 600 torr when P_{ether} has fallen to 150 torr from the initial value $P_0 = 300$ torr. Equation 16-2 is

$$P = P_0 e^{-kt}$$

In our case $P/P_0 = 0.5$. Hence we want to find the value of t at which $e^{-kt} = 0.5$. The function e^{-x} has the value 0.5 when $x = 0.693$. (We have $-x = \ln 0.5$ or $x = \ln 2 = 2.303 \log 2 = 2.303 \times 0.301 = 0.693$.) Hence $kt = 0.693$ with $k = 0.00135$, or $t = 0.693/0.00135 = 513$ s.

We may derive this answer in another way, using the statement in the text that the half-life is 693 times the time required for 0.1% decomposition. (This statement is proved in the preceding paragraph, where it is shown that $e^{-x} = \frac{1}{2}$ when $x = 0.693$.) In 10 seconds the partial pressure of dimethyl ether decreases from 300 torr to a value 4.05 torr less, a decrease of 1.35%. Hence there is 0.1% decomposition in $10/13.5 = 0.741$ seconds. The half-life of the ether is hence $693 \times 0.741 = 513$ s, which is the answer.

Example 16-2.   The half-life of radium ($^{226}_{88}Ra$) is 1590 years. What is the value of the decay constant? What fraction decays in one year?

Solution.   From the argument of the preceding paragraph we write

$$k = \frac{0.693}{1590} = 0.000436 \text{ year}^{-1}$$

Hence the decay constant has the value 4.36×10^{-4} year^{-1}.

On expanding the exponential term in Equation 16-2 we obtain

$$e^{-kt} = 1 - kt + \tfrac{1}{2}k^2t^2 - \cdots$$

For t small only the linear term above needs to be retained. We see that in unit time the fraction k decays. Hence in one year the fraction 4.36×10^{-4}, which is 0.0436%, of the radium decays.

Example 16-3.   In Section 26-5 it is pointed out that the age of a piece of wood can be determined by measuring its carbon-14 activity. Carbon 14 has half-life 5760 years. The carbon 14 in fresh wood decomposes at the rate 15.3 atoms per minute per gram of carbon (this is the number of carbon-14 β-rays emitted per minute, as determined with a Geiger counter). Wood from trees that were buried by ash in the volcanic eruption of Mt. Mazama in southern Oregon has been found to give 6.90 carbon-14 beta counts per minute per gram of carbon. When did the eruption occur?

Solution.   By the method used in the preceding example we find for k the value

$$k = \frac{0.693}{5760\ \mathrm{Y}} = 1.204 \times 10^{-4}\ \mathrm{Y}^{-1}$$

The fraction of ^{14}C undecomposed is $6.90/15.3 = 0.451$. Hence we write

$$e^{-kt} = 0.451$$
$$kt = -\ln 0.451 = -2.303 \log 0.451 = 2.303 \times 0.347 = 0.800$$
$$t = 0.800/k = 0.800/(1.204 \times 10^{-4}\ \mathrm{year}^{-1}) = 6640\ \mathrm{years}$$

Hence the eruption of Mt. Mazama occurred about 6640 years ago.

Example 16-4.   Skeletons assigned to a form of early man called *Zinjanthropus* (East African man) have been found in the Olduvai Canyon in Tanganyika in association with volcanic ash containing potassium minerals. By use of a mass spectrograph the amount of argon 40 in the ash was measured; it was found to be 0.078% of the amount of potassium 40 present. (Potassium 40, half-life 15×10^8 years, is a radioactive isotope of potassium constituting 0.011% of natural potassium.) This argon 40 had been formed by beta decay of potassium 40, since the ash was deposited in a volcanic eruption; the older argon 40 would have been boiled out of the molten lava at the time of the eruption. How old are the skeletons?

Solution.   The rate constant k for decay of ^{40}K is $0.693/(15 \times 10^8\ \mathrm{Y}) = 4.6 \times 10^{-10}\ \mathrm{Y}^{-1}$. The time t required for 0.078% of the ^{40}K to decompose is given by the equation

$$kt = 7.8 \times 10^{-4}$$
$$t = 7.8 \times 10^{-4}/k = 7.8 \times 10^{-4}/4.6 \times 10^{-10}\ \mathrm{Y}^{-1} = 1.7 \times 10^6\ \mathrm{Y}$$

Hence the ash was laid down about 1,700,000 years ago; this is the presumable age of the skeletons found in association with it.

16-3. Reactions of Higher Order

If reaction occurs by collision and interaction of two molecules A and B, the rate of the reaction will be proportional to the number of collisions. The number of collisions in unit volume is seen from simple kinetic considerations to be proportional to the product of the concentrations of A and B. Hence we may write as the differential rate expression for this second-order reaction

$$\text{rate of reaction} = -\frac{d[A]}{dt} = -\frac{d[B]}{dt} = k[A][B] \qquad (16\text{-}3a)$$

Here $-d[A]/dt$ is the rate of decrease in concentration of $[A]$ and $-d[B]/dt$ is that of $[B]$; they are equal, the reaction being $A + B \longrightarrow$ products. The factor k is the *second-order rate constant*.

It must again be emphasized that *the stoichiometric equation for a reaction does not determine its rate*. Thus the oxidation of iodide ion by persulfate ion

$$S_2O_8{}^{--} + 2I^- \longrightarrow 2SO_4{}^{--} + I_2$$

might be a third-order reaction, with rate proportional to $[S_2O_8{}^{--}][I^-]^2$; it is in fact second-order, with rate proportional to $[S_2O_8{}^{--}][I^-]$. In this case the slow, rate-determining reaction is the reaction between one persulfate ion and one iodide ion,

$$S_2O_8{}^{--} + I^- \longrightarrow \text{products}$$

This is followed by a rapid reaction between the products and another iodide ion.*

For gases it is convenient to use partial pressures instead of concentrations of substances. For a bimolecular gas reaction we may write

$$\text{rate of reaction} = -\frac{dP_A}{dt} = -\frac{dP_B}{dt} = kP_AP_B \qquad (16\text{-}3b)$$

The second-order reaction-rate equation can be integrated. Let us consider the case when the two reactants are present in equal concentrations (or partial pressures), or are the same substance (for example, nitrogen dioxide, for which the reaction $2NO_2 \longrightarrow N_2O_4$ occurs as a bimolecular reaction). Let

$$x = [A] = [B] \qquad (\text{or } x = P_A = P_B)$$

Our equation in differential form is

$$-\frac{dx}{dt} = kx^2$$

---

*The slow reaction probably produces the hypoiodite ion, IO^-.

We rewrite and integrate:

$$-\frac{dx}{x^2} = k \, dt$$

$$\frac{1}{x} + \text{constant of integration} = kt$$

It is convenient to take the constant of integration as $-1/c$:

$$\frac{1}{x} - \frac{1}{c} = kt$$

We rearrange the terms and solve for x:

$$\frac{1}{x} = kt + \frac{1}{c} = \frac{ckt + 1}{c}$$

$$x = \frac{c}{ckt + 1} \qquad (16\text{-}4)$$

By placing $t = 0$, we see that c is the initial value of x.

A general reaction with rate-determining equation

$$a\text{A} + b\text{B} + c\text{C} \longrightarrow \text{products}$$

(this step requiring the interaction of a molecules of substance A, b of B, and c of C) would proceed at the rate

$$\text{rate of reaction} = -\frac{d[\text{A}]}{dt} = k[\text{A}]^a \, [\text{B}]^b \, [\text{C}]^c \qquad (16\text{-}5)$$

In this expression the concentration of A occurs to the a power, that of B to the b power, and that of C to the c power. The order of this reaction would be $a + b + c$.

This equation is valid for gases at ordinary or low partial pressures and substances in solution at low concentrations.

Example 16-5.   Hydrogen and iodine react by a bimolecular mechanism (Figure 16-3):

$$\text{H}_2(g) + \text{I}_2(g) \longrightarrow 2\text{HI}(g)$$

If at a certain temperature and pressure 1% of the substance present in the smaller amount had undergone reaction in 1 minute, how long would it take for 1% to react, at the same temperature, if the volume of the sample of gas were doubled?

Solution.   We see from Equation 16-3b that

$$-\frac{1}{P_\text{A}} \frac{dP_\text{A}}{dt} = kP_\text{B}$$

The fractional rate of removal of A is accordingly proportional to P_B. Since doubling the volume halves P_B, and hence the rate, it would take 2 minutes, twice as long, for the reaction to take place to the same extent.

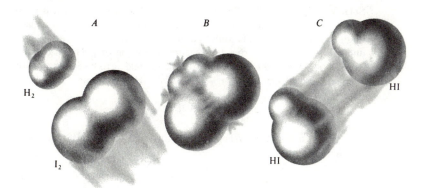

FIGURE 16-3
The mechanism of the reaction of hydrogen and iodine to form hydrogen iodide.

Example 16-6.   What is the time required for half the reactants to undergo second-order reaction, when they are present in equal amounts?

Solution.   We place $x = c/2$ in Equation 16-4 and solve for t:

$$\frac{c}{2} = \frac{c}{ckt' + 1}$$

$$ckt' + 1 = 2$$

$$ckt' = 1$$

$$t' \text{ (for half reaction)} = \frac{1}{ck}$$

We see that the time for half reaction is not determined alone by the reaction constant k, but is inversely proportional also to the initial concentration or pressure, c.

Example 16-7.   Esters, such as ethyl acetate, $CH_3COOC_2H_5$, react with hydroxide ion in aqueous solution:

$$CH_3COOC_2H_5 + OH^- \longrightarrow CH_3COO^- + C_2H_5OH$$

The rate is observed to be doubled by doubling either the ester concentration or the hydroxide ion concentration. In one experiment, at 25°C, 50 ml of solution 0.200 F in NaOH and 50 ml of solution 0.010 M in $CH_3COOC_2H_5$ were poured together and stirred. After 25 seconds, 50 ml of 0.200 F HCl was added; this stopped the reaction. On titrating the resultant solution to neutrality with 0.010 F NaOH, 10.0 ml was found to be needed. How much of the ester had been hydrolyzed? What is the value of the second-order reaction-rate constant for the reaction? At what time would 99.22% of the ester have reacted?

Solution. The solution after addition of HCl would have been neutral if no reaction had taken place. The amount of NaOH required to replace that used in the reaction is equivalent to 20% of the ester; hence 20% had undergone reaction. During the reaction the value of [OH⁻] decreased from 0.100 M to 0.099 M; we can accordingly take it to be constant, with average value 0.0995 M. Let $x = [CH_3COOC_2H_5]$. We write

$$-\frac{dx}{dt} = k \times 0.0995x = k'x$$

This is now a first-order reaction-rate expression. In the customary way (Section 16-3) we evaluate k':

$$0.80 = e^{-k't}, \qquad t = 25 \text{ s}$$

$$k' \times 25 \text{ s} = -\ln 0.80 = -2.303 \log 0.80 = 2.303 \times 0.097 = 0.223$$

$$k' = \frac{0.223}{25 \text{ s}} = 0.0089 \text{ s}^{-1}$$

The second-order reaction-rate constant hence has the value

$$k = \frac{k}{0.0995 \text{ mole liter}^{-1}} = \frac{0.0089 \text{ s}^{-1}}{0.0995 \text{ mole liter}^{-1}} = 0.089 \text{ s}^{-1} \text{mole}^{-1} \text{liter}$$

To determine the time required for 99.22% reaction of the ester, we note that the value of [OH⁻] would lie between 0.100 M and 0.095 M, and we multiply the average, 0.0975 mole liter⁻¹, by k, 0.089 s⁻¹ mole⁻¹ liter, to obtain $k' = 0.0975 \times 0.089 = 0.0087$ s⁻¹ for the pseudo-first-order reaction-rate constant. The time for half reaction is 0.693/0.0087 s⁻¹ = 80 s. For $(\frac{1}{2})^7 = \frac{1}{128} = 0.78\% = 100\% - 99.22\%$ to remain unreacted would require $7 \times 80 = 560$ s.

16-4. Mechanism of Reactions. Dependence of Reaction Rate on Temperature

It is everyday experience that chemical reactions are accelerated by increased temperature: the rate of many reactions at room temperature is approximately doubled for every 10°C increase in temperature.

This is a useful rule. It is only a rough rule. Reactions of large molecules, such as proteins, may have very large temperature factors; the rate of denaturation of ovalbumin (the process that occurs when an egg is boiled) increases about fiftyfold for a 10° rise in temperature.

We may obtain further insight into this question by considering the mechanism of reactions. For the reaction

$$H_2(g) + I_2(g) \longrightarrow 2HI(g)$$

at temperature T we may write the following rate equation:

$$- \frac{dP_{H_2}}{dt} = - \frac{dP_{I_2}}{dt} = k P_{H_2} P_{I_2}$$

The rate constant k is determined in part by the number of collisions at $P_{H_2} = P_{I_2} = 1$ atm (Figure 16-3). This number can be calculated by use of the kinetic theory of gases and certain assumptions about the sizes of the molecules. Only a small fraction of the collisions, however, are effective. If the molecules collide with small relative velocity, the van der Waals force of repulsion causes them to rebound elastically from one another.

Activation Energy

In order to react the pair of molecules must pass through a configuration (B in Figure 16-3) intermediate between the initial one (A) and the final one (C). This configuration is called the *activated complex*. In general this intermediate configuration, to which we may assign the structure

$$
\begin{array}{c}
\text{H---I} \\
\vdots \quad \vdots \\
\text{H---I}
\end{array}
$$

$\left(\text{resonance between } \begin{array}{cc} \text{H} & \text{I} \\ | & | \\ \text{H} & \text{I} \end{array} \text{ and } \begin{array}{c} \text{H—I} \\ \\ \text{H—I} \end{array} \right)$, has a larger energy (enthalpy) value

than the reactants. Let us use E_a to represent this energy difference, which is called the *activation energy* for the reaction. From the Boltzmann distribution law we know that the fractional number of collisions with this amount of energy involves the Boltzmann exponential factor (Section 9-5). We accordingly write

$$k = A \exp\left(- \frac{E_a}{RT}\right) \tag{16-6}$$

and

$$\ln k = - \frac{E_a}{RT} + \ln A \tag{16-7}$$

Here E_a is the activation energy per mole and A is a coefficient representing the collision number, the chance that the activated complex will break up to form the reactants rather than the products, and other factors affecting the rate.

These factors depend to some extent on the temperature, but as a good approximation we may take A to be a constant. We see from Equation 16-7 that $\ln k$ is a linear function of $1/T$, with slope $-E_a/R$:

$$\frac{d(\ln k)}{d(1/T)} = - \frac{E_a}{R} \tag{16-8}$$

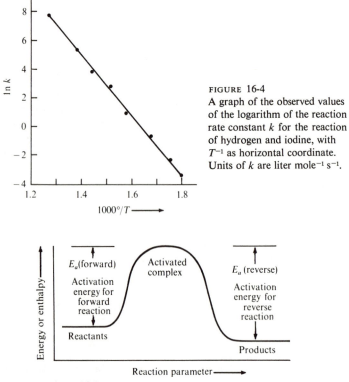

FIGURE 16-4
A graph of the observed values of the logarithm of the reaction rate constant k for the reaction of hydrogen and iodine, with T^{-1} as horizontal coordinate. Units of k are liter mole^{-1} s^{-1}.

FIGURE 16-5
Diagrammatic representation of the change in energy accompanying a chemical reaction. The top of the curve corresponds to the activated complex for the reaction.

Experimental values of k for the reaction of H_2 and I_2 over the range $T = 550°K$ to $800°K$ are shown in Figure 16-4. By differentiating Equation 16-7 with respect to T, we obtain the following expression for the temperature coefficient of the reaction-rate parameter k:

$$\frac{d \ln k}{dT} = \frac{E_a}{RT^2} \tag{16-9}$$

Equations 16-6 to 16-9 are called the *Arrhenius reaction-rate equations*. They were discovered in 1889 by the Swedish scientist Svante Arrhenius, who also discovered the existence of ions in electrolytic solutions.

Example 16-8. The observed reaction-rate curve for H_2 and I_2, Figure 16-4, has slope -21.6 in the units indicated. What is the value of the activation energy for the reaction?

Solution. With x-axis $1/T$ in place of $1000°/T$ the slope would be

$$\frac{d \ln k}{d(1/T)} = -21.6 \times 1000 = -21,600 \text{ deg}$$

From Equation 16-8 we see that

$$-\frac{E_a}{R} = -21,600 \text{ deg}$$

and hence
$$E_a = 21,600 \text{ deg} \times R = 21,600 \text{ deg} \times 8.315 \text{ J deg}^{-1} \text{ mole}^{-1}$$
$$= 180 \text{ kJ mole}^{-1}$$

The activation energy for the reaction is thus found to have the value 180 kJ mole^{-1}.

Activation Energy for the Reverse Reaction

In Figure 16-5 there is shown diagrammatically the change in energy accompanying a reaction. In order for the activated complex to be formed from the reactant molecules the activation energy E_a (forward) is needed. A detailed analysis shows that the activation energy E_a (reverse) differs from E_a (forward) by the difference in enthalpy of the reactants and the products. For the reactions $H_2 + I_2 \rightleftharpoons 2HI$ this enthalpy difference is 10 kJ mole^{-1} (Table 7-9); hence we can state that the value of E_a for the reaction $2HI \longrightarrow H_2 + I_2$ is $180 + 10 = 190$ kJ mole^{-1}.

The Relation between Activation Energy and Bond Energy

Values of the activation energy can often be estimated from bond-energy values (Table VII-1).

For example, the activation energy for the combination of two atoms, such as

$$I + I \longrightarrow I_2$$

is zero. Reaction may occur whenever the two atoms collide, with electron spins opposed, even at low relative velocity. (The bond energy may be converted to kinetic energy by collision with a third body.) The activation energy for the reverse reaction, the dissociation of a diatomic molecule, is just equal to the bond energy (enthalpy). For the dissociation of I_2 it is 151 kJ mole^{-1} (Table VIII-1).

A rough empirical rule can be formulated for reactions in which two bonds are broken and two are formed, such as

$$H_2 + I_2 \longrightarrow 2HI$$

and its reverse reaction

$$2HI \longrightarrow H_2 + I_2$$

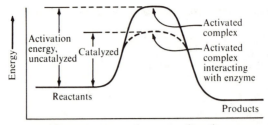

FIGURE 16-6

A diagram showing the possible effect of a catalyst in decreasing the value of the activation energy for a reaction, and thus increasing the rate of both the forward reaction and the reverse reaction.

It is found that the activation energy for the exothermic reaction (the first one of this pair) is approximately 30% of the sum of the bond energies of the two bonds that are broken (H—H and I—I, in this case).

For an exothermic reaction in which one bond is broken and another is formed, such as

$$F + H_2 \longrightarrow HF + H$$

or

$$D + H_2 \longrightarrow HD + H$$

the activation energy is approximately 8% of the energy of the bond that is broken.

16-5. Catalysis

The study of the factors that affect the rate of reactions has become more and more important with the continued great development of chemical industry. A modern method of manufacturing toluene, used for making the explosive trinitrotoluene and for other purposes, may be quoted as an example. The substance methylcyclohexane, C_7H_{14}, occurs in large quantities in petroleum. At high temperature and low pressure this substance should decompose into toluene, C_7H_8, and hydrogen. The reaction is so slow, however, that the process could not be carried out commercially until the discovery was made that a certain mixture of oxides increases the rate of reaction enough for the process to be put into practice. Many examples of catalysis (the process of accelerating a reaction by use of a catalyst) have already been mentioned, and others are mentioned in later chapters.

It is thought that catalysts speed up reactions by bringing the reacting molecules together and holding them in configurations favorable to reaction. Unfortunately so little is known about the fundamental nature of catalytic activity that the search for suitable catalysts is largely empirical. The test of a catalytic reaction, as of any proposed process, is made by trying it to see if it works.

Catalysis and Activation Energy

There is evidence that some catalysts have a structure that leads to strong interaction with the activated complex and only weak interaction with the reactants and products. The strong interaction with the activated complex leads to a decrease in the activation energy, as indicated in Figure 16-6, and hence to increase in the rate of the reaction.

16-6.   Kinetics of Enzyme Reactions

Most of the chemical reactions that take place in the human body and in other living organisms are catalyzed by certain protein molecules, which are called enzymes (Section 24-5). A reactant in an enzyme-catalyzed reaction is usually called the substrate. Many of these reactions take place at rates that can be described by a simple equation, the Michaelis-Menten equation (Equation 16-12).

Let us assume that reaction occurs through combination of the enzyme E and the substrate S to form the complex ES, which can then decompose in two ways, forming either E and S again, or E and the products P:

$$E + S \overset{K}{\rightleftarrows} ES \overset{k}{\longrightarrow} E + P$$

If the rate-constant k is small, the concentration of ES approximates closely the equilibrium value:

$$\frac{[ES]}{[E][S]} = K \qquad (16\text{-}10)$$

The rate of reaction is proportional to [ES], and hence to [E][S]:

$$-\frac{d[S]}{dt} = k[ES] = kK[E][S] \qquad (16\text{-}11)$$

The total enzyme concentration is $[E]_{total} = [E] + [ES]$. From Equation 16-10 we see that

$$[E] = \frac{[E]_{total}}{K[S] + 1}$$

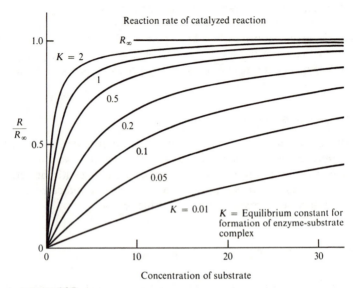

FIGURE 16-7
Curves showing calculated reaction rate R/R_∞ of a catalyzed reaction as function of the concentration of the substrate, for different values of the equilibrium constant K for formation of the enzyme-substrate complex.

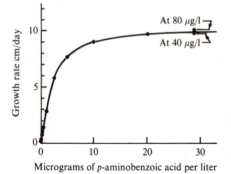

FIGURE 16-8
The observed rate of growth of a *p*-aminobenzoic-acid-requiring *Neurospora* mutant (Tatum and Beadle, 1942), as function of the concentration of the growth substance in the medium.

and hence Equation 16-11 may be rewritten as

$$\text{Rate} = -\frac{d[S]}{dt} = \frac{k[S][E]_{\text{total}}}{[S] + (1/K)} \tag{16-12}$$

From this equation we see that when [S] is small compared with K^{-1} the rate is proportional to [S], being $kK[S][E]_{\text{total}}$, and when [S] is so large as to saturate the enzyme the rate is independent of [S], being just $k[E]_{\text{total}}$. Some calculated curves are shown in Figure 16-7.

It is interesting to compare the curve for $K = 0.5$ with the experimental curve of Figure 16-8, which shows the rate of growth of a mutant of red bread mold along a glass tube containing the nutrient solution as a function of the concentration of a growth substance, para-aminobenzoic acid, in the solution. The similarity in shape indicates that an equilibrium corresponding to Equation 16-10 determines the rate of growth of the organism.

It is possible that the equilibrium involved is not that between an enzyme and a substrate but rather that between an apoenzyme (a protein without enzymatic activity) and a coenzyme (often a vitamin) to form the active enzyme:

$$A + C \rightleftharpoons E$$

$$\frac{[E]}{[A][C]} = K' \tag{16-13}$$

You can verify that the rate of the reaction is affected by the concentration of the coenzyme in the same way as by the concentration of the substrate.

Many diseases are known to involve gene mutations that decrease the activity of an important enzyme in the human body. One way in which this occurs is through the formation of an altered apoenzyme with a greatly decreased value of the combining constant K' with its coenzyme. One disease of this sort is cystathioninuria, which is recognized by the appearance of cystathionine, a derivative of the amino acid methionine (see Table 24-1), in the urine of the patient. The disease causes mental retardation and unpleasant physical manifestations. The enzyme involved in the metabolism of cystathionine has pyridoxine (vitamin B_6) as its coenzyme, which usually is ingested in the amount of about 1 mg per day. It has been found that ingestion of 100 times this amount leads to the disappearance of cystathionine from the urine and alleviation of other manifestations of the disease, presumably by shifting the equilibrium of Equation 16-13 in the direction of the active enzyme. This example illustrates the use of the principles of chemical equilibrium and chemical kinetics in the treatment of disease.

16-7.   Chain Reactions

When the reaction

$$H_2(g) + Br_2(g) \longrightarrow 2HBr(g)$$

was carefully investigated about 50 years ago, it was found that its rate is not proportional to the product $[H_2][Br_2]$, as expected (Equation 16-3). Instead, in its initial stages, with [HBr] small, the rate was observed to be proportional to $[H_2][Br_2]^{1/2}$:

$$-\frac{d[H_2]}{dt} = -\frac{d[Br_2]}{dt} = k[H_2][Br_2]^{1/2}$$

This observation was accounted for by the assumption that the reaction occurs in a series of steps. The first is the dissociation of a bromine molecule

$$Br_2 \longrightarrow 2Br$$

This is followed by a sequence of reactions, called a *chain*:

$$Br + H_2 \longrightarrow HBr + H$$
$$H + Br_2 \longrightarrow HBr + Br$$
$$Br + H_2 \longrightarrow HBr + H$$
$$H + Br_2 \longrightarrow HBr + Br$$
$$\cdots$$

The chain is ultimately broken by the reaction

$$2Br \longrightarrow Br_2$$

Explosions

A mixture of hydrogen and chlorine explodes when ignited. The overall reaction $H_2 + Cl_2 \longrightarrow 2HCl$, which takes place by the chain mechanism analogous to that for $H_2 + Br_2$, liberates so much heat that the gas may begin to increase in temperature, instead of dissipating the heat to the environment. The reaction then proceeds more rapidly, the temperature increases more rapidly, and a very rapid reaction, called *a thermal explosion*, results.

The explosion of a mixture of hydrogen and oxygen is different in kind; it is called a *branched-chain explosion*. An initial reaction with an atom of oxygen (or of hydrogen) is followed by two reactions, that one by three, then five, and so on:*

---

*Other reactions, involving the radical HO_2, also occur.

$$O + H_2 \longrightarrow OH + H$$
$$OH + H_2 \longrightarrow H_2O + O \quad H + O_2 \longrightarrow OH + O$$
$$O + H_2 \longrightarrow OH + H \quad OH + H_2 \longrightarrow H_2O + O \quad O + H_2 \longrightarrow OH + H$$
$$\cdots$$

Important chain reactions, which under some conditions lead to explosion, are the fission and fusion of atomic nuclei (Section 26-7).

Spontaneous Combustion

Reactions such as the combustion of fuels proceed very rapidly when combustion is begun, but fuels may remain indefinitely in contact with air without burning. In these cases the rate of reaction at room temperature is extremely small. The process of lighting a fire consists in increasing the temperature of part of the fuel until the reaction proceeds rapidly; the exothermic reaction then liberates enough heat to raise another portion of the fuel to the kindling temperature, and in this way the process is continued.

The oxidation of oil-soaked rags or other combustible material may occur rapidly enough at room temperature to produce heat enough to increase the temperature somewhat; this accelerates the oxidation, and causes further heating, until the mass bursts into flame. This process is called *spontaneous combustion.*

Exercises

16-1. By what factor would the percentage decomposition in given time of a sample of N_2O_5 (which decomposes unimolecularly) be increased by compressing to half the volume with temperature held constant?

16-2. In an investigation of the second-order gas reaction

$$2NO_2 \longrightarrow 2NO + O_2$$

the pressure changed from 100 to 101 torr in 12 s, starting from pure NO_2. How long would it take, later on, to change from 125 to 126 torr?

16-3. Azomethane gas decomposes according to the first-order reaction

$$CH_3NNCH_3 \longrightarrow C_2H_6 + N_2$$

(a) In a reaction at 287°C, with initial pressure (of azomethane) only 160 torr, what would the final total pressure be?

(b) When the total pressure reached 161.6 torr, after 100 seconds, what fraction of the azomethane had decomposed?

(c) Using the equation $\dfrac{-dp_{az}}{dt} = kp_{az}$ (p_{az} = partial pressure of azomethane), calculate k (in s^{-1}) at 287°C from the above data.

(d) How long would it take for half the azomethane to decompose?

(e) If the initial pressure were 80 torr, how long would it take for half the azomethane to decompose, at the same temperature?

16-4. The reduction of ferric ion by stannous ion in neutral solution was shown by A. A. Noyes to be third-order, corresponding to the reaction

$$2Fe^{+++} + Sn^{++} \longrightarrow 2Fe^{++} + Sn^{++++}$$

At certain concentrations 1% of the iron is reduced in 10 seconds.

(a) How long would it take to reduce 1% of the iron if $[Sn^{++}]$ were doubled?

(b) How long would it take to reduce 1% of the iron if $[Fe^{+++}]$ were doubled?

(c) About how long would it take, at the original concentrations, if the temperature were raised 30°?

16-5. Ramsperger reported the values 1.00×10^{-4} at 287.3°C and 20.8×10^{-4} at 327.4°C for the first-order rate constant of the azomethane decomposition.

(a) From these data determine the 10° rate-constant factor.

(b) Predict a value for the rate constant at 20°C.

(c) Calculate how long it would take for a sample to decompose to the extent of 1% at 20°C and at 327.4°C.

16-6. The half-life period of radium is 1,590 years. What is the reaction-rate constant for the reaction? What fraction decomposes in one day?

16-7. The half-life period of ^{11}C, made artificially and used in biological researches, is 20 min. How much of a sample remains after 3 hours?

16-8. A sample contains initially equal numbers of atoms of ^{11}C, with half-life 20 min., and of ^{14}C, with half-life about 5,568 years.
(*a*) What is the initial ratio of activities of ^{11}C to ^{14}C (that is, of rate of decomposition)?
(*b*) What is the ratio of activities after 6 hours? after 12 hours?

16-9. How long would it take before a mole of radium atoms disintegrates until only one or two radium atoms are left?

16-10. Calculate the half-life in years of the plutonium isotope of mass number 239 from the measurement that one microgram emits 140,000 alpha particles per minute. Each plutonium atom emits one alpha particle as it decomposes.

16-11. Automobile tires when stored age through oxidation and other reactions of the rubber. By about what factor would the safe period of storage be multiplied by lowering the temperature of the storage room 10°C?

16-12. Ozone gas decomposes according to the equation $2O_3 \longrightarrow 3O_2$. At one temperature, with ozone present at an initial pressure of 1.000 atm, the pressure rose to 1.012 atm in one minute. About how long would it take for the pressure to change from 1.000 to 1.012 atm at a temperature 15°C higher?

16-13. How does light cause the reaction between hydrogen and chlorine to take place? Would ultraviolet light with wavelength 1,000 Å, which causes the dissociation of hydrogen molecules into hydrogen atoms, initiate the reaction between hydrogen and chlorine?

16-14. What molecular processes can you imagine that would prevent a single quantum of light from causing all of the chlorine and all of the hydrogen in a vessel from combining?

16-15. We know that catalysts in the body cause carbohydrates (sugars), such as $C_6H_{12}O_6$, to react with oxygen to form carbon dioxide and water. In plants, because of the presence of chlorophyll, carbon dioxide and water are converted by a photochemical reaction to sugar and oxygen. Are these statements inconsistent with the statement that a catalyst can affect only the rate of a reaction and not its equilibrium? Why not?

16-16. The reaction

$$CF_3 + H_2 \longrightarrow CF_3H + H$$

is a second-order reaction with rate constant $k_2 = 4.5 \times 10^3$ liter mole^{-1} s^{-1} at 400°K. What is the electronic structure of the radical CF_3? If a small amount of this radical were to be introduced into hydrogen at 1 atm pressure and 400°K, how long would it take for half of it to be converted to methyl fluoride?

16-17. It was found by the American chemists D. P. Stevenson and D. O. Schissler, by use of a mass spectrograph to determine the composition of the reacting gas, that the reaction

$$Kr^+ + H_2 \longrightarrow KrH^+ + H$$

occurs at every collision between the reacting molecules, and that the activation energy for the reaction is zero.

(*a*) What electronic structure would you assign to KrH^+?

(*b*) Is the reaction exothermic or endothermic?

(*c*) What is the minimum value of the bond energy between Kr^+ and H in KrH^+? (Answer: (*c*) 435 kJ mole^{-1}.)

16-18. Stevenson and Schissler also observed the reaction

$$HCl^+ + HCl \longrightarrow H_2Cl^+ + Cl$$

This reaction has zero activation energy, and accordingly is exothermic. What is the electronic structure of H_2Cl^+? What value would you predict for the H—Cl—H bond angle? Can you suggest a reason for this reaction to be exothermic? (How would the electronegativity of Cl^+ compare with that of Cl?)

16-19. Ethyl chloride vapor when heated decomposes by the reaction

$$C_2H_5Cl \longrightarrow C_2H_4 + HCl$$

The reaction is observed to be of the first order, with rate constant $k = Ae^{-E*/TR}$, $A = 1.6 \times 10^{14}$ s^{-1}, $E* = 249$ kJ mole^{-1}. What is the value of k at 700°K? What fraction of the ethyl chloride would decompose in 10 minutes at this temperature? At what temperature is the reaction twice as fast? (Answer: 4.2×10^{-5} s^{-1}; 2.5%; 712°K.)

16-20. The reaction of the preceding Exercise might have as its initial step the breaking of a carbon-chlorine bond to form the radicals C_2H_5 and Cl, followed by a chain of reactions to give the final products. The investigators concluded that the observed value of the activation energy eliminates this mechanism. What is the nature of their argument? What value would be expected for the activation energy if this were the mechanism?

16-21. What structure would you assign to the activated ethyl chloride molecule if its decomposition involves its splitting directly into ethylene and hydrogen chloride? About what value would you expect the activation energy to have (Section 16-4; Table VIII-1)? (Answer: About 220 kJ mole^{-1}.)

16-22. An enzyme in the human body may increase the rate of a chemical reaction a millionfold. What energy of binding of the activated complex to the enzyme would be needed to account for this effect? (Answer: 36 kJ mole^{-1}.)

The Nature of Metals and Alloys

About eighty of the more than one hundred elementary substances are metals. A metal may be defined as a substance that has large conductivity of electricity and of heat, has a characteristic luster, called metallic luster, and can be hammered into sheets (is malleable) and drawn into wire (is ductile); in addition, the electric conductivity increases with decrease in temperature.*

---

*Sometimes there is difficulty in classifying an element as a metal, a metalloid, or a nonmetal. For example, the element tin can exist in two forms, one of which, the common form, called white tin, is metallic, whereas the other, gray tin, has the properties of a metalloid. The next element in the periodic table, antimony, exists in only one crystalline form, with metallic luster but with the electric properties of a metalloid, and it is brittle, rather than malleable and ductile. We shall consider tin to be a metal and antimony a metalloid.

17-1. The Metallic Elements

The metallic elements may be taken to include lithium and beryllium in the first short period of the periodic table, sodium, magnesium, and aluminum in the second short period, the thirteen elements from potassium to gallium in the first long period, the fourteen from rubidium to tin in the second long period, the twenty-nine from cesium to bismuth in the first very long period (including the fourteen rare-earth metals), as well as the eighteen from francium to khurchatovium.

The metals themselves and their alloys are of great usefulness to man, because of the properties characteristic of metals. Our modern civilization is based upon iron and steel, and valuable alloy steels are made that involve the incorporation with iron of vanadium, chromium, manganese, cobalt, nickel, molybdenum, tungsten, and other metals. The importance of these alloys is due primarily to their hardness and strength. These properties are a consequence of the presence in the metals of very strong bonds between the atoms. For this reason it is of especial interest to us to understand the nature of the forces that hold the metal atoms together in metals and alloys.

17-2. The Structure of Metals

In a nonmetal or metalloid the number of atoms that each atom has as its nearest neighbors is determined by its covalence. For example, the iodine atom, which is univalent, has only one other iodine atom close to it in a crystal of iodine: the crystal, like liquid iodine and iodine vapor, is composed of diatomic molecules. In a crystal of sulfur there are S_8 molecules, in which each sulfur atom has two nearest neighbors, to each of which it is attached by one of its two covalent bonds. In diamond the quadrivalent carbon atom has four nearest neighbors. On the other hand, the potassium atom in potassium metal, the calcium atom in calcium metal, and the titanium atom in titanium metal, which have one, two, and four outer electrons, respectively, do not have only one, two, and four nearest neighbors, but have, instead, eight or twelve nearest neighbors. We may state that one of the characteristic features of a metal is that each atom has a large number of neighbors; the number of small interatomic distances is greater than the number of valence electrons.

Most metals crystallize with an atomic arrangement in which each atom has surrounded itself with the maximum number of atoms that is geometrically possible. There are two common metallic structures that correspond to the closest possible packing of spheres of constant size. One of these structures, called the cubic closest-packed structure, has

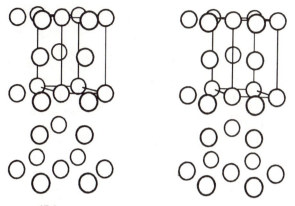

FIGURE 17-1
The hexagonal close-packed arrangement of spheres. Many metals crystallize with this structure.

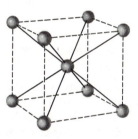

FIGURE 17-2
The atomic arrangement in α-iron (body-centered arrangement).

been described in Chapter 2. The other structure, called hexagonal closest packing, is represented in Figure 17-1. It is closely similar to the cubic closest-packed structure; each atom is surrounded by twelve equidistant neighbors, with, however, the arrangement of these neighbors slightly different from that in cubic closest packing. About fifty metals have the cubic closest-packed structure or the hexagonal closest-packed structure, or both.

Another common structure, assumed by about twenty metals, is the body-centered cubic structure. In this structure, shown as Figure 17-2, each atom has eight nearest neighbors, and six next-nearest neighbors. These six next-nearest neighbors are 15% more distant than the eight nearest neighbors; in discussing the structure it is difficult to decide whether to describe each atom as having ligancy 8 or ligancy 14.

The periodicity of properties of the elements, as functions of the atomic number, is illustrated by the observed values of the interatomic distances in the metals, as shown in Figure 17-3. These values are half of the directly determined interatomic distances for the metals with a cubic closest-

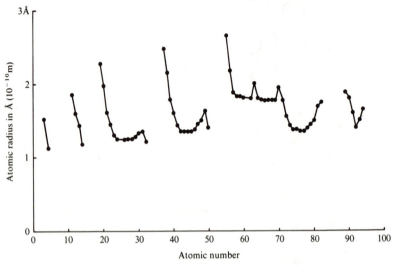

FIGURE 17-3
The atomic radii of metals, plotted against atomic number.

packed or hexagonal closest-packed structure. For other metals a small correction has been made; it has been observed, for example, that a metal such as iron, which crystallizes in a modification with a closest-packed structure and also a modification with the body-centered cubic structure, has contact interatomic distances about 3% less in the latter structure than in the former, and accordingly a correction of 3% can be made for body-centered cubic structures, to convert the interatomic distances to ligancy 12.

We may well expect that the strongest bonds would have the shortest interatomic distances, and it is accordingly not surprising that the large interatomic distances shown in Figure 17-3 are those for soft metals, such as potassium; the smallest ones, for chromium, iron, nickel, and others, refer to the hard, strong metals.

17-3.   The Nature of the Transition Metals

The long periods of the periodic system can be described as short periods with ten additional elements inserted. The first three elements of the long period between argon and krypton, which are the metals potassium, calcium, and scandium, resemble their congeners of the preceding short period, sodium, magnesium, and aluminum, respectively. Similarly the last four elements in the sequence, germanium, arsenic, selenium, and bromine, resemble their preceding congeners, silicon, phosphorus, sulfur, and chlorine, respectively. The remaining elements of the long period, titanium, vanadium, chromium, manganese, iron, cobalt, nickel, copper, zinc, and gallium, have no lighter congeners; they are not closely similar in their properties to any lighter elements.

The properties of these elements accordingly suggest that the long period can be described as involving the introduction of ten elements in the center of the series. The introduction of these elements is correlated with the insertion of ten additional electrons into the five $3d$ orbitals of the M shell, converting it from a shell of 8 electrons, as in the argon atom, to a shell of 18 electrons. It is convenient to describe the long period as involving ten *transition metals*, corresponding to the ten electrons. We shall consider the ten elements from titanium, group IVa, to gallium, group IIIb, as constituting the ten transition elements in the first long period, and shall take the heavier congeners of these elements as the transition elements in the later series.*

The chemical properties of the transition elements do not change so strikingly with change in atomic number as do those of the other elements. In the series potassium, calcium, scandium, the normal salts of the elements correspond to the maximum oxidation numbers given by the positions of the elements in the periodic system, 1 for potassium, 2 for calcium, and 3 for scandium; the sulfates, for example, of these elements are K_2SO_4, $CaSO_4$, and $Sc_2(SO_4)_3$. The fourth element, titanium, tends to form salts representing a lower oxidation number than its maximum, 4; although compounds such as titanium dioxide, TiO_2, and titanium tetrachloride, $TiCl_4$, can be prepared, most of the compounds of titanium represent lower oxidation states, $+2$ or $+3$. The same tendency is shown by the succeeding elements. The compounds of vanadium, chromium, and manganese that represent the maximum oxidation numbers $+5$, $+6$, and $+7$, respectively, are strong oxidizing agents, and are easily reduced to compounds in which these elements have oxidation numbers $+2$ or $+3$. The oxidation numbers $+2$ and $+3$ continue to be the important ones for the succeeding elements, iron, cobalt, nickel, copper, and zinc.

A striking characteristic of most of the compounds of the transition metals is their color. Nearly every compound formed by vanadium, chromium, manganese, iron, cobalt, nickel, and copper is strongly colored, the color depending not only on the atomic number of the metallic element but also on its state of oxidation, and, to some extent, on

---

*Some chemists take the ten elements scandium to zinc as the transition elements of the first long period. However, scandium and its congener yttrium rather closely resemble aluminum in their physical and chemical properties, whereas gallium and indium are quite dissimilar to aluminum; for this reason it seems wise to classify scandium and yttrium with aluminum, and gallium and indium among the transition elements.

the nature of the nonmetallic element or anion with which the metal is combined. It seems clear that the color of these compounds is associated with the presence of an incomplete M shell of electrons; that is, with an M shell containing less than its maximum number of electrons, 18. When the M shell is completed, as in the compounds of bipositive zinc ($ZnSO_4$ and others) and of unipositive copper (CuCl and others), the substances are in general colorless. Another property characteristic of incompleted inner shells is paramagnetism, the property of a substance of being attracted into a strong magnetic field. Nearly all the compounds of the transition elements in oxidation states corresponding to the presence of incompleted inner shells are strongly paramagnetic.

17-4. The Metallic State

The characteristic properties of hardness and strength of the transition metals and their alloys are a consequence of the presence in the metals of very strong bonds between the atoms. For this reason it is of especial interest to us to understand the nature of the forces that hold the metal atoms together in these metals and alloys.

Let us consider the first six metals of the first long period, potassium, calcium, scandium, titanium, vanadium, and chromium. The first of these metals, potassium, is a soft, light metal, with low melting point. The second metal, calcium, is much harder and denser, and has a much higher melting point. Similarly, the third metal, scandium, is still harder, still denser, and melts at a still higher temperature, and this change in properties continues through titanium, vanadium, and chromium. It is well illustrated in Figure 17-4, which shows a quantity called the ideal density, equal to 50/gram-atomic volume. This ideal density, which is inversely proportional to the gram-atomic volume of the metal, is the density that these metals would have if they all had the same atomic weight, 50. It is an inverse measure of the cube of the interatomic distances in the metals. We see that the ideal density increases steadily from its minimum value of about 1 for potassium to a value of about 7 for chromium, and many other properties of the metals, including hardness and tensile strength, show a similar steady increase through this series of six metals.

There is a simple explanation of this change in properties in terms of the electronic structure of the metals. The potassium atom has only one electron outside of its completed argon shell. It could use this electron to form a single covalent bond with another potassium atom, as in the diatomic molecules K_2 that are present, together with monatomic molecules K, in potassium vapor. In the crystal of metallic potassium each potassium atom has a number of neighboring atoms, at the same distance. It is held to these neighbors by its single covalent bond, which resonates among the

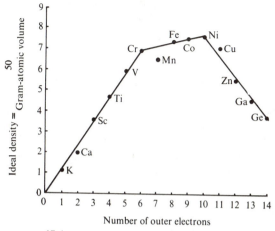

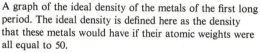

FIGURE 17-4

A graph of the ideal density of the metals of the first long period. The ideal density is defined here as the density that these metals would have if their atomic weights were all equal to 50.

neighbors. In metallic calcium there are *two* valence electrons per calcium atom, permitting each atom to form two bonds with its neighbors. These two bonds resonate among the calcium-calcium positions, giving a total bonding power in the metal twice as great as that in potassium. Similarly in scandium, with *three* valence electrons, the bonding is three times as great as in potassium, and so on to chromium, where, with *six* valence electrons, the bonding is six times as great.

This increase does not continue in the same way beyond chromium. Instead, the strength, hardness, and other properties of the transition metals remain essentially constant for the five elements chromium, manganese, iron, cobalt, and nickel, as is indicated by the small change in ideal density in Figure 17-4. (The low value for manganese results from the existence of this metal with an unusual crystal structure shown by no other element.) We can conclude that the metallic valence does not continue to increase, but remains at the value six for these elements. Then, after nickel, the metallic valence again decreases, through the series copper, zinc, gallium, and germanium, as is indicated by the rapid decrease in ideal density in Figure 17-4, and by a corresponding decrease in hardness, melting point, and other properties.

It is interesting to note that in the metallic state chromium has metallic valence 6, corresponding to the oxidation number +6 characteristic of the chromates and dichromates, rather than to the lower oxidation number +3 shown in the chromium salts, and that the metals manganese, iron, cobalt, and nickel also have metallic valence 6, although nearly all of their compounds represent the oxidation state $+2$ or $+3$. *The valuable physical properties of the transition metals are the result of the high metallic valence of the elements.*

Unsynchronized Resonance of Bonds in Metals

It is mentioned above that in potassium metal each atom, with one valence electron, can form one covalent bond, but that this bond is not between the atom and a single neighboring atom, but instead resonates among several positions. For four potassium atoms in a square, we might write two valence-bond structures:

$$
\begin{array}{cc}
\text{K} \quad \text{K} & \text{K—K} \\
\text{|} \quad \text{|} & \\
\text{K} \quad \text{K} & \text{K—K} \\
\text{I} & \text{II}
\end{array}
$$

These structures are analogous to the two Kekulé structures for the benzene molecule. Resonance between the two structures would stabilize the metal relative to a crystal composed of K_2 molecules, each with a fixed covalent bond.

There are other structures, involving a transfer of an electron from one atom to another, that might be considered:

$$
\begin{array}{cccc}
\text{K}^+ \ \text{K} & \text{K} \ \ \text{K}^+ & \text{K}^-\text{—K} & \text{K—K}^- \\
\ \ | & \ \ | & | & \ \ | \\
\text{K—K}^- & \text{K}^-\text{—K} & \text{K} \ \ \text{K}^+ & \text{K}^+ \ \text{K} \\
\text{III} & \text{IV} & \text{V} & \text{VI}
\end{array}
$$

Resonance among all six structures would lead to greater stabilization than resonance between only I and II above. Moreover, this unsynchronized resonance gives a simple explanation of the characteristic properties of metals—the large electric conductivity and the negative temperature coefficient of electric conductivity.

Let us consider a row of potassium atoms:

$$\text{K—K} \quad \text{K—K} \quad \text{K}^+ \quad \text{K—K}^-\text{—K} \quad \text{K—K} \quad \text{K—K}$$

In the presence of an electric field, produced by a cathode at the left and an anode at the right, the bonds would tend to shift in such a way as to move the positive charge toward the cathode and the negative charge toward the anode:

$$\text{K—K} \quad \text{K}^+ \ \text{K—K} \quad \text{K—K} \quad \text{K—K}^-\text{—K} \quad \text{K—K}$$

$$\text{K}^+ \ \text{K—K} \quad \text{K—K} \quad \text{K—K} \quad \text{K—K} \quad \text{K—K}^-\text{—K}$$
$$\text{etc.}$$

The unsynchronized resonance of the bonds corresponds to the transfer of electric charge (electrons) that leads to high electric conductivity. This conductivity is characteristic of the structure of the metal, and hence takes place most readily at very low temperatures, when the atoms are quite regularly arranged in the crystal. At higher temperatures the thermal oscillation of the atoms introduces some disorder in their arrangement, which interferes with the resonance of the bonds, and hence causes a decrease in conductivity (negative temperature coefficient).

The Metallic Orbital

In the valence-bond structure

$$K^+ \quad K$$
$$K\text{---}K^-$$

one atom, K^-, has assumed a second valence electron, permitting it to form two covalent bonds, rather than just one. Two bond orbitals are accordingly being utilized by this atom, one more than by the neutral (unicovalent) atoms. This extra orbital, called the *metallic orbital*, is needed to permit the unsynchronized valence-bond resonance characteristic of metals: *the metallic orbital, an extra orbital not occupied by an electron or electron pair in the neutral atom, is the characteristic structural feature of metals.*

Potassium has nine reasonably stable orbitals in its outer shell: one $4s$ orbital, three $4p$ orbitals, and five $3d$ orbitals. Only one (an spd hybrid) is used as a bond orbital by the unicovalent atom, and others are available to serve as the metallic orbital; hence potassium is a metal. In diamond, on the other hand, the four stable orbitals of the valence shell (the tetrahedral orbitals formed by hybridization of the $2s$ orbital and the three $2p$ orbitals, Chapter 6) are all occupied by bond electrons; there is no metallic orbital, and hence diamond is not a metal.

17-5.   Metallic Valence

It is mentioned in the preceding section that for the elements potassium, calcium, scandium, titanium, vanadium, and chromium the physical properties indicate that all of the electrons outside of the argon shell are used in forming bonds, and that the metallic valences for these elements are 1, 2, 3, 4, 5, and 6, respectively.

There are nine stable orbitals available for the transition elements (one $4s$, three $4p$, five $3d$), and, with one required as the metallic orbital, the metallic valence might be expected to continue to increase, and have the value 7 for manganese and 8 for iron. However, as mentioned above, the physical properties show that the metallic valence remains at the maximum of 6 for manganese, iron, cobalt, and nickel, and then begins to decrease at copper. The maximum value of 6 corresponds to the number of good bond orbitals that can be formed by hybridization of the s, p, and d orbitals. The decrease in metallic valence beginning at copper is caused by the limited number of orbitals, as shown by the example of tin.

Tin, element 50, has 14 electrons outside of the krypton shell, and nine stable orbitals ($4d$, $5s$, $5p$). The five $4d$ orbitals, which are more stable than the $5s$ and $5p$ orbitals, are occupied by five unshared electron pairs. The remaining four electrons may separately occupy the four tetrahedral $5s5p^3$ orbitals, and be used in forming four bonds, tetrahedrally directed. In fact, gray tin, one of the two allotropic forms of the element, has the diamond structure. The tin atoms in gray tin are quadrivalent, as are the carbon atoms in diamond. They have no metallic orbital, and gray tin is not a metal, but is a metalloid.

If the tin atom were to retain one of its orbitals for use as a metallic orbital, it would be bivalent rather than quadrivalent:

| Gray tin: | $5s$ | $5p$ | $5p$ | $5p$ | No metallic orbital |
| | ↑ | ↑ | ↑ | ↑ | Quadrivalent |

| White tin: | $5s$ | $5p$ | $5p$ | $5p$ | One metallic orbital |
| | ↑↓ | ↑ | ↑ | Metallic orbital | Bivalent |

The ordinary allotropic form of tin, white tin, has metallic properties. The observed bond lengths indicate that the valence of tin in this form is about 2.5.

The value 2.5 can be accounted for in the following way. The magnetic properties of the iron-group elements and their alloys indicate that the number of metallic orbitals per atom in a metal is 0.72, rather than 1 (see the discussion of magnetic properties in the next section). This fractional value can be explained by the reasonable assumption that the metal contains 28% M^+, 44% M, and 28% M^-. The ions M^- do not need a metallic orbital, because they cannot accept another electron (M^{--} would be unstable, according to the electroneutrality principle, Section 6-12). The structure of white tin can accordingly be represented in the following way:

	$5s$	$5p$	$5p$	$5p$		Contribution to valence
28% Sn^+	↑	↑	↑	Metallic		$3 \times 0.28 = 0.84$
44% Sn	↑↓	↑	↑	Metallic		$2 \times 0.44 = 0.88$
28% Sn^-	↑↓	↑	↑	↑		$3 \times 0.28 = 0.84$

Metallic valence of tin   2.56

The same argument leads to the following values for the metallic valence of the elements from copper to germanium:

Cu	Zn	Ga	Ge
5.56	4.56	3.56	2.56

Copper, zinc, and gallium are metals, with properties compatible with these values of the valence. Germanium under ordinary pressure is a metalloid, with the diamond structure and valence 4. At high pressure it is converted into another form, with greatly increased electric conductivity and density corresponding to the white tin structure and valence 2.56.

Ferromagnetism and Metallic Valence

Iron, cobalt, and nickel are ferromagnetic metals. The ferromagnetism of iron corresponds to 2.2 electrons with unpaired spins per atom. The alloys of iron with a small amount of cobalt are more strongly ferromagnetic than pure iron. The ferromagnetism increases to a maximum value at about 28% cobalt, and then decreases, reaching the value corresponding to 1.7 unpaired electrons per atom for pure cobalt.

The maximum ferromagnetism for the alloy of 72% iron and 28% cobalt can be interpreted in the following way. The atoms in this alloy have the average atomic number 26.28, and hence have 8.28 electrons outside of the argon shell. These electrons may occupy nine orbitals: the five $3d$ orbitals, the $4s$ orbital, and the three $3p$ orbitals. But if all nine orbitals were available for occupancy by the electrons (6 for bond formation and the others contributing to the ferromagnetism) the number of unpaired electrons would be expected to continue to increase beyond 28% cobalt and to reach its maximum at pure cobalt, which has nine electrons outside the argon shell. The fact that the maximum ferromagnetism is reached at 28% cobalt (8.28 electrons outside of the argon shell) indicates that only 8.28 of the nine orbitals are available for occupancy. The remaining 0.72 orbital per atom is interpreted as the metallic orbital of 72% of the atoms, as discussed above.

Example 17-1.   The ferromagnetism of alloys of nickel and copper decreases from the value corresponding to 0.6 unpaired electron per atom for pure nickel to 0 for the alloy with 56% copper. How is this fact interpreted?

Solution.   In these alloys 8.28 orbitals are available for occupancy by the electrons outside of the argon shell. The alloy of 44% nickel and 56% copper has an average of 10.56 such electrons per atom. Of these, 6 are bonding electrons, which occupy 6 of the 8.28 orbitals. The remaining 4.56 electrons occupy the remaining 2.28 orbitals; since the electron/orbital ratio is 2, these electrons are all paired. Hence there are no unpaired electrons in this alloy, and it is not ferromagnetic.

Example 17-2.   What is the metallic valence of zinc?

Solution.   Zinc has 12 electrons outside the argon shell, and 8.28 orbitals for them to occupy. We place 8.28 electrons with positive spin in these orbitals, and the remaining $12 - 8.28 = 3.72$ electrons with negative spin in 3.72 of the orbitals. Hence 3.72 orbitals per atom are occupied by electron pairs, and the remaining $8.28 - 3.72 = 4.56$ orbitals per atom are occupied by single electrons. These 4.56 electrons can be used in forming bonds. Hence the metallic valence of zinc is 4.56, as stated above.

17-6.   The Free-electron Theory of Metals

A theory of metals in which the valence electrons are treated as moving freely in the field of the ions was developed by the Dutch physicist Hendrik Antoon Lorentz (1853–1928) and, in quantum mechanical form, the Austrian physicist Wolfgang Pauli (1900–1958). The density of energy levels for a particle in a box is given by Equation 9-18. The Pauli exclusion principle permits two electrons (with opposed spins) to occupy each orbital. Hence the density $\rho(E)$ for free electron states for a metal is the following, obtained from Equation 9-18 by introducing the factor 2:

$$\rho(E) = \frac{8\pi 2^{1/2} m^{3/2} V E^{1/2}}{h^3} \tag{17-1}$$

At the absolute zero each of the lowest spin-orbit levels is occupied by an electron, as shown in Figure 17-5. At a higher temperature some electrons are excited to levels with larger energy, leaving a "hole" (one electron of an electron pair) in the orbital. The amount of excitation is proportional to kT.

The maximum energy of an occupied level for a metal with z valence electrons and molar volume V is found from Equation 9-17 to be

$$E_{\max} = \frac{h^2 (3zN)^{2/3}}{8m(\pi V)^{2/3}} \tag{17-2}$$

This is called the energy at the Fermi surface.

When the Boltzmann distribution law is applied to this problem it is found that the electronic heat capacity at low temperatures is proportional to T:

$$C_V \text{ (electronic)} = R\, \frac{\pi^2 kT}{2E_{\max}} \tag{17-3}$$

(Here C_V (electronic) refers to one mole of metal atoms.) We see that the value of C_V is just the equipartition value $3/2R$ for the fraction $\pi^2 kT/3E_{\max}$ of one mole of electrons.

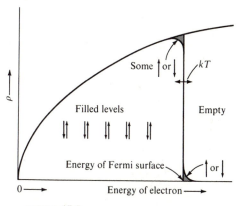

FIGURE 17-5
Density of free-electron spin-orbit levels, $\rho(E)$, as a function of the energy E. The curve and the vertical line (called the Fermi surface) include the filled levels at $T = 0°K$, and the hatched areas show excitation at a higher temperature.

The Heat Capacity of Metals

For a normal metal the heat capacity at low temperature can be written

$$C_V = \gamma T + \alpha T^3 \tag{17-4}$$

Here the first term is the electronic heat capacity and the second is the Debye heat capacity, with α proportional to Θ^{-3} (Section 10-16). The quantity C_V/T is a linear function of T^2, with γ the intercept and α the slope:

$$\frac{C_V}{T} = \gamma + \alpha T^2 \tag{17-5}$$

Experimental values for an alloy are shown in Figure 17-6. The intercept gives the value $\gamma = 0.0131$ J deg^{-2} mole^{-1}.

It is seen that at very low temperatures the electronic contribution to the heat capacity is the principal one for metals.

The free-electron theory is a rather crude approximation. A more refined treatment of metals (energy-band theory) gives a distribution function for electron energy levels differing from that shown in Figure 17-5. The value of γ is determined by the value of the density of levels ρ at the Fermi surface. The experimental values of γ often show disagreement with the free-electron theory, but good agreement with the refined theory.

FIGURE 17-6
Observed values of C_V/T (in J deg^{-2} mole^{-1}) for a chromium-iron alloy, as a function of T^2.

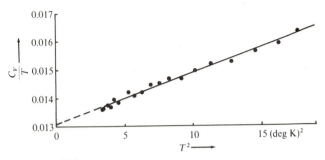

FIGURE 17-7
An alloy of gold and copper. The alloy consists of small crystals, each crystal being made of gold atoms and copper atoms in an orderly array, but with the atoms of the two different kinds distributed essentially at random among the atomic positions.

17-7.   The Nature of Alloys

An *alloy* is a metallic material containing two or more elements. It may be homogeneous, consisting of a single phase, or heterogeneous, consisting of a mixture of phases. An example of a homogeneous alloy is coinage gold. An ordinary sample of coinage gold consists of small crystal grains, each of which is a solid solution of copper and gold, with structure of the sort represented in Figure 17-7. An example of a different kind of homogeneous alloy is the very hard metallic substance tantalum carbide, TaC. It is a compound, with the same structure as sodium chloride (Figure 6-18). Each tantalum atom has twelve tantalum atoms as neighbors. In addition, carbon atoms, which are relatively small, are present in the interstices between the tantalum atoms and serve to bind them together. Each carbon atom is bonded to the six tantalum atoms that surround it. The bonds are $\frac{2}{3}$ bonds—the four covalent bonds resonate among the six positions about

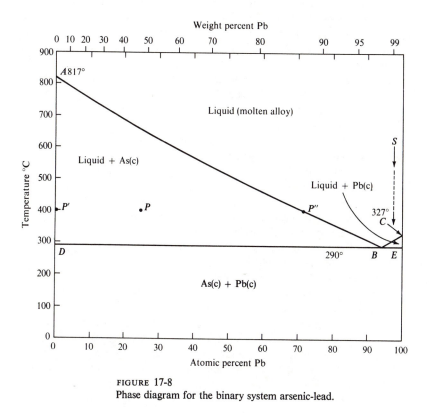

FIGURE 17-8
Phase diagram for the binary system arsenic-lead.

the carbon atom. Each tantalum atom is bonded not only to the adjacent carbon atoms but also to the twelve tantalum atoms surrounding it. The large number of bonds (nine valence electrons per TaC, as compared with five per Ta, occupying, in metallic tantalum, nearly the same volume) explains the greater hardness of the compound than of tantalum itself.

The Binary System Arsenic-Lead

The phase diagram for the binary system arsenic-lead is shown as Figure 17-8. In this diagram the vertical coordinate is the temperature, in degrees centigrade. The diagram corresponds to the pressure 1 atm. The horizontal coordinate is the composition of the alloy, represented along the bottom of the diagram in atomic percentage of lead, and along the top in weight percentage of lead. The diagram shows the temperature and composition corresponding to the presence in the alloy of different phases.

The range of temperatures and compositions represented by the region above the lines *AB* and *BC* is a region in which a single phase is present, the liquid phase, consisting of the molten alloy. The region included in the triangle *ADB* represents two phases, a liquid phase and a solid phase consisting of crystals of arsenic. The triangle *BEC* similarly represents a two-phase region, the two phases being the liquid and crystalline lead. The range below the horizontal line *DBE* consists of the two phases crystalline arsenic and crystalline lead, the alloy being a mixture of small grains of the two elements.

Let us apply the phase rule to an alloy in the one-phase region above the line *ABC*. Here we have a system of two components, and, in this region, one phase; the phase rule states that the variance should be three. The three quantities describing the system that may be varied in this region are the pressure (taken arbitrarily in this diagram as 1 atm, but capable of variation), the temperature, which may be varied through the range permitted by the boundaries of the region, and the composition of the molten alloy, which may similarly be varied through the range of compositions permitted by the boundaries of the region.

An alloy in the region *ADB*, such as that represented by the point *P*, at 35 atomic percent lead and 400°C, lies in a two-phase region, and the variance is accordingly stated by the phase rule to be two. The pressure and the temperature are the two variables; the phase rule hence states that it is not possible to vary the composition of the phases present in the alloy. The phases are crystalline arsenic, represented by the point *P′* directly to the left of *P*, and the molten alloy, with the composition *P″* directly to the right of *P*. The composition of the molten alloy in equilibrium with crystalline arsenic at 400°C and 1 atm pressure is definitely fixed at *P″*; it cannot be varied.

The only conditions under which three phases can be in equilibrium with one another at the arbitrary pressure 1 atm are represented by the point *B*. With three phases in equilibrium with one another for this two-component system, the phase rule requires that there be only one arbitrary variable, which we have used in fixing the pressure arbitrarily at 1 atm. Correspondingly we see that the composition of the liquid is fixed at that represented by the point *B*, 93 atomic percent lead, and the composition of the two solid phases is fixed, these phases being pure arsenic and pure lead. The temperature is also fixed, at the value 290°C, corresponding to the point *B*. This point is called the *eutectic point*, and the corresponding alloy is called the *eutectic alloy*, or simply the *eutectic*. The word eutectic means melting easily; the eutectic has a sharp melting point. When a liquid alloy with the eutectic composition is cooled, it crystallizes completely on reaching the temperature 290°C, forming a mixture of very small grains of pure arsenic and pure lead, with a fine texture. When this alloy is slowly heated, it melts sharply at the temperature 290°C.

The lines in the phase diagram are the boundaries separating a region in which one group of phases are present from a region in which another group of phases are present. A line such as *AB* is called the *freezing-point curve*, *liquidus curve*, or *liquidus*, and a line such as *DB* is called the *melting-point curve*, *solidus curve*, or *solidus*. These boundary lines can be located by various experimental methods, including the method of thermal analysis, discussed in Section 17-8.

If a molten alloy of arsenic and lead with the eutectic composition is cooled, the temperature drops at a regular rate until the eutectic temperature, 290°C, is reached; the liquid then crystallizes into the solid eutectic alloy, the temperature remaining constant until crystallization is complete. The eutectic has a constant melting point, just as has either one of the pure elementary substances.

The effect of the phenomenon of depression of the freezing point (Section 13-7) in causing the eutectic melting point to be lower than the melting point of the pure metals can be intensified by the use of additional components. Thus an alloy with eutectic melting point 70°C can be made by melting together 50 weight percent bismuth (m.p. 271°C), 27% lead (m.p. 327.5°C), 13% tin (m.p. 232°C), and 10% cadmium (m.p. 321°C), and the melting point can be reduced still further, to 47°C, by the incorporation in this alloy of 18% of its weight of indium (m.p. 155°C).

It is possible, in terms of this phase diagram, to discuss the following phenomenon. A small amount, about $\frac{1}{2}$% by weight, of arsenic is added to lead used to make lead shot, in order to increase the hardness of the shot and also to improve the properties of the molten material. Lead shot is made by dripping the molten alloy through a sieve. The fine droplets freeze during their passage through the air, and are caught in a tank of water after they have solidified. If pure lead were used, the falling drops would solidify rather suddenly on reaching the temperature 327°C. A falling drop tends not to be perfectly spherical, but to oscillate between prolate and oblate ellipsoidal shapes, as you may have noticed by observing drops of water dripping from a faucet; and hence the shot made of pure lead might be expected not to be perfectly spherical in shape. But the alloy containing $\frac{1}{2}$% arsenic by weight, represented by the arrow *S* (Figure 17-8), would begin to freeze on reaching the temperature 320°C, and would continue to freeze, forming small crystals of pure lead, until the eutectic temperature 290°C is reached. During this stage of its history the drop would consist of a sludge of lead crystals in the molten alloy, and this sluggish sludge would be expected to be drawn into good spherical shape by the action of the surface-tension forces of the liquid.

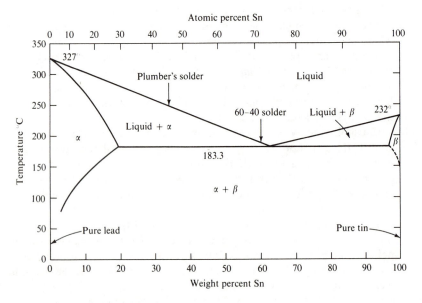

FIGURE 17-9
Phase diagram for the binary system lead-tin.

FIGURE 17-9
Phase diagram for the binary system lead-tin.

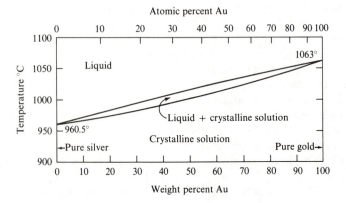

FIGURE 17-10
Phase diagram for the binary system silver-gold, showing the
formation of a complete series of crystalline solutions.

The Binary System Lead-Tin

The phase diagram for the lead-tin system of alloys is shown as Figure 17-9. This system rather closely resembles the system arsenic-lead, except that there is an appreciable solubility of tin in crystalline lead and a small solubility of lead in crystalline tin. The phase designated α is a solid solution of tin in lead, the solubility being 19.5 weight percent at the eutectic temperature and dropping to 2% at room temperature. The phase β is a solid solution of lead in tin, the solubility being about 2% at the eutectic temperature and extremely small at room temperature. The eutectic composition is about 62 weight percent tin, 38 weight percent lead.

The composition of *solder* is indicated by the two arrows, corresponding to ordinary plumbers' solder and to 60:40 solder. The properties of solder are explained by the phase diagram. The useful property of plumbers' solder is that it permits a wiped-joint to be made. As the solder cools it forms a sludge of crystals of the α phase in the liquid alloy, and the mechanical properties of this sludge are such as to permit it to be handled by the plumber in an effective way. The sludge corresponds to transition through the region of the phase diagram in which liquid and the α phase are present together. For plumbers' solder the temperature range involved is about 70°, from 250°C to 183°C, the eutectic temperature. The 60:40 tin-lead solder is preferred for electrical work because it has the lowest melting point, and is the least likely to lead to overheating the transistors.

The Binary System Silver-Gold

The metals silver and gold are completely miscible with one another not only in the liquid state but also in the crystalline state. A solid alloy of silver and gold consists of a single phase, homogeneous crystals with the cubic closest-packed structure, described for copper in Chapter 2, with gold and silver atoms occupying the positions in this lattice essentially at random (Figure 17-7). The phase diagram shown as Figure 17-10 represents this situation. It is seen that the addition of a small amount of gold to pure silver does not depress the freezing point, in the normal way, but instead causes an increase in the temperature of crystallization.

The alloys of silver and gold, usually containing some copper, are used in jewelry, in dentistry, and as a gold solder.

The Binary System Silver-Strontium

A somewhat more complicated binary system—that formed by silver and strontium—is represented in Figure 17-11. It is seen that four intermetallic compounds are formed, their formulas being Ag_5Sr, Ag_5Sr_3,

AgSr, and Ag_2Sr_3. These compounds and the pure elements form a series of eutectics; for example, the alloy containing 25 weight percent strontium is the eutectic mixture of Ag_5Sr and Ag_5Sr_3.

Some other binary systems are far more complicated than this one. As many as a dozen different phases may be present, and these phases may involve variation in composition, resulting from the formation of solid solutions. Ternary alloys (formed from three components) and alloys involving four or more components are of course still more complex.

It is seen that the formulas of intermetallic compounds, such as Ag_5Sr, do not correspond in any simple way to the usually accepted valences of the element. Compounds such as Ag_5Sr can be described by saying that the strontium atom uses its two valence electrons in forming bonds with the silver atoms that surround it, and that the silver atoms then use their remaining electrons in forming bonds with other silver atoms. Some progress has been made in developing a valence theory of the structure and properties of intermetallic compounds and of alloys in general, but this field of chemistry is still in a primitive state.

17-8. Experimental Methods of Studying Alloys

About 100 years ago a metallurgical technique called *metallography* was developed as a way of investigating the phases present in alloys. This technique consists of grinding and polishing the surface of a metallic specimen, sometimes etching it with reagents (such as nitric acid or picric acid) to emphasize grain boundaries and to help to distinguish between different phases, and then examining the surface with use of an optical microscope, with a method of illumination from above. In this way the sizes and shapes of crystal grains can be studied, and the presence of grains of two or more phases can be determined in alloys that to the unaided eye appear to be homogeneous. The polished and etched surface of a piece of copper is shown in Figure 2-2, and other photomicrographs of alloys, showing different phases, can be found in Chapter 20.

During recent years much use has been made of the electron microscope in the study of metals and alloys. Very thin foils of the metal are made, sometimes by dissolving the surface of the specimen with acid until holes develop; the region adjacent to a hole may be thin enough to permit penetration by the electron beam. The structure of individual crystal grains can also be determined by observing the electron diffraction pattern from a beam of electrons transmitted through a single grain. Changes in structure that occur with time, perhaps at elevated temperature, may be followed in this way.

These techniques for studying phase transitions, although powerful, are

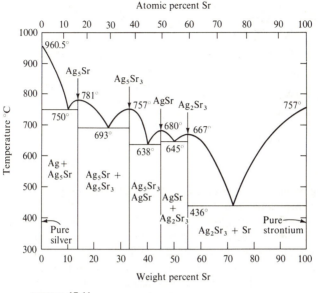

FIGURE 17-11
Phase diagram for the binary system silver-strontium,
showing the formation of four intermetallic compounds.

time-consuming and difficult. A simple and easily applied technique,
called *thermal analysis*, has been used for more than a century. Phase
transitions are characterized by the absorption or emission of a heat of
transition. The way in which the heat of transition is involved in the
technique of thermal analysis can be illustrated by discussion of a simple
experiment.

Samples of arsenic-lead alloys (Figure 17-8) may be made, corresponding to different compositions, from pure arsenic to pure lead. A sample of
pure arsenic is placed in a crucible in a furnace and heated to above the
melting point. One of the junctions of a thermocouple is inserted in the
sample, to permit measurement of the temperature. When the furnace is
turned off, the temperature of the sample begins to decrease, because of
conduction and radiation from the sample to the furnace and from the
furnace to the surrounding environment. The temperature decreases with
time, as indicated by the first curve in Figure 17-12. (These curves are
called *cooling curves.*) At the freezing point of arsenic, 817°C, there occurs
a discontinuity in the slope of the temperature-time curve, which begins to
follow a horizontal course, with slope zero. During a period of time the
heat of crystallization serves to balance the heat loss to the surrounding
furnace. The temperature of the horizontal section of the cooling curve is

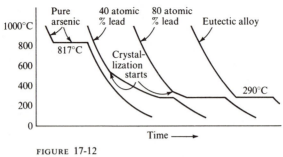

FIGURE 17-12
Cooling curves for samples of arsenic-lead alloys.

the melting point of the substance. After completion of crystallization the temperature begins to drop.

The second cooling curve in Figure 17-12, corresponding to 40 atomic percent lead, shows a decrease in temperature and then a change in slope, which represents the beginning of crystallization of arsenic from the molten alloy. The temperature continues to drop, even though arsenic is crystallizing, because the composition of the liquid alloy changes as arsenic is removed from it. The curve continues with changed slope until the temperature 290°C is reached. and then its slope becomes zero. The horizontal section represents the simultaneous crystallization of arsenic and lead, as separate phases, at the eutectic temperature.

The next curve, for the alloy with 80 atomic percent lead, shows a lower temperature at which crystallization of arsenic begins (that is, a lower liquidus temperature) and then a longer horizontal section, representing crystallization of a larger amount of the eutectic mixture of the two crystalline phases.

The next curve corresponds to the eutectic composition, with 93 atomic percent lead (point B on Figure 17-8). This curve is qualitatively similar to the curve for a pure substance.

The alloys in the region between the eutectic composition and pure lead give cooling curves that are like those for the alloys in the other half of the phase diagram.

The method of thermal analysis can be improved and refined by the use of two samples of metal or alloy and measurement of the temperature difference between the two samples; this is called *differential thermal analysis*. For example, suppose that an alloy of iron and cobalt has been made and that it is thought that the transition* between the α phase

---

*This transition, which does not involve a change in crystal structure but only a change in relative orientations of the magnetic moments of the atoms, occurs over a range of temperature, roughly 1°. It is called a *second-order transition*.

(ferromagnetic) and the β phase (paramagnetic, with the same body-centered crystal structure as the α phase) probably lies between 700°C and 900°C. A sample of the alloy is prepared and also a sample of a similar metal, such as copper, that does not have a phase transition in this temperature range. The samples are adjusted in weight so as to have approximately equal average values of their total heat capacity. A thermocouple is placed in each sample, and they are connected with a recording apparatus in such a way that the difference in temperature between the two samples can be continuously recorded as a function of the temperature of the first sample. The two samples are placed in a furnace and heated to 1000°C, the furnace is then turned off, and the temperature recordings are made as the samples cool. A change in slope of the differential temperature indicates the transition temperature, which can in this way be determined with considerable accuracy (0.1° to 1°).

Another important and useful method of investigating alloys is by preparing samples of different compositions and making x-ray photographs of them (especially powder photographs, which are the diffraction patterns given by a large number of small crystals in random orientation). By the analysis of the diffraction patterns the number of phases present can be determined. For example, the samples of silver-strontium alloys, with phase diagram represented in Figure 17-11, are found to give characteristic diffraction patterns at six compositions: pure silver, pure strontium, and the four compositions indicated by the arrows in Figure 17-11. For an alloy with intermediate composition the diffraction pattern shows the lines characteristic of two phases, with relative intensities proportional to the relative amounts of the two phases. Moreover, it is often possible by the analysis of the diffraction pattern to determine the structure of the crystal, and thus to verify the composition. It is in this way that the compound Ag_5Sr was identified.

Electron Microprobe X-ray Analysis

In the study of alloys the composition of phases can be determined by chemical analysis of samples shown by metallographic techniques to be single phases, or by the preparation of alloys of various compositions and determination of phase diagrams, as described above. During recent years another method, electron microprobe x-ray spectrometry, has been found to be very valuable for the investigation of heterogeneous substances, including rocks as well as alloys. This method makes possible a complete chemical analysis of very small portions of a sample, and it is nondestructive, leaving the specimen intact. A high-voltage electron beam, similar to the beam in an electron microscope, is brought to a focus about 1 μm in diameter on the polished surface of a specimen to be analyzed

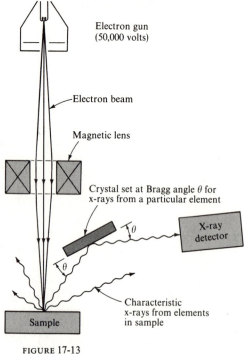

Electron gun
(50,000 volts)

Electron beam

Magnetic lens

Crystal set at Bragg angle θ for
x-rays from a particular element

X-ray
detector

θ

θ

Characteristic
x-rays from elements
in sample

Sample

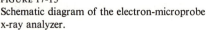

FIGURE 17-13
Schematic diagram of the electron-microprobe
x-ray analyzer.

(Figure 17-13). The electrons bombarding the metal surface excite x-rays in the same way as at the target of an x-ray tube, each type of atom in the sample emitting its characteristic x-ray wavelengths (Section 4-1). The wavelengths are measured by means of a crystal spectrometer similar to the Bragg x-ray spectrometer (Figure 3-23). From the wavelengths observed, the elements present in the micron-sized region bombarded by the electron beam can be identified, and from the intensities of the emitted x-rays, suitably calibrated, the amounts of the various elements in this region can be quantitatively measured. Moreover, the electron beam can be scanned across the sample in a raster (the way a television picture is scanned), with simultaneous measurement of the excited x-rays from a given element, and the measured x-ray intensity can be shown on an oscilloscope scanned synchronously with the electron beam. This produces a picture of the spatial distribution of the element (Figure 17-14), bright and dark portions of the image representing areas in the sample that are respectively rich and poor in the element measured. In this way it is found

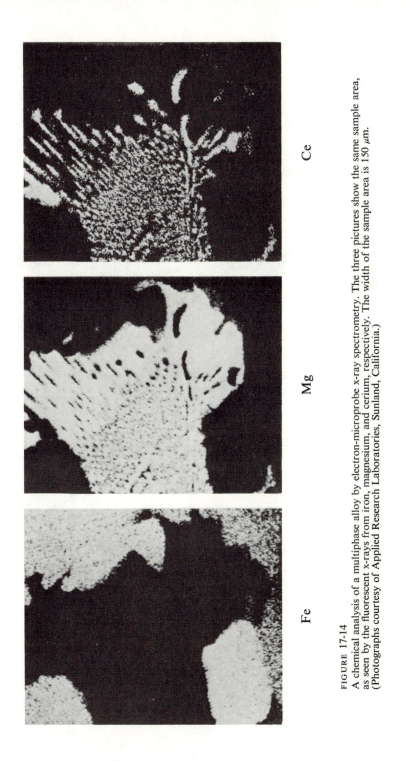

Fe   Mg   Ce

FIGURE 17-14

A chemical analysis of a multiphase alloy by electron-microprobe x-ray spectrometry. The three pictures show the same sample area, as seen by the fluorescent x-rays from iron, magnesium, and cerium, respectively. The width of the sample area is 150 μm. (Photographs courtesy of Applied Research Laboratories, Sunland, California.)

FIGURE 17-15
Polished surface of an iron-nickel meteorite, showing large crystal grains in parallel orientation. About 40% of natural size—the meteorite is about ten inches long. (Courtesy of Griffith Observatory.)

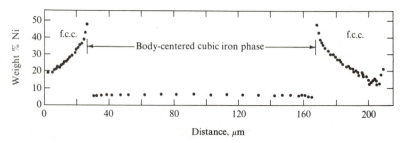

FIGURE 17-16
Chemical-composition traverse across an exsolution lamella in an iron meteorite, measured with the electron-microprobe x-ray analyzer. The face-centered cubic phase is labeled "f.c.c." (Courtesy of Pergamon Press: Geochimica et Cosmochimica Acta, Vol. 31, p. 1002, 1967.)

that many apparently homogeneous materials are in fact chemically inhomogeneous on a small scale, containing several phases of different composition or phases that vary in composition from place to place.

A natural example that illustrates the usefulness of the electron microprobe method of chemical analysis for studying metals and alloys is the internal structure of iron meteorites (Figure 17-15). These objects consist of a nickel-iron solid solution, from which lamellae of nearly pure iron have separated ("exsolved") during cooling of the alloy from an original high temperature of formation. The nickel-iron solid solution has a face-centered cubic structure, whereas the nearly pure iron phase in the lamellae has a body-centered cubic structure. The nickel content of an exsolved iron lamella and the adjacent nickel-iron host is shown in Figure 17-16, as measured with the electron microprobe in a traverse across the lamella. There is a gradient in the nickel concentration away from the lamella, representing the nickel expelled from the lamella in its conversion to the body-centered cubic structure. Diffusion of nickel atoms through the face-centered cubic structure was not fast enough to equalize the nickel content of this phase during the cooling time of the alloy. From the "frozen in" concentration gradients it is possible to calculate the rate of cooling that occurred, and this in turn allows the size of the original planetary body, from which the meteorite was derived, to be calculated.

17-9. Interstitial Solid Solutions and Substitutional Solid Solutions

Two clearly distinct types of solid solutions (crystalline solutions) have been recognized. In solid solutions of one type, called *interstitial solid solutions*, atoms of one element are inserted into some of the interstices in the crystal lattice formed by the atoms of a second element. Usually this results in a small increase in the lattice constant of the crystal; but usually the increase in lattice constant is not large enough to compensate for the mass of the inserted atom, and the density of the interstitial solid solution becomes larger than that of the substance without the interstitial atoms. The second sort of solid solutions, called *substitutional solid solutions*, involves the replacement of atoms of one kind in the crystal lattice by atoms of a second kind.

An example of an interstitial solid solution is provided by martensite, a solid solution of carbon in iron. The structure shown in Figure 17-17 is an ideal structure, corresponding to one carbon atom for every two iron atoms. It is seen from the figure that the iron atoms are arranged approximately as in α-iron—that is, in the body-centered arrangement. The

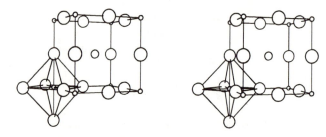

FIGURE 17-17
The structure of martensite, a carbide of iron that is present in steel. The atoms of carbon (small spheres) are in the centers of the horizontal squares formed by the iron atoms, which are in a body-centered arrangement.

carbon atoms are inserted in the centers of the horizontal faces of the unit of structure. The presence of carbon atoms in these faces, and not in the lateral faces, causes the crystal to be tetragonal in symmetry, rather than cubic. The vertical edge of the unit of structure shown in the figure is about 3% larger than the two horizontal edges. If there are not so many carbon atoms present in the phase, some of the interstitial positions in the horizontal faces are unoccupied. When the number of carbon atoms becomes smaller, the horizontal faces and the lateral faces are occupied at random by carbon atoms, and the interstitial solution becomes cubic, rather than tetragonal, in symmetry.

Interstitial solid solutions are usually formed when a substance with small atoms is dissolved in a substance with large atoms—the covalent radius of carbon is 0.77 Å, and the metallic radius of iron is 1.26 Å. Solid solutions composed of atoms of nearly the same size usually are substitutional solid solutions. For example, iron and nickel form substitutional solid solutions, having atoms of iron and nickel distributed at random over the positions of the body-centered structure (for compositions between 0 and about 25 atomic percent nickel) or over the face-centered positions (for compositions between 25 and 100 atomic percent nickel). The irregularity in the lattice produced by the presence of atoms of different sizes makes itself evident in an increased electric resistance; solid solutions are not such good conductors of electricity as the pure metals.

17-10. Physical Metallurgy

In recent years emphasis has been placed on the study of a branch of metallurgy called *physical metallurgy*. In this branch of metallurgy an attempt is made to explain the physical properties, such as tensile strength, hardness, ductility, electrical and thermal conductivity, and heat capacity, of pure metals and alloys in terms of their atomic and electronic structure. One of the ultimate aims of the physical metallurgist is to be able to design alloys with any desired set of properties.

Mechanical Properties of Metals

Most metals are malleable and ductile. Instead of being smashed into splinters when struck by a hammer, a piece of metal is flattened into a sheet of foil. A crystal of a metal must hence be able to deform itself without breaking.

If a crystal of sodium chloride is deformed in such a way that the ions are moved about one ionic diameter relative to one another, then sodium ions become adjacent to sodium ions and chloride ions to chloride ions, and the repulsion of the ions of like sign causes the crystal to break into pieces. In a metal, however, the atoms are all of the same kind, and any atom can form bonds with any other atom. Moreover, the valence bonds, which resonate easily from one position to another in the crystal, can still form between neighboring atoms even if the crystal is deformed, and accordingly a crystal of a metal remains strong during deformation.

The way in which a crystal of a metal changes its shape is by *slip along glide planes*. For example, the metal zinc has the hexagonal closest-packed structure indicated in Figure 17-1. The distance between the hexagonal layers of atoms is somewhat larger than for ideal closest packing—the distance between neighboring zinc atoms in the same hexagonal layer is 2.66 Å, whereas that between atoms in adjacent layers is 2.91 Å. Accordingly, we might expect it to be easy for a hexagonal layer to slip over another hexagonal layer. If a single crystal of zinc is made in the form of a round wire, with the hexagonal layers at an angle, as shown in the upper part of Figure 17-18, and the ends of the wire are pulled, the wire stretches out into a ribbon, through slip along the hexagonal planes, as illustrated on the right side of Figure 17-18. Photomicrographs of a metal that has been subjected to strain often show traces of these glide planes.

The slip along a glide plane does not occur by the simultaneous motion of a whole layer of atoms relative to an adjacent layer. Instead, the atoms move one at a time. There is a flaw in the structure, where an atom is missing. The atom to one side of this flaw (which is called a *dislocation*) moves to occupy the space, and leaves a space where it was; that is, the dislocation moves in the opposite direction to the atom. When the dislocation has moved all the way across the crystal grain, the whole row of atoms has moved, and the lower part of the crystal has slipped one atomic diameter in the direction of the strain. A description of some kinds of dislocations is given below.

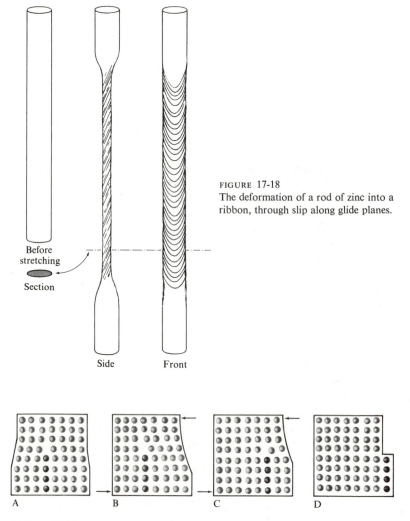

FIGURE 17-18
The deformation of a rod of zinc into a ribbon, through slip along glide planes.

Before stretching

Section

Side   Front

FIGURE 17-19
(A) An edge dislocation in a metal. The shaded atoms indicate the extra part-layer of atoms. (B) The crystal grain under stress. (C) Motion of the dislocation to the right; different atoms (shades) now constitute the dislocation. (D) The dislocation has reached the edge of the grain, forming a step; by motion of the dislocation the upper part of the crystal has moved to the left relative to the lower part.

Lattice Vacancies

One type of imperfection found in crystals is the *lattice vacancy* or *point imperfection*: an atom is missing at the place in the crystal lattice that is normally occupied by an atom, and the surrounding atoms have moved slightly toward this position. Lattice vacancies are formed by thermal agitation to an extent given by the Boltzmann distribution function (with the number of vacancies per unit volume in the metal about equal to the number of atoms per unit volume in the vapor in equilibrium with the metal). They may also be produced in larger numbers by bombardment of the metal with high-energy particles or x-rays.

Interstitial Atoms

Another type of point defect consists of an extra atom of a metal occupying a position that in the perfect crystal would be vacant. This extra atom may be a foreign atom, usually smaller than the atoms of the metal itself, such as hydrogen, carbon, nitrogen, or oxygen in iron. Larger foreign atoms may substitute for the atoms of the metal itself. It is found by experiment that a small amount of impurity in a metal may make it brittle. For example, copper containing sulfur or arsenic is brittle, rather than malleable and ductile. One way in which the foreign atoms may produce brittleness is by interfering with the motion of dislocations through the crystal; when the dislocation reaches a sulfur atom or other foreign atom in the copper crystal, it may be stopped, and the slip may thus be prevented from continuing.

Dislocations

The most important imperfections, so far as the mechanical properties of crystals are concerned, are the various imperfections called dislocations. The ease with which dislocations move through a crystal determine to a large extent its ranges of elastic and plastic deformation under an applied stress and its ultimate yield point—that is, the stress under which the crystal fractures. One kind of dislocation, called an *edge dislocation*, is shown in Figure 17-19. An edge dislocation can be described as involving removal of one-half of a plane of atoms from the crystal.

The *screw dislocation*, shown in Figure 17-20, has an axis that is either right-handed or left-handed. A crystal containing one screw dislocation is not made up of layers of atoms parallel to one another; instead it consists of a single layer of atoms, distorted about the screw axis into a helicoid or spiral ramp.

Screw axis

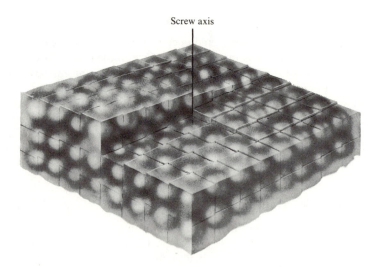

FIGURE 17-20
A screw dislocation emerging on the face of a metal crystal.

A dislocation can move through a crystal by a succession of processes, each of which involves the motion of a single atom from one position in the crystal to an adjacent position. The activation energy for this process may be small enough to permit it to occur at a rapid rate, and in consequence plastic deformation of metals under stress can take place.

Many of the mechanical properties can be understood in terms of the motion of dislocations. If a stress is applied in the right way to a metal, the metal bends. When the stress is removed, the metal either returns to its original shape or is permanently deformed. In the first case the dislocations have not moved or have moved in a reversible way; that is, the applied stress has not carried them over any foreign inclusion in the lattice or caused too many of them to collide. In the second case some dislocations have moved irreversibly, so that they do not return to their original positions upon removal of the applied stress. If the applied stress is very large, dislocations will move until many of them become piled up against some barrier, such as a foreign inclusion or the boundary between adjacent crystalline grains. In the region of a dislocation pile-up the applied stress strains the bonds in such a way as to initiate failure of the material.

A boundary between crystal grains can serve as a barrier to the motion of dislocations, and in this way can decrease the plasticity and increase the hardness of a metal. If a piece of copper is hammered until the large crystal grains are broken up into small crystal grains, the crystal boundaries may interfere with slip by stopping the motion of the dislocations.

This is the mechanism of the hardening of copper and other metals by *cold work* (by hammering them or otherwise working them in the cold). Heating the work-hardened metal to the temperature at which recrystallization occurs (growth of the small strained crystals to large unstrained crystals) restores plasticity; this process is called *annealing*. The recrystallization temperature is usually about one-third to one-half of the melting point of the metal (both on the absolute temperature scale).

Pure aluminum is a soft, malleable, and ductile metal. For some purposes an alloy of aluminum that is stronger, tougher, and less ductile is needed. Aluminum alloys of this sort can be made by incorporating small amounts of other metals, such as copper and magnesium. An alloy containing about 4% copper and 0.5% magnesium may strengthen the aluminum through the formation of hard, brittle crystals of the intermetallic compound $MgCu_2$. These minute crystals, interspersed through the crystals of aluminum, can serve to key the glide planes of aluminum so effectively as to improve the mechanical properties of the alloy significantly over those of the pure metal.

Exercises

17-1. Aluminum crystallizes in cubic closest packing. How many nearest neighbors does each atom have? Predict its metallic valence from its position in the periodic table. Would you predict it to have greater or less tensile strength than magnesium? Why?

17-2. Discuss the metallic valence of the elements rubidium, strontium, and yttrium. What would you predict about change in hardness, density, strength, and melting point in this series of elementary metals?

17-3. Compare the metallic valences of sodium, magnesium, and aluminum with their oxidation numbers in their principal compounds.

17-4. Describe the structure of tantalum carbide, TaC. Can you explain why it has much greater strength and hardness than tantalum itself?

17-5. Define alloy, intermetallic compound, phase, variance, eutectic, triple point.

17-6. State the phase rule, and give an application of it.

17-7. Cadmium (m.p. 321°C) and bismuth (m.p. 271°C) do not form solid solutions or compounds with one another. Their eutectic point lies at 61 weight percent bismuth and 146°C. Sketch their phase diagram, and label each region to show what phases are present.

17-8. Describe the alloy that would be obtained by cooling a melt of silver containing 8 atomic percent strontium. (See Figure 17-11.)

17-9. Describe the alloy that would be obtained by cooling a melt of silver and strontium containing 50 atomic percent Ag. Would it be homogeneous or heterogeneous? Would it melt sharply, at one temperature, or over a range of temperatures?

17-10. What is the lowest temperature at which an alloy of silver and strontium can remain liquid? What is the composition of this alloy? Is the solid alloy homogeneous or heterogeneous? Does it have a sharp melting point?

17-11. Why does the Ag-Sr alloy with 75 atomic percent Sr have a lower melting point than pure strontium?

17-12. From Figure 17-11 it is seen that the silver-strontium alloy containing 1 atomic percent Sr begins to freeze at a temperature 11° less than the freezing point of pure silver. What is the weight-molar freezing-point constant of silver? (See Section 13-7.) Silver and silicon have a phase diagram resembling that shown in Figure 17-9; neither element is soluble in the other in the crystalline state. At what temperature would the Ag-Si alloy containing 1 atomic percent Si begin to freeze? (Answer: 11° below the freezing point of silver.)

17-13. The x-ray examination of the crystalline compound of silver and strontium containing about 15% Sr has shown that it is hexagonal. The unit of structure has two edges with length 5.67 Å at the angle 120° with one another and a third edge with length 4.62 Å at right angles to the other two.
(*a*) What is the volume of the unit? (Answer: 128.7 Å³.)
(*b*) What is the volume of a mole of unit cells? (Answer: 76.85 cm³.)
(*c*) The density of the substance has been determined to be 8.16 g cm⁻³. What is the mass of a mole of unit cells? (Answer: 627 g.)
(*d*) What are the possible formula weights of the compound?
(*e*) How many atoms of strontium are there in one formula?
(*f*) How many atoms of silver?

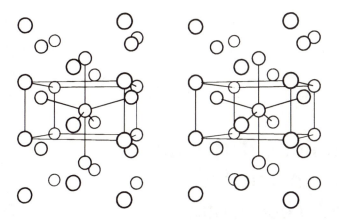

FIGURE 17-21
The structure of white tin (stereoscopic drawing; see Exercise 17-14).

17-14. White tin is tetragonal. Its unit of structure has $a = b = 5.819$ Å, $c = 3.175$ Å, 4 Sn at 000, $\frac{1}{2} 0 \frac{1}{4}$, $0 \frac{1}{2} \frac{3}{4}$, $\frac{1}{2} \frac{1}{2} \frac{1}{2}$.

(a) What value of the density of the metal is given by this information?

(b) Each atom has two neighboring atoms at 3.175 Å and four at a somewhat smaller distance. What is the value of this smaller distance? (See Figure 17-21.)

(c) A hypothetical alternative structure, based on ligancy 6, is the simple cubic structure (Figure 2-6), with a equal to the average of the six bond lengths in white tin. To what density does this correspond? Can you suggest a reason for the instability of this structure relative to white tin?

17-15. The enthalpy of formation of the intermetallic compound $Mg_2Sn(c)$ (which has the fluorite structure, Figure 18-3) from $Mg(c)$ and $Sn(gray)$ is -74 kJ mole^{-1}, and that for $Mg_2Si(c)$ is -80 kJ mole^{-1}. Note that these substances have the composition of normal-valence compounds. To what values of the electronegativity difference do these enthalpy values correspond? (Answer: about 0.45.)

17-16. The intermetallic compounds $Ba_2Sn(c)$ and $BaSn_3(c)$ have enthalpy of formation -379 and -189 kJ mole^{-1}, respectively, from $Ba(c)$ and $Sn(gray)$. Discuss these values in relation to the number of bonds between unlike atoms and the electronegativity difference.

18

Lithium, Beryllium Boron, and Silicon and Their Congeners

In this chapter we shall discuss the metals and metalloids of groups I, II, III, and IV of the periodic table, and their compounds.*

The alkali metals, group I, are the most strongly electropositive elements—the most strikingly metallic. Many of their compounds have been mentioned in earlier chapters. The alkaline-earth metals are also strongly electropositive.

Boron, silicon, and germanium are metalloids, with properties intermediate between those of metals and those of nonmetals. The electric conductivity† of boron, for example, is 1×10^{-6} mho/cm; this value is intermediate between the values for metals ($4 + 10^5$ mho/cm for aluminum, for example), and those for nonmetals (2×10^{-13} for diamond, for example). They have a corresponding tendency to form oxygen acids, rather than to serve as cations in salts.

---

*Compounds of carbon are discussed in Chapters 23 and 24.

†The electric conductivity, in mho/cm, is the current in amperes flowing through a rod with cross-section 1 cm² when there is an electric potential difference between the ends of the rod of 1 volt per cm length of the rod.

Silicon (from Latin *silex*, flint) is the second element in group IV, and is hence a congener of carbon. Silicon plays an important part in the inorganic world, similar to that played by carbon in the organic world. Most of the rocks that constitute the earth's crust are composed of the silicate minerals, of which silicon is the most important elementary constituent.

The importance of carbon in organic chemistry results from its ability to form carbon-carbon bonds, permitting complex molecules, with the most varied properties, to exist. The importance of silicon in the inorganic world results from a different property of the element—a few compounds are known in which silicon atoms are connected to one another by covalent bonds, but these compounds are relatively unimportant. The characteristic feature of the silicate minerals is the existence of chains and more complex structures (layers, three-dimensional frameworks) in which the silicon atoms are not bonded directly to one another but are connected by oxygen atoms. The nature of these structures is described briefly in later sections of this chapter.

18-1.   The Electronic Structures of Lithium, Beryllium, Boron, and Silicon and Their Congeners

The electronic structures of the elements of groups I, II, III, and IV are given in Table 5-5. The distribution of the electrons among the orbitals is the same in this table as in the energy-level chart, Figure 5-11, with one exception: the normal state of the lanthanum atom has been found by the study of the spectrum of lanthanum to correspond to the presence of one electron in the $5d$ orbital, rather than in the $4f$ orbital, as indicated in the energy-level chart.

The Russell-Saunders symbol for lithium and its congeners in the normal state is $^2S_{1/2}$, that of beryllium and its congeners is $^1S_0$, that of boron and its congeners is $^2P_{1/2}$, and that of carbon and its congeners is $^3P_0$.

The elements of group I have one more electron than the preceding noble gas, those of group II have two more, and those of group III have three more. The outermost shell of each of these noble-gas atoms is an octet of electrons, two electrons in the s orbital and six in the three p orbitals of the shell. The one, two, or three outermost electrons of the metallic elements are easily removed with formation of the cations Li^+, Na^+, K^+, Rb^+, Cs^+, Be^{++}, Mg^{++}, Ca^{++}, Sr^{++}, Ba^{++}, Al^{+++}, Sc^{+++}, Y^{+++}, and La^{+++}. Each of these elements forms only one principal series of compounds, in which it has oxidation number $+1$ for group I, $+2$ for group II, or $+3$ for group III. The metalloid boron also forms compounds in which its oxidation number is $+3$, but the cation B^{+++} is not stable.

Whereas carbon is adjacent to boron in the sequence of the elements, and also silicon to aluminum, the succeeding elements of group IV of the periodic table, germanium, tin, and lead, are widely separated from the corresponding elements of group III, scandium, yttrium, and lanthanum. Germanium is separated from scandium by the ten elements of the iron transition series, tin from yttrium by the ten elements of the palladium transition series, and lead from lanthanum by the ten elements of the platinum transition series, and also the fourteen lanthanons.*

Each of the elements of group IV has four valence electrons, which occupy *s* and *p* orbitals of the outermost shell. The maximum oxidation number of these elements is +4. All of the compounds of silicon correspond to this oxidation number. Germanium, tin, and lead form two series of compounds, representing oxidation number +4 and oxidation number +2, the latter being more important than the former for lead.

18-2. Radius Ratio, Ligancy, and the Properties of Substances

Some of the properties of substances can be discussed in a useful way in terms of the sizes of ions or atoms. Many of the substances mentioned in the later sections of this chapter and in the following chapters are compounds of metals, with small electronegativity, and nonmetals, with large electronegativity. The bonds between these atoms may have a sufficiently large amount of ionic character to justify the discussion of the substance as composed of cations and anions. Such a discussion may be helpful even for substances in which the bonds have a large amount of covalent character.

For example, let us consider again the fluorides of the elements of the second short period of the periodic table (see Section 11-4). Their formulas, melting points, boiling points, heats of fusion, and heats of vaporization (or sublimation) are the following:

	NaF	MgF$_2$	AlF$_3$	SiF$_4$	PF$_5$	SF$_6$	
M.p.	995°	1263°	>1257°	−90°	−94°	−51°C	
B.p.	1704°	2227°	1257°*	−95°*	−85°	−64°*	
Heat of fusion	33	58	—	7	12.8	5	kJ mole^{-1}
Heat of vaporization	209	272	322†	19‡	17.1	17.1	kJ mole^{-1}

*Temperature of sublimation of crystal at 1 atm pressure.
†Heat of sublimation of crystal.
‡Heat of vaporization at 1.74 atm.

---

*There is some disagreement among chemists about nomenclature of the groups of the periodic system. We have described the transition elements as coming between groups III and IV in the long periods of the periodic table. An alternative that has found about as wide acceptance is to place them between groups II and III.

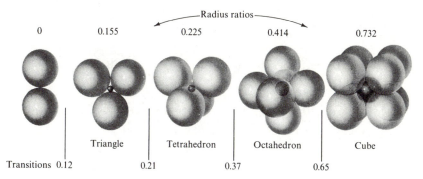

Radius ratios

| 0 | 0.155 | 0.225 | 0.414 | 0.732 |

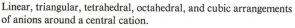

| | Triangle | Tetrahedron | Octahedron | Cube |

Transitions 0.12   0.21   0.37   0.65

FIGURE 18-1
Linear, triangular, tetrahedral, octahedral, and cubic arrangements
of anions around a central cation.

The first three substances are crystalline solids at room temperature, with
high melting and boiling points and large heats of fusion and vaporization,
and the other three are gases at room temperature, with low melting and
boiling points and small heats of fusion and vaporization. The pronounced change in properties between AlF_3 and SiF_4 cannot be attributed
in an obvious way to the change in oxidation number, composition (number of fluorine atoms per second-row atom), or electronegativity of the
second-row atom. It can, however, be accounted for by consideration of
the relative sizes of the atoms.

In Figure 18-1 there are shown several ways of arranging two or more
large anions, such as the fluoride ion, around a cation. For each arrangement there is given the ratio of radius of cation to radius of anion (the
radius ratio) corresponding to closest packing; that is, contact of anions
with one another as well as with the cation, both anions and cation being
considered to be spheres.

Thus for the planar triangular structure MX_3 the distances $r_M + r_X$ and
$2r_X$ have the relative values $1:\sqrt{3}$, from which we calculate $r_M/r_X =
2/\sqrt{3} - 1 = 0.155$. In a similar way the values 0.225 for ligancy 4 (a
tetrahedron of anions about the cation), 0.414 for ligancy 6 (octahedron),
and 0.645 for ligancy 8 (cube) are obtained.

Of substances MX_2, silicon dioxide (radius ratio 0.29) forms crystals
with tetrahedral coordination of four oxygen ions about each silicon ion
(Figure 18-8), magnesium fluoride (radius ratio 0.48) and stannic oxide
(radius ratio 0.51) form crystals with octahedral coordination of six
anions around each cation (the rutile structure, Figure 18-2), and calcium
fluoride (radius ratio 0.73) forms crystals with cubic coordination of eight
anions around each cation (the fluorite structure, Figure 18-3). The
ligancy (coordination number) increases with increase in the radius ratio,
as indicated in Figure 18-1.

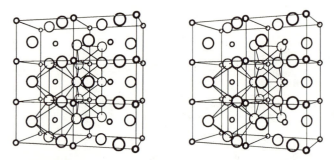

FIGURE 18-2
The structure of magnesium fluoride (stereo); this substance has
high melting point and boiling point. (This structure is usually
called the rutile structure; it is the structure of the mineral
rutile, TiO_2.)

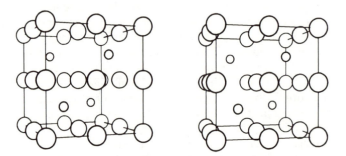

FIGURE 18-3
The structure of the fluorite crystal, CaF_2 (stereo).

The increase in stability (decrease in energy) with increase in ligancy is
easy to understand. Let us consider two ionic molecules, M^+X^-, with
M^+—X^- distance r. The electrostatic interaction energy of the electric
charge $+e$ and that $-e$ in one molecule is $-e^2/r$, and for two molecules it
is $-2e^2/r$. Now if the ligancy changes from 1 to 2, through the formation
of a square,

$$\begin{array}{c} M^+\!\!-\!\!X^- \\ |\quad\ | \\ X^-\!\!-\!\!M^+ \end{array}$$

and if the M^+—X^- distance retains the value r, each of the four M^+—X^-
interactions contributes $-e^2/r$ and each of the two repulsions across the
diagonals contributes $e^2/(\sqrt{2}\,r)$. The total electrostatic energy for the
square is then $(-4 + \sqrt{2})e^2/r = -2.59e^2/r$. The square arrangement is
thus 29% more stable than two separate molecules, with respect to the
electrostatic interactions.

Similar calculations show that, for constant cation-anion distance, the rutile structure (ligancy 6) is 8% more stable than the quartz structure (ligancy 4), and the fluorite structure (ligancy 8) is 5% more stable than the rutile structure.

If, however, the radius ratio is less than the value given in Figure 18-1, the anions come into contact with one another, the cation-anion distance becomes larger than the contact distance, and the structure becomes unstable relative to the structure with smaller ligancy. The approximate values of the radius ratio at which the transitions occur are shown in Figure 18-1.

Germanium dioxide is an interesting example. Its radius ratio (Table 6-2) is 0.53 Å/1.40 Å = 0.38. This value is very near the transition value, 0.37, from tetrahedral to octahedral coordination, and GeO_2 is in fact dimorphous, with one crystalline form having the quartz structure (ligancy 4) and the other having the rutile structure (ligancy 6).

We can now discuss the melting points and boiling points of the second-row fluorides. The ionic radii of the cations (the radius of F^- is 1.36 Å) and the radius ratios are the following:

	NaF	MgF_2	AlF_3	SiF_4	PF_5	SF_6
Radius of cation	0.95 Å	0.65 Å	0.50 Å	0.41 Å	0.34 Å	0.29 Å
Radius ratio	0.70	0.48	0.37	0.30	0.25	0.21
Expected ligancy of cation	6 or 8	6	4 or 6	4	4	4 or 3

We see that for silicon tetrafluoride the expected ligancy of silicon, 4, corresponds exactly to the formula of the molecule, SiF_4. Hence we conclude that in the crystal and liquid as well as the gas the substance consists of SiF_4 molecules. The structure of the crystal as determined by x-ray diffraction is shown in Figure 18-4. The crystal is an arrangement of tetrahedral SiF_4 molecules held together only by the van der Waals forces discussed in Section 11-4. The heat of fusion and heat of vaporization are correspondingly small, and the substance accordingly melts and boils (in fact, sublimes at 1 atm pressure) at a low temperature. For PF_5 and SF_6 the expected ligancy is less than the number of fluorine atoms. There is accordingly some strain in the molecules—the fluorine atoms in contact with one another are under compression and the P—F and S—F bonds are stretched; but the crystals, like those of SiF_4, consist of molecules held together only by van der Waals forces, and the melting and boiling points and heats of fusion and vaporization are close to those of silicon tetrafluoride.

In AlF_3, on the other hand, the expected ligancy of aluminum is 4 or 6, and the x-ray diffraction study of the crystals has shown the ligancy to be 6. Each aluminum atom is surrounded octahedrally by six fluorine atoms,

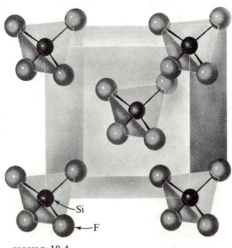

FIGURE 18-4
The structure of a molecular crystal, silicon
tetrafluoride. The tetrahedral SiF_4 molecules are
arranged in a body-centered cubic arrangement.

and each fluorine atom is ligated to two aluminum atoms.* The crystal
MgF_2 has the rutile structure (Figure 18-2) corresponding to the expected
ligancy 6 for magnesium and with each fluorine atom ligated to three
magnesium atoms, and NaF has the sodium chloride structure, in which
both sodium and fluorine have ligancy 6.

Fusion and vaporization of these three substances involves not just
overcoming the van der Waals attractive forces, as for SiF_4, but rather the
breaking of some Al—F, Mg—F, or Na—F bonds. For this reason the
heats of fusion and vaporization are large and the melting points and
boiling points are high.

The foregoing discussion has been based on the relative sizes of cations
and anions. A closely parallel discussion could be presented based on
covalent bond radii and van der Waals radii of atoms (Section 6-13). The
van der Waals radius of the fluorine atom, 1.35 Å (Table 6-7), is nearly
equal to the ionic radius of the fluoride ion, 1.36 Å, and the sums of
covalent radii are approximately equal to the corresponding sums of ionic
radii.

---

*The ligancy 6 rather than 4 for aluminum in the crystal is seen to be reasonable from
the following argument. With ligancy 4 for aluminum, the composition AlF_3 would require
an average $\frac{4}{3}$ aluminum atoms about each fluorine atom; that is, some fluorine atoms
ligated to one aluminum atom and some to two. Such a structure would be less stable
than the one with ligancy 6 for aluminum.

The Electrostatic Valence Rule

The description of a crystal or molecule in terms of cations and anions permits us to describe the bonds in a simple and useful way. The strength of each of the bonds formed by a cation can be defined as the electric charge of the cation divided by its ligancy. Thus in the crystal AlF_3, in which Al^{3+} has ligancy 6 (has six F^- ions coordinated about it), each of the six bonds formed by the aluminum ion has strength $\frac{3}{6} = \frac{1}{2}$. In the molecule SiF_4 each of the four bonds formed by SiF_4 has strength $\frac{4}{4} = 1$.

The electrostatic valence rule states that *the most stable structures of crystals and molecules are those in which the sum of the strengths of the bonds reaching each anion is just equal to its negative charge.*

For example, in the AlF_3 crystal each F^- is held to Al^{3+} by two bonds with strength $\frac{1}{2}$, and in the SiF_4 molecule it is held to Si^{4+} by one bond with strength 1; in each case the sum of the bond strengths equals the negative charge of the fluoride ion.*

Many of the properties of substances can be explained by the consideration of the relative sizes of ions or atoms, as illustrated above and in the following examples. Structural inorganic chemistry is, however, a new subject, and as yet far from precise. You may find it worth while to attempt to explain in terms of structure some of the properties of substances mentioned in later sections of this chapter and in the following chapters, but you must not become discouraged if you are unsuccessful. The fault may lie not with you but with the chemists of the present generation and earlier generations, who have not yet succeeded in the task of developing a really powerful theory of structural inorganic chemistry. If you become a chemist, you yourself may make a major contribution to the solution of this problem.

Example 18-1. Magnesium oxide and sodium fluoride have the same crystal structure, that of sodium chloride (ligancy 6). But their melting points are very different: 2800°C for MgO and 995°C for NaF. Explain.

Solution. In NaF the Na^+—F^- bonds have strength $\frac{1}{6}$, and in MgO the Mg^{++}—O^{--} bonds have strength $\frac{1}{3}$, twice as great. Because of the similarity in structure of the two substances, we may assume that the process of melting involves breaking the same fraction of the bonds. A higher temperature would be needed to break the stronger bonds in MgO than to break the weaker bonds in NaF; hence the melting point of MgO is higher than that of NaF.

---

*These arguments can also be presented in terms of covalent bonds with partial ionic character.

Example 18-2.   The mineral periclase, MgO, has hardness 5.8 on the Mohs scale (Section 7-1), whereas villaumite, NaF, has much smaller hardness, 3.5. Explain.

Solution.   As discussed in the answer to Example 18-1, the bonds in MgO are stronger than those in the NaF. In scratching a crystal to determine its hardness on the Mohs scale some of the bonds are broken, and for two crystals with the same structure the process of scratching can be expected to involve breaking the same number of bonds; hence for these two crystals the one with the stronger bonds should have the greater hardness.

Example 18-3.   The crystals NaF, NaCl, and KCl, with the same structure and the same ionic valences, have Mohs hardness 3.5, 2.5, and 2.0, respectively. Explain.

Solution.   The bond strengths (ratio of cation charge to ligancy) have the same value, $\frac{1}{6}$, for the three crystals, but the electrostatic forces become weaker in the sequence NaF, NaCl, KCl because the interionic distances (cation-anion bond lengths) increase: 2.31 Å, 2.76 Å, 3.14 Å (sum of radii, Table 6-2). Hence the crystals decrease in hardness in this sequence.

The Mohs hardness values are reliable only to about ± 0.3. Note that the empirical expression $19/(r_+ + r_-)^2$ gives the values 3.56, 2.50, and 1.93, respectively, for the hardness of the three substances. The use of the inverse square might be supported by an argument based on Coulomb's law.

Example 18-4.   The boiling (subliming) point of AlF_3, 1257°C, is much lower than those of NaF, 1704°C, and MgF_2, 2227°C. Both the increase in charge of the cation and the decrease in interionic distance (bond length) should lead to an increase in boiling point. What is the explanation?

Solution.   It was pointed out above in the text that the value of the radius ratio for Al^{+++} and F^-, 0.37, is near to the transition value between ligancy 6 and 4, and that ligancy 6 is preferred for the AlF_3 crystal because it permits all the fluoride ions to be bonded in the same way. Aluminum chloride exists in the gas phase as molecules Al_2Cl_6, with the structure of two tetrahedra sharing an edge:

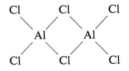

We may assume that aluminum fluoride also forms dimers Al_2F_6 with this structure, and that the stability of these molecules in the gas phase leads to the low boiling point, relative to MgF_2 and NaF.

TABLE 18-1
Some Properties of the Alkali Metals

	Z	Melting Point	Boiling Point	Density (g/cm^3)	Metallic Radius*	Ionic Radius†	Bond Energy‡
Li	3	186°C	1336°C	0.530	1.55 Å	0.60 Å	115
Na	11	97.5°	880°	.963	1.90	.95	75
K	19	62.3°	760°	.857	2.35	1.33	51
Rb	37	38.5°	700°	1.594	2.48	1.48	48
Cs	55	28.5°	670°	1.992	2.67	1.69	45

*For ligancy 12.

†For singly charged cation (Na^+, for example), with ligancy 6, as in the sodium chloride crystal.

‡Bond energy of $M_2(g)$, in kJ $mole^{-1}$.

18-3. The Alkali Metals and Their Compounds

The elements of the first group—lithium, sodium, potassium, rubidium, and cesium*—are soft, silvery-white metals with great chemical reactivity. These metals are excellent conductors of electricity. Some of their physical properties are given in Table 18-1. It can be seen from the table that they melt at low temperatures—four of the five metals melt below the boiling point of water. Lithium, sodium, and potassium are lighter than water. The vapors of the alkali metals are mainly monatomic, with a small concentration of diatomic molecules, such as Li_2, in which the two atoms are held together by a covalent bond.

The alkali metals are made by electrolysis of the molten hydroxides or chlorides (Chapter 15). Because of their reactivity, the metals must be kept in an inert atmosphere or under oil. The metals are useful chemical reagents in the laboratory, and they find industrial use (especially sodium) in the manufacture of organic chemicals, dyestuffs, and lead tetraethyl (a constituent of "ethyl gasoline"). Sodium is used in sodium-vapor lamps, and, because of its large heat conductivity, in the stems of valves of air-plane engines, to conduct heat away from the valve heads. A sodium-potassium alloy is used as a cooling liquid in nuclear reactors. Cesium is used in vacuum tubes, to increase electron emission from cathodes.

Compounds of sodium are readily identified by the yellow color that

---

*The sixth alkali metal, francium (Fr), element 87, has been obtained only in minute quantities, and no information has been published about its properties.

they give to a flame. Lithium causes a carmine coloration of the flame, and potassium, rubidium, and cesium cause a violet coloration. These elements may be tested for in the presence of sodium by use of a blue filter of cobalt glass.

The Discovery of the Alkali Metals

The alchemists had recognized many compounds of sodium and potassium. The metals themselves were isolated by Sir Humphry Davy in 1807 by electrolyzing their hydroxides. Compounds of lithium were recognized as containing a new element by the Swedish chemist Johan August Arfwedson, in 1817. The metal itself was first isolated in 1855. Rubidium and cesium were discovered in 1860 by the German chemist Robert Wilhelm Bunsen (1811–1899), by use of the spectroscope. Bunsen and the physicist Kirchhoff had invented the spectroscope just the year before, and cesium was the first element to be discovered by the use of this instrument. The spectrum of cesium contains two bright lines in the blue region and the spectrum of rubidium contains two bright lines in the extreme red.

Compounds of Lithium

Lithium occurs in the minerals* *spodumene*, $LiAlSi_2O_6$, *amblygonite*, $LiAlPO_4F$, and *lepidolite*, $K_2Li_3Al_5Si_6O_{20}F_4$. Lithium chloride, LiCl, is made by fusing (melting) a mineral containing lithium with barium chloride, $BaCl_2$, and extracting the fusion with water. It is used in the preparation of other compounds of lithium.

Compounds of lithium have found use in the manufacture of glass and of glazes for dishes and porcelain objects.

Compounds of Sodium

The most important compound of sodium is sodium chloride (common salt), NaCl. It crystallizes as colorless cubes, with melting point 801°C, and it has a characteristic salty taste. It occurs in seawater to the extent of 3%, and in solid deposits and concentrated brines (salt solutions) that are pumped from wells. Many million tons of the substance are obtained from these sources every year. It is used mainly for the preparation of other compounds of sodium and of chlorine, as well as of sodium metal and chlorine gas. Blood plasma and other body fluids contain about 0.9 g of sodium chloride per 100 ml.

Sodium hydroxide (caustic soda), NaOH, is a white hygroscopic

---

*Only specialists try to remember complicated formulas, such as that of lepidolite.

(water-attracting) solid, which dissolves readily in water. Its solutions have a smooth, soapy feeling, and are very corrosive to the skin (this is the meaning of "caustic" in the name caustic soda). Sodium hydroxide is made either by the electrolysis of sodium chloride solution or by the action of calcium hydroxide, $Ca(OH)_2$, on sodium carbonate, Na_2CO_3:

$$Na_2CO_3 + Ca(OH)_2 \longrightarrow CaCO_3 + 2NaOH$$

Calcium carbonate is insoluble, and precipitates out during this reaction, leaving the sodium hydroxide in solution. Sodium hydroxide is a useful laboratory reagent and a very important industrial chemical. It is used in industry in the manufacture of soap, the refining of petroleum, and the manufacture of paper, textiles, rayon and cellulose film, and many other products.

Compounds of Potassium

Potassium chloride, KCl, forms colorless cubic crystals, resembling those of sodium chloride. There are very large deposits of potassium chloride, together with other salts, at Stassfurt, Germany, and near Carlsbad, New Mexico. Potassium chloride is also obtained from Searles Lake in the Mojave Desert in California.

Potassium hydroxide, KOH, is a strongly alkaline substance, with properties similar to those of sodium hydroxide. Other important salts of potassium, which resemble the corresponding salts of sodium, are potassium sulfate, K_2SO_4, potassium carbonate, K_2CO_3, and potassium hydrogen carbonate, $KHCO_3$.

Potassium hydrogen tartrate (*cream of tartar*), $KHC_4H_4O_6$, is a constituent of grape juice; sometimes crystals of the substance form in grape jelly. It is used in making baking powder.

The principal use of potassium compounds is in *fertilizers*. Plant fluids contain large amounts of potassium ion, concentrated from the soil, and potassium salts must be present in the soil in order for plants to grow. A fertilizer containing potassium sulfate or some other salt of potassium must be used if the soil becomes depleted in this element.

The compounds of rubidium and cesium resemble those of potassium closely. They do not have any important uses.

Enthalpy of Formation of Compounds of the Alkali Metals

Values of the enthalpy of formation of some compounds of the alkali metals are given in Table 18-2. These values with their signs changed are the heats evolved on formation of the compounds from the elementary substances in their standard states.

TABLE 18-2
Standard Enthalpy of Formation of Compounds of Alkali Metals at 25°C (kJ mole⁻¹)

	M = Li	Na	K	Rb	Cs
M(g)	155	109	90	86	79
M^+(g)	681	611	515	495	461
M^+(aq)*	−278	−240	−251	−246	−248
M_2(g)	199	142	129	124	113
M_2O(c)	−596	−416	−361	−330	−318
MH(g)	128	125	126	138	121
MH(c)	−90	−57	−57	−59	−84
MF(c)	−612	−569	−563	−549	−531
MCl(c)	−409	−411	−436	−431	−433
MBr(c)	−350	−360	−392	−389	−395
MI(c)	−271	−288	−328	−328	−337
M_2S(c)		−373	−418	−348	−339
M_2Se(c)	−381	−264	−332		

*Relative to assumed value 0 for H^+(aq).

TABLE 18-3
Some Properties of the Alkaline-earth Metals

Symbol	Atomic Number	Atomic Weight	Melting Point*	Density (g/cm³)	Metallic Radius	Ionic Radius†
Be	4	9.0122	1350°C	1.86	1.12 Å	0.31 Å
Mg	12	24.312	651°	1.75	1.60	.65
Ca	20	40.08	810°	1.55	1.97	.99
Sr	38	87.62	800°	2.60	2.15	1.13
Ba	56	137.34	850°	3.61	2.22	1.35
Ra	88	226.04	960°	(4.45)‡	(2.46)‡	

*The boiling points of these metals are uncertain; they are about 600° higher than the melting points.
†For doubly charged cation with ligancy 6.
‡Estimated.

The small values of the electronegativity of the alkali metals (0.7 to 1.0) are reflected in the large values of the heats of formation of their compounds with the nonmetallic elements. These values (530 to 610 kJ mole⁻¹ for fluorides, 410 to 435 kJ mole⁻¹ for chlorides, and so on) are in rough agreement with the values calculated with use of the electronegativity expression, Equation 6-1.

TABLE 18-4

Standard Enthalpy of Formation of Compounds of Alkaline-earth Metals at 25°C (kJ mole⁻¹)

	M = Be	Mg	Ca	Sr	Ba
M(g)	321	150	193	164	176
M⁺(g)	1226	894	789	719	684
M⁺⁺(g)	2989	2351	1940	1790	1656
M⁺⁺(aq)*	−389	−462	−543	−546	−538
MO(c)	−611	−602	−636	−590	−558
MF₂(c)		−1102	−1215	−1215	−1200
MCl₂(c)	−512	−642	−795	−828	−860
MBr₂(c)	−370	−518	−675	−716	−755
MI₂(c)	−212	−360	−535	−567	−602
MS(c)	−234	−347	−482	−473	−485
MSe(c)			−313	−329	−310
MTe(c)		−209			
MH₂(c)			−195	−177	−172

*Relative to assumed value 0 for H⁺(aq).

The difference in enthalpy of formation of MH(g) and MH(c) is the heat of sublimation of the crystal. For the alkali hydrides it is a large quantity of energy—for LiH its value is $128 - (-90) = 218$ kJ mole⁻¹. This value is far larger than the van der Waals energy of attraction of molecules LiH, and we are required to conclude that the crystal is not a molecular crystal, but is instead an ionic crystal, with structure similar to that of the alkali halides. The alkali hydrides have been shown by x-ray diffraction to contain the hydride ion, H^-; they have the sodium chloride crystal structure, with ligancy 6 for M^+ and H^-. On electrolysis of the molten hydrides molecular hydrogen is liberated at the anode.

18-4.   The Alkaline-earth Metals and Their Compounds

The metals of group II of the periodic table—beryllium, magnesium, calcium, strontium, barium, and radium—are called the alkaline-earth metals. Some of their properties are listed in Table 18-3. These metals are much harder and less reactive than the alkali metals. The compounds of all the alkaline-earth metals are similar in composition; they all form oxides MO, hydroxides M(OH)₂, carbonates MCO₃, sulfates MSO₄, and other compounds (M = Be, Mg, Ca, Sr, Ba, or Ra). Enthalpy values are given in Table 18-4.

A Note on the Alkaline-earth Family

The early chemists gave the name "earth" to many nonmetallic substances. Magnesium oxide and calcium oxide were found to have an alkaline reaction, and hence were called the *alkaline earths*. The metals themselves (magnesium, calcium, strontium, and barium) were isolated in 1808 by Humphry Davy. Beryllium was discovered in the mineral beryl ($Be_3Al_2Si_6O_{18}$) in 1798 and was isolated in 1828.

Beryllium

Beryllium is a light, silvery white metal, which can be made by electrolysis of a fused mixture of beryllium chloride, $BeCl_2$, and sodium chloride. The metal is used for making windows for x-ray tubes (x-rays readily penetrate elements with low atomic number, and beryllium metal has the best mechanical properties of the very light elements). It is also used as a constituent of special alloys. About 2% of beryllium in copper produces a hard alloy especially suited for use in springs.

The principal ore of beryllium is *beryl*, $Be_3Al_2Si_6O_{18}$. *Emeralds* are beryl crystals containing traces of chromium, which give them a green color. *Aquamarine* is a bluish-green variety of beryl.

The compounds of beryllium have little special value, except that beryllium oxide, BeO, is used in the uranium reactors in which plutonium is made from uranium (Chapter 26).

Compounds of beryllium are very poisonous. Even the dust of the powdered metal or its oxide may cause very serious illness.

Magnesium

Magnesium metal is made by electrolysis of fused magnesium chloride, and also by the reduction of magnesium oxide by carbon or by ferrosilicon (an alloy of iron and silicon). Except for calcium and the alkali metals, magnesium is the lightest metal known; and it finds use in lightweight alloys, such as *magnalium* (10% magnesium, 90% aluminum).

Magnesium reacts with boiling water, to form magnesium hydroxide, $Mg(OH)_2$, an alkaline substance:

$$Mg + 2H_2O \longrightarrow Mg(OH)_2 + H_2$$

The metal burns in air with a bright white light, to form magnesium oxide, MgO, the old name of which is *magnesia*:

$$2Mg + O_2 \longrightarrow 2MgO$$

Magnesium oxide suspended in water is used in medicine (as "milk of magnesia"), for neutralizing excess acid in the stomach and as a laxative. Magnesium sulfate, "Epsom salt," $MgSO_4 \cdot 7H_2O$, is used as a cathartic.

Magnesium carbonate, $MgCO_3$, occurs in nature as the mineral *magnesite*. It is used as a basic lining for copper converters and open-hearth steel furnaces (Chapter 20).

Calcium

Metallic calcium is made by the electrolysis of fused calcium chloride, $CaCl_2$. The metal is silvery white in color, and is somewhat harder than lead. It reacts with water, and burns in air when ignited, forming a mixture of calcium oxide, CaO, and calcium nitride, Ca_3N_2.

Calcium has a number of practical uses—as a deoxidizer (substance removing oxygen) for iron and steel and for copper and copper alloys, as a constituent of lead alloys (metal for bearings, or the sheath for electric cables) and of aluminum alloys, and as a reducing agent for making other metals from their oxides.

Calcium reacts with cold water to form calcium hydroxide, $Ca(OH)_2$, and burns readily in air, when ignited, to produce calcium oxide, CaO.

Calcium sulfate occurs in nature as the mineral *gypsum*, $CaSO_4 \cdot 2H_2O$. Gypsum is a white substance, which is used commercially for fabrication into wallboard, and conversion into *plaster of Paris*. When gypsum is heated a little above 100°C it loses three-quarters of its water of crystallization, forming the powdered substance $CaSO_4 \cdot \frac{1}{2}H_2O$, which is called plaster of Paris. (Heating to a higher temperature produces anhydrous $CaSO_4$, which reacts more slowly with water.) When mixed with water the small crystals of plaster of Paris dissolve and then crystallize as long needles of $CaSO_4 \cdot 2H_2O$. These needles grow together, and form a solid mass, with the shape into which the wet powder was molded.

Strontium

The principal minerals of strontium are strontium sulfate, *celestite*, $SrSO_4$, and strontium carbonate, *strontianite*, $SrCO_3$.

Strontium nitrate, $Sr(NO_3)_2$, is made by dissolving strontium carbonate in nitric acid. It is mixed with carbon and sulfur to make red fire for use in fireworks, signal shells, and railroad flares. Strontium chlorate, $Sr(ClO_3)_2$, is used for the same purpose. The other compounds of strontium are similar to the corresponding compounds of calcium. Strontium metal has no practical uses.

Barium

The metal barium has no significant use. Its principal compounds are barium sulfate, $BaSO_4$, which is only very slightly soluble in water and dilute acids, and barium chloride, $BaCl_2 \cdot 2H_2O$, which is soluble in water. Barium sulfate occurs in nature as the mineral *barite*.

TABLE 18-5
Some Physical Properties of Elements of Groups III and IV

	Atomic Number	Atomic Weight	Density (g/cm³)	Melting Point	Atomic Radius*	Ionic Radius†
B	5	10.811	2.54	2300°C	0.80 Å	0.20 Å
Al	13	26.9815	2.71	660°	1.43	.50
Sc	21	44.956	3.18	1200°	1.62	.81
Y	39	88.905	4.51	1490°	1.80	.93
La	57	138.91	6.17	826°	1.87	1.15
C‡	6	12.01115	3.52	3500°	0.77	–
Si	14	28.086	2.36	1440°	1.17	0.41
Ge	32	72.59	5.35	959°	1.22	.53
Sn	50	118.69	7.30	232°	1.62	.71
Pb	82	207.19	11.40	327°	1.75	.84

*Single-bond covalent radius for B, C, Si, and Ge; metallic radius (ligancy 12) for the others.
†Section 6-8.
‡Diamond.

Barium, like all elements with large atomic number, absorbs x-rays strongly, and a thin paste of barium sulfate and water is swallowed as a "barium meal" to obtain contrasting x-ray photographs and fluoroscopic views of the alimentary tract. The solubility of the substance is so small that the poisonous action of most barium compounds is avoided.

Barium nitrate, $Ba(NO_3)_2$, and barium chlorate, $Ba(ClO_3)_2$, are used for producing green fire in fireworks.

Radium

Compounds of radium are closely similar to those of barium. The only important property of radium and its compounds is its radioactivity, which will be discussed further in Chapter 26.

18-5.   Boron

Boron can be made by heating potassium tetrafluoroborate, KBF_4, with sodium in a crucible lined with magnesium oxide:

$$KBF_4 + 3Na \longrightarrow KF + 3NaF + B$$

The element can also be made by heating boric oxide, B_2O_3, with powdered magnesium:

$$B_2O_3 + 3Mg \longrightarrow 3MgO + 2B$$

Boron forms brilliant transparent crystals, nearly as hard as diamond. Some of its properties are given in Table 18-5.

Boron forms a compound with carbon, B_4C. This substance, *boron carbide*, is one of the hardest substances known and it has found extensive use as an abrasive and for the manufacture of small mortars and pestles for grinding very hard substances. The cubic form of boron nitride, BN, with a tetrahedral structure like that of diamond, has about the same hardness.

Boric acid, H_3BO_3, occurs in the volcanic steam jets of central Italy. The substance is a white crystalline solid, which is sufficiently volatile to be carried along with a stream of steam. Boric acid can be made by treating borax with an acid.

The principal source of compounds of boron is the complex borate minerals, including *borax*, sodium tetraborate decahydrate, $Na_2B_4O_7 \cdot 10H_2O$; *kernite*, sodium tetraborate tetrahydrate, $Na_2B_4O_7 \cdot 4H_2O$ (which gives borax when water is added); and *colemanite*, calcium hexaborate pentahydrate, $Ca_2B_6O_{11} \cdot 5H_2O$. The main deposits of these minerals are in California.

Borax is used in making certain types of enamels and glass (such as Pyrex glass, which contains about 12% of B_2O_3), for softening water, as a household cleanser, and as a flux in welding metals. The last of these uses depends upon the power of molten borax to dissolve metallic oxides, forming borates.

Values of Enthalpy of Formation of Boron Compounds

In Table 18-6 there are given values of the enthalpy of formation of some compounds of boron, aluminum, scandium, yttrium, and lanthanum. It is seen that for a series of corresponding compounds the heats of formation increase in the sequence B, Al, Sc, Y, La, corresponding to the decrease in electronegativity in this sequence.

18-6.   The Boranes. Electron-deficient Substances

The reaction of magnesium boride, Mg_3B_2, with water would be expected from simple valence theory to result in the production of molecules of boron trihydride, BH_3. Instead, the substance diborane, B_2H_6, is produced:

$$Mg_3B_2 + 6H_2O \longrightarrow 3Mg(OH)_2 + B_2H_6(g)$$

Diborane is a gas under ordinary conditions (m.p. $-165.5°C$, b.p. $-92.5°C$).

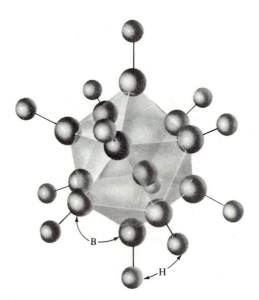

FIGURE 18-5
The structure of the icosahedral dodecaborane
ion, $B_{12}H_{12}^{--}$.

TABLE 18-6
*Standard Enthalpy of Formation of Compounds of Boron, Aluminum, and Their Congeners
at 25°C (kJ mole^{-1})*

	M = B	Al	Sc	Y	La
M(g)	407	314	389	431	368
M$^+$(g)	1213	897	1028	1067	916
M^{++}(g)	3646	2720	2278	2269	2025
M^{+++}(g)	7312	5468	4674	4252	3881
M^{+++}(aq)*		−525	−623	−703	−737
M$_2$O$_3$(c)	−1264	−1670			−1916
MF$_3$	−1110(g)	−1301(c)			
MCl$_3$	−418(l)	−695(c)	−924(c)	−982(c)	−1103(c)
MBr$_3$	−221(l)	−526(c)	−751(c)		
MI$_3$(c)		−315		−599	−700
M$_2$S$_3$(c)	−238	−509			−1284

*Relative to assumed value 0 for H$^+$(aq).

Many other boranes are known; those that have been the most thoroughly investigated have formulas B_4H_{10}, B_5H_9, B_5H_{10}, and $B_{10}H_{14}$. They have found some use as rocket fuels.

The B_2H_6 molecule has the following structure:

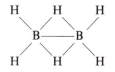

Each boron atom has ligancy 5, and two of the hydrogen atoms have ligancy 2. The bond lengths B—B = 1.77 Å, B—H = 1.33 Å (for bridging hydrogen atoms) and 1.19 Å (for outer hydrogen atoms) indicate that the bonds are fractional bonds, each involving less than one electron pair. There are six pairs of valence electrons in the molecule, and nine bonds; hence on the average each bond is two-thirds of a single bond. The bond lengths indicate that the six electron pairs resonate among the nine positions in such a way that the five central bonds involve about five valence electrons and the four outer bonds involve about seven.

Some borane ions are also known, such as $B_4H_{10}^{--}$ in $Na_2B_4H_{10}$ and $B_{12}H_{12}^{--}$ in $K_2B_{12}H_{12}$. The $B_{12}H_{12}^{--}$ ion has an interesting structure: the twelve boron atoms lie at the corners of a regular icosahedron, as shown in Figure 18-5, and each boron atom forms six bonds, five to the adjacent boron atoms in the icosahedron and one, directed outward radially, to a hydrogen atom. The bond lengths indicate bond numbers about 0.5 for the thirty B—B bonds and 0.83 for the twelve B—H bonds.

$B_{10}H_{14}$ and some other borane molecules have also been found to have structures based on the icosahedron, with some of the corners not occupied by boron atoms, and with some bridging hydrogen atoms, as in diborane. The B_{12} icosahedron is also present in elementary boron (Figure 18-6) and in the hard substance B_4C.

These substances can be described as *electron-deficient substances*. Electron-deficient substances are substances in which some or all of the atoms have more stable orbitals than electrons in the valence shell. The boron atom has four orbitals in its valence shell, and three valence electrons.

A characteristic feature of the structure of most electron-deficient substances is that the atoms have ligancy that is not only greater than the number of valence electrons but is even greater than the number of stable orbitals. Thus most of the boron atoms in the tetragonal form of crystalline boron have ligancy 6. Also, lithium and beryllium, with four stable orbitals and only one and two valence electrons, respectively, have structures in which the atoms have ligancy 8 or 12. All metals can be considered to be electron-deficient substances.

Another generalization that may deserve to be called a structural principle is that an electron-deficient atom causes adjacent atoms to increase their ligancy to a value greater than the orbital number. For example, in the boranes some of the hydrogen atoms, adjacent to the electron-deficient atoms of boron, have ligancy 2.

The boron-boron bond lengths, 1.80 Å, are 0.18 Å greater than the single-bond value, 1.62 Å; we may say that in tetragonal boron each atom with ligancy 6 uses its three electrons to form six half-bonds, rather than three electron-pair bonds.

Bridging Methyl Groups

In some substances an electron-deficient atom causes an adjacent carbon atom to increase its ligancy from four to five. An example is the dimer of aluminum trimethyl, which has the following structure:

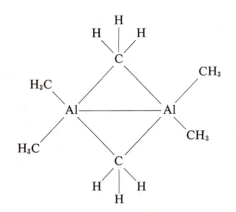

The electron deficiency of the aluminum atoms has increased the ligancy of both themselves and the bridging carbon atoms to 5. Another example is beryllium dimethyl, which forms infinite linear polymers in which each beryllium atom is surrounded tetrahedrally by four bridging methyl groups.

18-7. Aluminum and Its Congeners

Some of the physical properties of aluminum and its congeners are given in Table 18-5. Aluminum is only about one-third as dense as iron, and some of its alloys, such as duralumin (described below), are as strong as mild steel; it is this combination of lightness and strength, together with

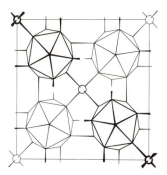

FIGURE 18-6

Structure of tetragonal boron as viewed in the direction of the c axis. One unit cell is shown. Two of the icosahedral B_{12} groups (light lines) are centered at $z = \frac{1}{4}$, and the other two (heavy lines) at $z = \frac{3}{4}$. The interstitial boron atoms, which have ligancy 4 (open circles), are at $(0, 0, 0)$ and $(\frac{1}{2}, \frac{1}{2}, \frac{1}{2})$. All of the extra-icosahedral bonds are shown with the exception of those parallel to the c axis from each icosahedron to the icosahedra in cells directly above and below. (Reprinted from Linus Pauling, *The Nature of the Chemical Bond.* Copyright 1939 and 1940 by Cornell University. Third edition © 1960 by Cornell University. Used by permission of Cornell University Press.)

low cost, that has led to the extensive use of aluminum alloys. Aluminum is also used, in place of copper, as a conductor of electricity; its electric conductivity is about 80% that of copper.* Its metallurgy has been discussed in Chapter 15.

The metal is reactive (note its position in the electromotive-force series), and when strongly heated it burns rapidly in air or oxygen. Aluminum dust forms an explosive mixture with air. Under ordinary conditions, however, aluminum rapidly becomes coated with a thin, tough layer of aluminum oxide, which protects it against further corrosion.

Some of the alloys of aluminum are very useful. *Duralumin* or *dural* is an alloy (containing about 94.3% aluminum, 4% copper, 0.5% manganese, 0.5% magnesium, and 0.7% silicon), which is stronger and tougher than pure aluminum. It is less resistant to corrosion, however, and often is protected by a coating of pure aluminum. Plate made by rolling a billet of

---

*The conductivity refers to the conductance of electricity by a wire of unit cross-sectional area. The density of aluminum is only 30% of that of copper; accordingly, an aluminum wire with the same weight as a copper wire with the same length conducts 2.7 times as much electricity as the copper wire with the same transmission loss.

dural sandwiched between and welded to two pieces of pure aluminum is called alclad plate.

Aluminum oxide (*alumina*), Al_2O_3, occurs in nature as the mineral *corundum*. Corundum and impure corundum (*emery*) are used as abrasives. Pure corundum is colorless. The precious stones *ruby* (red) and *sapphire* (blue or other colors) are transparent crystalline corundum containing small amounts of other metallic oxides (chromic oxide, titanium oxide). Artificial rubies and sapphires can be made by melting aluminum oxide (m.p. 2050°C) with small admixtures of other oxides, and cooling the melt in such a way as to produce large crystals. These stones are indistinguishable from natural stones, except for the presence of characteristic rounded microscopic air bubbles. They are used as gems, as bearings ("jewels") in watches and other instruments, and as dies through which wires are drawn.

Aluminum sulfate, $Al_2(SO_4)_3 \cdot 18H_2O$, may be made by dissolving aluminum hydroxide in sulfuric acid:

$$2Al(OH)_3 + 3H_2SO_4 + 12H_2O \longrightarrow Al_2(SO_4)_3 \cdot 18H_2O$$

It is used in water purification and as a mordant in dyeing and printing cloth (a *mordant* is a substance that fixes the dye to the cloth, rendering it insoluble). Both of these uses depend upon its property of producing a gelatinous precipitate of aluminum hydroxide, $Al(OH)_3$, when it is dissolved in a large amount of neutral or slightly alkaline water. The reaction that occurs is the reverse of the above reaction. In dyeing and printing cloth the gelatinous precipitate aids in holding the dye onto the cloth. In water purification it adsorbs dissolved and suspended impurities, which are removed as it settles to the bottom of the reservoir.

A solution containing aluminum sulfate and potassium sulfate, K_2SO_4, forms, on evaporation, beautiful colorless cubic (octahedral) crystals of *alum*, $KAl(SO_4)_2 \cdot 12H_2O$. Similar crystals of ammonium *alum*, $NH_4Al(SO_4)_2 \cdot 12H_2O$, are formed with ammonium sulfate. The alums also are used as mordants in dyeing cloth, in water purification, and in weighting and sizing paper (by precipitating aluminum hydroxide in the meshes of the cellulose fibers).

Aluminum chloride, $AlCl_3$, is made by passing dry chlorine or hydrogen chloride over heated aluminum:

$$2Al + 3Cl_2 \longrightarrow 2AlCl_3$$
$$2Al + 6HCl \longrightarrow 2AlCl_3 + 3H_2$$

The anhydrous salt is used in many chemical processes, including a cracking process for making gasoline.

Scandium, Yttrium, Lanthanum, and the Lanthanons

Scandium, yttrium, and lanthanum,* the congeners of boron and aluminum, form colorless compounds similar to those of aluminum, their oxides having the formulas Sc_2O_3, Y_2O_3, and La_2O_3. These elements and their compounds have not yet found any important use.

Scandium, yttrium, and lanthanum usually occur in nature with the fourteen lanthanons, cerium (atomic number 58) to lutetium (atomic number 71).† All of these elements except promethium (which is made artificially) occur in nature in very small quantities, the principal source being the mineral *monazite*, a mixture of phosphates containing also some thorium phosphate.

The metals themselves are very electropositive, and are accordingly difficult to prepare. Electrolytic reduction of a fused oxide-fluoride mixture may be used. An alloy containing about 70% cerium and smaller amounts of other lanthanons and iron gives sparks when scratched. This alloy is widely used for cigarette lighters and gas lighters.

These elements are usually terpositive, forming salts such as $La(NO_3)_3 \cdot 6H_2O$. Cerium forms also a well-defined series of salts in which it is quadripositive. This oxidation state corresponds to its atomic number, 4 greater than that of xenon. Praseodymium, neodymium, and terbium form dioxides, but not the corresponding salts.

The bipositive europium(II) ion is stable, and europium forms a series of europium(II) salts as well as of europium(III) salts. Ytterbium and samarium have a somewhat smaller tendency to form salts representing the +2 state of oxidation.

The ions of several of the lanthanons have characteristic colors. A special glass containing lanthanon ions is used in glassblowers' goggles and in optical instruments.

Many of the lanthanon compounds are strongly paramagnetic. Crystalline compounds of gadolinium, especially gadolinium sulfate octahydrate, $Gd_2(SO_4)_3 \cdot 8H_2O$, are used in the magnetic method of obtaining extremely low temperatures.

The sulfides cerium monosulfide, CeS, and thorium monosulfide, ThS, and related sulfides have been found valuable as refractory substances. The melting point of cerium monosulfide is 2450°C.

---

*Actinium, the heaviest member of group III, is a radioactive element that occurs in minute quantities in uranium ores.

†Lanthanum is often considered as one of the rare-earth elements (lanthanons). For convenience, the convention is adopted here of including lanthanum as a member of group III, leaving fourteen elements in the lanthanon group.

TABLE 18-7

Standard Enthalpy of Formation of Compounds of Silicon, Germanium, Tin, and Lead at 25°C (kJ mole⁻¹)

	M = Si	Ge	Sn*	Pb
M(g)	368	328	301	194
MO	−113(g)	−95(g)	−286(c)	−219(c)
MO$_2$(c)	−859	−537	−581	−277
MH$_4$(g)	−62			
MF$_2$(c)				−663
MF$_4$	−1548(g)			−930(c)
MCl$_2$(c)			−350	−359
MCl$_4$(l)	−640	−544	−545	
MBr$_2$(c)			−266	−277
MBr$_4$	−398(l)		−406(c)	
MI$_2$(c)			−144	−175
MI$_4$(c)	−132			
MS(c)			−78	−94

*White tin is the standard state; the value for gray tin is 2.5 kJ mole⁻¹.

18-8.   Silicon and Its Simpler Compounds

Elementary Silicon and Silicon Alloys

Silicon is a brittle steel-gray metalloid. Some of its physical properties are given in Table 18-5. Enthalpy values are given in Table 18-7. It can be made by the reduction of silicon tetrachloride by sodium:

$$SiCl_4 + 4Na \longrightarrow Si + 4NaCl$$

The element has the same crystal structure as diamond, each silicon atom forming single covalent bonds with four adjacent silicon atoms, which surround it tetrahedrally. It is used in transistors, especially for service at elevated temperatures (Section 18-15).

Silicon contaminated with carbon can be obtained by reduction of silica, SiO_2, with carbon in an electric furnace. An alloy of iron and silicon, called *ferrosilicon*, is obtained by reducing a mixture of iron oxide and silica with carbon.

Ferrosilicon, which has composition approximately FeSi, is used in the manufacture of acid-resisting alloys, such as *duriron*, which contains about 15% silicon. Duriron is used in chemical laboratories and manufacturing plants. A mild steel containing a few percent of silicon may be made which has a high magnetic permeability, and is used for the cores of electric transformers.

Silicides

Many metals form compounds with silicon, called silicides. These compounds include Mg_2Si, Fe_2Si, $FeSi$, $CoSi$, $NiSi$, $CaSi_2$, $Cu_{15}Si_4$, and $CoSi_2$. Ferrosilicon consists largely of the compound $FeSi$. Calcium silicide, $CaSi_2$, is made by heating a mixture of lime, silica, and carbon in an electric furnace. It is a powerful reducing agent, and is used for removing oxygen from molten steel in the process of manufacture of steel.

Silicon Carbide

Silicon carbide, SiC, is made by heating a mixture of carbon and sand in a special electric furnace:

$$SiO_2 + 3C \longrightarrow SiC + 2CO$$

The structure of this substance is similar to that of diamond, with carbon and silicon atoms alternating; each carbon atom is surrounded by a tetrahedron of silicon atoms, and each silicon atom by a tetrahedron of carbon atoms. The covalent bonds connecting all of the atoms in this structure make silicon carbide very hard. The substance is used as an abrasive.

18-9. Silicon Dioxide

Silicon dioxide (*silica*), SiO_2, occurs in nature in three different crystal forms: as the minerals *quartz* (hexagonal), *cristobalite* (cubic), and *tridymite* (hexagonal). Quartz is the most widespread of these minerals; it occurs in many deposits as well-formed crystals, and also as a crystalline constituent of many rocks, such as granite. It is a hard, colorless substance. Its crystals may be identified as right-handed or left-handed by their face development and also by the direction in which they rotate the plane of polarization of polarized light.

The structure of quartz is closely related to that of silicic acid, H_4SiO_4. In this acid silicon has ligancy 4, the silicon atom being surrounded by a tetrahedron of four oxygen atoms, with one hydrogen atom attached to each oxygen atom. Silicic acid, which is a very weak acid, has the property of undergoing condensation very readily, with elimination of water. If each of the four hydroxyl groups of a silicic acid molecule condenses with a similar hydroxyl group of an adjacent molecule, eliminating water, a structure is obtained in which the silicon atom is bonded to four surrounding silicon atoms by silicon-oxygen-silicon bonds. This process leads to a condensation product with formula SiO_2, since each silicon atom is surrounded by four oxygen atoms, and each oxygen atom serves as a neighbor to two silicon atoms. The structure of quartz and of the other forms of silica may be described as consisting of SiO_4 tetrahedra, with each oxygen atom serving as the corner of two of these tetrahedra. In order to break a crystal of quartz it is necessary to break some silicon-oxygen bonds. In this way the structure of quartz accounts for the hardness of the mineral.

Cristobalite and tridymite are similarly made from SiO_4 tetrahedra fused together by sharing oxygen atoms, with, however, different arrangements of the tetrahedra in space from that of quartz. Tridymite resembles ordinary ice (Figure 12-7) in structure, with silicon atoms in the oxygen-atom positions; cristobolite similarly resembles cubic ice. Three other crystalline modifications of silica have been discovered since 1956—keatite, coesite, and stishovite. Keatite and coesite contain SiO_4 tetrahedra somewhat distorted from the regular configuration (bond angles different from 109°28'). Coesite was discovered in the laboratory by subjecting silica to high pressure (about 30,000 atm), and was later found at Meteor Crater, Arizona, and other places where large meteorites have struck the earth. Stishovite was first made in 1961 by use of pressures of about 120,000 atm. It has a structure (like that of rutile, TiO_2; Figure 18-2) in which each silicon atom is octahedrally surrounded by six oxygen atoms. The presence of coesite and stishovite in rocks near a crater is evidence that the crater was formed by impact of a meteorite.

Silica Glass

If any form of silica is melted (m.p. about 1600°C) and the molten material is then cooled, it usually does not crystallize at the original melting point, but the liquid becomes more viscous as the temperature is lowered, until, at about 1500°C, it is so stiff that it cannot flow. The material obtained in this way is not crystalline, but is a supercooled liquid, or glass. It is called *silica glass* (or sometimes *quartz glass* or *fused quartz*). Silica glass does not have the properties of a crystal—it does not cleave, nor form crystal faces, nor show other differences in properties in different directions. The reason for this is that the atoms that constitute it are not arranged in a completely regular manner in space, but show a randomness in arrangement similar to that of the liquid.

The structure of silica glass is very similar in its general nature to that of quartz and the other crystalline forms of silica. Nearly every silicon atom is surrounded by a tetrahedron of four oxygen atoms, and nearly every oxygen atom serves as the common corner of two of these tetrahedra. The arrangement of the framework of tetrahedra in the glass is not regular, however, as it is in the crystalline forms of silica, but is irregular, so that a very small region may resemble quartz, and an adjacent region may resemble cristobalite or tridymite, in the same way that liquid silica, above the melting point of the crystalline forms, would show some resemblance to the structures of the crystals.

Silica glass is used for making chemical apparatus and scientific instruments. The coefficient of thermal expansion of silica glass is very small, so that vessels made of the material do not break readily on sudden heating or cooling. Silica is transparent to ultraviolet light, and because of this property it is used in making mercury-vapor ultraviolet lamps and optical instruments for use with ultraviolet light.

Ordinary glass is discussed in Section 18-12.

Cryptocrystalline Silica

A cryptocrystalline mineral is one in which the crystal grains are so small that they cannot be seen, even with a microscope. Silica, sometimes partially hydrated, occurs in nature in many cryptocrystalline varieties, distinguished from one another mainly by their color (usually arising from impurities). Among these are *chalcedony* (waxy luster, transparent or translucent, white, grayish, blue, brown, black), *carnelian* (a clear red or red-brown chalcedony), *chrysoprase* (an apple-green chalcedony, containing bipositive nickel), *agate* (a variegated chalcedony, either banded or cloudy), *onyx* (agate with layers in planes), *sardonyx* (onyx with some layers of carnelian, which is also called sard), *flint* (resembling chalcedony, but opaque and dull in color, usually gray, smoky-brown, or brownish-black), and *jasper* (more dull and opaque than flint; often red in color from iron(III) oxide, or yellow, gray-blue, brown-black).

The foregoing minerals are varieties of quartz. *Opal* is a cryptocrystalline variety of cristobalite, somewhat hydrated. The striking colors of opal are caused by Bragg diffraction of visible light by spherulites of cristobalite about 3000 Å in diameter that have settled into a close-packed array and have been cemented together by a silicious cement with index of refraction different from that of the spherulites.

18-10. Sodium Silicate and Other Silicates

Silicic acid (orthosilicic acid), H_4SiO_4, cannot be made by the hydration of silica. The sodium and potassium salts of silicic acid are soluble in water, however, and can be made by boiling silica with a solution of sodium hydroxide or potassium hydroxide, in which it slowly dissolves. A concentrated solution of sodium silicate, called water glass, is available commercially and is used for fireproofing wood and cloth, as an adhesive, and for preserving eggs. This solution is not sodium orthosilicate, Na_4SiO_4, but is a mixture of the sodium salts of various condensed silicic acids, such as $H_6Si_2O_7$, $H_4Si_3O_8$, and $(H_2SiO_3)_\infty$.

A gelatinous precipitate of condensed silicic acids $(SiO_2 \cdot xH_2O)$ is obtained when an ordinary acid, such as hydrochloric acid, is added to a solution of sodium silicate. When this precipitate is partially dehydrated it forms a porous product called silica gel. This material has great powers of adsorption for water and other molecules and is used as a drying agent and decolorizing agent.

Except for the alkali silicates, most silicates are insoluble in water. Many occur in nature, as ores and minerals.

18-11. The Silicate Minerals

Most of the minerals that constitute rocks and soil are silicates, which usually also contain aluminum. Many of these minerals have complex formulas, corresponding to the complex condensed silicic acids from which they are derived. These minerals can be divided into three principal classes: the *framework minerals* (hard minerals similar in their properties to quartz), the *layer minerals* (such as mica), and the *fibrous minerals* (such as asbestos).

The Framework Minerals

Many silicate minerals have tetrahedral framework structures in which some of the tetrahedra are AlO_4 tetrahedra instead of SiO_4 tetrahedra. These minerals have structures somewhat resembling that of quartz, with additional ions, usually alkali or alkaline-earth ions, introduced in the larger openings in the framework structure. Ordinary *feldspar* (*orthoclase*), $KAlSi_3O_8$, is an example of a tetrahedral aluminosilicate mineral. The aluminosilicate tetrahedral framework, $(AlSi_3O_8^-)_\infty$, extends throughout the entire crystal, giving it hardness nearly as great as that of quartz. Some other aluminosilicate minerals with tetrahedral framework structures are the following:

Kaliophilite	$KAlSiO_4$	Analcite	$NaAlSi_2O_6 \cdot H_2O$
Leucite	$KAlSi_2O_6$	Natrolite	$Na_2Al_2Si_3O_{10} \cdot 2H_2O$
Albite	$NaAlSi_3O_8$	Chabazite	$CaAl_2Si_4O_{12} \cdot 6H_2O$
Anorthite	$CaAl_2Si_2O_8$	Sodalite	$Na_4Al_3Si_3O_{12}Cl$

A characteristic feature of these tetrahedral framework minerals is that the number of oxygen atoms is just twice the sum of the number of aluminum and silicon atoms. In some of these minerals the framework is an open one, through which corridors run that are sufficiently large to permit ions to move in and out. The *zeolite minerals*, used for softening water, are of this nature. As the hard water, containing Ca^{++} and Fe^{+++} ions, passes around the grains of the mineral, these cations enter the mineral, replacing an equivalent number of sodium ions (Section 12-1).

Some of the zeolite minerals contain water molecules in the corridors and chambers within the aluminosilicate framework, as well as alkali and alkaline-earth ions. When a crystal of one of these minerals, such as chabazite, $CaAl_2Si_4O_{12} \cdot 6H_2O$, is heated, the water molecules are driven out of the structure. The crystal does not collapse, however, but retains essentially its original size and shape, the spaces within the framework formerly occupied by water molecules remaining unoccupied. This dehydrated chabazite has a strong attraction for water molecules and for molecules of other vapors, and can be used as a drying agent or absorbing agent for them. The structure of silica gel, mentioned above as a drying agent, is similar.

Some of the important minerals in soil are aluminosilicate minerals that have the property of base exchange, and that, because of this property, serve a useful function in the nutrition of the plant.

An interesting framework mineral is *lazurite*, or *lapis lazuli*, a mineral with a beautiful blue color. When ground into a powder, this mineral constitutes the pigment called *ultramarine*. Lazurite has the formula $Na_8Al_6Si_6O_{24}(S_x)$. It consists of an aluminosilicate framework in which there are sodium ions (some of which neutralize the charge of the framework) and anions S_x^{--}, such as S_2^{--} and S_3^{--}. These polysulfide ions are responsible for the color of the pigment. It was discovered at the beginning of the eighteenth century that a synthetic ultramarine can be made by melting together a suitable sodium aluminosilicate mixture with sulfur. Similar stable pigments with different colors can also be made by replacing the sulfur by selenium and the sodium ion by other cations.

Minerals with Layer Structures

By a condensation reaction involving three of the four hydroxyl groups of each silicic acid molecule, a condensed silicic acid can be made, with composition $(H_2Si_2O_5)_\infty$, which has the form of an infinite layer, as shown in Figure 18-7. The mineral *hydrargillite*, $Al(OH)_3$, has a similar layer structure, which involves AlO_6 octahedra (Figure 18-8). More complex layers, involving both tetrahedra and octahedra, are present in other layer minerals, such as *talc*, *kaolinite* (clay), and *mica*.

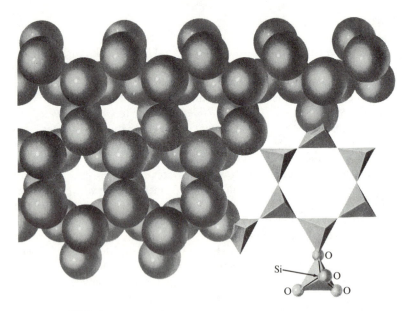

FIGURE 18-7
A portion of an infinite layer of silicate tetrahedra, as present in talc and other minerals with layer structures.

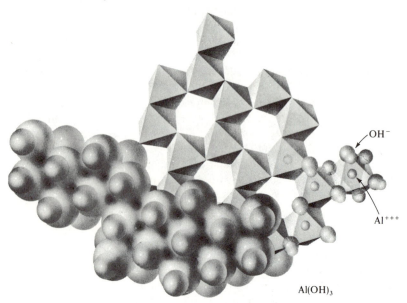

FIGURE 18-8
The crystal structure of aluminum hydroxide, $Al(OH)_3$. This substance crystallizes in layers, consisting of octahedra of oxygen atoms (hydroxide ions) about the aluminum atom. Each oxygen atom serves as a corner for two aluminum octahedra.

In talc and kaolinite, with formulas $Mg_3Si_4O_{10}(OH)_2$ and $Al_2Si_2O_5(OH)_4$, respectively, the layers are electrically neutral, and they are loosely superimposed on one another to form the crystalline material. These layers slide over one another very readily, which gives to these minerals their characteristic properties (softness, easy cleavage, soapy feel). In mica, $KAl_3Si_3O_{10}(OH)_2$, the aluminosilicate layers are negatively charged, and positive ions, usually potassium ions, must be present between the layers in order to give the mineral electric neutrality. The electrostatic forces between these positive ions and the negatively charged layers make mica considerably harder than kaolinite and talc, but its layer structure is still evident in its perfect basic cleavage, which permits the mineral to be split into very thin sheets. These sheets of mica are used for windows in stoves and furnaces, and for electric insulation in machines and instruments.

Other layer minerals, such as *montmorillonite*, with formula approximately $AlSi_2O_5(OH) \cdot xH_2O$, are important constituents of soils, and have also found industrial uses, as catalysts in the conversion of long-chain hydrocarbons into branched-chain hydrocarbons (to make high-octane gasoline), and for other special purposes.

The Fibrous Minerals

The fibrous minerals contain very long silicate ions in the form of tetrahedra condensed into a chain. These crystals can be cleaved readily in directions parallel to the silicate chains, but not in the directions that cut the chains. Accordingly crystals of these minerals show the extraordinary property of being easily unraveled into fibers. The principal minerals of this sort, *tremolite*, $Ca_2Mg_5Si_8O_{22}(OH)_2$, and *chrysotile*, $Mg_6Si_4O_{11}(OH)_6 \cdot H_2O$, are called *asbestos*. Deposits of these minerals are found, especially in South Africa, in layers several inches thick. These minerals are shredded into fibers, which are then spun or felted into asbestos yarn, fabric, and board for use for thermal insulation and as a heat-resistant structural material.

18-12. Glass

Silicate materials with important uses include glass, porcelain, glazes and enamels, and cement. Ordinary glass is a mixture of silicates in the form of a supercooled liquid. It is made by melting a mixture of sodium carbonate (or sodium sulfate), limestone, and sand, usually with some scrap glass of the same grade to serve as a flux. After the bubbles of gas have been expelled, the clear melt is poured into molds or stamped with dies, to produce pressed glassware, or a lump of the semifluid material on the end of a hollow tube is blown, sometimes in a mold, to produce hollow ware, such as bottles and flasks. *Plate glass* is made by pouring liquid glass onto a flat table and rolling it into a sheet. The sheet is then ground flat and polished on both sides. *Safety glass* consists of a sheet of tough plastic sandwiched between two sheets of glass.

Ordinary glass (soda-lime glass, soft glass) contains about 10% sodium, 5% calcium, and 1% aluminum, the remainder being silicon and oxygen. It consists of an aluminosilicate tetrahedral framework, within which are embedded sodium ions and calcium ions and some smaller complex anions. Soda-lime glass softens over a range of temperatures beginning at a dull-red heat, and can be conveniently worked in this temperature range.

Boric acid easily forms highly condensed acids, similar to those of silicic acid, and borate glasses are similar to silicate glasses in their properties. *Pyrex glass*, used for chemical glassware and baking dishes, is a boro-alumino-silicate glass containing only about 4% of alkali and alkaline-earth metal ions. This glass is not as soluble in water as is soft glass, and it also has a smaller coefficient of thermal expansion than soft glass, so that it does not break readily when it is suddenly heated or cooled.

Glazes on chinaware and pottery and *enamels* on iron kitchen utensils and bathtubs consist of easily fusible glass containing pigments or white fillers such as titanium dioxide and tin dioxide.

18-13. Cement

Portland cement is an aluminosilicate powder that sets to a solid mass on treatment with water. It is usually manufactured by grinding limestone and clay to a fine powder, mixing with water to form a slurry, and burning the mixture, with a flame of gas, oil, or coal dust, in a long rotary kiln. At the hot end of the kiln, where the temperature is about 1500°C, the alumino-silicate mixture is sintered together into small round marbles, called "clinker." The clinker is ground to a fine powder in a ball mill (a rotating cylindrical mill filled with steel balls) to produce the final product.

Portland cement before treatment with water consists of a mixture of calcium silicates, mainly Ca_2SiO_4 and Ca_3SiO_5, and calcium aluminate, $Ca_3Al_2O_6$. When treated with water the calcium aluminate hydrolyzes, forming calcium hydroxide and aluminum hydroxide, and these substances react further with the calcium silicates to produce calcium aluminosili-cates, in the form of intermeshed crystals.

Ordinary *mortar* for laying bricks is made by mixing sand with slaked lime (Section 8-5). This mortar slowly becomes hard through reaction with carbon dioxide of the air, forming calcium carbonate. A stronger mortar is made by mixing sand with Portland cement. The amount of cement needed for a construction job is greatly reduced by mixing sand and crushed stone or gravel with the cement, forming the material called *concrete*. Concrete is a very valuable building material. It does not require carbon dioxide from the air in order to harden, and it will set under water and in very large masses.

18-14.   The Silicones

When we consider the variety of structures represented by the silicate minerals, and their resultant characteristic and useful properties, we might well expect chemists to synthesize many new and valuable silicon compounds. In recent years this has been done; many silicon compounds, especially those of the class called *silicones*, have been found to have valuable properties.

The simplest silicones are the methyl silicones. These substances exist as oils, resins, and elastomers (rubberlike substances). Methyl silicone oil consists of long molecules, each of which is a silicon-oxygen chain with methyl groups attached to the silicon atoms. A short silicone molecule would have the following structure:

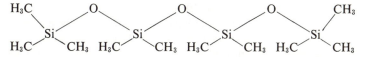

A *silicone oil* for use as a lubricating oil or in hydraulic systems contains molecules with an average of about 10 silicon atoms per molecule.

The valuable properties of the silicone oils are their very low coefficient of viscosity with temperature, ability to withstand high temperature without decomposition, and chemical inertness to metals and most reagents. A typical silicone oil increases only about sevenfold in viscosity on cooling from 100°F to −35°F, whereas a hydrocarbon oil with the same viscosity at 100°F increases in viscosity about 1800-fold at −35°F.

Resinous silicones can be made by polymerizing silicones into crosslinked molecules. These resinous materials are used for electric insulation. They have excellent dielectric properties and are stable at operating temperatures at which the usual organic insulating materials decompose rapidly. The use of these materials permits electric machines to be operated with increased loads.

Silicones may be polymerized to molecules containing 2000 or more $(CH_3)_2SiO$ units, and then milled with inorganic fillers (such as zinc oxide or carbon black, used also for ordinary rubber), and vulcanized, by heating to cause cross-links to form between the molecules, bonding them into an insoluble, infusible three-dimensional framework.

Similar silicones with ethyl groups or other organic groups in place of the methyl groups are also used.

The coating of materials with a water-repellent film has been achieved by use of the *methylchlorosilanes*. A piece of cotton cloth exposed for a second or two to the vapor of trimethylchlorosilane, $(CH_3)_3SiCl$, becomes coated with a layer of trimethylsilicyl groups, through reaction with hydroxyl groups of the cellulose:

$$(CH_3)_3SiCl + HOR \longrightarrow (CH_3)_3SiOR + HCl$$

The exposed methyl groups repel water in the way that a hydrocarbon film such as lubricating oil would. Paper, wool, silk, glass, procelain, and other materials can be treated in this way. The treatment has been found especially useful for ceramic insulators.

18-15.　Germanium

The chemistry of germanium, a moderately rare element, is similar to that of silicon. Most of the compounds of germanium correspond to oxidation number $+4$; examples are germanium tetrachloride, $GeCl_4$, a colorless liquid with boiling point 83°C, and germanium dioxide, GeO_2, a colorless crystalline substance melting at 1086°C.

The compounds of germanium have found little use. The element itself, a gray metalloid, is now extensively used in electronic devices.

The electrical properties of a single crystal of germanium (or silicon) can be drastically changed by alloying the element with very small amounts of other elements. As explained in the following paragraphs, these effects are the basis of the operation of the semiconductor junction rectifier, the transistor, and integrated circuits.

The electrical conductivity of pure germanium at very low temperatures is close to zero. The crystal, like that of diamond, contains atoms with ligancy 4 and a pair of electrons for every bond. We may use a two-dimensional representation:

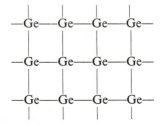

The electrons are restricted to the bond regions, and are not free to move when an electric field is applied. At higher temperatures an electron may be promoted to an excited orbit ($5s$, for example), leaving one electron in a bond where there should be two:

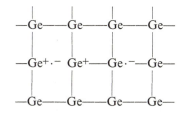

The electron left alone in the bond between two germanium atoms each with a positive charge, $Ge^{+\cdot-}Ge^{+}$, is called a *hole*. The hole can contribute to the electric conductance by the motion to it of an electron from an adjacent bond, as in the following sequence:

$$-Ge^{+\cdot-}\ Ge^{+}-\!\!-Ge-\!\!-Ge-\!\!-Ge-\!\!-$$

$$-Ge-\!\!-Ge^{+\cdot-}\ Ge^{+}-\!\!-Ge-\!\!-Ge-\!\!-$$

$$-Ge-\!\!-Ge-\!\!-Ge^{+\cdot-}\ Ge^{+}-\!\!-Ge-\!\!-$$

$$-Ge-\!\!-Ge-\!\!-Ge-\!\!-Ge^{+\cdot-}\ Ge^{+}\ -$$

The promoted electron also contributes to the conductance; it moves in the opposite direction to the hole.

A crystal of germanium containing some atoms of arsenic (germanium doped with arsenic) has extra electrons in the excited orbitals, because each arsenic atom contributes not only the four electrons needed for the tetrahedral bonds but also a fifth electron. Such a crystal has greater conductivity than pure germanium, and the conductivity is of the *n* type (carried by the negative electrons).

A crystal containing some atoms of aluminum, each contributing only three valence electrons, has for every aluminum atom a hole in the set of bonding electron pairs, and has conductivity of the *p* type (carried by the positive holes, which, of course, move in one direction as electrons jump in the opposite direction into them from adjacent bonding pairs).

A *p-n* junction rectifier is made by placing a *p* crystal and an *n* crystal in contact with one another, as shown in Figure 18-9. Each crystal is attached at the other end to a metal plate carrying the terminals. Both holes and electrons transfer readily to the metal plates and across the junction, and a steady current is carried when a potential is applied in such a direction as to cause both holes and electrons to move toward the junction. When the potential is reversed, however, the holes and electrons move away from the junction (bottom of Figure 18-9). There is no mechanism

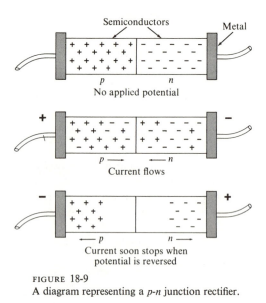

FIGURE 18-9
A diagram representing a *p-n* junction rectifier.

for rapidly producing new holes and promoted electrons at the junction—this process requires the energy to raise an electron from a bond orbital (for germanium the $4s4p^3$ tetrahedral hybrid orbitals) to an excited orbital, $5s$, and its rate is determined by the temperature (the Arrhenius exponential rate factor, Section 16-4). In consequence, the region near the junction becomes depleted of carriers and the current ceases to flow.

An *n-p-n* transistor can be made by sandwiching a *p* crystal between two *n* crystals and attaching terminals in such a way that one applied potential depletes or augments the supply of carriers for another. In this way a current in one circuit can be caused to produce a proportional current in another circuit at a higher power level.

18-16.   Tin

Tin is a silvery white metal, with great malleability, permitting it to be hammered into thin sheets, called tin foil. Ordinary *white tin*, which has metallic properties, slowly changes at temperatures below 18°C to a nonmetallic allotropic modification, *gray tin*, which has the diamond structure. (The physical properties given in Table 18-5 pertain to white tin.) At very low temperatures, around −40°C, the speed of this conversion is sufficiently great that metallic tin objects sometimes fall into a powder of gray tin. This phenomenon has been called the "tin pest."

Tin finds extensive use as a protective layer for mild steel. Tin plating is done by dipping clean sheets of mild steel into molten tin, or by electrolytic deposition. Copper and other metals are sometimes also coated with tin.

The principal alloys of tin are *bronze* (tin and copper), *soft solder* (tin and lead), *pewter* (75% tin and 25% lead), and *britannia metal* (tin with small amounts of antimony and copper).

Bearing metals, used as the bearing surfaces of sliding-contact bearings, are usually alloys of tin, lead, antimony, and copper. They contain small, hard crystals of a compound such as SnSb embedded in a soft matrix of tin or lead. The good bearing properties result from orientation of the hard crystals to present flat faces at the bearing surface.

Tin is reactive enough to displace hydrogen from dilute acids, but it does not tarnish in moist air. It reacts with warm hydrochloric acid to produce stannous chloride, $SnCl_2$, and hydrogen, and with hot concentrated sulfuric acid to produce stannous sulfate, $SnSO_4$, and sulfur dioxide, the equations for these reactions being

$$Sn + 2HCl \longrightarrow SnCl_2 + H_2$$

and

$$Sn + 2H_2SO_4 \longrightarrow SnSO_4 + SO_2 + 2H_2O$$

With cold dilute nitric acid it forms stannous nitrate, and with concentrated nitric acid it is oxidized to a hydrated stannic acid, H_2SnO_3.

Compounds of Tin

Stannous chloride, made by solution of tin in hydrochloric acid, forms colorless crystals, $SnCl_2 \cdot H_2O$, on evaporation of the solution. In neutral solution the substance hydrolyzes, forming a precipitate of stannous hydroxychloride, Sn(OH)Cl. The hydrolysis in solution may be prevented by the presence of an excess of acid. Stannous chloride solution is used as a mordant in dyeing cloth.

The stannous ion is an active reducing agent, which is easily oxidized to stannic chloride, $SnCl_4$, or, in the presence of excess chloride ion, to the complex chlorostannate ion, $SnCl_6^{--}$.

Stannic chloride, $SnCl_4$, is a colorless liquid (boiling point 114°C), which fumes very strongly in moist air, producing hydrochloric acid and stannic acid, $H_2Sn(OH)_6$. Sodium stannate, $Na_2Sn(OH)_6$, contains the octahedral hexahydroxystannate ion (stannate ion). This complex ion is similar in structure to the chlorostannate ion. Sodium stannate is used as a mordant and in preparing fireproof cotton cloth and weighting silk. The cloth is soaked in the sodium stannate solution, dried, and treated with ammonium sulfate solution. This treatment causes hydrated stannic oxide to be deposited in the fibers.

Stannous hydroxide, $Sn(OH)_2$, is formed by adding dilute sodium hydroxide solution to stannous chloride. It is readily soluble in excess alkali, producing the stannite ion, $Sn(OH)_3^-$.

Stannous sulfide, SnS, is obtained as a dark brown precipitate by addition of hydrogen sulfide or sulfide ion to a solution of a stannous salt. Stannic sulfide, SnS_2, is formed in the same way from stannic solution; it is yellow in color. Stannic sulfide is soluble in solutions of ammonium sulfide or sodium sulfide, producing the sulfostannate ion, SnS_4^{----}. Stannous sulfide is not soluble in sulfide solution, but is easily oxidized in the presence of polysulfide solutions to the sulfostannate ion. These properties are used in some schemes of qualitative analysis.

18-17. Lead

Lead is a soft, heavy, dull gray metal with low tensile strength. It is used in making type, for covering electric cables, and in many alloys. The organic lead compound lead tetraethyl, $Pb(C_2H_5)_4$, is added to gasoline to prevent knock in automobile engines.

Lead forms a thin surface layer of oxide in air. This oxide slowly changes to a basic carbonate. Hard water forms a similar coating on lead, which protects the water from contamination with soluble lead compounds. Soft water dissolves appreciable amounts of lead, which is poisonous; for this reason lead pipes should not be used to carry drinking water.

There are several oxides of lead, of which the most important are lead monoxide (*litharge*), PbO, minium or red lead, Pb_3O_4, and lead dioxide, PbO_2.

Litharge is made by heating lead in air. It is a yellow powder or yellowish-red crystalline material, used in making lead glass and for preparing compounds of lead. Red lead, Pb_3O_4, can be made by heating lead in oxygen. It is used in glass making, and for making a red paint for protecting iron and steel structures. Lead dioxide, PbO_2, is a brown substance made by oxidizing a solution of sodium plumbite, $Na_2Pb(OH)_4$, with hypochlorite ion, or by anodic oxidation of lead sulfate. It is soluble in sodium hydroxide and potassium hydroxide, forming the hexahydroxyplumbate ion, $Pb(OH)_6^{--}$. The principal use of lead dioxide is in the lead storage battery.

Lead nitrate, $Pb(NO_3)_2$, is a white crystalline substance made by dissolving lead, lead monoxide, or lead carbonate in nitric acid. Lead carbonate, $PbCO_3$, occurs in nature as the mineral *cerussite*. It appears as a precipitate when a solution containing the hydrogen carbonate ion, HCO_3^-, is added to lead nitrate solution. With a more basic carbonate solution a basic carbonate of lead, $Pb_3(OH)_2(CO_3)_2$, is deposited. This basic salt, called *white lead*, is used as a white pigment in paint. For this use it is manufactured by methods involving the oxidation of lead by air, the formation of a basic acetate by interaction with vinegar or acetic acid, and the decomposition of this salt by carbon dioxide. Lead chromate, $PbCrO_4$, is also used as a pigment, under the name *chrome yellow*.

Lead sulfate, $PbSO_4$, is a white, nearly insoluble substance. Its precipitation is used as a test for either lead ion or sulfate ion in analytical chemistry.

Exercises

18-1. Compare the properties of elements of groups I, II, III, and IV with their electronegativities (Table 6-4). What electronegativity value separates the metals from the metalloids?

18-2. Beryllium hydroxide is essentially insoluble in water, but is soluble both in acids and in alkalis. What do you think the products of its reaction with sodium hydroxide solution are? Discuss these properties of the substance in relation to the position of beryllium in the periodic table and in the electronegativity scale.

18-3. Discuss the electronic structure of potassium fluoroborate, KBF_4. Its solution in water contains the ion BF_4^-.

18-4. What is the electronic structure of the aluminum atom? How does it explain the fact that almost all the compounds of aluminum correspond to oxidation number $+3$?

18-5. One of the crystalline forms of silicon carbide is cubic, with $a = 4.358$ Å; 4C at 0 0 0, 0 $\frac{1}{2}$ $\frac{1}{2}$, $\frac{1}{2}$ 0 $\frac{1}{2}$, $\frac{1}{2}$ $\frac{1}{2}$ 0; 4Si at $\frac{1}{4}$ $\frac{1}{4}$ $\frac{1}{4}$, $\frac{1}{4}$ $\frac{3}{4}$ $\frac{3}{4}$, $\frac{3}{4}$ $\frac{1}{4}$ $\frac{3}{4}$, $\frac{3}{4}$ $\frac{3}{4}$ $\frac{1}{4}$. What nearest neighbors does a carbon atom have? A silicon atom? At what distances? What are the values of the bond angles? Can you suggest an explanation of the great hardness of the substance?

18-6. From the electronegativities of the elements calculate a value for the standard enthalpy of formation of SiC(c). The experimental value is -111 kJ mole^{-1}. (Answer: -96 kJ mole^{-1}.)

18-7. The compound AlP has a tetrahedral structure resembling that of SiC. Would you expect it to be a possible substitute for germanium in a p-n junction rectifier? How could AlP be doped to give a *p*-crystal and to give an *n*-crystal?

18-8. The trichloride of boron has m.p. $-107°C$ and boiling point $12.5°C$, whereas that of its congener lanthanum has m.p. $870°C$ and very high b.p. What is the explanation of these greatly different physical properties?

18-9. Of the two crystal structures that might be expected for BaF$_2$, the MgF$_2$ structure and the CaF$_2$ structure, which would you assign to it on the basis of values of the ionic radii?

18-10. Lithium vapor is an equilibrium mixture of molecules Li and Li$_2$. What is the electronic structure of Li$_2$? Calculate the Li—Li bond energy from the values of enthalpy of formation given in Table 18-2.

18-11. What is the value of the Li—H bond energy in LiH(g)? (See Table 18-2.)

18-12. When liquid NaH is electrolyzed, hydrogen is evolved at the anode. Explain in terms of the electronic structure of the substance. What volume of H$_2$ (standard conditions) would be produced per faraday?

18-13. Do you think that lanthanum metal could be made from lanthanum oxide by reaction with aluminum powder (see Table 18-6)?

18-14. Why is the presence of coesite or stishovite near a crater taken as evidence that the crater was formed by impact of a meteorite? In what way is the principle of Le Chatelier involved in your answer?

18-15. Using the Mulliken relation (Exercise 7-19), evaluate the electronegativities of the five alkali metals. Values of the enthalpy of ionization can be obtained from Table 18-2, and the electron affinity of the alkali-metal atoms (for which there are no experimental values) can be taken as zero. (Answer: 0.99, 0.95, 0.80, 0.77, 0.72.)

18-16. From values given in Table 18-4, calculate the enthalpy of removal of the first electron and that of the second electron from each of the alkaline-earth metals. What are the values of the ratio of the second to the first? Can you suggest a simple explanation of the approximation to the value 2?

18-17. Using the method of Exercise 18-15, find the divisor of the first ionization enthalpy (preceding Exercise) for the five alkaline-earth elements that gives for the sum of their electronegativities the same value as in Table 6-4, and calculate the electronegativities. (Answer: 591 kJ mole^{-1}; 1.53, 1.26, 1.01, 0.94, 0.86.)

18-18. The amount of covalent character of their bonds is such that in the compounds of the alkaline-earth metals the metal atoms have electric charge approximately $+1$. The Mulliken method of evaluating the electronegativity might accordingly be applied by assuming proportionality to the sum of the first and second ionization energies. Using the results of Exercise 18-16, calculate the appropriate divisor and the corresponding electronegativity values. Can you explain the close agreement with the values obtained in Exercise 18-17? (Answer: 1736 kJ mole^{-1}, 1.54, 1.27, 1.01, 0.94, 0.85.)

18-19. On the assumptions that Equation 6-1 gives the enthalpy of formation of a compound in its standard state from the elements in their standard states and that the electronegativity of chlorine is 3.00, use the values of enthalpy of formation of dichlorides in Table 18-4 to evaluate the electronegativities of the alkaline-earth metals. (Answer: 1.37, 1.18, 0.97, 0.93, 0.89.)

18-20. Apply the method of Exercise 18-18 to obtain electronegativity values for B, Al, Sc, Y, and La (Table 18-6). Note that a different divisor (1577 kJ mole^{-1}) is to be used; the need for a different divisor results from the difference in electronic interactions for atoms with different numbers of valence electrons and different Russell-Saunders states. (Answer: 2.06, 1.53, 1.20, 1.17, 1.15.)

18-21. Apply the method of Exercise 18-19 to the trichlorides of Table 18-6 to obtain electronegativity values. (Answer: 1.84, 1.45, 1.21, 1.16, 1.05.)

18-22. Assign electronic structures to Si_2Cl_6 and Si_2Cl_6O. Would you expect both of these molecules to have an electric dipole moment differing from zero?

19

Inorganic Complexes and the Chemistry of the Transition Metals

19-1. The Nature of Inorganic Complexes

An inorganic molecule that contains several atoms, including one or more metal atoms, is called an *inorganic complex*. An example is nickel tetracarbonyl, $Ni(CO)_4$. An inorganic complex with an electric charge is called a *complex ion*. Familiar examples of complex ions are the ferrocyanide ion, $Fe(CN)_6^{----}$, the ferricyanide ion, $Fe(CN)_6^{---}$, the hydrated aluminum ion, $Al(H_2O)_6^{+++}$, and the deep blue cupric ammonia complex ion, $Cu(NH_3)_4^{++}$, which is formed by adding ammonium hydroxide to a solution of cupric salt. Complex ions are important in the methods of separation used in qualitative and quantitative chemical analysis and in various industrial processes.

The formation of complexes constitutes an especially important part of the chemistry of the transition metals. The special feature of the electronic structure of the transition metals that leads to their formation of stable complexes is the availability of *d* orbitals, as well as *s* and *p* orbitals, for bond formation, as discussed in the following section.

19-2. Tetrahedral, Octahedral, and Square Bond Orbitals

The transition metals have in their outer shells electrons occupying d, s, and p orbitals. Thus for the elements from potassium to krypton the outer-shell electrons may occupy the five $3d$ orbitals, the $4s$ orbital, and the three $4p$ orbitals, and in the succeeding sequences of transition metals the available orbitals are similar, but with increase of the total quantum number by 1 or 2.

The different transition metals have different numbers of d orbitals available for hybridization with the s orbitals and the three p orbitals of the valence shell, to form bond orbitals, and the nature of the bonds formed by the metal atom depends upon the number of d orbitals available. With no d orbitals available, tetrahedral sp^3 bond orbitals of the type described in Chapter 6 may be formed. An example is provided by the zinc ion, Zn^{++}. The zinc ion has ten electrons outside of the argon shell. These ten electrons can occupy the five $3d$ orbitals in pairs, leaving the $4s$ orbital and the three $4p$ orbitals available for hybridization to form four tetrahedral bond orbitals. It is in fact found by experiment that bipositive zinc has ligancy four, forming complexes in which four atoms or groups of atoms are tetrahedrally bonded to it. Among the complexes of this sort that are discussed in following sections of the chapter and later chapters are $Zn(NH_3)_4^{++}$, $Zn(OH)_4^{--}$, and $Zn(CN)_4^{--}$.

Octahedral Orbitals

The doubly charged iron cation, Fe^{++}, has six electrons outside of the argon shell. These six electrons can be placed in three of the five $3d$ orbitals, in pairs. The ion would then have two $3d$ orbitals available to hybridize with the $4s$ orbital and the three $4p$ orbitals, to form six bond orbitals. These d^2sp^3 hybrid bond orbitals have been found to constitute a set of six orbitals with their maxima directed in the six octahedral directions (along the $+x$, $-x$, $+y$, $-y$, $+z$, and $-z$ directions in a set of Cartesian coordinates); that is, toward the corners of a regular octahedron. Bipositive iron might accordingly be expected to use these orbitals in forming an octahedral complex, and in fact the complex ion $Fe(CN)_6^{----}$ has been shown by x-ray diffraction of ferrocyanide crystals to have the octahedral structure. Other examples of octahedral complexes are described in later sections of this chapter.

By Hund's first rule (Section 5-3) the electronic structure that would be expected for an isolated Fe^{++} ion is the one in which four of the $3d$ orbitals are occupied by single electrons, with parallel spin, and one is occupied by a pair of electrons. The ion with this structure would have a magnetic moment corresponding to four unpaired electron spins in parallel orientation. It is found by experiment that the hydrated ferrous ion,

$Fe(H_2O)_6^{++}$, has a magnetic moment with this value, whereas the ferrocyanide ion has no magnetic moment. The conclusion can be drawn that the bonds in these two complex ions are different in character: in the hydrated ferrous ion the bonds, which have a large amount of ionic character, are formed with use of the $4s$ orbital and the three $4p$ orbitals, whereas in the ferrocyanide ion the orbitals are d^2sp^3 covalent bonds. Investigation of the magnetic properties of a complex can in many cases permit a decision to be made as to the nature of the bond orbitals used by the metal atom. It has been found by use of this magnetic criterion that complexes of metals with strongly electronegative atoms or groups are usually essentially ionic in character (without the $3d$ orbitals used in bonding), whereas those with less electronegative atoms or groups are covalent in character (with use of $3d$ orbitals in the hybrid bond orbitals).

Square Bond Orbitals

The bipositive nickel ion, Ni^{++}, has eight electrons outside of the argon shell. These electrons may be introduced in the five $3d$ orbitals in two ways: either by placing three electron pairs in three of the $3d$ orbitals and an odd electron in each of the other two, with their spins parallel, or by placing four electron pairs in four of the $3d$ orbitals, leaving one $3d$ orbital available for bond formation. Complexes in which bipositive nickel has the first electronic structure would have a magnetic moment, leading to paramagnetism, whereas those in which bipositive nickel has the second structure would have zero magnetic moment.

It has been found by study of the magnetic properties of different compounds of bipositive nickel that some of them, such as the hydrated nickel ion, are paramagnetic, and accordingly form bonds in which the $3d$ orbitals do not participate. Others, such as the nickel tetracyanide ion, $Ni(CN)_4^{--}$, have no magnetic moment, and the bonds may be considered to be formed by bond orbitals involving one $3d$ orbital.

The hybrid bond orbitals that can be formed by one $3d$ orbital, one $4s$ orbital, and the set of $3p$ orbitals are four bond orbitals that lie in a plane and are directed toward the corners of a square. (The third p orbital is not involved in this set of bond orbitals.) X-ray examination of crystals has shown that bipositive nickel, palladium, and platinum form complexes of this square planar type.

The Discovery of Octahedral and Square Complexes

The concept of the coordination of ions or groups of atoms in a definite geometric arrangement about a central metal atom was developed shortly after the beginning of the present century by the Swiss chemist A. Werner (1866–1919) to account for the existence and properties of compounds such as K_2SnCl_6, $Co(NH_3)_6I_3$, and so on. Before Werner's work was

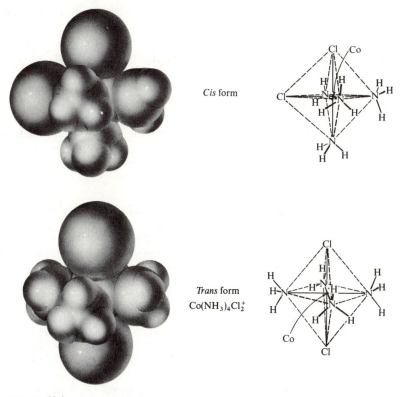

Cis form

Trans form
$Co(NH_3)_4Cl_2^+$

FIGURE 19-1

The *cis* and *trans* isomers of the cobaltic tetramminedichloride ions, $Co(NH_3)_4Cl_2^+$. In the *cis* form the two chlorine atoms occupy adjacent corners of the coordination octahedron about the cobalt atom, and in the *trans* form the two chlorine atoms occupy opposite corners.

carried out these compounds had been assigned formulas such as $SnCl_4 \cdot 2KCl$ and $CoI_3 \cdot 6NH_3$, and had been classed as "molecular compounds," of unknown nature. Werner showed that the properties of many complexes formed by various transition metals could be explained by the postulate that the metal atoms have ligancy 6, with the six attached groups arranged about the central atom at the corners of a circumscribed regular octahedron.

One important property that Werner explained in this way is the existence of *isomers of inorganic complexes*. For example, there are two complexes with the formula $Co(NH_3)_4Cl_2^+$, one of which is violet in color and one green. Werner identified these two complexes with the *cis* and *trans* structures shown in Figure 19-1. In the *cis* form the chloride ions are in adjacent positions, and in the *trans* form in opposite positions. Werner identified the violet complex with the *cis* configuration through the observation that it could be made easily from the carbonate-ammonia complex $Co(NH_3)_4CO_3^+$, for which only the *cis* form is possible. Werner also discovered square coordination and identified the square *cis* and *trans* isomers.

In recent years a great amount of information about the structure of inorganic complexes has been gathered by the methods of x-ray diffraction, measurement of magnetic susceptibility, magnetic resonance spectroscopy, Mössbauer spectroscopy, and other techniques. This information about the structure of complexes has been correlated with their chemical properties in such a way as to bring reasonable order into this field of chemistry.

Configurations about Atoms with Unshared Electron Pairs

It has been found by experiment that in most molecules an unshared pair of electrons seems to occupy one of the corners of a coordination polyhedron, with approximately the same configuration as for a molecule with a bond in place of the unshared pair.

For example, molecules such as NH_3 and PCl_3 have trigonal pyramidal configurations that may be described as involving bonds directed approximately toward three corners of a tetrahedron, with the fourth corner occupied by the unshared electron pair. A similar description can be given to the water molecule, in which two bonds are directed approximately toward two corners of a tetrahedron and the two unshared electron pairs of the oxygen atom can be described as occupying the other two corners.

The bond angles are, however, a few degrees different from the tetrahedral angle, because of the larger amount of p character for the bond orbitals than for tetrahedral bond orbitals when unshared pairs are present, which occupy the s orbital.

For a molecule in which the central atom forms five bonds and has one unshared pair an octahedral arrangement is to be expected, with the bonds directed toward the five corners of a square pyramid and the unshared pair occupying the sixth octahedral corner. The molecule BrF_5 has been shown to have this configuration. The bromine atom lies about 0.15 Å below the base of the pyramid, so that the F—Br—F bond angles (from the apical fluorine atom to the basal atoms) are about 86°. Accordingly the unshared electron pair occupies a somewhat larger volume about the bromine atom than the shared pairs occupy. Similarly, in ammonia, water, and related molecules each unshared pair can be described as occupying a larger solid angle than a shared pair.

The molecule PCl_5, with five shared pairs in the outer shell of the phosphorus atom, has the configuration of a trigonal bipyramid. The molecule

TeCl$_4$ is similar to PCl$_5$, except that one unshared electron pair replaces one of the bonds. The unshared pair lies in the basal plane, rather than in one of the apical positions. Bromine trifluoride, BrF$_3$, can also be described as being based upon the trigonal bipyramid. The three fluorine atoms lie in the same plane as the bromine atom, the bond angles having the value 86°. The molecule can be described as forming bonds toward the two apical positions of the trigonal bypyramid and toward one of the three equatorial positions, with unshared electron pairs occupying the two other equatorial positions.

19-3. Ammonia Complexes

A solution of a cupric salt is blue in color. This blue color is due to the absorption of yellow and red light, and consequent preferential transmission of blue light. The molecular species that absorbs the light is the *hydrated copper ion*, probably Cu(H$_2$O)$_4^{++}$. Crystalline hydrated cupric salts such as CuSO$_4 \cdot$5H$_2$O are blue, like the aqueous solution, whereas anhydrous CuSO$_4$ is white.*

When a few drops of sodium hydroxide solution are added to a cupric solution a blue precipitate is formed. This is cupric hydroxide, Cu(OH)$_2$, which precipitates when the ion concentration product [Cu^{++}][OH$^-$]^2 reaches the value corresponding to a saturated solution of the hydroxide. (Here the symbol Cu^{++} is used, as is conventional, for the ion species Cu(H$_2$O)$_4^{++}$.) Addition of more sodium hydroxide solution leads to no further change, other than the formation of more precipitate.

If ammonium hydroxide is added in place of sodium hydroxide the same precipitate of Cu(OH)$_2$ is formed. On addition of more ammonium hydroxide, however, the precipitate dissolves, giving a clear solution with a deeper and more intense blue color than the original cupric solution.†

The solution of the precipitate cannot be attributed to increase in hydroxide-ion concentration, because sodium hydroxide does not cause it, nor to ammonium ion, because ammonium salts do not cause it. There remains undissociated NH$_4$OH or NH$_3$, which might combine with the cupric ion. It has in fact been found that the new deep blue ion species formed by addition of an excess of ammonium hydroxide is the *cupric ammonia complex* Cu(NH$_3$)$_4^{++}$, similar to the hydrated cupric ion except that the four water molecules have been replaced by ammonia molecules.

---

*The crystal structure of CuSO$_4 \cdot$5H$_2$O shows that in the crystal four water molecules are attached closely to the cupric ion, and the fifth is more distant.

†In describing color the adjective deep refers not to intensity but to shade; deep blue tends toward indigo.

This complex is sometimes called the *cupric tetrammine complex*, the word *ammine* meaning an attached ammonia molecule.

Salts of this complex ion can be crystallized from ammonia solution. The best known one is *cupric tetrammine sulfate monohydrate*, $Cu(NH_3)_4SO_4 \cdot H_2O$, which has the same deep blue color as the solution.

The reason that the precipitate of cupric hydroxide dissolves in an excess of ammonium hydroxide can be given in the following way. A precipitate of cupric hydroxide is formed because the concentration of cupric ion and the concentration of hydroxide ion are greater than the values corresponding to the solubility product of cupric hydroxide. If there were some way for copper to be present in the solution without exceeding the solubility product of cupric hydroxide, then precipitation would not occur. In the presence of ammonia, copper exists in the solution not as the cupric ion (that is, the hydrated cupric ion), but principally as the cupric ammonia complex, $Cu(NH_3)_4^{++}$. This complex is far more stable than the hydrated cupric ion. The reaction of formation of the cupric ammonia complex is

$$Cu^{++} + 4NH_3 \rightleftharpoons Cu(NH_3)_4^{++}$$

We see from the equation for the reaction that the addition of ammonia to the solution causes the equilibrium to shift to the right, more of the cupric ion being converted into cupric ammonia complex as more and more ammonia is added to the solution. When sufficient ammonia is present a large amount of copper may exist in the solution as cupric ammonia complex, at the same time that the cupric ion concentration is less than that required to cause precipitation of cupric hydroxide. When ammonia is added to a solution in contact with the precipitate of cupric hydroxide, the cupric ion in the solution is converted to cupric ammonia complex, causing the solution to be unsaturated with respect to cupric hydroxide. The cupric hydroxide precipitate then dissolves, and if enough ammonia is present the process continues until the precipitate has dissolved completely.

The process of *solution of a slightly soluble substance through formation of a complex by one of its ions* is the basis of some of the most important practical applications of complex formation. Several examples are mentioned later in this chapter.

The nickel ion forms two rather stable ammonia complexes. When a small amount of ammonium hydroxide solution is added to a solution of a nickel salt (green in color) a pale green precipitate of nickel hydroxide, $Ni(OH)_2$, is formed. On addition of more ammonium hydroxide solution this dissolves to give a blue solution, which with still more ammonium hydroxide changes color to light blue-violet.

The light blue-violet complex is shown to be the *nickel hexammine ion*,

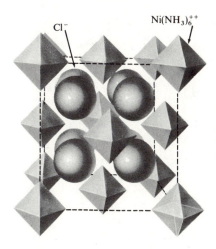

Cl^- $Ni(NH_3)_6^{++}$

FIGURE 19-2
The structure of crystalline nickel
hexammine chloride, $Ni(NH_3)_6Cl_2$.
The crystal contains octahedral nickel
hexammine ions and chloride ions.

$Ni(NH_3)_6^{++}$, by the facts that the same color is shown by crystalline $Ni(NH_3)_6Cl_2$ and other crystals containing six ammonia molecules per nickel ion, and that x-ray studies have revealed the presence in these crystals of octahedral complexes in which the six ammonia molecules are situated about the nickel ion at the corners of a regular octahedron. The structure of crystalline $Ni(NH_3)_6Cl_2$ is shown in Figure 19-2.

The blue complex is probably the *nickel tetramminedihydrate ion,* $Ni(NH_3)_4(H_2O)_2^{++}$. Careful studies of the change in color with increasing ammonia concentration indicate that the ammonia molecules are added one by one and that all the complexes $Ni(H_2O)_6^{++}$, $Ni(H_2O)_5NH_3^{++}$, $Ni(H_2O)_4(NH_3)_2^{++}$, $Ni(H_2O)_3(NH_3)_3^{++}$, $Ni(H_2O)_2(NH_3)_4^{++}$, $Ni(H_2O)(NH_3)_5^{++}$, and $Ni(NH_3)_6^{++}$ exist.

Several metal ions form ammonia complexes with sufficient stability to put the hydroxides into solution. Others, such as aluminum and iron, do not. The formulas of the stable complexes are given below. There is no great apparent order about the stability or composition of the complexes, except that often the unipositive ions add two, the bipositive ions four, and the terpositive ions six ammonia molecules.

The *silver ammonia complex,* $Ag(NH_3)_2^+$, is sufficiently stable for ammonium hydroxide to dissolve precipitated silver chloride by reducing the concentration of silver ion, $[Ag^+]$, below the value required for precipitation by the solubility product of AgCl. A satisfactory test for silver ion is the formation with chloride ion of a precipitate that is soluble in ammonium hydroxide. Ammonia complexes in general are decomposed by acid, because of formation of ammonium ion; for example, as in the reaction

$$Ag(NH_3)_2^+ + Cl^- + 2H^+ \longrightarrow AgCl + 2NH_4^+$$

Stable Ammonia Complexes

$Cu(NH_3)_2{}^+$	$Cu(NH_3)_4{}^{++}$	$Co(NH_3)_6{}^{+++}$
$Ag(NH_3)_2{}^+$	$Zn(NH_3)_4{}^{++}$	$Cr(NH_3)_6{}^{+++}$
$Au(NH_3)_2{}^+$	$Cd(NH_3)_4{}^{++}$	
	$Hg(NH_3)_2{}^{++}$	
	$Hg(NH_3)_4{}^{++}$	
	$Ni(NH_3)_4{}^{++}$	
	$Ni(NH_3)_6{}^{++}$	
	$Co(NH_3)_6{}^{++}$	

Notes: 1. Cobaltous ammonia ion is easily oxidized by air to cobaltic ammonia ion.
2. Chromic ammonia ion forms only slowly, and is decomposed by boiling, to give chromium hydroxide precipitate.

19-4. Cyanide Complexes

Another important class of complex ions includes those formed by the metal ions with cyanide ion. The common cyanide complexes are given in the following table.

Cyanide Complexes

$Cu(CN)_2{}^-$	$Zn(CN)_4{}^{--}$	$Fe(CN)_6{}^{---}$	$Au(CN)_4{}^-$
$Ag(CN)_2{}^-$	$Cd(CN)_4{}^{--}$	$Co(CN)_6{}^{---}$	
$Au(CN)_2{}^-$	$Hg(CN)_4{}^{--}$	$Mn(CN)_6{}^{----}$	
		$Fe(CN)_6{}^{----}$	
		$Co(CN)_6{}^{----}$	

Some of these complexes are very stable—the stability of the *argento-cyanide ion*, $Ag(CN)_2{}^-$, for example, is so great that addition of iodide ion does not cause silver iodide to precipitate, even though the solubility product of silver iodide is very small. The *ferrocyanide ion*, $Fe(CN)_6{}^{----}$, *ferricyanide ion*, $Fe(CN)_6{}^{---}$, and *cobalticyanide ion*, $Co(CN)_6{}^{---}$, are so stable that they are not appreciably decomposed by strong acid. The others are decomposed by strong acid, with the formation of hydrocyanic acid, HCN.

An illustration of the stability of the ferrocyanide complex is provided by the old method of making potassium ferrocyanide, $K_4Fe(CN)_6$, by strongly heating nitrogenous organic material (such as dried blood and hides) with potassium carbonate and iron filings.

The *cobaltocyanide ion*, $Co(CN)_6{}^{----}$, like the cobaltous ammonia complex, is a very strong reducing agent; it is able to decompose water, liberating hydrogen, as it changes into cobalticyanide ion.

Cyanide solutions are used in the *electroplating* of gold, silver, zinc, cadmium, and other metals. In these solutions the concentrations of uncomplexed metal ions are very small, and this favors the production of a uniform fine-grained deposit. Other complex-forming anions (tartrate, citrate, chloride, hydroxide) are also used in plating solutions.

19-5. Complex Halides and Other Complex Ions

Nearly all anions can enter into complex formation with metal ions. Thus stannic chloride, $SnCl_4$, forms with chloride ion the stable *hexachlorostannate ion*, $SnCl_6^{--}$, which with cations crystallizes in an extensive series of salts. Various complexes of this kind are discussed below.

Many chloride complexes are known; representative are the following:

$$CuCl_2(H_2O)_2, CuCl_3(H_2O)^-, CuCl_4^{--}$$
$$AgCl_2^-, AuCl_2^-$$
$$HgCl_4^{--}$$
$$CdCl_4^{--}, CdCl_6^{----}$$
$$SnCl_6^{--}$$
$$PtCl_6^{--}$$
$$AuCl_4^-$$

The cupric chloride complexes are recognizable in strong hydrochloric acid solutions by their green color. The crystal $CuCl_2 \cdot 2H_2O$ is bright green, and x-ray studies have shown that it contains the complex molecule $CuCl_2(H_2O)_2$. The ion $CuCl_3(H_2O)^-$ is usually written $CuCl_3^-$; it is highly probable that the indicated water molecule is present, and, indeed, the ion $Cu(H_2O)_3Cl^+$ very probably also exists in solution.

The stability of the *tetrachloroaurate ion*, $AuCl_4^-$, is responsible for the ability of *aqua regia*, a mixture of nitric and hydrochloric acids, to dissolve gold, which is not significantly soluble in the acids separately. Nitric acid serves as the oxidizing agent that oxidizes gold to the terpositive state, and the chloride ions provided by the hydrochloric acid further the reaction by combining with the auric ion to form the stable complex:

$$Au + 6H^+ + 4Cl^- + 3NO_3^- \longrightarrow AuCl_4^- + 3NO_2 + 3H_2O$$

The solution of platinum in aqua regia likewise results in the stable *hexachloroplatinate ion*, $PtCl_6^{--}$.

The bromide and iodide complexes closely resemble the chloride complexes, and usually have similar formulas.

Fluoride ion is more effective than the other halide ions in forming complexes. Important examples are the *tetrafluoroborate ion*, BF_4^-, the *hexafluorosilicate ion*, SiF_6^{--}, the *hexafluoroaluminate ion*, AlF_6^{---}, and the *ferric hexafluoride ion*, FeF_6^{---}.

A useful complex is that formed by thiosulfate ion, $S_2O_3^{--}$, and silver ion. Its formula is $Ag(S_2O_3)_2^{---}$, and its structure is

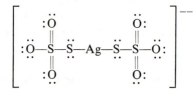

This complex ion is sufficiently stable to cause silver chloride and bromide to be soluble in thiosulfate solutions, and this is the reason that sodium thiosulfate solution ("hypo") is used after development of a photographic film or paper to dissolve away the unreduced silver halide, which if allowed to remain in the emulsion would in the course of time darken through long exposure to light.

Of the nitrite complexes that with cobaltic ion, $Co(NO_2)_6^{---}$, called the *cobaltinitrite ion* or *hexanitrocobaltic ion*, is the most familiar. *Potassium cobaltinitrite*, $K_3Co(NO_2)_6$, is one of the least soluble potassium salts, and its precipitation on addition of sodium cobaltinitrite reagent is commonly used as a test for potassium ion.

Ferric ion and thiocyanate ion combine to give a product with an intense red color; this reaction is used as a test for ferric ion. The red color seems to be due to various complexes, ranging from $Fe(H_2O)_5NCS^{++}$ to $Fe(NCS)_6^{---}$. The azide ion, NNN^-, gives a similar color with ferric ion.

The Chromic and Cobaltic Complexes

Terpositive chromium and cobalt combine with cyanide ion, nitrite ion, chloride ion, sulfate ion, oxalate ion, water, ammonia, and many other ions and molecules to form a very great number of complexes, with a wide range of colors that are nearly the same for corresponding chromic and cobaltic complexes. Most of these complexes are stable, and are formed and decomposed slowly. Representative are the members of the series

$Cr(NH_3)_6^{+++}$	$Cr(NH_3)_5Cl^{++}$	$Cr(NH_3)_4Cl_2^+$
Yellow	Purple	Green

$Cr(NH_3)_3Cl_3$	$Cr(NH_3)_2Cl_4^-$
Violet	Orange-red

and

$Co(NH_3)_6^{+++}$	$Co(NH_3)_5H_2O^{+++}$	$\cdots$	$Co(H_2O)_6^{+++}$
Yellow	Rose-red		Purple

A group such as oxalate ion, $C_2O_4^{--}$, or carbonate ion, CO_3^{--}, may occupy two of the six coordination places in an octahedral complex; examples are $Co(NH_3)_4CO_3^+$ and $Cr(C_2O_4)_3^{---}$.

The often puzzling color changes shown by chromic solutions are caused by reactions of these complexes. Solutions containing chromic ion, $Cr(H_2O)_6^{+++}$, are purple in color; on heating they become green, because of the formation of complexes such as $Cr(H_2O)_4Cl_2^+$ and $Cr(H_2O)_5SO_4^+$. At room temperature these green complexes slowly decompose, again forming the purple solution.

19-6. Hydroxide Complexes

If sodium hydroxide is added to a solution containing zinc ion a precipitate of zinc hydroxide is formed:

$$Zn^{++} + 2OH^- \rightleftharpoons Zn(OH)_2$$

This hydroxide precipitate is of course soluble in acid; it is also soluble in alkali. On addition of more sodium hydroxide the precipitate goes back into solution, this process occurring at hydroxide-ion concentrations about 0.1 M to 1 M.

To explain this phenomenon we postulate the formation of a complex ion. The complex ion that is formed is the zincate ion, $Zn(OH)_4^{--}$, by the reaction

$$Zn(OH)_2 + 2OH^- \rightleftharpoons Zn(OH)_4^{--}$$

The ion is closely similar to other complexes of zinc, such as $Zn(H_2O)_4^{++}$, $Zn(NH_3)_4^{++}$, and $Zn(CN)_4^{--}$, with hydroxide ions in place of water or ammonia molecules or cyanide ions. The ion $Zn(H_2O)(OH)_3^-$ is also formed to some extent.

The molecular species that exist in zinc solutions of different pH values are the following:

Acidic solution $\begin{cases} Zn(H_2O)_4^{++} \\ Zn(H_2O)_3(OH)^+ \end{cases}$

Neutral solution $\quad Zn(H_2O)_2(OH)_2 \rightleftharpoons Zn(OH)_2(c)$

Basic solution $\begin{cases} Zn(H_2O)(OH)_3^- \\ Zn(OH)_4^{--} \end{cases}$

The conversion of each complex into the following one occurs by removal of a proton from one of the four water molecules of the tetrahydrated zinc ion. Each complex except $Zn(H_2O)_4^{++}$ and $Zn(OH)_4^{--}$ is amphiprotic.

The principal common amphiprotic hydroxides and their anions are the following:

$Zn(OH)_2$	$Zn(OH)_4^{--}$,	zincate ion
$Al(OH)_3$	$Al(OH_2)_2(OH)_4^-$,	aluminate ion
$Cr(OH)_3$	$Cr(OH_2)_2(OH)_4^-$,	chromite ion
$Pb(OH)_2$	$Pb(OH)_3^-$,	plumbite ion
$Sn(OH)_2$	$Sn(OH)_3^-$,	stannite ion

In addition, the following hydroxides evidence acidic properties by combining with hydroxide ion to form complex anions:

$Sn(OH)_4$	$Sn(OH)_6^{--}$,	stannate ion
$As(OH)_3$	$As(OH)_4^-$,	arsenite ion
$Sb(OH)_3$	$Sb(OH)_4^-$,	antimonite ion
$Sb(OH)_5$	$Sb(OH)_6^-$,	antimonate ion

The hydroxides listed above form hydroxide complex anions to a sufficient extent to make them soluble in moderately strong alkali. Other common hydroxides have weaker acidic properties: $Cu(OH)_2$ and $Co(OH)_2$ are only slightly soluble in very strong alkali, and $Cd(OH)_2$, $Fe(OH)_3$, $Mn(OH)_2$, and $Ni(OH)_2$ are effectively insoluble. The common analytical method of separation of Al^{+++}, Cr^{+++}, and Zn^{++} from Fe^{+++}, Mn^{++}, Co^{++}, and Ni^{++} with use of sodium hydroxide is based on these facts.

19-7. Sulfide Complexes

Sulfur, which is directly below oxygen in the periodic table of the elements, has many properties similar to those of oxygen. One of these is the property of combining with another atom to form complexes; there exist *sulfo acids* (thio acids) of many elements similar to the oxygen acids. An example is *sulfophosphoric acid*, H_3PS_4, which corresponds exactly in formula to phosphoric acid, H_3PO_4. This sulfo acid is not of much importance; it is unstable, and hydrolyzes in water to phosphoric acid and hydrogen sulfide:

$$H_3PS_4 + 4H_2O \longrightarrow H_3PO_4 + 4H_2S$$

But other sulfo acids, such as *sulfarsenic acid*, H_3AsS_4, are stable, and are of use in analytical chemistry and in chemical industry.

All of the following arsenic acids are known:

$$H_3AsO_4 \qquad H_3AsO_3S \qquad H_3AsO_2S_2 \qquad H_3AsOS_3 \qquad H_3AsS_4$$

The structure of the five complex anions AsO_4^{---}, AsO_3S^{---}, $AsO_2S_2^{---}$, $AsOS_3^{---}$, and AsS_4^{---} is the same: an arsenic atom surrounded tetrahedrally by four other atoms, oxygen or sulfur.

Some metal sulfides are soluble in solutions of sodium sulfide or ammonium sulfide because of formation of a complex sulfo anion. The important members of this class are HgS, As_2S_3, Sb_2S_3, As_2S_5, Sb_2S_5, and SnS_2, which react with sulfide ion in the following ways:

$$HgS + S^{--} \rightleftarrows HgS_2^{--}$$
$$As_2S_3 + 3S^{--} \rightleftarrows 2AsS_3^{---}$$
$$Sb_2S_3 + 3S^{--} \rightleftarrows 2SbS_3^{---}$$
$$As_2S_5 + 3S^{--} \rightleftarrows 2AsS_4^{---}$$
$$Sb_2S_5 + 3S^{--} \rightleftarrows 2SbS_4^{---}$$
$$SnS_2 + S^{--} \rightleftarrows SnS_3^{--}$$

Mercuric sulfide is soluble in a solution of sodium sulfide and sodium hydroxide (to repress hydrolysis of the sulfide, which would decrease the sulfide-ion concentration), but not in a solution of ammonium sulfide and ammonium hydroxide, in which the sulfide-ion concentration is smaller. The other sulfides listed are soluble in both solutions. CuS, Ag_2S, Bi_2S_3, CdS, PbS, ZnS, CoS, NiS, FeS, MnS, and SnS are not soluble in sulfide solutions, but most of these form complex sulfides by fusion with Na_2S or K_2S. Although SnS is not soluble in Na_2S or $(NH_4)_2S$ solutions, it dissolves in solutions containing both sulfide and disulfide, Na_2S_2 or $(NH_4)_2S_2$, or sulfide and peroxide. The disulfide ion, S_2^{--}, or peroxide oxidizes the tin to the stannic level, and the sulfostannate ion is then formed:

$$SnS + S_2^{--} \rightleftarrows SnS_3^{--}$$

Many schemes of qualitative analysis involve separation of the copper-group sulfides (PbS, Bi_2S_3, CuS, CdS) from the tin-group sulfides (HgS, As_2S_3, As_2S_5, Sb_2S_3, Sb_2S_5, SnS, SnS_2) by treatment with Na_2S-Na_2S_2 solution, which dissolves only the tin-group sulfides.

19-8.   The Quantitative Treatment of Complex Formation

The quantitative theory of chemical equilibrium, as discussed in earlier chapters, can be applied in a straightforward manner to problems involving the formation of complexes. Some of the ways in which this can be done are exemplified in the following paragraphs.

Example 19-1. Ammonium hydroxide is added to a cupric solution until a precipitate is formed, and the addition is continued until part of the precipitate has dissolved to give a deep blue solution. What would be the effect of dissolving some ammonium chloride in the solution?

TABLE 19-1
Ammonia Concentrations Producing 50% Conversion of Metal Ions to Complexes

Metal Ion	Complex Ion	Ammonia Concentration
Cu^+	$Cu(NH_3)_2^+$	5×10^{-6}
Ag^+	$Ag(NH_3)_2^+$	2×10^{-4}
Zn^{++}	$Zn(NH_3)_4^{++}$	5×10^{-3}
Cd^{++}	$Cd(NH_3)_4^{++}$	5×10^{-2}
	$Cd(NH_3)_6^{++}$	10
Hg^{++}	$Hg(NH_3)_2^{++}$	2×10^{-9}
	$Hg(NH_3)_4^{++}$	2×10^{-1}
Cu^{++}	$Cu(NH_3)_4^{++}$	5×10^{-4}
Ni^{++}	$Ni(NH_3)_4^{++}$	5×10^{-2}
	$Ni(NH_3)_6^{++}$	5×10^{-1}
Co^{++}	$Co(NH_3)_6^{++}$	1×10^{-1}
Co^{+++}	$Co(NH_3)_6^{+++}$	1×10^{-6}

TABLE 19-2
Ion Concentrations Producing 50% Conversion of Metal Ions to Complexes

Metal Ion	Complex Ion	Ion Concentration
Cu^+	$Cu(CN)_2^-$	1×10^{-8}
	$CuCl_2^-$	4×10^{-3}
Ag^+	$Ag(CN)_2^-$	3×10^{-11}
	$AgCl_2^-$	3×10^{-3}
	$Ag(NO_2)_2^-$	4×10^{-2}
	$Ag(S_2O_3)_2^{---}$	3×10^{-7}
Zn^{++}	$Zn(CN)_4^{--}$	1×10^{-4}
Cd^{++}	$Cd(CN)_4^{--}$	6×10^{-5}
	CdI_4^{--}	3×10^{-2}
Hg^{++}	$Hg(CN)_4^{--}$	5×10^{-11}
	$HgCl_4^{--}$	9×10^{-5}
	$HgBr_4^{--}$	4×10^{-6}
	HgI_4^{--}	1×10^{-8}
	$Hg(SCN)_4^{--}$	3×10^{-6}

Solution. The weak base NH_4OH is partially ionized and is in equilibrium with dissolved ammonia:

$$NH_3 + H_2O \rightleftarrows NH_4OH \rightleftarrows NH_4^+ + OH^-$$

Addition of NH_4Cl would increase $[NH_4]^+$, which would shift the equilibrium to the left, producing more NH_3 and decreasing the hydroxide-ion concentration. The precipitate $Cu(OH)_2$ is in equilibrium with the solution according to the reaction

$$Cu(OH)_2(c) + 4NH_3 \rightleftarrows Cu(NH_3)_4^{++} + 2OH^-$$

Both the increase of $[NH_3]$ and the decrease of $[OH^-]$ caused by addition of NH_4Cl to the solution would shift this reaction to the right; hence more of the precipitate would dissolve.

In Tables 19-1 and 19-2 there are given values of equilibrium constants or equivalent constants for the reactions of formation of some complexes. The values of equilibrium constants must be used with some caution in making calculations. Thus for the reaction

$$Cu^{++} + 4NH_3 \rightleftarrows Cu(NH_3)_4^{++}$$

we would write

$$K = \frac{[Cu(NH_3)_4^{++}]}{[Cu^{++}][NH_3]^4}$$

as the equilibrium constant, and expect the concentration ratio

$$[Cu(NH_3)_4^{++}]/[Cu^{++}]$$

to vary with the fourth power of the ammonia concentration. This is true, however, only as an approximation, because of the fact that the reaction is more complicated than this. Actually the ammonia molecules attach themselves to the copper ion one at a time (replacing water molecules), and an accurate treatment would require that there be considered the four successive equilibria

$$Cu(H_2O)_4^{++} + NH_3 \rightleftarrows Cu(H_2O)_3NH_3^{++} + H_2O$$
$$Cu(H_2O)_3NH_3^{++} + NH_3 \rightleftarrows Cu(H_2O)_2(NH_3)_2^{++} + H_2O$$
$$Cu(H_2O)_2(NH_3)_2^{++} + NH_3 \rightleftarrows CuH_2O(NH_3)_3^{++} + H_2O$$
$$CuH_2O(NH_3)_3^{++} + NH_3 \rightleftarrows Cu(NH_3)_4^{++} + H_2O$$

The consequence of the existence of these intermediate complexes is that the formation of the final product takes place over a larger range of values of the ammonia concentration than it would otherwise. If the complex were formed in one step the change from 1% to 99% conversion would require only a tenfold increase in $[NH_3]$; it is found by experiment, however, that the ammonia concentration must be increased 10,000-fold to produce this conversion, as indicated by the color change.

19-9.   Polydentate Complexing Agents

In analytical chemistry and industrial chemistry extensive use is made of complexing agents with more than one atom capable of attachment to the central metal atom in a complex. Such a complexing agent is called a *polydentate ligand* or a *chelating agent* (pronounced kee'lating; from the Greek *chēlē*, claw).

An example is triaminotriethylamine (tren), with the following structural formula:

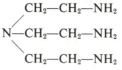

All four nitrogen atoms of this molecule can coordinate with a metal atom. Thus the Zn^{++} ion forms a complex with tren in which each of the four nitrogen atoms uses its unshared pair of electrons to form a bond with the zinc atom and the nitrogen atoms are arranged approximately tetrahedrally about the central atom. The formation constant $[Zn(tren)^{++}]/[Zn^{++}][tren]$ for the complex between tren and Zn^{++}, 4.5×10^{14}, is over 400,000 times larger than that $[Zn(NH_3)_4^{++}]/[Zn^{++}][NH_3]^4$ of the reaction between the zinc ion and four ammonia molecules. The large value of the formation constant for the $Zn(tren)^{++}$ complex is primarily the result of the entropy factor (the fact that the four nitrogen atoms are not free to move about in the solution independently of one another, but are bonded to one another at approximately the same distance apart as in the complex).

Another polydentate complexing agent that forms complexes with many metal ions is EDTA (ethylenediaminetetra-acetic acid), with formula

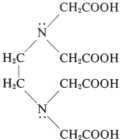

The anion of EDTA has a quadruple negative charge. The four carboxylate ion groups and also the two nitrogen atoms can form bonds with a metal atom; the anion is accordingly a hexadentate complexing agent. In the stable complexes that it forms with many metal ions the two nitrogen atoms and four oxygen atoms, one of each carboxylate ion group, are

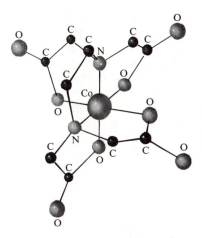

FIGURE 19-3
The structure of the complex between
tripositive cobalt and the anion of
EDTA.

approximately octahedrally arranged around the central ion. The structure of this complex with tripositive cobalt as determined by x-ray diffraction of a crystal containing the complex is shown in Figure 19-3.

EDTA (also called versene) is used in analytical chemistry and also in
chemical industries. In many industrial processes, such as dyeing and the
manufacture of soaps and detergents, even very small concentrations of
heavy metal ions in the water interfere with the reactions. EDTA and
similar agents (sequestering agents) convert the metal ions into complexes, which may not have the harmful properties of the metal ions.

19-10.   The Structure and Stability of Carbonyls and Other Covalent Complexes of the Transition Metals

The problem of the stability of the complexes of the transition metals was
for many years a puzzling one. Why is the cyanide group so facile in the
formation of complexes with these elements, whereas the carbon atom in
other groups, such as the methyl group, does not form bonds with them?
Why do the transition metals and not other metals (beryllium, aluminum,
and so on) form cyanide complexes? In the ferrocyanide ion, $Fe(CN)_6^{----}$,
for example, the iron atom has a formal charge of 4^-, on the assumption
that it forms six covalent bonds with the six ligands; how can this large
negative charge be made compatible with the tendency of metals to lose
electrons and form positive ions?

The answers to these questions and other questions about the cyanide
and carbonyl transition-metal complexes can be derived from the idea
that the cyanide and carbonyl groups form double bonds with the transition metal atom.

Nickel tetracarbonyl, $Ni(CO)_4$, is a volatile liquid with freezing point

−25°C and boiling point 43°C. It plays an important part in the refining of nickel (Section 20-6). The electron-diffraction pattern of the gas has shown that the molecule has a tetrahedral configuration with bond lengths Ni—C = 1.82 Å and C—O = 1.16 Å. These bond lengths show that the Ni—C bond has a large amount of double-bond character and the C—O bond a considerable amount of triple-bond character.* The structures A and B are suggested for the molecule by the electroneutrality principle (plus the structures that correspond to the partial ionic character of the Ni—C bonds):

A

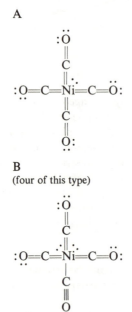

B
(four of this type)

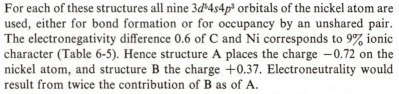

For each of these structures all nine $3d^54s4p^3$ orbitals of the nickel atom are used, either for bond formation or for occupancy by an unshared pair. The electronegativity difference 0.6 of C and Ni corresponds to 9% ionic character (Table 6-5). Hence structure A places the charge −0.72 on the nickel atom, and structure B the charge +0.37. Electroneutrality would result from twice the contribution of B as of A.

Iron forms the pentacarbonyl $Fe(CO)_5$, a liquid with freezing point −21°C and boiling point 103°C. It has the trigonal bipyramidal configuration, with Fe—C bond length 1.84 Å, and principal electronic

---

*Section 6-13. The observed bond length in NiH(g) is 1.47 Å, which leads to 1.17 Å for the single-bond radius of Ni.

structure

Chromium hexacarbonyl is a crystalline substance. It is less stable than nickel carbonyl and iron carbonyl, and decomposes at about 110°C. Its bond length, Cr—C = 1.92 Å, is compatible with the electronic structure

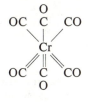

Many other carbonyl compounds are formed by the transition metals. For example, the molecules $Co(CO)_3NO$ and $Fe(CO)_2(NO)_2$, which are isoelectronic with nickel tetracarbonyl, have structures of the same type, the interatomic distances being observed to have the values Co—C = 1.83, Fe—C = 1.84, Co—N = 1.76, Fe—N = 1.77, C—O = 1.15, and N—O = 1.11 Å. Similar tetrahedral structures have also been found for the isoelectronic iron and cobalt carbonyl hydrides $HCo(CO)_4$ and $H_2Fe(CO)_4$, with interatomic distance Co—C = 1.81 Å, Fe—C = 1.81 Å, and C—O = 1.15 Å. These bond lengths indicate that the bonds are similar to those in nickel tetracarbonyl. There is evidence that the hydrogen atoms are held to the metal atoms by covalent bonds. The formulas of all of these substances correspond to structures in which all nine of the outer orbitals of the metal atom are occupied by shared or unshared electron pairs.

The Cyanide Complexes of the Transition Elements

The structural formula usually written for the ferrocyanide ion,

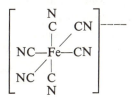

with single covalent bonds from the iron atom to each of the six carbon atoms, is seen to be surprising in that it places a charge of 4− on the iron atom, whereas iron tends to assume a positive charge, as in the ferrous ion, and not a negative charge. As suggested by the foregoing discussion of the carbonyl compounds, we assign to the complex a structure involving some iron-carbon double bonds. The structure

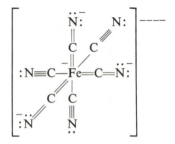

places the formal charge −1 on the iron atom (obtained by dividing the bond electrons equally between the bonded atoms), which becomes +0.08 on correction for the 12% ionic character for iron-carbon bonds indicated by the electronegativity difference. This structure thus agrees with the electroneutrality principle.

19-11. Polynuclear Complexes

The transition metals form polynuclear complexes, containing two or more transition-metal atoms, as well as the simple complexes discussed in the first part of this chapter. Cobalt(III), for example, forms many octahedral complexes, including $[Co(NH_3)_6]^{+++}$, which is yellow, and $[Co(NH_3)_5Cl]^{++}$, which is purple-red in color. It also forms the bright-blue binuclear ion $[(NH_3)_5CoNH_2Co(NH_3)_5]^{5+}$, in which each atom is octahedrally ligated to five ammino groups and one immino group, NH_2, which occupies the shared corner of each of the two octahedra. This ion in a solution containing hydrochloric acid reacts with H^+ and Cl^- to form the hexammino complex and pentamminochloro complex mentioned above. The OH group can also serve to link two octahedra together.

Vanadium, niobium, molybdenum, and tungsten form many polynuclear complexes, including some in which other metal atoms are incorporated. An example is the enneamolybdomanganese(IV) anion*, $[MnMo_9O_{32}]^{6-}$, which is formed when an acidic solution of manganese(II)

---

*The prefix ennea means nine (Greek).

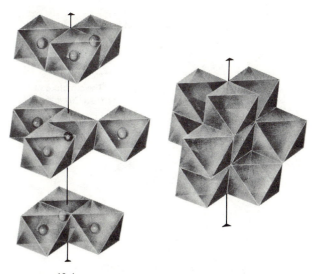

FIGURE 19-4
The structure of the $(MnMo_9O_{32})^{6-}$ anion; it is composed of
octahedra of oxygen atoms with a metal atom in the center of
each octahedron. Left: an exploded view of the ion, showing the
three layers of octahedra. The double circle is manganese and
the single circles are molybdenum. Right: the oxygen framework
of the ion as a whole (metal atoms omitted), showing how the
three layers are joined by the sharing of twelve oxygen atoms.

ion and a molybdate, containing also an oxidizing agent such as the
peroxysulfate ion, $S_2O_8^{--}$, is heated. The enneamolybdomanganese(IV)
ion has the structure shown in Figure 19-4. The manganese atom, at the
center of the complex, is surrounded by an octahedron of six oxygen
atoms, each of which is shared with three molybdenum atoms. Each
molybdenum atom also shows octahedral ligation. Each ion has the
shape of a propeller, either right-handed or left-handed.

In some polynuclear transition-metal complexes the metal atoms are
bonded directly to one another. An example is provided by the yellow
substance with composition corresponding to the simple formula $MoCl_2$.
This substance was discovered in 1859 by C. W. Blomstrand, who pointed
out that it has the surprising property that, when it is dissolved in water
and a solution of silver nitrate is added, only one third of the chlorine is
precipitated as silver chloride. X-ray study of the crystals has shown that
they contain the complex ion $[Mo_6Cl_8]^{4+}$, with the structure shown in
Figure 19-5. The six molybdenum atoms form an octahedron; each atom
is bonded to four molybdenum atoms by single bonds (Mo—Mo distance

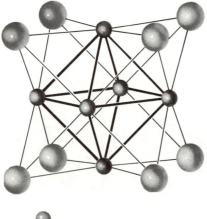

FIGURE 19-5
The structure of the complex ion $(Mo_6Cl_8)^{++++}$.

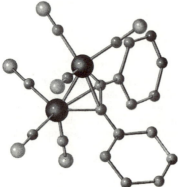

FIGURE 19-6
The structure of dicobalt hexacarbonyl diphenylacetylene, $Co_2(CO)_6C_2(C_6H_5)_2$. Large circles represent cobalt atoms, small circles carbon atoms, and circles of intermediate size oxygen atoms.

2.63 Å, which is less than in the metal, 2.73 Å for ligancy 8), as well as to four bridging chlorine atoms. The cationic complex can also add six anions, such as chloride or hydroxide, in the six positions directly out from each metal atom. $MoBr_2$, WCl_2, and WBr_2 also contain similar polynuclear complexes.

Many metal carbonyls and related substances are polynuclear. A representative example is dicobalt hexacarbonyl diphenylacetylene, the structure of which, determined by x-ray diffraction, is shown in Figure 19-6. The carbon-carbon triple bond has been replaced by a carbon-carbon single bond and four carbon-cobalt single bonds. Each cobalt atom forms a single bond to the other cobalt atom, two single bonds to the acetylenic carbon atoms, and a double bond to each of its attached carbonyl groups, thus using all of its nine outer electrons and nine outer orbitals in bond formation. In some polynuclear carbonyl complexes there are bridging carbonyl groups, in which the carbonyl carbon atom forms single bonds with two metal atoms, in addition to the double bond to the oxygen atom.

Exercises

19-1. Discuss the effects of adding to three portions of a cupric solution (*a*) NH_4OH, (*b*) NaOH, (*c*) NH_4Cl. Write equations for the reactions.

19-2. To three portions of a solution containing Ni^{++} and Al^{+++} there are added (*a*) NaOH, (*b*) NH_4OH, (*c*) NaOH + NH_4OH. What happens in each case?

19-3. Is silver iodide more or less soluble in 1 *F* NH_4OH than in a solution 1 *F* in NH_4I and 1 *F* in NH_4OH? Why? [Note that there are two opposing effects—one resulting from the change in degree of ionization of NH_4OH and the other from the increase in concentration of iodide ion (Sections 19-8, 13-5). Which of these effects is the larger?]

19-4. Write the equation for the principal chemical reaction involved in fixing a photographic film.

19-5. Write the chemical equation for the solution of platinum in aqua regia. Explain why platinum dissolves in aqua regia but not in either hydrochloric acid or nitric acid alone.

19-6. Would sodium cyanide be an effective substitute for sodium thiosulfate as a photographic fixer? (See Table 19-2 for data.)

19-7. Perchlorate ion is generally found to be the weakest complexing reagent of the common anions. Which solution is more acidic, 0.2 *F* $Zn(ClO_4)_2$ or 0.2 *F* $ZnCl_2$?

19-8. How many structural isomers of the octahedral complex $Co(NH_3)_3Cl_3$ are there?

19-9. How many isomers of the tetrahedral complex $Zn(NH_3)_2Cl_2$ are there? Of the planar, square complex $Pt(NH_3)_2Cl_2$?

19-10. The silicon hexafluoride ion, SiF_6^{--}, is octahedral. What orbitals of the silicon atom are occupied by unshared electron pairs? By bonding electron pairs? What is the electric charge on the silicon atom and each fluorine atom, as calculated from electronegativity values?

19-11. The compounds K_2SiF_6, K_2SnF_6, and K_2SnCl_6 are known, but not K_2SiCl_6. Can you explain this fact?

19-12. The aqueous solution made by dissolving crystals of ferric nitrate, $Fe(NO_3)_3 \cdot 6H_2O$ (which has a violet color), is yellow. When an equal volume of strong nitric acid is added to the solution, it assumes a pale violet color. What reaction do you suggest to be responsible for the change in color?

19-13. How much triaminotriethylamine (Section 19-9) would need to be added to a 10,000-liter tank filled with a solution containing 1 mg of zinc ion per liter in order to decrease the concentration of Zn^{++} to 1 ng l^{-1}?

20

Iron, Cobalt, Nickel and the Platinum Metals

In this chapter and in the two following chapters we shall discuss further the chemistry of the transition metals—the elements that occur in the central region of the periodic table. These elements and their compounds have great practical importance. Their chemical properties are complex and interesting.

We shall begin the discussion of the transition metals with iron, cobalt, nickel, and the platinum metals, which lie in the middle of the transition-metal region in the periodic table. The following chapter will be devoted to the elements that lie to the right of these metals; these are copper, zinc, and gallium and their congeners. Chapter 22 will deal with the chemistry of titanium, vanadium, chromium, and manganese and other elements of groups IVa, Va, VIa, and VIIa of the periodic table.

20-1. The Electronic Structures and Oxidation States of Iron, Cobalt, Nickel, and the Platinum Metals

The electronic structures of iron, cobalt, nickel, and the platinum metals are given in Table 5-5, as represented in the energy-level diagram of Figure 5-11. It is seen that each of the atoms has two outermost electrons, in the 4s orbital for iron, cobalt, and nickel, the 5s orbital for ruthenium, rhodium, and palladium, and the 6s orbital for osmium, iridium, and platinum. The next inner shell is incomplete, the 3d subshell (or 4d, or 5d) contains only six, seven, or eight electrons, instead of the full complement of ten.

The Russell-Saunders symbol for iron and its congeners in the normal state is $^5D_4$, that for cobalt and its congeners is $^4F_{9/2}$, and that for nickel and its congeners is $^3F_4$.

It might be expected that the two outermost electrons would be easily removed, to form a bipositive ion. In fact, iron, cobalt, and nickel all form important series of compounds in which the metal is bipositive. These metals also have one or more higher oxidation states. The platinum metals form covalent compounds representing various oxidation states between $+2$ and $+8$.

Iron can assume the oxidation states $+2$, $+3$, and $+6$, the last being rare, and represented by only a few compounds, such as potassium ferrate, K_2FeO_4. The oxidation states $+2$ and $+3$ correspond to the ferrous ion, Fe^{++}, and ferric ion, Fe^{+++}, respectively. The ferrous ion has six electrons in the incomplete 3d subshell, and the ferric ion has five electrons in this orbital. The magnetic properties of the compounds of iron and other transition elements are due to the presence of a smaller number of electrons in the 3d subshell than required to fill this subshell. For example, ferric ion can have all five of its 3d electrons with spins oriented in the same direction, because there are five 3d orbitals in the 3d subshell, and the Pauli principle permits parallel orientation of the spins of electrons so long as there is only one electron per orbital. The ferrous ion is easily oxidized to ferric ion by air or other oxidizing agents. Both bipositive and terpositive iron form complexes, such as the ferrocyanide ion, $Fe(CN)_6^{----}$, and the ferricyanide ion, $Fe(CN)_6^{---}$, but they do not form complexes with ammonia.

Cobalt(II) and cobalt(III) compounds are known; the cobalt(II) ion, Co^{++}, is more stable than the cobalt(III) ion, Co^{+++}, which is a sufficiently powerful oxidizing agent to oxidize water, liberating oxygen. But the covalent cobalt(III) complexes, such as the cobalticyanide ion, $Co(CN)_6^{---}$, are very stable, and the cobalt(II) complexes, such as the cobaltocyanide ion, $Co(CN)_6^{----}$, are unstable, being strong reducing agents.

TABLE 20-1
Standard Enthalpy of Formation of Compounds of Iron, Cobalt, and Nickel at 25°C (kJ mole^{-1})

	M = Fe	Co	Ni
M(c)	0	0	0
M(g)	416	425	430
M^+(g)	1174	1188	1173
M^{++}(g)	2659	2863	2937
M^{++}(aq)	−88	−67	−64
M^{+++}(g)	5678		
M^{+++}(aq)	−48		
MO(c)	−267*	−239	−240
M_2O_3(c)	−823		
M_3O_4(c)	−1120	−854	
MF_2(c)		−665	−667
MF_2(aq)	−744	−726	−718
MF_3	−1017(aq)	−782(c)	
MCl_2(c)	−341	−326	−316
MCl_3(c)	−405		
MBr_2(c)	−251	−232	−227
MI_2(c)	−125	−102	−86
MS(c)	−95	−85	−73
M_2S_3(c)		−213	
MS_2(c)	−178 (pyrite)		
	−154 (marcasite)		
MSe(c)	−69	−42	−42
MTe(c)	−78	−38	−38
M_3C(c)	21	40	46
MP(c)	−117	−146	

*For $Fe_{0.95}$ O.

TABLE 20-2
Some Physical Properties of Iron, Cobalt, and Nickel

	Atomic Number	Atomic Weight	Density (g/cm^3)	Melting Point	Boiling Point	Metallic Radius*	Heat of Sublimation at 25°C
Iron	26	55.847	7.86	1535°C	3000°C	1.26 Å	405 kJ mole^{-1}
Cobalt	27	58.9332	8.93	1480	2900	1.25	439
Nickel	28	58.71	8.89	1452	2900	1.24	425

*For ligancy 12.

Nickel forms only one series of salts, containing the nickel ion, Ni^{++}. A few compounds of nickel with higher oxidation number are known; of these the nickel(IV) oxide, NiO_2, is important.

Some standard enthalpy values are given in Table 20-1. They show the close similarity of the three metals. The most striking difference is that of 266 kJ mole^{-1} between the enthalpies of Fe_3O_4 and Co_3O_4. This difference can be attributed to the instability of Co(III) in forming largely ionic bonds.

As was mentioned in Chapter 17, iron, cobalt, and nickel are sexivalent in the metals and their alloys. This high metallic valence causes the bonds to be especially strong, and confers valuable properties of strength and hardness on the alloys.

Some physical properties of iron, cobalt, and nickel are given in Table 20-2.

20-2. Iron

Pure iron is a bright silvery-white metal, which tarnishes (rusts rapidly) in moist air or in water containing dissolved oxygen. It is soft, malleable, and ductile, and is strongly magnetic ("ferromagnetic"). Its melting point is 1535°C, and its boiling point 3000°C. Ordinary iron (alpha-iron) has the atomic arrangement shown in Figure 17-2 (the body-centered arrangement—each atom is in the center of a cube formed by the eight surrounding atoms). At 912°C alpha-iron undergoes a transition to another allotropic form, gamma-iron, which has the face-centered arrangement described for copper in Chapter 2 (Figure 2-3 and 2-4). At 1400°C another transition occurs, to delta-iron, which has the same body-centered structure as alpha-iron.

Pure iron, containing only about 0.01% of impurities, can be made by electrolytic reduction of iron salts. It has little use.

Metallic iron is greatly strengthened by the presence of a small amount of carbon, and its mechanical and chemical properties are also improved by moderate amounts of other elements, especially other transition metals. Wrought iron, cast iron, and steel are described in the following sections.

The Ores of Iron

The chief ores of iron are its oxides *hematite*, Fe_2O_3, and *magnetite*, Fe_3O_4, and its carbonate *siderite*, $FeCO_3$. The hydrated ferric oxides such as *limonite* are also important. The sulfide *pyrite*, FeS_2, is used as a source of sulfur dioxide, but the impure iron oxide left from its roasting is not satisfactory for smelting iron, because the remaining sulfur is a troublesome impurity.

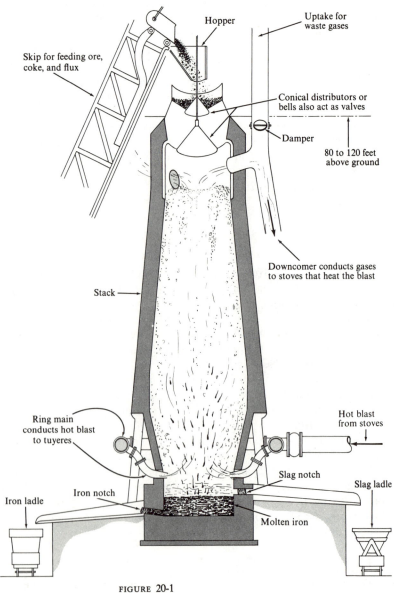

FIGURE 20-1
A blast furnace for smelting iron ore.

The Metallurgy of Iron

The ores of iron are usually first roasted, in order to remove water, to decompose carbonates, and to oxidize sulfides. They are then reduced with coke, in a structure called a *blast furnace* (Figure 20-1). Ores containing limestone or magnesium carbonate are mixed with an acidic flux (containing an excess of silica), such as sand or clay, in order to make a liquid slag. Limestone is used as flux for ores containing an excess of silica. The mixture of ore, flux, and coke is introduced at the top of the blast furnace, and preheated air is blown in the bottom through *tuyeres*. As the solid materials slowly descend they are converted into gases, which escape at the top and two liquids, molten iron and slag, which are tapped off at the bottom. The parts of the blast furnace where the temperature is highest are water-cooled, to keep the lining from melting.

The important reactions that occur in the blast furnace are the combustion of coke to carbon monoxide, the reduction of iron oxide by the carbon monoxide, and the combination of acidic and basic oxides (the impurities of the ore and the added flux) to form slag:

$$2C(c) + O_2(g) \longrightarrow 2CO(g) \qquad \Delta H° = -221 \text{ kJ mole}^{-1}$$

$$3CO(g) + Fe_2O_3(g) \longrightarrow 2Fe(c) + 3CO_2(g) \quad \Delta H° = -27 \text{ kJ mole}^{-1}$$

$$CaCO_3(c) \longrightarrow CaO(c) + CO_2(g) \quad \Delta H° = 178 \text{ kJ mole}^{-1}$$

$$CaO(c) + SiO_2(c) \longrightarrow CaSiO_3(c) \qquad \Delta H° = -89 \text{ kJ mole}^{-1}$$

The values of $\Delta H°$ are for 298°K. The third reaction is endothermic. Because of the formation of a gas, however, the entropy change is large (Section 11-10): $\Delta S° = 161$ J deg^{-1} mole^{-1} at 298°K. The term $-T\Delta S°$ in the expression for the free-energy change causes $\Delta G°$ to be negative at the temperature of the furnace, about 1500°K.

The slag is a glassy silicate mixture of complex composition, idealized as calcium metasilicate, $CaSiO_3$, in the above equation.

The hot exhaust gases, which contain some unoxidized carbon monoxide, are cleaned of dust and then are mixed with air and burned in large steel structures filled with fire brick. When one of these structures (called *stoves*) has thus been heated to a high temperature the burning exhaust gas is shifted to another stove and the heated stove is used to preheat the air for the blast furnace.

The amount of coke required in smelting iron can be reduced by using air that has been enriched by addition of oxygen to about 23.5%. In blast furnaces using oxygen-enriched air, steam is also added to the air, in order to prevent the temperature in the blast furnace from becoming too high.

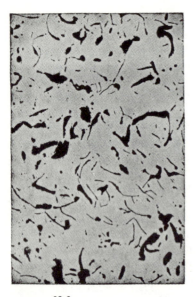

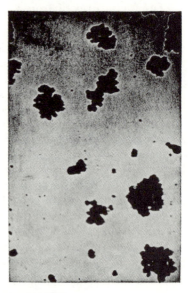

FIGURE 20-2
A photomicrograph of gray cast iron,
unetched. The white background is
ferrite, and the black particles are
flakes of graphite. Magnification 100×.
(From Malleable Founders' Society.)

FIGURE 20-3
A photomicrograph of malleable
cast iron, showing ferrite (background)
and globular particles of graphite.
Unetched. Magnification 100×.
(From Malleable Founders' Society.)

Cast Iron

The molten iron from the blast furnace, having been in contact with
coke in the lower part of the furnace, contains several percent of dissolved carbon (usually about 3 or 4%), together with silicon, manganese,
phosphorus, and sulfur in smaller amounts. These impurities lower its
melting point from 1535°C, that of pure iron, to about 1200°C. This
iron is often cast into bars called *pigs;* the cast iron itself is called *pig iron*.

When cast iron is made by sudden cooling from the liquid state it is
white in color, and is called *white cast iron*. It consists largely of the compound *cementite*, Fe_3C, a hard, brittle substance.

Gray cast iron, made by slow cooling, consists of crystalline grains of
pure iron (called *ferrite*) and flakes of graphite (Figure 20-2). Both white
cast iron and gray cast iron are brittle, the former because its principal
constituent, cementite, is brittle, and the latter because the tougher
ferrite in it is weakened by the soft flakes of graphite distributed through it.

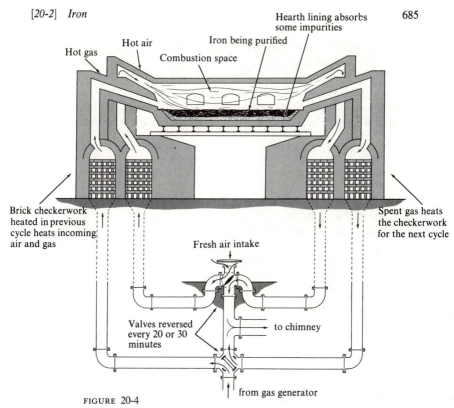

FIGURE 20-4

Reverberatory furnace, used for making wrought iron and steel.

Malleable cast iron, which is tougher and less brittle than either white or ordinary gray cast iron, is made by heat treatment of gray cast iron of suitable composition. Under this treatment the flakes of graphite coalesce into globular particles, which, because of their small cross-sectional area, weaken the ferrite less than do the flakes (Figure 20-3).

Cast iron is the cheapest form of iron, but its usefulness is limited by its low strength. A great amount is converted into steel, and a smaller amount into wrought iron.

Wrought Iron

Wrought iron is nearly pure iron, with only 0.1 or 0.2% carbon and less than 0.5% of all impurities. It is made by melting cast iron on a bed of iron oxide in a reverberatory furnace, in which the flame is reflected by the roof onto the material to be heated (Figure 20-4). As the molten cast iron is stirred, the iron oxide oxidizes the dissolved carbon to carbon monoxide, and the sulfur, phosphorus, and silicon are also oxidized and pass into the slag. As the impurities are removed, the melting point of the iron rises, and the mass becomes pasty. It is then taken out of the furnace and beaten under steam hammers to force out the slag.

Wrought iron is a strong, tough metal that can be readily welded and forged. In past years it was extensively used for making chains, wire, and similar objects. It has now been largely displaced by mild steel.

20-3.   Steel

Steel is a purified alloy of iron, carbon, and other elements that is manufactured in the liquid state. Most steels are almost free from phosphorus, sulfur, and silicon, and contain between 0.1 and 1.5% of carbon. *Mild steels* are low-carbon steels (less than 0.2%). They are malleable and ductile, and are used in place of wrought iron. They are not hardened by being quenched (suddenly cooled) from a red heat. *Medium steels*, containing from 0.2 to 0.6% carbon, are used for making rails and structural elements (beams, girders, and so on). Mild steels and medium steels can be forged and welded. *High-carbon steels* (0.75 to 1.50% carbon) are used for making razors, surgical instruments, drills, and other tools. Medium steels and high-carbon steels can be hardened and tempered (see below, The Properties of Steel).

Steel is made from pig iron chiefly by the *open-hearth process* (by which over 80% of that produced in the United States is made), the *Bessemer process*, and the *oxygen top-blowing process*. In each process either a basic or an acidic lining may be used in the furnace or converter. A basic lining (lime, magnesia, or a mixture of the two) is used if the pig iron contains elements, such as phosphorus, that form acidic oxides, and an acidic lining (silica) if the pig iron contains base-forming elements.

The Open-hearth Process

Open-hearth steel is made in a reverberatory furnace (Figure 20-4). Cast iron is melted with scrap steel and some hematite in a furnace heated with gas or oil fuel. The fuel and air (sometimes enriched with oxygen) are preheated by passage through a lattice of hot brick at one side of the furnace, and a similar lattice on the other side is heated by the hot outgoing gases. From time to time the direction of flow of gas is reversed. The carbon and other impurities in the molten iron are oxidized by the hematite and by excess air in the furnace gas. Analyses are made during the run, which requires about 8 hours, and when almost all the carbon is oxidized the amount desired for the steel is added as coke or as a high-carbon alloy, usually ferromanganese or spiegeleisen (Section 22-7). The

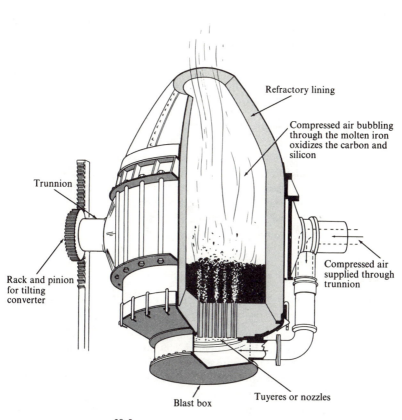

FIGURE 20-5
Bessemer converter, used for making steel from pig iron.

molten steel is then cast into ingots. Open-hearth steel of very uniform quality can be made, because the process can be closely checked by analyses during the several hours of the run.

The Bessemer Process

The Bessemer process of making steel was invented by an American, William Kelly, in 1852 and independently by an Englishman, Henry Bessemer, in 1855. Molten pig iron is poured into an egg-shaped converter (Figure 20-5). Air is blown up through the liquid from tuyeres in the bottom, oxidizing silicon, manganese, and other impurities and finally the carbon. In about ten minutes the reaction is nearly complete, as is seen from the change in character of the flame of burning carbon monoxide from the mouth of the converter. High-carbon alloy is then added, and the steel is poured.

The Bessemer process is inexpensive, but the steel is not as good as open-hearth steel.

The Oxygen Top-blowing Process

Since 1955 an increasing fraction of the steel produced in the United States has been made by a new process, the oxygen top-blowing process. Iron is placed in a converter resembling the Bessemer converter (Figure 20-5), but without the tuyeres at the base. Pure oxygen (99.5%) is then blown onto the surface of the molten metal through a long water-cooled copper lance, to oxidize carbon and phosphorus. The treatment of the charge of 50 to 250 tons is completed in 40 to 50 minutes. This process gives steel of high quality.

The Properties of Steel

When high-carbon steel is heated to bright redness and slowly cooled it is comparatively soft. If it is rapidly cooled, by quenching in water, oil, or mercury, it becomes harder than glass, and brittle instead of tough. This hardened steel can be "tempered" by suitable reheating, to give a product with the desired combination of hardness and toughness. Often the tempering is carried out in such a way as to leave a very hard cutting edge backed up by softer, tougher metal.

The amount of tempering can be estimated roughly by the interference colors of the thin film of oxide formed on a polished surface of the steel during reheating: a straw color (230°C) corresponds to a satisfactory temper for razors, yellow (250°C) for pocket knives, brown (260°C) for scissors and chisels, purple (270°C) for butcher knives, blue (290°C) for watch springs, and blue-black (320°C) for saws.

These processes of hardening and tempering can be understood by consideration of the phases that can be formed by iron and carbon. Carbon is soluble in gamma-iron, the form stable above 912°C. If the steel is quenched from above this temperature there is obtained a solid solution of carbon in gamma-iron. This material, called *martensite*, is very hard and brittle. It confers hardness and brittleness upon hardened high-carbon steel. Martensite is not stable at room temperature, but its rate of conversion to more stable phases is so small at room temperature as to be negligible, and hardened steel containing martensite remains hard so long as it is not reheated.

When hardened steel is tempered by mild reheating, the martensite

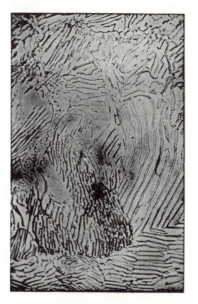

FIGURE 20-6
A photomicrograph of pearlite, showing lamellae of ferrite and cementite. Magnification 1000×. (From Dr. D. S. Clark.)

FIGURE 20-7
A photomicrograph of hypo-eutectoid steel, showing grains of pearlite. Carbon content of steel 0.38%. Magnification 500×. (From Dr. D. S. Clark.)

undergoes transformation to more stable phases. The changes that it undergoes are complex, but result ultimately in a mixture of grains of alpha-iron (ferrite) and the hard carbide Fe_3C, cementite. Steel containing 0.9% carbon (*eutectoid steel*) changes on tempering into *pearlite*, which is composed of extremely thin alternating layers of ferrite and cementite (Figure 20-6). Pearlite is strong and tough. Steel containing less than 0.9% carbon (*hypo-eutectoid steel*) changes on tempering into a microcrystalline metal consisting of grains of ferrite and grains of pearlite (Figure 20-7), whereas that containing more than 0.9% carbon (*hyper-eutectoid steel*) on tempering yields grains of cementite and grains of pearlite.

Steel intended to withstand both shock and wear must be tough and strong and must also present a very hard surface. Steel objects with these properties are made by a process called *case-hardening*. Medium-carbon steel objects are heated in contact with carbon or sodium cyanide until a thin surface layer is converted into high-carbon steel, which can be hardened by suitable heat treatment. Some alloy steels are case-hardened by formation of a surface layer of metal nitrides, by heating the objects in an atmosphere of ammonia.

Alloy Steels

Many alloy steels, steel containing considerable amounts of metals other than iron, have valuable properties and extensive industrial uses. Manganese steel (12 to 14% Mn) is extraordinarily hard, and crushing and grinding machines and safes are made of it. Nickel steels have many special uses. Chromium-vanadium steel (5 to 10% Cr, 0.15% V) is tough and elastic, and is used for automobile axles, frames, and other parts. Stainless steels usually contain chromium; a common composition is 18% Cr, 8% Ni. Molybdenum and tungsten steels are used for high-speed cutting tools.

20-4. Compounds of Iron

Iron is an active metal, which displaces hydrogen easily from dilute acids. It burns in oxygen to produce ferrous-ferric oxide, Fe_3O_4. This oxide is also made by interaction with superheated steam. One method of preventing rusting involves the production of an adherent surface layer of this oxide on iron.

Iron becomes *passive* when it is dipped in very concentrated nitric acid. It then no longer displaces hydrogen from dilute acids. However, a sharp blow on the metal produces a change that spreads over the surface from the point struck, the metal once more becoming active. This production of passivity is due to the formation of a protective layer of oxide, and the passivity is lost when the layer is broken. Passivity is also produced by other oxidizing agents, such as chromate ion; safety razor blades kept in a solution of potassium chromate remain sharp much longer than blades kept in air.

When exposed to moist air, iron becomes oxidized, forming a loose coating of rust, which is a partially hydrated ferric oxide.

Ferrous Compounds

The ferrous compounds, containing bipositive iron, are usually green in color. Most of the ferrous salts are easily oxidized to the corresponding ferric salts through the action of atmospheric oxygen.

Ferrous sulfate, $FeSO_4 \cdot 7H_2O$, is made by dissolving iron in sulfuric acid, or by allowing pyrite to oxidize in air. The green crystals of the substance are efflorescent, and often have a brown coating of a ferric hydroxide-sulfate, produced by atmospheric oxidation. Ferrous sulfate is used in dyeing and in making ink. To make ink, a solution of tannic acid—a complex organic acid obtained by extraction of nut-galls—is mixed with ferrous sulfate, producing ferrous tannate. On oxidation by the air a fine black insoluble pigment is produced.

Ferrous chloride, $FeCl_2 \cdot 4H_2O$, is made by dissolving iron in hydrochloric acid. It is pale green in color. *Ferrous hydroxide*, $Fe(OH)_2$, is formed as a nearly white precipitate on addition of alkali to a ferrous solution. The precipitate rapidly becomes a dirty green, and finally brown, by oxidation by air. *Ferrous sulfide*, FeS, is a black compound made by heating iron filings with sulfur. It is used in making hydrogen sulfide. Ferrous sulfide is also obtained as a black precipitate by the action of sulfide ion on a ferrous salt in solution.

Ferrous carbonate, $FeCO_3$, occurs in nature as a mineral, and can be obtained as a white precipitate by the action of carbonate ion on ferrous ion in the absence of dissolved oxygen. Like calcium carbonate, ferrous carbonate is soluble in acidic waters. Hard waters often contain ferrous or ferric ion.

Ferric Compounds

The hydrated ferric ion, $Fe(H_2O)_6^{+++}$, is pale violet in color. The ion loses protons readily, however, and ferric salts in solution usually are yellow or brown, because of the formation of hydroxide complexes. *Ferric nitrate*, $Fe(NO_3)_3 \cdot 6H_2O$, exists as pale violet deliquescent crystals. Anhydrous *ferric sulfate*, $Fe_2(SO_4)_3$, is obtained as a white powder by evaporation of a ferric sulfate solution. *Iron alum*, $KFe(SO_4)_2 \cdot 12H_2O$, forms pale violet octahedral crystals.

Ferric chloride, $FeCl_3 \cdot 6H_2O$, is obtained as yellow deliquescent crystals by evaporation of a solution made by oxidation of ferrous chloride with chlorine. Solutions of ferric ion containing chloride ion are more intensely colored, yellow or brown, than nitrate or sulfate solutions because of the formation of ferric chloride complexes. Anhydrous ferric chloride, Fe_2Cl_6, can be made by passing chlorine over heated iron.

Ferric ion in solution can be reduced to ferrous ion by treatment with metallic iron or by reduction with hydrogen sulfide or stannous ion.

Ferric hydroxide, $Fe(OH)_3$, is formed as a brown precipitate when alkali is added to a solution of ferric ion. When it is strongly heated ferric hydroxide is converted into *ferric oxide*, Fe_2O_3, which, as a fine powder, is called *rouge* and, as a pigment, *Venetian red*.

Complex Cyanides of Iron

Cyanide ion added to a solution of ferrous or ferric ion forms precipitates, which dissolve in excess cyanide to produce complex ions. Yellow crystals of *potassium ferrocyanide*, $K_4Fe(CN)_6 \cdot 3H_2O$, are made by heating organic material, such as dried blood, with iron filings and potassium carbonate. The mass produced by the heating is extracted with warm water, and the crystals are made by evaporation of the solution. *Potassium ferricyanide*, $K_3Fe(CN)_6$, is made as red crystals by oxidation of ferrocyanide. These substances contain the complexes *ferrocyanide ion*, $Fe(CN)_6^{----}$, and *ferricyanide ion*, $Fe(CN)_6^{---}$, respectively, and the ferrocyanides and ferricyanides of other metals are easily made from them. Their structure has been discussed in Section 19-10.

The pigments *Turnbull's blue* and *Prussian blue* are made by addition of ferrous ion to a ferricyanide solution or ferric ion to a ferrocyanide solution. The pigments that precipitate have the approximate composition $KFeFe(CN)_6 \cdot H_2O$. They have a brilliant blue color. Ferrous ion and ferrocyanide ion produce a white precipitate of $K_2FeFe(CN)_6$, whereas ferric ion and ferricyanide ion form only a brown solution.

20-5. Cobalt

Cobalt occurs in nature in the minerals *smaltite*, $CoAs_2$, and *cobaltite*, CoAsS, usually associated with nickel. The metal is obtained by reducing the oxide with aluminum.

Metallic cobalt is silvery-white, with a slight reddish tinge. It is less reactive than iron, and displaces hydrogen slowly from dilute acids. It is used in special alloys, including *Alnico*, a strong ferromagnetic alloy of aluminum, nickel, cobalt, and iron, which is used for making permanent magnets.

Cobalt ion, $Co(H_2O)_6^{++}$, in solution and in hydrated salts is red or pink in color. *Cobalt chloride*, $CoCl_2 \cdot 6H_2O$, forms red crystals, which when dehydrated change into a deep blue powder. Writing made with a dilute solution of cobalt chloride is almost invisible, but becomes blue when the paper is warmed, dehydrating the salt. *Cobalt oxide*, CoO, is a black substance that dissolves in molten glass, to give it a blue color (*cobalt glass*).

Tripositive cobalt ion is unstable, and an attempt to oxidize Co^{++} usually leads to the precipitation of *cobalt(III) hydroxide*, $Co(OH)_3$. The covalent compounds of cobalt(III) are very stable. The most important of these are *potassium cobaltinitrite*, $K_3Co(NO_2)_6$, and *potassium cobalticyanide*, $K_3Co(CN)_6$.

20-6. Nickel

Nickel occurs, with iron, in meteorites. Its principal ores are *niccolite*, NiAs, *millerite*, NiS, and *pentlandite*, (Ni,Fe)S. The metal is produced as an alloy containing iron and other elements, by roasting the ore and reducing with carbon. In the purification of nickel by the Mond process the compound *nickel carbonyl*, $Ni(CO)_4$, is manufactured and then decomposed. The ore is reduced with hydrogen to metallic nickel under conditions such that the iron oxide is not reduced. Carbon monoxide is then passed through the reduced ore at room temperature; it combines with the nickel to form nickel carbonyl:

$$Ni + 4CO \longrightarrow Ni(CO)_4$$

Nickel carbonyl is a gas. It is passed into a decomposer heated to 150°C; the gas decomposes, depositing pure metallic nickel, and the liberated carbon monoxide is returned to be used again.

Nickel is a white metal, with a faint tinge of yellow. It is used in making alloys, including the copper-nickel alloy (75% Cu, 25% Ni) used in coinage. Iron objects are plated with nickel by electrolysis from an ammoniacal solution. The metal is still less reactive than cobalt, and displaces hydrogen only very slowly from acids.

The hydrated salts of nickel such as *nickel sulfate*, $NiSO_4 \cdot 6H_2O$, and *nickel chloride*, $NiCl_2 \cdot 6H_2O$, are green in color. *Nickel(II) hydroxide*, $Ni(OH)_2$, is formed as an apple-green precipitate by addition of alkali to a solution containing nickel ion. When heated it produces the insoluble green substance *nickel(II) oxide*, NiO. Nickel(II) hydroxide is soluble in ammonium hydroxide, forming ammonia complexes such as $Ni(NH_3)_4(H_2O)_2^{++}$ and $Ni(NH_3)_6^{++}$.

In alkaline solution nickel(II) hydroxide can be oxidized to a hydrated *nickel(IV) oxide*, $NiO_2 \cdot xH_2O$. This reaction is used in the *Edison storage cell*. The electrodes of this cell are plates coated with $NiO_2 \cdot xH_2O$ and metallic iron, which are converted on discharge of the cell into nickel(II) hydroxide and ferrous hydroxide, respectively. The electrolyte in this cell is a solution of sodium hydroxide.

20-7. The Platinum Metals

The congeners of iron, cobalt, and nickel are the *platinum metals*—ruthenium, rhodium, palladium, osmium, iridium, and platinum. Some properties of these elements are given in Table 20-3.

The platinum metals are noble metals, chemically unreactive, which are found in nature as native alloys, consisting mainly of platinum.

Ruthenium and *osmium* are iron-gray metals, the other four elements being whiter in color. Ruthenium can be oxidized to RuO_2, and even to the octavalent compound RuO_4. Osmium unites with oxygen to form osmium tetroxide ("osmic acid"), OsO_4, a white crystalline substance melting at 40°C and boiling at about 100°C. Osmium tetroxide has an irritating odor similar to that of chlorine. It is a very poisonous substance.

TABLE 20-3
Some Physical Properties of the Platinum Metals

	Atomic Number	Atomic Weight	Density (g/cm³)	Melting Point	Heat of Sublimation at 25°C
Ru	44	101.07	12.36	2450°C	670 kJ mole⁻¹
Rh	45	102.905	12.48	1985°	577
Pd	46	106.4	12.09	1555°	390
Os	76	190.2	22.69	2700°	732
Ir	77	192.2	22.82	2440°	690
Pt	78	195.09	21.60	1755°	509

Its aqueous solution is used in histology (the study of the tissues of plants and animals); it stains tissues through its reduction by organic matter to metallic osmium, and also hardens the material without distorting it.

Ruthenium and osmium form compounds corresponding to various states of oxidation, such as the following: $RuCl_3$, K_2RuO_4, Os_2O_3, $OsCl_4$, K_2OsO_4.

Rhodium and *iridium* are very unreactive metals, not being attacked by aqua regia (a mixture of nitric acid and hydrochloric acid). Iridium is alloyed with platinum to produce a very hard alloy, which is used for the tips of gold pens, surgical tools, and scientific apparatus. Representative compounds are Rh_2O_3, K_3RhCl_6, Ir_2O_3, K_3IrCl_6, and K_2IrCl_6.

Palladium is the only one of the platinum metals that is attacked by nitric acid. Metallic palladium has an unusual ability to absorb hydrogen. At 1000°C it absorbs enough hydrogen to correspond to the formula $PdH_{0.6}$.

The principal compounds of palladium are the salts of chloropalladous acid, H_2PdCl_4, and chloropalladic acid, H_2PdCl_6. The chloropalladite ion, $PdCl_4^{--}$, is a planar ion, consisting of the palladium atom with four coplanar chlorine atoms arranged about it at the corners of a square. The chloropalladate ion, $PdCl_6^{--}$, is an octahedral covalent complex ion.

Platinum is the most important of the palladium and platinum metals. It is grayish-white in color, and is very ductile. It can be welded at a red heat, and melted in an oxyhydrogen flame. Because of its very small chemical activity it is used in electrical apparatus and in making crucibles and other apparatus for use in the laboratory. Platinum is attacked by chlorine and dissolves in a mixture of nitric and hydrochloric acids. It also interacts with fused alkalis, such as potassium hydroxide, but not with alkali carbonates.

The principal compounds of platinum are the salts of chloroplatinous acid, H_2PtCl_4, and chloroplatinic acid, H_2PtCl_6. These salts are similar in structure to the corresponding palladium salts. Both palladium and platinum form many other covalent complexes, such as the platinum(II) ammonia complex ion, $Pt(NH_3)_4^{++}$.

A finely divided form of metallic platinum, called *platinum sponge*, is made by strongly heating ammonium chloroplatinate, $(NH_4)_2PtCl_6$. *Platinum black* is a fine powder of metallic platinum made by adding zinc to chloroplatinic acid. These substances have very strong catalytic activity, and are used as catalysts in commercial processes, such as the oxidation of sulfur dioxide to sulfur trioxide. Platinum black causes the ignition of a mixture of illuminating gas and air or hydrogen and air as a result of the heat developed by the rapid chemical combination of the gases in contact with the surface of the metal.

Exercises

20-1. Compare the stability of the free cobalt(III) ion, Co^{+++}, with that of the cobalticyanide ion, $Co(CN)_6^{---}$, and explain in terms of electronic structure. What are the hybrid orbitals involved in bond formation? How many electrons with unpaired spins are there in the free ion? In the complex?

20-2. What are the oxidation states of iron in hematite, magnetite, and siderite?

20-3. Which do you predict would have the lower *p*H, an aqueous solution of ferric nitrate or an aqueous solution of ferric chloride?

20-4. Write an electronic structural formula for nickel carbonyl, and discuss the arrangement of the electrons around the nickel atoms in relation to the structure of krypton. Iron forms a carbonyl $Fe(CO)_5$, and chromium forms a carbonyl $Cr(CO)_6$; discuss the electronic structures of these substances.

20-5. Apply Equation 6-1 to the values of the standard enthalpy of formation of $FeCl_2(c)$, $FeBr_2(c)$, and $FeI_2(c)$ and the corresponding compounds of cobalt(II) and nickel(II) given in Table 20-1 to obtain values of the electronegativity of these metals in the bipositive state, assuming $x = 3.00$ for chlorine, 2.80 for bromine, and 2.50 for iodine. (Answer: averages 1.71, 1.76, 1.79.)

20-6. By the method of the preceding Exercise use the standard enthalpy of $FeCl_3(c)$ to calculate the electronegativity of iron(III). (Answer: 1.88.)

20-7. Assuming the dependence of electronegativity on oxidation number to be the same for nickel as for iron, estimate the standard enthalpy of $NiCl_3(c)$.

20-8. By use of heat-capacity measurements and the third law of thermodynamics the following values for the standard molar entropy at 298°K have been obtained: $Fe(c)$, 27.15 J deg^{-1} $mole^{-1}$; $S_8(c)$, 255.06 J deg^{-1} $mole^{-1}$; FeS, 67.36 J deg^{-1} $mole^{-1}$. What is the value of $\Delta S°$ at this temperature for the reaction of formation of $FeS(c)$ from the elements? (Answer: 8.33 J deg^{-1} $mole^{-1}$.)

20-9. With use of Table 20-1 and the result of the preceding Exercise calculate $\Delta H°$ and $\Delta G°$ for the reaction of formation of $FeS(c)$ from the elements at 298°K. Would you expect iron(II) sulfide to be formed by heating iron and sulfur? (Answer: -95 kJ $mole^{-1}$, -97 kJ $mole^{-1}$.)

20-10. The standard entropy at 298°K of cementite, $Fe_3C(c)$, is 107.5 J deg^{-1} $mole^{-1}$, and that of graphite, $C(c)$, is 5.69 J deg^{-1} $mole^{-1}$. Use these values and information given in Exercise 20-8 and Table 20-1 in discussing the statement in Section 20-3 that white cast iron (largely cementite) is made by sudden cooling from the liquid state and gray cast iron (iron and graphite) by slow cooling. What is the value of $\Delta G°$ at 298°K for the reaction of decomposition of cementite to ferrite and graphite?

20-11. As a rough approximation assume $\Delta H°$ and $\Delta S°$ to be constant for the reaction
$$Fe_3C(c) \rightleftarrows 3Fe(c) + C(c)$$
with the known values for 298°K (preceding Exercise). What does this approximation give as the temperature marking the limit between stability and instability of cementite? (Answer: 1031°K.)

20-12. The values of the molar heat capacity at 298° of cementite, iron, and graphite are $C_P = 105.86$, 25.64, and 8.65 J deg^{-1} $mole^{-1}$, respectively. Assuming the difference in heat capacity of cementite and its decomposition products to be constant from 298°K to 1031°K, calculate $\Delta H°$, $\Delta S°$, and $\Delta G°$ for the reaction of decomposition of cementite at 1031°K. From this calculation, is cementite stable or unstable at this temperature? (Answer: 36 kJ $mole^{-1}$, 33.5 J deg^{-1} $mole^{-1}$; 1.3 kJ $mole^{-1}$.)

20-13. Assuming $\Delta H°$ and $\Delta S°$ to be constant over a small temperature range near 1031°K, use the results of the preceding Exercise to calculate a more accurate value of temperature marking the limit between stability and instability of cementite. (Answer: 1070°K.)

Copper, Zinc, and Gallium and Their Congeners

In the preceding chapter we have begun the discussion of the chemistry of the transition metals through the consideration of iron, cobalt, nickel, and their congeners, the palladium and platinum metals. We shall now take up the chemistry of the elements that lie to the right of these elements in the periodic table.

The three metals copper, silver, and gold comprise group Ib of the periodic table. These metals all form compounds representing oxidation state $+1$, as do the alkali metals, but aside from this they show very little similarity in properties to the alkali metals. The alkali metals are very soft and light, and very reactive chemically, whereas the metals of the copper group are much harder and heavier and are rather inert, sufficiently so to occur in the free state in nature and to be easily obtainable by reducing their compounds, sometimes simply by heating. The metals zinc, cadmium, and mercury (group IIb) are also much different from the alkaline-earth metals (group II), as are gallium and its congeners (group IIIb) from the elements of group III.

In this chapter, in connection with the discussion of the compounds of silver, there is also a section on photography, including color photography (Section 21-5).

21-1.   The Electronic Structures and Oxidation States of Copper, Silver, and Gold

The electronic structures of copper, silver, and gold, as well as those of zinc and gallium and their congeners, are given in Table 5-5.

It is seen that copper has one outer electron, in the $4s$ orbital of the M shell, zinc has two outer electrons, in the $4s$ orbital, and gallium has three outer electrons, two in the $4s$ orbital and one in the $4p$ orbital. The congeners of these elements also have one, two, or three electrons in the outermost shell. The shell next to the outermost shell in each case contains 18 electrons; this is the M shell for copper, zinc, and gallium, the N shell for silver, cadmium, and indium, and the O shell for gold, mercury, and thallium. This shell is called an *eighteen-electron shell.*

The Russell-Saunders symbol for copper and its congeners in the normal state is $^2S_{1/2}$, that for zinc and its congeners is $^1S_0$, and that for gallium and its congeners is $^2P_{1/2}$.

The electrons in the outermost shell are held loosely, and can be easily removed. The resulting ions, Cu^+, Zn^{++}, Ga^{+++}, and so on, have an outer shell of eighteen electrons, and are called *eighteen-shell ions.* If these elements either lose their outermost electrons, forming eighteen-shell ions, or share the outermost electrons with other atoms, the resulting oxidation state is $+1$ for copper, silver, and gold, $+2$ for zinc, cadmium, and mercury, and $+3$ for gallium, indium, and thallium.

These are important oxidation states for all of these elements; there are, however, also some other important oxidation states. The cuprous ion, Cu^+, is unstable, and the cuprous compounds, except the very insoluble ones, are easily oxidized. The cupric ion, Cu^{++} (hydrated to $Cu(H_2O)_4^{++}$), occurs in many copper salts, and the cupric compounds are the principal compounds of copper. In the cupric ion the copper atom has lost two electrons, leaving it with only seventeen electrons in the M shell. In fact, the $3d$ electrons and the $4s$ electrons in copper are held by the atom with about the same energy—you may have noticed that the electronic structure given in Table 5-5 for copper differs from that given in the energy-level diagram, Figure 5-11, in that in the diagram copper is represented as having two $4s$ electrons and only nine $3d$ electrons.

The unipositive silver ion, Ag^+, is stable, and forms many salts. A very few compounds have also been made containing bipositive and tripositive silver. These compounds are very strong oxidizing agents. The stable oxidation state $+1$ shown by silver corresponds to the electronic structure of the element as given in Table 5-5. The Ag^+ ion is an eighteen-shell ion.

The gold(I) ion, Au^+, and the gold(III) ion, Au^{+++}, are unstable in aqueous solution. The stable gold(I) compounds and gold(III) compounds contain covalent bonds, as in the complex ions $AuCl_2^-$ and $AuCl_4^-$.

TABLE 21-1
Some Physical Properties of Copper, Silver, and Gold

	Atomic Number	Atomic Weight	Density (g/cm³)	Melting Point	Boiling Point	Metallic Radius	Color
Copper	29	63.54	8.97	1083°C	2310°C	1.28 Å	Red
Silver	47	107.870	10.54	960.5°	1950°	1.44	White
Gold	79	196.967	19.42	1063°	2600°	1.44	Yellow

The chemistry of zinc and cadmium is especially simple, in that these elements form compounds representing only the oxidation state $+2$. This oxidation state is closely correlated with the electronic structures shown in Table 5-5; it represents the loss or the sharing of the two outermost electrons. The ions Zn^{++} and Cd^{++} are eighteen-shell ions.

Mercury also forms compounds (the mercuric compounds) representing the oxidation state $+2$. The mercuric ion, Hg^{++}, is an eighteen-shell ion. In addition, mercury forms a series of compounds, the mercurous compounds, in which it has oxidation number $+1$. The electronic structure of the mercurous compounds is discussed in Section 21-10.

21-2. The Properties of Copper, Silver, and Gold

Copper is a red, tough metal with a moderately high melting point (Table 21-1). It is an excellent conductor of heat and of electricity when pure, and it finds extensive use as an electric conductor. Pure copper that has been heated is soft, and can be drawn into wire or shaped by hammering. This "cold work" (of drawing or hammering) causes the metal to become hard, because the crystal grains are broken into much smaller grains, with grain boundaries that interfere with the process of deformation and thus strengthen the metal. The hardened metal can be made soft by heating ("annealing"), which permits the grains to coalesce into larger grains.

Silver is a soft, white metal, somewhat denser than copper, and with a lower melting point. It is used in coinage, jewelry, and tableware, and as a filling for teeth.

Gold is a soft, very dense metal, which is used for jewelry, coinage, dental work, and scientific and technical apparatus. Gold is bright yellow by reflected light; very thin sheets are blue or green. Its beautiful color and fine luster, which, because of its inertness, are not affected by exposure to the atmosphere, are responsible for its use for ornamental purposes. Gold is the most malleable and most ductile of all metals; it can be hammered into sheets only 1/100,000 cm thick, and drawn into wires 1/5000 cm in diameter.

TABLE 21-2

Standard Enthalpy of Formation of Compounds of Copper, Silver, and Gold at 25°C (kJ mole⁻¹)

	M = Cu	Ag	Au
M(c)	0	0	0
M(g)	341	289	344
$M^+(g)$	1091	1026	1241
$M^+(aq)$	52	106	
$M^{++}(g)$	3055	3105	
$M^{++}(aq)$	64		
$M_2O(c)$	−167	−31	
MO(g)	146		
MO(c)	−155		
$M_2O_3(c)$			81
MH(g)	297	283	
$M_2F(c)$		−211	
MF(c)		−203	
$MF_2(c)$	−531	−370	
MCl(g)	134	97	
MCl(c)	−135	−127	−35
$MCl_2(c)$	−206		
$MCl_3(c)$			−118
$MCl_4^-(aq)$			−326
MBr(g)	159		
MBr(c)	−105	−99	−18
$MBr_2(c)$	−139		
$MBr_3(c)$			−54
MI(g)	259		
MI(c)	−68	−62	1
$MI_2(c)$	−7		
$M_2S(c)$	−79	−32	
MS(c)	−49		

Alloys of Copper, Silver, and Gold

The transition metals find their greatest use in alloys. Alloys are often far stronger, harder, and tougher than their constituent elementary metals. The alloys of copper and zinc are called *brass*, those of copper and tin are called *bronze*, and those of copper and aluminum are called *aluminum bronze.* Many of these alloys have valuable properties. Copper is a constituent also of other useful alloys, such as beryllium copper, coinage silver, and coinage gold.

Coinage silver in the United States contains 90% silver and 10% copper. This composition also constitutes *sterling silver* in the United States. British sterling silver is 92.5% silver and 7.5% copper.

Gold is often alloyed with copper, silver, palladium, or other metals. The amount of gold in these alloys is usually described in *carats*, the number of parts of gold in 24 parts of alloy—pure gold is 24 carat. American coinage gold is 21.6 carat and British coinage gold is 22 carat. *White gold*, used in jewelry, is usually a white alloy of gold and nickel.

21-3.   The Compounds of Copper

Values of the standard enthalpy of formation of some of the principal compounds of copper (and also of silver and gold) are given in Table 21-2. These values show that with the more electronegative nonmetals copper tends to form compounds of copper(II) (cupric compounds). For example, the heat of reaction of cuprous chloride with chlorine to form cupric chloride is positive:

$$CuCl(c) + \tfrac{1}{2}Cl_2(g) \longrightarrow CuCl_2(c) + 71 \text{ kJ mole}^{-1}$$

With sulfur and iodine, in which the bonds have little ionic character (electronegativity of copper, 1.9; of sulfur and iodine, 2.5), the cuprous compounds are the more stable.

Cupric Compounds

The hydrated *cupric ion*, $Cu(H_2O)_4{}^{++}$, is an ion with light blue color that occurs in aqueous solutions of cupric salts and in some of the hydrated crystals. The most important cupric salt is *copper sulfate*, which forms blue crystals, $CuSO_4 \cdot 5H_2O$. The metal copper is not sufficiently reactive to displace hydrogen ion from dilute acids (it is below hydrogen in the electromotive-force series, Chapter 15), and copper does not dissolve in acids unless an oxidizing agent is present. However, hot concentrated sulfuric acid is itself an oxidizing agent, and can dissolve the metal, and dilute sulfuric acid also slowly dissolves it in the presence of air:

$$Cu + 2H_2SO_4 + 3H_2O \longrightarrow CuSO_4 \cdot 5H_2O + SO_2$$

or

$$2Cu + 2H_2SO_4 + O_2 + 8H_2O \longrightarrow 2CuSO_4 \cdot 5H_2O$$

Copper sulfate, which has the common names *blue vitriol* and *bluestone*, is used in copper plating, in printing calico, in electric cells, and in the manufacture of other compounds of copper.

Cupric chloride, $CuCl_2$, can be made as yellow crystals by direct union of the elements. The hydrated salt, $CuCl_2 \cdot 2H_2O$, is blue-green in color, and its solution in hydrochloric acid is green. The blue-green color of the salt is due to its existence as a complex,

in which the chlorine atoms are bonded directly to the copper atom. The green solution contains ions $CuCl_3(H_2O)^-$ and $CuCl_4^{--}$. All of these ions are planar, the copper atom being at the center of a square formed by the four attached groups. The planar configuration is shown also by other complexes of copper, including the deep-blue ammonia complex, $Cu(NH_3)_4^{++}$ (Chapter 19).

Cupric bromide, $CuBr_2$, is a black solid obtained by reaction of copper and bromine or by solution of cupric oxide, CuO, in hydrobromic acid. It is interesting that cupric iodide, CuI_2, is unstable; when a solution containing cupric ion is added to an iodide solution there occurs an oxidation-reduction reaction, with precipitation of cuprous iodide, CuI:

$$2Cu^{++} + 4I^- \longrightarrow 2CuI(c) + I_2$$

This reaction occurs because of the extraordinary stability of cuprous iodide, which is discussed in the following section. The reaction is used in a method of quantitative analysis for copper, the liberated iodine being determined by titration with sodium thiosulfate solution.

Cupric hydroxide, $Cu(OH)_2$, forms as a pale blue gelatinous precipitate when alkali hydroxide or ammonium hydroxide is added to a cupric solution. It dissolves very readily in excess ammonium hydroxide, forming the deep-blue complex $Cu(NH_3)_4^{++}$. Cupric hydroxide is slightly amphoteric, and dissolves to a small extent in a very concentrated alkali, forming $Cu(OH)_4^{--}$.

The complex of cupric ion with tartrate ion, $C_4H_4O_6^{--}$, in alkaline solution is used as a test reagent (*Fehling's solution*) for organic reducing agents, such as certain sugars. This complex ion, $Cu(C_4H_4O_6)_2^{--}$, ionizes to give only a very small concentration of Cu^{++}, not enough to cause a precipitate of $Cu(OH)_2$ to form. The organic reducing agents reduce the copper to the unipositive state, and it then forms a brick-red precipitate of cuprous oxide, Cu_2O. This reagent is used in testing for sugar in the urine, in the diagnosis of diabetes.

Cuprous Compounds

Cuprous ion, Cu^+, is so unstable in aqueous solution that it undergoes auto-oxidation-reduction into copper and cupric ion:

$$2Cu^+ \longrightarrow Cu + Cu^{++}$$

Very few cuprous salts of oxygen acids exist. The stable cuprous compounds are either insoluble crystals containing covalent bonds or covalent complexes.

When copper is added to a solution of cupric chloride in strong hydrochloric acid a reaction occurs that results in the formation of a colorless solution containing cuprous chloride complex ions such as $CuCl_2^-$:

$$CuCl_4^{--} + Cu \longrightarrow 2CuCl_2^-$$

This complex ion involves two covalent bonds, its electronic structure being

$$\left[:\overset{..}{\underset{..}{Cl}}-Cu-\overset{..}{\underset{..}{Cl}}: \right]^-$$

Other cuprous complexes, $CuCl_3^{--}$ and $CuCl_4^{---}$, also exist.

If the solution is diluted with water a colorless precipitate of *cuprous chloride*, CuCl, forms. This precipitate also contains covalent bonds, each copper atom being bonded to four neighboring chlorine atoms and each chlorine atom to four neighboring copper atoms, with use of the outer electrons of the chloride ion. The structure, called the sphalerite structure, is closely related to that of diamond, with alternating carbon atoms replaced by copper and chlorine (Figure 21-3).

Cuprous bromide, CuBr, and *cuprous iodide*, CuI, are also colorless insoluble substances. The covalent bonds between copper and iodine in cuprous iodide are so strong as to make cupric iodide relatively unstable, as mentioned above.

Other stable cuprous compounds are the insoluble substances cuprous oxide, Cu_2O (red), cuprous sulfide, Cu_2S (black), cuprous cyanide, CuCN (white), and cuprous thiocyanate, CuSCN (white).

21-4.   The Compounds of Silver

Silver oxide, Ag_2O, is obtained as a dark-brown precipitate on the addition of sodium hydroxide to a solution of silver nitrate. It is slightly soluble, producing a weakly alkaline solution of silver hydroxide:

$$Ag_2O + H_2O \longrightarrow 2Ag^+ + 2OH^-$$

Silver oxide is used in inorganic chemistry to convert a soluble chloride, bromide, or iodide into the hydroxide. For example, cesium chloride solution can be converted into cesium hydroxide solution in this way:

$$2Cs^+ + 2Cl^- + Ag_2O + H_2O \longrightarrow 2AgCl + 2Cs^+ + 2OH^-$$

This reaction proceeds to the right because silver chloride is much less soluble than silver oxide.

The *silver halides*—AgF, AgCl, AgBr, and AgI—can be made by adding silver oxide to solutions of the corresponding halogen acids. Silver fluoride is very soluble in water, and the other halides are nearly insoluble. Silver chloride, bromide, and iodide form as curdy precipitates when the ions are mixed. They are respectively white, pale yellow, and yellow in color, and on exposure to light they slowly turn black, through photochemical decomposition. Silver chloride and bromide dissolve in ammonium hydroxide solution, forming the *silver ammonia complex* $Ag(NH_3)_2^+$ (Chapter 19); silver iodide does not dissolve in ammonium hydroxide. These reactions are used as qualitative tests for silver ion and the halide ions.

Other complex ions formed by silver, such as the silver cyanide complex $Ag(CN)_2^-$ and the silver thiosulfate complex $Ag(S_2O_3)_2^{---}$, have been mentioned in Chapter 19.

Silver nitrate, $AgNO_3$, is a colorless, soluble salt made by dissolving silver in nitric acid. It is used to cauterize sores. Silver nitrate is easily reduced to metallic silver by organic matter, such as skin or cloth, and is for this reason used in making indelible ink.

Silver ion is an excellent antiseptic, and several of the compounds of silver are used in medicine because of their germicidal power.

21-5. Photochemistry and Photography

Many chemical reactions are caused to proceed by the effect of light. For example, a dyed cloth may fade when exposed to sunlight because of the destruction of molecules of the dye under the influence of the sunlight. Reactions of this sort are called *photochemical reactions*. A very important photochemical reaction is the conversion of carbon dioxide and water into carbohydrate and oxygen in the leaves of plants, where the green substance chlorophyll serves as a photochemical catalyst.

One law of photochemistry, discovered by Grotthus in 1818, is that *only light that is absorbed is photochemically effective*. Hence a colored substance must be present in a system that shows photochemical reactivity with visible light. In the process of natural photosynthesis this substance is chlorophyll.

The second law of photochemistry, formulated in 1912 by Einstein, is that *one molecule of reacting substance may be activated and caused to react by the absorption of one photon.* In some systems, such as material containing rather stable dyes, many photons are absorbed by the molecules for each molecule that is decomposed; the fading of the dye by light is a slow and inefficient process in these materials. In some simple systems the absorption of one photon results in the reaction or decomposition of one molecule.

There are also chemical systems in which a *chain of reactions* may be set off by one light quantum. An example is the photochemical reaction of hydrogen and chlorine. A mixture of hydrogen and chlorine kept in the dark does not react at room temperature. When, however, it is illuminated with blue light, reaction immediately begins. Hydrogen is transparent to all visible light; chlorine, which owes its yellow-green color to its strong absorption of blue light, is the photochemically active constituent in the mixture. The absorption of a photon of blue light by a chlorine molecule splits the molecule into two chlorine atoms:

$$Cl_2 + h\nu \longrightarrow 2Cl$$

These chlorine atoms initiate a chain of reactions, as described in Section 16-7:

$$Cl + H_2 \longrightarrow HCl + H$$
$$H + Cl_2 \longrightarrow HCl + Cl$$

It may be observed that the mixture of hydrogen and chlorine explodes when exposed to blue light. The chain of reactions may be broken through the recombination of chlorine atoms to form chlorine molecules; this reaction occurs on the collison of two chlorine atoms with the wall of the vessel containing the gas or with another atom or molecule in the gas.

A photochemical reaction of much geophysical and biological importance is the formation of ozone from oxygen. Oxygen is practically transparent to visible light and to light in the near ultraviolet region, but it strongly absorbs light in the far ultraviolet region—in the region from 1600 Å to 2400 Å. Each photon that is absorbed dissociates an oxygen molecule into two oxygen atoms:

$$O_2 + h\nu \longrightarrow 2O$$

A reaction that does not require absorption of a photon then follows:

$$O + O_2 \longrightarrow O_3$$

Accordingly there are produced two molecules of ozone, O_3, for each photon absorbed. In addition, however, the ozone molecules can be destroyed by combining with oxygen atoms, or by a photochemical reaction. The reaction of combining with oxygen atom is

$$O + O_3 \longrightarrow 2O_2$$

The reactions of photochemical production of ozone and destruction of ozone lead to a photochemical equilibrium, which maintains a small concentration of ozone in the oxygen being irradiated. The layer of the atmosphere in which the major part of the ozone is present is about 24 km above the earth's surface; it is called the *ozone layer*.

The geophysical and biological importance of the ozone layer results from the absorption of light in the near ultraviolet region, from 2400 Å to 3600 Å, by the ozone. The photochemical reaction is

$$O_3 + h\nu \longrightarrow O + O_2$$

This reaction permits ozone to absorb ultraviolet light so strongly as to remove practically all of the ultraviolet light from the sunlight before it reaches the earth's surface. The ultraviolet light that it absorbs is photochemically destructive toward many of the organic molecules necessary in life processes, and if the ultraviolet light of sunlight were not prevented by the ozone layer from reaching the surface of the earth life in its present form could not exist.

Blueprint paper provides another interesting example of a photochemical reaction. Blueprint paper is made by treating paper with a solution of potassium ferricyanide and ferric citrate. Under action of light the citrate ion reduces the ferric ion to ferrous ion, which combines with ferricyanide to form the insoluble blue compound $KFeFe(CN)_6 \cdot H_2O$, Prussian blue. The unreacted substances are then washed out of the paper with water.

Photography

A photographic film is a sheet of cellulose acetate coated with a thin layer of gelatin in which very fine grains of silver bromide are suspended. This layer of gelatin and silver bromide is called the *photographic emulsion*. The silver halides are sensitive to light, and undergo photochemical decomposition. The gelatin increases this sensitivity, apparently because of the sulfur it contains.

When the film is briefly exposed to light some of the grains of silver bromide undergo a small amount of decomposition, perhaps forming a small particle of silver sulfide on the surface of the grain. The film can then be *developed* by treatment with an alkaline solution of an organic reducing agent, such as hydroquinone, the *developer*. This causes the silver bromide grains that have been sensitized to be reduced to metallic silver, whereas the unsensitized silver bromide grains remain unchanged. By this process the developed film reproduces the pattern of the light that exposed it. This film is called the *negative*, because it is darkest (with the greatest amount of silver) in the places that were exposed to the most light.

The undeveloped grains of silver halide are next removed, by treatment with a fixing bath, which contains thiosulfate ion, $S_2O_3^{--}$ (from sodium thiosulfate, "hypo," $Na_2S_2O_3 \cdot 5H_2O$). The soluble silver thiosulfate complex is formed:

$$AgBr + 2S_2O_3^{--} \longrightarrow Ag(S_2O_3)_2^{---} + Br^-$$

The fixed negative is then washed. Care must be taken not to transfer the negative from a used fixing bath, containing a considerable concentration of silver complex, directly to the wash water, as insoluble silver thiosulfate might precipitate in the emulsion:

$$2Ag(S_2O_3)_2^{---} \longrightarrow Ag_2S_2O_3(c) + 3S_2O_3^{--}$$

Since there are three ions on the right, and only two on the left, dilution causes the equilibrium to shift toward the right.

A positive print can be made by exposing print paper, coated with a silver halide emulsion, to light that passes through the superimposed negative, and then developing and fixing the exposed paper.

Sepia tones are obtained by converting the silver to silver sulfide, and gold and platinum tones by replacing silver by these metals.

Many other very interesting chemical processes are used in photography, especially for the reproduction of color.

The Chemistry of Color Photography

The electromagnetic waves of light of different colors have different wavelengths. In the visible spectrum these wavelengths extend from a little below 4000 Å (violet in color) to nearly 8000 Å (red in color). The sequence of colors in the visible region is shown in the next to the top diagram of Figure 21-1.

When white light (light containing all wavelengths in the visible region) is passed through a substance, light of certain wavelengths may be absorbed by the substance. The solar spectrum is shown in Figure 21-1. It consists of a background of white light, produced by the very hot gases in the sun, on which there are superimposed some dark lines, resulting from absorption of certain wavelengths by atoms in the cooler surface layers of the sun. It is seen that the yellow sodium lines, which occur as bright lines in the emission spectrum of sodium atoms, are shown as dark lines in the solar spectrum.

Molecules and complex ions in solution and in solid substances sometimes show sharp line spectra, but usually show rather broad absorption bands, as is indicated for the permanganate ion near the bottom of Figure 21-1. The permanganate ion has the power of absorbing light in the green region of the spectrum, permitting the blue-violet light and red light to pass through. The combination of blue-violet and red light appears magenta in color. We accordingly say that permanganate ion has a magenta color.

The human eye does not have the power of completely differentiating between light of one wavelength and that of another wavelength in the visible spectrum. Instead, it responds to three different wavelength regions in different ways. All of the colors that can be recognized by the eye can be composed from three fundamental colors. These may be taken as red-green (seen by the eye as yellow), which is complementary to blue-violet; blue-red, or magenta, which is complementary to green; and blue-green, or cyan, which is complementary to red. Three *primary colors*, such as these, need to be used in the development of any method of color representation.

An important modern method of color photography is the *Kodachrome method*, developed by the Kodak Research Laboratories. This method is illustrated in Figure 21-2. The film consists of several layers of emulsion, superimposed on a cellulose acetate base. The uppermost layer of photographic emulsion is the ordinary photographic emulsion, which is sensitive to blue and violet light. The second layer of photographic emulsion is a green-sensitive emulsion. It consists of a photographic emulsion that has been treated with a magenta-colored dye, which absorbs green light and sensitizes the silver bromide grains, thus making the emulsion sensitive to green light as well as to blue and violet light. The third photographic emulsion, red-sensitive emulsion, has been treated with a blue dye, which absorbs red light, making the emulsion sensitive to red light as well as to blue and violet (but not to green). Between the first layer and the middle layer there is a layer of yellow filter, containing a yellow dye, which during exposure prevents blue and violet light from penetrating to the lower layers. Accordingly, when such a film is exposed to light the blue-sensitive emulsion is exposed by blue light, the middle emulsion is exposed by green light, and the bottom emulsion is exposed by red light.

The exposure of the different layers of photographic emulsion in the film is illustrated diagrammatically as Process 1 in Figure 21-2.

The development of Kodachrome film involves several steps, which are represented as Processes 2 to 9 in Figure 21-2. First (Process 2) the Kodachrome film after exposure is developed with an ordinary black and white developer, which develops the silver negative in all three emulsions. Then after simple washing in water (not shown in the figure) the film is exposed through the back to red light, which makes the previously unexposed silver bromide in the red-sensitive emulsion capable of development (Process 3). The film then passes into a special developer, called cyan developer and coupler (Process 4). This mixture of chemical substances has the power of interacting with the exposed silver bromide grains in such a way as to deposit a cyan dye in the bottom layer, at the same time that the silver bromide grains are reduced to metallic silver. The cyan dye is deposited only in the regions occupied by the sensitized silver bromide

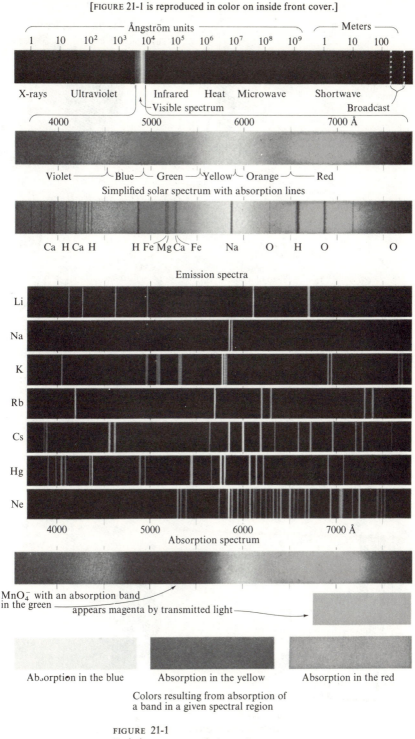

[FIGURE 21-1 is reproduced in color on inside front cover.]

Ångström units

1 10 10^2 10^3 10^4 10^5 10^6 10^7 10^8 10^9

Meters

1 10 100

X-rays   Ultraviolet   Infrared   Heat   Microwave   Shortwave

Visible spectrum   Broadcast

4000   5000   6000   7000 Å

Violet ——— Blue — Green — Yellow — Orange — Red

Simplified solar spectrum with absorption lines

Ca H Ca H   H Fe Mg Ca Fe   Na   O   H   O   O

Emission spectra

Li

Na

K

Rb

Cs

Hg

Ne

4000   5000   6000   7000 Å

Absorption spectrum

MnO_4^- with an absorption band
in the green —— appears magenta by transmitted light ——

Absorption in the blue   Absorption in the yellow   Absorption in the red

Colors resulting from absorption of
a band in a given spectral region

FIGURE 21-1
Emission spectra and absorption spectra.

[FIGURE 21-2 is reproduced in color on inside back cover.]

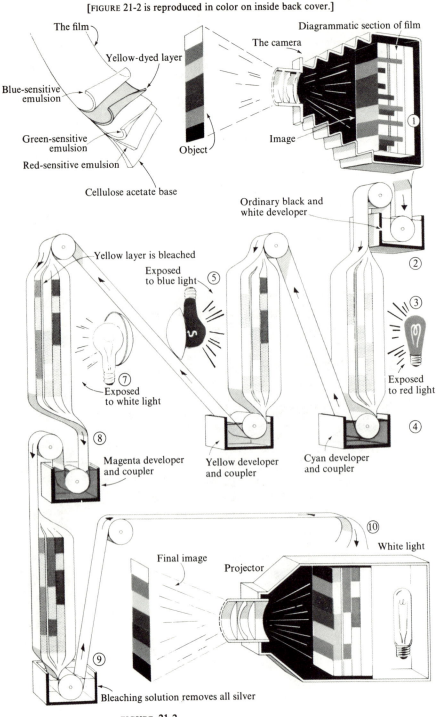

The film

Yellow-dyed layer

Blue-sensitive emulsion

Green-sensitive emulsion

Red-sensitive emulsion

Cellulose acetate base

The camera

Diagrammatic section of film

Object

Image

①

Ordinary black and white developer

②

Yellow layer is bleached

Exposed to blue light ⑤

③

Exposed to red light

Exposed to white light ⑦

Magenta developer and coupler ⑧

Yellow developer and coupler

Cyan developer and coupler ④

⑩

White light

Final image

Projector

⑨

Bleaching solution removes all silver

FIGURE 21-2

The Kodachrome process of color photography.

grains. The next process (Process 5) consists in exposure to blue light from the front of the negative. The blue light is absorbed by the yellow dye, and so affects only the previously unexposed grains in the first emulsion, the blue-sensitive emulsion. This emulsion is then developed in a special developer (Process 6), a yellow developer and coupler, which deposits a yellow dye in the neighborhood of these recently exposed grains. The film is then exposed to white light, to sensitize the undeveloped silver bromide grains in the middle emulsion, the yellow layer is bleached, the middle emulsion is developed with a magenta developer and coupler (Process 8), and the deposited metallic silver in all three solutions is removed by a bleaching solution (Process 9), leaving only a film containing deposited cyan, yellow, and magenta dyes in the three emulsion layers, in such a way that by transmitted light the originally incident colors are reproduced (Process 10).

21-6.   The Compounds of Gold

$KAu(CN)_2$, the potassium salt of the complex *gold(I) cyanide ion* $Au(CN)_2^-$, with electronic structure

$$[:N\equiv C—Au—C\equiv N:]^-$$

is an example of a gold(I) compound.* The *gold(I) chloride* complex $AuCl_2^-$ has a similar structure, and the *halides*, $AuCl$, $AuBr$, and AuI, resemble the corresponding halogenides of silver.

Gold dissolves in a mixture of concentrated nitric and hydrochloric acids to form *hydrogen aurichloride*, $HAuCl_4$. This acid contains the aurichloride ion, $AuCl_4^-$, a square planar complex ion:

$$\left[\begin{array}{c} :\ddot{Cl}: \\ | \\ :\ddot{Cl}—Au—\ddot{Cl}: \\ | \\ :\ddot{Cl}: \end{array} \right]^-$$

Hydrogen aurichloride can be obtained as a yellow crystalline substance, which forms salts with bases. When heated it forms *gold(III) chloride*, $AuCl_3$, and then gold(I) gold(III) chloride, Au_2Cl_4, and then gold(I) chloride, $AuCl$. On further heating all the chlorine is lost, and pure gold remains.

---

*The gold(I) and gold(III) compounds are often called *aurous* and *auric* compounds, respectively.

21-7. Color and Mixed Oxidation States

The gold halides provide examples of an interesting phenomenon—the *deep, intense color often observed for a substance that contains an element in two different oxidation states.* Gold(I) gold(III) chloride, Au_2Cl_4, is intensely black, although both gold(I) chloride and gold(III) chloride are yellow. Cesium gold(I) gold(III) bromide, $Cs_2^+[AuBr_2]^-[AuBr_4]^-$, is deep black in color and both $CsAuBr_2$ and $CsAuBr_4$ are much lighter. Black mica (biotite) and black tourmaline contain both ferrous and ferric iron. Prussian blue is ferrous ferricyanide; ferrous ferrocyanide is white, and ferric ferricyanide is light yellow. When copper is added to a light green solution of cupric chloride, a deep brownish-black solution is formed, before complete conversion to the colorless cuprous chloride complex.

The theory of this phenomenon is not understood. The very strong absorption of light is presumably connected with the transfer of an electron from one atom to another of the element that is present in two valence states.

21-8. The Properties and Uses of Zinc, Cadmium, and Mercury

Zinc is a bluish-white, moderately hard metal. It is brittle at room temperature, but is malleable and ductile between 100° and 150°C, and becomes brittle again above 150°C. It is an active metal, above hydrogen in the electromotive-force series, and it displaces hydrogen even from dilute acids. In moist air zinc is oxidized and becomes coated with a tough film of basic zinc carbonate, $Zn_2CO_3(OH)_2$, which protects it from further corrosion. This behavior is responsible for its principal use, in protecting iron from rusting. Iron wire or sheet iron is *galvanized* by cleaning with sulfuric acid or a sandblast, and then dipping in molten zinc; a thin layer of zinc adheres to the iron. Galvanized iron in some shapes is made by electroplating zinc onto the iron pieces.

Zinc is also used in making alloys, the most important of which is *brass* (the alloy with copper), and as a reacting electrode in dry cells and wet cells.

Cadmium is a bluish-white metal of pleasing appearance. It has found increasing use as a protective coating for iron and steel. The cadmium plate is deposited electrolytically from a bath containing the cadmium cyanide complex ion, $Cd(CN)_4^{--}$. Cadmium is also used in some alloys, such as the low-melting alloys needed for automatic fire extinguishers. *Wood's metal*, which melts at 65.5°C, contains 50% Bi, 25% Pb, 12.5% Sn, and 12.5% Cd. Because of the toxicity of compounds of elements of this

TABLE 21-3
Some Physical Properties of Zinc, Cadmium, and Mercury

	Atomic Number	Atomic Weight	Density (g/cm³)	Melting Point	Boiling Point	Metallic Radius	Color	Heat of Sublimation at 25°C
Zinc	30	65.37	7.14	419.4°C	907°C	1.38 Å	Bluish-white	131 kJ mole⁻¹
Cadmium	48	112.40	8.64	320.9°	767°	1.54	Bluish-white	113
Mercury	80	200.59	13.55	−38.89°	356.9°	1.57	Silvery-white	61

group, care must be taken not to use cadmium-plated vessels for cooking, and not to inhale fumes of zinc, cadmium, or mercury.

Mercury is the only metal that is liquid at room temperature (cesium melts at 28.5°C, and gallium at 29.8°C). It is unreactive, being below hydrogen in the electromotive-force series. Because of its unreactivity, fluidity, high density, and high electric conductivity it finds extensive use in thermometers, barometers, and many special kinds of scientific apparatus.

The alloys of mercury are called *amalgams*. Amalgams of silver, gold, and tin are used in dentistry.

The low melting points and small values of the heats of sublimation of zinc and its congeners (Table 21-3) are attributed to the fact that the gas atoms in the normal state contain only completed subshells of electrons (Russell-Saunders symbol $^1S_0$), and hence have no unpaired electrons that can be used to form chemical bonds. The first excited state of the zinc atom, ^3P, is less stable than the normal state by 385 kJ mole⁻¹. It has two unpaired electrons ($4s4p$), corresponding to bivalence.

21-9. Compounds of Zinc and Cadmium

Values of the standard enthalpy of formation of some compounds of zinc, cadmium, and mercury are given in Table 21-4. It is seen that there is a close similarity between zinc and cadmium, and that mercury differs considerably from its two lighter congeners.

The *zinc ion*, $Zn(H_2O)_4^{++}$, is a colorless ion formed by solution of zinc in acid. It is poisonous to man and to bacteria, and is used as a disinfectant. It forms tetraligated complexes readily, such as $Zn(NH_3)_4^{++}$, $Zn(CN)_4^{--}$, and $Zn(OH)_4^{--}$. The white precipitate of *zinc hydroxide*, $Zn(OH)_2$, which

TABLE 21-4

Standard Enthalpy of Formation of Compounds of Zinc, Cadmium, and Mercury at 25°C (kJ mole^{-1})

	M = Zn	Cd	Hg
M	0(c)	0(c)	0(l)
M(g)	130	113	61
M$^+$(g)	1043	987	1076
M^{++}(g)	2782	2624	2885
M^{++}(aq)	−152	−72	
M$_2$O(c)			−91
MO(c)	−348	−255	−91
MH(g)	228	262	243
MF(g)			58
MF$_2$(c)		−690	
MCl(g)	4	19	79
M$_2$Cl$_2$(c)			−265
MCl$_2$(c)	−416	−389	−230
MBr(g)		50	96
M$_2$Br$_2$(c)			−209
MBr$_2$(c)	−327	−314	−169
MI(g)	63	82	138
M$_2$I$_2$(c)			−121
MI$_2$(c)	−209	−201	−105
MS(g)	−59		13
MS(c)	−203	−144	−58
MSe(g)		7	67
MSe(c)	−142		−21
MTe(g)	126		
MTe(c)	−126	−102	

ïorms when ammonium hydroxide is added to a solution containing zinc ion, dissolves in excess ammonium hydroxide, forming the zinc ammonia complex. The zinc hydroxide complex, $Zn(OH)_4^{--}$, which is called *zincate ion*, is similarly formed on solution of zinc hydroxide in an excess of strong base; zinc hydroxide is amphoteric.

Zinc sulfate, $ZnSO_4 \cdot 7H_2O$, is used as a disinfectant and in dyeing calico, and in making *lithopone*, which is a mixture of barium sulfate and zinc sulfide used as a white pigment in paints:

$$Ba^{++}S^{--} + Zn^{++}SO_4^{--} \longrightarrow BaSO_4 + ZnS$$

Zinc oxide, ZnO, is a white powder (yellow when hot) made by burning zinc vapor or by roasting zinc ores. It is used as a pigment (zinc white), as a filler in automobile tires, adhesive tape, and other articles, and as an antiseptic (zinc oxide ointment).

Zinc sulfide, ZnS, is the only white sulfide (colorless in large crystals, if pure) among the sulfides of the common metals. It occurs in nature as the minerals sphalerite and wurtzite. Sphalerite has cubic (tetrahedral) symmetry, with 4Zn in the unit cell, at $0\ 0\ 0$, $0\ \frac{1}{2}\ \frac{1}{2}$, $\frac{1}{2}\ 0\ \frac{1}{2}$, and $\frac{1}{2}\ \frac{1}{2}\ 0$, and 4S at $\frac{1}{4}\ \frac{1}{4}\ \frac{1}{4}$, $\frac{1}{4}\ \frac{3}{4}\ \frac{3}{4}$, $\frac{3}{4}\ \frac{1}{4}\ \frac{3}{4}$, and $\frac{3}{4}\ \frac{3}{4}\ \frac{1}{4}$, as shown in Figure 21-3. Each atom is surrounded tetrahedrally by four of the other sort. The edge of the unit cube has the value 5.40 Å, which leads to 2.34 Å for the length of the Zn—S bond. Wurtzite is hexagonal; its structure, shown in Figure 21-4, is closely related to that of sphalerite, the relationship being similar to that between hexagonal and cubic closest packing of spheres. The bond length is the same as for sphalerite.

Many binary compounds crystallize with the sphalerite or the wurtzite structure. The assumption of these structures, with ligancy 4, rather than the NaCl structure ($L = 6$) or CsCl structure ($L = 8$) by BeO, BeS, BeSe, and BeTe may be attributed to the small size of the beryllium ion (Section 6-8). For many other substances with these tetrahedral structures the bonds are essentially covalent and the structures are determined by the availability of tetrahedral sp^3 bond orbitals for the atoms; these substances include CuF, CuCl, CuBr, CuI, ZnO, ZnS, ZnSe, ZnTe, GaN, GaP, GaAs, and GaSb, and many of the corresponding compounds of the heavier congeners of these metals.

The compounds of cadmium are closely similar to those of zinc. *Cadmium ion*, Cd^{++}, is a colorless ion, which forms complexes $(Cd(NH_3)_4^{++}$, $Cd(CN)_4^{--})$ similar to those of zinc. The cadmium hydroxide ion, $Cd(OH)_4^{--}$, is not stable, and *cadmium hydroxide*, $Cd(OH)_2$, is formed as a white precipitate by addition of sodium hydroxide to a solution containing cadmium ion. The precipitate is soluble in ammonium hydroxide or in a solution containing cyanide ion. *Cadmium oxide*, CdO, is a brown powder obtained by heating the hydroxide or burning the metal. *Cadmium sulfide*, CdS, is a bright yellow precipitate obtained by passing hydrogen sulfide through a solution containing cadmium ion; it is used as a pigment (*cadmium yellow*).

21-10. Compounds of Mercury

The mercuric compounds, in which mercury is bipositive, differ somewhat in their properties from the corresponding compounds of zinc and cadmium. The differences are due in part to the very strong tendency of the mercuric ion, Hg^{++}, to form covalent bonds. Thus the covalent crystal

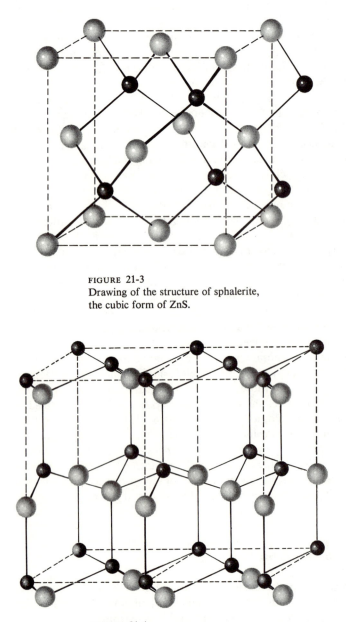

FIGURE 21-3
Drawing of the structure of sphalerite,
the cubic form of ZnS.

FIGURE 21-4
Drawing of the structure of wurtzite,
the hexagonal form of ZnS.

Mercuric   Mercurous   Mercuric   Mercurous
ion Hg⁺⁺   ion Hg₂⁺⁺   chloride HgCl₂   chloride Hg₂Cl₂

FIGURE 21-5
The structure of the mercuric ion, mercurous ion, mercuric chloride
molecule, and mercurous chloride molecule. In the mercurous ion and
the two molecules the atoms are held together by covalent bonds.

mercuric sulfide, HgS, is far less soluble than cadmium sulfide or zinc
sulfide.

Mercuric nitrate, $Hg(NO_3)_2$ or $Hg(NO_3)_2 \cdot \frac{1}{2}H_2O$, is made by dissolving
mercury in hot concentrated nitric acid:

$$Hg + 4HNO_3 \longrightarrow Hg(NO_3)_2 + 2NO_2 + 2H_2O$$

It hydrolyzes on dilution, unless a sufficient excess of acid is present, to
form basic mercuric nitrates, such as $HgNO_3OH$, as a white precipitate.

Mercuric chloride, $HgCl_2$, is a white crystalline substance usually made
by dissolving mercury in hot concentrated sulfuric acid, and then heating
the dry mercuric sulfate with sodium chloride, subliming the volatile
mercuric chloride:

$$Hg + 2H_2SO_4 \longrightarrow HgSO_4 + SO_2 + H_2O$$
$$HgSO_4 + 2NaCl \longrightarrow Na_2SO_4 + HgCl_2$$

A dilute solution of mercuric chloride (about 0.1%) is used as a disinfectant. Any somewhat soluble mercuric salt would serve equally well,
except for the tendency of mercuric ion to hydrolyze and to precipitate
basic salts. Mercuric chloride has only a small tendency to hydrolyze because its solution contains only a small concentration of mercuric ion, the
mercury being present mainly as un-ionized covalent molecules:

$$: \overset{..}{\underset{..}{Cl}} - Hg - \overset{..}{\underset{..}{Cl}} :$$

The electronic structure of these molecules, which have a linear configuration (Figure 21-5), is analogous to that of the gold(I) chloride complex,
$AuCl_2^-$. The ease of sublimation of mercuric chloride (melting point
275°C, boiling point 301°C) results from the stability of these molecules.

Mercuric chloride, like other soluble salts of mercury, is very poisonous when taken internally. The mercuric ion combines strongly with proteins; in the human body it acts especially on the tissues of the kidney, destroying the ability of this organ to remove waste products from the blood. Egg white and milk are swallowed as antidotes; their proteins precipitate the mercury in the stomach.

With ammonium hydroxide, mercuric chloride forms a white precipitate, $HgNH_2Cl$:

$$HgCl_2 + 2NH_3 \longrightarrow HgNH_2Cl(c) + NH_4^+ + Cl^-$$

Mercuric sulfide, HgS, is formed as a black precipitate when hydrogen sulfide is passed through a solution of a mercuric salt. It can also be made by rubbing mercury and sulfur together in a mortar. The black sulfide (which also occurs in nature as the mineral *metacinnabarite*) is converted by heat into the red form (cinnabar). Mercuric sulfide is the most insoluble of metallic sulfides. It is not dissolved even by boiling concentrated nitric acid, but it does dissolve in aqua regia, under the combined action of the nitric acid, which oxidizes the sulfide to free sulfur, and hydrochloric acid, which provides chloride ion to form the stable complex $HgCl_4^{--}$:

$$3HgS + 12HCl + 2HNO_3 \longrightarrow 3HgCl_4^{--} + 6H^+ + \tfrac{3}{8}S_8 + 2NO + 4H_2O$$

Mercuric oxide, HgO, is formed as a yellow precipitate by adding a base to a solution of mercuric nitrate or as a red powder by heating dry mercuric nitrate or, slowly, by heating mercury in air. The yellow and red forms differ only in grain size; it is a common phenomenon that red crystals (such as potassium dichromate or potassium ferricyanide) form a yellow powder when they are ground up. Mercuric oxide liberates oxygen when it is strongly heated.

Mercuric fulminate, $Hg(CNO)_2$, is made by dissolving mercury in nitric acid and adding ethyl alcohol, C_2H_5OH. It is a very unstable substance, which detonates when it is struck or heated, and it is used for making detonators and percussion caps.

Mercurous nitrate, $Hg_2(NO_3)_2$, is formed by reduction of a mercuric nitrate solution with mercury:

$$Hg^{++} + Hg \longrightarrow Hg_2^{++}$$

The solution contains the *mercurous ion*, Hg_2^{++}, a colorless ion that has a unique structure; it consists of two mercuric ions plus two electrons, which form a covalent bond between them (Figure 21-5):

$$2Hg^{++} + 2e^- \longrightarrow [Hg\text{—}Hg]^{++}$$

Mercurous chloride, Hg_2Cl_2, is an insoluble white crystalline substance obtained by adding a solution containing chloride ion to a mercurous nitrate solution:

$$Hg_2^{++} + 2Cl^- \longrightarrow Hg_2Cl_2(c)$$

It is used in medicine under the name *calomel*. The mercurous chloride molecule (Figure 21-5) has the linear covalent structure

$$:\overset{..}{\underset{..}{Cl}}-Hg-Hg-\overset{..}{\underset{..}{Cl}}:$$

The precipitation of mercurous chloride and its change in color from white to black on addition of ammonium hydroxide are used as the test for mercurous mercury in qualitative analysis. The effect of ammonium hydroxide is due to the formation of finely divided mercury (black) and mercuric aminochloride (white) by an auto-oxidation-reduction reaction:

$$Hg_2Cl_2 + 2NH_3 \longrightarrow Hg + HgNH_2Cl + NH_4^+ + Cl^-$$

Mercurous sulfide, Hg_2S, is unstable, and when formed as a brownish-black precipitate by action of sulfide ion on mercurous ion it immediately decomposes into mercury and mercuric sulfide:

$$Hg_2^{++} + S^{--} \longrightarrow Hg_2S \longrightarrow Hg + HgS$$

21-11. Gallium, Indium, and Thallium

The elements of group IIIb—gallium, indium, and thallium—are rare and have little practical importance. Their principal compounds represent oxidation state $+3$; thallium also forms compounds in which it has oxidation number $+1$. Gallium is liquid from 29°C, its melting point, to 1700°C, its boiling point. It has found use as the liquid in quartz-tube thermometers, which can be used to above 1200°C.

Exercises

21-1. What is the electronic structure of the Ag^+ ion? Of the Cu^{++} ion?

21-2. In what form does copper exist in a cupric sulfate solution? In a strong hydrochloric acid solution? In an ammoniacal solution? In a solution of potassium cyanide?

21-3. Under what conditions can dilute sulfuric acid dissolve copper? Write an equation for the reaction.

21-4. Write the equation for the formation of hydrogen aurichloride by solution of gold in a mixture of nitric and hydrochloric acids, assuming that nitric oxide, NO, is also produced.

21-5. If a solution containing cupric ion and a solution containing iodide ion are mixed, a precipitate of cuprous iodide is formed, and free iodine is liberated. Write the equation for this reaction, assuming that iodide ion is present in excess, leading to the formation of triiodide ion.

21-6. To what is the black color of biotite and black tourmaline attributed?

21-7. Assuming that Equation 6-1 can be applied to the formation of compounds in their standard states from elements in their standard states, discuss the values of enthalpy of formation of the monohalogenides in Table 21-2. Should silver be assigned an electronegativity slightly larger that that of copper? How much larger?

21-8. Describe the electronic structure of the mercurous ion, the mercuric ion, the mercurous chloride molecule, and the mercuric chloride molecule. Compare the total number of electrons surrounding each mercury atom with the number in the nearest noble gas. What hybrid orbitals are used in bond formation?

21-9. Write the equation for the reaction of zinc with hydrochloric acid. Would you expect zinc to dissolve in a concentrated solution of sodium hydroxide? If so, write the equation for this reaction.

21-10. The electroneutrality principle indicates that the bonds between gold and carbon in the ion $Au(CN)_2^-$ are double bonds. What is the electronic structure of the gold atom in this complex? What orbitals are occupied by unshared electron pairs? By shared electron pairs?

21-11. What is the electronic structure of the zincate ion, $Zn(OH)_4^{--}$?

21-12. Discuss some of the enthalpy values in Table 21-3 in relation to the electronegativities of the elements.

21-13. Can you advance a likely explanation of the low melting points and boiling points of zinc, cadmium, and mercury, based on their electronic structure?

21-14. The energy of dissociation of the oxygen molecule is 494 kJ mole^{-1}. Calculate the wavelength of the photon of longest wavelength with enough energy to dissociate the molecule, and compare with the statement in Section 21-5 that ultraviolet light with wavelength less than 2400 Å is able to effect this dissociation.

21-15. From values of the enthalpy of formation of ozone and atomic oxygen given in Table 7-1, calculate the maximum wavelength of a photon with enough energy to dissociate an ozone molecule into an oxygen molecule and an oxygen atom if it were absorbed by the ozone. (Answer: 11,180 Å.)

21-16. In December 1962 it was announced that xenon difluoride, XeF_2, had been made by a photochemical reaction. The first step in this reaction is presumably the dissociation of the fluorine molecule into two fluorine atoms through the absorption of a photon with sufficient energy to effect the dissociation. It has been found by experiment that the bond energy in the F_2 molecule is about 160 kJ mole^{-1}. What is the longest wavelength of light that would be predicted to produce this photochemical reaction, if it were absorbed?

21-17. If you were refining the electronegativity table (Table 6-4) by introducing different values of the electronegativity of an element in different oxidation states, what difference in values for copper(I) and copper(II) would you assign by applying Equation 6-1 to the standard enthalpy values for substances such as $CuCl(c)$ and $CuCl_2(c)$ in Table 21-2?

21-18. (*a*) Copper and zinc form stable compounds in which they are bipositive. In general, the values of $\Delta H°_f$ (Tables 21-2 and 21-4) are more negative for the compounds of zinc(II) than for the corresponding compounds of copper(II). (*b*) On the other hand, many stable compounds of copper(I) are known, whereas the corresponding compounds of zinc(I) do not exist. By what difference in atomic properties of copper and zinc would you explain these two facts?

22

Titanium, Vanadium, Chromium, and Manganese and Their Congeners

In the present chapter we shall conclude the discussion of the chemistry of the transition metals. This chapter deals with the chemistry of chromium and manganese and their congeners, of groups VIa and VIIa of the periodic table, and also the preceding elements titanium and vanadium and their congeners, of groups IVa and Va. These elements are not so well known nor so important as some other transition elements, especially iron and nickel, but their chemistry is interesting, and serves well to illustrate the general principles discussed in preceding chapters.

22-1. The Electronic Structures of Titanium, Vanadium, Chromium, and Manganese and Their Congeners

The electronic structures of the elements of groups IVa, Va, VIa, and VIIa as represented in the energy-level diagram (Figure 5-11), are given in Table 5-5. Each of the elements has either one electron or two electrons in the s orbital of the outermost shell. In addition, there are two, three, four, or five electrons in the *d* orbitals of the next inner shell. Reference to Figure 5-11 shows that the heaviest elements of these groups—thorium, protactinium, uranium, and neptunium—are thought to have the additional two to five electrons, respectively, in the 5*f* subshell, rather than the 6*d* subshell.

The Russell-Saunders symbol for titanium and its congeners in the normal state is $^3F_2$, that for vanadium and tantalum is $^4F_{3/2}$, that for niobium is $^6D_{1/2}$, that for chromium and molybdenum is $^7S_3$, that for tungsten is $^5D_0$, and that for manganese and its congeners is $^6S_{5/2}$.

The oxidation state $+2$, corresponding to the loss of two electrons, is an important one for all of these elements. In particular, the elements in the first long period form the ions Ti^{++}, V^{++}, Cr^{++}, and Mn^{++}. Several other oxidation states, involving the loss or sharing of additional electrons, are also represented by compounds of these elements. The maximum oxidation state is that corresponding to the loss or sharing of all of the electrons in the *d* orbitals of the next inner shell, as well as the two electrons in the outermost shell. Accordingly, the maximum oxidation numbers of titanium, vanadium, chromium, and manganese are $+4$, $+5$, $+6$, and $+7$, respectively.

Values of the standard enthalpy of formation of compounds of titanium and its congeners are given in Table 22-2, of vanadium and its congeners in Table 22-3, of chromium and its congeners in Table 22-5, and of manganese and rhenium in Table 22-6. For each set of congeners there is seen to be an increase in stability of compounds representing higher oxidation states with increase in atomic number. Many of the enthalpy values correspond satisfactorily to the electronegativity values of the elements, but there are some as yet unexplained deviations, such as the very large heats of formation of some compounds of uranium.

22-2. Titanium, Zirconium, Hafnium, and Thorium

The elements of group IVa of the periodic system are titanium, zirconium, hafnium, and thorium. Some of the properties of the elementary substances are given in Table 22-1.

Titanium occurs in the minerals *rutile*, TiO_2, and *ilmenite*, $FeTiO_3$. It forms compounds representing oxidation states $+2$, $+3$, and $+4$. Pure *titanium dioxide*, TiO_2, is a white substance. As a powder it has great power of scattering light, which makes it an important pigment. It is used in special paints and in face powders. Crystals of titanium dioxide (rutile) colored with small amounts of other metal oxides have been made recently for use as gems. *Titanium tetrachloride*, $TiCl_4$, is a molecular liquid at room temperature. On being sprayed into air it hydrolyzes,

TABLE 22-1
Some Properties of Titanium, Vanadium, Chromium, and Manganese and Their Congeners

	Atomic Number	Atomic Weight	Density (g cm^{-3})	Melting Point	Boiling Point	Metallic Radius*
Titanium	22	47.90	4.44	1800°C	3000°C	1.47 Å
Vanadium	23	50.942	6.06	1700°	3000°	1.34
Chromium	24	51.996	7.22	1920°	2330°	1.27
Manganese	25	54.9380	7.26	1260°	2150°	1.26
Zirconium	40	91.22	6.53	1860°		1.60
Niobium	41	92.906	8.21	2500°		1.46
Molybdenum	42	95.94	10.27	2620°	4700°	1.39
Hafnium	72	178.49	13.17	2200°		1.36
Tantalum	73	180.948	16.76	2850°		1.46
Tungsten	74	183.85	19.36	3382°	6000°	1.39
Rhenium	75	186.2	21.10	3167°		1.37
Thorium	90	232.038	11.75	1850°	3500°	1.80
Uranium	92	238.03	18.97	1690°		1.52

*For ligancy 12.

forming hydrogen chloride and fine particles of titanium dioxide; for this reason it is sometimes used in making smoke screens:

$$TiCl_4 + 2H_2O \longrightarrow TiO_2 + 4HCl$$

Titanium metal is very strong, light (density 4.44 g/cm³), refractory (melting point 1800°C), and resistant to corrosion. Since 1950 it has been produced in quantity, and has found many uses for which a light, strong metal with high melting point is needed; for example, it is used in airplane wings where the metal is in contact with exhaust flame or is subject to frictional heating. The availability of this metal has made the supersonic plane possible.

Zirconium occurs in nature principally as the mineral *zircon*, ZrSiO₄. Zircon crystals are found in a variety of colors—white, blue, green, and red—and because of its beauty and hardness (7.5) the mineral is used as a semiprecious stone. The principal oxidation state of zirconium is +4; the states +2 and +3 are represented by only a few compounds.

Hafnium is closely similar to zirconium, and natural zirconium minerals usually contain a few percent of hafnium. The element was not discovered until 1923, and it has found little use.

Thorium is found in nature as the mineral *thorite*, ThO₂, and in *monazite sand*, which consists of thorium phosphate mixed with the phosphates of the lanthanons (Section 18-7). The principal use of thorium is in the manufacture of gas mantles, which are made by saturating cloth fabric with thorium nitrate, Th(NO₃)₄, and cerium nitrate, Ce(NO₃)₄. When the

TABLE 22-2
Standard Enthalpy of Formation of Compounds of Titanium, Zirconium, Hafnium, and Thorium at 25°C (kJ mole⁻¹)

	M = Ti	Zr	Hf	Th
M(c)	0	0	0	0
M(g)	472	611	703	572
MO(c)	−519			
M_2O_3(c)	−1520			
MO_2(c)	−945	−1094	−1113	−1222
MF_2(c)	−828	−962		
MF_3(c)	−1318	−1464		
MF_4(c)	−1548	−1862		−1996
MCl(g)	510			
MCl_2(c)	−477	−607		
MCl_3(c)	−690	−870		
MCl_4	−761(g)	−975(c)		−1192(c)
MBr_2(c)	−397	−502		
MBr_3(c)	−552	−728		
MBr_4(c)	−649	−803		−950
MI_2(c)	−255	−377		
MI_3(c)	−335	−536		
MI_4(c)	−424	−544		−548
M_2S_3(c)				−1096
MN(c)	−338	−365		
MC(c)	−185	−188		

treated cloth is burned, there remains a residue of thorium dioxide and cerium dioxide, ThO_2 and CeO_2, which has the property of exhibiting a brilliant white luminescence when it is heated to a high temperature. Thorium dioxide is also used in the manufacture of laboratory crucibles, for use at temperatures as high as 2300°C. Thorium can be made to undergo nuclear fission, and it may become an important nuclear fuel (Chapter 26).

22-3.   Vanadium, Niobium, Tantalum, and Protactinium

Vanadium is the most important element of group Va. It finds extensive use in the manufacture of special steels. Vanadium steel is tough and strong, and is used in automobile crank shafts and for similar purposes. The principal ores of vanadium are *vanadinite*, $Pb_5(VO_4)_3Cl$, and *carnotite*, $K(UO_2)VO_4 \cdot \frac{3}{2}H_2O$. The latter mineral is also important as an ore of uranium.

TABLE 22-3
Standard Enthalpy of Formation of Compounds of Vanadium, Niobium, and Tantalum at 25°C ($kJ\ mole^{-1}$)

	M = V	Nb	Ta
M(c)	0	0	0
M(g)	515	772	782
MO(g)	230		
MO(c)	−418		
M_2O_3(c)	−1238		
MO_2(c)	−715	−812	
M_2O_5(c)	−1561	−1937	−2092
MCl_2(c)	−427		
MCl_3(c)	−598		
MCl_4(l)	−577		
$MOCl_3$(c)	−720	−887	
MBr_5(c)		−556	−598
MN(c)	−172		−243

TABLE 22-4
Superconduction Critical Temperatures of Elements

III	IVa	Va	VIa	VIIa	VIII	IIb	IIIb	IVb
Al 1.18°K								
Sc	Ti 0.4°	V 5.2°				Zn 0.86°	Ga 1.09°	
Y	Zr 0.54°	Nb 9.2°	Mo 0.9°	Tc 8.2°	Ru 0.5°	Cd 0.55°	In 3.40°	Sn 3.72°
La 6.1°	Hf 0.8°	Ta 4.4°	W 1.0°	Re 1.70°	Os 0.7°	Hg 4.15°	Tl 2.38°	Pb 7.21°

The chemistry of vanadium is very complex. The element forms compounds representing the oxidation states +2, +3, +4, and +5. The hydroxides of bipositive and terpositive vanadium are basic, and those of the higher oxidation states are amphiprotic. The compounds of vanadium are striking for their varied colors. The bipositive ion, V^{++}, has a deep violet color; the terpositive compounds, such as *potassium vanadium alum*, $KV(SO_4)_2 \cdot 12H_2O$, are green; the dark-green substance *vanadium dioxide*, VO_2, dissolves in acid to form the blue *vanadyl ion*, VO^{++}. *Vanadium(V) oxide*, V_2O_5, an orange substance, is used as a catalyst in the contact process for making sulfuric acid. *Ammonium metavanadate*, NH_4VO_3, which forms yellow crystals from solution, is used for making preparations of vanadium(V) oxide for the contact process.

Niobium (columbium) and *tantalum* usually occur together, as the minerals *columbite*, $FeNb_2O_6$, and *tantalite*, $FeTa_2O_6$. Niobium finds some use as a constituent of alloy steels. *Tantalum carbide*, TaC, a very hard substance, is used in making high-speed cutting tools.

Protactinium is a radioactive element (Chapter 26) that occurs in minute amounts in all uranium ores.

22-4. Superconductivity

In 1908 the Dutch physicist Heike Kamerlingh Onnes (1853–1926) succeeded in liquefying helium (normal boiling point 4.6°K). By boiling the liquid helium under reduced pressure he was able to reach temperatures as low as 1.15°K. While studying the properties of substances at these very low temperatures he discovered that mercury at about 4.1°K undergoes a transition to a state with properties different from those of the metal at higher temperatures. The most striking change is in the electric resistance, which drops sharply to zero. This state is called the *superconducting state*.

Many metallic elements are superconducting, as shown in Table 22-4, in which values of the critical temperature are given. The maximum values are for niobium and technetium, 9.2°K and 8.2°K, respectively. There is a minimum between groups Va and VIa, probably 0°K.

Many alloys, also, are superconductors. The highest superconduction transition temperature so far reported is 21°K, a little above the normal boiling point of hydrogen, for an alloy with composition $Nb_3Al_{0.75}Ge_{0.25}$. Such an alloy may be used, with cooling by liquid hydrogen, as the winding of the stator coils of an electric motor or generator, with a saving in both materials and power loss. At the present time other alloys cooled with liquid helium are used.

The theoretical explanation of superconductivity is that it results from an interaction of the conduction electrons at the Fermi surface (Section 17-6) and the phonons (vibrational waves; that is, sound waves) with the same wavelength in the metal. This interaction decreases the enthalpy for the superconducting state and produces a small gap at the Fermi surface; some levels, occupied by electrons, are stabilized, and the adjacent unoccupied levels are increased in energy. As a result of the existence of the gap the electrons no longer contribute to the heat capacity: γ is zero for the superconducting state (Section 17-6). Also, the coefficient α in the T^3

term is increased,* as shown in Figure 22-1. The larger value of α indicates a smaller value of the Debye characteristic temperature Θ for the superconducting state than for the normal metal, and hence smaller values of the Hooke's-law force constants for the bonds. It is in fact observed that a metal is more easily deformed in the superconducting state than in the normal state.

Some thermodynamic properties of superconductors are discussed in Example 22-1.

The metals with the largest electric conductivity at room temperature (Li, Be, Cu and their congeners) are not superconductors at temperatures above about 0.2°K. This observation provides support for the electron-phonon-interaction theory of superconductivity, inasmuch as the normal electric resistance of metals is attributed to scattering of the electron waves of the conducting electrons by phonons; low resistance results from small electron-phonon interaction, which results also in a decreased stability of the superconducting state.

Example 22-1. Discuss the entropy and enthalpy of the metal with heat capacity shown in Figure 22-1.

Answer. The heat capacity of a metal at low temperatures is given by the equation $C_P = \gamma T + \alpha T^3$. From Figure 2-1 we see that this metal has $\gamma = 0.012$ J deg^{-2} mole^{-1} and $\alpha = 0.00012$ J deg^{-4} mole^{-1} in the normal state, above the critical temperature $T_C = 10$°K, and $\alpha_{SC} = 0.00048$ J deg^{-4} mole^{-1} in the superconducting state, below 10°K.

The transition at T_C does not involve a heat of transition. The entropy of the superconducting metal should be equal to the entropy of the normal metal at this temperature, if the third law of thermodynamics applies; that is, if in each state the metal has no disorder. The entropy of the normal state at 10°K (assuming the validity of the expression $C_P = \gamma T + \alpha T^3$ in this region) is $\int_{0°}^{T_C} C_P dT/T = \gamma T_C + \alpha T_C^3/3 = 0.16$ J deg^{-1} mole^{-1}. The entropy of the superconducting state is similarly calculated to be $\alpha_{SC} T_C^3/3 = 0.16$ J deg^{-1} mole^{-1}, in agreement with the third law.

The value of the enthalpy difference $H(T_C) - H(0°K)$ for the normal state is $\int_{0°}^{T_C} (\gamma T + \alpha T^3) dT = \gamma T_C^2/2 + \alpha T_C^4/4 = 0.90$ J mole^{-1}, and that for the superconducting state is $\alpha_{SC} T_C^4/4 = 1.20$ J mole^{-1}. In order for the transition at T_C to be accompanied by no enthalpy of transition it is hence required that $H_{SC}(0°K) - H(0°K) = -0.30$ J mole^{-1}; that is, at 0°K the superconducting state is more stable than the nonsuperconducting state by this amount, which is the result of the electron-phonon interaction.

---

*For most superconductors the heat capacity curve deviates somewhat from the straight line shown in Figure 22-1.

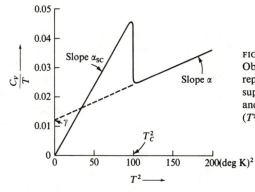

FIGURE 22-1
Observed values of C_P/T for a representative metal in the non-superconducting state ($T^2 > 100$) and the superconducting state ($T^2 < 100$).

22-5.   Chromium

The principal oxidation states of chromium are represented in the diagram below.

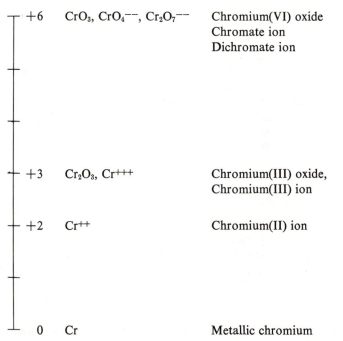

+6	CrO_3, CrO_4^{--}, $Cr_2O_7^{--}$	Chromium(VI) oxide Chromate ion Dichromate ion
+3	Cr_2O_3, Cr^{+++}	Chromium(III) oxide, Chromium(III) ion
+2	Cr^{++}	Chromium(II) ion
0	Cr	Metallic chromium

The maximum oxidation number, $+6$, corresponds to the position of the element in the periodic table.

The most important ore of chromium is *chromite*, $FeCr_2O_4$. The element was not known to the ancients, but was discovered in 1798 in lead chromate, $PbCrO_4$, which occurs in nature as the mineral *crocoite*.

The metal can be prepared by reducing chromic oxide with metallic aluminum (Chapter 15). Metallic chromium is also made by electrolytic reduction of compounds, usually chromic acid in aqueous solution.

Chromium is a silvery-white metal, with a bluish tinge. It is a very strong metal, with a high melting point, 1830°C. Because of its high melting point it resists erosion by the hot powder gases in big guns, the linings of which are accordingly sometimes plated with chromium.

Although the metal is more electropositive than iron, it easily assumes a passive (unreactive) state, by becoming coated with a thin layer of oxide, which protects it against further chemical attack. This property and its pleasing color are the reasons for its use for plating iron and brass objects, such as plumbing fixtures.

Ferrochrome, a high-chromium alloy with iron, is made by reducing chromite with carbon in the electric furnace. It is used for making alloy steels. The alloys of chromium are very important, especially the alloy steels. The chromium steels are very hard, tough, and strong. Their properties can be attributed to the high metallic valence (6) of chromium and to an interaction between unlike atoms that in general makes alloys harder and tougher than elementary metals. They are used for armor plate, projectiles, safes, and other articles. Ordinary *stainless steel* contains 14 to 18% chromium, and usually 8% nickel.

Chromium in its highest oxidation state (+6) does not form a hydroxide. The corresponding oxide, CrO_3, a red substance called *chromium(VI) oxide*, has acid properties. It dissolves in water to form a red solution of *dichromic acid*, $H_2Cr_2O_7$:

$$2CrO_3 + H_2O \longrightarrow H_2Cr_2O_7 \rightleftharpoons 2H^+ + Cr_2O_7^{--}$$

The salts of dichromic acid are called *dichromates*; they contain the dichromate ion, $Cr_2O_7^{--}$. Sexivalent chromium also forms another important series of salts, the *chromates*, which contain the ion CrO_4^{--}.

The chromates and dichromates are made by a method that has general usefulness for preparing salts of an acidic oxide—the method of *fusion with an alkali hydroxide or carbonate*. The carbonate functions as a basic oxide by losing carbon dioxide when heated strongly. Potassium carbonate is preferred to sodium carbonate because potassium chromate and potassium dichromate crystallize well from aqueous solution, and can be easily purified by recrystallization, whereas the corresponding sodium salts are deliquescent and are difficult to purify.

A mixture of powdered chromite ore and potassium carbonate slowly forms *potassium chromate*, K_2CrO_4, when strongly heated in air. The oxygen of the air oxidizes chromium to the sexipositive state, and also oxidizes the iron to ferric oxide:

$$4FeCr_2O_4 + 8K_2CO_3 + 7O_2 \longrightarrow 2Fe_2O_3 + 8K_2CrO_4 + 8CO_2$$

Sometimes the oxidation reaction is aided by the addition of an oxidizing agent, such as potassium nitrate, KNO_3, or potassium chlorate, $KClO_3$. The potassium chromate, a yellow substance, can be dissolved in water and recrystallized.

On addition of an acid, such as sulfuric acid, to a solution containing chromate ion, CrO_4^{--}, the solution changes from yellow to orange-red in color, because of the formation of dichromate ion, $Cr_2O_7^{--}$:

$$\underset{\text{Yellow}}{2CrO_4^{--}} + 2H^+ \rightleftarrows \underset{\text{Orange-red}}{Cr_2O_7^{--}} + H_2O$$

The reaction can be reversed by the addition of a base:

$$\underset{\text{Orange-red}}{Cr_2O_7^{--}} + 2OH^- \rightleftarrows \underset{\text{Yellow}}{2CrO_4^{--}} + H_2O$$

At an intermediate stage* both chromate ion and dichromate ion are present in the solution, in chemical equilibrium.

The chromate ion has a tetrahedral structure. The formation of dichromate ion involves the removal of one oxygen ion O^{--} (as water), by combination with two hydrogen ions, and its replacement by an oxygen atom of another chromate ion.

Both chromates and dichromates are strong oxidizing agents, the chromium being easily reduced from +6 to +3 in acid solution. *Potassium dichromate*, $K_2Cr_2O_7$, is a beautifully crystallizable bright-red substance used considerably in chemistry and industry. A solution of this substance or of chromium(VI) oxide, CrO_3, in concentrated sulfuric acid is a very strong oxidizing agent, which serves as a cleaning solution for laboratory glassware.

Large amounts of *sodium dichromate*, $Na_2Cr_2O_7 \cdot 2H_2O$, are used in the tanning of hides, to produce "chrome-tanned" leather. The chromium forms an insoluble compound with the leather protein.

Lead chromate, $PbCrO_4$, is a bright yellow, practically insoluble substance that is used as a pigment (*chrome yellow*).

Compounds of Terpositive Chromium

When ammonium dichromate, $(NH_4)_2Cr_2O_7$, a red salt resembling potassium dichromate, is ignited, it decomposes to form a green powder,

---

*There is also present in the solution some hydrogen chromate ion, $HCrO_4^-$:
$$H^+ + CrO_4^{--} \rightleftarrows HCrO_4^-$$

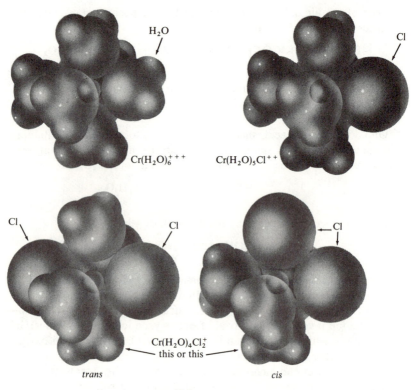

$Cr(H_2O)_6^{+++}$ $Cr(H_2O)_5Cl^{++}$

$Cr(H_2O)_4Cl_2^+$
← this or this →

trans   *cis*

FIGURE 22-2
Octahedral chromic complex ions.

chromium(III) oxide, Cr_2O_3:

$$(NH_4)_2Cr_2O_7 \longrightarrow N_2 + 4H_2O + Cr_2O_3$$

This reaction involves the reduction of the dichromate ion by ammonium ion. Chromium(III) oxide is also made by heating sodium dichromate with sulfur, and leaching out the sodium sulfate with water:

$$Na_2Cr_2O_7 + S \longrightarrow Na_2SO_4 + Cr_2O_3$$

It is a very stable substance, which is resistant to acids and has a very high melting point. It is used as a pigment (*chrome green*, used in the green ink for paper money).

Reduction of a dichromate in aqueous solution produces *chromium(III) ion*, Cr^{+++} (really the hexahydrated ion, $[Cr(H_2O)_6]^{+++}$), which has a violet color. The salts of this ion are similar in formula to those of aluminum. *Chrome alum*, $KCr(SO_4)_2 \cdot 12H_2O$, forms large violet octahedral crystals.

Chromium(III) chloride, $CrCl_3 \cdot 6H_2O$, forms several kinds of crystals, varying in color from violet to green, the solutions of which have similar colors. These different colors are due to the formation of stable complex ions (Figure 22-2):

$$[Cr(H_2O)_6]^{+++} \qquad \text{Violet}$$
$$[Cr(H_2O)_5Cl]^{++} \qquad \text{Green}$$
$$[Cr(H_2O)_4Cl_2]^{+} \qquad \text{Green}$$

In each of these complex ions there are six groups (water molecules and chloride ions) attached to the chromium ion. Chromium ion can be oxidized to chromate ion or dichromate ion by strong oxidizing agents, such as sodium peroxide in alkaline solution.

Chromium(III) hydroxide, $Cr(OH)_3$, is obtained as a pale grayish-green flocculent precipitate when ammonium hydroxide or sodium hydroxide is added to a chromium(III) solution. The precipitate dissolves in an excess of sodium hydroxide, forming the *chromite anion*, $Cr(OH)_4^-$ (bright green):

$$Cr(OH)_3 + OH^- \longrightarrow Cr(OH)_4^-$$

Chromium(III) solutions are reduced by zinc in acid solution or by other strong reducing agents to *chromium(II) ion*, Cr^{++} or $[Cr(H_2O)_6]^{++}$, which is blue in color. This solution and solid chromium(II) salts are very strong reducing agents, and must be protected from the air.

22-6. The Congeners of Chromium

The three heavier elements in group VIa—molybdenum, tungsten, and uranium—have all found important special uses.

Molybdenum

The principal ore of molybdenum is *molybdenite*, MoS_2, which occurs especially in a great deposit near Climax, Colorado. This mineral forms shiny black plates, closely similar in appearance to graphite.

Molybdenum metal is used to make filament supports in radio tubes and for other special uses. It is an important constituent of alloy steels.

The chemistry of molybdenum is complicated. It forms compounds corresponding to oxidation numbers $+6, +5, +4, +3$, and $+2$.

Molybdenum(VI) oxide, MoO_3, is a yellow-white substance made by roasting molybdenite. It dissolves in alkalis to produce molybdates, such as *ammonium molybdate*, $(NH_4)_6Mo_7O_{24} \cdot 4H_2O$. This reagent is used to precipitate orthophosphates, as the substance $(NH_4)_3PMo_{12}O_{40} \cdot 18H_2O$.

TABLE 22-5

Standard Enthalpy of Formation of Compounds of Chromium, Molybdenum, Tungsten, and Uranium at 25°C (kJ mole⁻¹)

	M = Cr	Mo	W	U
M(c)	0	0	0	0
M(g)	397	659	837	523
M^{++}(aq)	−139			
M^{+++}(aq)	−256			−515
M_2O_3(c)	−1141			
MO_2(c)		−544	−570	−1084
MO_3(c)	−579	−745	−840	−1218
MO_4^{--}(aq)	−863	−1064	−1115	
MF_2(c)	−757			
MF_3(c)	−1110			−1494
MF_4(c)				−1854
MF_5(c)				−2042
MF_6(g)			−1741	−2113
MCl_2(c)	−396	−184	−159	
MCl_3(c)	−563	−272		−891
MCl_4	−435(g)	−331(c)	−297(c)	−1051(c)
MCl_5(c)		−380	−351	−1097
MCl_6(c)		−377	−413	−1140
MBr_2(c)		−121	−79	
MBr_3(c)		−172		−712
MBr_4(c)		−188	−146	−823
MBr_5(c)		−213	−176	
MBr_6(c)			−184	
MI_2(c)	−227	−50	−4	
MI_3(c)		−63		−480
MI_4(c)		−75	0	−531
MI_5(c)		−75	113	
MS_2(c)		−232	−194	
MS_3(c)		−256		
MN(c)	−125			−335
MC(c)			−38	

Tungsten

Tungsten (also called *wolfram*) is a strong, heavy metal, with very high melting point (3370°C). It has important uses, as filaments in electric light bulbs, for electric contact points in spark plugs, as electron targets in x-ray tubes, and, in tungsten steel (which retains its hardness even when very hot), as cutting tools for high-speed machining.

The principal ores of tungsten are *scheelite*, $CaWO_4$, and *wolframite*, $(Fe,Mn)WO_4$.*

Tungsten forms compounds in which it has oxidation number $+6$ (tungstates, including the minerals mentioned above), $+5$, $+4$, $+3$, and $+2$. *Tungsten carbide*, WC, is a very hard compound that is used for the cutting edge of high-speed tools.

Uranium

Uranium is the rarest metal of the chromium group. Its principal ores are *pitchblende*, U_3O_8, and *carnotite*, $K_2U_2V_2O_{12} \cdot 3H_2O$. Its most important oxidation state is $+6$ (*sodium diuranate*, $Na_2U_2O(OH)_{12}$; *uranyl nitrate*, $UO_2(NO_3)_2 \cdot 6H_2O$; and others).

Before 1942 uranium was said to have no important uses—it was used mainly to give a greenish-yellow color to glass and glazes. In 1942, however, exactly one hundred years after the metal was first isolated, uranium became one of the most important of all elements. It was discovered in that year that uranium could be made a source of nuclear energy, liberated in tremendous quantity at the will of man.

Nuclear Fission

Ordinary uranium contains two isotopes,† ^{238}U (99.3%) and ^{235}U (0.7%). When a neutron collides with a ^{235}U nucleus it combines with it, forming a ^{236}U nucleus. This nucleus is unstable, and it immediately decomposes spontaneously by splitting into two large fragments, plus several neutrons. Each of the two fragments is itself an atomic nucleus, the sum of their atomic numbers being 92, the atomic number of uranium.

This nuclear fission is accompanied by the emission of a very large amount of energy—about 20×10^{12} J‡ per gram-atom of uranium decomposed (235 g of uranium). This is about 2,500,000 times the amount of heat evolved by burning the same weight of coal, and about 12,000,000 times that evolved by exploding the same weight of nitroglycerine. The large numbers indicate the very great importance of uranium as a source of energy; one ton of uranium (prewar price about $5000) could produce the same amount of energy as 2,500,000 tons of coal; and the use of uranium and other fissionable elements in place of coal may ultimately

---

*The formula $(Fe,Mn)WO_4$ means a solid solution of $FeWO_4$ and $MnWO_4$, in indefinite ratio.

†A minute amount, 0.006%, of a third isotope, ^{231}U, is also present.

‡This amount of energy weighs about 0.25 g, by the Einstein equation $E = mc^2$ ($E =$ energy, $m =$ mass, $c =$ velocity of light). The material products of the fission are 0.25 g lighter than the gram-atom of ^{235}U.

eliminate the disagreeable, but at present necessary, coal-mining industry.

The heavier uranium isotope, ^{238}U, also can be made to undergo fission, but by an indirect route—through the transuranium elements. These elements are discussed in Chapter 26.

22-7.   Manganese

The principal oxidation states of manganese are represented in the diagram below.

+7	MnO_4^-, Mn_2O_7	Permanganate ion Manganese heptoxide
+6	MnO_4^{--}	Manganate ion
+4	MnO_2	Manganese dioxide
+3	Mn_2O_3, Mn^{+++}	Manganese(III) oxide Manganese(III) ion
+2	Mn^{++}	Manganese(II) ion
0	Mn	Metallic manganese

The maximum oxidation number, +7, corresponds to the position of the element in the periodic table (group VIIa).

The principal ore of manganese is *pyrolusite*, MnO_2. Pyrolusite occurs as a black massive mineral and also as a very fine black powder. Less important ores are *braunite*, Mn_2O_3 (containing some silicate); *manganite*, $MnO(OH)$; and *rhodochrosite*, $MnCO_3$.

Impure manganese can be made by reducing manganese dioxide with carbon:

$$MnO_2 + 2C \longrightarrow Mn + 2CO$$

Manganese is also made by the aluminothermic process:

$$3MnO_2 + 4Al \longrightarrow 2Al_2O_3 + 3Mn$$

Manganese steels are usually made from special high-manganese alloys prepared by reducing mixed oxides of iron and manganese with coke in a blast furnace (see Chapter 20). The high-manganese alloys (70 to 80% Mn, 20 to 30% Fe) are called *ferromanganese*, and the low-manganese alloys (10 to 30% Mn) are called *spiegeleisen*.

Manganese is a silvery-gray metal, with a pinkish tinge. It is reactive, and displaces hydrogen even from cold water. Its principal use is in the manufacture of alloy steel.

Manganese dioxide (pyrolusite) is the only important compound of quadripositive manganese. This substance has many uses, most of which depend upon its action as an oxidizing agent (with change from Mn^{+4} to Mn^{+2}) or as a reducing agent (with change from Mn^{+4} to Mn^{+6} or Mn^{+7}).

Manganese dioxide oxidizes hydrochloric acid to free chlorine, and is used for this purpose:

$$MnO_2 + 2Cl^- + 4H^+ \longrightarrow Cl_2 + Mn^{++} + 2H_2O$$

Its oxidizing power also underlies its use in the ordinary dry cell (Chapter 15).

When manganese dioxide is heated with potassium hydroxide in the presence of air it is oxidized to *potassium manganate*, K_2MnO_4:

$$2MnO_2 + 4KOH + O_2 \longrightarrow 2K_2MnO_4 + 2H_2O$$

Potassium manganate is a green salt that can be dissolved in a small amount of water to give a green solution, containing potassium ion and the *manganate ion*, MnO_4^{--}. The manganates are the only compounds of Mn^{+6}. They are powerful oxidizing agents, and are used to a small extent as disinfectants.

The manganate ion can be oxidized to *permanganate ion*, MnO_4^-, which contains Mn^{+7}. The electron reaction for this process is

$$MnO_4^{--} \longrightarrow MnO_4^- + e^-$$

In practice this oxidation is carried out electrolytically (by anodic oxidation) or by use of chlorine:

$$2MnO_4^{--} + Cl_2 \longrightarrow 2MnO_4^- + 2Cl^-$$

The process of auto-oxidation-reduction is also used; manganate ion is stable in alkaline solution, but not in neutral or acidic solution. The

TABLE 22-6
Standard Enthalpy of Formation of Compounds of Manganese and Rhenium at 25°C (kJ mole^{-1})

Mn(c)	0	MnO_2(c)	−521	Re(c)	0
Mn(g)	281	MnF_2(c)	−795	Re(g)	791
Mn^{++}(aq)	−219	$MnCl_2$(c)	−482	ReO_3(c)	−347
Mn^{+++}(aq)	−100	$MnBr_2$(c)	−379	Re_2O_7(c)	−1245
MnO(g)	140	MnI_2(c)	−248	ReF_6(g)	−1142
MnO(c)	−385	MnS(c)	−204	ReS_2(c)	−185
Mn_2O_3(c)	−956	MnSe(c)	−117	$ReAs_2$(c)	4
$KMnO_4$(c)	−813	$MnCO_3$(c)	−895		

*No values have been reported for technetium.

addition of any acid, even carbon dioxide (carbonic acid), to a manganate solution causes the production of permanganate ion and the precipitation of manganese dioxide:

$$3MnO_4^{--} + 4H^+ \rightleftharpoons 2MnO_4^- + MnO_2 + 2H_2O$$
   Green   Magenta

When hydroxide is added to the mixture of the purple solution and the brown or black precipitate, a clear green solution is again formed, showing that the reaction is reversible.

This reaction serves as another example of Le Chatelier's principle: the addition of hydrogen ion, which occurs on the left side of the equation, causes the reaction to shift to the right.

Potassium permanganate, $KMnO_4$, is the most important chemical compound of manganese. It forms deep purple-red prisms, which dissolve readily in water to give a solution intensely colored with the magenta color characteristic of permanganate ion. The substance is a powerful oxidizing agent, which is used as a disinfectant. It is an important chemical reagent, especially in analytical chemistry.

On reduction in acidic solution the permanganate ion accepts five electrons, to form the manganese(II) ion:

$$MnO_4^- + 8H^+ + 5e^- \longrightarrow Mn^{++} + 4H_2O$$

In neutral or basic solution it accepts three electrons, to form a precipitate of manganese dioxide:

$$MnO_4^- + 2H_2O + 3e^- \longrightarrow MnO_2 + 4OH^-$$

A one-electron reduction to manganate ion can be made to take place in strongly basic solution:

$$MnO_4^- + e^- \longrightarrow MnO_4^{--}$$

Permanganic acid, $HMnO_4$, is a strong acid that is very unstable. Its anhydride, *manganese(VII) oxide*, can be made by the reaction of potassium permanganate and concentrated sulfuric acid:

$$2KMnO_4 + H_2SO_4 \longrightarrow K_2SO_4 + Mn_2O_7 + H_2O$$

It is an unstable, dark-brown oily liquid.

The manganese(III) ion, Mn^{+++}, is a strong oxidizing agent, and its salts are unimportant. The insoluble oxide, Mn_2O_3, and its hydrate, $MnO(OH)$, are stable. When manganese(II) ion is precipitated as hydroxide, $Mn(OH)_2$, in the presence of air, the white precipitate is rapidly oxidized to the brown compound $MnO(OH)$:

$$Mn^{++} + 2OH^- \longrightarrow Mn(OH)_2$$
White

$$4Mn(OH)_2 + O_2 \longrightarrow 4MnO(OH) + 2H_2O$$
Brown

Manganese(II) ion, Mn^{++} or $[Mn(H_2O)_6]^{++}$, is the stable cationic form of manganese. The hydrated ion is pale rose-pink in color. Representative salts are $Mn(NO_3)_2 \cdot 6H_2O$, $MnSO_4 \cdot 7H_2O$, and $MnCl_2 \cdot 4H_2O$. These salts and the mineral *rhodochrosite*, $MnCO_3$, are all rose-pink or rose-red. Crystals of rhodochrosite are isomorphous with calcite.

With hydrogen sulfide manganese(II) ion forms a light pink precipitate of *manganese sulfide*, MnS:

$$Mn^{++} + H_2S \longrightarrow MnS + 2H^+$$

22-8. Acid-forming and Base-forming Oxides and Hydroxides

Chromium and manganese illustrate the general rules about the acidic and basic properties of metallic oxides and hydroxides:

1. *The oxides of an element in its higher oxidation states tend to form acids.*
2. *The lower oxides of an element tend to form bases.*
3. *The intermediate oxides may be amphoteric; that is, they may serve either as acid-forming or as base-forming oxides.*

The highest oxide of chromium, chromium(VI) oxide, is acidic, and forms chromates and dichromates. The lowest oxide, CrO, is basic, forming the chromium(II) ion Cr^{++} and its salts. Chromium(III) hydroxide, $Cr(OH)_3$, representing the intermediate oxidation state, is amphoteric. With acids it forms the salts of chromium(III) ion, such as chromium(III) sulfate, $Cr_2(SO_4)_3$, and with strong bases it dissolves to form the chromite ion, $Cr(OH)_4^-$.

Similarly, the two highest oxidation states of manganese, $+7$ and $+6$, are represented by the anions MnO_4^-, and MnO_4^{--}, and the two lowest states are represented by the cations Mn^{++} and Mn^{+++}. The intermediate state $+4$ is unstable (except for the compound MnO_2), and is feebly amphoteric.

You should check the rules given above by considering the properties of oxides of other elements.

22-9. The Congeners of Manganese

Technetium

No stable isotopes of element 43 exist. Minute amounts of radioactive isotopes have been made, by Segrè and his collaborators, who have named the element technetium, symbol Tc.

Rhenium

The element rhenium, atomic number 75, was discovered by the German chemists Walter Noddack and Ida Tacke in 1925. The principal compound of rhenium is potassium perrhenate, $KReO_4$, a colorless substance. In other compounds all oxidation numbers from $+7$ to -1 are represented: examples are Re_2O_7, ReO_3, $ReCl_5$, ReO_2, Re_2O_3, $Re(OH)_2$.

Neptunium

Neptunium, element 93, was first made in 1940, by E. M. McMillan and P. H. Abelson, at the University of California, by the reaction of a neutron with ^{238}U, to form ^{239}U, and the subsequent emission of an electron from this nucleus, increasing the atomic number by 1:

$$^{238}_{92}U + {}_0n^1 \longrightarrow {}^{239}_{92}U$$
$$^{239}_{92}U \longrightarrow e^- + {}^{239}_{93}Np$$

Neptunium is important as an intermediate in the manufacture of plutonium (Chapter 26).

Exercises

22-1. Discuss the oxidation states of titanium, vanadium, chromium, and manganese in relation to their electronic structures. What electrons are removed in forming the bipositive ions? What electrons determine the highest oxidation states?

22-2. What reduction product is formed when dichromate ion is reduced in acidic solution? When permanganate ion is reduced in acidic solution? When permanganate ion is reduced in basic solution? Write the electron reactions for these three cases.

22-3. Write equations for the reduction of dichromate ion by (a) sulfur dioxide; (b) ethyl alcohol, C_2H_5OH, which is oxidized to acetaldehyde, H_3CCHO; (c) iodide ion, which is oxidized to iodine.

22-4. Write an equation for the chemical reaction that occurs on fusion of a mixture of chromite ($FeCr_2O_4$), potassium carbonate, and potassium chlorate (which forms potassium chloride).

22-5. Write the chemical equations for the preparation of potassium manganate and potassium permanganate from manganese dioxide, using potassium hydroxide, air, and carbon dioxide.

22-6. What chemical reactions are taking place when a violet solution of chrome alum on treatment with hydrochloric acid turns green in color?

22-7. The two most important oxidation levels of uranium are $+4$ and $+6$. Which of these levels would you expect to have the more acidic properties?

22-8. Assign an electronic structure to the hexahydrated chromium(III) ion. What orbitals of the chromium atom are used in bond formation? How many $3d$ orbitals are occupied by unpaired electrons? What are the electric charges on the various atoms, as indicated by the electronegativity differences?

22-9. To what property of the elements do you attribute the steady decrease in a row, from left to right, in values of the standard enthalpy of formation of compounds in Tables 22-2, 22-3, and 22-5? Use one sequence of three or four values as an illustration.

22-10. From the values in Table 22-2 calculate the standard enthalpy of formation of the crystalline diiodide, triiodide, and tetriodide of titanium and of zirconium. Compare with the values calculated by use of Equation 6-1.

22-11. Make the calculation and comparison of Exercise 22-10 for the bromides of titanium and zirconium.

22-12. From the results of Exercises 22-10 and 22-11, what do you conclude about the dependence of the electronegativity of titanium and zirconium on oxidation number?

22-13. Use the conclusions of Exercise 22-12 (extended to thorium) and the values for $ThBr_4(c)$ and $ThI_4(c)$ to predict rough values for the standard molar enthalpy of $ThBr_2(c)$, $ThBr_3(c)$, $ThI_2(c)$, and $ThI_3(c)$.

22-14. The molecule WCl_6 has the configuration of a regular octahedron, with the W—Cl distance 2.25 Å. What is the chlorine-chlorine distance? Compare with the van der Waals radius of chlorine, in relation to the statement in Section 6-13 about the decrease in van der Waals radius in the hemisphere around the bond direction. (Answer: 3.18 Å.)

22-15. The molecule $CrCl_2O_2$ has a tetrahedral structure, with bond lengths Cr—Cl = 2.12 Å and Cr—O = 1.57 Å. Discuss its electronic structure. The single-bond radius of chromium(VI) may be taken to be 1.19 Å. What do the observed bond lengths indicate about the character of the bonds in this molecule?

22-16. The high volatility of $CrCl_2O_2$, which has melting point $-96°C$ and boiling point 117°C, shows that the molecule has only a small electric dipole moment. By use of Table 6-5 calculate the amount of double-bond or triple-bond character of the bonds to chlorine and oxygen required to give zero electric charge to the atoms.

22-17. The observed value of the Cl—Cr—Cl bond angle in $CrCl_2O_2$ is 114°. What is the chlorine-chlorine distance? Compare with the van der Waals radius of chlorine. (Answer: 3.56 Å.)

22-18. What configuration and dimensions would you predict for the $CrCl_6$ molecule? Although $MoCl_6$ and WCl_6 are known stable compounds and many compounds of chromium(VI) are known, $CrCl_6$ has never been synthesized. Can you suggest a reason for its apparent instability? (Answer: Cr—Cl = 2.12 Å, Cl—Cl = 3.00 Å.)

23

Organic Chemistry

23-1. The Nature and Extent of Organic Chemistry

Organic chemistry is the chemistry of the compounds of carbon. It is a very great subject—over a million different organic compounds have already been reported and described in the chemical literature. Many of these substances have been isolated from living matter, and many more have been synthesized by chemists in the laboratory.

The occurrence in nature, methods of preparation, composition, structure, properties, and uses of some organic compounds (hydrocarbons, alcohols, chlorine derivatives of hydrocarbons, and organic acids) were discussed in Chapter 7. This discussion is continued in the following sections, with emphasis on natural products, especially the valuable substances obtained from plants, and on synthetic substances useful to man. Several large parts of organic chemistry will not be discussed at all; these include the methods of isolation and purification of naturally occurring compounds, the methods of analysis and determination of structure, and the methods of synthesis used in organic chemistry, except to the extent that they have been described in Chapter 7.

There are two principal ways in which organic chemists work. One of these ways is to begin the investigation of some natural material, such as a plant, that is known to have special properties. This plant might, for example, have been found by the natives of a tropical region to be beneficial in the treatment of malaria. The chemist then proceeds to make an extract from the plant, with use of a solvent such as alcohol or ether, and, by various methods of separation, to divide the extract into fractions. After each fractionation a study is made to see which fraction still contains the active substance. Finally this process may be carried so far that a pure crystalline active substance is obtained. The chemist then analyzes the substance, and determines its molecular weight, in order to find out what atoms are contained in the molecule of the substance. He next investigates the chemical properties of the substance, splitting its molecules into smaller molecules of known substances, in order to determine its molecular structure. When the structure has been determined, he attempts to synthesize the substance; if he is successful, the active material may be made available in large quantity and at low cost.

The other way in which organic chemists work involves the synthesis and study of a large number of organic compounds, and the continued effort to correlate the empirical facts by means of theoretical principles. Often a knowledge of the structure and properties of natural substances is valuable in indicating the general nature of the compounds that are worth investigation. The ultimate goal of this branch of organic chemistry is the complete understanding of the physical and chemical properties, and also the physiological properties, of substances in terms of their molecular structure. At the present time chemists have obtained a remarkable insight into the dependence of the physical and chemical properties of substances on the structure of their molecules. So far, however, only a small beginning has been made in attacking the great problem of the relation between structure and physiological activity. This problem remains one of the greatest and most important problems of science, challenging the new generation of scientists.

23-2. Petroleum and the Hydrocarbons

One of the most important sources of organic compounds is petroleum (crude oil). Petroleum, which is obtained from underground deposits that have been tapped by drilling oil wells, is a dark-colored, viscous liquid that is in the main a mixture of hydrocarbons (compounds of hydrogen and carbon; see Section 7-2). A very great amount of it, approximately one billion tons, is produced and used each year. Much of it is burned, for direct use as a fuel, but much is separated or converted into other materials.

The Refining of Petroleum

Petroleum may be separated into especially useful materials by a process of distillation, called *refining*. It was mentioned in Section 7-2 that petroleum ether, obtained in this way, is an easily volatile pentane-hexane-heptane (C_5H_{12} to C_7H_{16}) mixture that is used as a solvent and in the dry cleaning of clothes, gasoline is the heptane-to-nonane (C_7H_{16} to C_9H_{20}) mixture used in internal-combustion engines, kerosene is the decane-to-hexadecane ($C_{10}H_{22}$ to $C_{16}H_{34}$) mixture used as a fuel, and heavy fuel oil is a mixture of still larger hydrocarbon molecules.

The residue from distillation is a black, tarry material called *petroleum asphalt*. It is used in making roads, for asphalt composition roofing materials, for stabilizing loose soil, and as a binder for coal dust in the manufacture of briquets for use as a fuel. A similar material, *bitumen* or *rock asphalt*, is found in Trinidad, Texas, Oklahoma, and other parts of the world, where it presumably has been formed as the residue from the slow distillation of pools of oil.

It is thought that petroleum, like coal, is the result of the decomposition of the remains of plants that grew on the earth long ago (about 250 million years ago).

Cracking and Polymerizing Processes

As the demand for gasoline became greater, methods were devised for increasing the yield of gasoline from petroleum. The simple "cracking" process consists in the use of high temperature to break the larger molecules into smaller ones; for example, a molecule of $C_{12}H_{26}$ might be broken into a molecule of C_6H_{14} (hexane) and a molecule of C_6H_{12} (hexene, containing one double bond). There are now several rather complicated cracking processes in use. Some involve heating liquid petroleum, under pressure of about 50 atm, to about 500°C, perhaps with a catalyst such as aluminum chloride, $AlCl_3$. Others involve heating petroleum vapor with a catalyst such as clay containing some zirconium dioxide.

Polymerization is also used to make gasoline from the lighter hydrocarbons containing double bonds. For example, two molecules of ethylene, C_2H_4, can react to form one molecule of butylene, C_4H_8 (structural formula $CH_3—CH=CH—CH_3$).

Some gasoline is also made by the hydrogenation (reaction with hydrogen) of petroleum and coal. Many organic chemicals are prepared in great quantities from these important raw materials.

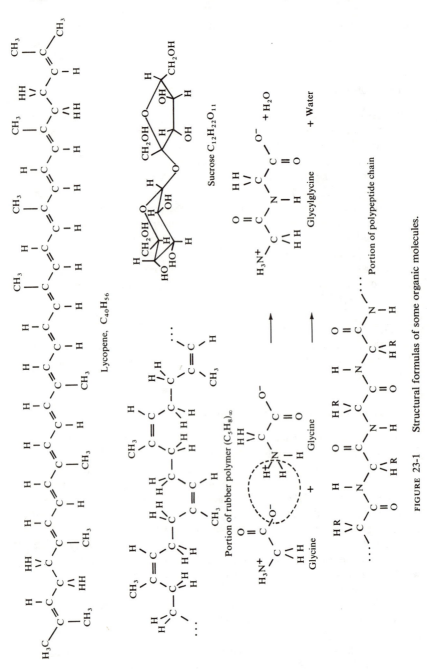

Lycopene, $C_{40}H_{56}$

Sucrose $C_{12}H_{22}O_{11}$

Portion of rubber polymer $(C_5H_8)_\infty$

Glycine + Glycine → Glycylglycine + H_2O

Portion of polypeptide chain

FIGURE 23-1   Structural formulas of some organic molecules.

Hydrocarbons Containing Several Double Bonds

The structure and properties of ethylene, a substance whose molecules contain a double bond, were discussed in Section 7-3. Some important natural products are hydrocarbons containing several double bonds. For example, the red coloring matter of tomatoes, called *lycopene*, is an unsaturated hydrocarbon, $C_{40}H_{56}$, with the structure shown in Figure 23-1. The molecule of this substance contains thirteen double bonds. It is seen that eleven of these double bonds are related to one another in a special way—they alternate regularly with single bonds. A regular alteration of double bonds and single bonds in a hydrocarbon chain is called a *conjugated system of double bonds*. The existence of this structural feature in a molecule confers upon the molecule special properties, such as the power of absorbing visible light, causing the substance to be colored.

Other yellow and red substances, isomers of lycopene, with the same formula $C_{40}H_{56}$, are called α-carotene, β-carotene, and similar names. These substances occur in butter, milk, green leafy vegetables, eggs, cod liver oil, halibut liver oil, carrots, tomatoes, and other vegetables and fruits. They are important substances because they serve in the human body as a source of Vitamin A (see Chapter 24).

Polycyclic Substances

Many important substances exist whose molecules contain two or more rings of atoms: these substances are called *polycyclic substances;* naphthalene, anthracene, and phenanthrene are examples of polycyclic aromatic hydrocarbons (Section 7-4). An example of a polycyclic aliphatic hydrocarbon is *pinene*, $C_{10}H_{16}$, which is the principal constituent of *turpentine*. Turpentine is an oil obtained by distilling a semifluid resinous material that exudes from pine trees. The pinene molecule has the following structure:

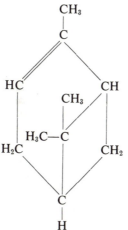

Another interesting polycyclic substance is *camphor*, obtained by steam distillation of the wood of the camphor tree, or, in recent years, by a synthetic process starting with pinene. The molecule of camphor is roughly spherical in shape—it is a sort of "cage" molecule:

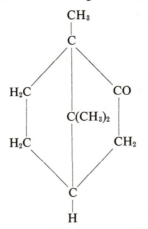

Camphor contains one oxygen atom, its formula being $C_{10}H_{16}O$. A hydrocarbon is obtained by replacing the oxygen atom by two hydrogen atoms, producing the substance called *camphane*. Camphor is used in medicine and in the manufacture of plastics. Ordinary *celluloid* consists of nitrocellulose plasticized with camphor.

Rubber

Rubber is an organic substance, obtained mainly from the sap of the rubber tree, *Hevea brasiliensis*. Rubber consists of very long molecules, which are polymers of *isoprene*, C_5H_8. The structure of isoprene is

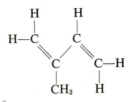

and that of the rubber polymer, as produced in the plant, is shown in Figure 23-1.

The characteristic properties of rubber are due to the fact that it is an aggregate of very long molecules, intertwined with one another in a rather random way. The structure of the molecules is such that they do not tend to align themselves side by side in a regular way—that is, to crystallize—but instead tend to retain an irregular arrangement.

It is interesting to note that the rubber molecule contains a large number of double bonds, one for each C_5H_8 residue. In natural rubber the configuration about the double bonds is the *cis* configuration, as shown in the structural formula in Figure 23-1. *Gutta percha*, a similar plant product that does not have the elasticity of rubber, contains the same molecules, with, however, the *trans* configuration around the double bonds. This difference in configuration permits the molecules of gutta percha to crystallize more readily than those of rubber.

Ordinary unvulcanized rubber is sticky, as a result of a tendency for the molecules to pull away from one another, a portion of the rubber thus adhering to any material with which it comes in contact. The stickiness is eliminated by the process of *vulcanization*, which consists in heating rubber with sulfur. During this process sulfur molecules, S_8, open up and combine with the double bonds of rubber molecules, forming bridges of sulfur chains from one rubber molecule to another rubber molecule. These sulfur bridges bind the aggregate of rubber molecules together into a large molecular framework, extending through the whole sample of rubber. Vulcanization with a small amount of sulfur leads to a soft product, such as that in rubber bands or (with a filler of carbon black or zinc oxide) in automobile tires. A much harder material, called vulcanite, is obtained by using a larger amount of sulfur.

The materials called *synthetic rubber* are not really synthetic rubber, since they are not identical with the natural product. They are, rather, substitutes for rubber—materials with properties and structure similar to but not identical with those of natural rubber. For example, the substance *chloroprene*, C_4H_5Cl, with the structure

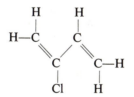

is similar to isoprene except for the replacement of a methyl group by a chlorine atom. Chloroprene polymerizes to a rubber called *chloroprene rubber*. It and other synthetic rubbers have found extensive uses, and are superior to natural rubber for some purposes.

23-3.   Alcohols and Phenols

The aliphatic alcohols have a hydroxyl group, —OH, attached to a carbon atom in place of one of the hydrogen atoms of an aliphatic hydrocarbon. The two simplest alcohols, methyl alcohol (methanol) and ethyl alcohol

TABLE 23-1
Physical Properties of Some Alcohols and Phenols

	Melting Point	Boiling Point	Density of Liquid
Methyl alcohol, CH₃OH	−97.8°C	64.7°C	0.796 g/ml
Ethyl alcohol, CH₃CH₂OH	−117.3°	78.5°	.789
Propyl alcohol, CH₃(CH₂)₂OH	−127°	97.2°	.804
Isopropyl alcohol, CH₃CHOHCH₃	−89°	82.3°	.785
Butyl alcohol, CH₃(CH₂)₃OH	−89°	117.7°	.810
sec-Butyl alcohol, CH₃CHOHCH₂CH₃	−89°	100°	.808
tert-Butyl alcohol, (CH₃)₃COH	25°	83°	.789
1-Pentanol, CH₃(CH₂)₄OH	−78°	138°	.814
Glycol, CH₂OHCH₂OH	−17°	197°	1.116
1,2-Propanediol, CH₂OHCHOHCH₃		189°	1.040
1,3-Propanediol, CH₂OHCH₂CH₂OH		214°	1.053
Glycerol, CH₂OHCHOHCH₂OH	17.9°	290°	1.260
Benzyl alcohol, C₆H₅CH₂OH	−15.3°	205°	1.050
Phenol, C₆H₅OH	41°	182°	1.072*
o-Cresol, CH₃C₆H₄OH	30°	192°	1.047*
m-Cresol, CH₃C₆H₄OH	11°	203°	1.034
p-Cresol, CH₃C₆H₄OH	36°	203°	1.035*

*Density of crystalline substance.

(ethanol), have been discussed in Section 7-6. The melting points, boiling points, and densities of some alcohols are given in Table 23-1.

Some of the heavier alcohols are made from the olefines that are obtained as by-products in the refining of petroleum. For example, propylene, $CH_2{=}CH{-}CH_3$, can be hydrated by addition of water vapor at high temperature and pressure in the presence of a catalyst:

$$CH_2{=}CH{-}CH_3 + H_2O \longrightarrow CH_3{-}\underset{\underset{OH}{|}}{C}H{-}CH_3$$

The product is called *isopropyl alcohol* or *2-propanol* (the number 2 means that the substituent is on the second carbon atom in the chain, and the suffix *ol* means that the substituent is the hydroxyl group). An alcohol of this kind, with formula

(R being a radical with a carbon atom forming the bond) is called a secondary alcohol. Isopropyl alcohol may be called *sec*-propanol.

An alcohol

is called a *primary alcohol;* examples are ethanol and 1-propanol, $CH_3CH_2CH_2OH$. *Tertiary alcohols* have the formula

The simplest example is *tert*-butyl alcohol $(CH_3)_3COH$. The propyl and butyl alcohols are used as solvents for lacquers and other materials.

The formation of hydrogen bonds by the hydroxyl groups causes the alcohols to have higher melting and boiling points and larger solubility in water than other organic compounds with corresponding molecular weight. The lower alcohols, including *tert*-butanol, are soluble in water in all proportions. The other butanols have limited solubility in water, presumably because their less compact —C_4H_9 groups fit less readily than the *tert*-butyl group into the water structure (see the discussion of crystalline hydrates in Chapter 12).

The Polyhydroxy Alcohols

Alcohols containing two or more hydroxyl groups attached to different carbon atoms can be made. They are called the *polyhydroxy alcohols* or *polyhydric* alcohols. *Glycol,*

$$CH_2OH$$
$$|$$
$$CH_2OH$$

is used as a solvent and as an antifreeze material for automobile radiators. *Glycerol* (glycerine), $C_3H_5(OH)_3$, is a trihydroxypropane, with the structure

$$
\begin{array}{c}
H \\
| \\
H-C-OH \\
| \\
H-C-OH \\
| \\
H-C-OH \\
| \\
H
\end{array}
$$

Glycerol is a viscous liquid that is used as an antifreeze material, as a humectant (moistening agent) for tobacco, and especially for use in manufacturing explosives. It reacts with a mixture of nitric acid and sulfuric acid to form the viscous liquid *glyceryl trinitrate* (common name, *nitroglycerine*):

$$
\begin{array}{l}
CH_2OH \\
| \\
CHOH \\
| \\
CH_2OH
\end{array}
\; + 3HONO_2 \;\xrightarrow[H_2SO_4]{}\;
\begin{array}{l}
CH_2ONO_2 \\
| \\
CHONO_2 \\
| \\
CH_2ONO_2
\end{array}
\; + 3H_2O
$$

Glyceryl trinitrate is a powerful and treacherous explosive. It was used extensively for blasting and mining in the decades about 1860, despite numerous fatal accidents. Then in 1867 the Swedish industrial chemist Alfred Nobel (1833–1896) discovered that the hazards of handling it would be greatly reduced by mixing it with an absorbent material such as diatomaceous earth, to form the product called *dynamite*. In the year 1876 Nobel also discovered the powerful detonating explosive blasting gelatin, which consists of cellulose nitrate (guncotton) that has soaked up glyceryl trinitrate, and in 1889 he developed the propellant ballistite, a plasticized mixture of cellulose nitrate and glyceryl trinitrate with composition such that it burns smoothly and rapidly and does not detonate.

The Aromatic Alcohols

An example of an aromatic alcohol is *benzyl alcohol*, $C_6H_5—CH_2OH$. In this substance the hydroxyl group is attached to the carbon atom of the alkyl group (methyl group) that is itself attached to the benzene ring. The properties of benzyl alcohol and other aromatic alcohols resemble those of the aliphatic alcohols.

The Phenols

A compound in which a hydroxyl group is attached directly to the carbon atom of a benzene ring (or of naphthalene or other aromatic ring system) is called a *phenol*. The simplest phenol is phenol (hydroxybenzene), C_6H_5OH. The three *cresols* (ortho, meta, and para) are 1-hydroxy-2-methylbenzene, 1-hydroxy-3-methylbenzene, and 1-hydroxy-4-methylbenzene, respectively:

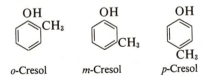

o-Cresol   *m*-Cresol   *p*-Cresol

They are obtained in the refining of coal tar, and are used as disinfectants and in the manufacture of plastics.

The properties of phenols differ considerably from those of the aliphatic and aromatic alcohols in ways that can be accounted for by the theory of resonance. The main difference is in acid strength: the alcohols (in aqueous solution) have acid constants about 1×10^{-16}, whereas the phenols are about a million times stronger, with acid constants about 1×10^{-10}.

The acid dissociation corresponds to the equilibrium reaction

$$ROH \rightleftharpoons RO^- + H^+$$

For an alcohol, such as methanol, the anion RO^- has the electronic structure $H_3C\!-\!\overset{\cdot\cdot}{\underset{\cdot\cdot}{O}}:^-$. For phenol, however, the phenolate ion can be assigned a structure that is the hybrid of several valence-bond structures:

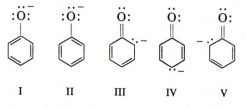

  I   II   III   IV   V

The resonance energy for these five structures stabilizes the phenolate ion more than the amount by which the undissociated phenol molecule is stabilized by resonance between the two Kekulé structures (with only small contributions by the other three, which involve a separation of charges). The extra stabilization of the anion increases the acid constant; the observed factor 10^6 corresponds to the reasonable value 33 kJ mole^{-1} for the extra resonance energy of the phenolate ion.

23-4.   Aldehydes and Ketones

The alcohols and ethers represent the first stage of oxidation of hydrocarbons. Further oxidation leads to substances called *aldehydes* and *ketones*. The aldehydes have the formula

and the ketones the formula

TABLE 23-2
Physical Properties of Some Aldehydes and Ketones

	Melting Point	Boiling Point	Density of Liquid
Formaldehyde, HCHO	$-92°C$	$-21°C$	0.82 g/ml
Acetaldehyde, CH_3CHO	$-124°$	$21°$	.782
Propionaldehyde, CH_3CH_2CHO	$-81°$	$49°$	.807
n-Butyraldehyde, $CH_3(CH_2)_2CHO$	$-98°$	$76°$	.817
Isobutyraldehyde, $(CH_3)_2CHCHO$	$-66°$	$62°$	.794
Glyoxal, OHCCHO	$15°$	$50°$	1.14
Acrolein, $CH_2{=}CHCHO$	$-88°$	$53°$	0.841
Benzaldehyde, C_6H_5CHO	$-26°$	$180°$	1.050
Acetone, CH_3COCH_3	$-95°$	$57°$	0.792
Methyl ethyl ketone, $CH_3COCH_2CH_3$	$-86°$	$80°$	.805
Methyl n-propylketone, $CH_3CO(CH_2)_2CH_3$	$-79°$	$102°$	.812
Diethyl ketone, $CH_3CH_2COCH_2CH_3$	$-42°$	$103°$	.815
Biacetyl, $CH_3COCOCH_3$		$88°$	.978
Acetylacetone, $CH_3COCH_2COCH_3$	$-23°$	$137°$	.976
Acetophenone, $CH_3COC_6H_5$	$20°$	$202°$	1.026
Benzophenone, $C_6H_5COC_6H_5$	$49°$	$306°$	1.098*

*Density of crystalline substance.

The group

is called the *carbonyl group*. The substance *formaldehyde*,

is also classed as an aldehyde. It can be made by passing methyl alcohol vapor and air over a heated metal catalyst:

$$2CH_3OH + O_2 \longrightarrow 2HCHO + 2H_2O$$

Formaldehyde is a gas with a sharp irritating odor. It is used as a disinfectant and antiseptic, and in the manufacture of plastics and of leather and artificial silk.

Acetaldehyde, CH_3CHO, is a similar substance made from ethyl alcohol.

The ketones are effective solvents for organic compounds, and are extensively used in chemical industry for this purpose. *Acetone*, $(CH_3)_2CO$, which is dimethyl ketone, is the simplest and most important of these substances. It is a good solvent for nitrocellulose.

Acrolein, $CH_2{=}CHCHO$, is the simplest unsaturated aldehyde. It is a liquid with the characteristic pungent odor of burning fat. It is produced when fats or oils are heated above 300°C, and it can be made by heating glycerol with a dehydrating agent:

$$C_3H_5(OH)_3 \xrightarrow[\text{KHSO}_4]{} CH_2CHCHO + 2H_2O$$

Many of the higher aldehydes and ketones have pleasant odors, and some of the aromatic aldehydes are used as flavors. An example is *vanillin*, the fragrant principle of the vanilla bean; its structural formula is

Vanillin is seen to be a phenol and an aromatic ether as well as an aldehyde. An example of a strongly fragrant ketone is *muscone*, which is obtained from the scent glands of the male musk deer and is used in perfumes. Its formula is

$$H_3C-CH-CH_2-C{=}O$$
$$\underset{\textstyle -(CH_2)_{12}-}{\rule{0pt}{1.2em}}$$

It contains an unusually large ring (15 carbon atoms).

Physical properties of some aldehydes and ketones are given in Table 23-2.

23-5. The Organic Acids and Their Esters

Acetic acid, CH_3COOH, was mentioned in Section 7-6 as an example of an organic acid. The simplest organic acid is *formic acid*, $HCOOH$. It can be made by distilling ants, and its name is from the Latin word for ants.

Properties of some of the organic acids are given in Table 23-3. It is seen that the acid constants for the monocarboxylic acids lie in the range 2×10^{-4} to 1×10^{-5} (*pK* 3.7 to 5). The explanation of the greater acid strength of the —OH group in the carboxylic acids than in the alcohols is given by the theory of resonance; it is similar to that already given

TABLE 23-3
Properties of Some Carboxylic Acids

	Melting Point	Boiling Point	Density of Liquid	pK_A
Formic, HCOOH	8°C	101°C	1.226	3.77
Acetic, CH_3COOH	17°	118°	1.049	4.76
Propionic, CH_3CH_2COOH	−22°	141°	0.992	4.88
Butyric, $CH_3(CH_2)_2COOH$	−6°	164°	.959	4.82
Isobutyric, $(CH_3)_2CHCOOH$	−47°	154°	.949	4.85
Valeric, $CH_3(CH_2)_3COOH$	−35°	187°	.942	4.81
Caproic, $CH_3(CH_2)_4COOH$	−1°	205°	.945	4.81
Palmitic, $CH_3(CH_2)_{14}COOH$	64°	380°	.853	
Stearic, $CH_3(CH_2)_{16}COOH$	69°	383°	.847	
Acrylic, $CH_2=CHCOOH$	12°	142°	1.062	4.26
Oleic, $CH_3(CH_2)_7CH=CH(CH_2)_7COOH$	14°	300°	0.895	
Lactic, $CH_3CHOHCOOH$	18°		1.248	3.87
Oxalic, HOOCCOOH	189°			1.46*
Malonic, $HOOCCH_2COOH$	136°		1.631†	2.80*
Succinic, $HOOC(CH_2)_2COOH$	185°		1.564†	4.17*
Benzoic, C_6H_5COOH	122°	249°	1.266†	4.17
Salicylic, $o\text{-}HOC_6H_4COOH$	159°		1.443†	3.00

*For first dissociation.
†Density of crystalline substance.

(Section 23-3) of the acid strength of the phenols. The dissociation of a carboxylic acid is represented by the equation

$$RCOOH \rightleftarrows RCOO^- + H^+$$

The anion RCOO⁻ can be assigned two electronic structures:

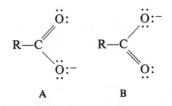

                              A   B

These two structures are equivalent, and the normal state of the anion can be described as a hybrid structure to which the two valence-bond structures A and B contribute equally. The anion is stabilized by the maximum amount of resonance energy, corresponding to complete resonance between the two valence-bond structures. For the undissociated acid the two valence-bond structures are A' and B':

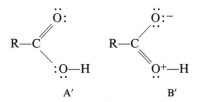

A'   B'

Structure B' is less stable than structure A' because it involves the separation of electric charge, and accordingly the normal state of the acid is a hybrid involving mainly A', with only a small contribution of B', and only a small amount of resonance stabilization. The anion is accordingly stabilized by resonance relative to the undissociated acid; this stabilization energy shifts the equilibrium to favor the ion, and thus increases the acid strength. The change in acid constant from about 1×10^{-16} (for alcohols) to 1×10^{-4} corresponds to about 67 kJ mole^{-1} greater resonance energy in the carboxylate anion than in the undissociated acid.

Formic acid and acetic acid are the first two members of a series of carboxylic acids, the *fatty acids*. The next two acids in the series are *propionic acid*, CH_3CH_2COOH, and *butyric acid*, $CH_3CH_2CH_2COOH$. Butyric acid is the principal odorous substance in rancid butter.

Some of the important organic acids occurring in nature are those in which there is a carboxyl group at the end of a long hydrocarbon chain. *Palmitic acid*, $CH_3(CH_2)_{14}COOH$, and *stearic acid*, $CH_3(CH_2)_{16}COOH$, have structures of this sort. *Oleic acid*, $CH_3(CH_2)_7CH{=}CH(CH_3)_7COOH$, is similar to stearic acid except that it contains a double bond between two of the carbon atoms in the chain.

Oxalic acid, $(COOH)_2$, is a poisonous substance that occurs in some plants. Its molecule consists of two carboxyl groups bonded together:

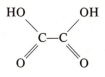

Lactic acid, having the structural formula

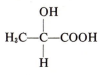

contains a hydroxyl group as well as a carboxyl group; it is a hydroxypropionic acid. It is formed when milk sours and when cabbage ferments, and it gives the sour taste to sour milk and sauerkraut. Tartaric acid, which occurs in grapes, is a dihydroxydicarboxylic acid, with the structural formula

Citric acid, which occurs in the citrus fruits, is a hydroxytricarboxylic acid, with the formula

Benzoic acid, C_6H_5COOH, is the simplest aromatic acid. It is used in medicine as an antiseptic (in benzoated lard). *Salicylic acid,* which is *o*-hydroxybenzoic acid, o-HOC_6H_4COOH, is also used in medicine.

Esters are the products of reaction of acids and alcohols or phenols. For example, ethyl alcohol and acetic acid react with the elimination of water to produce *ethyl acetate*:

$$C_2H_5OH + CH_3COOH \longrightarrow H_2O + CH_3COOC_2H_5$$

Ethyl acetate is a volatile liquid with a pleasing, fruity odor. It is used as a solvent, especially in lacquers.

Many of the esters have pleasant odors, and are used in perfumes and flavorings. The esters are the principal flavorful and odorous constituents of fruits and flowers. Butyl acetate, $CH_3COO(CH_2)_3CH_3$, and amyl acetate, $CH_3COO(CH_2)_4CH_3$, have the odor characteristic of bananas, methyl butyrate, $CH_3(CH_2)_2COOCH_3$, has the odor characteristic of pineapples, and amyl butyrate, $CH_3(CH_2)_2COO(CH_2)_4CH_3$, has that of apricots. Methyl salicylate, o-$OHC_6H_4COOCH_3$, is oil of wintergreen.

Aspirin, a valuable and widely used analgesic and antipyretic drug, is the acetate of salicyclic acid. Its structural formula is

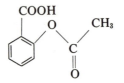

Its common chemical name is acetylsalicylic acid.

The natural *fats* and *oils* are also esters, principally of the trihydroxy alcohol glycerol. Animal fats consist mainly of the glyceryl esters of palmitic acid and stearic acid. *Glyceryl oleate*, the glyceryl ester of oleic acid, is found in olive oil, whale oil, and the fats of cold-blooded animals; these fats tend to remain liquid at ordinary temperatures, whereas *glyceryl palmitate* and *glyceryl stearate* form the solid fats.

Esters can be decomposed by boiling with strong alkali, such as sodium hydroxide. This treatment forms the alcohol and the sodium salt of the carboxylic acid. When fat is boiled with sodium hydroxide, glycerine and sodium salts of the fatty acids (sodium palmitate, sodium stearate, and sodium oleate) are formed. These sodium salts of the fatty acids are called *soap*.

23-6.   Amines and Other Organic Compounds of Nitrogen

The amines are derivatives of ammonia, NH_3, obtained by replacing one or more of the hydrogen atoms by organic radicals. The lighter amines, such as *methylamine*, CH_3NH_2, *dimethylamine*, $(CH_3)_2NH$, and *trimethylamine*, $(CH_3)_3N$, are gases.

Aniline is aminobenzene, $C_6H_5NH_2$. It is a colorless oily liquid, which on standing becomes dark in color because of oxidation to highly colored derivatives. It is used in the manufacture of dyes and other chemicals.

Many substances that occur in plant and animal tissues are compounds of nitrogen. Especially important are the proteins and nucleic acids, which are discussed in the following chapter. The principal product of the metabolism of proteins in the human body is *urea*, $(NH_2)_2CO$. It is the main nitrogenous constituent of urine.

Hexamethylene tetramine, $C_6H_{12}N_4$, is made by the condensation of formaldehyde with aqueous ammonia:

$$6CH_2O + 4NH_3 \longrightarrow C_6H_{12}N_4 + 6H_2O \qquad (23\text{-}1)$$

It forms colorless cubic crystals, which sublime at 280°C. It was the first organic compound for which a detailed determination of molecular structure was carried out (by R. G. Dickinson and A. L. Raymond, in 1922). The molecule was found to have tetrahedral symmetry, as shown in Figure 23-2, with a polycyclic structure, four six-atom rings. All bond angles have the tetrahedral value, to within 1°. The bond lengths, $C—N = 1.47$ Å and $C—H = 1.10$ Å, have the normal single-bond values. The high melting point, above the subliming point 280°C, is to be attributed to the large symmetry number, 12. The hydrocarbon adamantane, $C_{10}H_{16} = (CH_2)_6(CH)_4$, has the same structure, with CH replacing N. It melts at 268°C, as compared with, for example, -30°C for *n*-decane, $C_{10}H_{22}$.

Before the discovery of modern bacteriostatic drugs hexamethylene tetramine was used for treating infections of the kidney and bladder. Its doses were alternated with doses of sodium dihydrogen phosphate to reduce the *p*H of the urine to such an extent as to shift the equilibrium of the reaction of Equation 23-1 to the left, through the decrease in concentration of NH_3 in acid solution (conversion to NH_4^+), thus producing the bactericidal substance formaldehyde.

Heterocyclic Nitrogen Compounds. Purines and Pyrimidines

Heterocyclic compounds are cyclic compounds in which one or more atoms other than carbon (usually nitrogen, oxygen, or sulfur) are present in the ring. An example is *pyridine*, C_5H_5N, a colorless liquid with an unpleasant odor, which is among the products of distilling coal. The electronic structure of pyridine can be described as a hybrid of several valence-bond structures:

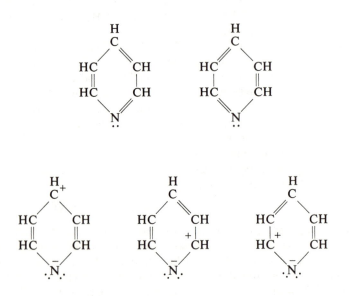

The resonance energy of pyridine, relative to one of the Kekulé-like structures, is 180 kJ mole^{-1}. Pyridine is a base; in acidic solution it adds a proton to the unshared electron pair of the nitrogen atom, forming the pyridonium ion, $C_5H_5NH^+$.

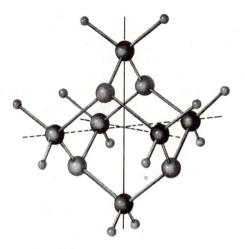

FIGURE 23-2
The molecular structure of hexamethylene
tetramine, $C_6H_{12}N_4$, as determined by x-ray
diffraction of the crystal and electron diffraction
of the vapor.

Six-membered rings containing two or more nitrogen atoms also exist. *Pyrimidine*, $C_4H_4N_2$, is an important example. It is a colorless substance with melting point 22°C and boiling point 124°C. The two nitrogen atoms are in the meta position in the ring. Its electronic structure is a hybrid of

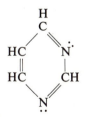

and other valence-bond structures similar to those shown above for pyridine. The partial double-bond character of all of the bonds in the ring requires that the molecules of pyridine and pyrimidine be planar.

The derivatives of pyrimidine, called the pyrimidines, include three substances, thymine, uracil, and cytosine, that are of great importance in the chemistry of heredity. They will be discussed in the following chapter.

The *barbiturates,* which include several important drugs used as seda-
tives (tranquilizers) and hypnotics (sleep-producers), are closely related
to the pyrimidines. The structural formulas of barbituric acid and two of
its derivatives are given below; in these formulas the distribution of the
hydrogen atoms between oxygen and nitrogen is uncertain, and only one
of several pertinent valence-bond structures is indicated. The mechanism
of the physiological action of the drugs is not known.

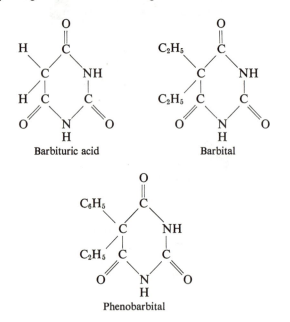

Barbituric acid   Barbital

Phenobarbital

The purines constitute another important class of nitrogen hetero-
cycles. They are the derivatives of the substance *purine,* $C_5H_4N_4$, a colorless crystalline substance with melting point 217°C. The purine molecule
is planar; its electronic structure is a hybrid of

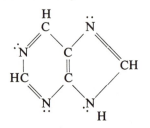

and several other valence-bond structures.

Two of the purines, adenine and guanine, are important in the chemistry of heredity, and will be discussed in the following chapter.

Caffeine, a stimulant found in coffee and tea, is a purine. It is a colorless, odorless substance with melting point 236°C. Its structural formula (showing only one of the several valence-bond distributions) is

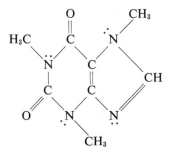

Alkaloids

Alkaloids are basic (that is, alkali-like) substances of plant origin that contain at least one nitrogen atom, usually in a heterocyclic ring. Most of the alkaloids are physiologically active, and many are useful in medicine. An example is *cocaine*, a powerful local anesthetic and stimulant obtained from coca leaves. Its formula is

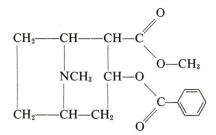

Nicotine, $C_{10}H_{14}N_2$, is the principal alkaloid in the tobacco plant. Its formula is

It is highly toxic and is used as an insecticide. In small quantities it acts as a stimulant and raises the blood pressure. The decreased life expectancy of cigarette smokers is thought to be due in some part to the effect of

the inhaled nicotine, which is absorbed into the blood stream, but for the most part to carcinogenic hydrocarbons and other harmful substances in the smoke.

23-7. Carbohydrates, Sugars, Polysaccharides

The *carbohydrates* are substances with the general formula $C_x(H_2O)_y$. They occur widely in nature. The simpler carbohydrates are called *sugars*, and the complex ones, consisting of very large molecules, are called *polysaccharides*.

A common simple sugar is D-*glucose* (also called *dextrose* and *grape sugar*), $C_6H_{12}O_6$. It occurs in many fruits, and is present in the blood of animals. Its structural formula (not showing the spatial configuration of bonds around the four central carbon atoms) is

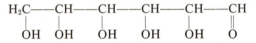

The molecule thus contains five hydroxyl groups and one aldehyde group.

Many other simple carbohydrates occur in nature. These include *fructose* (fruit sugar), *maltose* (malt sugar), and *lactose* (milk sugar).

Ordinary sugar, obtained from sugar cane and sugar beets, is *sucrose*, $C_{12}H_{22}O_{11}$. The molecules of sucrose have a complex structure, consisting of two rings (each containing one oxygen atom), held together by bonds to an oxygen atom as shown in Figure 23-1. It is a combination of glucose and fructose.

Important polysaccharides include *starch*, *glycogen*, and *cellulose*. Starch, $(C_6H_{10}O_5)_x$, occurs in plants, mainly in their seeds or tubers. It is an important constituent of foods. Glycogen, $(C_6H_{10}O_5)_x$, is a substance similar to starch, which occurs in the blood and the internal organs, especially the liver, of animals. Glycogen serves as a reservoir of readily available food for the body; whenever the concentration of glucose in the blood becomes low, glycogen is rapidly hydrolyzed into glucose.

Cellulose, which also has the formula $(C_6H_{10}O_5)_x$, is a stable polysaccharide that serves as a structural element for plants, forming the walls of cells. Like starch and glycogen, cellulose consists of long molecules that contains rings of atoms held together by oxygen atoms, in the way shown in Figure 23-1 for the two rings of sucrose.

The sugars have the properties of dissolving readily in water and of crystallizing in rather hard crystals. These properties are attributed to the presence of a number of hydroxyl groups in these molecules, which form hydrogen bonds with water molecules and (in the crystals) with each other.

23-8. Fibers and Plastics

Silk and wool are protein fibers, consisting of long polypeptide chains (see Chapter 24). Cotton and linen are polysaccharides (carbohydrates), with composition $(C_6H_{10}O_5)_x$. These fibers consist of long chains made from carbon, hydrogen, and oxygen atoms, with no nitrogen atoms present.

In recent years synthetic fibers have been made, by synthesizing long molecules in the laboratory. One of these, which has valuable properties, is *nylon*. It is the product of condensation of adipic acid and diaminohexane. These two substances have the following structures:

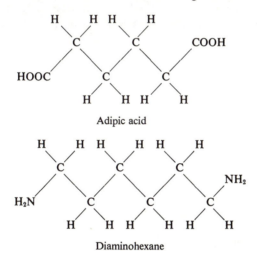

Adipic acid

Diaminohexane

Adipic acid is a chain of four methylene groups with a carboxyl group at each end, and diaminohexane is a similar chain of six methylene groups with an amino group at each end. A molecule of adipic acid can react with a molecule of diaminohexane in the following way:

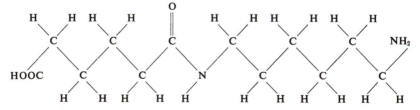

If this process is continued, a very long molecule can be made, in which the adipic acid residues alternate with the diaminohexane residues. Nylon is a fibrous material that consists of these long molecules in approximately parallel orientation.

Other artificial fibers and plastics are made by similar condensation reactions. A *thermolabile plastic* usually is an aggregate of long molecules of this sort that softens upon heating, and can be molded into shape. A *thermosetting plastic* is an aggregate of long molecules containing some reactive groups, capable of further condensation. When this material is molded and heated, these groups react in such a way as to tie the molecules together into a three-dimensional framework, producing a plastic material that cannot be further molded.

With a great number of substances available for use as his starting materials, the chemist has succeeded in making fibers and plastics that are for many purposes superior to natural materials. This field of chemistry, that of synthetic giant molecules, is now advancing rapidly, and we may look forward to further great progress in it in the coming years.

24

Biochemistry

Biochemistry is the study of the chemical composition and structure of the human body and other living organisms, of the chemical reactions that take place within these organisms, and of the drugs and other substances that interact with them.

During the past century biochemistry has developed into an important branch of science. We shall not be able in the limited space of the present chapter to give a general survey of this interesting subject, but shall instead have to content ourselves with a simple introductory discussion of a few of its aspects.

24-1. The Nature of Life

All of our ideas about life involve chemical reactions. What is it that distinguishes a living organism,* such as a man or some other animal or a plant, from an inanimate object, such as a piece of granite? We recognize

*The word *organism* is used to refer to anything that lives or has ever been living—we speak of dead organisms, as well as of living organisms.

that the plant or animal may have several attributes that are not possessed by the rock. The plant or animal has, in general, the power of *reproduction*—the power of having progeny, which are sufficiently similar to itself to be recognized as belonging to the same species of living organisms. The process of reproduction involves chemical reactions, the reactions that take place during the growth of the progeny. The growth of the new organism may occur only during a small fraction of the total lifetime of the animal, or may continue throughout its lifetime.

A plant or animal in general has the ability of ingesting certain materials, foods, subjecting them to chemical reactions, involving the release of energy, and secreting some of the products of the reactions. This process, by which the organism makes use of the food that it ingests by subjecting it to chemical reaction, is called *metabolism*.

Most plants and animals have the ability to respond to their environment. A plant may grow toward the direction from which a beam of light is coming, in response to the stimulus of the beam of light, and an animal may walk or run in a direction indicated by increasing intensity of the odor of a palatable food.

To illustrate the difficulty of defining a living organism, let us consider the simplest kinds of matter that have been thought to be alive. These are the *viruses*, such as the tomato bushy stunt virus, of which an electron micrograph has been shown as Figure 2-14. These viruses have the power of reproducing themselves when in the appropriate environment. A single particle (individual organism) of tomato bushy stunt virus, when placed on the leaf of a tomato plant, can cause the material in the cells of the leaf to be in large part converted into replicas of itself. This power of reproduction seems, however, to be the only characteristic of living organisms possessed by the virus. After the particles are formed, they do not grow. They do not ingest food nor carry on any metabolic processes. So far as can be told by use of the electron microscope and by other methods of investigation, the individual particles of the virus are identical with one another, and show no change with time—there is no phenomenon of aging, of growing old. The virus particles seem to have no means of locomotion, and seem not to respond to external stimuli in the way that large living organisms do. But they do have the power of reproducing themselves.

Considering these facts, should we say that a virus is a living organism, or that it is not? At the present time scientists do not agree about the answer to this question—indeed, the question may not be a scientific one at all, but simply a matter of the definition of words. If we were to define a living organism as a material structure with the power of reproducing itself, then we would include the plant viruses among the living organisms. If, however, we require that living organisms also have the property of carrying on some metabolic reactions, then the plant viruses would be described simply as molecules (with molecular weight of the order of magnitude of 10,000,000) that have such a molecular structure as to permit them to catalyze a chemical reaction, in a proper medium, leading to the synthesis of molecules identical with themselves.

24-2. The Structure of Living Organisms

Chemical investigation of the plant viruses has shown that they consist of the materials called *proteins* and *nucleic acids*, the nature of which is discussed in the following two sections. The giant virus particles or molecules, with molecular weight of the order of magnitude of 10,000,000, may be described as aggregates of smaller molecules, tied together in a definite way.

Many microorganisms, such as molds and bacteria, consist of single *cells*. These cells may be just big enough to be seen with an ordinary microscope, having diameter around 10,000 Å (10^{-4} cm), or they may be much bigger — as large as a millimeter or more in diameter. The cells have a well-organized structure, consisting of a *cell wall*, a few hundred Ångströms in thickness, within which is enclosed a semifluid material called *cytoplasm*, and often other structures that can be seen with the microscope. Other plants and animals consist largely of aggregates of cells, which may be of many different kinds in one organism. The muscles, blood vessel and lymph vessel walls, tendons, connective tissues, nerves, skin, and other parts of the body of a man consist of cells attached to one another to constitute a well-defined structure. In addition there are many cells that are not attached to this structure, but float around in the body fluids. Most numerous among these cells are the *red corpuscles* of the blood. The red corpuscles in man are flattened disks, about 75,000 Å in diameter and 20,000 Å thick. The number of red cells in a human adult is very large. There are about 5 million red cells per cubic millimeter of blood, and a man contains about 5 liters of blood, that is, 5 million cubic millimeters of blood. Accordingly there are 25×10^{12} red cells in his body. In addition, there are many other cells, some of them small, like the red cells, and some somewhat larger — a single nerve cell may be about 10,000 Å in diameter and 100 cm long, extending from the toe to the spinal cord. The total number of cells in the human body is about 5×10^{14}. The amount of *organization* in the human organism is accordingly very great.

The human body does not consist of cells alone. In addition there are the *bones*, which have been laid down as excretions of bone-making cells. The bones consist of inorganic constituents, calcium hydroxyphosphate, $Ca_5(PO_4)_3OH$, and calcium carbonate, and an organic constituent, *collagen*, which is a protein. The body also contains the body fluids blood and lymph, as well as fluids that are secreted by special organs, such as saliva and the digestive juices. Very many different chemical substances are present in these fluids.

The structure of cells is determined by their framework materials, which constitute the cell walls and, in some cases, reinforcing frameworks within the cells. In plants the carbohydrate cellulose, described in the preceding chapter, is the most important constituent of the cell walls. In animals the framework materials are proteins. Moreover, the cell contents consist largely of proteins. For example, a red cell is a thin membrane enclosing a medium that consists of 60% water, 5% miscellaneous materials, and 35% *hemoglobin*, an iron-containing protein, which has molecular weight 68,000, and has the power of combining reversibly with oxygen. It is this power that permits the blood to combine with a large amount of oxygen in the lungs, and to carry it to the tissues, making it available there for oxidation of foodstuffs and body constituents. It has been mentioned earlier in this section that the simplest forms of matter with the power of reproducing themselves, the viruses, consist in part of proteins, as do also the most complex living organisms.

24-3. Amino Acids and Proteins

Proteins may well be considered the most important of all the substances present in plants and animals. Proteins occur either as separate molecules, usually with very large molecular weight, ranging from about 10,000 to many millions, or as reticular constituents of cells, constituting their structural framework. The human body contains many thousands of different proteins, which have special structures that permit them to carry out specific tasks.

All proteins are nitrogenous substances, containing approximately 16% of nitrogen, together with carbon, hydrogen, oxygen, and often other elements such as sulfur, phosphorus, iron (four atoms of iron are present in each molecule of hemoglobin), and copper.

Amino Acids

When proteins are heated in acidic or basic solution they undergo hydrolysis, producing substances called amino acids. Amino acids are carboxylic acids in which one hydrogen atom has been replaced by an amino group, $-NH_2$. The amino acids that are obtained from proteins are *alpha* amino acids, with the amino group attached to the carbon atom next to the carboxyl group (this carbon atom is called the alpha carbon atom). The simplest of these amino acids is *glycine*, $CH_2(NH_2)COOH$. The other natural amino acids contain another group, usually called R, in place of one of the hydrogen atoms on the alpha carbon atom, their general formula thus being $CHR(NH_2)COOH$.

The amino group is sufficiently basic and the carboxyl group is sufficiently acidic that in solution in water the proton is transferred from the carboxyl group to the amino group. The carboxyl group is thus converted into a carboxyl ion, and the amino group into a substituted ammonium ion. The structure of glycine and of the other amino acids in aqueous solution is accordingly the following:

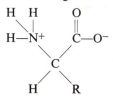

The amino groups and carboxyl groups of most amino acids dissolved in animal or plant liquids, which usually have pH about 7, are internally ionized in this way, to form an ammonium ion group and a carboxyl ion group within the same molecule.

There are twenty-three amino acids that have been recognized as important constituents of proteins. Their names are given in Table 24-1, together with the formulas of the characteristic group R. Some of the amino acids have an extra carboxyl group or an extra amino group. There is one double amino acid, *cystine*, which is closely related to a simple amino acid, *cysteine*. Four of the amino acids contain heterocyclic rings —rings of carbon atoms and one or more other atoms, in this case nitrogen atoms. Two of the amino acids given in the table, *asparagine* and *glutamine*, are closely related to two others, *aspartic acid* and *glutamic acid*, differing from them only in having the extra carboxyl group changed into an amide group,

Proteins are important constituents of food. They are digested by the digestive juices in the stomach and intestines, being split in the process of digestion into small molecules, probably mainly the amino acids themselves. These small molecules are able to pass through the walls of the stomach and intestines into the blood stream, by which they are carried around into the tissues, where they may then serve as building stones for the manufacture of the body proteins. Sometimes people who are ill and cannot digest foods satisfactorily are fed by the injection of a solution of amino acids directly into the blood stream. A solution of amino acids for this purpose is usually obtained by hydrolyzing proteins.

TABLE 24-1
The Principal Amino Acids Occurring in Proteins

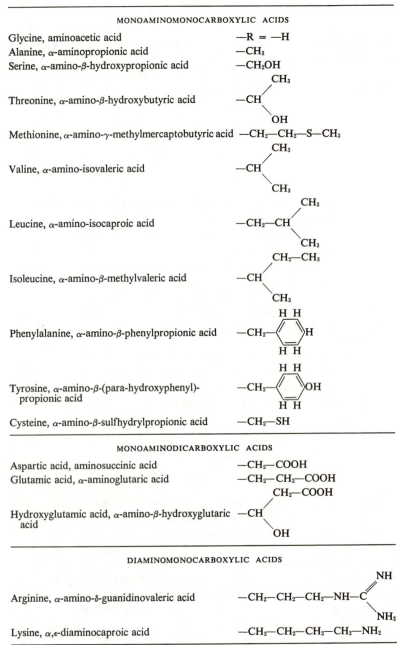

MONOAMINOMONOCARBOXYLIC ACIDS

Glycine, aminoacetic acid   —R = —H

Alanine, α-aminopropionic acid   —CH₃

Serine, α-amino-β-hydroxypropionic acid   —CH₂OH

Threonine, α-amino-β-hydroxybutyric acid

Methionine, α-amino-γ-methylmercaptobutyric acid   —CH₂—CH₂—S—CH₃

Valine, α-amino-isovaleric acid

Leucine, α-amino-isocaproic acid

Isoleucine, α-amino-β-methylvaleric acid

Phenylalanine, α-amino-β-phenylpropionic acid

Tyrosine, α-amino-β-(para-hydroxyphenyl)-propionic acid

Cysteine, α-amino-β-sulfhydrylpropionic acid   —CH₂—SH

MONOAMINODICARBOXYLIC ACIDS

Aspartic acid, aminosuccinic acid   —CH₂—COOH

Glutamic acid, α-aminoglutaric acid   —CH₂—CH₂—COOH

Hydroxyglutamic acid, α-amino-β-hydroxyglutaric acid

DIAMINOMONOCARBOXYLIC ACIDS

Arginine, α-amino-δ-guanidinovaleric acid

Lysine, α,ϵ-diaminocaproic acid   —CH₂—CH₂—CH₂—CH₂—NH₂

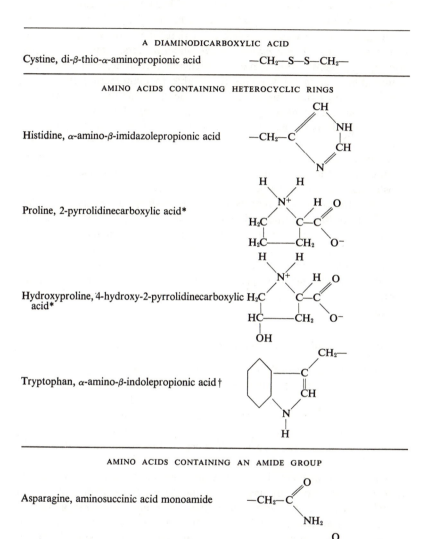

A DIAMINODICARBOXYLIC ACID

Cystine, di-β-thio-α-aminopropionic acid   —CH₂—S—S—CH₂—

AMINO ACIDS CONTAINING HETEROCYCLIC RINGS

Histidine, α-amino-β-imidazolepropionic acid

Proline, 2-pyrrolidinecarboxylic acid*

Hydroxyproline, 4-hydroxy-2-pyrrolidinecarboxylic acid*

Tryptophan, α-amino-β-indolepropionic acid †

AMINO ACIDS CONTAINING AN AMIDE GROUP

Asparagine, aminosuccinic acid monoamide

Glutamine, α-aminoglutaric acid monoamide

*The formulas given for proline and hydroxyproline are those of the complete molecules, and not just of the groups R.

†The hexagon represents a benzene ring.

Although all the amino acids listed in Table 24-1 are present in the proteins of the human body, not all of them need to be in the food. Experiments have been carried out which show that nine of the amino acids are essential to man. These *nine essential amino acids* are histidine, lysine, tryptophan, phenylalanine, leucine, isoleucine, threonine, methionine, and valine. The human body seems to be able to manufacture the others, which are called the nonessential amino acids. Some organisms that we usually consider to be simpler than man have greater powers than the human organisms in that they are able to manufacture all of the amino acids from inorganic constituents. The red bread mold, *Neurospora*, has this power. There is an evolutionary advantage for the organism in getting rid of the chemical machinery (enzymes) for manufacturing vital substances that are available in the food supply.

Protein foods for man may be classed as *good protein foods*, those that contain all of the essential amino acids, and *poor protein foods*, those that are lacking in one or more of the essential amino acids. *Casein*, the principal protein in milk, is a good protein, from this point of view, whereas *gelatin*, a protein obtained by boiling bones and tendons (partial hydrolysis of the insoluble protein collagen produces gelatin) is a poor protein. Gelatin contains no tryptophan, no valine, and little or no threonine.

Right-handed and Left-handed Amino-acid Molecules

It was pointed out in Section 6-3 that some substances exist in two isomeric (enantiomeric) forms, called L (levo) and D (dextro) forms, with molecules that are mirror images of one another. These two forms exist for every amino acid except glycine; they differ from one another in the arrangement in space of the four groups attached to the α-carbon atom. Figure 24-1 shows the two enantiomers of the amino acid alanine, in which R is the methyl group, CH_3.

A most extraordinary fact is that only one of the two enantiomers of each of the amino acids has been found to occur in plant and animal proteins, and that this enantiomer has the same configuration for all of these amino acids; that is, the hydrogen atom, carboxyl ion group, and ammonium ion group occupy the same position relative to the group R around the alpha carbon atom. This configuration is called the L configuration—*proteins are built entirely of L-amino acids*.

This is a very puzzling fact. Nobody knows why it is that we are built of L-amino acid molecules, rather than of D-amino acid molecules. All the proteins that have been investigated, obtained from animals and from plants, from higher organisms and from very simple organisms—bacteria, molds, even viruses—are found to have been made of L-amino acids.*

---

*Residues of D-amino acids are found in a few simple peptides in living organisms.

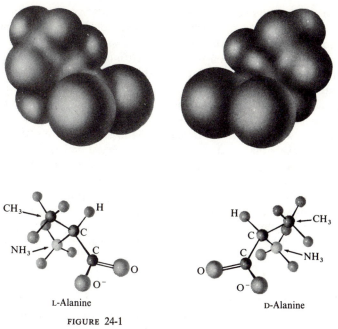

L-Alanine   D-Alanine

FIGURE 24-1
The two enantiomers of the amino acid alanine.

Right-handed molecules and left-handed molecules have exactly the same properties, so far as their interaction with ordinary substances is concerned—they differ in their properties only when they interact with other right-handed or left-handed molecules. The earth might just as well be populated with living organisms made of D-amino acids as with those made of L-amino acids. A man who was suddenly converted into an exact mirror image of himself would not at first know that anything had changed about him, except that he would write with his left hand, instead of his right, his hair would be parted on the right side instead of the left, his heartbeat would show his heart to be on the right side, and so on; he could drink water, inhale air and use the oxygen in it for combustion, exhale carbon dioxide, and carry on other bodily functions just as well as ever—so long as he did not eat any ordinary food. If he were to eat ordinary plant or animal food, he would find that he could not digest it.* He could be kept alive only on a diet containing synthetic D-amino acids, made in the chemical laboratory. He could not have any children, unless he could find a wife who had been subjected to the same process of reflection into a mirror image of her original self. We see that there is the possibility that the earth might have been populated with two completely

---

*Alice: "Perhaps Looking-glass milk isn't good to drink." In *Through the Looking-Glass*, by Lewis Carroll (Charles Lutwidge Dodgson), 1872.

independent kinds of life—plants, animals, human beings of two kinds, who could not use one another's food, could not produce hybrid progeny.

No one knows why living organisms are constructed of L-amino acids. We have no strong reason to believe that molecules resembling proteins could not be built up of equal numbers of right-handed and left-handed amino acid molecules. Perhaps the protein molecules that are made of amino acid molecules of one sort only are especially suited to the construction of a living organism—but if this is so, we do not know why.*

Nor do we know why it is that living organisms have evolved in the L-system rather than in the D-system. The suggestion has been made that the first living organism happened by chance to make use of a few molecules with the L configuration, which were present with D molecules in equal number; and that all succeeding forms of life that have evolved have continued to use L-amino acid molecules through inheritance of the character from the original form of life. Perhaps a better explanation than this can be found—but I do not know what it is.

The Primary Structure of Proteins

During the past century much effort has been devoted by scientists to the problem of the structure of proteins. This is a very important problem; when it is solved, we shall have a much better understanding than at present of the nature of physiological reactions, and the knowledge of the structure of protein molecules will help in the attack on important medical problems, such as the problems of the control of heart disease and cancer.

In the period between 1900 and 1910 strong evidence was obtained by the German chemist Emil Fischer (1852–1919) to indicate that the amino acids in proteins are combined into long chains, called *polypeptide chains.* For example, two molecules of glycine can be condensed together, with elimination of water, to form the double molecule glycylglycine, shown in Figure 23-1. The bond formed in this way is called a *peptide bond.* The process of forming these bonds can be continued, resulting in the production of a long chain containing many amino-acid residues, as shown in Figure 23-1.

Chemical methods have been developed to determine how many polypeptide chains there are in a protein molecule. These methods involve the use of a reagent (fluorodinitrobenzene) that combines with the free amino group of the amino acid residue at the end of the polypeptide chain to form a colored complex, which can be isolated and identified after the protein has been hydrolyzed into its constituent amino acids (and the

---

*A possible reason is that there would be serious overcrowding of the side chains for a mixture of D and L residues in the alpha helix (Figure 24-2).

end amino acid with the colored group attached). For example, the kind of hemoglobin molecule that is found in the red corpuscles of most adult human beings (called normal adult human hemoglobin or hemoglobin A) has been shown to contain four polypeptide chains. There are two chains of one kind, called the alpha chain, and two of the other, called the beta chain. The alpha chain begins with the sequence val-leu-···· (the abbreviations commonly used are the first three letters of the name of the amino acid), and the beta chain with the sequence val-his-leu-·····. The hemoglobin molecule has been shown by use of the ultracentrifuge, x-ray diffraction, and other methods of investigation to be approximately spherical in shape, with diameter about 40 Å. Hence the polypeptide chains cannot be stretched out, but must be folded back and forth, to produce the globular molecule.

The order of amino acid residues in the polypeptide chains (called the *primary structure*) was first determined for the protein insulin. The insulin molecule has molecular weight about 12,000. It consists of four polypeptide chains, of which two contain 21 amino acid residues apiece, and the other two contain 30. The sequence of amino acids in the short chains and in the long chains was determined in the years between 1945 and 1952 by the English biochemist F. Sanger (born 1918) and his collaborators. The four chains in the molecule are attached to one another by sulfur-sulfur bonds, between the halves of cystine residues (see Table 24-1). Sequence determinations have now been made by Sanger's method for the alpha and beta chains of normal adult human hemoglobin and many other proteins. The sequence for the beta chain of human hemoglobin A (146 amino acid residues) is the following: (Amino end or N-terminus) val-his-leu-thr-pro-glu-glu-lys-ser-ala-val-thr-ala-leu-try-gly-lys-val-aspNH$_2$-val-asp-glu-val-gly-gly-glu-ala-leu-gly-arg-leu-leu-val-val-tyr-pro-try-thr-gluNH$_2$-arg-phe-phe-glu-ser-phe-gly-asp-leu-ser-thr-pro-asp-ala-val-met-gly-aspNH$_2$-pro-lys-val-lys-ala-his-gly-lys-lys-val-leu-gly-ala-phe-ser-asp-gly-leu-ala-his-leu-asp-asp-leu-lys-gly-thr-phe-ala-thr-leu-ser-glu-leu-his-cys-asp-lys-leu-his-val-asp-pro-glu-aspNH$_2$-phe-arg-leu-leu-gly-aspNH$_2$-val-leu-val-cys-val-leu-ala-his-his-phe-gly-lys-glu-phe-thr-pro-pro-val-gluNH$_2$-ala-ala-tyr-gluNH$_2$-lys-val-val-ala-gly-val-ala-aspNH$_2$-ala-leu-ala-his-lys-tyr-his (carboxyl end or C-terminus). The sequence for the alpha chain (141 residues) has some similarity to that for the beta chain: about 75 amino acid residues occur in essentially the same places in the chains. The alpha chain of gorilla hemoglobin differs from that of human hemoglobin in two substitutions of one amino acid residue for another, and the gorilla and human beta chains differ by only one substitution. The difference between horse hemoglobin and human hemoglobin is about 18 substitutions per chain. These observations, and many similar ones for other proteins, provide strong independent support for the theory of the evolution of species.

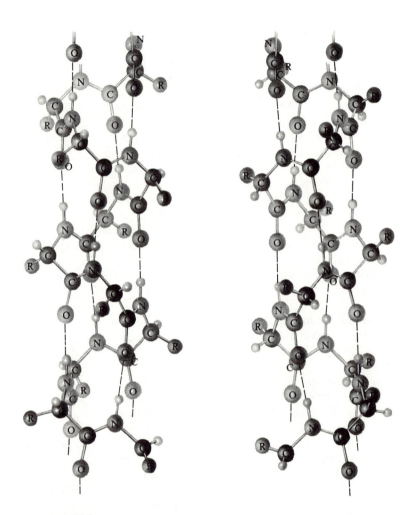

FIGURE 24-2

A drawing showing two possible forms of the α helix; the one on the left is a left-handed helix, and the one on the right is a right-handed helix. The right-handed helix of polypeptide chains is found in many proteins. The amino-acid residues have the L configuration in each case. The circles labeled R represent the side chains of the various residues.

The Denaturation of Proteins

Proteins such as insulin and hemoglobin have certain special properties that make them valuable to the organism. Insulin is a hormone that assists in the process of oxidation of sugar in the body. Hemoglobin has the power of combining reversibly with oxygen, permitting it to attach oxygen molecules to itself in the lungs, and to liberate them in the tissues. These well-defined properties show that the protein molecules have very definite structures.

A protein that retains its characteristic properties is called a *native protein:* hemoglobin as it exists in the red cell or in a carefully prepared hemoglobin solution, in which it still has the power of combining reversibly with oxygen, is called native hemoglobin. Many proteins lose their characteristic properties very easily. They are then said to have been *denatured.* Hemoglobin can be denatured simply by heating its solution to 65°C. It then coagulates, to form a brick-red insoluble coagulum of denatured hemoglobin. Most other proteins are also denatured by heating to approximately this temperature. Egg white, for example, is a solution consisting mainly of the protein *ovalbumin,* with molecular weight 43,000. Ovalbumin is a soluble protein. When its solution is heated for a little while at about 65°C the ovalbumin is denatured, forming an insoluble white coagulum of denatured ovalbumin. This phenomenon is observed when an egg is cooked.

It is believed that the process of denaturation involves uncoiling the polypeptide chains from the characteristic structure of the native protein. In the coagulum of denatured hemoglobin or denatured ovalbumin the uncoiled polypeptide chains of different molecules of the protein have become tangled up with one another in such a way that they cannot be separated; hence the denatured protein is insoluble. Some chemical agents, including strong acid, strong alkali, and alcohol, are good denaturing agents.

The Secondary Structure of Proteins

A regular way in which the polypeptide chains of a protein molecule are arranged in space is called a *secondary structure* of the protein. During recent years much progress has been made in this field, especially by use of the x-ray diffraction method.

The principal type of secondary structure is shown in Figure 24-2. The polypeptide chain is folded into a helix. There are about 3.6 amino acid residues per turn of the helix—about 18 residues in 5 turns. Each residue is linked to residues in the preceding and following turns by hydrogen bonds between the N—H groups and the oxygen atom of the C=O

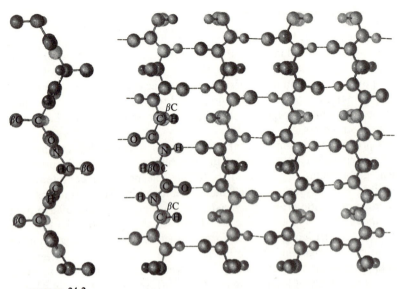

FIGURE 24-3
A drawing of the antiparallel-chain pleated sheet, a protein structure found for silk fibers.

group. The side chains R of the different residues project radially from the helix; there is plenty of room for them, so that the sequence of residues can be an arbitrary one. This configuration is called the alpha helix.

Many fibrous proteins, including hair, fingernails, horn, and muscle, consist of polypeptide chains with the configuration of the alpha helix, arranged approximately parallel to one another, with the axis of the helix in the direction of the fiber. In some of these proteins the polypeptide chains, with the configuration of the alpha helix, are twisted about one another to form cables or ropes. Hair and horn can be stretched out to over twice their normal length; this process involves breaking the hydrogen bonds of the alpha helix, and forcing the polypeptide chains into a stretched configuration. Silk fibers consist of polypeptide chains with the stretched configuration, attached to one another by hydrogen bonds that extend laterally, as shown in Figure 24-3.

Tertiary and Quaternary Structure

During the period 1946 to 1960 a complete determination of the structure of a globular protein, *myoglobin*, was carried out by the English scientist J. D. Kendrew and his collaborators. Myoglobin, which is found in muscle, is a protein rather similar to hemoglobin, but with only one poly-peptide chain in the molecule (molecular weight about 17,000). The myo-globin molecule has been shown by x-ray diffraction of myoglobin crys-tals to contain a polypeptide chain that does not form a single helix, but instead coils into eight short segments with the configuration of the alpha helix, which are connected by nonhelical sections. This aspect of the three-dimensional structure of the polypeptide chain, the arrangement in space of segments with a regular (repeating) structure (secondary struc-ture), is called the *tertiary structure* of the protein. The tertiary structure, like the secondary structure, is determined by the amino acid sequence (the primary structure).

The x-ray investigation of hemoglobin (by the English scientist Max Perutz and his collaborators) has shown that the molecule is an aggregate of four groups, each of which closely resembles a myoglobin molecule and contains one polypeptide chain. The combination of two or more polypeptide chains in this way is described as the quaternary structure of a protein.

24-4. Nucleic Acids. The Chemistry of Heredity

One of the most amazing and interesting aspects of the world is the existence of human beings and other living organisms who are able to have progeny, to whom they transmit many of their own characteristics. The mechanism by means of which a child develops in such a way as to resemble his parents has been under intensive study for a century, and the progress in understanding this phenomenon has been especially rapid during the last few years.

In 1866 the Abbot Gregor Johann Mendel (1822–1884) developed a simple theory of inheritance on the basis of experiments that he had carried out with peas in the garden of the Augustinian monastery at Brno, in Moravia (now Czechoslovakia). He found that his experimental results could be accounted for by assuming that each of the plants of the second generation receives from each of the two parent plants a determiner or factor (now called a *gene*) for each inherited character. The genes are now described as being arranged linearly in a larger structure, one of the chromosomes, which can be seen in the nuclei of cells.

Different genes that may occur at the same locus in a chromosome are called *alleles* or *allelomorphic genes*. For example, Mendel hybridized two strains of peas that differed from one another in that the seeds were round in one strain and wrinkled in the other. The first-generation hybrid progeny had round seeds. However, when they were allowed to become self-fertilized he found that about three-quarters of the second-generation progeny had round seeds and about one-quarter had wrinkled seeds. His explanation of this observation, and of many others like it, is that the peas of the first strain carry two alleles for roundness, and those of the second strain two alleles for wrinkledness. The hybrids of these two strains inherit one of each of these two alleles (one from each parent), and Mendel assumed that the allele for roundness is the *dominant* gene and that for wrinkledness is *recessive*, so that the possession of one each of the two allelomorphic genes leads to roundness (as does the possession of two genes for roundness). In the next generation, obtained by self-fertilization of the first-generation progeny, the allele for roundness or the allele for wrinkledness is inherited at random from the one parent, and also at random from the other parent. About one-quarter of the progeny would then be expected to have the genic composition RR (with R representing the dominant allele), one-half to have the genic constitution Rr or rR, and one-quarter to have the genic constitution rr. The progeny RR would have round seeds, the heterozygotes Rr and rR would also have round seeds, because of the assumed dominance of R, and the recessive homozygotes rr would have wrinkled seeds.

The theory of the gene was greatly developed in the years following 1910 as the result of work on the fruit fly, *Drosophila*, carried out by Thomas Hunt Morgan and his collaborators (especially A. H. Sturtevant, Calvin Bridges, and H. J. Muller), who were able to determine the order in which many genes are located in the chromosomes of this organism. Further progress was made by other investigators (G. W. Beadle and E. L. Tatum, in particular) with use of the red bread mold, *Neurospora*, and by J. Lederberg and others who have studied the genetics of bacteria.

An example of the relation between genes and protein molecules is provided by the different kinds of hemoglobin that have been found in the red cells of human beings. In 1949 it was discovered that some human beings, patients with the disease sickle-cell anemia, have in their red cells a form of hemoglobin (hemoglobin S) that is different from that in the red cells of most people (hemoglobin A). The difference is not great: the two alpha chains of the hemoglobin-S molecule are identical with those of the hemoglobin-A molecule, and each beta chain has one amino acid residue that is different. The beta chain of hemoglobin A has a residue of glutamic acid in the sixth position from the free amino end (see the sequence on p. 777), whereas the beta chain of hemoglobin S has in this position a residue of valine; all of the other amino acid residues are the same.

The abnormal hemoglobin in the red cells of the sickle-cell-anemia patients causes a very serious disease. Each of the two parents of a patient with this disease is found by experiment to have in his red cells a fifty-fifty mixture of hemoglobin A and hemoglobin S, and one-quarter of the children of such marriages are found, on the average, to be sickle-cell homozygotes, with the genic constitution SS and the disease sickle-cell anemia. It is evident that the two genes A and S carry out their functions essentially independently of one another; in a heterozygote, with genic constitution AS, each of the genes manufactures its own kind of hemoglobin, and each red cell contains a mixture of hemoglobin A and hemoglobin S.

About 25 years ago evidence was obtained showing that a gene is a molecule of *deoxyribonucleic acid* (usually abbreviated as *DNA*). The chemical nature of DNA has now been determined, and its molecular structure is known. The nature of this structure is such as to permit considerable insight to be obtained about the mechanism by means of which these molecules duplicate themselves, in order that the duplicates may be passed on to the progeny, or in order that the living organism may grow, through cell division, with each cell having its complement of genes.

DNA consists of units, called nucleotides (several hundred), that are held together by chemical bonds in a linear array, called a polynucleotide chain or a nucleic acid molecule. Each nucleotide consists of three parts: a molecule of phosphoric acid, a molecule of a sugar, *deoxyribose*, and a molecule of a nitrogen compound, called a nitrogen base. The molecules of sugar and molecules of phosphoric acid are condensed together to form long chains:

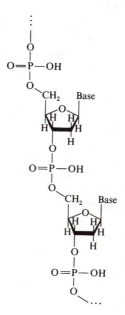

Deoxyribose is a pentose (sugar with formula $C_5H_{10}O_5$) that has lost one oxygen atom, giving it the formula $C_5H_{10}O_4$; its structural formula is

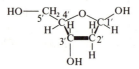

In DNA the two hydroxyl groups attached to carbon atoms 3' and 5' condense with hydroxyl groups of separate molecules of phosphoric acid, $OP(OH)_3$, to form the DNA chain. The nitrogen atom of the nitrogen base replaces the hydroxyl group attached to carbon atom 1'.

The nitrogen bases found in DNA comprise the two purines *adenine* and *guanine* and the two pyrimidines *thymine* and *cytosine*; in the formulas shown in Figure 24-4 the asterisk indicates the hydrogen atom that is replaced by the carbon atom of the sugar ring in DNA, and the double bonds correspond to only one of the several valence-bond structures for each molecule. The molecules are planar, because each of the bonds in the purine and pyrimidine rings has some double-bond character.

Chemical analysis of DNA from the nuclei of cells showed that, although the relative number of molecules of the two purines adenine and guanine varies from species to species, the molecular ratio adenine/thymine is unity and the ratio guanine/cytosine is unity. For example, the percentages in human sperm are 31% adenine, 19% guanine, 31% thymine, and 19% cytosine.

This experimental result was interpreted only when a theory of the structure of DNA had been developed. In 1953, making use of excellent x-ray diffraction patterns of DNA that had been made by M. H. F. Wilkins, the American biologist J. D. Watson and the British biophysicist F. H. C. Crick proposed that molecules of DNA consist of two chains wrapped about one another in a helical configuration, in such a way that at every level, 3.3 Å apart along the axis of the double helix, there occurs a residue of either adenine or guanine and one of either thymine or cytosine, and that these residues occur in complementary pairs: either as an adenine-thymine pair or as a guanine-cytosine pair. The explanation of this complementary pairing is shown in Figure 24-4. It is seen that adenine and thymine can form two hydrogen bonds with one another, whereas cytosine and guanine can form three.

According to the Watson-Crick proposal, the four bases adenine, thymine, guanine, and cytosine, which may be represented by the letters A, T, G, and C, occur in a characteristic sequence in one of the two polynucleotide chains of a gene and in the complementary sequence in the other polynucleotide chain. At each level there is one of the following four pairs of nitrogen bases: —A====T—, —T====A—, —G≡≡≡C—,

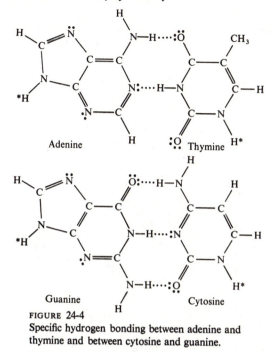

Adenine   Thymine

Guanine   Cytosine

FIGURE 24-4
Specific hydrogen bonding between adenine and
thymine and between cytosine and guanine.

—C≡≡≡G—. The dashes indicate either two or three hydrogen bonds,
as shown in Figure 24-4.

The sequence of bases in a gene constitute a code that determines the
nature of the character that is conferred upon the organism that has inherited the gene. It is believed that the sequence of bases in a gene usually
determines the sequence of amino acid residues in a polypeptide chain (a
protein) that is synthesized in the cell under the influence of the gene, as
described below.

In addition to controlling the manufacture of other molecules, the DNA
replicates itself. The Watson-Crick mechanism of reduplication of DNA
molecules in the course of cell division is postulated to be the following:
a double helix of complementary polynucleotides begins to uncoil into
the separate chains; new polynucleotide chains begin to be synthesized,
with the old ones as the templates, and the new chain that is being synthesized in approximation to each of the old chains is identical with the
other old chain, in order to preserve the complementariness. Thus when
the process is completed there are two identical double helixes, each consisting of one old chain and one newly synthesized chain (Figure 24-5).
The half-old half-new character of first-generation daughter molecules
of DNA in bacterial cultures has been verified by experiments with
tracer isotopes (^{15}N; Section 26-5).

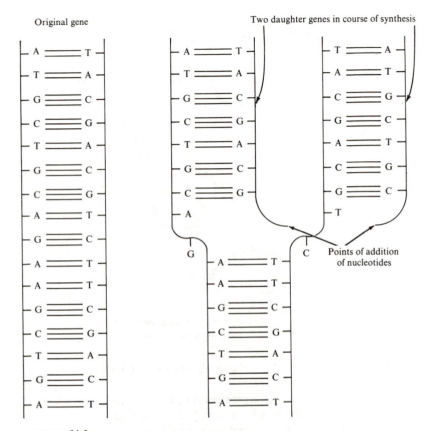

FIGURE 24-5
A diagram showing the postulated method of reduplication of the gene through
formation of a polynucleotide chain complementary to each of the two mutually
complementary chains of the original gene. The helical arrangement of the two
chains is not indicated in this diagram.

The mechanism of protein synthesis involves the transfer of information
from one of the chains of the DNA helix to a molecule of RNA (ribo-
nucleic acid) that is a complement of the DNA chain. RNA contains the
sugar ribose

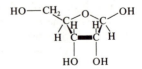

in place of the deoxyribose of DNA, and the pyrimidine base uracil (U)

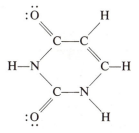

in place of thymine. The power of uracil to form hydrogen bonds (two to adenine) is the same as for thymine, from which it differs only in having hydrogen in place of a methyl group (see Figure 24-4). Each gene (molecule of DNA) can serve as the template for the synthesis of many molecules of messenger RNA, each of which carries the information stored in the gene. This information is then used, with the aid of other molecules, especially certain enzymes, in the synthesis of the polypeptide chains of proteins.

It has been found that three nucleotides select an amino acid for incorporation in the chain; we may say that the gene is a sequence of three-letter words (called *codons*) formed with a four-letter alphabet, A, T, G, C for DNA and the equivalent A, U, G, C for RNA. Thus 146 codons, 438 letters (plus a few to carry the messages to start and to stop the synthesis) are needed in the gene for the beta chain of hemoglobin, containing 146 amino acid residues. Each RNA molecule probably manufactures some tens of thousands of beta chains; there are about 100,000,000 hemoglobin molecules in the mature red cell.

The genetic code seems to be essentially the same in all organisms. It is given in Table 24-2. The code is redundant, in selecting among 20 amino acids; there are 64 three-letter words in the alphabet. The redundancy involves primarily the third letter.

How the code has been worked out is illustrated by the following experiment. An enzyme solution obtained from bacterial cells and added to a solution of all twenty amino acids produces a polypeptide chain consisting only of residues of the amino acid phenylalanine when provided with a synthetic RNA consisting of poly-uracil (that is, U-U-U-U- . . .). Hence UUU is the codon for phenylalanine, as shown in the table. Much of this work was done by the American scientists M. W. Nirenberg, H. G. Khorana, and R. H. Holley, and their collaborators, with use of enzymes that had been discovered by A. Kornberg and S. Ochoa.

TABLE 24-2
The Genetic Code

SECOND LETTER

		U	C	A	G	
F I R S T L E T T E R	**U**	UUU ⎤ Phe UUC ⎦ UUA ⎤ Leu UUG ⎦	UCU ⎤ UCC UCA ⎥ Ser UCG ⎦	UAU ⎤ Tyr UAC ⎦ UAA * UAG *	UGU ⎤ Cys UGC ⎦ UGA * UGG Trp	U C A G
	C	CUU ⎤ CUC CUA ⎥ Leu CUG ⎦	CCU ⎤ CCC CCA ⎥ Pro CCG ⎦	CAU ⎤ His CAC ⎦ CAA ⎤ Gln CAG ⎦	CGU ⎤ CGC CGA ⎥ Arg CGG ⎦	U C A G
	A	AUU ⎤ AUC ⎥ Ile AUA ⎦ AUG Met	ACU ⎤ ACC ACA ⎥ Thr ACG ⎦	AAU ⎤ Asn AAC ⎦ AAA ⎤ Lys AAG ⎦	AGU ⎤ Ser AGC ⎦ AGA ⎤ Arg AGG ⎦	U C A G
	G	GUU ⎤ GUC GUA ⎥ Val GUG ⎦	GCU ⎤ GCC GCA ⎥ Ala GCG ⎦	GAU ⎤ Asp GAC ⎦ GAA ⎤ Glu GAG ⎦	GGU ⎤ GGC GGA ⎥ Gly GGG ⎦	U C A G

T H I R D   L E T T E R

*May act as signals for terminating polypeptide chains.

24-5. Metabolic Processes. Enzymes and Their Action

The chemical reactions that take place in a living organism are called *metabolic processes* (Greek *metabole*, change). These reactions are of very many kinds. Let us consider what happens to food that is ingested. The food may contain complex carbohydrates, especially starch, that are split up into simple sugars in the process of digestion, and then pass through the walls of the digestive tract into the blood stream. The sugars may then be converted, in the liver, into glycogen (animal starch), which has the same formula as starch, $(C_6H_{10}O_5)_x$, where x is a large number. Glycogen and other polysaccharides constitute one of the important sources of energy for animals. They combine with oxygen to form carbon dioxide and water, with liberation of energy, part of which can be used for doing work, and part to keep the body warm.

We have mentioned before that proteins in foodstuffs are split in the stomach and intestines into amino acids or simple peptides, which pass through the walls into the blood stream, and then may be built up into the special proteins needed by the organism. A process of tearing down the proteins of the body also takes place. For example, red cells have a life-time of a few weeks, at the end of which they are destroyed, being replaced by newly formed red cells. The nitrogen of the protein molecules that are torn down is eliminated in the urine, as urea, $CO(NH_2)_2$.

Fats that are ingested are also decomposed in the process of digestion into simpler substances, which then are used by the body for fuel and as structural material.

Some of the chemical reactions that take place in the body can also be made to take place in beakers or flasks in the laboratory. For example, a protein can be decomposed into amino acids in the laboratory by adding strong acids to it and boiling for a long time. Similarly, sugar can be oxidized to carbon dioxide and water; if a little cigarette ash or other solid material is rubbed onto a cube of sugar, the sugar can be lighted by a match, and it will then burn in air, producing carbon dioxide and water:

$$C_{12}H_{22}O_{11} + 12O_2 \longrightarrow 12CO_2 + 11H_2O$$

However, it has not been found possible to cause these chemical reactions to take place in the laboratory at the temperature of the human body, except in the presence of special substances obtained from plants or animals. These substances, enzymes, are proteins that have a catalytic power for certain reactions. Thus the saliva contains a special protein, an enzyme called *salivary amylase* or *ptyalin*, which has the power of catalyzing the decomposition of starch into a sugar, maltose, $C_{12}H_{22}O_{11}$.

The reaction that is catalyzed by salivary amylase is

$$(C_6H_{10}O_5)_x + \frac{x}{2} H_2O \longrightarrow \frac{x}{2} C_{12}H_{22}O_{11}$$

Saliva is mixed with a food, such as potato, while the food is being chewed, and during the first few minutes that the food is in the stomach the salivary amylase causes the conversion of the starch into maltose to take place.

Similarly, there is an enzyme in the stomach, *pepsin*, that has the power of serving as a very effective catalyst for the reaction of hydrolysis of proteins into amino acids—that is, for splitting the peptide bond by reaction with water, to form an amino group and a carboxyl group. Pepsin does its work most effectively in a somewhat acidic solution. Gastric juice is in fact rather strongly acidic, its *p*H being about 0.8—it is hence somewhat more strongly acidic than 0.1 *F* hydrochloric acid.

The stomach also contains an enzyme, *rennin*, that assists in the digestion of milk, and another enzyme, *lipase*, that catalyzes the decomposition of fats into simpler substances. Additional enzymes involved in the digestion of polysaccharides, proteins, and fats take part in the continuation of the digestion in the intestines; these enzymes are contained in the intestinal juice, pancreatic juice, and bile.

The chemical reactions that take place in the blood and in the cells of the body are also in general catalyzed by enzymes. For example, the process of oxidation of sugar is a complicated one, involving a number of steps, and it is believed that a special enzyme is present to catalyze each of these steps. It has been estimated that there may be twenty thousand or thirty thousand different enzymes in the human body, each constructed in such a way as to permit it to serve as an effective catalyst for a particular chemical reaction useful to the organism.

In recent years many enzymes have been isolated and purified. Many, indeed, have been crystallized. A great deal of work has been done in an effort to discover the mechanism of the catalytic activity of enzymes. This general problem is one of the most important of all of the problems of biochemistry.

Heat Values of Foods

One important use of foods is to serve as a source of energy, permitting work to be done, and of heat, keeping the body warm. Foods serve in this way through their oxidation within the body by oxygen that is extracted from the air in the lungs and is carried to the tissues by the hemoglobin of the blood. The ultimate products of oxidation of most of the hydrogen and carbon in foods are water and carbon dioxide.

Heats of combustion of foods and their relation to dietary requirements have been thoroughly studied. The food ingested daily by a healthy
man of average size doing a moderate amount of muscular work should
have a total heat of combustion of about 3000 kcal. About 90% of this
is made available as work and heat by digestion and metabolism of the
food.

Fats and carbohydrates are the principal sources of energy in foods.
Pure fat has a caloric value (heat of combustion) of 4080 kcal per pound,
and pure carbohydrate (sugar) a caloric value of about 1860 kcal per
pound. The caloric values of foods are obtained by use of a bomb calorimeter, just as was described for fuels. The third main constituent of food,
protein, is needed primarily for growth and for the repair of tissues.
About 50 g of protein is the daily requirement for an adult of average
size. Usually about twice this amount of protein is ingested. This amount,
100 g, has a caloric value of only about 400 kcal, the heat of combustion
of protein being about 2000 kcal per pound. Accordingly, fat and carbohydrate must provide about 2600 kcal of the 3000 kcal required daily.

24-6. Vitamins

Man requires nine amino acids in his diet, in order to keep in good
health. It is not enough, however, that the diet contain proteins that provide these nine amino acids, and a sufficient supply of carbohydrates
and fats to provide energy. Other substances, both inorganic and organic,
are also essential to health.

Among the inorganic constituents that must be present in foods in
order that a human being be kept in good health we may mention sodium
ion, chloride ion, potassium ion, calcium ion, magnesium ion, iodide ion,
phosphorus (which may be ingested as phosphate), and several of the
transition metals. Iron is necessary for incorporation in hemoglobin and
some other protein molecules in the body that serve as enzymes; in the
absence of sufficient iron in the diet, anemia will develop. Copper is also
required; it seems to be involved in the process of manufacture of hemoglobin and the other iron-containing compounds in the body.

The organic compounds other than the essential amino acids that are
required for health are called *vitamins*. Man is known to require at least
thirteen vitamins: vitamins A, B_1 (thiamine), B_2 (riboflavin), B_6 (pyridoxine), B_{12}, C (ascorbic acid), D, K, niacin, pantothenic acid, inositol,
para-aminobenzoic acid, and biotin.

Although it has been recognized for over a century that certain diseases
occur when the diet is restricted, and can be prevented by additions to
the diet (such as lime juice for the prevention of scurvy), the identification

of the essential food factors as chemical substances was not made until a few decades ago. Progress in the isolation of these substances and in the determination of their structure has been rapid in recent years, and many of the vitamins are now being made synthetically, for use as dietary supplements. It is possible for a diet to be obtained that provides all of the essential food substances in satisfactory amounts, but it is usually wise to have the diet supplemented by vitamin preparations, in order to achieve the best of health.

Vitamin A has the formula $C_{20}H_{29}OH$, and the structure

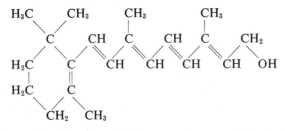

It is a yellow, oily substance, which occurs in nature in butterfat and fish oils. Lack of vitamin A in the diet causes a scaly condition of the eyes, and similar abnormality of the skin in general, together with a decreased resistance to infection of the eyes and skin. In addition there occurs a decreased ability to see at night, called *night blindness*. There are two mechanisms for vision, one situated in the cones of the retina of the eye, which are especially concentrated in the neighborhood of the fovea (the center of vision), and the other situated in the rods of the retina. Color vision, which is the ordinary vision, used when the intensity of light is normal, involves the retinal cones. Night vision, which operates when the intensity of light is very small, involves the rods; it is not associated with a recognition of color. It has been found that a certain protein, *visual purple*, which occurs in the rods, takes part in the process of night vision. Vitamin A is the prosthetic group of the visual purple molecule, and a deficiency in this vitamin leads for this reason to a decrease in the ability to see at night.

A protein, such as visual purple, that has a characteristic chemical group other than the amino acid residues as part of its structure is called a *conjugated protein*. Such a characteristic group in a conjugated protein is called a *prosthetic group* (Greek *prosthesis*, an addition). Hemoglobin is another example of a conjugated protein. Each hemoglobin molecule consists of a protein called globin to which there are attached four prosthetic groups called *heme groups*. The formula of the heme group is $C_{34}H_{32}O_4N_4Fe$.

It is not essential that vitamin A itself be present in food in order to prevent the vitamin A deficiency symptoms. Certain hydrocarbons, the *carotenes*, with formula $C_{40}H_{56}$ (similar in structure to lycopene, Figure 23-1), can be converted into vitamin A in the body. These substances, which are designated by the name *provitamin A*, are red and yellow substances that are found in carrots, tomatoes, and other vegetables and fruits, as well as in butter, milk, green leafy vegetables, and eggs.

Thiamine, Vitamin B_1, has the following formula (that shown is for thiamine chloride):

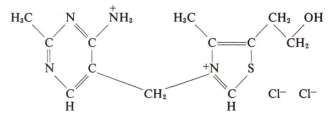

A lack of thiamine in the diet causes the disease beri-beri, a nerve disease that in past years was common in the Orient. Just before 1900 it was found by Eijkman in Java that beri-beri occurred as a consequence of a diet consisting largely of polished rice, and that it could be cured by adding the rice polishings to the diet. In 1911 Casimir Funk assumed that beri-beri and similar diseases were due to lack of a substance present in a satisfactory diet and missing from a deficient diet, and he attempted to isolate the substance whose lack was responsible for beri-beri. He coined the name vitamin for substances of this sort (he spelled it vitamine because he thought that the substances were amines). The structure of vitamin B_1, thiamine, was determined by R. R. Williams, E. R. Buchman, and their collaborators in 1936.

Thiamine seems to be important for metabolic processes in the cells of the body, but the exact way in which it operates is not known. There is some evidence that it is the prosthetic group for an enzyme involved in the oxidation of carbohydrates. The vitamin is present in potatoes, whole cereals, milk, pork, eggs, and other vegetables and meats.

Riboflavin, Vitamin B_2, has the following structure:

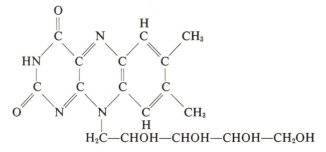

It seems to be essential for growth and for a healthy condition of the skin. Riboflavin is known to be the prosthetic group of an enzyme, called *yellow enzyme*, that catalyzes the oxidation of glucose and certain other substances in the animal body.

Vitamin B₆ (pyridoxine) has the formula

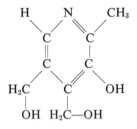

It is present in yeast, liver, rice polishings, and other plant and animal foods, and is also produced synthetically. It has the power of stimulating growth, and of preventing skin eruptions (dermatitis).

Vitamin B₁₂ is involved in the manufacture of the red corpuscles of the blood. It can be used for the treatment of pernicious anemia, and it is perhaps the most potent substance known in its physiological activity: 1 microgram per day (1×10^{-6} g) of vitamin B_{12} is effective in the control of the disease. The vitamin can be isolated from liver tissue, and is also produced by molds and other microorganisms. Each molecule of vitamin B_{12} contains one cobalt atom. This is the only compound of cobalt that is known to be present in the human body.

Ascorbic acid, Vitamin C, is a water-soluble vitamin of great importance. A deficiency of vitamin C in the diet leads to scurvy, a disease characterized by loss of weight, general weakness, hemorrhagic condition of the gums and skin, loosening of the teeth, and other symptoms. Sound tooth development seems to depend upon a satisfactory supply of this vitamin, and a deficiency is thought to cause a tendency to incidence of a number of diseases.

The formula of ascorbic acid is the following:

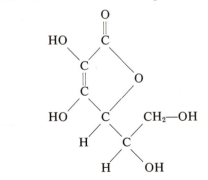

The vitamin is present in many foods, especially fresh green peppers, turnip greens, parsnip greens, spinach, orange juice, and tomato juice. The daily requirement of vitamin C is about 60 mg. Larger amounts, 1000 to 5000 mg per day, seem to be valuable in preventing or ameliorating the common cold.

Vitamin D is necessary in the diet for the prevention of rickets, a disease involving malformation of the bones and unsatisfactory development of the teeth. There are several substances with antirachitic activity. The form that occurs in oils from fish livers is called vitamin D_3; it has the following chemical structure:

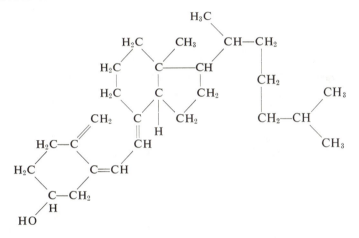

Only a very small amount of vitamin D is necessary for health— approximately 0.01 mg per day. The vitamin is a fat-soluble vitamin, occurring in cod-liver oil, egg yolks, milk, and in very small amounts in other foods. Cereals, yeast, and milk acquire an added vitamin D potency when irradiated with ultraviolet light. The radiation converts a fatty substance (a *lipid*) that is present in the food (called *ergosterol*) into another substance, *calciferol* (vitamin D_2), that has vitamin D activity. The structure of calciferol is closely related to that of vitamin D_3.

Whereas most vitamins are harmless even when large quantities are ingested, vitamin D is harmful when taken in large amounts.

Vitamin E, while not necessary for health, seems to be required for the reproduction and lactation of animals. Pantothenic acid, inositol, *p*-aminobenzoic acid, and biotin are substances involved in the process of normal growth. Vitamin K is a vitamin that prevents bleeding, by assisting in the process of clotting of the blood.

It is interesting that many "simpler organisms" do not require so many substances for growth as does man. It was mentioned above that the red bread mold, *Neurospora*, can synthesize all the amino acids present in proteins, whereas man is unable to synthesize nine of them, but must obtain them in his diet. The red bread mold is also able to manufacture other substances that man requires as vitamins. The only organic growth substance required by this organism is biotin. Similarly, the food requirements of the rat, while greater than those of *Neurospora*, are not so great as those of man. The rat, for example, does not require ascorbic acid (vitamin C) in its diet, but is able to synthesize this substance, which is present as an important constituent in the tissues of the animal.

24-7. Hormones

Another class of substances of importance in the activity of the human body consists of the *hormones*, which are substances that serve as messengers from one part of the body to another, moving by way of the body fluids. The hormones control various physiological processes. For example, when a man is suddenly frightened, a substance called *epinephrine* (also called adrenalin) is secreted by the suprarenal glands, small glands that lie just above the kidneys. The formula of epinephrine is

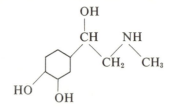

When epinephrine is introduced into the blood stream it speeds up the action of the heart, causes the blood vessels to contract, thus increasing the blood pressure, and causes glucose to be released from the liver, thus providing an immediate source of extra energy.

Thyroxin is a secretion of the thyroid gland that controls metabolism. *Insulin* is a secretion of the pancreas that controls the combustion of carbohydrates. Both of these hormones are proteins, thyroxin having a prosthetic group that contains iodine. Many other hormones are known, some of which are proteins and some simpler chemical substances.

It has been recognized that diseases (such as goiter) affecting the thyroid gland may arise from a deficient production of thyroxin, which can be remedied by the introduction of added iodide ion into the diet. The disease *diabetes mellitus*, characterized by the appearance of sugar in the urine and resulting from a deficient production of the hormone insulin, has in recent decades been treated by the injection of insulin, obtained from the pancreatic glands of animals. The hormones *cortisone* and *ACTH* (adrenocorticotropic hormone) have been shown to have strong therapeutic activity toward rheumatoid arthritis and some other diseases.

24-8. Chemistry and Medicine

From the earliest times chemicals have been used in the treatment of disease. The substances that were first used as drugs are natural products such as in the leaves, branches, and roots of plants. As the alchemists discovered or made new chemical substances, these substances were tried out to see if they had physiological activity, and many of them were introduced into early medical practice. For example, both mercuric chloride, $HgCl_2$, and mercurous chloride, Hg_2Cl_2, were used in medicine, mercuric chloride as an antiseptic, and mercurous chloride, taken internally, as a cathartic and general medicament.

The modern period of *chemotherapy*, the treatment of disease by use of chemical substances, began with the work of Paul Ehrlich (1854–1915). It was known at the beginning of the present century that certain organic compounds of arsenic would kill protozoa, parasitic microorganisms responsible for certain diseases, and Ehrlich set himself the task of synthesizing a large number of arsenic compounds, in an effort to find one that would be at the same time toxic (poisonous) to protozoa in the human body and nontoxic to the human host of the microorganism. After preparing many compounds he synthesized *arsphenamine*, which has the structure of a linear high polymer:

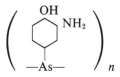

Arsphenamine has been found to be extremely valuable. Its greatest use is in the treatment of syphilis; the drug attacks the microorganism responsible for this disease, *Spirocheta pallida*. It has also been useful in the treatment of some other diseases. Now it has been superseded by penicillin (which we shall discuss below) in the treatment of syphilis.

Since Ehrlich's time there has been continual progress in the development of new chemotherapeutic agents. Thirty years ago the infectious diseases constituted the principal cause of death; now most of these diseases are under effective control by chemotherapeutic agents, some of

TABLE 24-3
Structural Formulas of Sulfa Drugs and Penicillin

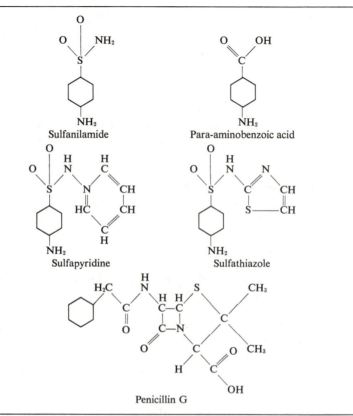

Sulfanilamide

Para-aminobenzoic acid

Sulfapyridine

Sulfathiazole

Penicillin G

which have been synthesized in the laboratory and some of which have been isolated from microorganisms. At the present time only a few of the infectious diseases constitute major hazards to the health of man, and we may confidently anticipate that the control of these diseases by chemotherapeutic agents will be achieved in a few years.

The recent period of rapid progress began with the discovery of the *sulfa drugs* by G. Domagk. In 1935 Domagk discovered that the compound prontosil, a derivative of sulfanilamide, was effective in the control of streptococcus infections. It was soon found by other workers that sulfanilamide itself is just as effective in the treatment of these diseases, and that it could be administered by mouth. The formula of sulfanilamide is given in Table 24-3. Sulfanilamide is effective against hemolytic streptococcic infections and meningococcic infections. As soon as the value of sulfanilamide was recognized chemists synthesized hundreds of related substances, and investigations were made of their usefulness as bacteriostatic agents (agents with the power of controlling the spread of bacterial infections). It was found that many of these related substances are valuable, and their use is now an important part of medical practice. Sulfapyridine has been found valuable for the control of pneumococcic pneumonia (pneumonia due to the *Pneumococcus* microorganisms), as well as of other pneumococcic infections and gonorrhea. Sulfathiazole is used for these infections and also for the control of staphylococcic infections, which occur especially in carbuncles and eruptions of the skin. These and other sulfa drugs are all derivatives of sulfanilamide itself, obtained by replacing one of the hydrogen atoms of the amide group (the NH_2 bonded to the sulfur atom) by some other group (Table 24-3).

The introduction of *penicillin* into medical treatment was the next great step forward. In 1929 Professor Alexander Fleming, a bacteriologist working in the University of London, noticed that bacteria that he was growing in a dish in his laboratory were not able to grow in the region immediately surrounding a bit of mold that had accidentally begun to develop. He surmised that the mold was able to produce a chemical substance that had bacteriostatic action, the power of preventing the bacteria from growing, and he made a preliminary investigation of the nature of this substance. Ten years later, perhaps spurred on by the successful use of the sulfa drugs in medicine, Professor Howard Florey and Dr. E. B. Chain of the University of Oxford decided to make a careful study of the antibacterial substances that had been reported in order to see whether they would be similarly useful in the treatment of disease. When they tested the bacteriostatic power of the liquid in which the mold *Penicillium notatum* that had been observed by Fleming was growing, they found it to be very great, and within a few months the new antibiotic substance penicillin was being used in the treatment of patients. Through the cooperative effort of many investigators in the United States and England rapid progress was made during the next two or three years in the determination of the structure of penicillin, the development of methods of manufacturing it in large quantities, and the investigation of the diseases that could be effectively treated by use of it. Within less than a decade this new antibiotic agent had become the most valuable of all drugs. It provides an effective therapeutic treatment of many diseases.

The structure of penicillin is shown in Table 24-3. The substance has been synthesized, but no cheap method of synthesizing it has been developed, and the large amount of penicillin that is being manufactured and used in the treatment of disease is made by growing the mold penicillium in a suitable medium and then extracting the penicillin from the medium. Important forward steps in the introduction of penicillin into medical treatment were the development of strains of the mold that produced the desired penicillin in large quantities, and the discovery of the best medium on which to grow the mold.

It is interesting that a number of slightly different penicillins are formed in nature by different strains of the mold. The formula in Table 24-3 represents benzyl penicillin (penicillin G), which is the product that is now manufactured and used.

The spectacular success of penicillin as a chemotherapeutic agent has led to the search for other antibiotic products of living organisms. *Streptomycin*, which is produced by the mold *Actinomyces griseus*, has been found to be valuable in the treatment of diseases that are not effectively controlled by penicillin, and some other bacteriostatic agents also have been found to have significant value.

Another very great step forward has been made since 1955 by the discovery of substances that can control the development of viral infections. Penicillin, streptomycin, and the sulfa drugs are effective against bacteria but not against viruses. It has recently been found, however, that *chloramphenicol* (Chloromycetin) and *aureomycin*, both of which are substances manufactured by molds (the molds *Streptomyces venezuele* and *Streptomyces aureofaciens* respectively), have the power of controlling certain viral infections.

The Relation between the Molecular Structure of Substances and Their Physiological Activity

No one knows what the relation between the molecular structure of substances and their physiological activity is. We know the structural formulas of many drugs, vitamins, and hormones—some of these formulas have been given in the preceding sections. It is probable, however, that most of these substances produce their physiological action by interacting with or combining with proteins in the human body or in the bacterium or virus that they counteract; and we do not yet know the structure of any of these proteins.

A suggestion has been made about the way in which the sulfa drugs exercise their bacteriostatic action. It seems probable that this suggestion is essentially correct. It has been found that a concentration of sulfanilamide or other sulfa drug that would prevent bacterial cultures from growing under ordinary circumstances loses this power when some para-aminobenzoic acid is added. The amount of para-aminobenzoic acid required to permit the bacteria to increase in number is approximately proportional to the excess of the amount of the sulfa drug over the minimum that would produce bacteriostatic action. This *competition* between the sulfa drug and para-aminobenzoic acid can be given a reasonable explanation. Let us assume that the bacteria need to have some para-aminobenzoic acid in order to grow; that is, that para-aminobenzoic acid is a vitamin for the bacteria. Probably it serves as a vitamin by combining with a protein to form an essential enzyme; presumably it serves as the prosthetic group of this enzyme. It is likely that the bacterium synthesizes a protein molecule that has a small region, a cavity, on one side of itself into which the para-aminobenzoic acid molecule just fits.

The sulfanilamide molecule is closely similar in structure to the para-aminobenzoic molecule (see Table 24-3). Each of the molecules contains a benzene ring, an amino group ($-NH_2$) attached to one of the carbon atoms of the benzene ring, and another group attached to the opposite carbon atom. It seems not unlikely that the sulfanilamide molecule can fit into the cavity on the protein, thus preventing the para-aminobenzoic molecule from getting into this place. If it is further assumed that the sulfanilamide molecule is not able to function in such a way as to make the complex with the protein able to act as an enzyme, then the explanation of the action of sulfanilamide is complete. It is thought that the protein fits tightly around the benzene ring and the amino group, but not around the other end of the molecule. The evidence for this is that derivatives of sulfanilamide in which various other groups are attached to the sulfur atom are effective as bacteriostatic agents, whereas compounds in which other groups are attached to the benzene ring or the amino group are not effective.

25

The Chemistry of
the Fundamental Particles

During recent years there has been a great increase in our knowledge of the world. Atoms have been found to consist of electrons and nuclei, and the atomic nuclei have been found to consist of protons and neutrons. Moreover, in addition to the electron, the proton, and the neutron, many other particles classed as fundamental have been discovered.

The field of science dealing with the nature and the reactions of the fundamental particles is developing very rapidly at the present time. Work in this field of science has been carried out largely by physicists, but the reactions by means of which the fundamental particles are created, converted into others, and destroyed are in a general way similar to chemical reactions, and we may be justified in considering the study of these reactions and the properties of the fundamental particles themselves as constituting the field of the chemistry of the fundamental particles.

In the following paragraphs we shall describe 34 particles as fundamental particles. This number includes 6 (the photon, the graviton, two neutrinos, and two antineutrinos) that move only with the speed of light, and 28 that move only at speeds less than the speed of light. In accordance with the theory of relativity, the particles that move only at the speed of light have zero rest-mass, whereas the others have finite rest-mass.

Much of the knowledge about the fundamental particles has been obtained during the last decade. The scientists who have been working in this field have made many completely unexpected discoveries, which are changing our ways of thinking about the world. Just as the discoveries in the field of atomic and molecular science, discussed in earlier chapters, and the field of nuclear science, to be discussed in the following chapter, have had profound effects upon our daily lives, changing the nature of our civilization and especially the methods of waging war, so may we expect that the new knowledge about fundamental particles will in the course of time have equally profound effects upon our lives.

25-1.   The Classification of the Fundamental Particles

At the present time it is convenient to classify the thirty-four fundamental particles in the following way:

> 8 baryons (the proton, the neutron, and six heavier particles)
> 8 antibaryons
> 8 mesons and antimesons
> 8 leptons and antileptons
> The photon
> The graviton

Most of the fundamental particles can be described as constituting either *matter* or *antimatter*. The existence of these two kinds of matter was predicted in 1928, on the basis of relativistic quantum mechanics, by P. A. M. Dirac (born 1902), the English theoretical physicist who first developed a theory of quantum mechanics compatible with the theory of relativity. His prediction has been thoroughly confirmed by experiment. Every electrically charged particle has a counterpart that is identical with it in some properties and opposite to it in others: the masses and spins are identical, but the electric charges are opposite. For example, the electron, which constitutes a part of ordinary matter, and the positron, which is the antielectron, have opposite electric charges, $-e$ and $+e$, respectively; their masses are the same; and each has a spin represented by the spin quantum number $\frac{1}{2}$, which permits two ways of orienting the spinning particle in a magnetic field. Some neutral particles have antiparticles and some are their own antiparticles. Whenever a particle and the corresponding antiparticle come together they annihilate each other. Their masses are totally converted into high-energy light waves or, in some cases, into lighter particles moving with great speeds. The Einstein equation $E = mc^2$ gives the amount of energy that is released when a particle and its antiparticle annihilate one another with formation of radiant energy. The neutral particles that are their own antiparticles decay very rapidly.

Antimatter does not exist except fleetingly on earth. Particles of antimatter are created by collisions, as described in the following section, and the antiparticles are then rapidly destroyed as they react with particles of ordinary matter with which they collide.

There is the possibility that some regions of the universe, perhaps some nebulae, are composed of antimatter. The hydrogen atom in such a region consists of a positron moving about an antiproton. The collision between an antimatter nebula and a nebula composed of ordinary matter would result in the liberation of a tremendous amount of radiant energy, and might be recognized by astronomers.

Fermions and Bosons

The elementary particles may be divided into two classes on the basis of the magnitude of their spin. The electron was described in Chapter 3 as having spin $\frac{1}{2}$. It has an angular momentum determined by the spin quantum number $\frac{1}{2}$, and in a magnetic field it can orient its angular momentum with component either $+\frac{1}{2}$ or $-\frac{1}{2}$ in the direction of the field (the unit of angular momentum is the Bohr unit $h/2\pi$). It was also mentioned in Chapter 5 that two electrons cannot occupy the same orbital in an atom unless they have opposite orientations of their spin; that is, they cannot be in exactly the same quantum state, as they would be if they occupied the same orbital and both had positive orientation of the spin. This is the expression of the Pauli exclusion principle.

Particles that have spin $\frac{1}{2}$ are called fermions, named after the physicist Enrico Fermi. In accordance with the Pauli exclusion principle, no two identical fermions can be in exactly the same quantum state.

The baryons, antibaryons, leptons, and antileptons are all fermions with spin $\frac{1}{2}$. According to theory, particles with spins $\frac{3}{2}, \frac{5}{2}, \cdots$ would also be fermions.

Particles with integral spin $(0, 1, 2, \cdots)$ are called bosons, named after the Indian physicist S. N. Bose. They interact with one another in a way that permits two or more particles to be in exactly the same quantum state. The photon, the graviton, and the mesons are bosons. The mesons all have spin 0. The photon has spin 1. The graviton, which is the quantum of the gravitational field, is expected to have spin 2.

The *photon* or light quantum is now accepted as one of the fundamental particles. Its nature has been discussed in Chapter 3.

The symbol used for the photon is γ (the Greek letter gamma). This symbol was originally used for γ-rays.

The value 1 for the spin of the photon is connected with the polarization of light, discussed in Chapter 6. Right-handed circularly polarized light corresponds to a component $+1$ of the spin in the direction of motion of light, and left-handed circularly polarized light to the component -1.

Photons may be emitted or absorbed by an oscillating electric dipole, such as a negatively charged electron rotating around a positively charged proton. It might be thought that a system of two masses, such as the earth and the moon, rotating about their common center would emit gravitational quanta. These gravitational quanta are called gravitons. Only in 1969 have any reports been published on the experimental detection of gravitational waves, and the existence of the graviton, a quantized gravitational wave, has not yet been verified by any experiment.

25-2. The Discovery of the Fundamental Particles

The *electron* has been discussed throughout this book. It was the first of the fundamental particles to be recognized, having been discovered by J. J. Thomson in 1897. It is present in ordinary matter, and is easily separated from the atomic nuclei to which it is ordinarily attached.

The *proton* was observed as positively charged rays in a discharge tube in 1886 by the German physicist E. Goldstein. The nature of the rays was not at first understood. In 1898 the German physicist W. Wien made a rough determination of their ratio of charge to mass, and accurate measurements of this sort, which verified the existence of protons as independent particles in a discharge tube containing ionized hydrogen at low pressure, were made by J. J. Thomson in 1906.

The next particle to be discovered (aside from the photon) was the *positron* (the antielectron), found in 1932 by the American physicist Carl D. Anderson (born 1905). The positrons were found among the particles produced by the interaction of cosmic rays with matter. They are identical with electrons except that their electric charge is $+e$ instead of $-e$.

The mass of the electron corresponds, according to the Einstein equation $E = mc^2$, to the energy 510,976 electron volts (0.510976 MeV). Hence the annihilation of an electron and a positron liberates 1.022 MeV of energy, which might be in the form of two photons, each with the energy 0.511 MeV and corresponding wavelength 0.02426 Å.

A rapidly moving electron that strikes the anode in an x-ray tube is suddenly slowed down, and much of its energy is converted into a photon of x-radiation. If its kinetic energy is greater than 1.022 MeV, this amount of energy may be converted into an electron-positron pair. Electron-positron pair production can be carried out in this way in the laboratory with use of particles that have been given large amounts of kinetic energy in a particle accelerator, as described later in this section. The positrons that were first observed by Anderson were produced, together with electrons, by the impingement of cosmic-ray particles against particles of ordinary matter. Cosmic rays are described later in this section.

The discovery of the neutron has been discussed in Section 4-2.

The existence of the *antiproton* was verified in 1955 by Segrè, Chamberlain, Wiegand, and Ypsilantis, by use of a particle accelerator (the Berkeley synchrotron) that could generate particles with energy 6 GeV. The mass of the proton-antiproton pair is 1836 times that of the electron-positron pair, and accordingly 1836×1.022 MeV $= 1876$ MeV of energy is needed to produce this pair of heavier particles.* The antiproton has negative electric charge, mass equal to the mass of the proton, and spin $\frac{1}{2}$.

The discovery of some of the other fundamental particles will be described in later sections.

Cosmic Rays

Cosmic rays are particles of very high energy that reach the earth from interstellar space or other parts of the cosmos or that are produced in the earth's atmosphere by the rays from outer space. The discovery that some ionizing radiation on the earth's surface comes from outer space was made by the Austrian physicist Victor Hess (1883–1964), who made measurements of the amount of ionization in the earth's atmosphere during balloon ascents to a height of 5000 m in 1911 and 1912. Many of the fundamental particles in addition to the positron were discovered in the course of studies of cosmic rays.

Cosmic rays that impinge on the outer part of the atmosphere consist of protons and the nuclei of heavier atoms moving with great speed. The cosmic rays that reach the earth's surface consist in large part of mesons, positrons, electrons, and protons produced by reaction of the fast protons and other atomic nuclei with atomic nuclei in the atmosphere.

---

*This is the amount of energy needed for proton-antiproton pair production by collision of two similar particles moving with equal speeds in opposite directions in the coordinate system of the laboratory. A much larger amount of energy—nearly 6 GeV—must be imparted to a particle in order that a pair may be produced when it collides with a stationary particle. It is for this reason that an accelerator with a reverse-direction storage ring, such that two beams of particles might be directed against one another, is now being built.

Some of the phenomena produced by cosmic rays can be explained only if it is assumed that particles are present with energy in the range from 10^{15} to 10^{20} eV. The great accelerators that have been or are being built (following section) produce or will produce particles with energies in the range 10^6 to 10^{12} eV. There is no way known at present to accelerate particles to energies as great as those of the fastest particles in cosmic rays, and accordingly the study of cosmic rays will probably continue to yield information about the universe that cannot be obtained in any other way.

Particle Accelerators

In recent years great progress has been made in the laboratory production of high-speed particles. The first efforts to accomplish this involved the use of transformers. Different investigators built transformers and vacuum tubes operating to voltages as high as three million volts, in which protons, deuterons, and helium nuclei could be accelerated. In 1931 an electrostatic generator was developed by R. J. Van de Graaff, an American physicist, involving the carrying of electric charges to the high-potential electrode on a moving insulated belt. Van de Graaff generators have been built and operated to produce potential differences up to fifteen million volts.

The *cyclotron* was invented by the American physicist Ernest Orlando Lawrence (1901–1958) in 1929. In the cyclotron positive ions (protons, deuterons, or other light nuclei) are given successive accelerations by repeatedly falling through a potential difference of a few thousand volts. The charged particles are caused to move in circular paths by a magnetic field, produced by a large magnet between whose pole pieces the apparatus is placed (Figure 25-1). Cyclotrons can be used to accelerate particles to about 100 MeV, but the relativistic change in mass of the particle then causes it to get out of phase with the alternating electric field, so that higher energies cannot be obtained.

An accelerator, the *synchrotron*, in which a number of the particles are injected and the frequency of the alternating field is adjusted to compensate for the relativistic change in mass, was proposed by the Russian physicist V. Veksler and independently by the American physicist E. M. McMillan in 1945. By use of the synchrotron principle particles have now been accelerated to about 100 GeV, and plans are at present being made for intercontinental cooperation in the construction of a giant accelerator to produce particles in the range 300 GeV to 1000 GeV.

The reactions of particles can be observed by the study of the tracks of the particles in a cloud chamber or a bubble chamber. The *cloud chamber*,

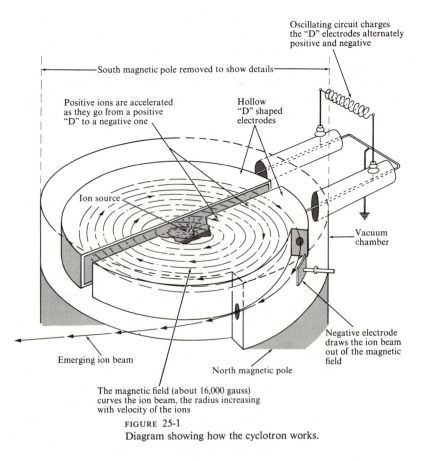

Oscillating circuit charges
the "D" electrodes alternately
positive and negative

←————————South magnetic pole removed to show details————————→

Positive ions are accelerated
as they go from a positive
"D" to a negative one

Hollow
"D" shaped
electrodes

Ion source

Vacuum
chamber

Negative electrode
draws the ion beam
out of the magnetic
field

Emerging ion beam

North magnetic pole

The magnetic field (about 16,000 gauss)
curves the ion beam, the radius increasing
with velocity of the ions

FIGURE 25-1
Diagram showing how the cyclotron works.

which was invented by the English physicist C. T. R. Wilson (1869–1959) in 1911, is a chamber containing air saturated with water vapor. When the air is suddenly expanded by increasing the volume of the chamber by moving a piston, the air is cooled and becomes supersaturated, so that droplets of water form. These droplets tend to form around the ions that are produced as high-energy electrically charged particles traverse the gas, and thus the droplets define the paths of the particles. Neutral particles do not form paths, but their presence can sometimes be detected by the presence of paths radiating from a point where the neutral particle underwent a reaction that produced high-energy charged particles. The *bubble chamber*, invented by the American physicist D. A. Glaser (born 1926) in 1952, has found extensive use in recent years. It is a chamber containing a liquid held at a temperature slightly above its boiling point. The ions

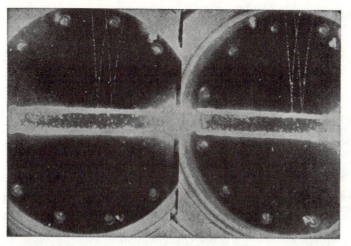

FIGURE 25-2
A direct view and a mirror view of two electron-positron pairs produced
by a cosmic-ray photon near the nucleus of a lead atom in a lead plate
1 cm thick in a cloud chamber. This photograph was made about 1934
by Carl D. Anderson, the discoverer of the positron. There is a magnetic
field present, which causes the paths of the electron and the positron to
curve in opposite directions.

formed by high-energy particles traversing the liquid serve as centers of
formation of small vapor bubbles, which define the tracks of the particles.

A cloud-chamber photograph is shown as Figure 25-2 and a bubble-
chamber photograph as Figure 25-3.

Another instrument for defining the tracks of high-energy particles is
the *spark chamber*. This instrument contains a gas, and a series of metal
plates that can be electrically charged to such a potential between alternate
plates to cause a spark to pass from one plate to an adjacent one. The
spark, which can be photographed, follows the tracks of the ions formed
by high-energy particles. In the 1962 neutrino experiment (Section 25-5) a
spark chamber 10 feet by 6 feet by 4 feet was used, containing 90 aluminum
plates 4 feet square and 1 inch thick, $\frac{1}{2}$ inch apart, with neon as the gas.

25-3. The Forces between Nucleons.
Strong Interactions

In 1932, when the neutron was discovered, it was recognized that the
heavier atomic nuclei can be described as being built of protons and
neutrons, with the electric charge equal to the number of protons and the

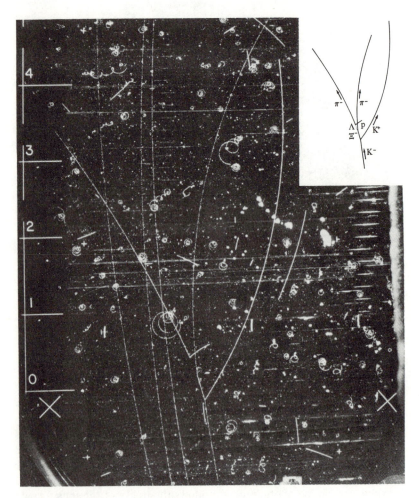

FIGURE 25-3
An event recorded in the 72-inch liquid hydrogen bubble chamber of the University
of California (L. W. Alvarez and coworkers). The incident particle is a negative
kaon, in a beam of these particles. By collision with a proton it forms a positive kaon
and a negative xion. The negative xion then decomposes to form a lambda particle
and a negative pion. The lambda particle, which is neutral, produces no track.
It is shown as decomposing to form a proton and a negative pion.

A   ●———π———●

B   ●————π————⊃   ●

$|$-1.4 × 10^{-15} m-$|$

FIGURE 25-4
A diagram illustrating the range of
internucleonic forces.

(Nucleons and pion messenger)

mass number equal to the sum of the number of protons and the number of neutrons—that is, equal to the number of nucleons, with a nucleon either a proton or a neutron. The question immediately arose as to the nature of the forces holding the neutrons and protons together. If electrostatic forces were the only forces operating between nucleons the heavier nuclei would break up, because of the electrostatic repulsion between the protons.

It was evident that the force of attraction between nucleons must be a strong force at small distances, stronger than the repulsion due to the positive electric charges on the protons, and a weak force at large distances, weaker than the electrostatic repulsion. Careful studies of the size of the heavier nuclei and the scattering of nucleons from one another led to the discovery that two nucleons attract one another with approximately a constant force when they are less than 1.4 fm apart, and that the internucleonic force, other than electrostatic repulsion of protons, drops rapidly to zero at distances greater than 1.4 fm.

The idea of action at a distance is not a satisfying one; instead, physicists have developed a *quantum theory of force fields*, in which the field at some distance from its source is thought of as carried to that point by a messenger or quantum of the field. For electrostatic attraction and repulsion these messengers are the photons, and for gravitational attraction they are described as being gravitons. In 1935 the Japanese physicist Hideki Yukawa (born 1907) proposed an answer to the question of the mechanism of the force of attraction between nucleons. He pointed out that, whereas messengers that have zero rest-mass, such as photons and gravitons, can extend their influence to infinity, a messenger with finite rest-mass could reach only a limited distance from the particle. He suggested that messengers of this sort are involved in the interaction of nucleons, and from the known range of internucleonic force, 1.4 fm, he calculated that the rest-mass of these particles should be about 274 times the electronic mass. These particles, which are intermediate in mass between electrons and nucleons, are called *mesons* (Greek *mesos*, middle).

Let us consider two nucleons a small distance apart, less than 1.4×10^{-15} m, as shown in A in Figure 25-4. One of the nucleons produces or emits a messenger particle, a meson, which travels with a speed close to that of light and is destroyed in the neighborhood of the second particle. This process of production and destruction of the messenger particle gives rise to the force of attraction.

If, however, the two particles are farther apart than 1.4×10^{-15} m, as in B, the emitted messenger particle is not able to traverse the distance between the particles, but instead turns back and disappears. There is accordingly no interaction between the nucleons at the larger distance. The reason that the range of the messenger particles is restricted can be understood by consideration of the uncertainty principle.

Let us consider the reaction

$$p^+ \rightleftarrows p^+ + \pi^0$$

Here we use π^0 to represent a messenger particle; the mesons responsible in the main for internuclear forces are called *pions*. This reaction, the reaction of a proton to form a proton plus a pion, violates the principle of conservation of mass-energy. Until the uncertainty principle was discovered no reaction of this sort would ever have been considered.

However, because of the uncertainty relation between energy and time we may consider a reaction such as this, which violates the principle of conservation of mass-energy, provided that the length of time during which we consider the reaction to be taking place is less than the time Δt given by the uncertainty principle. We make use of the equation $\Delta E \cdot \Delta t = \hbar$ given in Section 3-10. The length of time in which we are interested is the time required for a particle moving with a speed close to that of light to move the distance 1.4×10^{-15} m. The corresponding value of ΔE, the uncertainty for mass-energy, is $\hbar$ divided by this time Δt, $1.4 \times 10^{-15}/3 \times 10^8$ (the speed of light), which is $\Delta t = 0.47 \times 10^{-23}$ s. The value of ΔE is accordingly $1.05 \times 10^{-34}/0.47 \times 10^{-23} = 2.23 \times 10^{-11}$ J, and the corresponding value in mass units, obtained by dividing by c^2 (since $E = mc^2$), is 2.48×10^{-25} g, which is 274 times the mass of the electron. Yukawa accordingly stated that the short range of internucleonic forces could be explained by assuming that the interactions are carried out by particles with mass about 274 times the mass of the electron. No such particles were known at that time.

In 1936 particles with mass 207 times the mass of the electron and with either a positive or a negative charge were discovered by Anderson and Neddermeyer and independently by Street and Stevenson in the course of cosmic-ray experiments. These particles, which are now called *muons*, were at first thought to be the Yukawa particles. However, if they were responsible for the internucleonic forces they would interact strongly with nucleons. This strong interaction should cause them to react within a period of time about 10^{-23} s when in the neighborhood of a nucleon. Muons were found to decompose in free space, their half-life being about 10^{-6} s, and their rate of decomposition was found not to be greatly changed when a beam of muons was passed through solid substances and the muons were thus subjected to the influence of nucleons; hence they could not be the Yukawa particles.

When this last experiment was carried out, in 1945, the physicists were again at a loss to account for internucleonic forces, but not for long, because the strongly interacting mesons, which were named pions, were soon discovered. Cosmic-ray experiments with use of stacks of photographic emulsions to detect the tracks of the charged particles, carried out in 1947 by the British physicist C. F. Powell (born 1903) and his coworkers, led to the discovery of three particles, the positive pion, neutral pion, and negative pion, with masses 273.3 for π^+ and π^- and 264.3 for π^0 and with the properties of strong interaction with nucleons that had been predicted by Yukawa. There is now no doubt that the internucleonic forces that operate in atomic nuclei involve pions. It has been shown by experiment that charged pions as well as neutral pions are involved in the internucleonic forces. The equations for the charged-pion forces are

$$p^+ \rightleftarrows n + \pi^+$$
$$n \rightleftarrows p^+ + \pi^-$$

Other particles, especially the rho and omega particles (Section 25-10), are probably also involved in the internucleonic forces.

25-4.   The Structure of Nucleons

The proton and the neutron are closely similar in properties except that the proton has positive charge and the neutron is electrically neutral. The mass of the neutron is only about 0.1% greater than that of the proton. Both particles have spin $\frac{1}{2}$. The proton-proton, proton-neutron, and neutron-neutron internucleonic forces at small distances are essentially the same. Because of these facts, the idea arose some years ago that the proton and the neutron are simply two states of one particle, the nucleon.

In Chapter 3 we have pointed out that an electron in an atom may have two orientations of its spin relative to the direction of a magnetic field or of the angular momentum vector produced by its orbital motion. These two directions, represented by $+\frac{1}{2}$ and $-\frac{1}{2}$, respectively, are said to give rise to a doublet. The doublet is associated with the spin quantum number $\frac{1}{2}$. This suggested that the proton and the neutron may constitute an *electric-charge doublet*. It has been suggested that the nucleon has an intrinsic electric charge with magnitude $+\frac{1}{2}$ (in units e) and an electric-charge vector with magnitude $\frac{1}{2}$ that can assume two orientations (not in ordinary, three-dimensional space, but in some undefined space) such as to contribute either $+\frac{1}{2}$ to the resultant charge, to produce the proton, or to contribute $-\frac{1}{2}$, to produce the neutron. The proton and the neutron, according to this picture, constitute the two states of the electric-charge doublet of a nucleon with intrinsic charge $+\frac{1}{2}$ and electric-charge vector $\frac{1}{2}$. Similarly, the antiproton and the antineutron constitute the two states of the corresponding type of antimatter, the antinucleon, with intrinsic charge $-\frac{1}{2}$ and electric-charge vector $\frac{1}{2}$.

In 1961 some experimental results providing support for this picture of the proton and the neutron were reported by Robert Hofstadter and his coworkers at Stanford University and by a group of investigators at Cornell University. These physicists studied the scattering of high-speed electrons by protons and neutrons, and were able to interpret their experiments to determine the distribution of electric charge within the proton and the neutron.

They reported that both the proton and the neutron can be described as involving a central ball of positive charge, somewhat less than $0.5e$, extending to the radius about 0.3 fm. Surrounding the ball is a shell, extending to about 1 fm, and with positive charge $+\frac{1}{2}e$ for the proton and negative charge $-\frac{1}{2}e$ for the neutron. In addition, there is a fringe of positive electricity in both the proton and the neutron, amounting to about $0.15e$ and extending to about 1.5 fm.

It is possible that the fringe represents ephemeral mesons that constitute the mechanism of production of the strong internucleonic interactions. Except for the cloud of mesons surrounding it, the nucleon can be described, in its two states, the proton and the neutron, as consisting of a central ball of positive charge, $+\frac{1}{2}e$, which may be identified with the intrinsic charge of the nucleon and a shell, $+\frac{1}{2}e$ for the proton and $-\frac{1}{2}e$ for the neutron, representing the component of the electric-charge vector.

These results about the structure of the nucleon give exciting promise of great future developments in the understanding of the fundamental nature of the universe.

Several other charge doublets, corresponding to the two aspects of an electric-charge vector $\frac{1}{2}$ (also called *isotopic spin*), are known. In addition, as will be seen in the tables given in the following sections, there are several charge triplets that are known, groups of three particles with closely similar properties except for their electric charge, $+1$, 0, and -1. These charge triplets can be described as the three states of a single particle with electric-charge vector 1, which can have the component $+1$, 0, or -1. The three pions, π^+, π^0, π^-, constitute such a charge triplet (Section 25-9).

25-5. Leptons and Antileptons

We begin the tabulation of the fundamental particles by discussing the leptons and antileptons. There are eight of these particles known. Some of their properties are given in Table 25-1. Except for the muon and antimuon, they are stable particles. The word lepton is from the Greek *leptos*, small.

The muon, μ^-, was the first particle with mass intermediate between the

TABLE 25-1
*Leptons and Antileptons**

Name	Electric Charge			Mass	Xenicity (strangeness)	Spin
	+1	0	−1			
Electron			e^-	0.511 MeV	0	$\frac{1}{2}$
Muon			μ^-	105.66	0	$\frac{1}{2}$
Electron neutrino		ν		0	0	$\frac{1}{2}$R †
Muon neutrino		ν'		0	0	$\frac{1}{2}$R †
Positron	$\bar{e}^+$			0.511	0	$\frac{1}{2}$
Antimuon	$\bar{\mu}^+$			105.66	0	$\frac{1}{2}$
Electron antineutrino		$\bar{\nu}$		0	0	$\frac{1}{2}$L †
Muon antineutrino		$\bar{\nu}'$		0	0	$\frac{1}{2}$L †

*The electron, muon, and neutrino have lepton number +1; the positron, antimuon, and antineutrino have lepton number −1; all other particles have lepton number 0.
†The spin of the neutrinos corresponds to a right-handed screw, that of the antineutrinos to a left-handed screw.

electron and the proton to be discovered. It is present in cosmic rays. It is made by the following reaction:

$$\bar{\pi}^- \longrightarrow \mu^- + \bar{\nu}'$$

The positive muon, the antimuon ($\bar{\mu}^+$), is made by a similar reaction from the positive pion. Both the positive pion and the negative pion are present in cosmic rays. They decompose rapidly, with half-life about 2.56×10^{-8} s, to form muons. The muon and the antimuon themselves decompose, to form an electron (or positron), a neutrino, and an antineutrino:

$$\mu^- \longrightarrow e^- + \nu + \bar{\nu}$$
$$\bar{\mu}^+ \longrightarrow \bar{e}^+ + \nu + \bar{\nu}$$

The muon and the antimuon have no significance with respect to internucleonic forces. Their nature is uncertain. It is possible that they represent an excited state of the electron and positron. A striking indication that they are closely related to the electron and positron is provided by the observed values of the magnetic moment of the muon. The electron has been found by magnetic resonance experiments to have values ±1.00116 Bohr magnetons for the component of its magnetic moment in the direction of a magnetic field. (The deviation from unity is attributed to the photon field surrounding the electron.) The proton and neutron, which have a complex structure (Section 25-4), have magnetic moments that are not related in a simple way to the Bohr magneton. But the observed components of the muon in the direction of a magnetic field are $\pm(1.0015 \pm 0.00002)$ muonic Bohr magnetons. (The muonic Bohr magneton is the Bohr magneton multiplied by the ratio of electron mass to muon mass.) The identity of this value with the value for the electron shows that the muon and the electron have closely similar structures, and the close approximation of each value to unity is strong evidence that their structure is simple in comparison with that of the proton and neutron.

In 1962 P. A. M. Dirac published a theory of the muon in which it is described as an excited vibrational state of the electron. The postulated vibration is spherically symmetric; it is the rhythmic increase and decrease of size of the sphere of negative electricity constituting the particle.

Neutrinos and Antineutrinos. Weak Interactions

The neutrino is a particle with zero rest-mass and spin $\frac{1}{2}$; it differs from the photon primarily in the value of the spin (the photon has spin 1). The existence of the neutrino was proposed in 1927 by W. Pauli, in order to account for the apparent lack of conservation of energy in the process of emission of a β particle (an electron) by a radioactive nucleus, as discussed in Chapter 26. It had been observed that all radioactive nuclei of the same kind that emitted an α particle, such as radium 226 (Figure 26-1), shoot out their particles with the same energy, as expected from the law of conservation of mass-energy, but that, on the other hand, radioactive atoms that emit β particles, such as ^{214}Pb, emit the β particles with varying energies. Pauli, and later Fermi, suggested that another particle, with small or zero rest-mass, is also emitted when the nucleus undergoes radioactive decay with emission of a β particle, and that the energy of the reaction is divided between the β particle and the other particle, which Fermi named the neutrino.

In 1934 Fermi developed his theory of β decay, in order to explain the puzzling observation that some radioactive nuclides shoot out an electron in the course of radioactive decomposition, although they were supposed to be composed only of protons and neutrons. He pointed out that atoms emit photons when they change from one quantum state to another, although it is not believed that the atoms contain the photons; instead, it is accepted that the photon is created at the time when it is emitted. Fermi suggested that the electrons, the β particles, are created when the radioactive nucleus undergoes decomposition, and that at the same time one of the neutrons inside the nucleus becomes a proton, and a neutrino (or, rather, an antineutrino) is emitted.

The fundamental reaction of the Fermi theory is

$$n \longrightarrow p^+ + e^- + \bar{\nu}$$

This is the reaction of decomposition of the free neutron (Table 25-4). The free neutron decomposes with a half-life of 1040 s. In many nuclei the neutron is made stable by interaction with other nucleons, but in some nuclei it remains unstable, and this reaction takes place.

Neutrinos interact only very weakly with other particles, and the existence of the neutrino was not verified by experiment until 1956. In that year the American physicists F. Reines and C. L. Cowan, Jr., showed that neutrinos from a nuclear reactor passing through a liquid-hydrogen bubble chamber cause a reaction to take place that is approximately the reverse of the decay of a neutron:

$$\bar{\nu} + p^+ \longrightarrow n + \bar{e}^+$$

The decay of a neutron into a proton, an electron, and a neutrino cannot be explained by strong interactions (Section 25-3) or by electromagnetic forces. Fermi assumed that another kind of interaction, called weak interaction, occurs among some particles. It is about 10^{-15} times as strong as the strong interactions that occur between nucleons and similar particles, and it leads to reaction times of the order of 10^{-8} s, instead of the time 10^{-23} s that applies to strong interactions.

Neutrinos and antineutrinos have spin $\frac{1}{2}$, but they have an extraordinary property that was discovered in 1957 as a result of the work of the Chinese physicists Tsung-Dao Lee (born 1926) and Chen Ning Yang (born 1922), working in the United States. These theoretical physicists and the experimental physicists whom they inspired found that the neutrino, which has spin $\frac{1}{2}$, always orients its spin in the direction of its motion, so that it moves through space with the speed of light as though it were a right-handed propeller. The antineutrino always orients its spin in the opposite direction, and moves as though it were a left-handed propeller.

In 1960 it was proposed by several physicists, in order to explain a number of experimental observations in a simple way, that there are two neutrinos and two antineutrinos, with somewhat different properties. It was postulated that one neutrino (ν) and one antineutrino ($\bar{\nu}$) have a close relation of some sort to the electron and positron, and the other neutrino (ν') and antineutrino ($\bar{\nu}'$) have a similar relation to the muon and antimuon. Experimental verification of this hypothesis was obtained in 1962 by a difficult experiment carried out by a group of Columbia University and Brookhaven National Laboratory scientists. As mentioned above, Reines and Cowan had shown that a neutrino produced by a reaction involving electrons reacts with a proton to produce a neutron and an

TABLE 25-2
*Mesons and Antimesons**

Name	Electric Charge			Mass	Intrinsic Charge	Charge Spin	Xenicity (strange-ness)	Spin
	+1	0	−1					
Eta		η^0		500 MeV	0	0	0	0
Kaons		K^0	K^-	497.7, 493.8	$-\frac{1}{2}$	$\frac{1}{2}$	−1	0
Antikaons	$\overline{K}^+$	$\overline{K}^0$		493.8, 497.7	$+\frac{1}{2}$	$\frac{1}{2}$	+1	0
Pions	π^+	π^0	π^-	139.6, 135.0, 139.6	0	1	0	0

*The muon was originally named the meson, and then the μ meson, but it is now placed in the lepton class. Mesons and antimesons have baryon number 0 and lepton number 0. All the particles listed in this table have spin 0 (zero angular momentum). The positive pion and the negative pion are the antiparticles of one another. The neutral pion is its own antiparticle, and the eta is its own antiparticle.

TABLE 25-3
*Baryons and Antibaryons**

Name	Electric Charge			Mass	In-trinsic Charge	Charge Spin	Xenicity (strange-ness)
	+1	0	−1				
Xi particles		Ξ^0	Ξ^-	1315, 1321.2 MeV	$-\frac{1}{2}$	$\frac{1}{2}$	−2
Sigma particles	Σ^+	Σ^0	Σ^-	1189.5, 1192.6, 1197.4	0	1	−1
Lambda particle		Λ		1115.6	0	0	−1
Proton, neutron	p^+	n		938.2, 939.5	$+\frac{1}{2}$	$\frac{1}{2}$	0
Xi antiparticles	$\overline{\Xi}^+$	$\overline{\Xi}^0$			$+\frac{1}{2}$	$\frac{1}{2}$	+2
Sigma antiparticles	$\overline{\Sigma}^+$	$\overline{\Sigma}^0$	$\overline{\Sigma}^-$	Same as for particles	0	1	+1
Lambda antiparticle		$\overline{\Lambda}$			0	0	+1
Antineutron, antiproton		$\overline{n}$	$\overline{p}^-$		$-\frac{1}{2}$	$\frac{1}{2}$	0

*Baryons have baryon number +1. Antibaryons have baryon number −1. Both have lepton number 0. All have spin $\frac{1}{2}$.

electron. In the 1962 experiment it was shown that neutrinos produced by the decomposition of muons react with protons to produce only muons, and not electrons:

$$\bar{v}' + p^+ \longrightarrow n + \bar{\mu}^+$$

We shall call the two neutrinos the *electron neutrino*, v, and the *muon neutrino*, v'. At the present time nothing can be said about their nature, to explain the difference in their properties in terms of a difference in structure.

25-6.   Mesons and Antimesons

The known mesons and antimesons, eight in number, are listed in Table 25-2. The kaons are the antiparticles of the antikaons, and the two charged pions are antiparticles of one another. The neutral pion is its own antiparticle, and the eta particle is its own antiparticle. All of the mesons are unstable; their decay reactions will be discussed in Section 25-8.

The pions and kaons were discovered in experiments with cosmic rays, and their properties have been determined by use both of cosmic rays and of high-energy particles produced by particle accelerators. The pions were discovered by Powell and his collaborators, as mentioned in an earlier section. The kaons were discovered about 1950 by many investigators.

25-7.   Baryons and Antibaryons

The baryons include the nucleons and heavier particles. There are eight baryons and eight antibaryons known, as listed in Table 25-3. The word baryon is from the Greek *barys*, heavy. The word hyperon (Greek *hyper*, beyond) is also used; it refers to the baryons other than the proton and the neutron.

The baryons other than the proton and the neutron were discovered in the period between 1950 and 1960 by use of cosmic rays and particle accelerators. Their masses range from 1115 to 1318 MeV. All baryons have spin $\frac{1}{2}$ and are fermions, obeying the Pauli exclusion principle.

25-8.   The Decay Reactions of the Fundamental Particles

Most of the fundamental particles decompose spontaneously. The exceptions, the stable particles, comprise the proton, the antiproton, the electron, the positron, and the particles that move with the speed of light.

TABLE 25-4
Main Reactions of Decay of Particles

	Reaction	Ratio (%)	Half-Life (seconds)
Baryons:	$\Xi^0 \longrightarrow \Lambda + \pi^0$		$\sim 2 \times 10^{-10}$
	$\Xi^- \longrightarrow \Lambda + \pi^-$		2×10^{-10}
	$\Sigma^+ \longrightarrow p^+ + \pi^0$	46 ± 6	0.8×10^{-10}
	$n + \pi^+$	54 ± 6	
	$\Sigma^0 \longrightarrow \Lambda + \gamma$		$\sim 10^{-20}$
	$\Sigma^- \longrightarrow n + \pi^-$		1.6×10^{-10}
	$\Lambda \longrightarrow p^+ + \pi^-$	63 ± 3	2.4×10^{-10}
	$n + \pi^0$	37 ± 3	
	$n \longrightarrow p^+ + e^- + \bar{\nu}$		1040
Mesons:	$\eta^0 \longrightarrow \pi^+ + \pi^0 + \pi^-$, etc.		$\sim 10^{-20}$
	$K_1^0 \longrightarrow \pi^+ + \pi^-$	78 ± 6	$1.0 \times 10^{-10*}$
	$\pi^0 + \pi^0$	21 ± 6	
	$K_2^0 \longrightarrow \pi^+ + \pi^-$	78 ± 6	6×10^{-8}
	$\pi^0 + \pi^0$	22 ± 6	
	$K^- \longrightarrow \mu^- + \bar{\nu}'$	59 ± 2	1.22×10^{-8}
	$\pi^0 + \pi^-$	26 ± 2	
	$\pi^+ + \pi^- + \pi^-$	5.7 ± 0.3	
	$\pi^0 + \pi^0 + \pi^-$	1.7 ± 0.3	
	$e^- + \bar{\nu} + \pi^0$	4.2 ± 0.4	
	$\mu^- + \bar{\nu}' + \pi^0$	4.0 ± 0.8	
	$\pi^+ \longrightarrow \bar{\mu}^+ + \nu'$	100	2.56×10^{-8}
	$\bar{e}^+ + \nu$	0.013	
	$\pi^0 \longrightarrow \gamma + \gamma$		2×10^{-15}
Leptons:	$\mu^- \longrightarrow e^- + \nu' + \bar{\nu}$		10^{-6}

*In a beam of neutral kaons K^0 and antikaons $\bar{K}^0$ the particles decompose at two rates, to give the same products. This behavior is explained by saying that the beam contains particles K_2^0 that are in the quantum state corresponding to symmetric resonance of K^0 and $\bar{K}^0$ and also particles K_1^0 that are in the quantum state corresponding to antisymmetric resonance of K^0 and $\bar{K}^0$. The kaon-antikaon pair is the only pair known to have this property.

Even though many of the fundamental particles were discovered only a few years ago, a tremendous amount of information has been gained about their properties and the reactions by which they are produced, changed into other forms of matter, and destroyed. The principal reactions by which the unstable particles decay are listed in Table 25-4, which also gives the values of the half-life. All of these decay reactions are unimolecular reactions, the nature of which has been discussed in Chapter 16.

Conservation Principles

By analyzing the tracks produced by individual particles in cloud chambers, stacks of photographic emulsions, and bubble chambers, and by other methods of detecting particles, the decay of individual particles has been studied, and it has been found that in every case there is conservation of mass-energy and conservation of momentum. Other conservation principles have also been found to be adhered to rigorously:

> Conservation of mass-energy
> Conservation of momentum
> Conservation of angular momentum
> Conservation of electric charge
> Conservation of baryon number
> Conservation of lepton number

The principle of conservation of electric charge is illustrated by the decay reactions given in Table 25-4. For example, the lambda particle, which is a hyperon, with mass somewhat greater than that of a nucleon, can decompose either to form a proton and a negative pion or to form a neutron and a neutral pion. In the first case the lambda particle, which is neutral, forms a positively charged particle and a negatively charged particle; in the second case it forms two neutral particles.

A more complicated example, also given in Table 25-4, is the decomposition of the negative kaon. This particle has been observed to decompose in six different ways. Five of the reactions of decomposition lead to the formation of a negatively charged particle and one or two neutral particles. The sixth reaction leads to the formation of a positively charged particle, a positive pion, and two negatively charged particles, negative pions. Hence in each of the six reactions there is conservation of electric charge.

There is also conservation of the baryon number in every reaction. The baryons have baryon number $+1$ and the antibaryons have baryon number -1; all other particles have baryon number 0. In the various processes of formation of baryons and antibaryons they are always formed in pairs, one baryon and one antibaryon. Similarly, the decomposition of a baryon always leads to the formation of another baryon, plus other particles with 0 baryon number. Thus the negative xi particle is observed to decompose to form a lambda particle, which has baryon number $+1$, and a negative pion, which has baryon number 0.

Leptons, which include the electron, the neutrino, and the muon, have lepton number $+1$, and antileptons have lepton number -1; all other particles have lepton number 0. There is rigorous conservation of the lepton number in all reactions. For example, the neutron, which has lepton number 0, decomposes to form a proton, also with lepton number 0, an electron, and an antineutrino. The lepton numbers of the electron and the antineutrino add up to 0, so that in this reaction, as in all others listed in Table 25-4, there is conservation of the lepton number.

There are also some conservation principles that are observed to hold for strong interactions but not for weak interactions. This matter is discussed in the following section.

25-9. Strangeness (Xenicity)

A great contribution to the understanding of the nature of the fundamental particles was made in the period between 1953 and 1956 by the American physicist Murray Gell-Mann and the Japanese physicist K. Nishijima, working independently. The classification of the fundamental particles given in Tables 25-1, 25-2, and 25-3 is in considerable part due to their efforts. This classification is based upon the concept of charge multiplets and the concept of strangeness. Neither of these concepts can be said to be thoroughly understood at the present time, and it is likely that some additional great contributions will be made in the near future.

In Section 25-4 it was pointed out that the close similarity in properties of the neutron and the proton, except for electric charge, suggests that these two particles represent two aspects of the same particle, the nucleon. The nucleon may be said to have intrinsic electric charge $+\frac{1}{2}$ and an electric-charge vector $\frac{1}{2}$, which can have the component $+\frac{1}{2}$ or $-\frac{1}{2}$ in charge space, leading to the resultant electric charge $+1$ for the proton and 0 for the neutron. The proton and neutron can then be described as a charge doublet.*

The diagram in Figure 25-5 shows that the 24 particles represented in the diagram constitute three charge singlets, six doublets, and three triplets. The charge singlets have electric-charge vector equal to 0, and intrinsic charge 0. The doublets all have electric-charge vector equal to $\frac{1}{2}$; the nucleon, antinucleon, kaons, and xions have intrinsic charge $+\frac{1}{2}$ or $-\frac{1}{2}$. A charge doublet can thus have electric charges either 0 and $+1$ or 0 and -1. The triplets, with electric-charge vector 1 and intrinsic charge 0, have electric charges $+1$, 0, and -1, corresponding to the three orientations of the charge vector.

The idea of strangeness was introduced by Gell-Mann and Nishijima to explain in a rough way the rates of decay reactions. Some of the unstable

---

*The idea of charge multiplets was introduced into physics by the American physicists B. Cassen and E. U. Condon in 1936.

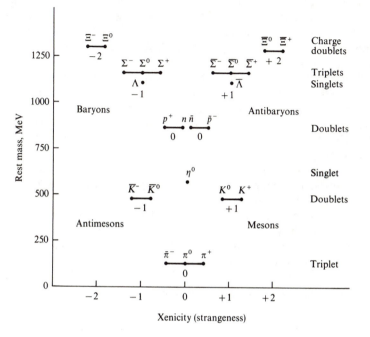

FIGURE 25-5
A diagram representing the masses and xenicities of some of the
elementary particles.

particles are expected to decay by virtue of the strong interactions (Section 25-3), and this decomposition should be very rapid, with half-lives of the order of 10^{-23} s. An example is the decay of the η^0 particle, to form three pions; its half-life is about 10^{-23} s.

Many other particles, however, are observed to have much longer half-lives, of the order of 10^{-9} s. These particles accordingly live 10^{14} times as long as predicted for them on the basis of the theory of strong interactions. Because of this deviation from the expected behavior, these particles were called strange particles.

Gell-Mann and Nishijima suggested that a characteristic property, which is called strangeness, should be assigned to the particles, such that there is conservation of strangeness for reactions involving strong interactions but violation of conservation of strangeness for weak interactions. *Xenicity** may be a better name for this property than strangeness.

---

*From Greek *xenos*, stranger.

The values of the xenicity are shown in Figure 25-5. Pions, the eta particle, nucleons, and antinucleons have xenicity 0. The kaons, anti-lambda particle, and antisigma particles have xenicity $+1$, and the anti-kaons, lambda particle, and sigma particles have xenicity -1. The anti-xions have xenicity $+2$, and the xions have xenicity -2.

The conservation principle is that for strong interactions there must be conservation of xenicity; the sum of the xenicities for the reactants equals the sum of the xenicities for the products. Reactions in which the sum of the xenicities changes by one unit can occur as a result of weak interactions, but these reactions are slow. Reactions in which the sum changes by two units are very slow.

The eta particle has xenicity 0, and the pions have xenicity 0. There is, accordingly, no change in xenicity accompanying the decay of the eta particle, and the reaction is very fast.

Table 25-4 contains many examples of reactions in which there is a change in xenicity. The negative antikaon, $\bar{K}^-$, has xenicity -1. It can decay in six ways, to form pions and leptons (the muon, the electron, an antineutrino), all of which have xenicity 0. The total half-life for these various reactions is 1.22×10^{-8}, far longer than the half-life for the eta decomposition, and this long half-life is attributed to the change in xenicity.

25-10. Resonance Particles and Complexes

In 1952 it was found by Enrico Fermi and his coworkers, who were study-ing the scattering of a beam of pions by protons, that the scattering is much larger when the pions have about 200 MeV of kinetic energy than for smaller or greater amounts of kinetic energy. This observation was interpreted as showing that there is a strong interaction between the pion and the proton, which can be described as corresponding to the formation of a short-lived particle or complex, to which the symbol Δ has been assigned:

$$\pi + p \rightleftarrows \Delta$$

The mass of Δ is about 1236 MeV. Its half-life is about 10^{-23}; and, in ac-cordance with the uncertainty relation between energy and time (Section 3-10), the mass (energy) of the particle is not well defined, but has an un-certainty of about ± 60 MeV. (The half-life 10^{-23} s is in fact calculated from the observed distribution function for the mass of the Δ complex.)

Since 1952 about a score of these short-lived particles or complexes have been discovered. They are called *resonance particles* or *resonance complexes*. One of them, η^0, has been included in our listing of the mesons

(Table 25-2). It is produced by reaction of a pion and a neutron (within a deuteron):

$$\pi^+ + d^+ \longrightarrow \eta^0 + p^+ + p^+$$

It is the lightest of the resonance particles, with mass 550 MeV. It decomposes, with half-life 10^{-20} s, by the reactions

$$\eta^0 \longrightarrow \gamma + \gamma \qquad\qquad 31\%$$
$$\eta^0 \longrightarrow \pi^\circ + \gamma + \gamma \qquad\quad 21\%$$
$$\eta^0 \longrightarrow 3\pi^0 \qquad\qquad\quad 21\%$$
$$\eta^0 \longrightarrow \pi^+ + \pi^0 + \pi^- \qquad 22\%$$
$$\eta^0 \longrightarrow \pi^+ + \pi^- + \gamma \qquad 5\%$$
$$\eta^0 \longrightarrow \pi^0 + e^+ + e^- \qquad 0.1\%$$
$$\eta^0 \longrightarrow \pi^+ + \pi^- + e^+ + e^- \qquad 0.1\%$$

The next lightest resonance particles are the ρ particles, ρ^+ and ρ^- with mass 778 MeV and ρ^0 with mass 770 MeV. They are formed by the following reactions:

$$\pi^+ + p^+ \longrightarrow \rho^+ + p^+$$
$$p^+ + \bar{p}^- \longrightarrow \rho^0 + \pi^+ + \pi^-$$
$$\pi^- + p^+ \longrightarrow \rho^- + p^+$$

They decompose, with half-life about 10^{-23} s, to give two pions. Except for having spin 1, they resemble the pions. Other particles with spin 1 are ω^0, which has mass 783.4 MeV, φ^0, with mass 1019 MeV, and K^{*+}, K^{*0}, $\bar{K}^{*0}$, and $\bar{K}^{*-}$, with mass 892 MeV for the charged and 888 MeV for the uncharged particles. The relation of these nine mesons with spin 1 to the eight mesons with spin 0 listed in Table 25-2 is discussed in Section 25-11.

The Δ particles mentioned above constitute a baryon charge quartet, Δ^{++}, Δ^+, Δ^0, and Δ^-, with spin $\frac{3}{2}$ and mass ranging from 1236 MeV for Δ^{++} to 1246 MeV for Δ^-. A charge triplet with xenicity 1 and spin $\frac{3}{2}$ exists: Σ^{*+} (1382 MeV), Σ^{*0} (1385 MeV), and Σ^{*-} (1388 MeV); also a charge doublet with xenicity 2 and spin $\frac{3}{2}$: Ξ^{*0} (1530 MeV) and Ξ^{*-} (1534 MeV), and a charge singlet with xenicity 3 and spin $\frac{3}{2}$: Ω^- (mass 1674 MeV). The existence of Ω^- was predicted by M. Gell-Mann, with use of ideas about structure resembling those presented in the next section.

25-11.   The Structure of the Fundamental Particles. Quarks

It is likely that in the course of time a structural theory of the fundamental particles and resonances will be developed. A set of at least three protogons (Greek *protos*, first, and *gone*, that which generates) and three

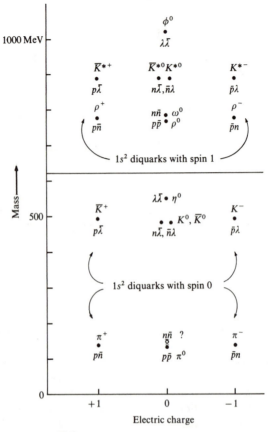

FIGURE 25-6
Diagram of mesons (diquarks) with spin 0 and spin 1.

antiprotogons seems to be needed. At the present time the most promising idea is that mesons and baryons are made of *quarks*, each meson being a diquark (a compound of a quark and an antiquark) and each baryon being a triquark. This idea was developed independently in 1964 by M. Gell-Mann and George Zweig.

The three quarks are the positive quark, p, the negative quark, n, and the strange quark, λ. (Note that we use p and n for the quarks, p and n for the nucleons.) All three are fermions, with spin $\frac{1}{2}$; p has electric charge $+\frac{2}{3}$, and n and λ have charge $-\frac{1}{3}$. The antiquarks $\bar{p}$, $\bar{n}$, and $\bar{\lambda}$ have charges $-\frac{2}{3}$, $+\frac{1}{3}$, and $+\frac{1}{3}$, respectively. Each quark has baryon number $+\frac{1}{3}$ (each antiquark $-\frac{1}{3}$); λ has xenicity 1 and $\bar{\lambda}$ has xenicity -1.

The failure so far to observe any single quarks among the products of high-energy reactions indicates that they have very large mass, several thousand MeV. The effective mass of λ in diquarks is about 145 MeV greater than that of n and p; that of n is about 4 MeV greater than that of p.

Quark Structure of Mesons

Let us consider the diquarks with baryon number 0—that is, the compounds of a quark and an antiquark. The most stable diquarks are expected to be those in which both particles are in a $1s$ orbital, as they move about their common center of mass. The quarks and antiquarks are different particles; hence the Pauli exclusion principle does not forbid parallel spins for a quark and its antiquark, and a $1s^2$ diquark can have resultant spin 0 or resultant spin 1. The possible $1s^2$ diquarks are the following: $p\bar{p}$, $p\bar{n}$, $\bar{p}n$, $n\bar{n}$, $p\bar{\lambda}$, $\bar{p}\lambda$, $n\bar{\lambda}$, $\bar{n}\lambda$, and $\lambda\bar{\lambda}$. The nine corresponding diquarks with spin 0 and the nine with spin 1 are shown in Figure 25-6, with the symbols of the corresponding known particles. There is only one particle missing: the $n\bar{n}$ neutral particle with spin 0 and mass about 150 MeV. There is the possibility that it has been observed but misidentified as π^0.

We see that the spin-spin interaction of a quark and an antiquark has the opposite sign to that of two electrons (Section 5-3); the state of a $1s^2$ diquark with spin 0 (antiparallel spins) is more stable by about 500 MeV than the state with spin 1 (parallel spins).

Quark Structure of Baryons

In the quark theory baryons consist of three quarks (giving baryon number $3 \times \frac{1}{3} = 1$), and antibaryons of three antiquarks. The most stable baryons are expected to be those with the three quarks in a $1s$ orbital. The $1s^3$ triquarks with resultant spin $\frac{1}{2}$ allowed by the Pauli exclusion principle are p^2n, pn^2, $p^2\lambda$, $pn\lambda(2)$, $n^2\lambda$, $p\lambda^2$, and $n\lambda^2$. In addition there is a $1s^3$ $pn\lambda$ state with spin $\frac{3}{2}$, as shown by the following argument.

With three different quarks, p, n, and λ, in the $1s$ orbital, each quark may have positive or negative orientation of its spin. With all three positive the component of the total spin I in the field direction is $+\frac{3}{2}$; hence $I = \frac{3}{2}$, and $M_I = +\frac{3}{2}, +\frac{1}{2}, -\frac{1}{2},$ and $-\frac{3}{2}$. The other four values of M_I are $+\frac{1}{2}, +\frac{1}{2}, -\frac{1}{2},$ and $-\frac{1}{2}$; they correspond to two $pn\lambda$ states with $I = \frac{1}{2}$.

The eight $1s^3$ triquarks with spin $\frac{1}{2}$ correspond to the eight most stable baryons (listed in Table 25-3), as shown in Figure 25-7. The eight $1s^3$ triantiquarks with spin $\frac{1}{2}$ correspond to the eight most stable antibaryons.

The ten particles Δ^{++}, Δ^+, Δ^0, Δ^-, Σ^{*+}, Σ^{*0}, Σ^{*-}, Ξ^{*0}, Ξ^{*-}, and Ω^-

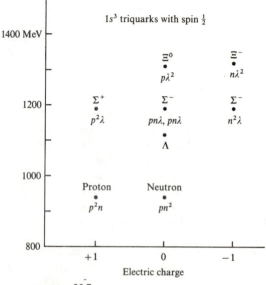

FIGURE 25-7
Diagram of baryons (triquarks) with spin 1/2.

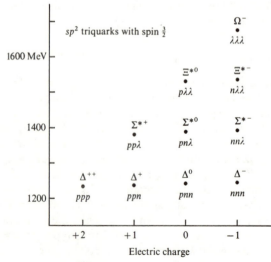

FIGURE 25-8
Diagram of baryons (triquarks) with spin 3/2.

mentioned in Section 25-10 constitute an interesting set. (There is also a corresponding set of ten antiparticles.) These particles have spin $\frac{3}{2}$, as would result from three quarks with parallel spins. The ten particles may be assumed to be the ten triquarks ppp to $\lambda\lambda\lambda$ shown in Figure 25-8. Three identical quarks, such as ppp, with parallel spin require three different orbitals, which we take to be the trigonal hybrid sp^2 orbitals (Appendix VII). We see that the effective mass of the p quark in these spin-$\frac{3}{2}$ triquarks is 412 MeV, that of the n quark is 416 MeV, and that of the λ quark is 558 MeV. The corresponding values for the diquarks with spin 1 are 385, 389, and 510 MeV, respectively, and those for the spin-$\frac{1}{2}$ baryons and spin-0 mesons are only about half as great, indicating stronger binding.

Scores of heavier mesons and baryons are known, with spin as large as $\frac{19}{2}$. No detailed assignment of structures to them (with use, for example, of $d, f, g, \ldots$ orbitals) has yet been made, but it seems likely that they can be assigned structures somewhat resembling the electronic structures of atoms and molecules.

The nature of the strange quark λ is not as yet understood. One possibility is that λ bears the same structural relation to n that the muon bears to the electron; this relation, however, is also not yet clear.

Whether or not we become nuclear chemists or fundamental-particle physicists, we may all await with eagerness the increase in knowledge about the nature of the universe that will surely be obtained through the efforts of scientists all over the world during the coming years.

25-12. Positronium, Muonium, Mesonic Atoms

In 1953 it was observed that a positron and an electron combine to form a pseudo-atom, somewhat similar to the hydrogen atom. In the hydrogen atom the electron can be described as moving around an essentially stationary nucleus, the proton. In the pseudo-atom formed by a positron and an electron, which has been given the name *positronium*, the two particles have the same mass, so that they carry out similar motions about their center of mass, the point midway between the two.

It was found by the American physicist Martin Deutsch that there are two kinds of positronium. The kind in which the spin of the positron is antiparallel to that of the electron is called parapositronium, and that in which the two spins are parallel is called orthopositronium. Parapositronium decomposes with destruction of the positron and the electron and production of two photons, its half-life being 0.9×10^{-10} s. Ortho-positronium decomposes with production of three photons, and half-life

1.0×10^{-7} s. The existence of positronium was detected by the observation of a delay between its production (by decomposition of sodium 22, which emits positrons) and its annihilation. The time of delay was found to correspond to the sum of two first-order reactions, with the values of the half-life given above.

Muonium, a pseudo-atom involving a negative muon moving about a proton, has also been observed. Other mesonic atoms, having structures similar to ordinary atoms but with a muon or other meson replacing one of the electrons, have also been observed. For example, muonic neon is a neon atom with a negative muon in place of an electron.

A muonic molecule ion, $[H^+\mu^-D^+]^+$, in which a proton and a deuteron are held together by a negative muon, has also been made. The proton and the deuteron are sufficiently close together, about 0.003 Å apart, to permit reaction between them, liberating the muon and producing a helium-3 nucleus plus an additional muon, with release of 5.4 MeV of energy. The use of a mesonic molecule of this sort might possibly permit the controlled release of energy through nuclear fusion.

26

Nuclear Chemistry

The field of nuclear chemistry deals with the reactions that involve changes in atomic nuclei. This field began with the discovery of radioactivity and the work of Pierre and Marie Curie on the chemical nature of the radioactive substances. After some decades, during which natural radioactivity was rather thoroughly investigated, a great increase in knowledge resulted through the discovery of artificial radioactivity.

Nuclear chemistry has now become a large and important branch of science. About 920 radioactive nucleides have been made in the laboratory, whereas only about 272 stable nucleides and 55 unstable (radioactive) nucleides have been detected in nature. The use of radioactive isotopes as "tracers" has become a valuable technique in scientific and medical research. The controlled release of nuclear energy promises to lead us into a new world, in which the achievement of man is no longer severely limited by the supply of energy available to him.

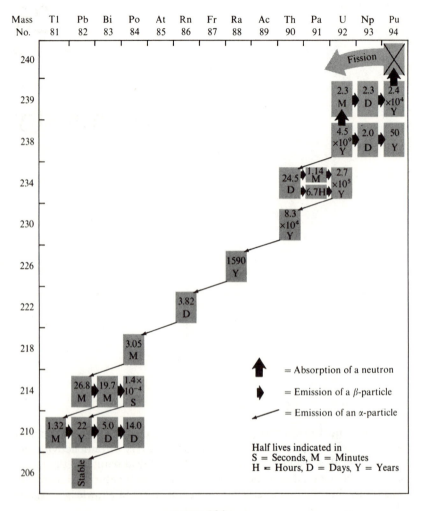

FIGURE 26-1
The uranium-radium series.

26-1. Natural Radioactivity

After their discovery of polonium and radium in 1898 (Chapter 3), the Curies found that radium chloride could be separated from barium chloride by fractional precipitation of the aqueous solution by addition of alcohol, and by 1902 Madame Curie had prepared 0.1 g of nearly pure radium chloride, with radioactivity about 3,000,000 times that of uranium. Within a few years it had been found that natural radioactive materials emit three kinds of rays capable of sensitizing the photographic plate (Chapter 3). These rays—alpha rays, beta rays, and gamma rays—are affected differently by a magnetic field (Figure 3-12). Alpha rays are the nuclei of helium atoms, moving at high speeds; beta rays are electrons, also moving at high speeds; and gamma rays are photons, with very short wavelengths.

It was soon discovered that the rays from radium and other radioactive elements cause regression of cancerous growths. These rays also affect normal cells, "radium burns" being caused by overexposure; but often the cancerous cells are more sensitive to radiation than normal cells, and can be killed by suitable treatment without serious injury to normal tissues. The medical use in the treatment of cancer is the main use for radium. Since about 1950, considerable use has also been made of the artificial radioactive nuclide cobalt 60 as a substitute for radium.

Through the efforts of many investigators the chemistry of the radioactive elements of the uranium series and the thorium series was unraveled during the first two decades of the twentieth century, and that of the neptunium series during a few years from 1939 on.

The Uranium Series of Radioactive Disintegrations

When an alpha particle (He^{++}) is emitted by an atomic nucleus the nuclear charge decreases by two units; the element hence is transmuted into the element two columns to the left in the periodic table. Its mass number (atomic weight) decreases by 4, the mass of the alpha particle. When a beta particle (an electron) is emitted by a nucleus the nuclear charge is increased by one unit, with no change in mass number (only a very small decrease in atomic weight); the element is transmuted into the element one column to its right. No change in atomic number or mass number is caused by emission of a gamma ray.

The nuclear reactions in the *uranium-radium series* are shown in Figure 26-1. The principal isotope of uranium, ^{238}U, constitutes 99.28% of the natural element. This isotope has a half-life of 4,500,000,000 years. It decomposes by emitting an alpha particle and forming ^{234}Th. This isotope

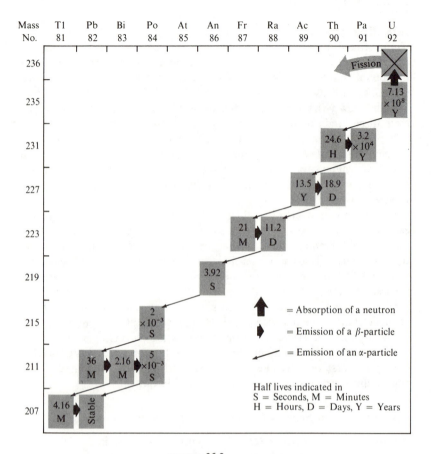

FIGURE 26-2
The uranium-actinium series.

of thorium undergoes decomposition with β-emission, forming ^{234}Pa, which in turn forms ^{234}U. Five successive α-emissions then occur, giving ^{214}Pb, which ultimately changes to ^{206}Pb, a stable isotope of lead.

The *uranium-actinium series*, shown in Figure 26-2, is a similar series beginning with ^{235}U, which occurs to the extent of 0.71% in natural uranium. It leads, through the emission of seven alpha particles and four beta particles, to the stable isotope ^{207}Pb.

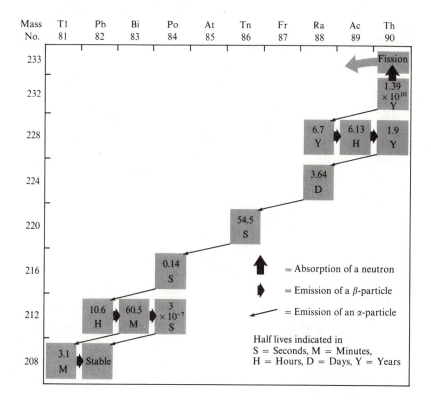

FIGURE 26-3
The thorium series.

The Thorium Series

The third natural radioactive series begins with the long-lived, naturally occurring isotope of thorium, ^{232}Th, which has half-life 1.39×10^{10} years (Figure 26-3). It leads to another stable isotope of lead, ^{208}Pb.

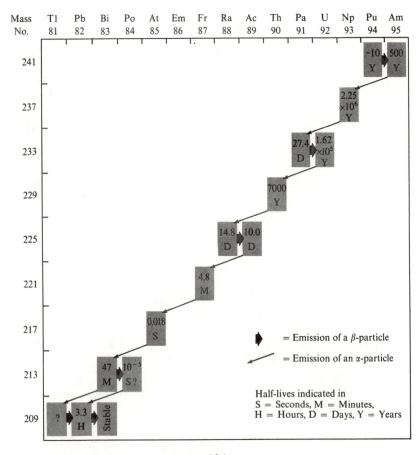

The Neptunium Series

The fourth radioactive series (Figure 26-4) is named after its longest-lived member, which is ^{237}Np.

The nature of radioactive disintegration within each of the four series—the emission of β-particles, with mass nearly zero, or of α-particles, with mass 4—is such that all the members of a series have mass numbers

differing by a multiple of 4. The four series can hence be classified as
follows (*n* being integral):

> The 4*n* series = the thorium series
> The 4*n* + 1 series = the neptunium series
> The 4*n* + 2 series = the uranium-radium series
> The 4*n* + 3 series = the uranium-actinium series

26-2. The Age of the Earth

Measurements made on rocks containing radioactive elements can be
interpreted to provide values of the age of the rocks, and hence of the
age of the earth; that is, the time that has elapsed since the oldest rocks
were laid down. For example, 1 g of ^{238}U would in its half-life of 4.5×10^9
years decompose to leave 0.5000 g of ^{238}U and to produce 0.0674 g of
helium and 0.4326 g of ^{206}Pb. (Each atom of ^{238}U that decomposes forms
eight atoms of helium, with total mass 32, leaving one atom of ^{206}Pb.)
If analyses showed that the nucleides were present in a rock in the ratios
of these numbers, the rock would be assumed to be 4.5×10^9 years old.
The $^{235}U/^{207}Pb$ ratio, the $^{232}Th/^{208}Pb$ ratio, the $^{40}K/^{40}Ar$ ratio, and the
$^{87}Rb/^{87}Sr$ ratio are also being used for determining the ages of rocks. Ages
around 3.0×10^9 years have been determined for rocks found in Finland,
Canada, and Africa, and about 4.5×10^9 years for meteorites. The
present estimate of the age of the earth and other planets in the solar
system is 4.5×10^9 years.

26-3. Artificial Radioactivity

Stable atoms can be converted into radioactive atoms by bombardment
with particles traveling at high speeds. In the early experimental work the
highspeed particles used were alpha particles from ^{214}Bi (called radium C).
The first nuclear reaction produced in the laboratory was that between
alpha particles and nitrogen, carried out by Lord Rutherford and his
collaborators in the Cavendish Laboratory at Cambridge in 1919. The
nuclear reaction that occurs when nitrogen is bombarded with alpha particles is the following:

$$^{14}_{7}N + {}^{4}_{2}He \longrightarrow {}^{17}_{8}O + {}^{1}_{1}H$$

In this reaction a nitrogen nucleus reacts with a helium nucleus, which
strikes it with considerable energy, to form two new nuclei, a ^{17}O nucleus
and a proton.

The ^{17}O nucleus is stable, so that this nuclear reaction does not lead to the production of artificial radioactivity. Many other elements, however, undergo similar reactions with the production of unstable nuclei, which then undergo radioactive decomposition.

Many nuclear reactions result from the interaction of nuclei and neutrons. The early experiments with neutrons were carried out by use of a mixture of radon and beryllium metal. The alpha particles from radon react with the beryllium isotope ^{9}Be to produce neutrons in the following ways:

$$_4^9\text{Be} + {}_2^4\text{He} \longrightarrow {}_6^{12}\text{C} + {}_0^1 n$$

$$_4^9\text{Be} + {}_2^4\text{He} \longrightarrow 3\,{}_2^4\text{He} + {}_0^1 n$$

Neutrons are also prepared by reactions in the cyclotron and in uranium reactors.

Manufacture of the Transuranium Elements

The first transuranium element to be made was a neptunium isotope, $_{93}^{239}$Np. This nucleide was made by E. M. McMillan and P. H. Abelson in 1940 by bombarding uranium with high-speed deuterons:

$$_{92}^{238}\text{U} + {}_1^2\text{H} \longrightarrow {}_{92}^{239}\text{U} + {}_1^1\text{H}$$

$$_{92}^{239}\text{U} \longrightarrow {}_{93}^{239}\text{Np} + e^-$$

The first isotope of plutonium to be made was ^{238}Pu, by the reactions

$$_{92}^{238}\text{U} + {}_1^2\text{H} \longrightarrow {}_{93}^{238}\text{Np} + 2\,{}_0^1 n$$

$$_{93}^{238}\text{Np} \longrightarrow {}_{94}^{238}\text{Pu} + e^-$$

The ^{238}Np decomposes spontaneously, emitting electrons. Its half-life is 2.0 days.

During and since World War II some quantity, of the order of one million kilograms, of the nucleide ^{239}Pu has been manufactured. This nucleide is relatively stable; it has a half-life of about 24,000 years. It slowly decomposes with the emission of alpha particles. It is made by the reaction of the principal isotope of uranium, ^{238}U, with a neutron, to form ^{239}U, which then undergoes spontaneous radioactive decomposition with emission of an electron to form ^{239}Np, which in turn emits an electron spontaneously, forming ^{239}Pu:

$$_{92}^{238}\text{U} + {}_0^1 n \longrightarrow {}_{92}^{239}\text{U}$$

$$_{92}^{239}\text{U} \longrightarrow {}_{93}^{239}\text{Np} + e^-$$

$$_{93}^{239}\text{Np} \longrightarrow {}_{94}^{239}\text{Pu} + e^-$$

Plutonium and the next four transuranium elements—americium, curium, berkelium, and californium—were discovered by G. T. Seaborg and his collaborators at the University of California in Berkeley. Americium has been made as ^{241}Am by the following reactions:

$$^{238}_{92}U + {}^4_2He \longrightarrow {}^{241}_{94}Pu + {}^1_0n$$

$$^{241}_{94}Pu \longrightarrow {}^{241}_{95}Am + e^-$$

This nucleide slowly undergoes radioactive decomposition, with emission of alpha particles. Its half-life is 500 years. Curium is made from plutonium 239 by bombardment with helium ions accelerated in the cyclotron:

$$^{239}_{94}Pu + {}^4_2He \longrightarrow {}^{242}_{96}Cm + {}^1_0n$$

The nucleide ^{242}Cm is an alpha-particle emitter, with half-life about 5 months. Other isotopes of curium have also been made. One is ^{240}Cm, made by bombarding plutonium, ^{239}Pu, with high-speed helium ions:

$$^{239}_{94}Pu + {}^4_2He \longrightarrow {}^{240}_{96}Cm + 3\,{}^1_0n$$

Using only very small quantities of the substances, Seaborg and his collaborators succeeded in obtaining a considerable amount of information about the chemical properties of the transuranium elements. They have found that, whereas uranium is similar to tungsten in its properties, in that it has a pronounced tendency to assume oxidation state $+6$, the succeeding elements are not similar to rhenium, osmium, iridium, and platinum, but show an increasing tendency to form ionic compounds in which their oxidation number is $+3$. This behavior is similar to that of the rare-earth metals.

26-4. The Kinds of Nuclear Reactions

Many different kinds of nuclear reactions have now been studied. Spontaneous radioactivity is a nuclear reaction in which the reactant is a single nucleus. Other known nuclear reactions involve a proton, a deuteron, an alpha particle, a neutron, or a photon (usually a gamma ray) interacting with the nucleus of an atom. The products of a nuclear reaction may be a heavy nucleus and a proton, an electron, a deuteron, an alpha particle, a neutron, two or more neutrons, or a gamma ray. In addition, there occurs a very important type of nuclear reaction in which a very heavy nucleus, made unstable by the addition of a neutron, breaks up into two parts of comparable size, plus several neutrons. This process of fission has been mentioned in Chapter 22 and is described in a later section of the present chapter.

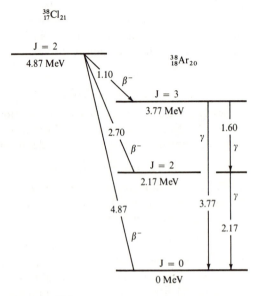

FIGURE 26-5
Beta emission by ^{38}Cl, showing the three associated gamma rays.

The following radioactive decompositions are examples of the different ways in which an unstable nucleus can decompose:

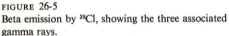

	Half-life
$^3_1H_2 \longrightarrow e^- + {}^3_2He_1 + \bar{\nu}$	12.26 y
$^{148}_{64}Gd_{84} \longrightarrow \alpha^{++} + {}^{144}_{62}Sm_{82}$	130 y
$^{15}_8O_7 \longrightarrow \bar{e}^+ + {}^{15}_7N_7 + \nu$	124 s
$^{37}_{18}Ar_{19} + e^- \longrightarrow {}^{37}_{17}Cl_{20} + \nu$	35.0 d
$^5_3Li_2 \longrightarrow p^+ + {}^4_2He_2$	10^{-21} s
$^5_2He_3 \longrightarrow n + {}^4_2He_2$	2×10^{-21} s

The first two reactions are found for many of the heavy radioactive nucleides. Alpha emission has been observed also for a number of the neutron-rich nucleides in the rare-earth region. The third reaction, positron emission, occurs for most neutron-rich nucleides, many of which also decompose by electron capture (the fourth reaction). (Electron capture is classed as a spontaneous decomposition because the electrons are always available in the atom for capture; it is the *s* electrons, principally

1s, that are captured; they are the only electrons with finite probability at the nucleus.) The last two reactions, proton and neutron emission, occur only rarely.

Most β decompositions (either e^+ or e^-) are accompanied (immediately followed) by the emission of γ rays. The β decomposition may take place to one or more of the excited states of the product nucleus, which then drops to the normal state by γ emission. A simple example is shown in Figure 26-5. A great amount of information about energy levels of nuclei has been obtained by measuring the wavelengths of the photons (the γ rays) and the maximum kinetic energy of the β rays (the maximum corresponds to zero energy for the neutrino).

Nuclear reactions are caused to take place by bombarding nuclei with photons, neutrons, protons, deuterons, tritons ($^3H^+$), trelions ($^3He^{++}$), helions (alpha particles), or heavier nuclei. An example is the production of ^{32}P by bombarding ordinary phosphorus, ^{31}P, with 10-MeV deuterons:

$$^{31}_{15}P_{16} + {}^2_1H_1 \longrightarrow {}^{32}_{15}P_{17} + {}^1_1H_0$$

Nuclear reactions are usually written in a shorter way:

$$^{31}P(d, p)^{32}P$$

The symbol (d, p) means that the bombarding (reacting) particle is a deuteron and the emitted particle is a proton. The following are examples of other nuclear reactions of this general type:

$$^6Li(n, \alpha)^3H \qquad\qquad {}^{11}Be(d, p)^{12}Be$$
$$^7Li(n, \gamma)^8Li \qquad\qquad {}^{11}Be(t, p)^{13}Be$$
$$^9Be(n, p)^9Li \qquad\qquad {}^{10}B(p, 2n)^9C$$
$$^9Be(d, 2p)^9Li \qquad\qquad {}^{10}B(p, \gamma)^{11}C$$
$$^6Li(d, n)^7Be \qquad\qquad {}^{10}B(\alpha, n)^{13}N$$
$$^{10}B(p, \alpha)^7Be \qquad\qquad {}^{23}Na(p, 3n)^{21}Mg$$
$$^6Li(^3He, n)^8B \qquad\qquad {}^{141}Pm(^{12}C, 4n)^{149}Tb$$

26-5.   The Use of Radioactive Elements as Tracers

A valuable technique for research is the use of both radioactive and non-radioactive isotopes as tracers.* By the use of these isotopes an element can be observed in the presence of large quantities of the same element. For example, one of the earliest uses of tracers was the experimental

---

*Analysis for nonradioactive isotopes used as tracers, such as ^{15}N, is carried out by use of a mass spectrograph.

determination of the rate at which lead atoms move around through a crystalline sample of the metal lead. This phenomenon is called *self-diffusion*. If some radioactive lead is placed as a surface layer on a sheet of lead, and the sample is allowed to stand for a while, it can then be cut up into thin sections parallel to the original surface layer, and the radio-activity present in each section can be measured. The presence of radio-activity in layers other than the original surface layer shows that lead atoms from the surface layer have diffused through the metal.

In the discussion of chemical equilibrium in Chapter 11 it was pointed out that a system in chemical equilibrium is not static, but that instead chemical reactions may be proceeding in the forward direction and the reverse direction at equal rates, so that the amounts of different substances present remain constant. At first thought it would seem to be impossible to determine experimentally the rates at which different chemical reactions are proceeding at equilibrium. It has now been found possible to make experiments of this sort, however, with the use of isotopes as tracers.

Perhaps the greatest use for radioactive nucleides as tracers will continue to be in the field of biology and medicine. The human body contains such large amounts of the elements carbon, hydrogen, nitrogen, oxygen, sulfur, and others, that it is difficult to determine the state of organic material in the body. An organic compound containing a radio-active nucleide, however, can be traced through the body. An especially useful radioactive nucleide for these purposes is carbon 14. This isotope of carbon has a half-life of about 5000 years. It undergoes slow decomposition with emission of beta rays, and the amount of the isotope present in a sample can be followed by measuring the beta activity. Large quantities of ^{14}C can be readily made in a nuclear reactor, by the action of slow neutrons on nitrogen:

$$^{14}_{7}N + ^{1}_{0}n \longrightarrow ^{14}_{6}C + ^{1}_{1}H$$

The process can be carried out by running a solution of ammonium nitrate into the nuclear reactor, where it is exposed to neutrons. The carbon that is made in this way is in the form of the hydrogen carbonate ion, HCO_3^-, and it can be precipitated as barium carbonate by adding barium hydroxide solution. The samples of radioactive carbon are very strongly radioactive, containing as much as 5% of the radioactive isotope.

The Unit of Radioactivity, the Curie

It has been found convenient to introduce a special unit in which to measure amounts of radioactive material. The unit of radioactivity is called the *curie*. One curie of any radioactive substance is an amount of the substance such that 3.70×10^{10} atoms of the substance undergo radioactive disintegration per second.

The curie is a rather large unit. One curie of radium is approximately one gram of the element. (The curie was originally defined in such a way as to make a curie of radium equal to one gram, but because of improvement in technique it has been found convenient to define it instead in the way given above.)

It is interesting to point out that in a disintegration chain of radioactive elements in a steady state all of the radioactive elements are present in the same radioactive amounts. For example, let us consider one gram of the element radium, in a steady state with the first product of its decomposition, radon (^{222}Rn), and the successive products of disintegration (see Figure 26-2). The rate at which radon is being produced is proportional to the amount of radium present, one atom of radon being produced for each atom of radium that undergoes decomposition. The number of atoms of radium that undergo decomposition in unit time is proportional to the number of atoms of radium present; the decomposition of radium is a unimolecular reaction. When the system has reached a steady state the number of atoms of radon present remains unchanged, so that the rate at which radon is itself undergoing radioactive decomposition must be equal to the rate at which it is being formed from radium. Hence the radon present in a steady state with one gram of radium itself amounts to one curie.

The amount of radon present in a steady state with one gram of radium can be calculated by consideration of the first-order reaction-rate equations discussed in Chapter 16. The reaction-rate constant for the decomposition of radium is inversely proportional to its half-life. Hence when a steady state exists, and the number of radium atoms undergoing decomposition is equal to the number of radon atoms undergoing decomposition, the ratio of the numbers of radon atoms and radium atoms present must be equal to the ratio of their half-lives.

26-6. Dating Objects by Use of Carbon 14

One of the most interesting recent applications of radioactivity is the determination of the age of carbonaceous materials by measurement of their carbon-14 radioactivity. This technique of radiocarbon dating, which was developed by an American physical chemist, Willard F. Libby, permits the dating of samples containing carbon with an accuracy of around 200 years. At the present time the method can be applied to materials that are not over about 50,000 years old.

Carbon 14 is being made at a steady rate in the upper atmosphere. Cosmic-ray neutrons transmute nitrogen into carbon 14, by the reaction given in the preceding section. The radiocarbon is oxidized to carbon dioxide, which is thoroughly mixed with the nonradioactive carbon dioxide in the atmosphere, through the action of winds. The steady-state concentration of carbon 14 built up in the atmosphere by cosmic rays is about one atom of radioactive carbon to 10^{12} atoms of ordinary carbon. The carbon dioxide, radioactive and nonradioactive alike, is absorbed by plants, which fix the carbon in their tissues. Animals that eat the plants also similarly fix the carbon, containing 1×10^{-12} part radiocarbon, in their tissues. When a plant or animal dies, the amount of radioactivity of the carbon in its tissues is determined by the amount of radiocarbon present, which is the amount corresponding to the steady state in the atmosphere. After 5,760 years (the half-life of carbon 14), however, half of the carbon 14 has undergone decomposition, and the radioactivity of the material is only half as great. After 11,520 years only one-quarter of the original radioactivity is left, and so on. Accordingly, by determining the radioactivity of a sample of carbon from wood, flesh, charcoal, skin, horn, or other plant or animal remains, the number of years that have gone by since the carbon was originally extracted from the atmosphere can be determined.

In applying the method of radiocarbon dating, a sample of material containing about 30 g of carbon is burned to carbon dioxide, which is then reduced to elementary carbon in the form of lamp black. The beta-ray activity of the elementary carbon is then determined, with the use of Geiger counters, and compared with the beta-ray activity of recent carbon, which is 15.3 ± 0.1 decompositions per minute per gram of carbon. The age of the sample is then calculated by the use of the equation for a first-order reaction (Chapter 16). The method was checked by measurement of carbon from the heartwood of a giant Sequoia tree, for which the number of tree rings showed that 2928 ± 50 years had passed since the wood was laid down. This check was satisfactory, as were also similar checks with other carbonaceous materials, such as wood in 1st Dynasty Egyptian tombs 4900 years old, whose dating was considered to be reliable.

The method of radiocarbon dating has now been applied to several thousand samples. One of the interesting conclusions that has been reached is that the last glaciation of the northern hemisphere occurred about 11,400 years ago. Specimens of wood from a buried forest in Wisconsin, in which all of the tree trunks are lying in the same direction as though pushed over by a glacier, were found to have an age of $11,400 \pm 700$ years. The age of specimens of organic materials laid down during the last period of glaciation in Europe was found to be $10,800 \pm 1200$ years. Many samples of organic matter, charcoal, and other carbonaceous material from human camp sites in the western hemisphere have been dated as extending to 11,400 years ago; a very few older ones (30,000 years) have been found.

The eruption of Mt. Mazama in southern Oregon, which formed the crater now called Crater Lake, was determined to have occurred 6453 ± 250 years ago, by the dating of charcoal from a tree killed by the eruption. Three hundred pairs of woven rope sandals found in Fort Rock Cave, Oregon, which had been covered by an earlier eruption, were found to be 9053 ± 350 years old. The Lascaux Cave near Montignac, France, contains some remarkable paintings made by prehistoric man; charcoal from camp fires in this cave was found to have the age $15,516 \pm 900$ years. Linen wrappings from the Dead Sea scrolls of the Book of Isaiah, recently found in a cave in Palestine and thought to be from about the first or second century B.C., were dated 1917 ± 200 years old.

26-7.   The Properties of Nucleides

The nucleides of the various elements show many interesting properties. Most of the known nucleides corresponding to the first ten elements are listed in Table 26-1.

For most elements other than those that form part of the natural radioactive series the distribution of nucleides for an element has been found to be the same for all natural occurrences. The average natural distribution is shown in the fourth column of the table.

Some striking regularities are evident, especially for the heavier elements. The elements of odd atomic number have only one or two natural nucleides, whereas those of even atomic number are much richer in nucleides, many having eight or more. It is also found that the odd elements are much rarer in nature than the even elements. The elements with no stable isotopes (technetium, atomic number 43; astatine, atomic number 85, promethium, atomic number 61) have odd atomic numbers.

Binding Energy

Consideration of the masses of the nucleides shows that they are not additive. Thus the mass of the ordinary hydrogen atom is $1.007825\ d$, and that of the neutron is $1.008665\ d$. If the helium atom were made from two hydrogen atoms and two neutrons without change in mass, its mass would be $4.032980\ d$, but it is in fact less, only $4.002604\ d$. The masses of the heavier atoms are also less than they would be if they were composed of hydrogen atom and neutrons without change in mass.

TABLE 26-1
Isotopes of the Lighter Elements

Z	Name	Mass Number	Mass*	Percent Abundance	Half-Life†	Radiation
0	Electron	0	0.0005486			
0	Neutron	1	1.008665		12 m	e^-
1	Proton	1	1.007276			
1	Hydrogen	1	1.007825	99.985		
		2	2.014102	0.015		
		3	3.014949		12.26 y	
2	Alpha	4	4.001507			e^-
2	Helium	3	3.016030	0.00013		
		4	4.002604	~100		
		5	5.012296		2×10^{-21} s	n
		6	6.018900		0.79 s	e^-
		7			60×10^{-6} s	e^-
3	Lithium	5	5.012541		~10^{-21} s	
		6	6.015126	7.42		
		7	7.016005	92.58		
		8	8.022488		0.85 s	
		9	9.027300		0.17 s	e^-
4	Beryllium	6	6.019780		$\geq 4 \times 10^{-21}$ s	
		7	7.016931		53 d	γ
		8	8.005308		~3×10^{-16} s	
		9	9.012186	100		
		10	10.013535		2.7×10^6 y	e^-
		11	11.021660		13.6 s	e^-, γ
5	Boron	8	8.024612		0.78 s	e^+
		9	9.013335		$\geq 3 \times 10^{-19}$ s	
		10	10.012939	19.6		
		11	11.009305	80.4		
		12	12.014353		0.020 s	e^-, γ
		13	13.017779		0.035 s	e^-

Z	Name	Mass Number	Mass*	Percent Abundance	Half-Life†	Radiation
6	Carbon	10	10.016830		19 s	e^+, γ
		11	11.011433		20.5 m	e^+
		12	12.000000	98.89		
		13	13.003354	1.11		
		14	14.003242		5760 y	e^-
		15	15.010600		2.25 s	e^-, γ
		16	16.014702		0.74 s	e^-
7	Nitrogen	12	12.018709		0.011 s	e^+
		13	13.005739		10.0 m	e^+
		14	14.003074	99.63		
		15	15.000108	0.37		
		16	16.006089		7.35 s	e^-, γ
		17	17.008449		4.14 s	e^-
8	Oxygen	14	14.008597		71 s	e^+, γ
		15	15.003072		124 s	e^+
		16	15.994915	99.759		
		17	16.999133	0.037		
		18	17.999160	0.204		
		19	19.003577		29 s	e^-, γ
		20	20.004071		14 s	e^-, γ
9	Fluorine	16	16.011707		$\sim 10^{-19}$ s	
		17	17.002098		66 s	e^+
		18	18.000950		111 m	e^+
		19	18.998405	100		
		20	19.999986		11 s	e^-, γ
		21	20.999972		5 s	e^-
10	Neon	18	18.005715		1.46 s	e^+, γ
		19	19.001892		18 s	e^+
		20	19.992440	90.92		
		21	20.993849	0.257		
		22	21.991384	8.82		
		23	22.994475		38 s	e^-, γ
		24	23.993597		3.38 m	e^-, γ

*Carbon-12 scale.
†s = second, m = minute, y = year.

The loss in mass accompanying the formation of a heavier atom from hydrogen atoms and neutrons shows that these reactions are strongly exothermic. A very large amount of energy is evolved in the formation of the heavier atoms from hydrogen atoms and neutrons, an amount given by the Einstein equation $E = mc^2$. The more stable the heavy nucleus, the larger is the decrease in mass from that of the neutrons and protons from which the nucleus may be considered to be made.

The decrease in mass accompanying the reaction of two hydrogen atoms and two neutrons to form a ^{4}He atom is 0.030376 d, which is equal to 28.294 MeV. It is customary to express the binding energy in MeV per nucleon (7.073 MeV per nucleon for ^{4}He). The binding energy per nucleon as a function of the number of nucleons (the mass number) is shown for some stable nuclides in Figure 26-6. It is seen that the elements of the first long period of the periodic table, between chromium and zinc, lie at the maximum of the curve, and can accordingly be considered to be the most stable of all the elements. If one of these elements were to be converted into other elements, the total mass of the other elements would be somewhat greater than that of the reactants, and accordingly energy would have to be added in order to cause the reaction to occur. On the other hand, either the heavier or the lighter elements could undergo a nuclear reaction to form the elements with mass numbers in the neighborhood of 60, and these nuclear reactions would be accompanied by the evolution of a large amount of energy.

Scientists have attempted to develop a theory of the origin of nuclear species on the basis of the extensive information now available about nuclear reactions. One idea is that the elements have been produced by synthesis from hydrogen by a succession of neutron captures interspersed where necessary by decrease in atomic number through β decay. There is convincing astronomical evidence that the universe is expanding. The light from distant galaxies contains spectral lines that can be identified, but their frequencies are not those observed in the laboratory; instead, there is a shift in wavelength to the red (the red shift). The same fractional shift in wavelength is observed for all of the spectral lines and for the continuum in the optical spectrum and also for all radio waves emitted by a particular galaxy. This fact is the basis for the belief that the red shift is due to the Doppler effect (the dependence of observed frequency on the relative velocities of emitter and observer) and that the distant galaxies are receding from us. It was discovered by the American astronomers Hubble and Humason about 40 years ago that the red shift is greatest for the most distant galaxies. The magnitude of the velocity of the galaxies as deduced from the red shift and distances of the galaxies fixes the time of creation of the universe at about 15×10^9 years ago.

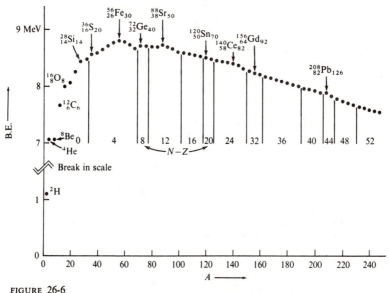

FIGURE 26-6

The binding energy per nucleon in stable even-even nuclei with A a multiple of 4.

The American scientist George Gamow postulated that at that time, the beginning, the universe consisted of a huge ball of neutrons bathed in radiation, which immediately began to expand because of its great internal energy. Some of the neutrons then began to decay to form protons, electrons, and neutrinos, liberating 0.78 MeV of energy per neutron. The protons could then capture neutrons to form deuterons, and, in the neutron-capture theory of the origin of nucleides, the process of neutron capture would continue, and would build up the distribution of nucleides that is observed.

There are, however, some difficulties about this theory. One is that there are no stable nucleides with mass 5 or mass 8, and thus no synthesis of the elements beyond these masses through neutron capture alone is possible; the synthesis stops when all of the hydrogen has been turned into helium 4.

An alternative theory of nucleide synthesis is that this synthesis has taken place and is still taking place in the center of stars. This theory has been supported principally by the British astrophysicist Fred Hoyle. The problem of the instability of the nucleides with mass 5 and mass 8 is overcome by way of reactions such as the following:

$$3\ ^4\text{He} \underset{\longleftarrow}{\overset{\longrightarrow}{\rightleftarrows}} {}^{12}_{6}\text{C}^* \longrightarrow {}^{12}_{6}\text{C} + \gamma$$

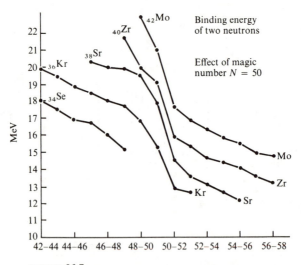

FIGURE 26-7
The energy of binding of two neutrons for nuclei in the
region 42 to 58 for N, showing the decreased binding
energy for $N > 50$.

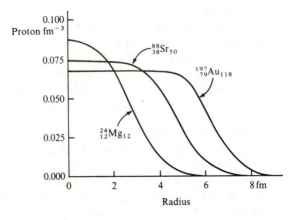

FIGURE 26-8
Proton density distribution for three nuclei, determined
from the observed scattering of high-energy electrons
(100 to 250 MeV). The neutron density function is nearly
the same (with the factor N/Z).

At a temperature of 100 million degrees and density of 10,000 g cm^{-3} in the center of a star, there is an equilibrium involving three alpha particles and an excited state of the carbon-12 nucleus, with energy 7.653 MeV greater than the normal state of the nucleus. The excited ^{12}C nucleus can change to the normal state by emission of a photon. Various other known nuclear reactions can then lead to the synthesis of all of the heavier nucleides.

Magic Numbers

In 1934 W. M. Elsasser, then working in France, discussed evidence showing that there is a special stability associated with certain numbers of protons and certain numbers of neutrons. These numbers, which are known as *magic numbers*, are 2, 8, 20, 28, 50, 82, and 126. The numbers can be correlated with the subshell numbers that have been developed in the discussion of the electrons in atoms. The magic number 2 is, of course, the number of fermions that can occupy a 1s orbital. The magic number 8 may be described as the number of fermions occupying a 1s orbital and the three 2p orbitals; in nuclei it is to be expected that a particle in a 2p orbital would be more stable than in a 2s orbital. The number 20 similarly corresponds to pairs of fermions occupying the 1s orbital, the three 2p orbitals, the 2s orbital, and the five 3d orbitals. The larger magic numbers presented a puzzle that was solved fourteen years later (Sections 26-8 and 26-9). The first evidence for the magic numbers was provided by the values of the binding energy and properties closely related to it. An example, for magic number 50, is shown in Figure 26-7, which represents the binding energy for pairs of neutrons in the neighborhood of $N = 50$. It is seen that the binding energy drops sharply at $N = 50$; the value for $50 \rightarrow 52$ is much less than for $48 \rightarrow 50$.

Radii of Nuclei

The results of the experiments on the scattering of helions (alpha particles) by gold foil were interpreted by Rutherford and his coworkers (Section 3-4) as showing that the interaction of the helion and the heavier nucleus shows no deviation from Coulomb repulsion at distances larger than about 10 fm. Other experiments have led to rather accurate values of the sizes of nuclei and to knowledge of the probability distribution function of nucleons within the nuclei. The scattering of high-energy electrons, studied especially by the American physicist Robert Hofstadter (born 1915) and his associates, has led to results such as those shown in Figure 26-8. The nucleonic density is found to be constant, with value about 0.17 nucleons per fm³, throughout the central region of every nucleus (except the lightest ones); it then drops to zero over the radial increment 2 fm (from density 90% to 10% of the maximum). The radius (measured to density 50% of maximum) is proportional to the cube root of the nucleon number:

$$R = 1.07 \, A^{1/3} \text{ fm} \tag{26-1}$$

The observed constant nucleonic density suggests that the nucleons are packed together in a nucleus in a way resembling the packing of molecules in a liquid, rather than in a gas (see Section 26-8).

Nuclear Spins and Magnetic Moments

The angular momentum of a nucleus has the value $\{I(I + 1)\}^{1/2} \hbar$, in which I is half-integral ($I = \frac{1}{2}, \frac{3}{2}, \ldots$) for A odd and integral ($I = 0, 1, 2, \ldots$) for A even. The component of angular momentum in the direction of a magnetic field has the allowed values $M_I \hbar$, with $M_I = -I, -I + 1, \ldots, + I$; M_I is called the nuclear spin magnetic quantum number. The quantum numbers I and M_I are analogous to J and M_J for an atom (Section 3-8; Chapter 5).

Values of I and of the parity are given for the normal states of some nuclei in Table 26-2. Positive parity is indicated by the superscript $+$ and negative parity by the superscript $-$. Positive parity corresponds to no change in the wave function of the nucleus when all nucleonic coordinates change sign, and negative parity to multiplication of the wave function by -1 when all change sign. (The wave function is said to be symmetric with respect to inversion about the center of mass of the nucleus for positive parity, and antisymmetric for negative parity.) The relation of parity to the shell model is discussed in the following section.

All even-even nuclei (Z and N both even) have $I = 0$ and positive parity in the normal state.

Observed values of the magnetic moment of some nuclei in their normal states are given in Table 26-2. Two values, μ and μ', are given, defined in the following way:

$$\mu = \{I(I + 1)\}^{1/2} \mu_N \, g \tag{26-2}$$
$$\mu' = I \mu_N \, g \tag{26-3}$$

Here μ_N, the *nuclear Bohr magneton*, is defined by the equation

$$\mu_N = \frac{2\pi \hbar e}{M_p} \times 10^{-7} = 6.347 \times 10^{-33} \text{ weber meter} \tag{26-4}$$

The nuclear Bohr magneton is defined in the same way as the electronic

TABLE 26-2
*Spin, Parity, and Magnetic Moment for Nuclei in their Normal States**

	I^p	μ	μ'	g
$^1_0n_1$	$\frac{1}{2}^+$	-3.31366	-1.91314	-3.82628
$^1_1p_0$	$\frac{1}{2}^+$	4.83722	2.79277	5.58554
$^2_1H_1$	1^+	1.21255	0.85740	0.85740
$^3_1H_2$	$\frac{1}{2}^+$	5.1594	2.9788	5.9576
$^3_2He_1$	$\frac{1}{2}^+$	-3.6849	-2.1275	-4.2550
$^6_3Li_3$	1^+	1.1625	0.8220	0.8220
$^7_3Li_4$	$\frac{3}{2}^-$	4.2039	3.2563	2.1709
$^9_4Be_5$	$\frac{3}{2}^-$	-1.5200	-1.1774	-0.7849
$^{10}_5B_5$	3^+	2.0792	1.8006	0.6002
$^{11}_5B_6$	$\frac{3}{2}^-$	3.4710	2.6886	1.7924
$^{13}_6C_7$	$\frac{1}{2}^-$	1.21656	0.70238	1.4048
$^{14}_7N_7$	1^+	0.5708	0.4036	0.4036
$^{15}_7N_8$	$\frac{1}{2}^-$	-0.49031	-0.28308	-0.56616
$^{17}_8O_9$	$\frac{5}{2}^+$	-2.24066	-1.8937	-0.75748
$^{19}_9F_{10}$	$\frac{1}{2}^+$	3.9194	2.6287	4.5257
$^{22}_{11}Na_{11}$	3^+	2.0150	1.745	0.5817
$^{23}_{11}Na_{12}$	$\frac{3}{2}^+$	2.8628	2.2175	1.4783
$^{24}_{11}Na_{13}$	4^+	1.89	1.69	0.423
$^{25}_{12}Mg_{13}$	$\frac{5}{2}^+$	-1.0118	-0.8551	-0.3420
$^{27}_{13}Al_{14}$	$\frac{5}{2}^+$	4.3086	3.6414	1.4566

*See text for meaning of μ and μ'.

Bohr magneton (Equation 3-18), except with the proton mass, M_p, replacing the electron mass, m. The magnetic moment μ is defined in the way that is customary in the discussion of the magnetic properties of substances (Appendix XIV). The usual practice of nuclear physicists is, however, to refer to μ', the maximum component of the nuclear magnetic moment in the field direction, as the magnetic moment of the nucleus, and it is μ', rather than μ, that is given in the tables of nuclear moments in reference books.

Nuclear spin was discovered in the course of the analysis of atomic spectra. It was observed, for example, that the line at 3596 Å emitted by ^{209}Bi has a hyperfine structure consisting of six lines with wavelengths from 3595.952 Å to 3596.256 Å. W. Pauli in 1924 suggested that these lines could result from the interaction of the angular momentum J of the electrons and the angular momentum I of the nucleus. The observed hyperfine structure was found to correspond to the nuclear spin $I = \frac{9}{2}$

for the ^{209}Bi nucleus. The spin $I = 1$ for the deuteron was discovered by
the American physicist I. I. Rabi by a molecular-beam method (a modification of the Stern-Gerlach experiment, Section 3-8), which permitted
also the determination of the value of the magnetic moment. Most of
the measured values of nuclear magnetic moments have been obtained
by the nuclear magnetic resonance technique (NMR). Nuclei are oriented
in a magnetic field (Figure 3-29), and the wavelength of photons in the
radio or microwave region with energy $h\nu$ enough to change M_I by one
unit is determined. The usual method is to keep the photon frequency
constant and to scan by varying the strength of the magnetic field.

The magnetic moment associated with the spin of the electron (Section
3-8) has the value corresponding to $g = 2.00232$. The value 2 is expected
for an elementary particle on the basis of the Dirac theory, and the small
deviation from this value has been accounted for as resulting from the
interaction of the electron and the electromagnetic (photon) field. It was
mentioned in Section 3-8 that the value of g for the proton (5.58554)
deviates greatly from the value 2 expected for a simple particle, and indicates that it has a complex structure, such as the triquark structure discussed in Section 25-11. The same conclusion is reached from the observed
deviation of g for the neutron from the value 0 expected for a simple
neutral particle.

26-8. The Shell Model of Nuclear Structure

In 1933 the American physicist J. H. Bartlett, Jr. (born 1904) suggested
that the protons and the neutrons in a nucleus could be assigned to orbitals
(about the center of mass) resembling the electron orbitals in an atom.
This orbital model of the nucleus accounts reasonably well for the smaller
magic numbers 2, 8, and 20: ^{4}He is assigned the configuration $1s^2$ for both
neutrons and protons, ^{16}O the configuration $1s^2 1p^6$, and ^{40}Ca the configuration $1s^2 1p^6 1d^{10} 2s^2$. (The principal quantum number is conventionally
taken as $n = 1, 2, 3, \ldots$ for each value of l, for nucleonic orbitals, rather
than $n = l + 1, l + 2, \ldots$, the convention for electronic orbitals.) The
other magic numbers, 28, 50, 82, and 126, were finally interpreted by a
refinement of the orbital model, called the *shell model*, that was developed
by the American physicist Maria Goeppert Mayer (born 1906) and the
German physicist J. Hans D. Jensen (born 1907) and his collaborators
in 1948.

The characteristic feature of the shell model is the assumption that each
nucleon couples its orbital angular momentum vector (with quantum
number l) and its spin vector (with quantum number $s = \frac{1}{2}$) to form a
resultant spin-orbit angular momentum vector, with quantum number j

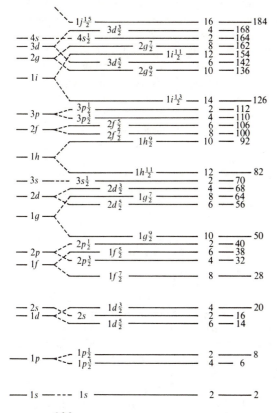

FIGURE 26-9
Sequence of energy levels for spin-orbit coupling,
$j = l + 1/2$ or $j = l - 1/2$, for protons and neutrons,
as given by the Mayer-Jensen shell model of nuclear
structure.

equal to $l + \frac{1}{2}$ or $l - \frac{1}{2}$. The total angular momentum of the nucleus, quantum number I, is the resultant of the j-vectors for all the nucleons. This sort of coupling of angular momenta is called jj coupling (compare Russell-Saunders coupling, Section 5-3). The nucleonic subsubshell with $j = l + \frac{1}{2}$ lies below that for $j = l - \frac{1}{2}$, as shown in Figure 26-9. It is seen that each of the larger magic numbers corresponds to completed shells plus a subsubshell, $(1f7/2)^8$, for example, for 28. There are, of course, $2j + 1$ protons or neutrons in a completed j subsubshell, corresponding to the $2j + 1$ values of m_j (from $-j$ to $+j$).

The shell model leads rather directly to the observed values of the spin and parity of most nuclei. The nuclei ${}^{17}_{8}O_9$ and ${}^{17}_{9}F_8$, for example, have an odd (unpaired) nucleon outside the completed-shell structure of ${}^{16}_{8}O_8$. From Figure 26-9 we assign this nucleon to $1d\frac{5}{2}$. Both ${}^{17}O$ and ${}^{17}F$ have $I^P = \frac{5}{2}^+$ for the normal state, and $\frac{1}{2}^+$ (corresponding to $2s$) as the first excited state. (Note that $s, d, g, \ldots$ orbitals have even parity and $p, f, \ldots$ have odd parity.) The nuclei ${}^{41}_{20}Ca_{21}$ and ${}^{41}_{21}Sc_{20}$, with an odd nucleon in $1f\frac{7}{2}$, have $I^P = \frac{7}{2}^-$ in the normal state.

Some properties are not satisfactorily accounted for by the simple shell model. An example is the magnetic moment. Thus for the normal state of ${}^{7}_{3}Li_4$ the value of the magnetic moment calculated (see Appendix VI) for a proton in a $1p\frac{3}{2}$ orbital is 4.90 nuclear magnetons, whereas the experimental value is 4.20 μ_N. The value calculated for a triton (${}^3H$) moving around a helion (next Section) is 4.27 μ_N, in much better agreement with experiment.

26-9. The Helion-Triton Model

Another useful model of nuclear structure, called the alpha-particle model or the helion-triton model, is based on the assumption that the nucleons in a nucleus can be considered to be grouped together into helions or tritons, occupying localized $1s$ orbitals. For ${}^{16}O$, for example, the 8 protons and the 8 neutrons could be described as forming four helions, which are arranged at the corners of a tetrahedron. In the shell model the protons and the neutrons would be described as occupying the $1s$ orbital and the three $1p$ orbitals. These four orbitals can be hybridized, as described in Appendix VII, to form four localized tetrahedral orbitals, each of which is concentrated about one of the four corners of a tetrahedron.

Many properties of nuclei can be simply explained on this basis. An example is the magnetic moment of ${}^7Li$, mentioned at the end of Section 26-8. The observed magnetic moment is approximately equal to the value expected for a triton moving about a helion (with charge-mass ratio 1/3) rather than for a moving proton (with charge-mass ratio 1/1). Another example is provided by ${}^{20}Ne$. The properties of this nucleus and related nuclei (such as ${}^{21}Ne$ and ${}^{21}Na$), indicate that the ${}^{20}Ne$ nucleus has a prolate deformation from spherical shape; that is, it is elongated. The structure expected for an aggregate of five helions is the trigonal bipyramid, five helions in positions corresponding to those of the fluorine atoms in phosphorus pentafluoride (Figure 11-5), in agreement with the observed prolate deformation.

Nuclei with a larger number of helions and tritons might be expected to have a structure in which one helion or triton is at the center of the nucleus, and the others are arranged in a layer about it. This structure would, of course, be expected for 13 spheres—a close-packing of 12 spheres around

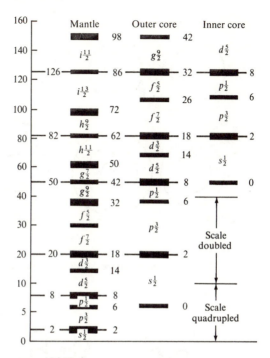

The sequence of nucleon energy levels (overlapping ranges), with assignment to successive layers (inner core, outer core, mantle) on the basis of the principal quantum number.

a central sphere. A layer structure can be assigned to nuclei, with the helions and tritons placed in an inner core, an outer core layer, and an outermost layer, called the mantle, by assigning the nucleonic orbitals to the successive layers in the way indicated in Figure 26-10. This assignment is based on the shell-model sequence of subsubshells, with the assumption that a subsubshell that occurs with only one value of the principal quantum number contributes to the outer layer of helions and tritons, one with two values of the principal quantum number (such as 1s and 2s) contributes to the mantle and the next inner layer, and one with three values of the principal quantum number contributes to the mantle, the outer core, and the inner core.

The shell-model diagrams for the magic numbers, with shells and subsubshells assigned to the different layers of the nucleus, are shown in Figure 26-11. The numbers of helions and tritons in these successive layers approximate the numbers for close-packed aggregates of spheres.

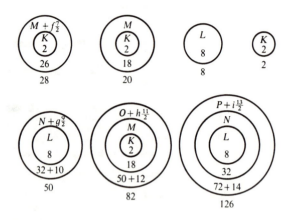

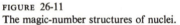

FIGURE 26-11
The magic-number structures of nuclei.

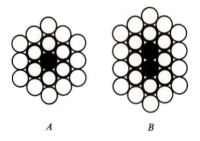

FIGURE 26-12
Two-dimensional representation of packing of
spheres around (A) one central sphere and (B) two
central spheres. In three dimensions icosahedral
packing of spheres of nearly the same size
(within 10%) places 12 in the second layer about
1 central sphere and 32 in the outer layer.

Prolate Deformation of Heavy Nuclei

The observed properties of many of the heavier nuclei have been interpreted as showing that the nuclei are not spherical but are permanently deformed. The principal ranges of deformation are from neutron number 90 to 116 and from 140 on. Most of the deformed nuclei are described as prolate ellipsoids of revolution, with major radii 20% to 40% larger than the minor radii.

The helion-triton model provides a simple explanation of the onset of prolate deformation at neutron number 90. This value corresponds to 45 helions and tritons. The magic number $N = 82$ corresponds to 41, which, from Figures 26-10 and 26-11, are assigned to the successive layers in the following way: 1 in the inner core, 9 in the outer core, and 31 in the mantle. The closest packing of spheres of approximately equal size is 1 sphere in the inner core, 12 in the outer core, and 32 in the mantle, a total of 45, as indicated in Figure 26-12A. An additional sphere would have to be placed on the surface of this nearly spherical aggregate, or be introduced into the inner core, as indicated in Figure 26-12B. We conclude that at neutron number 90 an extra pair of neutrons is introduced in the core (occupying a $p3/2$ orbital), causing the core to contain two tritons or helions, and thus leading to prolate deformation of the nucleus as a whole.

26-10. Nuclear Fission and Nuclear Fusion

The instability of the heavy elements relative to those of mass number around 60, as shown by the binding-energy curve, suggests the possibility of spontaneous decomposition of the heavy elements into fragments of approximately half-size. This fission has been accomplished.

It was reported on January 6, 1939, by the German physicists O. Hahn and F. Strassmann that barium, lanthanum, cerium, and krypton seemed to be present in substances containing uranium that had been exposed to neutrons. Within two months more than forty papers were then published on the fission of uranium. It was verified by direct calorimetric measurement that a very large amount of energy is liberated by fission—over 20×10^{12} J per mole. Since a kilogram of uranium contains 4.26 gramatoms, the complete fission of 1 kg of this element, or a similar heavy element, produces about 0.8×10^{14} J. This may be compared with the heat of combustion of 1 kg of coal, which is approximately 4×10^7 J. Thus uranium as a source of energy is 2 million times more valuable than coal.

Uranium 235 and plutonium 239, which can be made from ^{238}U, are

²³⁵U + neutron

²³⁶U

Fission products:

2 nuclei of mass
numbers 85 to 105
and 130 to 150 and
one to three neutrons

FIGURE 26-13
Nuclear fission of ²³⁵U.

capable of undergoing fission when exposed to slow neutrons. It was also shown by the Japanese physicist Y. Nishina in 1939 that the thorium isotope ²³²Th undergoes fission under the influence of fast neutrons.

It has been customary to use the *megaton* as a unit of energy released in nuclear fission (or fusion, discussed below). One megaton is equal to 8×10^{15} J, the energy of explosion of one million tons of the ordinary explosive TNT. The energy of fission of 100 kg of uranium or plutonium is one megaton.

Uranium and thorium have become important sources of heat and energy. There are large amounts of these elements available—the amount of uranium in the earth's crust has been estimated as 4 parts per million and the amount of thorium as 12 parts per million. The deposits are distributed all over the world.

The fission reactions can be chain reactions. These reactions are initiated by neutrons. A nucleus ²³⁵U, for example, may combine with a neutron to form ²³⁶U. This isotope is unstable, and undergoes spontaneous fission, into two smaller particles; the protons in the ²³⁶U nucleus are divided between the two daughter nuclei (Figure 26-13). These daughter nuclei also contain most of the neutrons originally present in the ²³⁶U nucleus. Since, however, the ratio of neutrons to protons is greater in the heavier nuclei than in those of intermediate mass, the fission is also accompanied by the liberation of a few free neutrons. The neutrons that are thus liberated may then combine with other ²³⁵U nuclei, forming additional ²³⁶U nuclei, which themselves undergo fission. A reaction of this sort, the products of which cause the reaction to continue, is called a chain reaction or an autocatalytic reaction.

The fission of some nuclei, such as those in the neighborhood of ^{208}Pb (gold, thallium, lead, bismuth), is symmetric. When one of these elements is bombarded with 20-MeV deuterons it undergoes fission to give two daughter nuclei with nearly the same values of Z and N, the distribution function having a half-width at half maximum of about 15 mass-number units. The heavier nuclei show unsymmetric fission, as illustrated in Figure 26-13 for ^{236}U.

If a mass of ^{235}U or ^{239}Pu weighing a few kilograms and in a suitable form, such as a hollow sphere, is suddenly compressed into a small volume, the autocatalytic fission of the nuclei occurs nearly completely, and the amount of energy 0.01 megaton per kilogram that undergoes fission is released. An ordinary *atomic bomb* (nuclear bomb) consists of a few kilograms of ^{235}U or ^{239}Pu and a mechanism for suddenly compressing the metal. Each of the atomic bombs exploded over Hiroshima and Nagasaki in August 1945 had explosive energy of about 0.020 megatons (20 kilotons). Modern nuclear weapons involving fission alone have explosive energy 0.001 megaton to 0.1 megaton.

The process of *nuclear fusion* also may liberate energy. From the binding-energy diagram we see that the fission of a very heavy nucleus converts about 0.1% of its mass into energy. Still larger fractions of the mass of very light nuclei are converted into energy by their fusion into heavier nuclei. The process 4H $\longrightarrow$ He, which is the principal source of the energy of the sun, involves the conversion of 0.7% of the mass into energy. The similar reaction of a deuteron and a triton to form a helium nucleus and a neutron is accompanied by the conversion of 0.4% of the mass into energy:

$$^{2}_{1}H + ^{3}_{1}H \longrightarrow ^{4}_{2}He + ^{1}_{0}n$$

It was found by experiment (in 1952) that these materials surrounding an ordinary atomic bomb undergo reaction at the temperature of many millions of degrees produced by the reaction. Tritium is, however, inconvenient and expensive to use because it is radioactive and unstable (half-life 12 years). In 1953 it was shown that a fission-fusion bomb could be made by placing some of the stable solid substance *lithium deuteride*, LiD, about an ordinary fission bomb. Some of the reactions that occur are the following:

$$^{2}D + ^{2}D \longrightarrow ^{3}He + n$$
$$^{2}D + ^{2}D \longrightarrow ^{3}H + p$$
$$n + ^{6}Li \longrightarrow ^{4}He + ^{3}H$$
$$^{3}H + D \longrightarrow ^{4}He + n$$
$$p + ^{7}Li \longrightarrow ^{4}He + ^{4}He$$

The amount of energy released in the fusion of lithium deuteride is about

60 megatons per ton of the material undergoing fusion, as compared with 10 megatons per ton of uranium undergoing fission. The largest nuclear bomb so far exploded, the Soviet bomb of November 1961, was a fission-fusion bomb with explosive energy of about 60 megatons, about 10 times the total for all bombs used in the Second World War.

The standard present-day nuclear weapons are three-stage fission-fusion-fission bombs (superbombs). An ordinary 20-megaton superbomb (50:50 fission-fusion) has as its first stage material (the detonator) a few kilograms of plutonium, with some ordinary explosive to compress it suddenly. The second-stage material is about 150 kg of lithium deuteride, which is surrounded by a shell of ordinary uranium metal (^{238}U, the third-stage material) weighing somewhat more than 1000 pounds. The ^{238}U undergoes fission through reaction with fast neutrons. The total weight of a 20-megaton bomb is about 1.5 tons.

The *manufacture of plutonium* is carried out by a controlled chain reaction. A piece of ordinary uranium contains 0.71% of ^{235}U. An occasional neutron strikes one of these atoms, causing it to undergo fission and release a number of neutrons. The autocatalytic reaction does not build up, however, if the piece of uranium is small, because the neutrons escape, and some of them may be absorbed by impurities, such as cadmium, the nuclei of which combine very readily with neutrons.

However, if a large enough sample of uranium is taken, nearly all of the neutrons that are formed by the fission remain within the sample of uranium, and either cause other ^{235}U nuclei to undergo fission, or are absorbed by ^{238}U, converting it into ^{239}U, which then undergoes spontaneous change to ^{239}Pu. This is the process used in practice for the manufacture of plutonium. A large number of lumps of uranium are piled together, alternately with bricks of graphite, in a structure called a reactor. The first uranium reactor ever constructed, built at the University of Chicago and put into operation on December 2, 1942, contained 5600 kg of uranium metal. Cadmium rods were held in readiness to be introduced into cavities in the reactor, and to serve to arrest the reaction by absorbing neutrons, whenever there was danger of its getting out of hand.

The large reactors that were put into operation in September, 1944, at Hanford, Washington, were of such size as to permit the fission reaction to proceed at the rate corresponding to an output of energy of 1,500,000 kilowatts.

The significance of the uranium reactors as a source of radioactive material can be made clear by a comparison with the supply of radium. About 1000 curies (1000 grams) of radium has been separated from its ores and put into use, mainly for medical treatment. The rate of operation mentioned above for the reactors at Hanford represents the fission of about 5×10^{20} nuclei per second, forming about 10×10^{20} radioactive atoms. The concentration of these radioactive atoms will build up until they are undergoing decomposition at the rate at which they are being formed. Since 1 curie corresponds to 3.70×10^{10} disintegrating atoms per second, these reactors develop a radioactivity of approximately 3×10^{10} curies—that is, about thirty million times the radioactivity of all the radium that has been so far isolated from its ores.

The foregoing calculation illustrates the great significance of the fissionable elements as a source of radioactive material. Their significance as a source of energy has also been pointed out, by the statement that 1 kg of uranium or thorium is equivalent to 2 million kg of coal. When we remember that uranium and thorium are not rare elements, but are among the more common elements—the amount of uranium and thorium in the the earth's crust being about the same as that of the common element lead—we begin to understand the promise of nuclear energy for the world of the future, and the possibility of its great contributions to human welfare, if civilization is not brought to an end by war. I believe that the discovery of the controlled fission of atomic nuclei and controlled release of atomic energy is the greatest discovery that has been made since the controlled use of fire was discovered by primitive man.

APPENDIX I
Units of Measurement

Basic IS Units

Physical quantity	Name of unit	Symbol for unit
length	meter	m
mass	kilogram	kg
time	second	s
electric current	ampere	A
thermodynamic temperature	degree Kelvin	K or °K
luminous intensity	candela	cd

*Supplementary Units**

plane angle	radian	rad
solid angle	steradian	sr

*These units are dimensionless.

*Fractions and Multiples**

Fraction	Prefix	Symbol	Multiple	Prefix	Symbol
10^{-1}	deci	d†	10	deka	da†
10^{-2}	centi	c†	10^{2}	hecto	h†
10^{-3}	milli	m	10^{3}	kilo	k
10^{-6}	micro	μ	10^{6}	mega	M
10^{-9}	nano	n	10^{9}	giga	G
10^{-12}	pico	p	10^{12}	tera	T
10^{-15}	femto	f			
10^{-18}	atto	a			

*Compound prefixes should not be used. Thus 10^{-9} meter is represented by 1 nm, not 1 mμm. The attaching of a prefix to a unit in effect constitutes a new unit, so that 1 km^2 = 1 (km)^2 = 10^6 m^2, not 1 k(m^2) = 10^3 m^2.

†To be restricted as much as possible.

Derived IS Units with Special Names

Physical quantity	Name of unit	Symbol for unit	Definition of unit
energy	joule	J	$kg\ m^2\ s^{-2} = N\ m$
force	newton	N	$kg\ m\ s^{-2} = J\ m^{-1}$
power	watt	W	$kg\ m^2\ s^{-3} = J\ s^{-1}$
electric charge	coulomb	C	A s
electric potential difference	volt	V	$kg\ m^2\ s^{-3}\ A^{-1} = J\ A^{-1}\ s^{-1}$
electric resistance	ohm	Ω	$kg\ m^2\ s^{-3}\ A^{-2} = V\ A^{-1}$
electric capacitance	farad	F	$A^2\ s^4\ kg^{-1}\ m^{-2} = A\ s\ V^{-1}$
magnetic flux	weber	Wb	$kg\ m^2\ s^{-2}\ A^{-1} = V\ s$
inductance	henry	H	$kg\ m^2\ s^{-2}\ A^{-2} = V\ s\ A^{-1}$
magnetic flux density	tesla	T	$kg\ s^{-2}\ A^{-1} = V\ s\ m^{-2}$
luminous flux	lumen	lm	cd sr
illumination	lux	lx	$cd\ sr\ m^{-2}$
frequency	hertz	Hz	cycle per second
customary temperature, t	degree Celsius	°C	$t\ °C = T\ °K - 273.15°$

Examples of Other Derived IS Units

Physical quantity	IS unit	Symbol for unit
area	square meter	m^2
volume	cubic meter	m^3
density	kilogram per cubic meter	$kg\ m^{-3}$
velocity	meter per second	$m\ s^{-1}$
angular velocity	radian per second	$rad\ s^{-1}$
acceleration	meter per second squared	$m\ s^{-2}$
pressure	newton per square meter	$N\ m^{-2}$
kinematic viscosity, diffusion coefficient	square meter per second	$m^2\ s^{-1}$
dynamic viscosity	newton second per square meter	$N\ s\ m^{-2}$
electric field strength	volt per meter	$V\ m^{-1}$
magnetic field strength	ampere per meter	$A\ m^{-1}$
luminance	candela per square meter	$cd\ m^{-2}$

Units to be Allowed in Conjunction with IS

Physical quantity	Name of unit	Symbol for unit	Definition of unit
length	parsec	pc	30.87×10^{15} m
area	barn	b	10^{-28} m^2
	hectare	ha	10^4 m^2
volume	liter	l	10^{-3} m^3 = dm^3
pressure	bar	bar	10^5 N m^{-2}
mass	tonne (metric ton)	t	10^3 kg = Mg
magnetic flux density (magnetic induction)	gauss	G	10^{-4} T
radioactivity	curie	Ci	37×10^9 s^{-1}
energy	electronvolt	eV	1.6021×10^{-19} J

The common units of time, such as hour or year, will persist, and also, in appropriate contexts, the angular degree.

Examples of Units Contrary to IS, with their Equivalents

Physical quantity	Name of unit	Equivalent
length	ångström	10^{-10} m
	inch	0.0254 m
	foot	0.3048 m
	mile	1.60934 km
area	square inch	645.16 mm^2
	square foot	0.092903 m^2
	acre	4046.9 m^2
volume	cubic inch	1.63871×10^{-5} m^3
	cubic foot	0.028317 m^3
mass	pound	0.4535924 kg
	ounce	28.3495 g
	grain	64.799 mg
density	pound/cubic inch	2.76799×10^4 kg m^{-3}
force	dyne	10^{-5} N
	poundal	0.138255 N
pressure	atmosphere	101.325 kN m^{-2}
	torr	133.322 N m^{-2}
energy	erg	10^{-7} J
	calorie (thermochemical)	4.184 J
	British thermal unit	1055.06 J
	kilowatt hour, kW h	3.6 MJ
power	horse power	745.700 W

Values of Some Physical and Chemical Constants
Based on the ^{12}C Scale

Avogadro's number	$N = 0.60229 \times 10^{24}$ mole^{-1}
Velocity of light	$c = 2.997925 \times 10^{8}$ m s^{-1}
Mass of electron	$m = 0.91083 \times 10^{-30}$ kg
Electronic charge	$e = 0.160206 \times 10^{-18}$ C
Electronic charge*	$\epsilon = 15.1880 \times 10^{-15}$ S
Faraday	$F = Ne = 96490$ C mole^{-1}
Dalton	$d = 1.66033 \times 10^{-27}$ kg
Planck's constant	$h = 0.66252 \times 10^{-33}$ J s
Angular-momentum quantum	$\hbar = h/2\pi = 0.105443 \times 10^{-33}$ J s
Mass of proton	$m_p = 1.67239 \times 10^{-27}$ kg
Mass of neutron	$m_n = 1.67470 \times 10^{-27}$ kg
Boltzmann constant	$k = 13.805 \times 10^{-24}$ J deg^{-1}
Gas constant	$R = Nk = 8.3146$ J deg^{-1} mole^{-1}
Gas constant	$R = 0.08206$ l atm deg^{-1} mole^{-1}
Standard molar gas volume	$273.15°\ R = 22.415$ l
Celsius temperature	$t°$ C $=$ T$°$K $- 273.15°$
Atmospheric pressure	1 atm $= 101.325$ kN m^{-2}
Electric dipole moment	1 ϵÅ $= 0.1602 \times 10^{-28}$ C m
	1 ϵÅ $= 4.8029$ Debye
Electron volt	1 eV $= 96.4905$ kJ mole^{-1}

*The stoney (S) is a unit of electric charge such that the force of repulsion of two 1-stoney charges one meter apart is one newton.

Relations Among Energy Quantities

$$1 \text{ eV} = 0.160206 \times 10^{-18} \text{ J}$$
$$1 \text{ eV} = 96.4905 \text{ kJ mole}^{-1}$$
$$1 \text{ eV} = 23.0618 \text{ kcal mole}^{-1}$$
$$1 \text{ erg} = 1 \times 10^{-7} \text{ J}$$
$$1 \text{ liter atm} = 9.869 \times 10^{-3} \text{ J}$$
$$1 \text{ cal} = 4.184 \text{ J}$$

The energy of a photon with wavelength 1Å is 12398 eV $= 1.1963 \times 10^{8}$ J mole^{-1}. The energy of a photon with wave-number 1 cm^{-1} (reciprocal of wavelength in cm) is 1.2398×10^{-4} eV $= 11.963$ J mole^{-1}.

A photon with wavelength 12398 Å, wave-number (λ^{-1}) 8066 cm^{-1}, frequency 2.418×10^{18} Hz has energy 1 eV.

Symmetry of Molecules and Crystals

Symmetry plays an important role in nature. Many structural features of molecules and crystals are governed by considerations of symmetry.

An object has symmetry if some operation can be carried out on it such that, when the operation is completed, the object appears unchanged. For example, a three-bladed propeller can be rotated about its axis through 120°, and it then appears unchanged from its original condition, if the three blades are identical. We shall consider here first the symmetry properties of individual molecules, and then those of aggregates of molecules, in crystals.

Point Symmetry

The symmetry operations that leave one point in space fixed are called *point symmetry* operations. The symmetry of an individual molecule is of this type, since other types of symmetry operations lead in general to successive duplications of the molecule and therefore to a crystal.

The point symmetry operations are (1) rotations around an axis by some fraction of 360°; (2) reflection across a plane; (3) inversion through a point; and (4) rotation around an axis, followed by inversion through a point on the axis. Operations 1, 2, and 4 are illustrated in Figure III-1.

In rotation symmetry, the symmetry element is called an *n-fold rotation axis*, where the symmetry operation consists in rotation through an angle $360°/n$, n being an integer. Linear molecules, such as CO_2, have an ∞-fold rotation axis along their length, meaning that they have complete rotational symmetry about this axis. In reflection symmetry, the symmetry element is called a *mirror plane* or a *plane of symmetry*. The symmetry operation—mirror-reflection across this plane—consists in replacing each atom on one side of the plane by an atom perpendicularly opposite on the other side, at the same distance from the plane. The operation of inversion consists in projecting each atom through a particular point in space, to a location on the opposite side, at the same distance. The point is called a *center of symmetry* if inversion through it leaves the molecule apparently unchanged. The *n-fold rotatory inversion axis* stands for the symmetry operation of rotation through an angle $360°/n$ about the axis, followed by inversion through a particular point on the axis.

All of the point symmetry elements of a molecule—rotation axes, inversion axes, mirror planes, and a center of symmetry, if present—must share a common point, at which the inversion center on any inversion

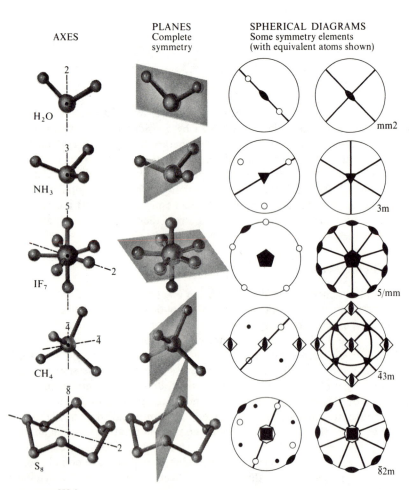

                                  PLANES   SPHERICAL DIAGRAMS
               AXES   Complete   Some symmetry elements
                                  symmetry   (with equivalent atoms shown)

FIGURE III-1
Symmetry elements of individual molecules. On the left, some of the rotation
axes, rotatory inversion axes, and mirror planes for each molecule are identified.
Rotation axes are labeled n (for n-fold), and rotatory inversion axes $\bar{n}$.
On the right, these symmetry elements are plotted on spherical diagrams,
drawn as viewed along the principal axes of each molecule (vertical axis in
drawings on the left). The center of the sphere is indicated by a dot in the
molecular drawings on the left, and atoms that do not lie at this center are
plotted as solid or open circles on the spherical diagrams. On the far right, the
complete collection of symmetry elements for each molecule is shown.

axes must also lie. This common point is, of course, the one that remains fixed under any of the point symmetry operations, as required by the definition of point symmetry.

The effect of the point symmetry on the molecule can best be understood by drawing a sphere around the common point, and plotting on the surface of this sphere a representative atom, plus all of the additional atoms generated from it by operation of the symmetry elements present. Such a spherical diagram is given alongside each of the molecular drawings in Figure III-1. Atoms on the upper hemisphere are represented as open circles, and atoms on the lower hemisphere as solid dots. The symmetry elements are represented where they intersect the surface of the sphere: mirror planes by heavy curves (great circles on the sphere), and rotation or rotatory inversion axes by special symbols placed at the point where each axis intersects the surface of the sphere (see Figure III-1). A center of symmetry cannot be shown in this manner, since it lies at the center of the sphere, but its presence can be inferred from the distribution of atoms on the surface of the sphere.

There are no restrictions on the point symmetry elements that individual molecules can have, except that, for any molecule, the symmetry elements taken together must form a *group* in the mathematical sense. In essence this requires that the collection of symmetry operations be internally consistent. For example, two mirror planes cannot lie at any arbitrary angle to one another, but only at certain angles. They can be perpendicular, and in this case a 2-fold rotation axis is necessarily present along their line of intersection, as in the water molecule (Figure III-1). The theory of the *symmetry groups* plays an important role in molecular spectroscopy and quantum theory, as well as in current ideas about the elementary particles (Chapter 25). The diagrams in the right-hand column of Figure III-1 are representations of the symmetry groups for the molecules shown.

Lattice Symmetry

The stacking of atoms or molecules side by side to build a crystal results in a second type of symmetry, called *translation symmetry*. Because the group of atoms contained in each unit cell is identical, a translation of the crystal by a vector displacement equal to any of the three crystal axes, or any combination of integral multiples of the axes, will leave the structure apparently unchanged, and hence is a symmetry operation.

Translation symmetry is also called *lattice symmetry*, because it is the symmetry shown by the *crystal lattice*. The crystal lattice is the array of points at the corners of all of the unit cells in the crystal, as illustrated in Figure III-2. It depends on the size and shape of the unit cell, but not on the unit-cell contents. There is thus a sharp distinction between the crystal *lattice*, which is simply an array of points showing the translation sym-

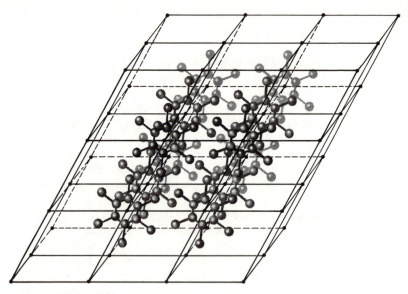

FIGURE III-2

Relationship between crystal lattice and crystal structure in a crystal of hexamethyl benzene, $C_6(CH_3)_6$. Lattice points are shown as dots. Each lattice point lies at the center of a hexamethyl benzene molecule, some of which are shown. Only the carbon atoms of each molecule are depicted. The lattice points are centers of symmetry for the crystal, and hence for the individual molecules. The crystal is triclinic, space group P1.

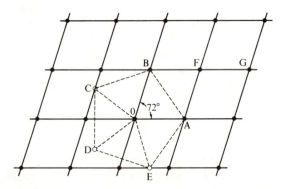

FIGURE III-3

Incompatibility of five-fold rotation axis and lattice symmetry. Starting from point A, the operation of a five-fold axis through O generates points B, C, D, E. Starting from points O, A, and B, the operation of lattice symmetry generates the lattice points (solid dots) at F, G, etc. Points C, D, and E fail to occur as lattice points, and the lattice fails to show five-fold rotation symmetry about O.

metry of the crystal, and the crystal *structure*, which is the complete array of atoms in the crystal. The lattice is important in the theory of x-ray diffraction (Appendix IV).

Crystallographic Point Groups

In crystals, the combined presence of point symmetry and lattice symmetry places restrictions on both. The way the restrictions arise can be illustrated by noting that if a crystal is to have point symmetry elements located at the lattice points, then each point symmetry operation must be a symmetry operation not only for the local grouping of atoms around the lattice points, but also for the complete array of lattice points themselves. This restricts the possible point symmetry operations to those that can be shown by lattices.

A lattice automatically has centers of symmetry at the lattice points, and it can have mirror symmetry. It can show 1-, 2-, 3-, 4-, and 6-fold rotation and rotatory-inversion axes, but no others.

The way in which the other types of axes are excluded is illustrated in the case of the 5-fold axis in Figure III-3. Exclusion of the 5-fold rotation axis and the 8-fold rotation-inversion axis does not mean that molecules such as IF_7 and S_8 (Figure III-1) cannot be incorporated into crystals, but only that when they are, the molecules may become distorted by interaction with surrounding molecules in the crystal, so that their 5- or 8-fold axes will not be rigorously retained.

When these restrictions on the possible symmetry axes are applied, there remain only thirty-two point-symmetry groups that are compatible with lattice symmetry and can therefore occur in crystals. They are called the *crystallographic point groups*. Each group is a self-consistent collection of the allowed point symmetry elements—rotation or rotatory inversion axes of multiplicity 2, 3, 4, or 6, mirror planes, and the inversion center. They are also called the *crystal classes*, because when crystals are classified according to the symmetry shown by their external form (the kinds and numbers of crystal faces), they separate into these thirty-two groups. Some examples from the thirty-two crystallographic point groups are shown in Figure III-4, with the help of spherical diagrams of the type introduced in Figure III-1.

To provide identifying names for the point groups, a system of symbols is used that indicates the basic symmetry elements present. The point-group symbol is listed for each point group alongside the corresponding diagram in Figure III-4, and also in Figure III-1 (where two of the point groups are, however, noncrystallographic). The symbol is constructed as follows. The rotation axes are designated 2, 3, 4, and 6, in accordance with their multiplicity. The rotatory inversion axes are designated $\bar{2}$, $\bar{3}$, $\bar{4}$, and $\bar{6}$; $\bar{1}$ represents a center of symmetry; a mirror plane is *m*. For each point group, the rotation or rotatory inversion axis of highest multiplicity is called the principal axis of the point group, and is listed first in the symbol, followed by secondary axes and mirror planes. A mirror plane parallel to a 4-fold principal axis is indicated by $4m$, and similarly for other axes. A mirror plane perpendicular to a 4-fold axis is designated $4/m$. Except for the cubic point groups, the only secondary axes possible are 2-fold axes, and they always are perpendicular to the principal axis. In the cubic point groups four 3-fold axes are always present, along the diagonals of a cube, and they are regarded as secondary to the three 4-fold or three 2-fold axes parallel to the cube edges. Any symmetry elements not listed in the symbol, but present as part of the point group symmetry, are implied by the symmetry elements listed, and can be determined by constructing the spherical diagram (Figure III-1 or III-4) based on these.

Lattice Types and Crystal Systems

In order that the crystal lattice conform to the point symmetry of the crystal, there are restrictions on the shape of the unit cell, and hence on the cell axes. Crystals are classified into six *crystal systems* on the basis of these restrictions, as follows (see Figure III-4):

1. If no symmetry, or only a center of symmetry, is present there are no restrictions on the axes, and the crystal is *triclinic*. The interaxial angles α, β, and γ are not 90°, and a, b, and c are unequal.
2. A single mirror plane, or a single 2-fold rotation axis, requires one of the crystal axes (conventionally taken as b) to be perpendicular to the mirror plane or parallel to the 2-fold axis. a and c must lie in the plane perpendicular to b, but they can make an arbitrary angle β with each other. The crystal is *monoclinic*.
3. A second mirror plane perpendicular to the first, or likewise a second 2-fold axis, makes the crystal *orthorhombic*. The a, b, and c axes are mutually perpendicular, but may have any lengths.
4. A 4-fold axis makes the crystal *tetragonal*. The 4-fold axis is conventionally taken as c. The cell axes are mutually perpendicular, and a and b are equal.
5. A 3- or 6-fold axis makes the crystal hexagonal. a and b are equal and lie at 120°, while c is perpendicular to them, lying along the 3- or 6-fold axis.
6. Crystals with four 3-fold axes and three 2-fold or three 4-fold axes are *cubic*. There are three equal cell axes of length a, mutually perpendicular, and aligned with the 4-fold or 2-fold rotation axes.

The representative point groups shown in Figure III-4 are classified in accordance with the crystal systems.

Unit cells with axes as specified above satisfy the point-symmetry requirements of the crystal systems, but there are some additional types of cells that do this also. These are conveniently described by adding extra lattice points to cells of the standard shapes described above. These extra lattice points can appear at the center of the cell, or at the center of one or all of the faces (Figure III-4). If there are lattice points at the corners only (as assumed previously), the cell is called *primitive* (designated *P*). If a lattice point appears at the center of the cell, it is called *body-centered* (*I*). If one face of the cell is centered, the cell is *end-centered* (*C*), and if all faces are centered, it is *face-centered* (*F*). In the hexagonal system, a cell can occur having three equal axes *a* making equal angles α with one another. It is called the rhombohedral cell (*R*), and can be derived from the conventional hexagonal cell by adding two extra lattice points, equally spaced along one of the long diagonals of the cell. The complete collection of fourteen possible unit cell types is shown in Figure III-4.

Note that the structure of copper (Figure 2-3) is based on a face-centered cubic cell, with a copper atom at each lattice point.

Space Symmetry

If the point symmetry of a molecule is the same as one of the crystallographic point groups, such molecules can be stacked together in a lattice to form a crystal of the same symmetry. There will be one molecule around each lattice point, which will be the point common to the point symmetry elements of the molecule (see Figure III-2). At the same time, these symmetry elements will be valid for the entire crystal. The symmetry represented by such an array of symmetry elements, comprising lattice symmetry plus certain rotations, reflections, etc., about the lattice points, is known as *space symmetry*. The three-dimensional array of symmetry elements itself is known as a *space group*, the group concept signifying, again, that the collection of symmetry elements is complete and internally consistent. Figure III-5a shows the array of symmetry elements that results from combining point symmetry 2 with a lattice of type *P*. The space group is identified symbolically by combining the symbol for the lattice type with the point group symbol, to give the space-group symbol *P*2 for the group represented in Figure III-5a. When the thirty-two point groups are combined in this way with the allowed types of space lattice there result seventy-three space groups.

The possible types of space symmetry are, however, more numerous than this, owing to the existence of other *space symmetry operations* beside the pure translations and pure rotations, reflections, and so forth, considered up to this point. These additional operations are ones that combine translation with rotation or reflection, and are called *screw axes* and *glide planes*.

FIGURE III-4
The crystal systems.

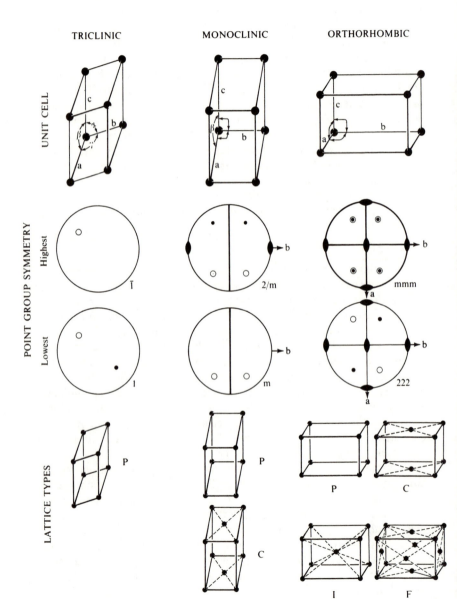

TETRAGONAL HEXAGONAL CUBIC

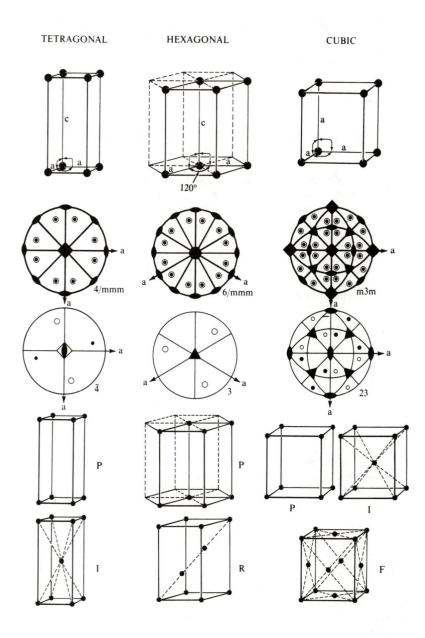

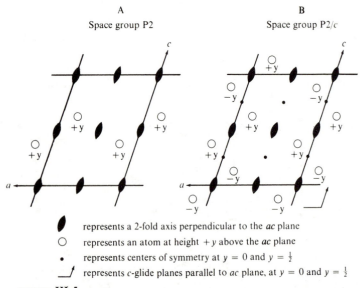

A
Space group P2

B
Space group P2/c

represents a 2-fold axis perpendicular to the *ac* plane

○ represents an atom at height $+y$ above the *ac* plane

• represents centers of symmetry at $y = 0$ and $y = \frac{1}{2}$

represents *c*-glide planes parallel to *ac* plane, at $y = 0$ and $y = \frac{1}{2}$

FIGURE III-5
Diagrams of space symmetry. The unit cells (outlined) are monoclinic, viewed in the direction of the *b* axis. The diagrams show the location of symmetry elements in the cell, and an array of atoms related to one another by operation of the symmetry elements.

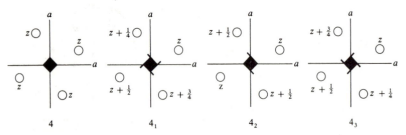

FIGURE III-6
Atoms related by four-fold rotation axes and screw axes. The axes are normal to the plane of the diagram (*aa* plane of a tetragonal crystal), and the heights of the atoms above this plane are listed, as fractions of the *c*-axis length.

Operation of a screw axis involves a rotation of $360°/n$ around the axis, coupled with a translation of Lt/n parallel to the axis, L being a lattice vector parallel to the screw axis (L is usually one of the axes of the unit cell); t and n are integers, and the screw axis is labeled with the symbol n_t. Thus a 4_2 axis parallel to the *c* axis of a tetragonal crystal implies that a rotation of 90° around this axis, coupled with a translation of $c/2$ along the axis, is a symmetry operation. Figure III-6 illustrates the symmetries resulting from the three types of 4-fold screw axis, 4_1, 4_2, and 4_3, as well as the pure rotation axis, 4. Figure III-7 shows the 3_1 screw axes that occur in the structure of selenium. A complete listing of the types of axes that can occur in space symmetry is as follows:

$$6, 6_1, 6_2, 6_3, 6_4, 6_5, \bar{6}$$
$$4, 4_1, 4_2, 4_3, \bar{4}$$
$$3, 3_1, 3_2, \bar{3}$$
$$2, 2_1, \bar{2} \ (= m)$$
$$1 \ (= \text{no symmetry}), \bar{1} \ (= \text{symmetry center})$$

The limitation to 6-, 4-, 3-, and 2-fold screw axes arises for reasons similar to those given earlier for pure rotation axes.

A glide plane operates by a reflection across the plane, coupled with a translation of $L/2$ parallel to the plane, L being a lattice vector parallel to the plane. L is usually one of the crystal axes, in which case the glide plane is identified by the corresponding symbol a, b, or c. When a c-glide plane is added to the space symmetry of Figure III-5a, there results the space group shown in Figure III-5b.

The space group of Figure III-5b is designated with the symbol $P2/c$ by combining the lattice symbol P with the symbols for the basic space symmetry elements present, following the same format explained previously for the point groups. In the same way, symbols for all of the possible space groups are formulated.

When the space symmetry elements (rotation axes, rotatory inversion axes, screw axes, mirror planes, glide planes, and symmetry centers) are combined in all possible arrays consistent with lattice symmetry, using lattices of the fourteen Bravais types,* it is found that there result, in total, 230 distinct space groups. Every crystal structure conforms to one or another of these space groups. A listing of the 230 space groups, together with diagrams like Figures III-5 showing the spatial array of symmetry elements, is found in the *International Tables for X-ray Crystallography*. To give an idea of the variety that arises when all of the possible space symmetry elements are taken into consideration, the thirteen space groups of the monoclinic system are listed below.

Point Group	Space Groups
m	Pm, Pc, Cm, Cc
2	$P2$, $P2_1$, $C2$
$2/m$	$P2/m$, $P2_1/m$, $C2/m$,
	$P2/c$, $P2_1/c$, $C2/c$

---

*The fourteen space lattices are called Bravais types after their discoverer, the French physicist A. Bravais (1811–1863).

FIGURE III-7
Structure of selenium. Along the
edges of the unit cell parallel to the
c axis (vertical) are 3_1 screw axes,
and around each of these is a helical
molecule of selenium atoms.

Description of a Crystal Structure

A complete description contains:

(1) Crystal system.
(2) Unit cell axial lengths and interaxial angles.
(3) Space group symbol.
(4) The kinds and numbers of atoms in the unit cell.
(5) The coordinates x, y, z of those atoms in a unique list that is
    sufficient, in combination with the space symmetry elements, to
    specify the positions of all atoms in the unit cell.

Helical Symmetry

Repeated duplication of a group of atoms by a screw axis produces a
pattern called a *helix*. If the atoms are joined by chemical bonds in a
continuous chain from one group to the next, the result is a helical molecule that extends the length of the crystal. This occurs in the crystal
structures of selenium and tellurium, which contain helical molecules of
3_1 or 3_2 symmetry (space group $P3_121$ or $P3_221$), shown in Figure III-7.
It is also possible to construct helical molecules by a symmetry operation
similar to a screw axis, except that the angle of rotation from one group
to the next is not an integral fraction of 360°. This is the most general
type of space symmetry operation—an arbitrary rotation coupled with
an arbitrary translation. Some molecules of great biological importance
have helical symmetry of this type: in particular, the α-helix of proteins
(Figure 24-2) and the helical backbone of the DNA molecule.

X-rays and Crystal Structure

Much of the detailed knowledge that we now have about the atomic architecture of chemical substances has been obtained by studying crystals with x-rays. Early x-ray studies revealed the structures of many elements and simple compounds, and established important basic principles, such as the distinctions between molecular and nonmolecular crystals. They thereby showed that there were great differences in the chemical-bonding forces that hold atoms together in different types of substances. With progress in developing the techniques of x-ray structure analysis it has become possible to determine the structures of more and more complex substances, culminating in the determination of the structures of protein molecules, built up of thousands of atoms. Along with this advance has come increased accuracy in determining the precise distances and angular relations among the atoms in molecules, which has allowed accurate principles of molecular architecture to be formulated.

This appendix presents the basic ideas involved in x-ray structure analysis. To find out the structure of a crystal, there are two fundamental steps: first, to identify the crystal lattice and associated space symmetry (Appendix III); second, to determine the atomic contents of the unit cell.

Diffraction and the Crystal Lattice

Figure IV-1 is a diffraction photograph made by placing a crystal in the beam of x-rays from a high voltage x-ray tube, and recording the diffracted beams on a photographic film placed behind the crystal, in the way shown in Figure IV-2. It is called a *Laue photograph*, because it is made in the way originally suggested by von Laue. As explained in Chapter 3 (Figure 3-24), the diffraction can be interpreted as reflection of the x-ray beam from planes of atoms in the crystal. Since the reflection is "specular" (equal angles of incidence and reflection), the position of each reflected spot on the film in Figure IV-1 depends only on the orientation of the corresponding atomic planes in the crystal. The orientation of these planes in turn depends on the geometry of the crystal lattice (Appendix III).

Each atom in the unit cell, when repeated by translation symmetry in all of the other unit cells of the crystal, forms an array of atoms lying at the points of a lattice identical to the crystal lattice, and each of these identical lattices has an identical array of atomic planes. The geometry of the diffraction pattern thus depends only on the crystal lattice, whereas,

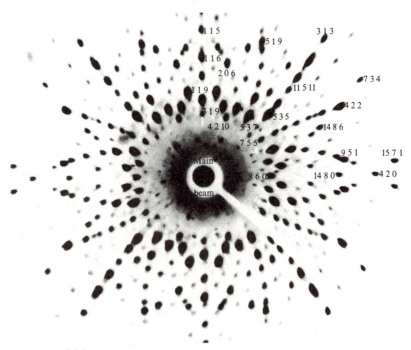

FIGURE IV-1

A Laue x-ray diffraction photograph of a cubic crystal (the mineral zunyite, $Al_{13}Si_5O_{20}(OH, F)_{18} Cl$). The photograph is taken with the incident beam nearly parallel to a two fold axis, making the intersection of two planes of symmetry. (The spots near the center are caused by very small crystallites on the surface of the main crystal.) Crystallographic indices of planes producing some of the reflections are indicated.

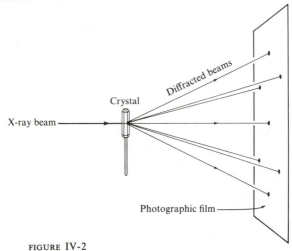
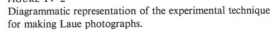

FIGURE IV-2

Diagrammatic representation of the experimental technique for making Laue photographs.

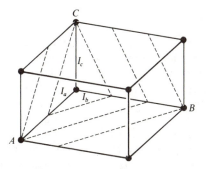

FIGURE IV-3
A unit cell with the traces of
equivalent parallel planes with
indices 231.

in contrast, the intensities of the diffracted beams depend on the number
and arrangement of atoms within the unit cell; that is, on the complete
crystal structure. In analyzing the structure of a crystal by x-rays the
geometry of the diffraction pattern is first used to deduce the size, shape,
and symmetry of the unit cell, and then the diffracted intensities are used
to locate the atoms within the cell.

Geometry of the Lattice Planes

Each reflection in Figure IV-1 can be labeled with a triplet of integers
called *indices*, which identify the corresponding lattice plane. The meaning
of the indices can be understood from Figure IV-3. This shows a group
of parallel, equidistant planes (dashed) cutting the sides of the unit cell
defined by axes OA, OB, and OC. The dashed planes will represent a
set of lattice planes if the lattice points O, A, B, and C fall on the planes,
because the pattern will then repeat in every unit cell, and all of the
lattice points in the crystal will therefore fall on the planes. This condition is met if the intercepts I_a, I_b, I_c, into which the lattice planes sub-
divide the crystal axes, are integral fractions of the axial lengths. Thus
$I_a = a/h$, $I_b = b/k$, and $I_c = c/l$, where h, k, and l are integers. These
integers are the indices of the set of lattice planes, and the x-ray reflection
from these planes is identified with the triplet hkl ($= 231$ in the example
in Figure IV-3).

If the indices h, k, l contain a common factor, say n, then only every
nth plane in the complete array of parallel planes actually contains lattice
points, and the reflection hkl represents the nth order reflection from the
set of actual lattice planes. The n here is the same as the n that occurs in
Equation 3-8. The lattice plane spacing d that occurs in Equation 3-8
can be calculated from a formula derived with the help of vector analysis.
For crystals in which the axes are not mutually perpendicular the formula
is somewhat complex, but for crystals with orthogonal axes it simplifies to

$$\frac{n^2}{d^2} = \frac{h^2}{a^2} + \frac{k^2}{b^2} + \frac{l^2}{c^2} \tag{IV-1}$$

Measurement of Unit Cell Constants

To measure the cell dimensions of a crystal with x-rays a monochromatic x-ray beam is used; that is, a beam of known, single wavelength, such as the characteristic x-rays of wavelength $\lambda = 1.5407$ Å emitted from an x-ray tube having a copper target (see Section 4-1). The Bragg angle θ is measured for a number of reflections from the crystal, and the corresponding lattice spacings d are calculated from Equation 3-8. For an orthorhombic crystal the d values for three reflections of known indices hkl are necessary to determine the three lattice constants a, b, c from Equation IV-1. For a triclinic crystal six reflections are necessary to determine the axial lengths and angles.

Symmetry

The first step in seeking the complete space symmetry of a crystal is to find its point group (Appendix III). The point group of a well-formed crystal can be deduced by examining the array of crystal faces. When the external shape of the crystal is not well enough developed, it is necessary to find the internal symmetry, with x-rays, and this is usually done anyway, as a check. The symmetry elements of the crystal can be found with Laue photographs, such as Figure IV-1. On a Laue photograph any symmetry element of the crystal that coincides with the axis of the x-ray beam will show up as a symmetrical pattern in the array of spots on the film. The example in Figure IV-1 shows that a 2-fold axis and two mirror planes are parallel to the x-ray beam. To find all of the symmetry elements it is necessary to adjust the orientation of the crystal until each axis or plane becomes parallel to the beam and can be identified. In this way the complete array of symmetry elements, representing one of the point groups, is found. A significant obstacle to doing this is the fact that all crystals appear centrosymmetric to x-rays, because a reflection from one side of a set of lattice planes is normally indistinguishable from a reflection from the other side. Methods have been devised for circumventing this difficulty; the simplest is to use the external form of the crystal to decide whether or not a center of symmetry is present. It is remarkable that a simple macroscopic observation can here add significant information to what is obtainable by the powerful method of x-ray diffraction.

Once the point group is known, the space group is determined by finding, first, the type of space lattice, and, second, the types of space symmetry elements that correspond to the macroscopically observed point symmetry elements (see Appendix III). To find the space-lattice type one must distinguish between a primitive cell and a cell with some type of centering. (Figure III-6), and to find the space symmetry elements one must distinguish between pure rotation axes and screw axes, or between mirror planes and glide planes. All of these distinctions can be made from the diffraction pattern, because a symmetry element involving a component of translation causes reflections of certain types to be absent systematically. For example, if the lattice is of type I (body centered), then all reflections hkl for which $h + k + l$ is odd will be absent. Again, if there is a c-glide plane perpendicular to the b axis, all reflections of type $h0l$ will be absent for which l is odd. The reason for these systematic absences will be noted later. By observing them, the lattice type and space symmetry elements can be deduced, and thus the space group determined.

Contents of the Unit Cell

The numbers of atoms of different kinds in the unit cell can be calculated from the density and chemical formula of the substance, once the cell dimensions are known. If ρ is the density and M the molecular weight or formula weight, the number Q of molecules or formula units of the substance in the unit cell is

$$Q = \frac{N_0 \rho V}{M} \tag{IV-2}$$

where N_0 is Avogadro's number, and V is the volume of the unit cell. V is calculated from the cell constants as follows:

$$V = abc \, [1 + 2 \cos \alpha \cos \beta \cos \gamma$$
$$- \cos^2 \alpha - \cos^2 \beta - \cos^2 \gamma]^{1/2} \tag{IV-3}$$

Q must be an integer, since the number of atoms in the unit cell is necessarily an integer. If an integer is not obtained when Q is calculated from Equation IV-2, an error is indicated in the density, in the cell volume, or in the chemical formula, or the crystal has some sort of random structure, such as a random arrangement of atoms and vacancies.

The number of atoms of each kind in the cell is Q times the number of atoms of that kind in the chemical formula of the substance. If the chemical formula is not well established, or if an integer value of Q is not obtained from Equation IV-2, then only an approximate number of atoms of each kind in the cell can be estimated. In this case a successful x-ray analysis of the structure will reveal the exact number of atoms of each kind in the cell and will thus allow the correct chemical formula for the substance to be written. The formulas of many complex substances have been established in this way; for example, the formulas of many silicates.

Atomic Arrangement in the Unit Cell

When the number of atoms of a particular kind in the unit cell is known the space symmetry may place restrictions on where these atoms can lie, and in simple cases this may lead directly to the complete atomic arrangement. With copper, for example, the unit cell contains four atoms, and there is only one arrangement of four atoms that satisfies the space symmetry ($Fm3m$); this is the arrangement shown in Figure 2-3. With more atoms in the unit cell, or with fewer space symmetry elements, many choices for the atomic positions are possible. Normally some of the atoms will occupy positions where their coordinates x, y, z are not fixed by symmetry requirements, and must therefore be determined. The essential task of crystal structure determination is to obtain at this stage an approximately correct version of the atomic arrangement. The primary data available for doing this, in addition to the unit cell size, symmetry, and cell content, are the intensities of the x-ray beams diffracted from the various lattice planes.

For reasons given below, it is usually not possible to proceed in a straightforward way from the x-ray data to the atomic arrangement. Direct methods for doing this have recently been developed, but the methods are elaborate, require extensive calculations, and work only under their own favorable circumstances. One such method, applicable to crystallized proteins, is described later.

Often a more or less indirect method is used to discover the atomic arrangement. Using his intuition and ingenuity, the chemist devises possible arrangements that conform to the constraints imposed by cell size, symmetry, and cell content. Sometimes certain features of the x-ray data, such as the fact that a particular reflection is outstandingly intense, can be used to help suggest possible atomic arrangements. When something is known about the kinds of structural features to be expected for a substance of the type under investigation, atomic arrangements with unreasonable features (such as having some atoms too close together) can be rejected. This test is often a powerful one for finding the correct structure among alternatives, because many reliable principles of molecular architecture have now been established by x-ray studies.

The validity of an atomic arrangement devised by any of these methods, direct or indirect, is tested by calculating from it the intensities of the x-ray reflections to be expected, and comparing them with the observed values. If reasonable agreement is obtained, methods are available for improving the agreement further by making small adjustments in the atomic coordinates, until the agreement is as good as can be expected in view of unavoidable experimental errors in the intensity measurements. Accurate values of the atomic coordinates are obtained by this means.

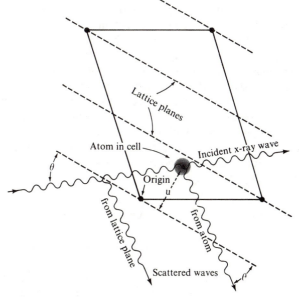

FIGURE IV-4
Diagram illustrating the calculation of the phases
of x-rays scattered by atoms in the unit cell.

Diffracted Intensities

In x-ray diffraction from a given set of lattice planes the waves scattered by each of the atoms in the crystal combine to form the complete diffracted wave. When the diffraction condition 3-8 is satisfied the scattered waves from all of the unit cells of the crystal are exactly in phase, and interfere constructively. In computing the amplitude and phase of the complete diffracted wave it is therefore sufficient to examine closely the wave scattered from a single unit cell.

The amplitude of the wave scattered from a given atom is proportional to the scattering factor, f, for that atom. The x-ray scattering by an atom is accomplished by the electrons, which act to a good approximation as free electrons distributed in accordance with the electronic charge cloud of the atom; hence the scattering factor f is taken as the equivalent number of free electrons.

For reflection from a given set of lattice planes the phase ϕ of the scattered wave from an atom is determined by the position of the atom in relation to the lattice planes (Figure IV-4). Because the angles of incidence and reflection are equal (the Bragg angle θ), displacement of the atom parallel to the lattice planes has no effect on ϕ, and only displacement perpendicular to the planes has an effect (Figure IV-4). Let u be the perpendicular distance measured from the given atom to the lattice plane that passes through the origin of the unit cell, and let the phase of a wave scattered from the origin be taken as 0. As u increases from 0 to d (the interplanar spacing), the phase of the scattered wave increases from 0 to $2\pi n$ radians (where n is the order of reflection from the lattice plane), because the wave scattered from one lattice plane is retarded by n wavelengths relative to the wave from the next plane, according to the diffraction condition in Equation 3-8. At intermediate values of u, the phase is proportionally intermediate, and thus is $\phi = 2\pi nu/d$.

A combination of scattered cosine waves with amplitudes $f_1, f_2, f_3, \ldots$ and phases $\phi_1, \phi_2, \phi_3, \ldots$ has the form

$$E = f_1 \cos\left(2\pi\frac{w}{\lambda} + \phi_1\right) + f_2 \cos\left(2\pi\frac{w}{\lambda} + \phi_2\right)$$
$$+ f_3 \cos\left(2\pi\frac{w}{\lambda} + \phi_3\right) + \cdots \quad \text{(IV-4)}$$

Here w is a coordinate measured along the scattered x-ray beam, and λ is the x-ray wavelength. It is convenient to rewrite Equation IV-4 in terms of complex numbers ($i = \sqrt{-1}$) as follows:

$$E = \text{Real part of}\left(\sum_{p=1}^{N} f_p e^{i\left(2\pi\frac{w}{\lambda}+\phi_p\right)}\right)$$
$$= \text{Real part of}\left[\left(\sum_{p=1}^{N} f_p e^{i\phi_p}\right)e^{2\pi i\frac{w}{\lambda}}\right] \quad \text{(IV-5)}$$

The summation in Equation IV-5 is over all N atoms in the unit cell. The factor $e^{2\pi i\frac{w}{\lambda}}$ simply describes a cosine wave of amplitude unity and phase zero; hence the amplitude and phase of the scattered wave are contained in the complex number F given by

$$F = \sum_{p=1}^{N} f_p e^{i\phi_p} \quad \text{(IV-6)}$$

As discussed above, each phase ϕ_p is given by

$$\phi_p = 2\pi nu_p/d \quad \text{(IV-7)}$$

where u_p is the perpendicular distance for the pth atom. If we write F in the form

$$F = |F|e^{i\phi} \quad \text{(IV-8)}$$

then $|F|$, the absolute value of the complex number F, represents the amplitude of the composite scattered wave, and ϕ its phase.

The intensity I of each x-ray reflection hkl is the square of the above amplitude, $|F|^2$, times a proportionality factor K that depends on the number of unit cells in the crystal, the scattering power of a single electron, and certain geometrical factors that depend on how the intensity I is measured experimentally:

$$I(hkl) = K|F(hkl)|^2 \tag{IV-9}$$

The important quantity F is called the *structure factor* for reflection hkl: it describes the effect of the atomic arrangement, or structure, on the intensity of the x-ray reflection. The notation $F(hkl)$ is used to show explicitly that the value of the structure factor depends on the indices of the reflection in question.

The diffracted intensities $I(hkl)$ can be measured from the brightness of the spots in an x-ray photograph, such as Figure IV-1, or with an ionization chamber or geiger counter, in the way shown in Figure 3-23. In modern methods of counter measurement the crystal and counter are set into the reflecting orientation by means of a system of arcs controlled by a high-speed computer, which calculates the required arc settings from the lattice constants and indices hkl, and then carries out automatically the measurement of intensities for a large number of reflections.

The structure factors can be computed from the measured intensities by use of Equation IV-9, since the proportionality factor K can be evaluated for the particular experimental conditions used. Note, however, that what is obtained in this way is only the absolute value, $|F|$, called the structure *amplitude*; the phase ϕ is not measured.

Use of Structure Factors to Find the Atomic Arrangement

Each structure factor $F(hkl)$ provides one piece of information about the atomic arrangement; namely, the particular combination of contributions in Equation IV-6 which depends on the distances u_p measured perpendicular to the lattice planes hkl. To utilize this information in finding the atomic coordinates we need to express u_p in terms of the coordinates x, y, z for each atom. With the help of vector analysis, it can be shown that the following simple relationship holds:

$$\frac{nu}{d} = hx + ky + lz \tag{IV-10}$$

When IV-10 and IV-7 are introduced into IV-6 the formula for the structure factor becomes

$$F = \sum_{p=1}^{N} f_p e^{2\pi i(hx_p + ky_p + lz_p)} \tag{IV-11}$$

where, again, the summation is over all N atoms in the unit cell. Equation

IV-11 is the fundamental formula of x-ray diffraction, on the basis of which the diffracted intensities are used to obtain the atomic coordinates in a crystal. Equation IV-11 can also be used to show why systematic absences ($F = 0$) occur when atoms are duplicated by screw axes, glide planes, or body-centering or face-centering translations.

If the structure factors F could be measured both in amplitude and in phase (see Equation IV-8) a direct solution of any crystal structure would be possible from measured quantities. In practice, the phase of F cannot be measured, and there is no simple way to go directly from the $|F|$ values to atomic coordinates x_p, y_p, z_p by use of Equation IV-11.

In the indirect approach described earlier, possible atomic arrangements are devised on the basis of cell geometry, symmetry, and chemical reasoning as to plausible structural features; these arrangements are then tested by means of Equation IV-11, and, finally, the coordinates of the correct model are improved by small adjustments to give the best agreement between calculated and observed $|F|$ values. In addition, a number of ingenious methods have been developed for overcoming the lack of structure-factor phases, and extracting information about the atomic arrangement directly from the structure amplitude values alone. These methods can reduce the amount of guesswork and chemical intuition required in solving a crystal structure. They require extensive computations, for which a high-speed computer is essential.

The Fourier Method

The approach most adaptable to complex structures makes use of the important method of *Fourier synthesis*. The basis for this method is contained essentially in the derivation of Equation IV-11 if, rather than summing over scattering centers localized at the individual atoms, we instead sum over contributions from the individual electrons as scatterers, distributed over the whole unit cell in accordance with an electron probability density function $\rho(x, y, z)$. The number of electrons contained in a small volume $V\Delta x\Delta y\Delta z$ in the neighborhood of the point x, y, z is $\rho(x, y, z)V\Delta x\Delta y\Delta z$. The contribution of these electrons to the scattered wave, in accordance with IV-11, is $e^{2\pi i(hx+ky+lz)} \rho(x, y, z)V\Delta x\Delta y\Delta z$. Here no f factor appears, because the contribution is localized from a small volume of space, rather than being the net contribution from an entire atom, of extended volume, as are the contributions in IV-11. These localized contributions are then summed for points x, y, z spanning the entire unit cell. As Δx, Δy, and Δz tend to zero, this sum tends to an integral over the cell:

$$F(hkl) = V\int_0^1\int_0^1\int_0^1 \rho(x, y, z)e^{2\pi i(hx+ky+lz)}dx\,dy\,dz \qquad \text{(IV-12)}$$

(The range of each coordinate from 0 to 1 represents a traverse of the unit cell from one side to the other, in the direction of the corresponding cell axis.)

Equation IV-12, which relates the structure factors to the electron density function $\rho(x, y, z)$, is identical to a standard formula in the theory of Fourier analysis, by which a periodic function is analyzed into its frequency components. The function $\rho(x, y, z)$ is periodic in three dimensions, because of the translation symmetry of the crystal, and its frequency components are therefore three-dimensional also. The integers h, k, and l act as frequencies in the directions of each of the three crystal axes. Equation IV-12 shows that the structure factors are the three-dimensional frequency components of the electron density in the crystal.

When the frequency components of a periodic function are known, the function can be synthesized by combining sine waves of the appropriate amplitude and phase. This is the process of *Fourier synthesis*. It is the inverse of Equation IV-12, in which the function is analyzed into its frequency components. The synthesis is a sum of sine waves, one for each frequency component:

$$\rho(x, y, z) = \frac{1}{V} \sum_h \sum_k \sum_l F(hkl)e^{-2\pi i(hx+ky+lz)} \qquad \text{(IV-13)}$$

Each sine wave is represented, as before, by the function $e^{-2\pi i(hx+ky+lz)}$, which describes a wave with crests aligned along the lattice planes hkl. The amplitude and phase of the wave are contained in the factor $F(hkl)$. Thus each x-ray reflection contributes to the Fourier series IV-13 a wave with the geometrical configuration of the lattice planes, and with amplitude and phase given by the structure factor for that reflection. The complete electron density is the sum of all such waves, crisscrossing in all directions. The sum extends over all values of the indices h, k, and l, ranging in principle from $-\infty$ to $+\infty$, but in practice over the range of observable structure factors.

As an example of the results of this method, the electron density calculated by Fourier synthesis for a crystal of nickel phthalocyanine is shown in Figure IV-5. The complete three-dimensional electron-density function is difficult to depict in two dimensions, and Figure IV-5 shows instead a projection of the complete electron density onto one face of the unit cell. The individual atoms appear as peaks in the distribution of electron density. The number of electrons contained in each peak is the atomic number of the corresponding atom, so that each type of atom can be identified. The location of the nucleus of each atom is at the center of the corresponding peak, and the atomic coordinates can thus be read directly from the electron density map. The Fourier method enables us in effect to "see" the complete structure.

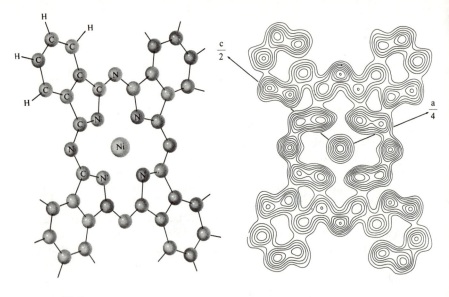

FIGURE IV-5
Contour lines showing the electron density for a molecule of nickel
phthalocyanine, calculated by Fourier synthesis from the observed intensities
of x-ray diffraction maxima for the crystal.

If it were possible to measure both amplitude and phase of the structure factors, then the complete structure could be calculated directly by the Fourier series, IV-13. In practice, successful use of the Fourier method in analyzing crystal structures is a matter of first ascertaining the phases by more or less indirect methods. If an approximate model of the structure, or of some part of the structure, can be devised by the methods described earlier, then approximate phases can be calculated from Equation IV-11. These can be combined with the observed amplitudes, by Equation IV-8, to give structure factors that can then be used in the Fourier series IV-13 to calculate the electron density. If the original model is nearly enough correct this procedure will result in an improvement in the originally assumed atomic positions, and sometimes in the appearance of additional atoms whose presence had not been included in the original model. The addition of these atoms, and the improvement in the atomic positions, leads to better values of the phases and thus to further improvement in the structure. The method then converges to give the complete structure.

The Fourier method is particularly useful for complex organic compounds, and the structures of many biochemically important molecules have been determined in this way. One of the outstanding examples is vitamin B-12, whose structure, as found by Fourier methods, is shown in Figure IV-6. The molecule, $C_{63}H_{84}N_{14}O_{14}PCo$, contains the 99 atoms shown in the figure, plus 84 hydrogen atoms not shown.

An important method for determining structures by Fourier synthesis is known as the method of *isomorphous substitution*. Two crystals are isomorphous if they have the same spatial arrangement of atoms, although the types of atoms differ (Section 4-10). An isomorphous substitution of one atom for another in a structure is a substitution that leaves the spatial arrangement of atoms unchanged. Isomorphous substitutions make possible a direct determination of the phases of the structure factors.

The method is easiest to use in centrosymmetric crystals. The molecule phthalocyanine, $C_{32}H_{18}N_8$, is centrosymmetric, and the molecules stack to form centrosymmetric crystals with the center point of each molecule at a center of symmetry. The molecule can also combine with certain metals to form stable derivatives, such as $C_{32}H_{16}N_8Ni$, whose structure is shown in Figure IV-5. The metal atom occupies the originally vacant center point of the molecule, forming covalent bonds to the four nearby nitrogen atoms. (A similar feature occurs also in the vitamin B-12 molecule.) Molecules of $C_{32}H_{18}N_8$ and $C_{32}H_{16}N_8Ni$ form isomorphous crystals, the only significant difference between them being the substitution of the nickel atom at the center of the molecule. The diffracted intensities from the two crystals thus differ only in the contribution from the nickel atom, a single additional term in Equation IV-11. (The contributions from the two hydrogen atoms displaced by nickel in the isomorphous substitution are too weak to have an observable effect.) Since there is a center of symmetry, there is for every atom at x, y, z in the unit cell also an atom of the same kind at $-x$, $-y$, $-z$, and the contributions from these pairs of atoms combine in Equation IV-12 to to give a real (rather than complex) term:

$$f(e^{2\pi i(hx+ky+lz)} + e^{-2\pi i(hx+ky+lz)}) = 2f\cos 2\pi(hx + ky + lz)$$

In the case of centrosymmetry, therefore, the structure factors $F(hkl)$ are all real quantities, and their phases are either 0 (meaning $F > 0$) or π (meaning $F < 0$). The contribution to Equation IV-11 from the nickel atom, at position 0, 0, 0, is necessarily positive. Hence if the structure amplitude decreases upon isomorphous substitution of nickel the original structure was negative (phase π) while if the amplitude increases the original structure factor was either positive (phase 0), or only slightly negative, the two possibilities being distinguishable by the magnitude of the amplitude increase. In this way, by comparing the intensities of corresponding reflections in the two crystals, the phases of all of the structure factors can be determined, and the Fourier series IV-13 can then be summed to obtain the complete structure. This is the way in which Figure IV-5 was derived.

Recently it has been found possible to extend the Fourier method to vastly more complex molecules. It was discovered that heavy atoms such as mercury or gold can be attached chemically to protein molecules at certain specific sites, which allows the method of isomorphous substitution

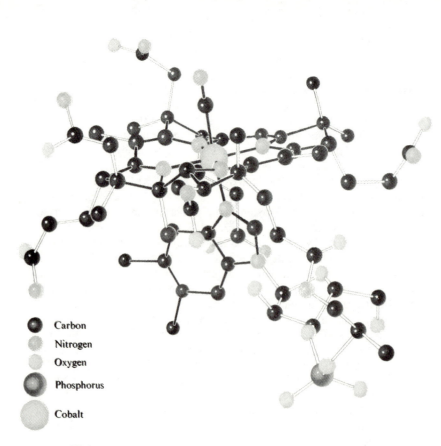

Carbon

Nitrogen

Oxygen

Phosphorus

Cobalt

to be used in determining the structure-factor phases. At least two iso-morphous substitutions are necessary to determine the unknown phases, when the restriction to phase 0 or π no longer applies. To use the intensity changes in determining the phases, it is first necessary to locate the sub-stituted heavy atoms in the unit cell. Fortunately, the effects of a heavy atom on the diffraction intensities from a crystal containing many light atoms are such that the position of the heavy atom can be determined directly, without a knowledge of the complete structure.

The first protein molecule to be analyzed by these methods was myo-globin, a relative of hemoglobin, containing about 2500 atoms. The structures of about half a dozen protein molecules, including hemoglobin itself, have now been determined in this way, and the results are providing valuable information on the biochemical functioning of these molecules.

The wave functions for a state of a hydrogenlike atom described by the quantum numbers n (total quantum number), l (azimuthal quantum number), and m (magnetic quantum number) are usually expressed in terms of the polar coordinates, r, θ, and ϕ. The orbital wave function is a product of three functions, each depending on one of the coordinates:

$$\psi_{nlm}(r, \theta, \phi) = R_{nl}(r)\Theta_{lm}(\theta)\Phi_m(\phi) \tag{V-1}$$

In this equation the functions Φ, Θ, and R have the forms

$$\Phi_m(\phi) = \frac{1}{\sqrt{2\pi}} e^{im\phi} \tag{V-2}$$

$$\Theta_{lm}(\theta) = \left\{ \frac{(2l+1)(l-|m|)!}{2(l+|m|)!} \right\}^{1/2} P_l^{|m|}(\cos\theta) \tag{V-3}$$

and

$$R_{nl}(r) = -\left[\left(\frac{2Z}{na_0} \right)^3 \frac{(n-l-1)!}{2n\{(n+l)!\}^3} \right]^{1/2} e^{-\rho/2} \rho^l L_{n+l}^{2l+1}(\rho) \tag{V-4}$$

in which

$$\rho = \frac{2Z}{na_0} r \tag{V-5}$$

and

$$a_0 = \frac{h^2}{4\pi^2 m\epsilon^2} = 0.530\ \text{Å} \tag{V-6}$$

The functions $P_l^{|m|}(\cos\theta)$ are the associated Legendre functions, and the functions $L_{n+l}^{2l+1}(\rho)$ are the associated Laguerre polynomials.

The wave functions are normalized, so that

$$\int_0^\infty \int_0^\pi \int_0^{2\pi} \psi_{nlm}^*(r, \theta, \phi)\psi_{nlm}(r, \theta, \phi)r^2 \sin\theta\, d\phi\, d\theta\, dr = 1 \tag{V-7}$$

ψ^* is the complex conjugate of ψ. The functions in r, θ, and ϕ are separately normalized to unity:

$$\int_0^{2\pi} \Phi_m(\phi)\Phi_m(\phi)d\phi = 1 \tag{V-8}$$

$$\int_0^\pi \{\Theta_{lm}(\theta)\}^2 \sin\theta\, d\theta = 1 \tag{V-9}$$

$$\int_0^\infty \{R_{nl}(r)\}^2 r^2\, dr = 1 \tag{V-10}$$

In Tables V-1, V-2, and V-3 there are given the expressions for the three component parts of the hydrogenlike wave functions for all values of the quantum that relate to the normal states of atoms. The expressions for $\Phi_m(\phi)$ are given in both the complex form and the real form.

TABLE V-1.
The functions $\Phi_m(\phi)$

$$\Phi_0(\phi) = \frac{1}{\sqrt{2\pi}} \qquad \text{or} \qquad \Phi_0(\phi) = \frac{1}{\sqrt{2\pi}}$$

$$\Phi_1(\phi) = \frac{1}{\sqrt{2\pi}} e^{i\phi} \qquad \text{or} \qquad \Phi_{1\ cos}(\phi) = \frac{1}{\sqrt{\pi}} \cos\phi$$

$$\Phi_{-1}(\phi) = \frac{1}{\sqrt{2\pi}} e^{-i\phi} \qquad \text{or} \qquad \Phi_{1\ sin}(\phi) = \frac{1}{\sqrt{\pi}} \sin\phi$$

$$\Phi_2(\phi) = \frac{1}{\sqrt{2\pi}} e^{i2\phi} \qquad \text{or} \qquad \Phi_{2\ cos}(\phi) = \frac{1}{\sqrt{\pi}} \cos 2\phi$$

$$\Phi_{-2}(\phi) = \frac{1}{\sqrt{2\pi}} e^{-i2\phi} \qquad \text{or} \qquad \Phi_{2\ sin}(\phi) = \frac{1}{\sqrt{\pi}} \sin 2\phi$$

TABLE V-2.
The Wave Functions $\theta_{lm}(\theta)$

$l = 0$, s orbitals:
$$\Theta_{00}(\theta) = \frac{\sqrt{2}}{2}$$

$l = 1$, p orbitals:
$$\Theta_{10}(\theta) = \frac{\sqrt{6}}{2} \cos\theta$$

$$\Theta_{1\pm1}(\theta) = \frac{\sqrt{3}}{2} \sin\theta$$

$l = 2$, d orbitals:
$$\Theta_{20}(\theta) = \frac{\sqrt{10}}{4} (3\cos^2\theta - 1)$$

$$\Theta_{2\pm1}(\theta) = \frac{\sqrt{15}}{2} \sin\theta\cos\theta$$

$$\Theta_{2\pm2}(\theta) = \frac{\sqrt{15}}{4} \sin^2\theta$$

$l = 3$, f orbitals:
$$\Theta_{30}(\theta) = \frac{3\sqrt{14}}{4} \left(\frac{5}{3}\cos^3\theta - \cos\theta\right)$$

$$\Theta_{3\pm1}(\theta) = \frac{\sqrt{42}}{8} \sin\theta(5\cos^2\theta - 1)$$

$$\Theta_{3\pm2}(\theta) = \frac{\sqrt{105}}{4} \sin^2\theta\cos\theta$$

$$\Theta_{3\pm3}(\theta) = \frac{\sqrt{70}}{8} \sin^2\theta$$

TABLE V-3. *The Hydrogen Radial Wave Functions*

$n = 1$, K shell:

$l = 0$, $1s$ $\qquad R_{10}(r) = (Z/a_0)^{3/2} \cdot 2e^{-\rho/2}$

$n = 2$, L shell:

$l = 0$, $2s$ $\qquad R_{20}(r) = \dfrac{(Z/a_0)^{3/2}}{2\sqrt{2}}\,(2 - \rho)e^{-\rho/2}$

$l = 1$, $2p$ $\qquad R_{21}(r) = \dfrac{(Z/a_0)^{3/2}}{2\sqrt{6}}\,\rho e^{-\rho/2}$

$n = 3$, M shell:

$l = 0$, $3s$ $\qquad R_{30}(r) = \dfrac{(Z/a_0)^{3/2}}{9\sqrt{3}}\,(6 - 6\rho + \rho^2)e^{-\rho/2}$

$l = 1$, $3p$ $\qquad R_{31}(r) = \dfrac{(Z/a_0)^{3/2}}{9\sqrt{6}}\,(4 - \rho)\rho e^{-\rho/2}$

$l = 2$, $3d$ $\qquad R_{32}(r) = \dfrac{(Z/a_0)^{3/2}}{9\sqrt{30}}\,\rho^2 e^{-\rho/2}$

$n = 4$, N shell:

$l = 0$, $4s$ $\qquad R_{40}(r) = \dfrac{(Z/a_0)^{3/2}}{96}\,(24 - 36\rho + 12\rho^2 - \rho^3)e^{-\rho/2}$

$l = 1$, $4p$ $\qquad R_{41}(r) = \dfrac{(Z/a_0)^{3/2}}{32\sqrt{15}}\,(20 - 10\rho + \rho^2)\rho e^{-\rho/2}$

$l = 2$, $4d$ $\qquad R_{42}(r) = \dfrac{(Z/a_0)^{3/2}}{96\sqrt{5}}\,(6 - \rho)\rho^2 e^{-\rho/2}$

$l = 3$, $4f$ $\qquad R_{43}(r) = \dfrac{(Z/a_0)^{3/2}}{96\sqrt{35}}\,\rho^3 e^{-\rho/2}$

$n = 5$, O shell:

$l = 0$, $5s$ $\qquad R_{50}(r) = \dfrac{(Z/a_0)^{3/2}}{300\sqrt{5}}\,(120 - 240\rho + 120\rho^2 - 20\rho^3 + \rho^4)e^{\,\rho/2}$

$l = 1$, $5p$ $\qquad R_{51}(r) = \dfrac{(Z/a_0)^{3/2}}{150\sqrt{30}}\,(120 - 90\rho + 18\rho^2 - \rho^3)\rho e^{\,\rho/2}$

$l = 2$, $5d$ $\qquad R_{52}(r) = \dfrac{(Z/a_0)^{3/2}}{150\sqrt{70}}\,(42 - 14\rho + \rho^2)\rho^2 e^{\,\rho/2}$

$l = 3$, $5f$ $\qquad R_{53}(r) = \dfrac{(Z/a_0)^{3/2}}{300\sqrt{70}}\,(8 - \rho)\rho^3 e^{\,\rho/2}$

$l = 4$, $5g$ $\qquad R_{54}(r) = \dfrac{(Z/a_0)^{3/2}}{900\sqrt{70}}\,\rho e^{\,\rho/2}$

$n = 6$, P shell:

$l = 0$, $6s$ $\qquad R_{60}(r) = \dfrac{(Z/a_0)^{3/2}}{2160\sqrt{6}}\,(720 - 1800\rho + 1200\rho^2 - 300\rho^3 + 30\rho^4 - \rho^5)e^{\rho/2}$

$l = 1$, $6p$ $\qquad R_{61}(r) = \dfrac{(Z/a_0)^{3/2}}{432\sqrt{210}}\,(840 - 840\rho + 252\rho^2 - 28\rho^3 + \rho\ )\rho e^{-\rho/2}$

Russell-Saunders States of Atoms Allowed by the Pauli Exclusion Principle

In Section 5-3 it was pointed out that the allowed Russell-Saunders states of an atom with two electrons with different quantum numbers can be found by combining the electron spins to produce a resultant spin, corresponding to the total spin quantum number S (in this case 0 and 1), combining the orbital angular momenta of the electrons to produce the values of the total angular momentum quantum number L that are permitted by the magnitudes of the individual orbital angular momenta of the electrons, and then combining the total spin angular momentum vector and the total orbital angular momentum vector in all of the ways permitted by the magnitudes of the vectors that correspond to the total angular momentum quantum number J, with J having integral values when S is integral (an even number of electron spins) and half-integral values $(1/2, 3/2, \cdots)$ when S has half-integral values (corresponding to an odd number of electrons). It was also mentioned that the Pauli exclusion principle introduces a restriction in case that the two electrons have the same value of the principal quantum number and the azimuthal quantum number. In particular, the normal state of the helium atom corresponds to the electron configuration $1s^2$, with each electron having $n = 1$, $l = 0$, $m_l = 0$, and, of course, $s = 1/2$; the Pauli exclusion principle requires that one of the electrons have $m_s = +1/2$ and the other have $m_s = -1/2$, so that the resultant spin angular momentum is zero, and the state must be a singlet state, $^1S_0$. The corresponding triplet state, $^3S_1$, is excluded by the exclusion principle, and in fact does not exist.

The application of the Pauli exclusion principle is necessary for the understanding of the normal states of atoms. There is a simple way of determining the allowed Russell-Saunders states for an atom with two or more electrons in the same subgroup (same values of n and l).

Sometimes the allowed states can be discovered by a very simple argument. For example, let us discuss the normal state of the nitrogen atom. The nitrogen atom, with seven electrons, has $1s^2 2s^2 2p^3$ as its most stable electron configuration. By the argument given above, the two $1s$ electrons contribute nothing to the spin angular momentum or the orbital angular momentum of the atom, as do also the two $2s$ electrons. The values of the quantum numbers S, L, and J for the normal state of the atom can accordingly be found by consideration of the three $2p$ electrons. The three electrons might give rise to one or more quartet states, with the spin quantum number $S = 3/2$, and one or more doublet states, with the spin quantum number $S = 1/2$. By Hund's first rule (Section

5-3) the quartet states will be more stable than the doublet states, and accordingly we may discuss the quartet states, in the search for the normal state. Each of the two p electrons has $l = 1$, and the possible values of the resultant angular momentum quantum number are hence $L = 0, 1, 2$, and 3. The quartet state might accordingly be ^4S, ^4P, ^4D, or ^4F. In order to obtain a quartet state, with $S = 3/2$, the spins of the three $2p$ electrons must be parallel. These three electrons have accordingly the same values of the quantum numbers n, l, s, and m_s, equal respectively to 2, 1, 1/2, and $+1/2$ (for orientation of the resultant spins in the positive direction). The Pauli exclusion principle requires that they differ from one another; accordingly the remaining quantum number, m_l, must have the values $+1$, 0, and -1 for the three electrons, respectively, and hence the resultant orbital angular momentum must be zero ($L = 0$). Accordingly the one quartet state allowed by the exclusion principle for the configuration $2p^3$ is $^4S_{3/2}$. The normal state of the nitrogen atom is thus found, in agreement with experiment, to be $1s^2 2s^2 2p^3$ $^4S_{3/2}$, as given in Table 5-5.

A somewhat more extended argument is needed to show that the allowed doublet states for this configuration are $^2D_{3/2}$, $^2D_{1/2}$, and $^2S_{1/2}$. The method used to reach this conclusion will be illustrated by application to a simpler case, that of two equivalent p electrons (two electrons with the same value of n and with $l = 1$).

The Zeeman Effect

It was discovered in 1896 by the Dutch physicist Pieter Zeeman (1865-1943) that spectral lines may be split into two or more components when a magnetic field is applied to the atoms that are emitting or absorbing the radiation. This effect is called the Zeeman effect. The splitting of the spectral lines is due to a splitting of the energy levels into two or more components as a result of the interaction of the magnetic moments associated with the spin of the electrons and their orbital motion with the magnetic field.

We may use the configuration $2p3p$ as an example. The Russell-Saunders states corresponding to this configuration, beginning with the most stable one, are $^3D_1$, $^3D_2$, $^3D_3$, $^3P_0$, $^3P_1$, $^3P_2$, $^3S_1$, $^1D_2$, $^1P_1$, and $^1S_0$. There are accordingly ten energy levels. However, when a magnetic field is applied, all of these levels except those corresponding to $J = 0$ are split into several levels by the interaction of the magnetic moments with the magnetic field. For example, the states with $J = 1$ are split into three levels, corresponding to the total magnetic quantum number $M_J = -1$, 0, and $+1$, and those with $J = 2$ are split into five levels, corresponding to $M_J = -2, -1, 0, +1$, and $+2$ (Figure VI-1). In general a state with given value of J is split into $2J + 1$ levels. No further splitting can be

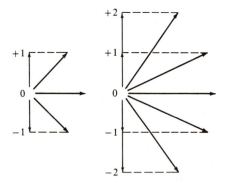

FIGURE VI-1
A diagram showing the orientation in a vertical magnetic field of the total angular
momentum vector corresponding to the value 1 for the angular momentum quantum
number J, and also to the value 2 for this quantum number. For $J = 1$ there are three
orientations, corresponding to the values -1, 0, and $+1$ for the total magnetic quantum
number M_J, and for $J = 2$ there are five orientations. This diagram also represents the
orientation of the total spin angular momentum and the total orbital angular momentum
for the Paschen-Back effect for the ^3D states, with quantum numbers $S = 1$ and $L = 2$.
The diagram at the left represents the orientation of the spin vector and that at the right
the independent orientation of the orbital vector in the vertical magnetic field.

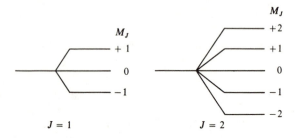

FIGURE VI-2
Energy levels for states with total angular momentum quantum number
$J = 1$ (left) and 2 (right) in the Zeeman effect. The degenerate energy level
is split into three or five components by the magnetic field, corresponding
to different values of the magnetic quantum number M_J.

obtained.* The energy level in the absence of the magnetic field is said to be *degenerate*, the degeneracy of the state being $2J + 1$; the Russell-Saunders state $^3D_1$ is in fact three states, which in the absence of a magnetic field happen to coincide in energy. The application of the magnetic field is said to remove the degeneracy.

By adding the values of $2J + 1$ for the ten Russell-Saunders states listed above we see that there are in fact 36 states based upon the configuration $2p3p$ and the application of a magnetic field gives rise to 36 energy levels.

The change in energy due to the magnetic field is equal to

$$\Delta E = M_J g \mu_B H \qquad (\text{VI-1})$$

In this equation M_J is the total magnetic quantum number, g is a factor that will be discussed later, μ_B is the Bohr magneton, and H is the strength of the magnetic field. The splitting of energy levels into equally separated components is illustrated in Figure VI-2.

The Paschen-Back Effect

It was then discovered by F. Paschen and E. Back that when the magnetic field becomes so strong that the Zeeman splitting of the energy levels of a Russell-Saunders state becomes approximately as great as the separation of the states with different values of J, such as $^3D_3$, $^3D_2$, and $^3D_1$, the nature of the pattern of energy levels changes. In a strong magnetic field the coupling between the orbital angular momentum and the spin angular momentum to give a resultant angular momentum represented by J is broken; instead, the orbital angular momentum, represented by L, and the spin angular momentum, represented by S, orient independently in the magnetic field, in the ways determined by the orbital magnetic quantum number M_L and the spin magnetic quantum number M_S. This situation is illustrated for the multiplet $^3D_1$, $^3D_2$, and $^3D_3$ in Figure VI-1. In this figure the spin angular momentum is shown to be oriented in three ways in the magnetic field, corresponding to $M_S = -1$, 0, and $+1$, and the orbital angular momentum in five ways, corresponding to $M_L = -2$, -1, 0, $+1$, and $+2$. The orientations of the spin and the orbital angular momenta are independent of one another, and accordingly 15 quantum states are represented. Similarly, the Paschen-Back effect for $^3P_0$, $^3P_1$, and $^3P_2$ gives rise to nine quantum states; the other Russell-Saunders states, with either $S = 0$ or $P = 0$, do not show a Paschen-Back effect; they correspond to a total of 12 quantum states,

---

*An exception is the hyperfine splitting that arises if the nucleus of the atom has a spin, as mentioned in Section 26-7. This hyperfine splitting does not invalidate the above argument.

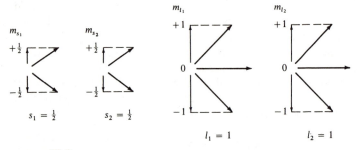

FIGURE VI-3

Diagram representing orientations of the spin vectors of the two electrons and the orbital angular momentum vectors of the two electrons in the extreme Paschen-Back effect for an atom with two $2p$ electrons. The two spins orient themselves separately in the vertical magnetic field, as do also the two orbital angular momentum vectors. Each electron spin can assume orientations such that the component of angular momentum along the field direction is represented by the quantum number $m_s = +\frac{1}{2}$ or $-\frac{1}{2}$, and each orbital angular momentum may orient itself in such a way that the component of the orbital angular momentum along the field direction is represented by the quantum number $m_1 = +1$, 0, or -1.

TABLE VI-1.
Allowed States for Two Equivalent p Electrons

m_{s_1}	m_{s_2}	m_{l_1}	m_{l_2}	$M_S = m_{s_1} + m_{s_2}$	$M_L = m_{l_1} + m_{l_2}$
+1/2	+1/2	+1	0	+1	+1
		+1	−1	+1	0
		0	−1	+1	−1
+1/2	−1/2	+1	+1	0	+2
		+1	0	0	+1
		+1	−1	0	0
		0	+1	0	+1
		0	0	0	0
		0	−1	0	−1
		−1	+1	0	0
		−1	0	0	−1
		−1	−1	0	−2
−1/2	−1/2	+1	0	−1	+1
		+1	−1	−1	0
		0	−1	−1	−1

so that the total for the configuration remains 36, as discussed for the Zeeman case, above. The application of a magnetic field of gradually increasing strength cannot lead to the destruction of quantum states or the formation of new ones, but only to the change in their energy values.

The Extreme Paschen-Back Effect

If the magnetic field is very strong, the interactions that cause the spins of the electrons to combine to a resultant spin and the orbital moments to combine to a resultant orbital moment are broken. Then each electron orients its spin independently in the magnetic field, having two possible values, $+1/2$ and $-1/2$. Similarly, each orbital moment orients itself independently in the magnetic field, there being only one orientation ($m_l = 0$) for an s electron, three ($m_l = -1, 0, +1$) for a p electron, and so on. For the configuration $2p3p$ there are two orientations of the spin for each electron and three orientations for the orbital moment, as shown in Figure VI-3. These orientations are independent of one another; accordingly, the extreme Paschen-Back effect for this configuration gives rise to $2 \times 2 \times 3 \times 3 = 36$ quantum states. These quantum states are equal in number to those for the ten Russell-Saunders states listed above, and also to those for the Paschen-Back states.

Two Equivalent p Electrons

If the two electrons do not have different values of the total quantum numbers, then some of the extreme Paschen-Back states indicated in Figure VI-3 are excluded by the exclusion principle. For example, the electrons cannot both have $m_s = +1/2$ and $m_l = +1$; this is an excluded state. The allowed states for two equivalent p electrons are seen by inspection to be 15 in number; they are listed in Table VI-1. Note that these allowed states require that the quantum numbers m_{s_1} and m_{l_1} for the first electron be not identical with m_{s_2} and m_{l_2} for the second electron; moreover, two assignments of quantum numbers that differ only in the interchange of the two electrons are not counted as two states, but only as one.

The correlation of the extreme Paschen-Back states with Russell-Saunders states can be made by considering the Paschen-Back effect. By adding the electron magnetic spin quantum numbers and electron magnetic orbital quantum number values of the resultant magnetic spin and orbital quantum numbers M_S and M_L are obtained. These may be interpreted at once in terms of Russell-Saunders states. The presence of $M_S = +1$ and -1 (as well as 0) requires that there be some triplet states, with $S = 1$. The values $M_S = +1$ or -1 are correlated with $M_L = +1$, 0, and -1, but not $+2$ or -2; hence there are no ^3D states, but only ^3P.

TABLE VI-2.
Allowed Russell-Saunders States for Equivalent Electrons

Equivalent *s* electrons
$s - {}^2S$
$s^2 - {}^1S$

Equivalent *p* electrons

$p^1 -$	${}^2P$				
$p^2 - {}^1S$		${}^1D$		${}^3P$	
$p^3 -$	${}^2P$		${}^2D$		${}^4S$
$p^4 - {}^1S$		${}^1D$		${}^3P$	
$p^5 -$	${}^2P$				
$p^6 - {}^1S$					

Equivalent *d* electrons

$d^1 -$	${}^2D$					
$d^2 - {}^1(SDG)$		${}^3(PF)$				
$d^3 -$	${}^2D$		${}^2(PDFGH)$	${}^4(PF)$		
$d^4 - {}^1(SDG)$		${}^3(PF)$	${}^1(SDFGI)$	${}^3(PDFGH)$	${}^5D$	
$d^5 -$	${}^2D$		${}^2(PDFGH)$	${}^4(PF)$	${}^4(SDFGI)$	${}^4(DG)$	${}^6S$
$d^6 - {}^1(SDG)$		${}^3(PF)$	${}^1(SDFGI)$	${}^3(PDFGH)$	${}^5D$	
$d^7 -$	${}^2D$		${}^2(PDFGH)$	${}^4(PF)$		
$d^8 - {}^1(SDG)$		${}^3(PF)$				
$d^9 -$	${}^2D$					
$d^{10} - {}^1S$						

Equivalent *f* electrons

f^1	${}^2F$	
f^2	${}^1(SDGI)$	${}^3(PFH)$
f^{12}	${}^1(SDGI)$	${}^3(PFH)$
f^{13}	${}^2F$	
f^{14}	${}^1S$	

When the nine values of M_S and M_L corresponding to ${}^3P$ are crossed out, there remain only the values $M_S = 0$ together with $M_L = +2, +1, 0, 0, -1$, and -2; it is seen that these correspond to the states ${}^1D$ and ${}^1S$. Accordingly the Russell-Saunders states ${}^3P_0$, ${}^3P_1$, ${}^3P_2$, ${}^1D_2$, and ${}^1S_0$ are allowed for two equivalent *p* electrons.

In Table VI-2 are listed the allowed Russell-Saunders states for equivalent *s*, *p*, *d*, and some cases of *f* electrons.

The g-factor

The magnetic moment of an atom can be expressed in a simple way in terms of its angular momentum. An electron moving in an orbit with x units of angular momentum has orbital magnetic moment equal to x Bohr magnetons.

However, the magnetic moment of electron spin bears a different relation to the spin angular momentum; it is approximately twice as great. This can be expressed by saying that the g-factor for orbital motion of the electron is 1 and that for spin of the electron is 2. The g-factor for an atom is the ratio of the magnetic moment of the atom in Bohr magnetons and the angular momentum of the atom in Bohr units $\hbar$.

It is possible to calculate the g-factor for an atom in a Russell-Saunders state by considering the angles between the vectors S and L and the vector J. The total angular momentum in units $\hbar$ is $\sqrt{J(J+1)}$. The magnetic moment in the direction of the angular momentum vector (the component perpendicular to the angular momentum vector cancels out) is equal to the sum of the components of the magnetic moment along S and that along L in the direction of J. The value can be calculated by trigonometry, the magnitudes of the vectors S and L being taken to be equal to $\sqrt{S(S+1)}$ and $\sqrt{L(L+1)}$, respectively. The equation obtained in this way is

$$g(J) = \frac{3J(J+1) + S(S+1) - L(L+1)}{2J(J+1)} \qquad \text{(VI-1)}$$

The general equation for the case g_S and g_L arbitrary is

$$g(J) = \frac{1}{2}(g_S + g_L) + \frac{1}{2}(g_S - g_L)\frac{S(S+1) - L(L+1)}{J(J+1)} \qquad \text{(VI-2)}$$

This equation may be used, for example, in the treatment of nuclear magnetic moments described in Section 26-7. For a moving proton the value of g_L is 1, and for a moving triton it is 1/3. The values of g_S for the proton and triton are given in the last column of Table 26-2.

Hybrid Bond Orbitals

It was mentioned in Section 5-3 that an s orbital is spherically symmetrical and each p orbital is concentrated along an axis: p_x along the x axis, p_y along the y axis, and p_z along the z axis, as shown in Figure 5-10. In Section 6-4 it was stated that the $2s$ orbital and the three $2p$ orbitals of the carbon atom (or $3s$ and the three $3p$'s for silicon, and so on) can be hybridized (combined) to form four equivalent bond orbitals, directed toward the four corners of a tetrahedron, and that these tetrahedral orbitals are the best sp hybrid bond orbitals that can be formed. The angular dependence of a tetrahedral sp hybrid orbital is shown in Figure VII-1.

Derivation of the Tetrahedral Orbitals

The tetrahedral bond orbitals mentioned above are derived in the following way. We assume that the radial parts of the wave functions ψ_s and ψ_{p_x}, ψ_{p_y}, ψ_{p_z} (Appendix V) are so closely similar that their differences can be neglected. The angular parts are

$$
\begin{aligned}
s &= 1 \\
p_x &= \sqrt{3} \sin \theta \cos \phi \\
p_y &= \sqrt{3} \sin \theta \sin \phi \\
p_z &= \sqrt{3} \cos \theta
\end{aligned}
\tag{VII-1}
$$

θ and ϕ being the angles used in spherical polar coordinates. These functions are *normalized* to 4π, the integral

$$
\int_0^{2\pi} \int_0^{\pi} f^2 \sin \theta d\theta d\phi
$$

of the square of the function taken over the surface of a sphere having the value 4π. The functions are *mutually orthogonal*, the integral of the product of any two of them (sp_z, say) over the surface of a sphere being zero.

Now we ask whether a new function

$$
f = as + bp_x + cp_y + dp_z \tag{VII-2}
$$

normalized to 4π (this requiring that $a^2 + b^2 + c^2 + d^2 = 1$) can be formed which has a larger bond strength than 1.732, and, if so, what

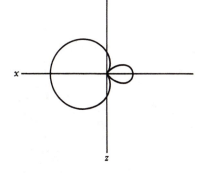

FIGURE VII-1
The angular dependence of a tetrahedral orbital with bond direction
along the x axis.

function of this type has the maximum bond strength. The direction of the bond is immaterial; let us choose the z axis. It is easily shown that p_x and p_y do not increase the strength of a bond in this direction, but decrease it, so they are ignored, the function thus assuming the form

$$f_1 = as + \sqrt{1 - a^2}\, p_z \qquad \text{(VII-3)}$$

in which d is replaced by $\sqrt{1 - a^2}$ for normalization. The value of this in the bond direction $\theta = 0$ is, on substituting the expressions for s and p_z,

$$f_1(\theta = 0) = a + \sqrt{3(1 - a^2)}$$

This is made a maximum by differentiating with respect to a, equating to zero, and solving, the value $a = \frac{1}{2}$ being obtained. Hence the best bond orbital in the z direction is

$$f_1 = \frac{1}{2}s + \frac{\sqrt{3}}{2}p_z = \frac{1}{2} + \frac{3}{2}\cos\theta. \qquad \text{(VII-4)}$$

This orbital has the form shown in Figure VII-1. Its strength is seen to be 2 by placing $\theta = 0$, $\cos\theta = 1$.

We now consider the function

$$f_2 = as + bp_x + dp_z$$

which is orthogonal to f_1, satisfying the requirement

$$\int_0^{2\pi} \int_0^{\pi} f_1 f_2 \sin\theta\, d\theta\, d\phi = 0,$$

and which has the maximum value possible in some direction. (This direction will lie in the xz plane; i.e., $\theta = 0$, since p_y has been left out.) It is found on solving the problem that the function is

$$f_2 = \frac{1}{2}s + \frac{\sqrt{2}}{\sqrt{3}}p_x - \frac{1}{2\sqrt{3}}p_z \qquad \text{(VII-5)}$$

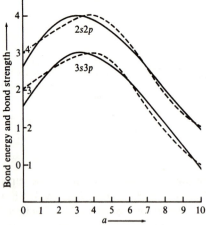

FIGURE VII-2
Square of bond strength (dashed curves) and calculated bond-energy values (full curves) for hybrid sp orbitals varying from pure p orbitals ($a = 0$, left) to pure s orbitals ($a = 10$, right). The upper pair of curves are for L orbitals ($2s$ and $2p$), and the lower, with shifted vertical scale, for M orbitals ($3s$ and $3p$).

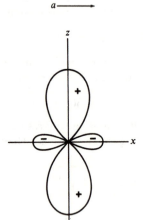

FIGURE VII-3
The angular dependence of the d_{z^2} orbital.

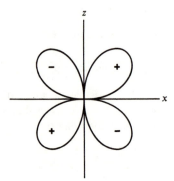

FIGURE VII-4
The angular dependence of the d_{xz} orbital.

This function is seen on examination to be identical with f_1 except that it is rotated through $109°28'$ from f_1. In the same way two more functions can be constructed, each identical with f_1 except for orientation.

An equivalent set of tetrahedral bond orbitals, differing from these only in orientation, is

$$
\left.
\begin{aligned}
t_{111} &= \tfrac{1}{2}(s + p_x + p_y + p_z) \\
t_{1\bar{1}\bar{1}} &= \tfrac{1}{2}(s + p_x - p_y - p_z) \\
t_{\bar{1}1\bar{1}} &= \tfrac{1}{2}(s - p_x + p_y - p_z) \\
t_{\bar{1}\bar{1}1} &= \tfrac{1}{2}(s - p_x - p_y + p_z)
\end{aligned}
\right\}
\qquad \text{(VII-6)}
$$

The strength of an *s-p* hybrid orbital increases with increase in the amount of p involved from 1 (pure s) to a maximum value of 2 (tetrahedral orbital) and then decreases to 1.732 (pure p), in the way shown by the dashed curves in Figure VII-2, in which the square of the bond strength (that is, the product of the strengths of equivalent orbitals of two atoms forming a bond) is shown as a function of the nature of the orbitals. That the strength of the orbital is a measure of its bond-forming power is shown by the approximation of these curves to the full curves, which represent the calculated energy of a one-electron bond as a function of the nature of the bond orbitals.

Octahedral Orbitals

The five d orbitals, in their angular dependence, are

$$
\left.
\begin{aligned}
d_{z^2} &= \sqrt{5/4}\,(3\cos^2\theta - 1) \\
d_{yz} &= \sqrt{15}\,\sin\theta\cos\theta\cos\phi \\
d_{xz} &= \sqrt{15}\,\sin\theta\cos\theta\sin\phi \\
d_{xy} &= \sqrt{15/4}\,\sin^2\theta\sin 2\phi \\
d_{x^2+y^2} &= \sqrt{15/4}\,\sin^2\theta\cos 2\phi
\end{aligned}
\right\}
\qquad \text{(VII-7)}
$$

The orbital d_{z^2} has the angular dependence shown in Figure VII-3. It is cylindrically symmetrical about the z axis and consists of two positive lobes extending in the directions $+z$ and $-z$ and a negative belt about the xy plane. The nodal zones are at $54°44'$ and $125°16'$ with the z direction. The strength of the orbital is 2.236, which is $\sqrt{5}$.

The four other d orbitals described in Equation VII-7 differ in shape from d_{z^2}. The four are equivalent except for spatial orientation. The angular dependence of one of them ($d_{x^2+y^2}$) is shown in Figure VII-4. It has four equivalent lobes, with extrema in the directions $+x$ and $-x$

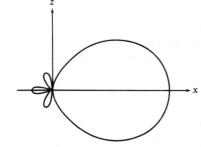

FIGURE VII-5
The angular dependence of an octahedral d^2sp^3 bond orbital with bond
direction along the x axis.

(positive lobes) and $+y$ and $-y$ (negative lobes). The strength (value in these directions) is 1.936. The five d orbitals (unlike the three p orbitals) are therefore not equivalent in shape.

It is found on analysis of the problem that when only two d orbitals are available for combination with the s and p orbitals six equivalent bond orbitals of strength 2.923 can be formed, and that these six orbitals have their bond directions toward the corners of a regular octahedron (Figure VII-5). The set of six equivalent octahedral orbitals formed from two d orbitals, the s orbital, and the three p orbitals is

$$\psi_1 = \frac{1}{\sqrt{6}}s + \frac{1}{\sqrt{2}}p_z + \frac{1}{\sqrt{3}}d_{z^2}$$

$$\psi_2 = \frac{1}{\sqrt{6}}s - \frac{1}{\sqrt{2}}p_z + \frac{1}{\sqrt{3}}d_{z^2}$$

$$\psi_3 = \frac{1}{\sqrt{6}}s + \frac{1}{\sqrt{2}}p_x - \frac{1}{\sqrt{12}}d_{z^2} + \frac{1}{2}d_{x^2+y^2}$$

$$\psi_4 = \frac{1}{\sqrt{6}}s - \frac{1}{\sqrt{2}}p_x - \frac{1}{\sqrt{12}}d_{z^2} + \frac{1}{2}d_{x^2+y^2}$$

$$\psi_5 = \frac{1}{\sqrt{6}}s + \frac{1}{\sqrt{2}}p_y - \frac{1}{\sqrt{12}}d_{z^2} - \frac{1}{2}d_{x^2+y^2}$$

$$\psi_6 = \frac{1}{\sqrt{6}}s - \frac{1}{\sqrt{2}}p_y - \frac{1}{\sqrt{12}}d_{z^2} - \frac{1}{2}d_{x^2+y^2}$$

(VII-8)

Square Bond Orbitals

In a covalent complex of bivalent nickel such as the nickel cyanide ion $[Ni(CN)_4]^{--}$ the 26 inner electrons of the nickel atom can be placed in pairs in the $1s$, $2s$, three $2p$, $3s$, three $3p$, and four of the $3d$ orbitals. This leaves available for use in bond formation the fifth $3d$ orbital as well as the 4*s* and three 4*p* orbitals. It is found on hybridizing these orbitals that four strong bond orbitals directed to the corners of a square can be formed. The four orbitals (written with the bonds directed along $+x$, $-x$, $+y$, and $-y$) are

$$\left. \begin{array}{l} \psi_1 = \dfrac{1}{2}s + \dfrac{1}{\sqrt{2}}p_x + \dfrac{1}{2}d_{xy} \\[2mm] \psi_2 = \dfrac{1}{2}s - \dfrac{1}{\sqrt{2}}p_x + \dfrac{1}{2}d_{xy} \\[2mm] \psi_3 = \dfrac{1}{2}s + \dfrac{1}{\sqrt{2}}p_y - \dfrac{1}{2}d_{xy} \\[2mm] \psi_4 = \dfrac{1}{2}s - \dfrac{1}{\sqrt{2}}p_y - \dfrac{1}{2}d_{xy} \end{array} \right] \qquad \text{(VII-9)}$$

They have the bond strength 2.694, much greater than that of sp^3 tetra-hedral orbitals (2.000). These four square orbitals are formed with use of only two of the 4*p* orbitals; the other *p* orbital might accordingly also be used by the nickel atom to form another (rather weak) bond.

Complexes involving octahedral and square bond orbitals are discussed in Chapter 19 and Appendix XIV.

Bond Energy and Bond-dissociation Energy

In Section 6-11 it was mentioned that a set of bond-energy values can be found such that their sum over all the bonds of a molecule that can be satisfactorily represented by a single valence-bond structure is equal to the heat of formation of the gaseous substance from atoms of the elements. Such a set of bond-energy values is given in Table VIII-1. Most of the values are reliable to about 3 kJ mole^{-1}, and calculations made with their use can be trusted to within about 3 kJ mole^{-1} per bond (see the examples at the end of this Appendix).

Values for single bonds not given in the table can be estimated roughly from the relation to the electronegativity difference:

$$E(A\!-\!B) = \tfrac{1}{2}\{E(A\!-\!A) + E(B\!-\!B)\}$$
$$+ 100(x_A - x_B)^2 - 6.5(x_A - x_B)^4 \text{ kJ mole}^{-1} \qquad \text{(VIII-1)}$$

For example, as yet no determination has been made of the heat of formation of Br_2O, which is an unstable substance for which the structure

$$: \overset{\displaystyle \cdot\cdot}{Br} :$$
$$\big|$$
$$: \overset{\displaystyle \cdot\cdot}{O} \! - \! \overset{\displaystyle \cdot\cdot}{Br} :$$

may be written with confidence. From Equation VIII-1 and the electronegativity values (Table 6-4) we obtain the value 215 kJ mole^{-1} for the Br—O bond energy:

$$E(Br\!-\!O) = \tfrac{1}{2}(193 + 143) + 100(3.5 - 2.8)^2 - 6.5(3.5 - 2.8)^4$$
$$= 168 + 49 - 2 = 215 \text{ kJ mole}^{-1}$$

Hence we obtain 430 kJ mole^{-1} for the heat of formation of $Br_2O(g)$ from atoms. By use of the values of the enthalpy of the atoms relative to the elements in their standard states (Table VIII-3) we can then obtain a predicted value of the standard heat of formation:

$$2Br(g) + O(g) \longrightarrow Br_2O(g) + 430 \text{ kJ mole}^{-1}$$
$$Br_2(l) \longrightarrow 2Br(g) - 224 \text{ kJ mole}^{-1}$$
$$\tfrac{1}{2}O_2(g) \longrightarrow O(g) - 249 \text{ kJ mole}^{-1}$$
$$Br_2(l) + \tfrac{1}{2}O_2(g) \longrightarrow Br_2O(g) - 43 \text{ kJ mole}^{-1}$$

The heat of formation is thus predicted to have a small negative value. The negative value may explain the ease with which the substance decomposes.

TABLE VIII-1
Bond-energy Values for Single Bonds

H—H	436 kJ mole^{-1}	C—H	415 kJ mole^{-1}	Si—Cl	396 kJ mole^{-1}
Li—Li	111	Si—H	295	Si—Br	289
Na—Na	75	N—H	391	Si—I	213
K—K	55	P—H	322	Ge—Cl	408
Rb—Rb	52	As—H	245	N—O	175
Cs—Cs	45	O—H	463	N—F	270
B—B	225	S—H	368	N—Cl	200
C—C	344	Se—H	277	P—O	360
Si—Si	187	Te—H	241	P—F	486
Ge—Ge	157	H—F	563	P—Cl	317
Sn—Sn	143	H—Cl	432	P—Br	266
N—N	159	H—Br	366	P—I	218
P—P	217	H—I	299	As—O	311
As—As	134	B—C	312	As—F	466
Sb—Sb	126	B—O	460	As—Cl	288
Bi—Bi	105	B—S	276	As—Br	236
O—O	143	B—F	582	As—I	174
S—S	266	B—Cl	388	O—F	212
Se—Se	184	B—Br	310	O—Cl	210
Te—Te	168	C—Si	290	O—Br	217
F—F	158	C—N	292	O—I	241
Cl—Cl	243	C—O	350	S—Cl	277
Br—Br	193	C—S	259	S—Br	239
I—I	151	C—F	441	Se—Cl	243
Li—H	245	C—Cl	328	Cl—F	251
Na—H	202	C—Br	276	Br—F	249
K—H	182	C—I	240	Br—Cl	218
Rb—H	167	Si—O	432	I—F	281
Cs—H	175	Si—S	227	I—Cl	210
B—H	331	Si—F	590	I—Br	178

TABLE VIII-2
Bond-energy Values for Multiple Bonds

C=C	615 kJ mole^{-1}	(−81)*	C≡C	812 kJ mole^{-1}	(−232)
N=N	418	(+96)	N≡N	946	(+460)
O=O†	402	(+124)	C≡N	890	(+15)
C=N	615	(+31)	P≡P	490	(−160)
C=O	725	(+25)			
C=S	477	(−41)			

*Values in parentheses are A=B − 2(A—B) and (A≡B) − 3(A—B). A positive value shows that a structure with double or triple bonds is more stable than one with single bonds.

†This value is for the first excited state of O_2, a singlet state that may be considered to involve a double bond.

TABLE VIII-3
Enthalpy (in kJ mole⁻¹) of Monatomic Gases of Elements Relative to the Elements in their Standard States

H	218.0								
Li	160.7	C	715.0	N	472.7	O	249.2	F	78.9
Na	107.8	Si	443.5	P	333.9	S	279	Cl	121.0
K	89.2	Ge	328	As	254	Se	202	Br	111.9
Rb	86	Sn	301	Sb	254	Te	199	I	106.9
Cs	79	Pb	194	Bi	208				

Bond energy values for multiple bonds are given in Table VIII-2.

The values in Table VIII-3, which are the values of ΔH for the reaction X(standard state) $\longrightarrow$ X(g), are useful in converting values of the heat of formation from atoms.

Bond-dissociation Energy

The bond-dissociation energy of a bond in a molecule is the energy required to break that bond alone—that is, to split the molecule into the two parts that were previously connected by that bond. For example, the bond-dissociation energy of the C—C bond in ethane, H_3C—CH_3, is the enthalpy of dissociation of ethane into two methyl radicals, $\cdot CH_3$.

For diatomic molecules the bond energy and the bond-dissociation energy are the same. For polyatomic molecules they are in general different. The sum of successive bond-dissociation energies is, of course, equal to the sum of the bond energies.

The differences between values of bond-dissociation energies and bond energies can usually be interpreted in terms of known features of the electronic structure of the products of dissociation, as illustrated by the following example.

Example 1. What are the values of the O—H bond dissociation energy for H_2O and for OH? To what structural feature is the difference between the two values to be ascribed?

Solution. From the enthalpy values of Table 7-1 we obtain the following heats of reaction:

$$\frac{1}{2}H_2(g) + \frac{1}{2}O_2(g) \longrightarrow OH(g) - 42 \text{ kJ mole}^{-1}$$
$$H(g) \longrightarrow \frac{1}{2}H_2(g) + 218$$
$$O(g) \longrightarrow \frac{1}{2}O_2(g) + 248$$
$$\overline{H(g) + O(g) \longrightarrow OH(g) + 424 \text{ kJ mole}^{-1}}$$

and

$$H_2(g) \; + \tfrac{1}{2}O_2(g) \longrightarrow \; H_2O(g) + 242 \text{ kJ mole}^{-1}$$
$$OH(g) \longrightarrow \; \tfrac{1}{2}H_2(g) + \tfrac{1}{2}O_2(g) + 42$$
$$H(g) \longrightarrow \; \tfrac{1}{2}H_2(g) + 218$$
$$\overline{H(g) \; + \; OH(g) \longrightarrow \; H_2O(g) + 502 \text{ kJ mole}^{-1}}$$

Thus we find that on breaking the two O—H bonds in the water molecule in succession the bond-dissociation energy for the first is 502 kJ mole⁻¹ and that for the second is only 424 kJ mole⁻¹. (The average of the two, 463 kJ mole⁻¹, is the value given in Table VIII-1 for the O—H bond energy.)

To answer the question about what structural feature the difference is to be ascribed to, we examine the electronic structures of H_2O, OH, and O (the hydrogen atom has the same structure, $1s\ ^2S$, whether it is the first one or the second one removed from the molecule, and hence it does not determine the difference in the two-bond-dissociation energies). The water molecule contains two unshared electron pairs on the oxygen atom, one occupying the $2s$ orbital and one a $2p$ orbital. The other two $2p$ orbitals of the oxygen atom are used in forming the two O—H bonds. Each bond consists of two electrons, with opposite spins; they may be described as interchanging their places between the two atoms.

The OH radical, produced when the first hydrogen atom is removed, has the structure :Ȯ—H. Again there are two unshared pairs, a $2s$ pair and a $2p$ pair, on the oxygen atom; of the other two $2p$ orbitals, one, as before is involved in forming an O—H bond and the other is occupied by an odd electron. This odd electron may have its spin oriented in either a positive or negative direction, with no difference in energy, because the other electrons in the molecule are paired.

When the second hydrogen atom is removed an oxygen atom, :Ö·, remains. It has a $2s$ pair, a $2p$ pair, and two odd electrons in the other two $2p$ orbitals.

When these two odd electrons were serving as bonding electrons, the spins were essentially equally often parallel and opposed. But in the oxygen atom the parallel orientation of the spins gives rise to a triplet state, ^3P, and the opposed orientation gives rise to two singlet states, ^1D and ^1S (see Russell-Saunders coupling, Section 5-3). The interaction between electrons in an atom is such as to make the state with parallel spins (maximum multiplicity) more stable than the states with anti-parallel spin. Hence the normal state of the oxygen atom is the triplet state, ^3P, which we may represent as :Ö↑. It is the stabilizing interaction of the two parallel electron spins in the oxygen atom that causes the second bond-dissociation energy of the water molecule to be smaller than the first.

The Vapor Pressure of Water at Different Temperatures

Temperature (°C)	Vapor Pressure
−10 (ice)	0.0028 atm
−5 ″	.0042
0	.0060
5	.0086
10	.0121
15	.0168
16	.0179
17	.0191
18	.0204
19	.0217
20	.0231
21	.0245
22	.0261
23	.0277
24	.0294
25	.0313
26	.0332
27	.0352
28	.0373
29	.0395
30	.0419
35	.0555
40	.0728
45	.0946
50	.1217
60	.1965
70	.3075
80	.4672
90	.6917
100	1.0000
110	1.414
150	4.698
200	15.340
300	84.774

An Alternative Derivation of the Boltzmann Distribution Law

The derivation of the Boltzmann distribution law given in Section 9-6 is a novel one. The derivation usually given in textbooks of statistical mechanics makes use of Stirling's approximation and of the Lagrange method of undetermined multipliers. Because of the importance of the subject and the additional insight that the treatment may give, this derivation is carried out in the following paragraphs.

We begin with Equation 9-22:

$$\text{Number of states of identical molecules} = \frac{n^N}{\overset{\infty}{\underset{j=1}{\Pi} N_j!}} \qquad \text{(X-1)}$$

We want to find the values of N_j that make this function a maximum; this maximum corresponds to the most probable distribution of the molecules among their quantum states. It is convenient to use the natural logarithm of the function:

$$\text{Natural logarithm of the number of states} = N \ln n - \sum_{j=1}^{\infty} \ln (N_j!) \qquad \text{(X-2)}$$

The logarithm, of course, has a maximum when the function has its maximum. This expression can be simplified by using Stirling's approximation (discussed at the end of this appendix) for the logarithm of a factorial, which is

$$\ln x! = x \ln x - x \qquad \text{(X-3)}$$

This is a very close approximation for large values of x. We thus obtain the expression

$$\text{Natural logarithm of number of states}$$
$$= N \ln n - \sum_{j=1}^{\infty} N_j \ln N_j + \sum_{j=1}^{\infty} N_j \qquad \text{(X-4)}$$

Our problem is to find the maximum of this function with respect to the variables, N_1, N_2, and so on, with the restraints that the total number of molecules be constant; that is, $\sum N_j = N$, and that the energy of the system be constant; that is, $\sum N_j E_j = E_{\text{total}}$.

Fortunately, there is a simple way of handling a problem involving constraints on the variables. It was discovered by the great French mathematician Joseph Louis Lagrange (1736–1813), and it is called the Lagrange method of undetermined multipliers.

Let us consider the function

$$F = N \ln n - \sum_{j=1}^{\infty} N_j \ln N_j + \sum_{j=1}^{\infty} N_j + \alpha \sum_{j=1}^{\infty} N_j - \beta \sum_{j=1}^{\infty} N_j E_j \qquad \text{(X-5)}$$

This function differs from the function X-4 by having the terms $\alpha \sum N_j$ and $-\beta \sum N_j E_j$ added (α and β are constants to which we shall later assign values). These two terms are constant, however, since $\sum N_j$ is just the total number of molecules and $\sum N_j E_j$ is the total energy of the system; hence the maximum for F comes at the same place as the maximum of the function X-4.

We now take the partial derivative of F with respect to N_j and equate it to zero:

$$\frac{\partial F}{\partial N_j} = -\ln N_j - 1 + 1 + \alpha - \beta E_j = 0$$

From this equation we obtain

$$\ln N_j = \alpha - \beta E_j$$

and hence

$$N_j = \exp \alpha \exp (-\beta E_j) \qquad \text{(X-6)}$$

The number N_j is the number of molecules occupying the n one-molecule states with energy equal to (or very close to) E_j. Hence N_j/n is the probability that a single quantized state with this energy is occupied. Let us call this probability N_i. Its value is

$$N_i = N C \exp (-\beta E_i) \qquad \text{(X-7)}$$

with NC written in place of $(\exp \alpha)/n$. The value of C is such that the sum of all values of N_i is equal to N, the total number of molecules; hence we obtain

$$C = \frac{1}{\sum_{i=1}^{\infty} \exp (-\beta E_i)} \qquad \text{(X-8)}$$

Equations X-7 and X-8 are identical with 9-28 and 9-29 in the text; with $\beta = \dfrac{1}{kT}$ they give the Boltzmann distribution law in its usual form.

An approximate value for $\ln N!$ can be derived in the following way. Consider the definite integral $\int_{3/2}^{N+1/2} \ln x \, dx$. As dx covers the range $\frac{3}{2}$ to $\frac{5}{2}$ the integrand $\ln x$ has the approximate average value $\ln 2$ (the midpoint value); similarly $\ln 3$ for the range $\frac{5}{2}$ to $\frac{7}{2}$, and so on to $\ln N$ for the range $N - \frac{1}{2}$ to $N + \frac{1}{2}$. But $\ln 2 + \ln 3 + \ldots + \ln N$ is $\ln N!$ Hence $\ln N! \cong \int_{3/2}^{N+1/2} \ln x \, dx$. (The foregoing method of approximating an integral by a sum is called Newton's method; it was discovered by Isaac Newton.) The integral $\int_{3/2}^{N+1/2} \ln x \, dx$ equals $(N + \frac{1}{2}) \ln(N+\frac{1}{2}) - (N+\frac{1}{2}) - \frac{3}{2} \ln \frac{3}{2} + \frac{3}{2}$. The Scottish mathematician James Stirling (1692–1770) derived a better approximation:

$$\ln N! = (N + \tfrac{1}{2}) \ln N - N + \tfrac{1}{2} \ln 2\pi \qquad \text{(X-9)}$$

For $N = 15$, for example, $\ln N!$ is 27.899, Stirling's approximation gives 27.894, and the Newton's-method approximation gives 27.875. For large values of N the approximation $\ln N! = N \ln N - N$ is usually used.

The Boltzmann Distribution Law in Classical Mechanics

In classical mechanics the state of a particle can be described by giving the values of its coordinates, x, y, and z, and of the corresponding components of its velocity, v_x, v_y, and v_z. It is convenient in classical statistical mechanics to use the components of linear momentum, $p_x = mv_x$, $p_y = mv_y$, and $p_z = mv_z$, in place of the components of the velocity; here m is the mass of the particle.

We ask: For the system consisting of a single particle in a rectangular box with volume V in thermodynamic equilibrium with its environment at absolute temperature T, what is the probability that the particle will be in the state in which its x coordinate lies between x and $x + dx$, its y coordinate between y and $y + dy$, its z coordinate between z and $z + dz$, and the three components of its momentum between p_x and $p_x + dp_x$, p_y and $p_y + dp_y$, and p_z and $p_z + dp_z$, respectively? It was shown by Maxwell and Boltzmann that the answer is the following:

$$\text{Probability} = C \exp\left(-E/kT\right) dx\,dy\,dz\,dp_x\,dp_y\,dp_z \qquad \text{(XI-1)}$$

Here C is a normalizing factor. The expression is similar to that in the numerator or denominator of Equation 9-30, with $dx\,dy\,dz\,dp_x\,dp_y\,dp_z$ in place of the quantum weight 1 of a state in the corresponding quantum-theory equation. In classical theory the number of states is infinite, but proportional to the six-dimensional differential volume $dx\,dy\,dz\,dp_x\,dp_y\,dp_z$. The Boltzmann exponential factor is the same for classical theory as for quantum theory.

In classical theory the energy E is the sum of the kinetic energy and the potential energy. The kinetic energy $\frac{1}{2}mv^2$ is a function of the momenta only,

$$\frac{p_x^2}{2m} + \frac{p_y^2}{2m} + \frac{p_z^2}{2m}$$

and the potential energy is a function of the coordinates x, y, and z only. [For example, if the particle is in the gravitational field of the earth the potential energy is mgz, in which z is the vertical coordinate (height above the surface of the earth) and g is the gravitational constant.] With $E = E_{\text{kin}} + E_{\text{pot}}$, the exponential factor $\exp\left[-(E_{\text{kin}} + E_{\text{pot}})/kT\right]$ can be written as the product of two factors, $\exp\left(-E_{\text{kin}}/kT\right)\exp\left(-E_{\text{pot}}/kT\right)$, and the probability expression of Equation XI-1 becomes the product of a function of x, y, and z and a function of p_x, p_y, and p_z. This form means that the distribution of probabilities in coordinate space x, y, z, which depends upon the potential energy function $E_{\text{pot}}(x, y, z)$, and the distribution of probabilities in momentum space p_x, p_y, p_z, which depends upon the kinetic energy function $E_{\text{kin}}(p_x, p_y, p_z) = (p_x^2 + p_y^2 + p_z^2)/2m$, can be discussed independently of one another. The corresponding expressions are

Probability in coordinate space $= A \exp(-E_{\text{pot}}/kT)\, dxdydz$ (XI-2)

Probability in momentum space $= B \exp(-E_{\text{kin}}/kT)\, dp_x dp_y dp_z$ (XI-3)

A and B are normalizing factors. Equation XI-3 leads directly to the Maxwell-Boltzmann distribution law for molecular velocities, which has been discussed in Section 9-5.

The Entropy of a Perfect Gas

The entropy S of a system has been defined in Equation 10-4 in terms of the multiplicity W of the system (the number of quantum states compatible with the description of the macroscopic state of the system) by the Boltzmann equation

$$S = k \ln W \qquad \text{(XII-1)}$$

We now consider a system composed of N identical monatomic molecules. Let w be the molecular multiplicity. A simple case is one in which there are w quantum states with the same energy accessible to any molecule. We ask how many quantum states W there are for the system; that is, how many ways can N identical molecules be distributed among w quantum states. The answer is

$$W = \frac{w^N}{N!} \qquad \text{(XII-2)}$$

Each of the N molecules can occupy any one of the w quantum states, giving w^N distributions, which must, however, be divided by $N!$ because interchange of identical molecules does not give a different state of the system. (We are treating the case of w very large compared with N.) With use of Stirling's approximation for $N!$ we obtain for the entropy the equation

$$S = k \ln W = R \ln w - R \ln N + R \qquad \text{(XII-3)}$$

In this derivation each of the w molecular quantum states has equal probability of occupancy. But we have shown (Section 9-6) that the probability of occupancy of the molecular quantum states at temperature T is proportional to the Boltzmann factor $\exp(-E_i/kT)$.

The average molecule has energy $\bar{E}$. The molecular quantum states with energy approximately $\bar{E}$ can be expected to contribute fully to the molecular multiplicity. This expectation and the Boltzmann distribution law are both satisfied by the following definition of the molecular multiplicity w:

$$w = \exp(\bar{E}/kT) \sum_i \exp(-E_i/kT) \qquad \text{(XII-4)}$$

For a monatomic gas at equipartition temperatures we know that $\bar{E} = \frac{3}{2}kT$. We also know the density of quantum states in energy (Equation 9-18):

$$dN = \frac{4\pi 2^{1/2} m^{3/2} V}{h^3} E^{1/2} dE \qquad \text{(XII-5)}$$

We introduce this quantity in Equation XII-4, replacing the sum by an integral, to obtain the result

$$w = \frac{4\pi 2^{1/2} m^{3/2} V}{h^3} \exp{(3/2)} \int_0^\infty \exp{(-E/kT)} E^{1/2} dE \qquad \text{(XII-6)}$$

The definite integral $\int_0^\infty \exp{(-x/a)} x^{1/2} dx$ is equal to $\pi^{1/2} a^{3/2}/2$; hence w is found to be

$$w = \exp{(3/2)} \frac{(2\pi k T m)^{3/2} V}{h^3}$$

Introduction of this value for w in Equation XII-3 gives the following expression for the molar entropy of a monatomic perfect gas in the equipartition temperature range:

$$S = \tfrac{3}{2} R \ln m + \tfrac{3}{2} R \ln T + R \ln V$$
$$- R \ln N + \tfrac{5}{2} R + \tfrac{3}{2} R \ln{(2\pi k/h^2)} \qquad \text{(XII-7)}$$

This equation is called the Sackur-Tetrode equation; it was discovered in 1912 by the German scientists O. Sackur and H. Tetrode, by use of the old quantum theory. Its discovery and experimental verification played an important part in the development of the third law of thermodynamics.

Replacement of m by the molecular weight M and introduction of the values of the constants gives the equation

$$S = \tfrac{3}{2} R \ln M + \tfrac{3}{2} R \ln T + R \ln V + 11.11 \text{ J deg}^{-1} \text{ mole}^{-1} \quad \text{(XII-8)}$$

Substitution of RT/P for V leads to

$$S = \tfrac{3}{2} R \ln M + \tfrac{5}{2} R \ln T - R \ln P - 9.69 \text{ J deg}^{-1} \text{ mole}^{-1} \qquad \text{(XII-9)}$$

(V in liter mole^{-1}, P in atm). These equations are given in the text as Equations 10-6 and 10-7.

The Molecular Partition Function

In most textbooks of statistical mechanics use is made of a quantity q, which is usually called the *molecular partition function* (sometimes the *sum-over-states*). It is defined by the equation

$$q = \sum_i \exp{(-E_i/kT)} \qquad \text{(XII-10)}$$

We see that it is closely related to the molecular multiplicity (Equation XII-4), which differs from it by the factor $\exp{(\bar{E}/kT)}$.

The relation of q to the average energy of a molecule is obtained in the following way. First we evaluate the derivative of $\ln q$ with respect to T:

$$\frac{d \ln q}{dT} = \frac{1}{q} \frac{dq}{dT} = \frac{1}{kT^2 q} \sum_i E_i \exp{(-E_i/kT)} \qquad \text{(XII-11)}$$

Since $\exp(-E_i/kT)/q$ is the probability that the molecule is in the state with energy E_i, we see that the term on the right side is the average energy divided by kT^2. Accordingly we rewrite Equation XII-11 as

$$\bar{E} = kT^2 \frac{d \ln q}{dT} \qquad \text{(XII-12)}$$

The heat capacity per mole can be obtained by differentiating $\bar{E}$ with respect to T and multiplying by N. The entropy per mole is given by Equation XII-3 by replacing w by $\exp(\bar{E}/kT)q$; it is

$$S = \frac{\bar{E}}{T} + R \ln q - R \ln N + R \qquad \text{(XII-13)}$$

Equations XII-12 and XII-13 are given in textbooks of statistical mechanics.

Electric Polarizabilities
and Electric Dipole Moments

It was mentioned in Section 6-9 that a great deal of information has been obtained about the structure of molecules by the study of the electric properties of substances. A sample of a substance placed in an electric field undergoes a change in structure that is described as *electric polarization*. In general this change in structure involves motion of the electrons relative to the adjacent atomic nuclei, and also motion of the atomic nuclei relative to one another. Theoretical treatments have been developed that permit the observed polarization to be related to the properties of the atoms, ions, or molecules that compose the substance.

Electric Polarization and Dielectric Constant

Under the influence of an electric field E acting upon a gas, liquid, or cubic crystal (the restriction to cubic crystals is made because of the complication introduced for other crystals by the dependence of their properties upon direction relative to the crystal axes), the positively and negatively charged particles that make up the substance undergo some relative motion, producing an induced average electric moment. Let P be the induced average electric moment per unit volume. The electric moment is defined as the product of the charge, positive and negative, by the distance of separation, for example, the electric moment of a pair of ions, with charge $+e$ and $-e$, the distance d apart being de. In electromagnetic theory the *electric induction D* is defined as

$$D = E + 4\pi P \qquad \text{(XIII-1)}$$

and the dielectric constant ϵ is defined as

$$\epsilon = D/E = 1 + 4\pi P/E \qquad \text{(XIII-2)}$$

The dielectric constant of a substance may be measured by determining the ratio of the capacity of a capacitor filled with the substance and the capacity of the empty capacitor. The electrical apparatus involves the capicitor whose capacity is to be determined in parallel with a calibrated variable capacitor, in a tuned resonant circuit; the determination is made by adjusting the variable capacitor to keep the resonance frequency constant, and this requires that the sum of the capacities of the two capacitors be constant.

Let us first consider the dielectric constant of a dilute gas consisting of molecules without a permanent electric dipole moment. We assume that the molecules are far enough apart for them to contribute independently to the polarization and that the electric field E induces an electric dipole moment αE in each molecule, in which α is the electronic polarizability of the molecule (Section 11-4). The number of moles per unit volume of gas is the density ρ divided by the molecular weight M, and the number of molecules in unit volume is this ratio multiplied by Avogadro's number N. Hence the polarization of the gas (the induced dipole moment per unit volume) is given by the following equation:

$$P = N \frac{\rho}{M} \alpha E \qquad \text{(XIII-3)}$$

Combining this equation with Equation XIII-2, we obtain

$$(\epsilon - 1) \frac{M}{\rho} = 4\pi N\alpha \qquad \text{(XIII-4)}$$

This equation is not valid for liquids or crystals, but only for substances for which the dielectric constant is very close to unity, as for gases. For other substances an equation derived by consideration of the effect of the induced moments of neighboring molecules upon the molecule undergoing polarization must be considered. In a polarized medium each molecule is affected by the electric field in the region occupied by the molecule, called the local field. For many substances the local field is satisfactorily represented by the Clausius-Mossotti expression, derived in 1850. Each molecule is considered to occupy a spherical cavity. The part of the substance outside the spherical cavity undergoes polarization in the applied field. A simple calculation shows that the shift of positive charges and negative charges corresponding to the polarization P produces inside the cavity the field $(4\pi/3)P$, in addition to the applied field E; hence the local field is given by the equation

$$E_{\text{local}} = E + \frac{4\pi}{3} P \qquad \text{(XIII-5)}$$

The polarization in unit volume is accordingly given by the expression

$$P = N \frac{\rho}{M} \alpha E_{\text{local}} = N\rho \frac{\alpha}{M} \left(E + \frac{4\pi}{3} P \right) \qquad \text{(XIII-6)}$$

This equation, together with the definition of the dielectric constant (Equation XIII-2), leads at once to the equation

$$\frac{(\epsilon - 1)M}{(\epsilon + 2)\rho} = \frac{4\pi N\alpha}{3} \qquad \text{(XIII-7)}$$

This equation, called the Lorenz-Lorentz equation (see Equation 11-19), was derived in 1880 by combining the Clausius-Mossotti expression for the local field with the idea of molecular polarizability.

The principal interaction of an electromagnetic wave, such as visible light, with a substance is that of the electric field of the wave and the electric charges of the substance. The dielectric constant of the substance determines the magnitude of this interaction; in fact, it is equal to the square of the index of refraction:

$$\epsilon = n^2 \tag{XIII-8}$$

The amount of polarization of the medium by the electric field of the electromagnetic wave is a function of the frequency; for example, the dielectric constant of water is 80 when the frequency is very low or zero (static field) and falls to 1.78 for visible light. The reason for the difference is that in a static electric field or the field of an electromagnetic wave with very low frequency the molecules of water, which have a permanent electric dipole moment, are able to orient themselves in the field, and thus to produce a great increase in polarization of the liquid; whereas in the electric field of high frequency of visible light the molecular orientation cannot occur, and the only polarization that contributes to the dielectric constant is electronic polarization. A detailed discussion of the contribution of orientation of permanent molecular electric dipoles to the dielectric constant is given in a following section.

The Lorenz-Lorentz equation for the index of refraction is

$$R = \frac{(n^2 - 1)MW}{(n^2 + 2)\rho} = \frac{4\pi}{3} N\alpha \tag{XIII-9}$$

The quantity R is called the mole refraction.

Values of α for some atoms, ions, and molecules are given in Tables 11-4 and 11-5.

The Debye Equation for Dielectric Constant

The Debye equation for the dielectric constant of a gas whose molecules have a permanent electric moment μ_0 (Section 6-9) is

$$P = \frac{4\pi(\epsilon - 1)MN}{3(\epsilon + 2)\rho} \left(\frac{\mu_0^2}{3kT} + \alpha \right) \tag{XIII-10}$$

This equation can be derived from Equation XIII-4 by including in the expression for the polarization the contribution due to preferential orientation of the permanent dipole moments μ_0 in the field direction. The component of the dipole moment of a molecule in the field direction is $\mu_0 \cos \theta$, where θ is the polar angle between the dipole-moment vector and the field direction, and the energy of interaction is $-\mu_0 E \cos \theta$. The relative probability of orientation in volume element $\sin \theta d\theta d\phi$ (in polar coordinates) is given by the Boltzmann principle as $e^{\mu_0 E \cos\theta / kT} \sin \theta d\theta d\phi$. The average value of the component is hence given by the expression

$$\mu_{av} = \frac{\int_0^{2\pi} \int_0^\pi \mu_0 \cos\theta \exp(\mu_0 E \cos\theta / kT) \sin\theta d\theta d\phi}{\int_0^{2\pi} \int_0^\pi \exp(\mu_0 E \cos\theta / kT) \sin\theta d\theta d\phi} \tag{XIII-11}$$

(The integral in the denominator normalizes the probability.) The integrals are easily evaluated by expanding the exponential functions and retaining the first nonvanishing term:

$$\mu_{av} = \frac{\mu_0^2 E}{kT} \frac{\int_0^{2\pi} \int_0^\pi \cos^2\theta \sin\theta d\theta d\phi}{4\pi} \tag{XIII-12}$$

The integral (with the divisor 4π) is just the mean value of $\cos^2\theta$ over the surface of a sphere. Its value is $\frac{1}{3}$. (The same value is found in quantum mechanics, as the average of $M_J^2/J(J+1)$, with $M_J = J, J-1, \ldots,$ $-J$, for J either integral or half-integral.) Hence we obtain $\mu_{av} = \mu_0^2 E/3kT$. This expression for the contribution of the permanent dipole moments of the molecule leads to the first term on the right in Equation XIII-10. The second term, α, includes the electronic polarizability of the molecule and also the so-called atom polarization, the small change in relative positions of the nuclei caused by the field. This term in independent of the temperature.

Hydrogen chloride, for example, has dielectric constant that decreases from 1.0055 at 200°K to 1.0028 at 500°K, for constant density, corresponding to 1 atm at 0°C. When values of the polarization P (proportional to $\epsilon - 1$) are plotted against $1/T$, the slope leads to the value 0.225 ϵÅ for μ.

Extensive tables of values of dipole moments for gases and solute molecules can be found in reference books. These values are usually given with the debye ($1D = 1 \times 10^{-18}$ statcoulomb cm) as the unit. Conversion to ϵÅ is made by dividing by 4.803 (1ϵÅ $= 4.803$ D).

The relation of electric dipole moment to the partial ionic character of bonds is discussed in Section 6-9.

The Magnetic Properties of Substances

The principal types of interaction of a substance with a magnetic field are called diamagnetism, paramagnetism, ferromagnetism, antiferromagnetism, and ferrimagnetism. They are useful in providing information about the electronic structures of the substances.

Diamagnetism

It was discovered by Faraday that most substances when placed in a magnetic field develop a magnetic moment opposed to the field. Such a substance is said to be diamagnetic. Substances that develop a moment parallel to the field are called paramagnetic substances.

A sample of a diamagnetic substance placed in an inhomogeneous magnetic field is acted on by a force that tends to push it away from the strong-field region. This force is proportional to the diamagnetic susceptibility of the substance, which is defined as the ratio of the induced moment, μ, to the field strength, H:

$$\mu = \chi H \qquad \text{(XIV-1)}$$

The common methods of determining the magnetic susceptibility involve measuring this force.

There is a common misapprehension that a bar of a paramagnetic substance in a uniform magnetic field sets itself parallel to the lines of force of the field and that a bar of diamagnetic substance sets itself perpendicular to the lines of force; in fact, a bar of substance either paramagnetic or diamagnetic sets itself parallel to the lines of force in a uniform field.

Let us consider a metal wire in the form of a circle. If a magnetic field is applied perpendicularly to the plane of the circle a current is induced in the wire. Corresponding to this current there is a magnetic field, resembling that of a magnetic dipole with orientation opposed to the field.

The effect of the application of a magnetic field to an atom or monatomic ion is to cause the electrons to assume an added rotation about an axis parallel to the field direction and passing through the nucleus. This rotation, called the Larmor precession, has the angular velocity $eH/2mc$. The angular momentum of an electron with cylindrical radius ρ about the field axis and with the angular velocity $eH/2mc$ is $eH\rho^2/2c$, and its magnetic moment is related to its angular momentum by the factor $-e/2mc$

and therefore has the value $-e^2\rho^2 H/4mc^2$. Hence the molar diamagnetic susceptibility is

$$\chi_{\text{molar}} = -\frac{Ne^2}{4mc^2} \sum_i \overline{\rho_i^2} \tag{XIV-2}$$

Here $\overline{\rho_i^2}$ is the average value of ρ^2 for the ith electron, and the sum is to be taken over all the electrons in the atom. For spherically symmetrical atoms, with $\rho^2 = x^2 + y^2$ and $r^2 = x^2 + y^2 + z^2$, where r is the distance of the electron from the nucleus, $\overline{\rho^2}$ is equal to $\frac{2}{3}\overline{r^2}$, and hence we may write

$$\chi_{\text{molar}} = -\frac{Ne^2}{6mc^2} \sum_i \overline{r_i^2} \tag{XIV-3}$$

Measured values of the diamagnetic susceptibility of the noble gases correspond to reasonable values of $\sum \overline{r_i^2}$. For polyatomic molecules the interpretation of the diamagnetic susceptibility in terms of structural features is rather uncertain, and this property has not been found to be very valuable in structural chemistry.

Some diamagnetic crystals (graphite, bismuth, naphthalene and other aromatic substances) show pronounced diamagnetic anisotropy. The observed anisotropies of crystals of benzene derivatives correspond to the molar diamagnetic susceptibility -54×10^{-6} with the field direction perpendicular to the plane of the benzene ring and -37×10^{-6} with it in the plane. This molecular anisotropy has been found to be of some use in determining the orientation of the planes of aromatic molecules in crystals.

Diamagnetic susceptibility (per mole or per gram) is in general independent of the temperature. The value for water at 25°C is -0.719×10^{-6} cgsu per gram. Values between -0.55 and -0.75×10^{-6} cgsu per gram are found for most organic compounds.

Paramagnetism

It is customary to restrict the use of the word paramagnetism to substances that in a magnetic field of ordinary strength develop a magnetic moment in the field direction that is proportional to the strength of the field. (This usage excludes ferromagnetic substances.) Most paramagnetic substances have susceptibilities a hundred or a thousand times as great as the average diamagnetic susceptibilities, and with opposite sign (mass susceptibility [per g] $+10^{-4}$ or 10^{-3} cgsu g^{-1}, as compared with about -1×10^{-6} for diamagnetic substances). They also have, of course, a diamagnetic contribution to the total susceptibility.

It was shown by Pierre Curie (1859–1906) in 1895 that paramagnetic susceptibility is strongly dependent on temperature, and for many substances is inversely proportional to the absolute temperature. The equation

$$\chi_{\text{molar}} = \frac{C_{\text{molar}}}{T} + D \qquad \text{(XIV-4)}$$

is called Curie's law, and the constant C_{molar} is called the molar Curie constant. D represents the diamagnetic contribution (it is negative).

The German physicist Wilhelm Weber (1804–1891) in 1854 had attributed paramagnetism to the orientation of little permanent magnets in the substance (and diamagnetism to induced currents, as discussed above). A quantitative treatment was developed by the French physicist Paul Langevin (1872–1946) in 1895 by application of the Boltzmann principle. The theory is the same as for the orientation of electric dipoles (see Appendix XIII). It leads to the equation

$$C_{\text{molar}} = \frac{N\mu^2}{3k} \qquad \text{(XIV-5)}$$

in which μ is the value of the magnetic dipole moment per atom or molecule.

The magnetic moment μ is related to the molar Curie constant by the equation

$$\mu \text{ (in Bohr magnetons)} = 2.824 C_{\text{molar}}^{1/2} \qquad \text{(VIX-6)}$$

Curie's equation applies to gases, solutions, and some crystals. For other crystals a more general equation, the Weiss equation, may be used (derived by P. Weiss in 1907). Weiss assumed that the local magnetic field orienting the dipoles is equal to the applied field plus an added field proportional to the magnetic volume polarization M:

$$H_{\text{local}} = H + aM \qquad \text{(XIV-7)}$$

Application of the Boltzmann distribution law leads to the equation

$$M = \frac{N\rho\mu^2}{3kTW} (H + aM) \qquad \text{(XIV-8)}$$

in which ρ is the density and W the molecular weight. The molar susceptibility is defined as

$$\chi_{\text{molar}} = WM/\rho H \qquad \text{(XIV-9)}$$

These equations lead to the Weiss equation:

$$\chi_{\text{molar}} = C_{\text{molar}}/(T - \theta) \qquad \text{(XIV-10)}$$

with θ, the Curie temperature, given by the expression

$$\theta = N\rho\mu^2 a/3kW \qquad \text{(XIV-11)}$$

and C_{molar} by Equation XIV-5.

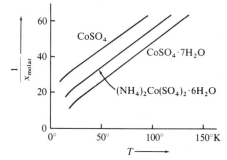

FIGURE XIV-1
Curves showing the reciprocal of the molar magnetic susceptibility of three compounds of cobalt(II).

In a graph of $1/\chi_{molar}$ against T, the points lie on a straight line if the Weiss equation is valid. Measurements for three salts of cobalt(II) are shown in Figure XIV-1. It is seen that the curves are straight lines except at very low temperatures. Their slopes are the same; the slope is the reciprocal of the Curie constant, and accordingly the cobalt(II) atom has the same magnetic moment in the three substances.

The Magnetic Criterion for Bond Type

It has been found that the magnetic moments of complexes can be discussed in a generally satisfactory way by assigning the atomic electrons (the electrons that are not involved in bond formation) to the stable orbitals that are not used as bond orbitals. The assignment is made in the way corresponding to maximum stability, as given by Hund's rules for atoms (Section 5-3); in particular, electrons are introduced into equivalent orbitals in such a way as to give the maximum number of unpaired electron spins compatible with the Pauli exclusion principle. Observed values of the magnetic moment can often be used in selecting one from among several alternative electronic structures for a complex. Application of this magnetic criterion to octahedral and square complexes is made in the following paragraphs.

The Magnetic Moments of Octahedral Complexes

There are two main kinds of electronic structures that may be expected for the octahedral complexes MX_6 of the iron-group transition elements (and also for those of the palladium and platinum groups).

The first kind is that in which no $3d$ orbitals are involved in bond formation; the bonds may be formed with use of the $4s$ orbital and the three $4p$ orbitals (four sp^3 bonds resonating among the six positions), or with use of these four orbitals and two $4d$ orbitals. For this structure all five $3d$ orbitals of M are available for occupancy by the atomic electrons,

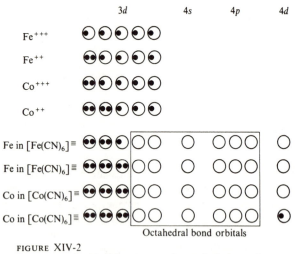

FIGURE XIV-2

Occupancy of orbitals by electrons in octahedral complexes of iron and cobalt.

and the expected magnetic moment is close to that for the monatomic ion M^{+z}. We shall here refer to complexes with this kind of structure as *hypoligated* complexes (the ligands are bonded less strongly than in complexes with the other structures).

Octahedral structures of the second kind are those in which d^2sp^3 bonds are formed with use of two of the $3d$ orbitals, leaving only three for occupancy by atomic electrons. Complexes with these structures were formerly described as essentially covalent; here we shall describe them as *hyperligated* complexes (complexes with strong bonds).

The two kinds of complexes are illustrated in Figure XIV-2.

The way in which the magnetic criterion can be used to distinguish between hypoligated and hyperligated octahedral complexes can be described for iron(II). The Fe^{++} ion has six electrons outside the argon shell. For hypoligated complexes five $3d$ orbitals are available, and the stable disposition of the six electrons among the five orbitals leaves four unpaired, with one orbital occupied by a pair, as shown in Figure XIV-2; the corresponding magnetic moment due to the spins of four electrons is 4.90 magnetons. The hydrated iron(II) ion, $[Fe(OH)_2)_6]^{++}$, is observed to have $\mu = 5.25$, and hence this ion is a hypoligated octahedral complex. On the other hand, the hyperligated octahedral complexes of iron(II) must place the six electrons in three orbitals, and hence must have $\mu = 0$, as is observed for $[Fe(CN)_6]^{----}$.

The magnetic moments predicted for the normal states of the monatomic ions Fe^{++}, Co^{++}, and so on are due in part to spin and in part to

TABLE XIV-1
Magnetic Moments of Iron-Group Ions in Aqueous Solution

Ion	Number of $3d$ Electrons	Number of Unpaired Electrons	Calculated Spin Moment*	Observed Moment*
K^+, Ca^{++}, Sc^{+++}, Ti^{4+}	0	0	0.00	0.00
Ti^{+++}, V^{4+}	1	1	1.73	1.78
V^{+++}	2	2	2.83	2.80
V^{++}, Cr^{+++}, Mn^{4+}	3	3	3.88	3.7–4.0
Cr^{++}, Mn^{+++}	4	4	4.90	4.8–5.0
Mn^{++}, Fe^{+++}	5	5	5.92	5.9
Fe^{++}	6	4	4.90	5.2
Co^{++}	7	3	3.88	5.0
Ni^{++}	8	2	2.83	3.2
Cu^{++}	9	1	1.73	1.9
Cu^+, Zn^{++}	10	0	0.00	0.00

*In Bohr magnetons.

orbital motion. Their values may be calculated for the predicted stable Russell-Saunders state (Section 5-3 and Appendix VI) as $g\sqrt{J(J+1)}$, where J is the total angular momentum quantum number and g is the Landé g-factor appropriate to the Russell-Saunders state. For example, the normal state of Fe^{++} is $^5D_4$, for which $g = 1.500$ and $\mu = 6.70$. In complexes, however, the orbital magnetic moment of the complex is in large part quenched, and the moment approaches the value due to the spin alone, which is $\sqrt{n(n+2)}$, in which n is the number of electrons with unpaired spins. For $n = 4$ the spin moment is 4.90, as mentioned above. The experimental value for the hexahydrated iron(II) ion in solution and in several crystals is 5.25, showing that the orbital moment is largely quenched.

The value of the spin moment for iron-group ions rises to a maximum of 5.92, corresponding to five unpaired electrons, and then decreases, as shown in Table XIV-1. The observed values for the iron-group ions in aqueous solution are seen from the table to agree reasonably well with the theoretical values. The deviations observed can be explained as resulting from contributions of the orbital moments of the electrons.

In many crystalline salts of these elements values of μ are observed that are close to those for the aqueous ions; some of these are given in Table XIV-2.

Values of the observed magnetic moment of some complexes of metals of the iron group, the palladium group, and the platinum group are given in Table XIV-3. Octahedral complexes of iron with fluoride and water are

TABLE XIV-2
Magnetic Moments of Iron-Group Ions in Solid Compounds

Substance	Calculated Spin Moment*	Observed Moment*
$Cr_2O_3 \cdot 7H_2O$	3.88	3.85
$CrSO_4 \cdot 6H_2O$	4.90	4.82
$MnCl_2$	5.92	5.75
$MnSO_4 \cdot 4H_2O$		5.87
$(NH_4)_3FeF_6$		5.88
$FeCl_3$		5.84
$FeCl_2$	4.90	5.23
$(NH_4)_2Fe(SO_4)_2 \cdot 6H_2O$		5.25
$CoCl_2$	3.88	5.04
$(NH_4)_2Co(SO_4)_2 \cdot 6H_2O$		5.00
$Co(N_2H_4)_2Cl_2$		4.93
$NiCl_2$	2.83	3.3
$Ni(N_2H_4)_2(NO_2)_2$		2.80
$Ni(NH_3)_4SO_4$		2.63
$CuCl_2$	1.73	2.02
$(Cu(NH_3)_4(NO_2)_2$		1.82

*In Bohr magnetons.

seen to be hypoligated, whereas those with cyanide and nitrite are hyperligated. All of the complexes of cobalt(III) that have been investigated are hyperligated except that with fluorine, $[CoF_6]^{---}$, which is hypoligated. It is interesting that in the sequence $[Co(NH_3)_6]^{+++}$, $[Co(NH_3)_3F_3]$, $[CoF_6]^{---}$ the transition from hyperligation to hypoligation occurs between the second and third complex.

Bipositive cobalt forms hypoligating bonds with water and hyperligating bonds with nitrite groups.

All the octahedral complexes of the elements of the palladium and platinum groups that have been investigated are diamagnetic, showing the strong tendency of these elements to form hyperligating bonds.

The Magnetic Moments of
Tetrahedral and Square Coordinated Complexes

The bipositive nickel atom forming four covalent dsp^2 bonds has only four 3d orbitals available for the eight unshared 3d electrons, which must thus form four pairs, the square complex NiX_4 being diamagnetic. Bipositive nickel in a complex involving only the 4s and 4p orbitals (electro-

Table XIV-3

Observed Magnetic Moments of Octahedral Complexes of Transition Elements

	Hyperligated Complexes	
	μ Calculated	μ Observed
$K_4[Cr^{II}(CN)_6]$	2.83*	3.3*
$K_3[Mn^{III}(CN)_6]$		3.0
$K_4[Mn^{II}(CN)_6]$	1.73	2.0
$K_3[Fe^{III}(CN)_6]$		2.33
$K_4[Fe^{II}(CN)_6]$	0.00	0.00
$Na_3[Fe^{II}(CN)_5 \cdot NH_3]$		.00
$K_3[Co^{III}(CN)_6]$		.00
$[Co^{III}(NH_3)_3F_3]$		.00
$[Co^{III}(NH_3)_6]Cl_3$		.00
$K_2Ca[Co^{II}(NO_2)_6]$	1.73	1.9
$K_2[Pd^{IV}Cl_6]$	0.00	0.00
$[Pd^{IV}Cl_4(NH_3)_2]$		.00
$Na_3[Ir^{III}Cl_2(NO_2)_4]$		.00
$[Ir^{III}(NH_3)_3(NO_2)_3]$		.00
$K_2[Pt^{IV}Cl_6]$		.00
$[Pt^{IV}(NH_3)_6]Cl_4$		.00

	Hypoligated Complexes	
	μ Calculated	μ Observed
$[Mn^{II}(NH_3)_6]Br_2$	5.92*	5.9*
$(NH_4)_3[Fe^{III}F_6]$		5.9
$(NH_4)_2[Fe^{III}F_5 \cdot H_2O]$		5.9
$[Fe^{II}(H_2O)_6](NH_4SO_4)_2$	4.90	5.3
$K_3[Co^{III}F_6]$		5.3
$[Co^{II}(NH_3)_6]Cl_2$	3.88	4.96

*In Bohr magnetons.

static bonds or weak convalent bonds) distributes the eight $3d$ electrons among the five $3d$ orbitals in such a way as to leave two electrons unpaired, the complex having a magnetic moment of 2.83 Bohr magnetons. From this argument it is seen that the assignment of nickel complexes to the tetrahedral and square coplanar classes can be made by magnetic measurements.

The crystals $K_2Ni(CN)_4$ and $K_2Ni(CN)_4 \cdot H_2O$, shown by isomorphism to contain the planar complex $[Ni(CN)_4]^{--}$, are diamagnetic. Many other nickel complexes, some of which have been shown to be planar by x-ray diffraction, have been found to satisfy the magnetic criterion. All of the compounds of palladium(II) and platinum(II) are diamagnetic.

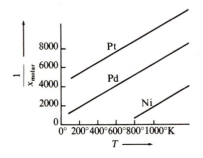

FIGURE XIV-3
Curves showing the reciprocal of the molar paramagnetic susceptibility of nickel, palladium, and platinum.

Ferromagnetism

Ferromagnetic substances assume a large magnetic polarization in weak fields, approaching a constant value (saturation) as the field strength increases. Many of them, including steel and magnetite (Fe_3O_4), retain their magnetization after the field is removed. The substances consist of domains, about 0.01 mm in diameter, that have their atomic moments parallel. In the absence of an external field different domains orient their moments in different directions (these directions are along cube edges for iron and along cube body diagonals for nickel). On application of a magnetic field the atomic moments of a domain reorient themselves.

Values of the low-temperature saturation magnetic moment of ferromagnetic substances represent the maximum component of the atomic magnetic moment in the field direction; for example, for spin alone the value in Bohr magnetons is $2S$, whereas the magnetic moment obtained from the paramagnetic susceptibility is $2\sqrt{S(S + 1)}$.

At higher temperatures thermal agitation disorients some of the atomic moments; at the ferromagnetic Curie temperature the substance becomes paramagnetic. The paramagnetic susceptibilities of nickel, palladium, and platinum are shown in Figure XIV-3. For all three substances the magnetic moment, as given by the slope of the lines, has approximately the value expected from the saturation moment for nickel in its ferromagnetic range, below 680°K. Palladium and platinum are not ferromagnetic.

The nature of the local field in the ferromagnetic metals is probably the interaction between the unpaired spins of atomic electrons and those of some electrons involved in forming one-electron bonds between the metal atoms.

Antiferromagnetism

Antiferromagnetic substances are paramagnetic substances with a characteristic temperature at which the magnetic susceptibility shows a pronounced maximum. This temperature is called the antiferromagnetic transition temperature or Néel temperature (after L. Néel, who first discussed the phenomenon). Above the Néel temperature the susceptibility depends on temperature in accordance with the Weiss equation (Equation XIV-10) with a negative value of the Curie temperature θ. Below the Néel temperature the susceptibility decreases toward zero with decreasing temperature.

All of these properties can be accounted for by the assumption (made first by Néel) that the magnetic moments of adjacent atoms are related by a resonance integral such that maximum stability is associated with alternating orientations of the moments, $\uparrow \downarrow \uparrow \downarrow \uparrow \cdots$, rather than parallel orientation, $\uparrow \uparrow \uparrow \uparrow \uparrow \cdots$, as in ferromagnetic substances. This interaction would make the Curie temperature θ negative, rather than positive. Moreover, at low temperatures the interaction could become cooperative, in such a way that nearly all the atomic magnetic moments would be held in the regular antiparallel arrangement, and the susceptibility would decrease rapidly toward zero.

It is possible to determine the arrangement of positive and negative spins in an antiferromagnetic crystal by neutron diffraction; their interaction with the magnetic moment of the neutron causes their scattering powers to be different. For example, in MnF_2, which has the rutile structure (Figure 18-2), the moments of the manganese atoms in a string of octahedra with shared edges have positive orientation, and those in the adjacent strings have negative orientation. The Néel transition temperature for MnF_2 is 72°K, and the Curie temperature θ is $-113°$.

Ferrimagnetism

Ferrimagnetic substances are substances in which there is an interaction between atomic magnetic moments such as to cause them to align themselves in antiparallel orientations, as in antiferromagnetic substances, but with a difference in the total moments in the two directions, so that the resultant moment is not zero. The properties of ferrimagnetic substances are similar qualitatively to those of ferromagnetic substances: there is a Curie transition temperature, above which the substance is paramagnetic and below which it is ferromagnetic. However, the total magnetic moment indicated in the paramagnetic region is much greater than that given by saturation in the ferromagnetic region.

For example, magnetite, the first ferromagnetic substance discovered, is in fact ferrimagnetic. The crystal has composition Fe_3O_4, with 8 iron atoms in one set of equivalent positions in the unit cube and 16 iron atoms in another set. The observed paramagnetic susceptibility above the Curie temperature is compatible with the magnetic moments 5.2 for iron(II) and 5.9 for iron(III), as found in hypoligated complexes. The sum of the maximum components for one iron(II) moment and two iron(III) moments is 14 Bohr magnetons (spin moment only). The observed value of the ferro-magnetic saturation moment is, however, only 4.2 Bohr magnetons per Fe_3O_4. This fact shows that 8Fe(II) plus 8Fe(III) moments align them-selves in parallel fashion and that the other 8Fe(III) moments in the unit cube have antiparallel orientation. The saturation moment of $MnFe_2O_4$, in which iron(II) is replaced by manganese(II), is 5.0 Bohr magnetons per $MnFe_2O_4$, and that of $NiFe_2O_4$ is 2.2 per $NiFe_2O_4$; these are the values expected for ferrimagnetism such that the iron(II) and iron(III) moments cancel. Substituted magnetites (spinels) of this sort, especially those with some zinc as well as manganese or nickel replacing iron(II), have great practical value in magnetic tape and other applications. They are called ferrites.

Values of Thermodynamic Properties of Some Substances at 25°C and 1 atm

Substance	$\Delta H°_f$ kJ mole^{-1}	$\Delta G°_f$ kJ mole^{-1}	$S°$ J deg^{-1} mole^{-1}	$C°_P$ J deg^{-1} mole^{-1}
Ag(c)	0.00	0.00	42.70	25.49
AgF(c)	−202.9	−185	84	
AgCl(c)	−127.0	−109.7	96.1	50.8
AgBr(c)	−99.5	−95.94	107.1	52.4
AgI(c)	−62.4	−66.3	114	54.4
Ag$_2$O(c)	−30.57	−10.82	121.7	65.6
Al(c)	0.00	0.00	28.32	24.34
AlF$_3$(c)	−1301	−1230	96	
AlCl$_3$(c)	−695	−637	167	89
AlBr$_3$(c)	−526	−505	184	103
AlI$_3$(c)	−315	−314	201	
Al$_2$O$_3$(c)	−1670	−1576	50.99	79.0
Au(c)	0.00	0.00	48	25
Au$_2$O$_3$(c)	81	163	125	
As(gray)	0.00	0.00	35	25
AsF$_3$(g)	−913	−898	289	
AsCl$_3$(l)	−336	−295	233	
As$_4$O$_6$(c)	−1314	−1152	214	191
Ba(c)	0.00	0.00	67	26
BaCl$_2$(c)	−860	−811	126	75
BaO(c)	−558	−529	70	47
Be(c)	0.00	0.00	9.5	17.8
BeO(c)	−611	−582	14.1	25.4
Bi(c)	0.00	0.00	57	26
BiCl$_3$(c)	−379	−319	190	
B(c)	0.00	0.00	6.5	12.0
BCl$_3$(g)	−395	−380	290	63
Br$_2$(l)	0.00	0.00	152	
C(graphite)	0.00	0.00	5.69	8.65
CO(g)	−110.52	−137.27	197.9	29.14
CO$_2$(g)	−393.51	−394.38	213.6	37.13
CH$_4$(g)	−74.85	−50.79	186.2	35.7
Ca(c)	0.00	0.00	41.6	26.3
CaO(c)	−636	−604	40	43
CaCO$_3$(c)	−1207	−1129	93	82

(continued)

(*continued*)

Substance	$\Delta H°_f$ kJ mole^{-1}	$\Delta G°_f$ kJ mole^{-1}	$S°$ J deg^{-1} mole^{-1}	$C°_P$ J deg^{-1} mole^{-1}
CaCl$_2$(c)	−795	−750	114	73
Cd(c)	0.00	0.00	51	26
CdO(c)	−255	−225	55	43
CdCl$_2$(c)	−389	−343	251	118
Cl$_2$(g)	0.00	0.00	222.9	33.9
Cu(c)	0.00	0.00	33.3	24.5
CuO(c)	−155	−127	44	44
Cu$_2$O(c)	−167	−146	101	70
F$_2$(g)	0.00	0.00	203	31.5
H$_2$(g)	0.00	0.00	130.6	28.8
H$_2$O(g)	−241.83	−228.60	188.7	33.6
HF(g)	−269	−271	174	29.1
HCl(g)	−92.31	−95.27	186.7	29.1
HBr(g)	−36.2	−53.2	198.5	29.1
HI(g)	25.9	1.3	206.3	29.2
Hg(l)	0.00	0.00	77	27.8
Hg$_2$Cl$_2$(c)	−265	−211	196	102
HgO(red)	−90.7	−58.5	72	46
I$_2$(c)	0.00	0.00	117	55
K(c)	0.00	0.00	64	29
KCl(c)	−435.86	−408.32	82.7	51.5
KClO$_3$(c)	−391.2	−289.9	143.0	100.2
KClO$_4$(c)	−433	−304	151	110
Li(c)	0.00	0.00	28.0	23.6
LiF(c)	−612	−584	36	42
Mg(c)	0.00	0.00	32.5	24
MgO(c)	−602	−570	27	37
MgCO$_3$(c)	−1113	−1029	66	76
N$_2$(g)	0.00	0.00	191.49	29.12
NH$_3$(g)	−46.2	−16.6	192.5	35.66

Substance	$\Delta H°_f$ kJ mole^{-1}	$\Delta G°_f$ kJ mole^{-1}	$S°$ J deg^{-1} mole^{-1}	$C°_P$ J deg^{-1} mole^{-1}
NH$_4$Cl(c)	−315.4	−203.9	95	84
NO(g)	90.37	86.69	210.6	29.9
NO$_2$(g)	33.85	51.84	240.5	38
N$_2$O$_4$(g)	9.66	98.29	304.3	79
O$_2$(g)	0.00	0.00	205.03	29.36
P(c, white)	0.00	0.00	44.4	23
P$_4$(g)	54.9	24.4	280	67
Rb(c)	0.00	0.00	69	30
RbClO$_3$(c)	−392	−292	152	103
RbClO$_4$(c)	−435	−306	224	161
S(c, rhombic)	0.00	0.00	31.9	22.6
S(c, monoclinic)	0.30	0.10	32.6	23.6
SO$_2$(g)	−297	−300	249	40
SO$_3$(g)	−395.2	−370.4	256	51
Sb(c)	0.00	0.00	44	25
SbCl$_3$(c)	−382	−325	186	
Si(c)	0.00	0.00	18.7	19.9
SiCl$_4$(g)	−610	−570	331	91
SiO$_2$(c, quartz)	−859	−805	42	44
Sn(c, white)	0.00	0.00	52	26.4
Sn(c, gray)	2.5	4.6	45	25.8
SnCl$_4$(l)	−545	−474	259	165
SnO(c)	−286	−257	56	44
SnO$_2$(c)	−581	−520	52	53
Ti(c)	0.00	0.00	30.3	25.1
TiCl$_4$(l)	−750	−674	253	157
TiO$_2$(c, rutile)	−912	−853	50	55
Zn(c)	0.00	0.00	41.6	25
ZnCl$_2$(c)	−416	−369	108	77
ZnO(c)	−348	−318	44	40

Selected Readings
From **SCIENTIFIC AMERICAN**

The *Scientific American* Offprints listed below by number are available from bookstores or from W. H. Freeman and Company, 660 Market Street, San Francisco, California 94104, and Warner House, Folkestone, Kent, England.

Index

A CATALOG OF SELECTED
DOVER BOOKS
IN SCIENCE AND MATHEMATICS

A CATALOG OF SELECTED
DOVER BOOKS
IN SCIENCE AND MATHEMATICS

QUALITATIVE THEORY OF DIFFERENTIAL EQUATIONS, V.V. Nemytskii and V.V. Stepanov. Classic graduate-level text by two prominent Soviet mathematicians covers classical differential equations as well as topological dynamics and ergodic theory. Bibliographies. 523pp. 5⅜ x 8½. 65954-2 Pa. $14.95

MATRICES AND LINEAR ALGEBRA, Hans Schneider and George Phillip Barker. Basic textbook covers theory of matrices and its applications to systems of linear equations and related topics such as determinants, eigenvalues and differential equations. Numerous exercises. 432pp. 5⅜ x 8½. 66014-1 Pa. $10.95

QUANTUM THEORY, David Bohm. This advanced undergraduate-level text presents the quantum theory in terms of qualitative and imaginative concepts, followed by specific applications worked out in mathematical detail. Preface. Index. 655pp. 5⅜ x 8½. 65969-0 Pa. $14.95

ATOMIC PHYSICS (8th edition), Max Born. Nobel laureate's lucid treatment of kinetic theory of gases, elementary particles, nuclear atom, wave-corpuscles, atomic structure and spectral lines, much more. Over 40 appendices, bibliography. 495pp. 5⅜ x 8½. 65984-4 Pa. $13.95

ELECTRONIC STRUCTURE AND THE PROPERTIES OF SOLIDS: The Physics of the Chemical Bond, Walter A. Harrison. Innovative text offers basic understanding of the electronic structure of covalent and ionic solids, simple metals, transition metals and their compounds. Problems. 1980 edition. 582pp. 6⅛ x 9¼. 66021-4 Pa. $16.95

BOUNDARY VALUE PROBLEMS OF HEAT CONDUCTION, M. Necati Özisik. Systematic, comprehensive treatment of modern mathematical methods of solving problems in heat conduction and diffusion. Numerous examples and problems. Selected references. Appendices. 505pp. 5⅜ x 8½. 65990-9 Pa. $12.95

A SHORT HISTORY OF CHEMISTRY (3rd edition), J.R. Partington. Classic exposition explores origins of chemistry, alchemy, early medical chemistry, nature of atmosphere, theory of valency, laws and structure of atomic theory, much more. 428pp. 5⅜ x 8½. (Available in U.S. only) 65977-1 Pa. $11.95

A HISTORY OF ASTRONOMY, A. Pannekoek. Well-balanced, carefully reasoned study covers such topics as Ptolemaic theory, work of Copernicus, Kepler, Newton, Eddington's work on stars, much more. Illustrated. References. 521pp. 5⅜ x 8½. 65994-1 Pa. $12.95

PRINCIPLES OF METEOROLOGICAL ANALYSIS, Walter J. Saucier. Highly respected, abundantly illustrated classic reviews atmospheric variables, hydrostatics, static stability, various analyses (scalar, cross-section, isobaric, isentropic, more). For intermediate meteorology students. 454pp. 6⅛ x 9¼. 65979-8 Pa. $14.95

HANDBOOK OF MATHEMATICAL FUNCTIONS WITH FORMULAS, GRAPHS, AND MATHEMATICAL TABLES, edited by Milton Abramowitz and Irene A. Stegun. Vast compendium: 29 sets of tables, some to as high as 20 places. 1,046pp. 8 x 10½. 61272-4 Pa. $26.95

MATHEMATICAL METHODS IN PHYSICS AND ENGINEERING, John W. Dettman. Algebraically based approach to vectors, mapping, diffraction, other topics in applied math. Also generalized functions, analytic function theory, more. Exercises. 448pp. 5⅜ x 8¼. 65649-7 Pa. $10.95

A SURVEY OF NUMERICAL MATHEMATICS, David M. Young and Robert Todd Gregory. Broad self-contained coverage of computer-oriented numerical algorithms for solving various types of mathematical problems in linear algebra, ordinary and partial, differential equations, much more. Exercises. Total of 1,248pp. 5⅜ x 8½. Two volumes.
Vol. I: 65691-8 Pa. $16.95
Vol. II: 65692-6 Pa. $16.95

TENSOR ANALYSIS FOR PHYSICISTS, J.A. Schouten. Concise exposition of the mathematical basis of tensor analysis, integrated with well-chosen physical examples of the theory. Exercises. Index. Bibliography. 289pp. 5⅜ x 8½. 65582-2 Pa. $8.95

INTRODUCTION TO NUMERICAL ANALYSIS (2nd Edition), F.B. Hildebrand. Classic, fundamental treatment covers computation, approximation, interpolation, numerical differentiation and integration, other topics. 150 new problems. 669pp. 5⅜ x 8½. 65363-3 Pa. $16.95

INVESTIGATIONS ON THE THEORY OF THE BROWNIAN MOVEMENT, Albert Einstein. Five papers (1905–8) investigating dynamics of Brownian motion and evolving elementary theory. Notes by R. Fürth. 122pp. 5⅜ x 8½. 60304-0 Pa. $5.95

CATASTROPHE THEORY FOR SCIENTISTS AND ENGINEERS, Robert Gilmore. Advanced-level treatment describes mathematics of theory grounded in the work of Poincaré, R. Thom, other mathematicians. Also important applications to problems in mathematics, physics, chemistry and engineering. 1981 edition. References. 28 tables. 397 black-and-white illustrations. xvii + 666pp. 6⅛ x 9¼. 67539-4 Pa. $17.95

AN INTRODUCTION TO STATISTICAL THERMODYNAMICS, Terrell L. Hill. Excellent basic text offers wide-ranging coverage of quantum statistical mechanics, systems of interacting molecules, quantum statistics, more. 523pp. 5⅜ x 8½. 65242-4 Pa. $12.95

STATISTICAL PHYSICS, Gregory H. Wannier. Classic text combines thermodynamics, statistical mechanics and kinetic theory in one unified presentation of thermal physics. Problems with solutions. Bibliography. 532pp. 5⅜ x 8½. 65401-X Pa. $12.95

ORDINARY DIFFERENTIAL EQUATIONS, Morris Tenenbaum and Harry Pollard. Exhaustive survey of ordinary differential equations for undergraduates in mathematics, engineering, science. Thorough analysis of theorems. Diagrams. Bibliography. Index. 818pp. 5⅜ x 8½. 64940-7 Pa. $18.95

STATISTICAL MECHANICS: Principles and Applications, Terrell L. Hill. Standard text covers fundamentals of statistical mechanics, applications to fluctuation theory, imperfect gases, distribution functions, more. 448pp. 5⅜ x 8½. 65390-0 Pa. $11.95

ORDINARY DIFFERENTIAL EQUATIONS AND STABILITY THEORY: An Introduction, David A. Sánchez. Brief, modern treatment. Linear equation, stability theory for autonomous and nonautonomous systems, etc. 164pp. 5⅜ x 8¼. 63828-6 Pa. $6.95

THIRTY YEARS THAT SHOOK PHYSICS: The Story of Quantum Theory, George Gamow. Lucid, accessible introduction to influential theory of energy and matter. Careful explanations of Dirac's anti-particles, Bohr's model of the atom, much more. 12 plates. Numerous drawings. 240pp. 5⅜ x 8½. 24895-X Pa. $7.95

THEORY OF MATRICES, Sam Perlis. Outstanding text covering rank, nonsingularity and inverses in connection with the development of canonical matrices under the relation of equivalence, and without the intervention of determinants. Includes exercises. 237pp. 5⅜ x 8½. 66810-X Pa. $8.95

GREAT EXPERIMENTS IN PHYSICS: Firsthand Accounts from Galileo to Einstein, edited by Morris H. Shamos. 25 crucial discoveries: Newton's laws of motion, Chadwick's study of the neutron, Hertz on electromagnetic waves, more. Original accounts clearly annotated. 370pp. 5⅜ x 8½. 25346-5 Pa. $10.95

INTRODUCTION TO PARTIAL DIFFERENTIAL EQUATIONS WITH APPLICATIONS, E.C. Zachmanoglou and Dale W. Thoe. Essentials of partial differential equations applied to common problems in engineering and the physical sciences. Problems and answers. 416pp. 5⅜ x 8½. 65251-3 Pa. $11.95

BURNHAM'S CELESTIAL HANDBOOK, Robert Burnham, Jr. Thorough guide to the stars beyond our solar system. Exhaustive treatment. Alphabetical by constellation: Andromeda to Cetus in Vol. 1; Chamaeleon to Orion in Vol. 2; and Pavo to Vulpecula in Vol. 3. Hundreds of illustrations. Index in Vol. 3. 2,000pp. 6⅛ x 9¼. 23567-X, 23568-8, 23673-0 Pa., Three-vol. set $44.85

CHEMICAL MAGIC, Leonard A. Ford. Second Edition, Revised by E. Winston Grundmeier. Over 100 unusual stunts demonstrating cold fire, dust explosions, much more. Text explains scientific principles and stresses safety precautions. 128pp. 5⅜ x 8½. 67628-5 Pa. $5.95

AMATEUR ASTRONOMER'S HANDBOOK, J.B. Sidgwick. Timeless, comprehensive coverage of telescopes, mirrors, lenses, mountings, telescope drives, micrometers, spectroscopes, more. 189 illustrations. 576pp. 5⅜ x 8¼. (Available in U.S. only) 24034-7 Pa. $11.95

CATALOG OF DOVER BOOKS

SPECIAL FUNCTIONS, N.N. Lebedev. Translated by Richard Silverman. Famous Russian work treating more important special functions, with applications to specific problems of physics and engineering. 38 figures. 308pp. 5⅜ x 8½. 60624-4 Pa. $9.95

OBSERVATIONAL ASTRONOMY FOR AMATEURS, J.B. Sidgwick. Mine of useful data for observation of sun, moon, planets, asteroids, aurorae, meteors, comets, variables, binaries, etc. 39 illustrations. 384pp. 5⅝ x 8¼. (Available in U.S. only) 24033-9 Pa. $8.95

INTEGRAL EQUATIONS, F.G. Tricomi. Authoritative, well-written treatment of extremely useful mathematical tool with wide applications. Volterra Equations, Fredholm Equations, much more. Advanced undergraduate to graduate level. Exercises. Bibliography. 238pp. 5⅜ x 8½. 64828-1 Pa. $8.95

POPULAR LECTURES ON MATHEMATICAL LOGIC, Hao Wang. Noted logician's lucid treatment of historical developments, set theory, model theory, recursion theory and constructivism, proof theory, more. 3 appendixes. Bibliography. 1981 edition. ix + 283pp. 5⅜ x 8½. 67632-3 Pa. $8.95

MODERN NONLINEAR EQUATIONS, Thomas L. Saaty. Emphasizes practical solution of problems; covers seven types of equations. ". . . a welcome contribution to the existing literature...."–*Math Reviews*. 490pp. 5⅜ x 8½. 64232-1 Pa. $13.95

FUNDAMENTALS OF ASTRODYNAMICS, Roger Bate et al. Modern approach developed by U.S. Air Force Academy. Designed as a first course. Problems, exercises. Numerous illustrations. 455pp. 5⅜ x 8½. 60061-0 Pa. $10.95

INTRODUCTION TO LINEAR ALGEBRA AND DIFFERENTIAL EQUATIONS, John W. Dettman. Excellent text covers complex numbers, determinants, orthonormal bases, Laplace transforms, much more. Exercises with solutions. Undergraduate level. 416pp. 5⅜ x 8½. 65191-6 Pa. $11.95

INCOMPRESSIBLE AERODYNAMICS, edited by Bryan Thwaites. Covers theoretical and experimental treatment of the uniform flow of air and viscous fluids past two-dimensional aerofoils and three-dimensional wings; many other topics. 654pp. 5⅜ x 8½. 65465-6 Pa. $16.95

INTRODUCTION TO DIFFERENCE EQUATIONS, Samuel Goldberg. Exceptionally clear exposition of important discipline with applications to sociology, psychology, economics. Many illustrative examples; over 250 problems. 260pp. 5⅜ x 8½. 65084-7 Pa. $8.95

LAMINAR BOUNDARY LAYERS, edited by L. Rosenhead. Engineering classic covers steady boundary layers in two- and three- dimensional flow, unsteady boundary layers, stability, observational techniques, much more. 708pp. 5⅜ x 8½. 65646-2 Pa. $18 95

LECTURES ON CLASSICAL DIFFERENTIAL GEOMETRY, Second Edition, Dirk J. Struik. Excellent brief introduction covers curves, theory of surfaces, fundamental equations, geometry on a surface, conformal mapping, other topics. Problems. 240pp. 5⅜ x 8½. 65609-8 Pa. $8.95

ROTARY-WING AERODYNAMICS, W.Z. Stepniewski. Clear, concise text covers aerodynamic phenomena of the rotor and offers guidelines for helicopter performance evaluation. Originally prepared for NASA. 537 figures. 640pp. 6⅛ x 9¼.
64647-5 Pa. $16.95

DIFFERENTIAL GEOMETRY, Heinrich W. Guggenheimer. Local differential geometry as an application of advanced calculus and linear algebra. Curvature, transformation groups, surfaces, more. Exercises. 62 figures. 378pp. 5⅜ x 8½.
63433-7 Pa. $9.95

INTRODUCTION TO SPACE DYNAMICS, William Tyrrell Thomson. Comprehensive, classic introduction to space-flight engineering for advanced undergraduate and graduate students. Includes vector algebra, kinematics, transformation of coordinates. Bibliography. Index. 352pp. 5⅜ x 8½.
65113-4 Pa. $9.95

A SURVEY OF MINIMAL SURFACES, Robert Osserman. Up-to-date, in-depth discussion of the field for advanced students. Corrected and enlarged edition covers new developments. Includes numerous problems. 192pp. 5⅜ x 8½.
64998-9 Pa. $8.95

ANALYTICAL MECHANICS OF GEARS, Earle Buckingham. Indispensable reference for modern gear manufacture covers conjugate gear-tooth action, gear-tooth profiles of various gears, many other topics. 263 figures. 102 tables. 546pp. 5⅜ x 8½.
65712-4 Pa. $14.95

SET THEORY AND LOGIC, Robert R. Stoll. Lucid introduction to unified theory of mathematical concepts. Set theory and logic seen as tools for conceptual understanding of real number system. 496pp. 5⅜ x 8¼.
63829-4 Pa. $12.95

A HISTORY OF MECHANICS, René Dugas. Monumental study of mechanical principles from antiquity to quantum mechanics. Contributions of ancient Greeks, Galileo, Leonardo, Kepler, Lagrange, many others. 671pp. 5⅜ x 8½.
65632-2 Pa. $14.95

FAMOUS PROBLEMS OF GEOMETRY AND HOW TO SOLVE THEM, Benjamin Bold. Squaring the circle, trisecting the angle, duplicating the cube: learn their history, why they are impossible to solve, then solve them yourself. 128pp. 5⅜ x 8½.
24297-8 Pa. $4.95

MECHANICAL VIBRATIONS, J.P. Den Hartog. Classic textbook offers lucid explanations and illustrative models, applying theories of vibrations to a variety of practical industrial engineering problems. Numerous figures. 233 problems, solutions. Appendix. Index. Preface. 436pp. 5⅜ x 8½.
64785-4 Pa. $11.95

CURVATURE AND HOMOLOGY, Samuel I. Goldberg. Thorough treatment of specialized branch of differential geometry. Covers Riemannian manifolds, topology of differentiable manifolds, compact Lie groups, other topics. Exercises. 315pp. 5⅜ x 8½.
64314-X Pa. $9.95

HISTORY OF STRENGTH OF MATERIALS, Stephen P. Timoshenko. Excellent historical survey of the strength of materials with many references to the theories of elasticity and structure. 245 figures. 452pp. 5⅜ x 8½.
61187-6 Pa. $12.95

DE RE METALLICA, Georgius Agricola. The famous Hoover translation of greatest treatise on technological chemistry, engineering, geology, mining of early modern times (1556). All 289 original woodcuts. 638pp. 6¾ x 11.   60006-8 Pa. $21.95

SOME THEORY OF SAMPLING, William Edwards Deming. Analysis of the problems, theory and design of sampling techniques for social scientists, industrial managers and others who find statistics increasingly important in their work. 61 tables. 90 figures. xvii + 602pp. 5⅜ x 8½.   64684-X Pa. $16.95

THE VARIOUS AND INGENIOUS MACHINES OF AGOSTINO RAMELLI: A Classic Sixteenth-Century Illustrated Treatise on Technology, Agostino Ramelli. One of the most widely known and copied works on machinery in the 16th century. 194 detailed plates of water pumps, grain mills, cranes, more. 608pp. 9 x 12.
28180-9 Pa. $24.95

LINEAR PROGRAMMING AND ECONOMIC ANALYSIS, Robert Dorfman, Paul A. Samuelson and Robert M. Solow. First comprehensive treatment of linear programming in standard economic analysis. Game theory, modern welfare economics, Leontief input-output, more. 525pp. 5⅜ x 8½.   65491-5 Pa. $14.95

ELEMENTARY DECISION THEORY, Herman Chernoff and Lincoln E. Moses. Clear introduction to statistics and statistical theory covers data processing, probability and random variables, testing hypotheses, much more. Exercises. 364pp. 5⅜ x 8½.   65218-1 Pa. $10.95

THE COMPLEAT STRATEGYST: Being a Primer on the Theory of Games of Strategy, J.D. Williams. Highly entertaining classic describes, with many illustrated examples, how to select best strategies in conflict situations. Prefaces. Appendices. 268pp. 5⅜ x 8½.   25101-2 Pa. $7.95

CONSTRUCTIONS AND COMBINATORIAL PROBLEMS IN DESIGN OF EXPERIMENTS, Damaraju Raghavarao. In-depth reference work examines orthogonal Latin squares, incomplete block designs, tactical configuration, partial geometry, much more. Abundant explanations, examples. 416pp. 5⅜ x 8¼.
65685-3 Pa. $10.95

THE ABSOLUTE DIFFERENTIAL CALCULUS (CALCULUS OF TENSORS), Tullio Levi-Civita. Great 20th-century mathematician's classic work on material necessary for mathematical grasp of theory of relativity. 452pp. 5⅜ x 8½.
63401-9 Pa. $11.95

VECTOR AND TENSOR ANALYSIS WITH APPLICATIONS, A.I. Borisenko and I.E. Tarapov. Concise introduction. Worked-out problems, solutions, exercises. 257pp. 5⅝ x 8¼.   63833-2 Pa. $8.95

THE FOUR-COLOR PROBLEM: Assaults and Conquest, Thomas L. Saaty and Paul G. Kainen. Engrossing, comprehensive account of the century-old combinatorial topological problem, its history and solution. Bibliographies. Index. 110 figures. 228pp. 5⅜ x 8½.   65092-8 Pa. $7.95

A CONCISE HISTORY OF MATHEMATICS, Dirk J. Struik. The best brief history of mathematics. Stresses origins and covers every major figure from ancient Near East to 19th century. 41 illustrations. 195pp. 5⅜ x 8½.   60255-9 Pa. $8.95

A SHORT ACCOUNT OF THE HISTORY OF MATHEMATICS, W.W. Rouse Ball. One of clearest, most authoritative surveys from the Egyptians and Phoenicians through 19th-century figures such as Grassman, Galois, Riemann. Fourth edition. 522pp. 5⅜ x 8½.   20630-0 Pa. $11.95

HISTORY OF MATHEMATICS, David E. Smith. Nontechnical survey from ancient Greece and Orient to late 19th century; evolution of arithmetic, geometry, trigonometry, calculating devices, algebra, the calculus. 362 illustrations. 1,355pp. 5⅜ x 8½.   20429-4, 20430-8 Pa., Two-vol. set $26.90

THE GEOMETRY OF RENÉ DESCARTES, René Descartes. The great work founded analytical geometry. Original French text, Descartes' own diagrams, together with definitive Smith-Latham translation. 244pp. 5⅜ x 8½.   60068-8 Pa. $8.95

THE ORIGINS OF THE INFINITESIMAL CALCULUS, Margaret E. Baron. Only fully detailed and documented account of crucial discipline: origins; development by Galileo, Kepler, Cavalieri; contributions of Newton, Leibniz, more. 304pp. 5⅜ x 8½. (Available in U.S. and Canada only)   65371-4 Pa. $9.95

THE HISTORY OF THE CALCULUS AND ITS CONCEPTUAL DEVELOPMENT, Carl B. Boyer. Origins in antiquity, medieval contributions, work of Newton, Leibniz, rigorous formulation. Treatment is verbal. 346pp. 5⅜ x 8½. 60509-4 Pa. $9.95

THE THIRTEEN BOOKS OF EUCLID'S ELEMENTS, translated with introduction and commentary by Sir Thomas L. Heath. Definitive edition. Textual and linguistic notes, mathematical analysis. 2,500 years of critical commentary. Not abridged. 1,414pp. 5⅜ x 8½.   60088-2, 60089-0, 60090-4 Pa., Three-vol. set $32.85

GAMES AND DECISIONS: Introduction and Critical Survey, R. Duncan Luce and Howard Raiffa. Superb nontechnical introduction to game theory, primarily applied to social sciences. Utility theory, zero-sum games, n-person games, decision-making, much more. Bibliography. 509pp. 5⅜ x 8½.   65943-7 Pa. $13.95

THE HISTORICAL ROOTS OF ELEMENTARY MATHEMATICS, Lucas N.H. Bunt, Phillip S. Jones, and Jack D. Bedient. Fundamental underpinnings of modern arithmetic, algebra, geometry and number systems derived from ancient civilizations. 320pp. 5⅜ x 8½.   25563-8 Pa. $8.95

CALCULUS REFRESHER FOR TECHNICAL PEOPLE, A. Albert Klaf. Covers important aspects of integral and differential calculus via 756 questions. 566 problems, most answered. 431pp. 5⅜ x 8½.   20370-0 Pa. $8.95

CATALOG OF DOVER BOOKS

CHALLENGING MATHEMATICAL PROBLEMS WITH ELEMENTARY SOLUTIONS, A.M. Yaglom and I.M. Yaglom. Over 170 challenging problems on probability theory, combinatorial analysis, points and lines, topology, convex polygons, many other topics. Solutions. Total of 445pp. 5⅜ x 8½. Two-vol. set.

Vol. I: 65536-9 Pa. $7.95
Vol. II: 65537-7 Pa. $7.95

FIFTY CHALLENGING PROBLEMS IN PROBABILITY WITH SOLUTIONS, Frederick Mosteller. Remarkable puzzlers, graded in difficulty, illustrate elementary and advanced aspects of probability. Detailed solutions. 88pp. 5⅜ x 8½.

65355-2 Pa. $4.95

EXPERIMENTS IN TOPOLOGY, Stephen Barr. Classic, lively explanation of one of the byways of mathematics. Klein bottles, Moebius strips, projective planes, map coloring, problem of the Koenigsberg bridges, much more, described with clarity and wit. 43 figures. 210pp. 5⅜ x 8½. 25933-1 Pa. $6.95

RELATIVITY IN ILLUSTRATIONS, Jacob T. Schwartz. Clear nontechnical treatment makes relativity more accessible than ever before. Over 60 drawings illustrate concepts more clearly than text alone. Only high school geometry needed. Bibliography. 128pp. 6⅛ x 9¼. 25965-X Pa. $7.95

AN INTRODUCTION TO ORDINARY DIFFERENTIAL EQUATIONS, Earl A. Coddington. A thorough and systematic first course in elementary differential equations for undergraduates in mathematics and science, with many exercises and problems (with answers). Index. 304pp. 5⅜ x 8½. 65942-9 Pa. $8.95

FOURIER SERIES AND ORTHOGONAL FUNCTIONS, Harry F. Davis. An incisive text combining theory and practical example to introduce Fourier series, orthogonal functions and applications of the Fourier method to boundary-value problems. 570 exercises. Answers and notes. 416pp. 5⅜ x 8½. 65973-9 Pa. $11.95

AN INTRODUCTION TO ALGEBRAIC STRUCTURES, Joseph Landin. Superb self-contained text covers "abstract algebra": sets and numbers, theory of groups, theory of rings, much more. Numerous well-chosen examples, exercises. 247pp. 5⅜ x 8½.

65940-2 Pa. $8.95

STARS AND RELATIVITY, Ya. B. Zel'dovich and I. D. Novikov. Vol. 1 of *Relativistic Astrophysics* by famed Russian scientists. General relativity, properties of matter under astrophysical conditions, stars and stellar systems. Deep physical insights, clear presentation. 1971 edition. References. 544pp. 5⅜ x 8½.

69424-0 Pa. $14.95

Prices subject to change without notice.

Available at your book dealer or write for free Mathematics and Science Catalog to Dept. GI, Dover Publications, Inc., 31 East 2nd St., Mineola, N.Y. 11501. Dover publishes more than 250 books each year on science, elementary and advanced mathematics, biology, music, art, literature, history, social sciences and other areas.